Springer Proceedings in Mathematics & Statistics

Volume 189

AF832187

Springer Proceedings in Mathematics & Statistics

This book series features volumes composed of selected contributions from workshops and conferences in all areas of current research in mathematics and statistics, including operation research and optimization. In addition to an overall evaluation of the interest, scientific quality, and timeliness of each proposal at the hands of the publisher, individual contributions are all refereed to the high quality standards of leading journals in the field. Thus, this series provides the research community with well-edited, authoritative reports on developments in the most exciting areas of mathematical and statistical research today.

More information about this series at http://www.springer.com/series/10533

Jan Kallsen · Antonis Papapantoleon
Editors

Advanced Modelling in Mathematical Finance

In Honour of Ernst Eberlein

Springer

Editors
Jan Kallsen
Department of Mathematics
Christian-Albrechts-Universität zu Kiel
Kiel
Germany

Antonis Papapantoleon
Institute of Mathematics
Technische Universität Berlin
Berlin
Germany

ISSN 2194-1009 ISSN 2194-1017 (electronic)
Springer Proceedings in Mathematics & Statistics
ISBN 978-3-319-83390-3 ISBN 978-3-319-45875-5 (eBook)
DOI 10.1007/978-3-319-45875-5

Mathematics Subject Classification (2010): 91Gxx, 91G10, 91G20, 91G30, 91G40, 91G60, 91G70, 91G80, 60G44, 60G48, 60G48

© Springer International Publishing Switzerland 2016
Softcover reprint of the hardcover 1st edition 2016
This work is subject to copyright. All rights are reserved by the Publisher, whether the whole or part of the material is concerned, specifically the rights of translation, reprinting, reuse of illustrations, recitation, broadcasting, reproduction on microfilms or in any other physical way, and transmission or information storage and retrieval, electronic adaptation, computer software, or by similar or dissimilar methodology now known or hereafter developed.
The use of general descriptive names, registered names, trademarks, service marks, etc. in this publication does not imply, even in the absence of a specific statement, that such names are exempt from the relevant protective laws and regulations and therefore free for general use.
The publisher, the authors and the editors are safe to assume that the advice and information in this book are believed to be true and accurate at the date of publication. Neither the publisher nor the authors or the editors give a warranty, express or implied, with respect to the material contained herein or for any errors or omissions that may have been made.

Printed on acid-free paper

This Springer imprint is published by Springer Nature
The registered company is Springer International Publishing AG
The registered company address is: Gewerbestrasse 11, 6330 Cham, Switzerland

Preface

Mathematical Finance from its modest beginnings just a few decades ago has turned into a thriving branch of applied mathematics, and in particular of probability and statistics. In Germany, Ernst Eberlein was among the first to recognise the potential and relevance of this field. He has contributed strongly to its development, not only in terms of research, but also through manifold contributions to the community as an organiser of conferences and workshops, by fostering international contacts and networks, serving in societies, promoting young researchers, assuming editorial duties, and so on. In appreciation of Ernst's merits, and on his 70th birthday, a conference titled *Advanced Modelling in Mathematical Finance* took place during 20–22 May 2015 in Kiel, where 19 talks were delivered by invited experts.

The Festschrift at hand is based on some of these presentations and on additional invited contributions. They all reflect the title of the conference and demonstrate the breadth of issues and methods that manifest in modern mathematical finance. First, we give Ernst Eberlein a chance to speak for himself. An interview with him is to be found right after the table of contents.

This volume consists of four major parts. The first one concerns the choice and properties of stochastic processes for models in finance and other applications. Hammerstein reviews popular classes of Lévy processes in detail before focusing on their tail behaviour. Lévy-driven processes with a flexible dependence structure are studied by Barndorff-Nielsen. Mandjes and Spreij, on the other hand, focus on a class of tractable Markov processes allowing, for example, for explicit representations of their characteristic function.

In the subsequent part, aspects of statistics and risk in the broad sense are analyzed in different respects. Geman and Liu discuss the recent evolution of energy markets and the consequences for modelling them. The performance of different models for the dependence of returns is studied in the contribution by Madan. Kimura and Yoshida consider estimating dependence as well, but from the point of view of statistical theory. Extreme value theory is applied by Beirlant, Schoutens, De Spiegeleer, Reynkens, and Herrmann in order to find out to what extent large losses in the financial crisis were beyond expectation. Lütkebohmert-Holtz and Xiao investigate the contribution of time-varying collaterals to default risk. The question of modelling risk and uncertainty is discussed by Stahl from a fundamental perspective.

Part III considers option pricing and hedging, as well as optimisation in the context of finance. Bayer and Schoenmakers study the computation of contingent claim prices in affine generalisations of Merton's jump diffusion model. Jahncke and Kallsen apply a small jump expansion in order to approximate the prices of options on the quadratic variation. Their approach is closely related to the one in the contribution by Grbac, Krief and Tankov in the last chapter. The error when hedging barrier options in exponential Lévy models is studied by Černý. Musiela, Sokolova, and Zariphopoulou study indifference pricing in the context of forward performance processes as an alternative to classical utility maximisation. Feodoria and Kallsen consider almost surely long-run optimal investment in the presence of transaction costs. The properties of optimal payoffs in the sense of Dybvig are reviewed in the contribution by Corcuera, Fajardo, and Pamen, whereas Rüschendorf and Wolf study this concept in the particular context of exponential Lévy models.

The final part concerns term structure models for interest rates and commodity prices. The basic question of absence of arbitrage is investigated by Klein, Schmidt, and Teichmann in this context. Different variants of discrete-tenor interest rate models are contrasted by Glau, Grbac, and Papapantoleon. A small jump approximation to option prices in such models is derived in the contribution by Grbac, Krief, and Tankov. Finally, Benth proposes a term structure model for cointegrated commodity markets, which allows for option pricing by integral transforms.

At the end of this short introduction, we want to thank those who made this Festschrift possible: first of all, the contributors and the anonymous referees, and in particular Gerhard Stahl from the Talanx group who not only—along with the German Science Foundation—provided generous financial support for the conference, but also encouraged us to put together this volume. Thanks are also due to the staff of Springer-Verlag for accepting this project and for their professional assistance. Last but not least, the editors thank Ernst himself for all his support over the years without which we would not be who we are now.

Kiel, Germany Jan Kallsen
Berlin, Germany Antonis Papapantoleon
July 2016

A Conference in Honour of Ernst Eberlein

Advanced Modelling in Mathematical Finance

A conference in honour of
Ernst Eberlein

May 20 - 22, 2015, Kiel, Germany
www.math.uni-kiel.de/ammf2015

Confirmed Speakers

- Christa Cuchiero (Vienna)
- Damir Filipovic (Lausanne)
- Hélyette Geman (London)
- Kathrin Glau (Munich)
- Zorana Grbac (Paris)
- Reinhard Höpfner (Mainz)
- Monique Jeanblanc (Paris)
- Paul Krühner (Vienna)
- Eva Lütkebohmert-Holtz (Freiburg)
- Dilip Madan (Maryland)
- Martijn Pistorius (London)
- Markus Porembski (Kiel)
- Thorsten Schmidt (Chemnitz)
- Wim Schoutens (Leuven)
- Albert Shiryaev (Moscow)
- Michael Sørensen (Copenhagen)
- Gerhard Stahl (Hannover)
- Peter Tankov (Paris)
- Josef Teichmann (Zürich)
- Nakahiro Yoshida (Tokyo)

Organisers

- Jan Kallsen (Kiel)
- Antonis Papapantoleon (Berlin)

Sponsors

DFG talanx.

CAU Christian-Albrechts-Universität - Online: www.math.uni-kiel.de/ammf2015 - Email: ammf@math.uni-kiel.de

Courtesy of Ernst Eberlein

Contents

Interview with Ernst Eberlein

Would you like to tell us about your youth? Where and how did you grow up?

I was born in Rothenburg ob der Tauber, a small town in Franconia, which is the northern half of Bavaria, where somewhat out of town in the Tauber valley my father ran a sawmill, that specialized in oak wood. Living outside of the town came with the benefit of being close to the river, hillsides and woods, which I enjoyed a lot. The disadvantage was that I had a longer way to school, which I did by bike in summer and by bus during the rest of the year. Shortly after I had entered high school, my father passed away unexpectedly. In this situation the expectation of the family was that I would replace him in the business as soon as I would be able to do it. As a consequence, I left high school after six years, despite the fact that the standard curriculum in Germany is nine years, in order to start a formal training as 'Industriekaufmann', a formation which I think can be translated as 'Industrial Business Manager'. I carried out the practical part of this training in a veneer mill 100 km away from home. Normally three years are necessary to obtain this qualification, but I got it in two. It was mainly the prospect of new intellectual challenges that led me at that point to try to go back to high school. I succeeded in recovering part of the time I had spent within professional training, as the ministry in Munich allowed me to skip one grade, on the basis of an exam. In summary, you can conclude that my youth ended somewhat earlier than for most of my contemporaries.

How did you develop your interest in mathematics?

This started certainly at high school (Gymnasium in German). Looking back I realize that I had excellent teachers in mathematics. Not that I did not do well in other subjects too, like writing essays for example, but that part was hard work for me, whereas doing the homework for mathematics was mostly pleasure. It was the right intellectual challenge.

You studied in Erlangen. What was the atmosphere like, in particular around Bauer & Jacobs?

Let me first tell you how it happened that I started in Erlangen. Of course I had not been aware of the great tradition which mathematics had at Erlangen University, but one day I got a letter from an assistant at the physics department who was working on his PhD. He had learned from a friend that I had the intention of studying either mathematics or physics. He sent me some information on the curriculum and recommended that I come to Erlangen. I followed his advice and we became long-term friends. In particular, for many years we met for jogging in the evening in the forest at the southern city limit.

When I started in the winter term of 1966/67, one of the two basic courses, namely Linear Algebra, was taught by Konrad Jacobs. He was the most impressive personality among the professors we got to know in the first year. When he lectured he used no notes. It looked so elegant and easy, but in fact it was not, neither for him nor for us. There was enormous respect in front of a professor. He invited us repeatedly to ask questions during the 15-minute break in the middle of the two-hour sessions, but nobody dared to approach him. We did not want to admit that we were having a hard time following his high speed at the blackboard. At the end of the first year, based on the results that we had obtained in the weekly exercises, he invited a small group of students for a special seminar that would start in the following semester. This 'by invitation only' seminar was continued in the subsequent terms and although at the beginning the topics where quite general like topology, number theory and combinatorics, his guidance, together with his course in probability theory, led towards stochastics. The surviving members of this group wrote their diploma thesis under his supervision and he accepted several of them as PhD students.

It was only in my fourth semester that I took a course taught by Heinz Bauer. His book on probability and measure theory which would soon become a standard reference was in press. His teaching style was very attractive. He was fond of giving every detail, so one had a good chance of following his arguments. I think his style of teaching had a strong influence on the teaching style that I developed myself later. There were plenty of other courses that we had to take, but on the basis of the courses these two young and dynamic professors were offering, it was difficult not to be hooked by probability theory and stochastic processes. Another attraction of following the courses of Jacobs and Bauer was that both from time to time had prominent visitors, who visited the department not only for research but also to lecture. Let me just mention Alexandra Ionescu-Tulcea (Bellow) and Robert Blumenthal.

As a student you spent a year in Paris, around 1968. What are your memories from this time?

After three years of studies in Erlangen I thought it would be a good idea to add a year abroad. Paris looked like the most attractive place. Jacobs introduced me with a letter to Jacques Neveu who signed the papers such that I was accepted as a troisième cycle student for the DEA. Neveu himself did not teach at this level during the academic year 1969/70. I attended courses by J. Azéma, P. Priouret and D. Revuz and besides that I worked on my diploma thesis, for which Jacobs had given me a very appealing subject. The diploma thesis would later become my first publication, 'Toeplitzfolgen und Gruppentranslationen', since Jacobs, without even informing me, submitted the manuscript right after I had it handed over to him for inspection. 1970 was also the year when the mathematics department moved from the Institut Henri Poincaré to the new building at Jussieu. The floor for the probabilists (couloir 56-66) was so spacious that they had reserved an office for foreign students which I shared with another student from Germany. That year in Paris was a wonderful period. Besides the productive time at the department I particularly enjoyed the international atmosphere at the Cité Universitaire.

Following your PhD, you spent time at IMPA in Rio de Janeiro and at the ETH in Zürich. You also spent a sabbatical at Stanford in California. How did these opportunities arise and how did they influence your career?

After I had received the PhD, a fundamental decision had to be made, either to continue with research or to look for a job in the industry. Jacobs had received an invitation from IMPA, but he did not want to go there himself. I think it was because of the political situation in South America at that time. So he asked me if I would be interested to spend a year at IMPA within the agreement on the exchange of researchers between Germany and Brazil. I did not hesitate too long. When I arrived in April 1973, the institute was not located where it is now, but downtown in a historical building which is now a cultural center. I lived in Copacabana and took a bus to the city center every morning. The probability group was headed by Pedro Fernandez, who had received his PhD under Lucien LeCam and had attracted other Berkeley alumni to IMPA. The first course I was asked to teach was an introductory course on probability theory. The second was a measure theory course which I taught in Portuguese!

Living in Rio in the early seventies cannot be compared to going there today. Already the flight required a stop in Casablanca for refueling. Making a phone call was complicated and too expensive. So I stayed in contact with family members and friends by writing letters. There were also some urgent matters for which I sent wires across the Atlantic. The colleagues at IMPA were very friendly. I remember in particular Lindolpho de Carvalho Dias, the director. On one occasion he told me that the Institute had acquired that magnificent piece of land at Jardim Botanico where the institute resides now. Needless to say, I enjoyed the beaches during sunny weekends. In particular the Barra da Tijuca, which is nowadays a part of the metropolis, and densely covered with buildings, was just nature at that time and a marvelous beach.

The reason why I went to Zürich was that Hans Föllmer had accepted a professorship at ETH and had an assistant position that he offered to me. As you know, in order to be able to apply for a professorship in Germany, a Habilitation was a necessary requirement at that time, in addition to a PhD. Zürich with its very active life in research in statistics, actuarial sciences and stochastics in general was a rather attractive place for such an endeavor. In January 1978, I submitted my Habilitationsschrift and the procedure was finished in the summer, after I had delivered a formal lecture to the ETH faculty. I had applied for the vacant position in probability at the Department of Mathematical Stochastics in Freiburg, accepted the offer which I got and I started to teach in October after the Dean had asked me to do so. It went so fast that the certificate of appointment, which had to be signed by the prime minister of the state, arrived only six weeks later. At that age you do not worry too much about legal risks. After four years of teaching one was allowed to ask for the first one-semester sabbatical, which I wanted to spend abroad. I had met Ingram Olkin who had been visiting our department. When I asked him if I could go to his department at Stanford, he agreed and so I spent the winter of 1982/83 in California. David Siegmund also helped greatly to make this visit successful.

During your PhD and early career, you were interested in dynamical systems and invariance principles. Would you like to describe your early results?

Ergodic theory attracted a lot of interest at that time. There was the famous Kolmogorov conjecture that entropy was supposed to be a complete isomorphism invariant. This conjecture was confirmed for Bernoulli shifts by Don Ornstein in a series of papers starting in 1970. As a student of Konrad Jacobs you were exposed to these developments and there was some probability that he would propose a topic from ergodic theory for your PhD. More technically, it would be a topic on measure-theoretic dynamical systems or its topological equivalent, but not on differential dynamical systems. What he proposed to me was a question about strict ergodicity in continuous-time dynamical systems. Strict ergodicity for a system with a topological structure means that it carries a unique transformation-invariant probability measure and the question was about conditions that guarantee this uniqueness. An equivalent way to characterize strict ergodicity is to require that, for any continuous function, if you take time-means along orbits of the transformation, you have to get in the limit the space mean with respect to the invariant measure. Jacobs had obtained a partial result by showing that a system that is weakly mixing is isomorphic to a strictly ergodic one. I was strongly motivated to work on this problem since in my diploma thesis I had already studied strictly ergodic systems in discrete time, namely those generated by Toeplitz sequences on the Bernoulli space of 0's and 1's. In order to tackle this problem I first proved a generator theorem that allowed me to embed aperiodic systems into a space of Lipschitz functions with the shift transformation. Once the system was embedded in a space with nice properties, in joint work with Manfred Denker we showed that ergodicity is enough to guarantee the isomorphism to a strictly ergodic one. This is the definite result since the opposite direction is evident. Any strictly ergodic system is ergodic. I continued for some time to look into the interplay between measure-theoretic and topological properties of dynamical systems by studying generators and in particular topological entropy, but then I got interested in a rather different object, namely random sheets, a two-dimensional analogue of the classical random walk. The study of random sheets led me to stochastic dependence structures, since the restriction that the values of a sheet can differ only by +1 or −1 between any neighboring points on the two-dimensional grid, entails a mixing structure for arrays of random variables once one uses the counting measure defined by the number of sheets. In the combinatorial part of this study I rediscovered the transfer matrix method, without being aware that this had been a fundamental tool used by Elliott Lieb in his paper on the six-vertex ice model. Random sheets are in fact isomorphic to six-vertex ice. Invariance principles and later strong approximation results under a variety of dependence assumptions were then an area that developed naturally out of this example. For several years, while I was an active member of the Probability in Banach Spaces group, I investigated the subtle balance between dependence assumptions and moment conditions in the context of probabilistic limit theorems. I tried to understand the borderline until which the limit theorems or strong approximation results are valid. I organized several conferences at the Oberwolfach Institute. The conference titles 'Dependence in Probability and Statistics' and 'High-Dimensional Probability' are in use still now.

How did you get interested in mathematical finance?

Well, with the family and education background that I indicated at the beginning, it is not really surprising that I was always interested in banking and finance in general and watched with increasing interest what was going on in the eighties after the Harrison and Pliska papers had appeared. Actually conversations with Michael Harrison played a key role. In 1987 I spent seven months, my second sabbatical, at UCSD. Murray Rosenblatt and Ron Getoor were my hosts in La Jolla. During that stay I went to Stanford for one week and had the opportunity for several discussions with Michael Harrison. On one occasion he gave me some handwritten notes on the derivation of the Black-Scholes formula. I was so impressed by the elegance with which he explained the subject that I studied further papers. The first conference on mathematical finance which I attended was held in July 1989 at Cornell, organized by David Heath and Bob Jarrow. It was the only occasion where I had a chance to listen to Fischer Black. Although modest, even shy, he was already at that time a sort of saint in the finance community. It was also during that conference that Stan Pliska proposed that we should start a journal for Black-Scholes theory which after a long discussion finally got the most natural name, namely *Mathematical Finance*. As early as August 1992 I myself organized a conference with the title *Mathematical Finance* at the Oberwolfach Institute, with Darrell Duffie and Stan Pliska as co-organizers. This Oberwolfach series was continued in later years by Hans Föllmer. Concerning your question allow me another remark. Your personal background always plays an important role in what you are interested in. I am convinced that Louis Bachelier's ingenious thesis, which his referees—none less than Appell, Poincaré and Boussinesq—classified in their report as being 'on a subject that is rather far removed from those usually treated by our candidates', could never have been written without his background in business and his profound knowledge about financial markets, which he had acquired through this.

You were one of the pioneers in the application of Lévy processes in finance. How did you start working in this direction?

This began with statistics and data analysis. In 1987 we had started an interdisciplinary seminar in our University where the people involved in statistics and probability met. The formal foundation of the Freiburg Center for Data Analysis and Modeling (FDM), which grew out of this seminar, had to wait until 1994. The talks in this seminar inspired me or, rather, put some pressure on me to contribute and to start with data analysis myself. It was clear that I wanted to analyze financial markets. I acquired daily stock price data from a data bank service. After consulting with Stan Pliska, I got access to some of the relevant literature and one of my talented students, Ulrich Keller, agreed to do the work on the computer. The results were ready to be presented in January 1994 in Paris at the first conference where our newly established European network met. Jean Jacod acted as head during the first four years of funding for this network. That was the reason why we met in Paris. Ole Barndorff-Nielsen was in the audience and when he saw our graphs of empirical return distributions from German stock price data he commented '*This looks very much like hyperbolic distributions*'. So, back at home, I read the papers on this class of probability distributions, which Ole had introduced in the seventies in the context of the so-called sand project. In this interdisciplinary project, people in Aarhus studied the drift of sand under the impact of waves and this sort of distribution turned out to be useful for the statistical description of the particle size. Preben Blaesild was kind enough to send us a program for parameter estimation. Once we had parameters, the question that I posed myself was: Is there a model such that the return distributions from this model—let us say at time 1—are hyperbolic? Being used to the Black-Scholes- or rather the Samuelson—setting, I tried it for several months with diffusion processes. To illustrate how difficult it sometimes is to abandon the thinking in which you were trained, let me tell you the following story. I presented the empirical findings again, but this time including parameter estimates, during a conference in Cortona in Italy in May, to which a good part of the then-élite of mathematical finance had been invited by Wolfgang Runggaldier. In the discussion following my talk one of the prominent members of the community commented '*This looks very interesting, but you will never be able to develop a suitable theory based on these distributions*'. I had gone through other difficulties and it would have taken more than that to discourage me. At some point I realized that diffusions do not and cannot work to reach the goal. With the exception of the simplest cases, one does not even know the distribution which is produced by a diffusion on a given time horizon. A diffusion equation or, equivalently, a stochastic exponential was just not the right starting point to get what I was looking for. Something more radical had to be done. Hyperbolic distributions are infinitely divisible and thus they generate Lévy processes, where you get the generating distribution back at time one. In Bauer's book in my student days these processes had just been called processes with stationary and independent increments. The name Lévy process came only later. Since we used log-returns for the statistics, one had to take the ordinary exponential of the corresponding Lévy process instead of a stochastic exponential, in order to get exactly the distribution that comes from the data. In the classical case of Brownian motion the difference between the ordinary and the stochastic exponential is not that crucial, but for jump-type Lévy processes it is. It was so simple once I had seen this point. The joint paper with Ulrich on the hyperbolic Lévy model appeared in October 1994 as No. 1 of the newly created FDM-preprint series and was published in the first volume of *Bernoulli* a year later. Let me finish by making one more remark. That meeting of the EU network in Paris helped showing the many avenues of research in this field, and many network members, none of whom had been involved in financial models before, were inspired by my talk to start working in stochastic finance. DYNSTOCH, as we called it later for the second funding period, headed by Michael Soerensen, became a very successful working platform and the group has continued with an annual meeting for more than twenty years already.

Later you considered Lévy processes in the context of term structure problems. Could you tell us about this?

Once we had introduced the exponential Lévy model for equity, it was rather natural to look into fixed income modeling. From the point of view of mathematics, interest rate models are a priori more challenging, because instead of the dynamics of a single value or a finite-dimensional vector, one has to model the movements of an infinite-dimensional object, i.e. the whole term structure as a function. We started with the instantaneous forward rate or Heath-Jarrow-Morton approach. In Sebastian Raible I had another talented student for this project. The paper on term-structure models appeared in the FDM preprint series as the output of my sabbatical in the summer of 1996, but was printed in '*Mathematical Finance*' only in January 1999 after a relatively slow refereeing process. By the way, Sebastian mentioned later to me that during his time as a PhD student he had had a number of very useful discussions on analytical issues with you, Jan. Maybe you contributed to this project without my becoming aware of it.

The next challenge in the program to develop a Lévy-driven finance theory was the Libor market model. This was done with Fehmi Özkan. Let me tell you how the breakthrough came. I had been scheduled to give a series of talks at a conference for central bankers in Mexico City. Only a few days before my departure, the BIS informed me that the conference had to be relocated to Rio de Janeiro six months later. Suddenly I had a free week. What a wonderful gift. At the end of that week in which I devoted all my time to working with Fehmi, we had the Lévy Libor model and in addition we had understood more, namely that it is more appealing to model the forward price processes instead of the Libor rates. With the former as basic objects, the analysis becomes much simpler as the quantity modeled is—up to the norming constant—also the density process needed in the backward induction to go from one forward measure to the next. One can forget about any approximation like the frozen drift and at the same time one can obtain negative rates in a natural way. Not only in the current market environment where the AAA Euro yield curve went negative up to ten years—this happened intraday on June 14, 2016, for the first time—but because of its superb calibration properties and the easier analysis mentioned before, the forward process approach should be the model of your choice. The quants in the banks are struggling currently with negative rates, since what is implemented there are models that produce positive rates only.

The Libor and the forward process model were by no means the end of the program for term-structure modeling. In the work with Wolfgang Kluge we included defaultable fixed-income instruments. This was extended with Zorana Grbac to a market model with a fully-fledged credit rating system. Currently we are working on Lévy-driven multiple curve models, which became necessary after the 2007–2009 financial crisis.

In the last decade, you did a lot of work on two-price economies. What was your motivation?

This came out of the recent financial crisis directly. In 2008 many of the major banks got into trouble. As a consequence their credit rating was downgraded, but surprisingly they reported windfall profits of the order of magnitude of hundreds of millions of US dollars in that year. Looking at their balance sheets it became clear where these profits came from. To finance their activities these banks issue bonds. The market value of this debt typically decreases once the issuer is downgraded. Since it is a position on the liability side of the balance sheet of such a bank, a decrease of its liabilities—other positions unchanged—produces profits. Following this line of thought further, it means that an issuer of bonds will always report large profits once its bankruptcy is looming. In discussions with Dilip Madan we came to the conclusion that something is wrong with the valuation of instruments and the current mark-to-market accounting rules. Entries in a balance sheet should actually give a realistic picture of the financial status of a corporation. This goal can be achieved if a financial instrument is marked with the bid or lower price on the asset side and the ask or higher price on the liability side. The price of an instrument depends then on the direction of the trade. How one can get two prices via Knightian uncertainty is a different story and the smart part of the project. The two-price theory has now reached a maturity where it applies to all markets in finance and not only to illiquid ones. By means of some portion of the spread it even provides an excellent measure for the liquidity of the corresponding market.

Would you like to describe the early days of mathematical finance? How has the field evolved over the years?

Well, it depends on what you mean by 'early days'. I can only comment on what I recollect from roughly 1990 onwards and from a European perspective. Actually mathematical finance started late on this side of the Atlantic. It was a rather small community. One has to be aware that the first exchange for derivatives in Germany, the DTB, started in 1990 only. Before that, derivatives were traded only OTC in this country. Some years later in order to motivate students for the subject, I even organised an excursion to the exchange in Frankfurt with those who attended my course. In 1998 the DTB mutated into the Eurex, which is now one of the biggest derivatives exchanges worldwide. I mentioned the Oberwolfach conference in 1992 earlier. What had certainly had an impact was a series of conferences which Wolfgang Runggaldier organized in Italy, first in 1992 in Erice in Sicily and later in Cortona. The major representatives of the field at that time where colleagues coming from areas where diffusions were used. To illustrate how difficult it was to present something that was not mainstream, let me tell you what happened with my first paper together with Jean Jacod. In 1995 I sent it to the freshly appointed editor of the new journal *Finance and Stochastics*. It is the paper in which we show that exponential Lévy models have a very rich structure in the sense that, under slight regularity assumptions, the range of values of options, using all possible equivalent martingale measures for the valuation, covers the whole no-arbitrage interval excluding the boundaries. I got the manuscript back within a few days with the comment that this approach is useless, since the model does not have a unique pricing measure. Do not misunderstand that I want to criticize Dieter Sondermann's work as editor. He did a marvelous job by establishing a journal which is now one of our best. I think his attitude had to do with the unfortunate choice of the notion of 'incompleteness' which insinuates that something is wrong with this model. In my opinion it is the opposite. On the basis of market data one can pick the most appropriate pricing operator via calibration. It is the same as extracting implied volatilities from price data in the classical diffusion world, but the parameter space is much richer. In any case, I had a hard time convincing the editor that the paper is worth publishing. It appeared in the first volume of the journal. I can give you another example of how strenuous it was not to swim with the current of diffusion models. I had initially submitted the paper with Keller and Prause, which later appeared in the *Journal of Business*, to another prominent journal whose editor at that time had been a well-known name in financial time series. He rejected it right away and wrote back that enough has been published already on Lévy models in finance. I think it is no exaggeration to say that, in 1997, the published papers on Lévy dynamics could be counted on the fingers of two hands...

Would you like to tell us about the perception of Mathematical Finance in the probability community in the early days and later?

First let me say that probability was not the only area from which the contributors came to the subject. Some had a background in functional analysis or PDE theory or even physics, but since the Harrison-Pliska papers had appeared, probability theory looked like the most natural starting point. In my opinion, academic positions in mathematical finance could be located mainly in departments of probability theory or statistics. A number of colleagues became interested in this applied area, but many gave up again after realizing that probability theory alone was not enough to do interesting work. Personally, I did not get any recognition from the probability community in Germany for working in mathematical finance. It was the opposite. There were several interesting vacancies for which I applied. The establishment in Germany used its position as referee to eliminate me from the list of candidates. I experienced a similarly hostile attitude at faculty level too. It is standard practice, and our faculty in Freiburg is definitely no exception to this rule, that in response to an offer from another University, the Dean works towards an increase of your compensation in order to improve the chance that you will stay. That did not happen when, due to increasing recognition from my colleagues abroad, I got an offer from a foreign university. Several years later I had a chance to respond in kind, when the faculty urged me to take over as Dean again. I had already served in this position back in the eighties. That had been a rather interesting and busy term, since it was during my term that the University of Freiburg initiated a curriculum in computer science. As Dean I became chairman of the recruitment committee for the new field. Until the establishment of the Faculty of Engineering, the new colleagues remained part of the Mathematical Institute for a number of years. Apparently the faculty had not been dissatisfied with the way I handled the job, but when, years later, they urged me to serve as Dean of Mathematics and Physics, I had to tell them that I needed to focus on research instead, in order to be able to reach a level of recognition where I could be treated in the same way as other colleagues. When you work with success in a very attractive area, it can also create some jealousy in the community and amongst your colleagues in the faculty. Meanwhile, mathematical finance has become a well-established part of a number of faculties in this country and maybe I contributed to this state of affairs.

Which persons have influenced your scientific career the most? And which have been very important companions over the years?

Without any doubt the most important person, as far as my career in academia is concerned, was Konrad Jacobs. Without his encouragement and guidance I would probably have left the academic world with a PhD or even earlier with a diploma degree. Konrad was not a supervisor in the sense that you would or could see him regularly and discuss the problems where you got stuck. Our contact with him was very different from that. His seminars were very broad and inspiring. When he became convinced that you were the right candidate, he guided you into a certain area by indicating the right papers. In some cases he identified a precise question on which you could work. From then on he expected that, without further discussions, you would submit a printable paper. From his way of doing mathematics I learned that you should look beyond the tip of your nose. In spite of the considerable distance between professor and student at the beginning, we became lifelong friends after my PhD. Unfortunately he passed away last year.

Albert Shiryaev visited our department for two months in the winter term 1981/82 and in a series of lectures gave a survey of semimartingale theory. It was through these lectures that I grew fascinated by this level of stochastic processes. This is somewhat strange because the architect of modern stochastic analysis, Paul-André Meyer, was in Strasbourg, just 'around the corner' from Freiburg and I had a chance to talk to him on a number of occasions. Let me share with you one of his statements. I was teaching Stochastic Processes that term, essentially using Michel Métivier's book on semimartingales and I asked Meyer: '*What do you think is the most elegant way to introduce the stochastic integral?*' He answered:

'You can do whatever you like, you will always need half of the semester for the integral.' Before Shiryaev's visit I had only become acquainted with semimartingales in terms of particular results and had not seen that they had reached the state of a mature theory. As a consequence of this visit, Albert and I became close friends. He visited our department quite often during the last 35 years. Some stays lasted several months. He must have spent a total of nearly two years in Freiburg in that time span. We had many inspiring discussions on stochastic analysis and later on financial mathematics and this is still going on.

During the last ten years I developed a very close cooperation and friendship with Dilip Madan. We had met at several conferences as we were both working on Lévy-driven models although using different processes. At a conference in Venice, while enjoying a magnificent sea bream for dinner, we decided to work together. Since then a considerable number of joint papers has appeared, some of them with further co-authors. I will mention only Hélyette Geman, Marc Yor and Wim Schoutens. To discuss with Dilip is a wonderful experience. He has extremely wide knowledge in finance. Another colleague with whom I had a very fruitful cooperation is Jean Jacod. You can only be very impressed by his speed of thinking and working.

Last but not least, let me emphasize that I had quite a number of extraordinarily talented PhD students including both of you. This local strength was vital for what we achieved on the basis of Lévy processes and more general ones.

How was the Bachelier Finance Society created? What are your impressions after serving for almost a decade as its Executive Secretary?

The Bachelier Finance Society was created on June 4, 1996, in Aarhus. Jorgen Nielsen had organized a conference at the University of Aarhus and that day he asked twelve participants to join him for dinner at Kellers Gaard. The dinner took place in a special room in the first floor of the restaurant. The building where Kellers Gaard was operating at that time no longer exists. Dinner had not been the main reason for bringing this group of people together. Jorgen proposed to form an association for mathematical finance and presented statutes which he had already drawn up. He became the first President of the Society and Stan Pliska its Executive Secretary. At the beginning of 2001, Hélyette Geman, who had by then become the third President of the Society, approached me asking if I would be willing to take over from Stan. I agreed and served as Executive Secretary for the next ten years. What kept me busy during that time were the membership administration including the journal subscriptions, elections, the work of committees, the congresses which we hold every second year and a whole variety of other issues which kept the Society going. Even after deleting insignificant e-mails, I still had several thousand e-mails in various folders. Nowadays these tasks are spread across more shoulders than during my time. I would not like to have missed the many contacts which I had due to that position. I am proud that I was able to contribute to the success of our Society and hope it will help to promote the field of mathematical finance for many years to come.

You have served as Co-Editor for Mathematical Finance and as Associate Editor for several other journals. How much time do these activities take up? How does it differ from journal to journal?

With respect to the scientific quality of journals, there is little doubt that we, the people who do the research, are best qualified to serve as editors, associate editors or referees and to select what is accepted. This is time-consuming. Sometimes one receives contradictory opinions from referees. This means that you have to read the article yourself in order to come to a decision. How much time editorship consumes depends largely on the reputation of the journal. The latter determines the number of submissions which you have to manage. This varies a lot from journal to journal. Top journals easily receive ten times as many articles as can be printed within a limited page budget.

What is your opinion of the peer-review system, which is often criticized? What do you think about open access journals?

The peer-review system has its weaknesses. It requires a lot of integrity from the persons who act from a position of anonymity. On the other hand, criticism is very often justified and helps to improve a submitted article. Without the anonymity of the referee there would be less freedom to express criticism. Therefore I do not know a system that would guarantee more fairness. If an article is rejected, you are free—after taking care of justified criticism you received—to submit it to another journal. The question of the integrity of referees is much more severe in the case of applications for research funds at your science foundation and it is of fundamental importance when somebody is asked to rank candidates for a vacant position. In such a situation the applicants might not easily have another chance at that organization or university. In the case of a vacant position I therefore very much favour the rules applied in the United States where up to ten letters on a candidate are collected, while in Germany decisions are based on two or three opinions, or in the worst case on the opinion of just one super-referee. This gives too much power to a single person. I do not know any person who would possess such an overview as to justify this sort of dictatorship. The latter is not to the benefit of the community, but carries a high risk for misuse of a position. Of course as an editor of a journal you are usually already happy if two reports become available. More is not possible for this purpose.

Concerning the second question it is clear that our way of publishing will change fundamentally in the future. The traditional journals have become so expensive that in many departments almost the entire available budget has to be used for subscription fees. Other technical means for the dissemination of research results are already available.

What does it take for a successful academic career?

What is most important is that you like what you do. The frustrations that will inevitably occur can only be overcome if you are highly motivated. Another aspect that I want to point out is enthusiasm for teaching. An academic career should not be reduced to a career in research only. Teaching with enthusiasm attracts excellent students and allows you to establish a strong working group.

What would be your advice for a young financial mathematician?

Be flexible. Get interested in new developments. What you learned during your studies will no longer be sufficient for your job in ten or in twenty years.

You were involved in industrial projects and have been in contact with financial regulators. How was that experience?

Working in industrial projects and with the regulators widens your horizon enormously. You get a much clearer picture about what their needs are. You will realize that data issues, statistical aspects in calibration and efficient numerical procedures are as important as providing a mathematical theory.

During the recent financial crisis, mathematical finance was heavily criticized for its role in it, notably Michel Rocard (a former French Prime Minister) said that "mathematicians are responsible for crimes against humanity". What is your opinion on this?

Marc Yor has published an adequate answer to Mr. Rocard's allegations. Therefore I do not want to repeat his arguments here. That would need too much space. Nevertheless let me add that we should use state-of-the-art models in our work and not stick to methods that were appropriate thirty years ago. On quite a number of occasions I have heard people in the banks or in the regulatory agencies saying that the methods applied should be simple. First of all, what you consider to be simple is relative to your level of knowledge. Secondly, can you imagine that diagnosis and therapy methods should primarily be simple if you have a health problem? You will without any doubt expect that the doctors will try to help you by applying state-of-the-art methods. Why should this be different when millions or billions are invested?

What should financial mathematicians do differently in order to avoid another crisis?

In the past, mathematicians usually were far away from the C-level hierarchy positions where the key decisions, which lead to a crisis like the last one, are made. The situation is somewhat better nowadays. More direct access to top management should be used to pinpoint risks that have been identified. Some of the banks which were hit the hardest did not even have any mathematical expertise to understand the risks of those financial instruments in which they had heavily invested.

What is your opinion about the future of mathematical finance?

Mathematical finance is an attractive and challenging field of mathematics and it is applied on a large scale in a rapidly developing industry, thus there is an excellent job market. What more do you want? I cannot see any reason why it should not have an excellent future in academia as well as in the industry.

Part I
Flexible Levy-based Models

Tail Behaviour and Tail Dependence of Generalized Hyperbolic Distributions

Ernst August v. Hammerstein

Abstract Generalized hyperbolic distributions have been well established in finance during the last two decades. However, their application, in particular the computation of distribution functions and quantiles, is numerically demanding. Moreover, they are, in general, not stable under convolution which makes the computation of quantiles in factor models driven by these distributions even more complicated. In the first part of the present paper, we take a closer look at the tail behaviour of univariate generalized hyperbolic distributions and their convolutions and provide asymptotic formulas for the quantile functions that allow for an approximate calculation of quantiles for very small resp. large probabilities. Using the latter, we then analyze the dependence structure of multivariate generalized hyperbolic distributions. In particular, we concentrate on the implied copula and determine its tail dependence coefficients. Our main result states that the generalized hyperbolic copula can only attain the two extremal values 0 or 1 for the latter, that is, it is either tail independent or completely dependent. We provide necessary conditions for each case to occur as well as a simpler criterion for tail independence. Possible limit distributions of the generalized hyperbolic family are also included in our investigations.

Keywords Normal mean-variance mixture · GH distribution · Convolution tails · Tail dependence · Copula

1 Introduction

Almost forty years ago, generalized hyperbolic distributions (henceforth GH) have been introduced in [5] in connection with the modeling of aeolian sand deposits and dune movements. Eighteen years later, they were introduced in finance by [14] where the hyperbolic subclass was used as a more realistic model for stock returns. The

E.A.v. Hammerstein (✉)
Department of Quantitative Finance, University of Freiburg,
Platz der Alten Synagoge, 79098 Freiburg im Breisgau, Germany
e-mail: ernst.august.hammerstein@finance.uni-freiburg.de

© Springer International Publishing Switzerland 2016

J. Kallsen and A. Papapantoleon (eds.), *Advanced Modelling in Mathematical Finance*,
Springer Proceedings in Mathematics & Statistics 189,
DOI 10.1007/978-3-319-45875-5_1

normal inverse Gaussian subclass followed shortly after in [6], and the general case was then considered in [11, 18, 19]. Starting from early applications to stock price modeling and option pricing, GH distributions have been successfully used in various fields of finance during the last two decades, for example, in interest rate theory and the pricing of interest rate derivatives (see [15, 16, 20]), currency markets [17], and portfolio credit risk models [12].

There are mainly two reasons for the widespread applicability of GH distributions: First, they are infinitely divisible and therefore allow to make use of the extensive theory of Lévy processes in continuous-time models based on them. The second is their convenient tail behaviour. On the one hand, the tails of the GH probability densities have considerably more mass than the ones of normal distributions. This means that, for example, extreme price movements which are observed more often nowadays are much more likely under the assumption of GH distributed asset returns, whereas such events are severely underestimated in models based on the normal distribution. On the other hand, the GH densities asymptotically still decay exponentially and therefore possess a moment generating function on some non-degenerate interval around the origin. This is an inevitable requirement for the construction of financial models of exponential Lévy type since stock or bond price models having infinite expectations are obviously not very realistic. Moreover, the existence of a moment generating function allows for an easy way to explicitly determine a risk-neutral measure for derivative pricing via an Esscher transform.

Despite these advantages, GH distributions are, to some extent, computationally demanding in practical applications nevertheless because their distribution functions can neither be given in closed form, nor does there exist a well-known and quickly convergent series expansion for them. Therefore, computing values of the distribution and quantile functions is challenging and can only be done numerically. These procedures naturally become less stable and reliable if the arguments of the distribution function are extremely large resp. the probabilities inserted into the quantile function are very close to zero or one. The latter difficulty can occur in risk management, especially in credit risk, where one has to calculate the value at risk or expected shortfall for probabilities beyond 99 % and to deal with small default probabilities. In the first part of this paper, we analyze the tail behaviour of univariate GH distributions in detail and derive asymptotic formulas for the distribution and quantile functions of GH distributions and their convolutions that enable a simple calculation of approximate values of the latter.

The other major topic we are concerned with in the second part of the paper is the dependence structure of multivariate GH distributions. In practice, correlation still seems to be the predominant dependence measure although it only provides a complete characterization of dependencies in case of a multivariate normal distribution. The dependence structure of the latter is indeed linear and fully described by the corresponding correlation matrix. However, the picture changes significantly if one departs from the normal world. In general, zero correlation does not imply independence, and maximal dependence (co- or countermonotonicity) can already occur for correlations with absolute value strictly smaller than one. We will show that the latter also holds for multivariate GH distributions.

Another dependence concept that has gained increasing attention, especially in credit portfolio modeling, is tail dependence. Roughly speaking, the tail dependence coefficients give the asymptotic probabilities of joint extremal events which may be, for example, multiple defaults in a credit portfolio within the same time interval or severe losses of different stocks at the same trading day. Tail dependence is solely determined by the implied copula which is inherent in every multivariate distribution and—in contrast to correlation—completely characterizes the dependence structure of the latter. The implied copula of a multivariate normal distribution is known to be tail independent, that is, extreme marginal outcomes occur (asymptotically) independent from each other. In credit and insurance risk modeling, this property often is not realistic, therefore dependence models in this area are usually based on copulas possessing tail dependence coefficients greater than zero like the t- or grouped t-copula (see [10]). To see whether the implied copula of a multivariate GH distribution provides a suitable model in this context, we determine the potential range of its tail dependence coefficients. It turns out that only the two extremal values 0 or 1 can be obtained, implying that the GH copula either is tail independent or completely dependent. For both cases, we derive explicit conditions on the GH parameters as well as a simpler criterion for tail independence.

The paper is structured as follows: In the next section, we recall the definition of univariate GH distributions as normal mean-variance mixtures, state possible limit distributions and provide some useful facts on normal mean-variance mixtures in general as well as the mixing generalized inverse Gaussian distributions which will be required later on. Section 3 then is devoted to a thorough study of the tail behaviour of univariate GH distributions and their convolutions. Multivariate GH distributions and their weak limits are introduced in Sect. 4, where also the most important properties for the subsequent analysis of its dependence structure are discussed. The latter is done in Sect. 5. Section 6 concludes with some final remarks on possible generalizations of multivariate GH distributions that have been introduced in the recent literature.

2 Univariate GIG and GH Distributions and Some of Their Limits

Generalized hyperbolic distributions can be defined as normal mean-variance mixtures where the mixing distribution is a generalized inverse Gaussian (GIG) one. For the convenience of the reader, we first define normal mean-variance mixtures in general and provide some of their properties which might be of their own interest.

Definition 1 A real valued random variable X is said to have a *normal mean-variance mixture distribution* if

$$X \stackrel{d}{=} \mu + \beta Z + \sqrt{Z}W,$$

where $\mu, \beta \in \mathbb{R}$, $W \sim N(0,1)$ and $Z \sim G$ is a real-valued, non-negative random variable which is independent of W. Equivalently, a probability measure F on $(\mathbb{R}, \mathscr{B})$ is said to be a normal mean-variance mixture if

$$F(\mathrm{d}x) = \int_{\mathbb{R}_+} N(\mu + \beta y, y)(\mathrm{d}x)\, G(\mathrm{d}y),$$

where the mixing distribution G is a probability measure on $(\mathbb{R}_+, \mathscr{B}_+)$. We shall use the short hand notation $F = N(\mu + \beta y, y) \circ G$.

The most important facts about normal mean-variance mixtures are summarized in the following lemma. It especially shows that properties like stability under convolutions and weak convergence are inherited from the mixing distributions. A detailed proof can be found in [27, Lemmas 1.6 and 1.7].

Lemma 1 *Let $\mathbb{G}$ be a class of probability distributions on $(\mathbb{R}_+, \mathscr{B}_+)$ and suppose $G, G_1, G_2 \in \mathbb{G}$.*

(a) If G possesses a moment generating function $M_G(u) = \int_{\mathbb{R}_+} e^{ux} G(dx)$ on some open interval (a,b) with $a < 0 < b$, then $F = N(\mu + \beta y, y) \circ G$ also possesses a moment generating function, and $M_F(u) = e^{\mu u} M_G(\frac{u^2}{2} + \beta u)$, $a < \frac{u^2}{2} + \beta u < b$.

(b) If $G = G_1 * G_2 \in \mathbb{G}$, then $\big(N(\mu_1 + \beta y, y) \circ G_1\big) * \big(N(\mu_2 + \beta y, y) \circ G_2\big) = N(\mu_1 + \mu_2 + \beta y, y) \circ G$.

(c) If $(\mu_n)_{n\ge1}$ and $(\beta_n)_{n\ge1}$ are convergent sequences of real numbers with finite limits $\mu, \beta < \infty$, and $(G_n)_{n\ge1}$ is a weakly convergent sequence of mixing distributions with $G_n \xrightarrow{w} G$, then $N(\mu_n + \beta_n y, y) \circ G_n \xrightarrow{w} N(\mu + \beta y, y) \circ G$.

We now leave the general case and concentrate on a specific class $\mathbb{G}$ of mixing distributions, namely the generalized inverse Gaussian one mentioned above. This class was introduced more than 60 years ago (one of the first papers where its densities are mentioned is [26]) and rediscovered in [5, 40, 41]. An extensive survey with statistical applications can be found in [29]. The density of a GIG distribution is as follows:

$$d_{GIG(\lambda,\delta,\gamma)}(x) = \left(\frac{\gamma}{\delta}\right)^{\lambda} \frac{1}{2K_\lambda(\delta\gamma)}\, x^{\lambda-1} e^{-\frac{1}{2}(\delta^2 x^{-1} + \gamma^2 x)}\, \mathbb{1}_{(0,\infty)}(x), \tag{1}$$

where $K_\lambda(x)$ denotes the modified Bessel function of third kind with index λ. Permitted parameters are

$$\begin{aligned} &\delta \ge 0,\ \gamma > 0, \quad \text{if } \lambda > 0,\\ &\delta > 0,\ \gamma > 0, \quad \text{if } \lambda = 0,\\ &\delta > 0,\ \gamma \ge 0, \quad \text{if } \lambda < 0. \end{aligned}$$

Parametrizations with $\delta = 0$ or $\gamma = 0$ have to be understood as limiting cases. Using the asymptotic behaviour (cf. [2, Formula 9.6.8])

$$K_\lambda(x) \sim \frac{\Gamma(|\lambda|)}{2}\left(\frac{x}{2}\right)^{-|\lambda|}, \qquad x \downarrow 0,\ \lambda \neq 0, \tag{2}$$

of the Bessel functions where $\Gamma(x)$ denotes the Gamma function, the limit for $\lambda > 0$ is obtained as

$$\lim_{\delta \to 0} d_{GIG(\lambda,\delta,\gamma)}(x) = \left(\frac{\gamma^2}{2}\right)^{\lambda} \frac{x^{\lambda-1}}{\Gamma(\lambda)} e^{-\frac{\gamma^2}{2}x}\, \mathbb{1}_{(0,\infty)}(x) = d_{G(\lambda,\frac{\gamma^2}{2})}(x) \tag{3}$$

which is nothing but the density of a Gamma distribution $G(\lambda, \frac{\gamma^2}{2})$ with shape parameter λ and scale parameter $\frac{\gamma^2}{2}$. For $\lambda < 0$, we arrive at

$$\lim_{\gamma \to 0} d_{GIG(\lambda,\delta,\gamma)}(x) = \left(\frac{2}{\delta^2}\right)^{\lambda} \frac{x^{\lambda-1}}{\Gamma(-\lambda)} e^{-\frac{\delta^2}{2x}}\, \mathbb{1}_{(0,\infty)}(x) = d_{iG(\lambda,\frac{\delta^2}{2})}(x) \tag{4}$$

which equals the density of an inverse Gamma distribution $iG(\lambda, \frac{\delta^2}{2})$.

For $|\lambda| = \frac{1}{2}$, the Bessel function $K_\lambda(x)$ can be given in explicit form: We have $K_{-\frac{1}{2}}(x) = K_{\frac{1}{2}}(x) = \sqrt{\frac{\pi}{2x}}\, e^{-x}$ (cf. [2, Formula (9.6.6)] and [43, Eq. (8) on p. 79]), thus

$$d_{GIG(-\frac{1}{2},\delta,\gamma)}(x) = \frac{\delta}{\sqrt{2\pi x^3}} e^{-\frac{1}{2x}(\gamma x - \delta)^2}\, \mathbb{1}_{(0,\infty)}(x) \tag{5}$$

which is the density of an inverse Gaussian distribution $IG(\delta, \gamma)$, showing that the GIG distributions are, in fact, a natural extension of this subclass.

A distribution G on $(\mathbb{R}_+, \mathscr{B}_+)$ is completely characterized by its Laplace transform $\mathscr{L}_G(u) = \int_{\mathbb{R}_+} e^{-ux} G(dx)$ from which many properties of G can be easily derived. For the GIG class, we obtain the following representations (see [27, Proposition 1.9] for a proof).

Proposition 1 *The Laplace transforms of GIG distributions are given by*

$$\mathscr{L}_{GIG(\lambda,\delta,\gamma)}(u) = \left(\frac{\gamma}{\sqrt{\gamma^2 + 2u}}\right)^{\lambda} \frac{K_\lambda\big(\delta\sqrt{\gamma^2 + 2u}\big)}{K_\lambda(\delta\gamma)}, \qquad \delta, \gamma > 0,$$

$$\mathscr{L}_{G(\lambda,\frac{\gamma^2}{2})}(u) = \left(1 + \frac{2u}{\gamma^2}\right)^{-\lambda}, \qquad \lambda > 0,$$

$$\mathscr{L}_{iG(\lambda,\frac{\delta^2}{2})}(u) = \left(\frac{2}{\delta\sqrt{2u}}\right)^{\lambda} \frac{2K_\lambda\big(\delta\sqrt{2u}\big)}{\Gamma(-\lambda)}, \qquad \lambda < 0.$$

With help of the preceding proposition and the fact that $\mathscr{L}_{G_1}(u)\mathscr{L}_{G_2}(u) = \mathscr{L}_G(u)$ implies $G_1 * G_2 = G$, one can derive the subsequent convolution properties of GIG distributions:

$$\begin{aligned}
&\text{(a)}\quad IG(\delta_1,\gamma) * IG(\delta_2,\gamma) = IG(\delta_1+\delta_2,\gamma), \\
&\text{(b)}\quad IG(\delta_1,\gamma) * GIG\big(\tfrac{1}{2},\delta_2,\gamma\big) = GIG\big(\tfrac{1}{2},\delta_1+\delta_2,\gamma\big), \\
&\text{(c)}\quad GIG(-\lambda,\delta,\gamma) * G\big(\lambda,\tfrac{\gamma^2}{2}\big) = GIG(\lambda,\delta,\gamma), \qquad \lambda>0, \\
&\text{(d)}\quad G\big(\lambda_1,\tfrac{\gamma^2}{2}\big) * G\big(\lambda_2,\tfrac{\gamma^2}{2}\big) = G\big(\lambda_1+\lambda_2,\tfrac{\gamma^2}{2}\big), \qquad \lambda_1,\lambda_2>0.
\end{aligned} \tag{6}$$

Further observe that all $GIG(\lambda,\delta,\gamma)$-distributions with $\gamma > 0$ decay at an exponential rate for $x \to \infty$, so they possess moments of arbitrary order, and the moment generating functions are given by

$$M_{GIG(\lambda,\delta,\gamma)}(u) = \int_0^\infty e^{ux}\, d_{GIG(\lambda,\delta,\gamma)}(x)\,\mathrm{d}x = \mathscr{L}_{GIG(\lambda,\delta,\gamma)}(-u), \quad u \in \big(-\infty, \tfrac{\gamma^2}{2}\big). \tag{7}$$

After these preliminaries, we can now study the class of generalized hyperbolic distributions which have been introduced in the seminal paper [5], motivated by empirical statistics of aeolian sand deposits. The GH distributions are defined as normal mean-variance mixtures with a GIG mixing distribution as follows:

$$GH(\lambda,\alpha,\beta,\delta,\mu) := N(\mu+\beta y, y) \circ GIG\big(\lambda,\delta,\sqrt{\alpha^2-\beta^2}\big). \tag{8}$$

The parameter restrictions for GIG distributions immediately imply that the GH parameters have to fulfill the constraints

$$\lambda,\mu \in \mathbb{R} \quad \text{and} \quad \begin{array}{ll} \delta \geq 0,\ 0 \leq |\beta| < \alpha, & \text{if } \lambda > 0, \\ \delta > 0,\ 0 \leq |\beta| < \alpha, & \text{if } \lambda = 0, \\ \delta > 0,\ 0 \leq |\beta| \leq \alpha, & \text{if } \lambda < 0. \end{array}$$

As before, parametrizations with $\delta = 0$ and $|\beta| = \alpha$ have to be understood as limiting cases which by Lemma 1(c) equal normal mean-variance mixtures with the corresponding GIG limit distributions. We defer a more precise introduction of the latter and first concentrate on GH distributions with parameters $\delta > 0$ and $|\beta| < \alpha$. Their Lebesgue densities are given by

$$\begin{aligned}
d_{GH(\lambda,\alpha,\beta,\delta,\mu)}(x) &= \int_0^\infty d_{N(\mu+\beta y,y)}(x)\, d_{GIG\left(\lambda,\delta,\sqrt{\alpha^2-\beta^2}\right)}(y)\,\mathrm{d}y \\
&= a(\lambda,\alpha,\beta,\delta,\mu)\big(\delta^2+(x-\mu)^2\big)^{(\lambda-\frac{1}{2})/2} K_{\lambda-\frac{1}{2}}\Big(\alpha\sqrt{\delta^2+(x-\mu)^2}\Big)\, e^{\beta(x-\mu)}
\end{aligned} \tag{9}$$

with the norming constant

$$a(\lambda, \alpha, \beta, \delta, \mu) = \frac{(\alpha^2 - \beta^2)^{\frac{\lambda}{2}}}{\sqrt{2\pi}\, \alpha^{\lambda - \frac{1}{2}} \delta^{\lambda} K_\lambda\big(\delta\sqrt{\alpha^2 - \beta^2}\big)}. \tag{10}$$

A closer look at the densities reveals that the influence of the parameters is as follows: α determines the shape, β the skewness, μ is a location parameter, and δ serves for scaling. λ characterizes certain subclasses and considerably influences the size of mass contained in the tails. Setting $\lambda = -\frac{1}{2}$ leads to the subclass of normal inverse Gaussian distributions (NIG). By (8), these are the normal mean-variance mixtures arising from inverse Gaussian mixing distributions which explains their name. With the symmetry relation $K_{-\lambda}(x) = K_\lambda(x)$ and the aforementioned representation of $K_{-\frac{1}{2}}(x)$, its densities are obtained from (9) and (10) as

$$d_{NIG(\alpha,\beta,\delta,\mu)}(x) = \frac{\alpha\delta}{\pi} \frac{K_1\big(\alpha\sqrt{\delta^2 + (x-\mu)^2}\big)}{\sqrt{\delta^2 + (x-\mu)^2}}\, e^{\delta\sqrt{\alpha^2-\beta^2} + \beta(x-\mu)}. \tag{11}$$

From Lemma 1(a), Proposition 1, and Eq. (7) we conclude that all GH distributions with parameters $\delta > 0$ and $|\beta| < \alpha$ possess a moment generating function of the following form:

$$M_{GH(\lambda,\alpha,\beta,\delta,\mu)}(u) = e^{\mu u} \left(\frac{\alpha^2 - \beta^2}{\alpha^2 - (\beta + u)^2}\right)^{\frac{\lambda}{2}} \frac{K_\lambda\big(\delta\sqrt{\alpha^2 - (\beta+u)^2}\big)}{K_\lambda\big(\delta\sqrt{\alpha^2 - \beta^2}\big)}, \tag{12}$$

which is defined for all $u \in (-\alpha - \beta, \alpha - \beta)$. The characteristic functions of GH distributions are easily obtained via the relation

$$\phi_{GH(\lambda,\alpha,\beta,\delta,\mu)}(u) = \int_{\mathbb{R}} e^{iux} d_{GH(\lambda,\alpha,\beta,\delta,\mu)}(x)\, \mathrm{d}x = M_{GH(\lambda,\alpha,\beta,\delta,\mu)}(iu). \tag{13}$$

A detailed derivation of the weak limits of GH distributions can be found in [13, Section 3] from which we summarize the main results below. The limit distributions emerging in the case of $\lambda > 0$ and $\delta \to 0$ are also known as Variance Gamma distributions (VG). By Lemma 1(c) and Eq. (3), they are normal mean-variance mixtures of the following form:

$$VG(\lambda, \alpha, \beta, \mu) = N(\mu + \beta y, y) \circ G\big(\lambda, \tfrac{\alpha^2 - \beta^2}{2}\big). \tag{14}$$

Using the asymptotic relationship (2), the corresponding densities can be obtained as pointwise limits (for $x - \mu \neq 0$) of the GH densities:

$$
\begin{aligned}
d_{VG(\lambda,\alpha,\beta,\mu)}(x) &= \lim_{\delta\to 0} d_{GH(\lambda,\alpha,\beta,\delta,\mu)}(x) \\
&= \frac{(\alpha^2-\beta^2)^{\lambda}}{\sqrt{\pi}\,(2\alpha)^{\lambda-\frac{1}{2}}\,\Gamma(\lambda)}\,|x-\mu|^{\lambda-\frac{1}{2}}\,K_{\lambda-\frac{1}{2}}(\alpha|x-\mu|)\,e^{\beta(x-\mu)}.
\end{aligned}
\tag{15}
$$

This class was introduced in [31] (symmetric case $\beta = \theta = 0$) and [30] (general case), but with a different parametrization $VG(\sigma, \nu, \theta, \tilde{\mu})$. The latter is obtained by

$$
\sigma^2 = \frac{2\lambda}{\alpha^2-\beta^2}, \qquad \nu = \frac{1}{\lambda}, \qquad \theta = \beta\sigma^2 = \frac{2\beta\lambda}{\alpha^2-\beta^2}, \qquad \tilde{\mu} = \mu.
$$

Lemma 1(a), Proposition 1, and Eq. (7) imply that all VG distributions possess a moment generating function which is given by

$$
M_{VG(\lambda,\alpha,\beta,\mu)}(u) = e^{\mu u}\,\mathscr{L}_{GIG(\lambda,0,\sqrt{\alpha^2-\beta^2})}\big(-\tfrac{u^2}{2}-\beta u\big) = e^{\mu u}\left(\frac{\alpha^2-\beta^2}{\alpha^2-(\beta+u)^2}\right)^{\lambda} \tag{16}
$$

for all $u \in (-\alpha-\beta, \alpha-\beta)$.

For $\lambda < 0$, there are two possible limit cases. If $\alpha, \beta \to 0$, Lemma 1(c) and Eq. (4) imply that the limit distributions are normal mean-variance mixtures

$$
t(\lambda, \delta, \mu) = N(\mu, y) \circ iG\big(\lambda, \tfrac{\delta^2}{2}\big) \tag{17}
$$

which equal scaled and shifted t-distributions with $f = -2\lambda$ degrees of freedom (the usual Student's t-distribution is obtained with $\delta^2 \equiv -2\lambda$). The associated densities can again be obtained as pointwise limits of the GH densities:

$$
d_{t(\lambda,\delta,\mu)}(x) = \lim_{\alpha,\beta\to 0} d_{GH(\lambda,\alpha,\beta,\delta,\mu)}(x) = \frac{\Gamma\big(-\lambda+\frac{1}{2}\big)}{\sqrt{\pi\delta^2}\,\Gamma(-\lambda)}\left(1+\frac{(x-\mu)^2}{\delta^2}\right)^{\lambda-\frac{1}{2}}. \tag{18}
$$

The other class of limit distributions for $\lambda < 0$ is obtained by letting $|\beta| \to \alpha > 0$. Again by Lemma 1(c) and Eq. (4), these are normal mean-variance mixtures given by

$$
GH(\lambda, \alpha, \pm\alpha, \delta, \mu) = N(\mu \pm \alpha y, y) \circ iG\big(\lambda, \tfrac{\delta^2}{2}\big) \tag{19}
$$

and possessing the density

$$
\begin{aligned}
d_{GH(\lambda,\alpha,\pm\alpha,\delta,\mu)}(x) = {} & \frac{2^{\lambda+\frac{1}{2}}}{\sqrt{\pi}\,\alpha^{\lambda-\frac{1}{2}}\,\delta^{2\lambda}\,\Gamma(-\lambda)}\big(\delta^2+(x-\mu)^2\big)^{(\lambda-\frac{1}{2})/2} \\
& \times K_{\lambda-\frac{1}{2}}\big(\alpha\sqrt{\delta^2+(x-\mu)^2}\big)\,e^{\pm\alpha(x-\mu)}.
\end{aligned}
\tag{20}
$$

This type of distribution was called generalized hyperbolic skew Student t-distribution and applied to financial data in [1].

Let us close this section by noting that also the normal and GIG distributions themselves can emerge as potential limits of univariate GH distributions. But the tail behaviour and tail dependence of the former are already well-known, and for GIG distributions there does not seem to exist a natural multivariate version of which the tail dependence could be studied, therefore we tacitly ignore these two limiting cases here and in the following.

3 Tail Behaviour of GH Distributions and Their Convolutions

From the existence of a moment generating function one can already conclude that the tails of the GH densities with $0 \leq |\beta| < \alpha$ decay at an exponential rate. More precisely, for $|x| \to \infty$ we have $\delta^2 + (x-\mu)^2 \sim x^2$, and the asymptotic behaviour of the Bessel functions (cf. [2, formula (9.7.2)])

$$K_\lambda(x) \sim \sqrt{\frac{\pi}{2x}} e^{-x}, \quad x \to \infty, \tag{21}$$

further implies $K_{\lambda-\frac{1}{2}}\left(\alpha\sqrt{\delta^2 + (x-\mu)^2}\right) \sim \sqrt{\frac{\pi}{2\alpha}}\, |x|^{-1/2}\, e^{-\alpha|x|}$, so we obtain from Eq. (9)

$$d_{GH(\lambda,\alpha,\beta,\delta,\mu)}(x) \sim c\, |x|^{\lambda-1}\, e^{-\alpha|x|+\beta x}, \quad x \to \pm\infty, \tag{22}$$

where $c = \sqrt{\frac{\pi}{2\alpha}} a(\lambda, \alpha, \beta, \delta, \mu)$, and $a(\lambda, \alpha, \beta, \delta, \mu)$ is the norming constant from (10). Completely analogously, we infer from Eqs. (15) and (21) that

$$d_{VG(\lambda,\alpha,\beta.\mu)}(x) \sim \tilde{c}\, |x|^{\lambda-1}\, e^{-\alpha|x|+\beta x}, \quad x \to \pm\infty, \tag{23}$$

where $\tilde{c} = \frac{(\alpha^2-\beta^2)^\lambda}{(2\alpha)^\lambda \Gamma(\lambda)}$. The GH and VG densities can thus be regarded more generally as subsets of a class of probability densities f with support $\mathbb{R}$ whose tail behaviour is given by

$$f(x) \sim c_1\, |x|^{a_1}\, e^{-b_1|x|}, \quad x \to -\infty, \quad \text{and} \quad f(x) \sim c_2\, x^{a_2}\, e^{-b_2 x}, \quad x \to +\infty, \tag{24}$$

for some constants $a_1, a_2 \in \mathbb{R}$ and $b_1, b_2, c_1, c_2 > 0$. It can be easily deduced that every probability distribution F having a Lebesgue density f that fulfills Eq. (24) also possesses a moment generating function which is defined at least on the open interval $(-b_1, b_2)$. In case of GH and VG distributions we have $a_1 = a_2 = \lambda - 1$, $b_1 = \alpha + \beta$, $b_2 = \alpha - \beta$ and $c_1 = c_2 = c$ resp. $c_1 = c_2 = \tilde{c}$.

A remarkable and probably surprising property of such densities is that the tail behaviour of the corresponding distribution functions is the same up to a multiplicative constant, which is shown in the next proposition.

Proposition 2 *Let f be a probability density fulfilling* (24) *for some constants* $a_1, a_2, b_1, b_2, c_1, c_2$, $F(x) := \int_{-\infty}^{x} f(y)\,dy$ *be the associated distribution function and* $\bar{F}(x) := 1 - F(x)$. *Then* $f(x) \sim b_1 F(x)$ *as* $x \to -\infty$ *and* $f(x) \sim b_2 \bar{F}(x)$ *as* $x \to +\infty$.

Proof Let us consider the right tail $\bar{F}(x)$ first. From the assumptions we get, using partial integration,

$$\bar{F}(x) = \int_x^{\infty} f(y)\,\mathrm{d}y \sim c_2 \int_x^{\infty} y^{a_2} e^{-b_2 y}\,\mathrm{d}y = \frac{c_2}{b_2} x^{a_2} e^{-b_2 x} + \frac{c_2 a_2}{b_2} \int_x^{\infty} y^{a_2 - 1} e^{-b_2 y}\,\mathrm{d}y.$$

The claim now follows if we can show that $\left(\int_x^{\infty} y^{a_2-1} e^{-b_2 y} \mathrm{d}y\right)\left(x^{a_2} e^{-b_2 x}\right)^{-1} \to 0$ as $x \to \infty$. But the latter quotient equals

$$\frac{1}{x} \int_x^{\infty} \left(\frac{y}{x}\right)^{a_2-1} e^{-b_2(y-x)} \mathrm{d}y = \frac{1}{x} \int_0^{\infty} \left(\frac{y+x}{x}\right)^{a_2-1} e^{-b_2 y}\,\mathrm{d}y$$

and thus converges to zero as $x \to \infty$ because the existence of an integrable majorant ensures that the integral on the right hand side remains bounded. Possible majorants are $g(y) = (y+1)^{a_2-1} e^{-b_2 y}$ if $a_2 > 1$ and $g(y) = e^{-b_2 y}$ if $a_2 \le 1$. Using the change of variables $z = -y$ we see that for $x \to -\infty$

$$F(x) \sim c_1 \int_{-\infty}^{x} |y|^{a_1} e^{-b_1 |y|}\,\mathrm{d}y = c_1 \int_{|x|}^{\infty} z^{a_1} e^{-b_1 z}\,\mathrm{d}z,$$

hence the assertion for the left tail immediately follows from what we have proven above. □

The tail behaviour of the t-distributions, however, can be derived more easily. The asymptotics of the corresponding densities are easily seen from (18) to equal $d_{t(\lambda,\delta,\mu)}(x) \sim \bar{c}\,|x|^{2\lambda-1}$, $x \to \pm\infty$. Hence, $F_{t(\lambda,\delta,\mu)}(x) \sim \mathrm{sgn}(x) \frac{\bar{c}}{2\lambda} |x|^{2\lambda} + \mathbb{1}_{(0,\infty)}(x)$. The knowledge of the tail behaviour allows to derive the asymptotic behaviour of the associated quantile functions as well. This is of particular importance for GH- and VG-distributions whose distribution functions cannot be given in closed form, and a reliable and rapidly convergent series expansion for these is not known either. To determine quantiles of the former, one therefore has to resort to numerical integration of their densities in most cases (an alternative way to numerically compute the quantile function of the hyperbolic distribution with given precision is described in [36]). This may—depending on the quality of the integration routine used—lead to more or less inaccurate and unstable results for p-quantiles if p is very close to 0 or 1. The quantile asymptotics are summarized in the following lemma which is a slightly modified version of [4, Lemma 3.1]. Due to its importance for the derivation of the tail dependence coefficients in Sect. 5, we also provide a short proof here (a more general result on asymptotics of inverse functions can be found in the short note [23]).

Lemma 2 *Suppose* $F : \mathbb{R} \to [0, 1]$ *is a continuous and strictly increasing distribution function.*

(a) If $F(x) \sim c_1|x|^{-a_1}$ *as* $x \to -\infty$ *and* $1 - F(x) \sim c_2 x^{-a_2}$ *as* $x \to \infty$ *for some* $a_1, a_2, c_1, c_2 > 0$*, then* $F^{-1}(u) \sim -\left(\frac{c_1}{u}\right)^{\frac{1}{a_1}}$ *and* $F^{-1}(1-u) \sim \left(\frac{c_2}{u}\right)^{\frac{1}{a_2}}$ *for* $u \downarrow 0$*.*

(b) If $F(x) \sim c_1|x|^{a_1}e^{-b_1|x|}$ *as* $x \to -\infty$ *and* $1 - F(x) \sim c_2 x^{a_2} e^{-b_2 x}$ *as* $x \to \infty$ *for some* $a_1, a_2 \in \mathbb{R}$ *and* $b_1, b_2, c_1, c_2 > 0$*, then* $F^{-1}(u) \sim \frac{\log(u)}{b_1}$ *and* $F^{-1}(1-u) \sim -\frac{\log(u)}{b_2}$ *for* $u \downarrow 0$*.*

Proof (a) If $1 - F(x) \sim c_2 x^{-a_2}$ as $x \to \infty$, then for any $r > 0$

$$\lim_{u \downarrow 0} \frac{1 - F\left(r\left(\frac{c_2}{u}\right)^{\frac{1}{a_2}}\right)}{u} = r^{-a_2}.$$

For $r < 1$, the right hand side of the above equation is greater than one, so we conclude that in this case $1 - F\left(r\left(\frac{c_2}{u}\right)^{\frac{1}{a_2}}\right) > u$ for sufficiently small u and hence $F^{-1}(1-u) > r\left(\frac{c_2}{u}\right)^{\frac{1}{a_2}}$ (note that the assumptions on F imply $F^{-1}(F(y)) = y$ for all $y \in \mathbb{R}$). If $r > 1$, then we similarly obtain $1 - F\left(r\left(\frac{c_2}{u}\right)^{\frac{1}{a_2}}\right) < u$ and thus $F^{-1}(1-u) < r\left(\frac{c_2}{u}\right)^{\frac{1}{a_2}}$ for sufficiently small u. This proves the assertion for $F^{-1}(1-u)$, and the asymptotic behaviour of $F^{-1}(u)$ for $u \downarrow 0$ can be shown analogously.

(b) If $1 - F(x) \sim c_2 x^{a_2} e^{-b_2 x}$ as $x \to \infty$, then we have

$$\lim_{u \downarrow 0} \frac{1 - F\left(-\frac{r \log(u)}{b_2}\right)}{u} = \lim_{u \downarrow 0} c_2 \left(-\frac{r \log(u)}{b_2}\right)^{a_2} u^{r-1} = \begin{cases} \infty, & r < 1, \\ 0, & r > 1. \end{cases}$$

With the same reasoning as before we conclude $F^{-1}(1-u) \sim -\frac{\log(u)}{b_2}$ for $u \downarrow 0$, and the corresponding result for $F^{-1}(u)$ is easily obtained along the same lines. □

Observe that the tails of the densities $d_{GH(\lambda,\alpha,\pm\alpha,\delta,\mu)}(x)$ of the generalized hyperbolic skew Student t-distribution behave completely differently for large arguments. If $\beta = \alpha$, then by (21) the asymptotic behaviour is as follows:

$$\begin{aligned} d_{GH(\lambda,\alpha,\alpha,\delta,\mu)}(x) &\sim \tilde{c}_1 \, |x|^{\lambda-1} \, e^{-2\alpha|x|}, & x \to -\infty, \\ d_{GH(\lambda,\alpha,\alpha,\delta,\mu)}(x) &\sim \tilde{c}_2 \, |x|^{\lambda-1}, & x \to +\infty, \end{aligned} \tag{25}$$

and the other way round if $\beta = -\alpha$. Hence, they have one semi-heavy and one heavy (power) tail, so the asymptotic behaviour of their distribution functions and quantiles is obtained by combining the corresponding results above. Further, it is easily seen from Eq. (1) that the GIG densities possess a right tail satisfying (24) with parameters $a_2 = \lambda - 1$, $b_2 = \frac{\gamma^2}{2}$, and $c_2 = \frac{\gamma^\lambda}{2\delta^\lambda K_\lambda(\delta\gamma)}$, so the above lemma and proposition can also be applied here.

But not only the tail behaviour of single GH distributions, also that of convolutions of the latter is of practical interest in finance. Think, for example, of factor models for credit portfolios where for each portfolio constituent a state variable $X_i = \sqrt{\rho}M + \sqrt{1-\rho}Z_i$, $0 \le \rho \le 1$, with a systematic factor M and an independent idiosyncratic factor Z_i is defined. The portfolio loss distribution derived from this approach then entails the quantile function $F_{X_i}^{-1}(p_d)$ of the distribution of X_i which has to be evaluated for typically very small default probabilities p_d. If the factor distributions F_M and F_{Z_i} are not stable under convolution, the distribution of X_i is usually unknown, therefore the quantiles $F_{X_i}^{-1}(p_d)$ can only be determined by either time-consuming simulations or advanced numerical methods. Precisely this is the case if one assumes the factors to be GH distributed. From Lemma 1(b) and Eq. (6), one can deduce the following convolution properties of the GH family:

$$\begin{aligned}
&\text{(a)}\ NIG(\alpha,\beta,\delta_1,\mu_1) * NIG(\alpha,\beta,\delta_2,\mu_2) = NIG(\alpha,\beta,\delta_1+\delta_2,\mu_1+\mu_2),\\
&\text{(b)}\ NIG(\alpha,\beta,\delta_1,\mu_1) * GH\big(\tfrac{1}{2},\alpha,\beta,\delta_2,\mu_2\big) = GH\big(\tfrac{1}{2},\alpha,\beta,\delta_1+\delta_2,\mu_1+\mu_2\big),\\
&\text{(c)}\ GH(-\lambda,\alpha,\beta,\delta,\mu) * VG\big(\lambda,\alpha,\beta,\mu_2\big) = GH(\lambda,\alpha,\beta,\delta,\mu_1+\mu_2),\ \lambda>0,\\
&\text{(d)}\ VG(\lambda_1,\alpha,\beta,\mu_1) * VG(\lambda_2,\alpha,\beta,\mu_2) = VG(\lambda_1+\lambda_2,\alpha,\beta,\mu_1+\mu_2),\ \lambda_1,\lambda_2>0.
\end{aligned} \tag{26}$$

Inspecting the Laplace transforms of GIG distributions given in Proposition 1 more closely, one can deduce that the list of GIG convolution formulas (6) is complete, that is, no other convolution of two GIG distributions will yield a distribution that itself is contained in the GIG class. Consequently, there do not exist more than the four convolution formulas above for the GH family either. In particular, a convolution of two GH distributions with different parameters α and/or β cannot be GH distributed itself. This fact makes the application of generalized hyperbolic factor models computationally demanding, therefore some (approximate) formulas for $F_{X_i}^{-1}(p_d)$, at least for small probabilities p_d, which are faster and easier to evaluate would be desirable here. For a more thorough introduction to GH factor models, we refer to [12] and [27, Chap. 3]; there the quantiles of the convolution were calculated with help of Fourier inversion.

The behaviour of GH convolution tails is described in Proposition 3 below. The latter applies, in fact, to an even slightly more general framework where both factors belong to $\mathscr{L}_{a,b}$, the class of distributions with exponential tails with rates a and b, which we define as follows:

Definition 2 A distribution function F is said to have *exponential tails* with rates $a > 0$ and $b > 0$ ($F \in \mathscr{L}_{a,b}$) if for all $y \in \mathbb{R}$

$$\lim_{x\to-\infty} \frac{F(x-y)}{F(x)} = e^{-ay} \quad \text{and} \quad \lim_{x\to\infty} \frac{\bar{F}(x-y)}{\bar{F}(x)} = e^{by}.$$

Note that most definitions of exponential tails only use one index which characterizes the behaviour of the right tail $\bar{F}(x)$. This is due to the fact that these arose from extreme value theory or more generally actuarial science where one typically works with probability distributions on $\mathbb{R}_+$. The above is a natural generalization to distributions having support $\mathbb{R}$ we are concerned with. The class $\mathcal{L}_{a,b}$ is closely related to the class $\mathcal{R}_p$ of regularly varying functions to be introduced in

Definition 3 A measurable function g is *regularly varying* with exponent $p \in \mathbb{R}$ ($g \in \mathcal{R}_p$) if $\lim_{t\to\infty} \frac{g(st)}{g(t)} = s^p$ for all $s > 0$.

We have $F \in \mathcal{L}_{a,b}$ iff $F(-\ln(x)) \in \mathcal{R}_{-a}$ and $\bar{F}(\ln(x)) \in \mathcal{R}_{-b}$. To see this, put $s = e^y$ and $t = e^{-x}$, then

$$e^{-ay} = \lim_{x\to-\infty} \frac{F(x-y)}{F(x)} \iff s^{-a} = \lim_{t\to\infty} \frac{F(-\ln(t) - \ln(s))}{F(-\ln(t))} = \lim_{t\to\infty} \frac{F(-\ln(st))}{F(-\ln(t))},$$

and the assertion for the right tails follows analogously with $s = e^{-y}$ and $t = e^x$.

From Proposition 2 it is immediately seen that for a probability distribution F that possesses a density f fulfilling (24) we have

$$\lim_{x\to-\infty} \frac{F(x-y)}{F(x)} = \lim_{x\to-\infty} \frac{f(x-y)}{f(x)} = \lim_{x\to-\infty} \left(\frac{|x-y|}{|x|}\right)^{a_1} e^{-b_1(|x-y|-|x|)} = e^{-b_1 y},$$

and an analogous limit is obtained for the right tails, hence $F \in \mathcal{L}_{b_1,b_2}$. In particular, we see that $GH(\lambda, \alpha, \beta, \delta, \mu)$- and $VG(\lambda, \alpha, \beta, \mu)$-distributions belong to the class $\mathcal{L}_{\alpha+\beta,\alpha-\beta}$. The asymptotic behaviour of the densities of the t-distributions, however, is easily seen from (18) to equal $d_{t(\lambda,\delta,\mu)}(x) \sim \bar{c}\,|x|^{2\lambda-1},\ x \to \pm\infty$, from which we immediately obtain that $F_{t(\lambda,\delta,\mu)}(-x), \bar{F}_{t(\lambda,\delta,\mu)}(x) \in \mathcal{R}_{2\lambda}$. We defer the latter for a moment and first consider convolutions of factors with exponential tails.

An easy solution occurs if the factors of the convolution have exponential tails which decay at different rates: the convolution tails are determined by the factor with the heavier left (respectively right) tail.

Proposition 3 *Suppose that $F_1 \in \mathcal{L}_{b_1,b_2}$, $F_2 \in \mathcal{L}_{\tilde{b}_1,\tilde{b}_2}$ with moment generating functions $M_{F_1}(u)$ and $M_{F_2}(u)$. If $b_1 < \tilde{b}_1$ and $b_2 < \tilde{b}_2$, then $F_1 * F_2 \in \mathcal{L}_{b_1,b_2}$ and*

$$\lim_{x\to-\infty} \frac{(F_1 * F_2)(x)}{F_1(x)} = M_{F_2}(-b_1), \qquad \lim_{x\to\infty} \frac{\overline{(F_1 * F_2)}(x)}{\bar{F}_1(x)} = M_{F_2}(b_2).$$

A detailed proof of this result can be found in [27, Proposition 1.16]. The assumption above that both tails of F_1 are heavier than those of F_2 was just made for notational convenience. As it is easily seen, in general we have $F_1 * F_2 = \mathcal{L}_{b_1\wedge\tilde{b}_1, b_2\wedge\tilde{b}_2}$, that is, one factor may determine the left tail of the convolution and the other one the right tail. In [21, Theorem 3 (b)] it has been shown that if the right tails of F_1 and F_2 are both exponential with the same rate a, then the right tail of $F_1 * F_2$ is also exponential with rate a, so we may conclude that $F_1 * F_2 = \mathcal{L}_{b_1\wedge\tilde{b}_1, b_2\wedge\tilde{b}_2}$ remains valid if $b_1 = \tilde{b}_1$ and/or $b_2 = \tilde{b}_2$. Summing up, we have the following

Corollary 1 *Let F_1, F_2 be the distribution functions of $GH(\lambda_1, \alpha_1, \beta_1, \delta_1, \mu_1)$ resp. $GH(\lambda_2, \alpha_2, \beta_2, \delta_2, \mu_2)$, and $F = F_1 * F_2$. If $\alpha_1 + \beta_1 \neq \alpha_2 + \beta_2$ and $\alpha_1 - \beta_1 \neq \alpha_2 - \beta_2$, then*

$$F(x) \sim M_{F^l_{max}}(-b_1) F^l_{max}(x), \; x \to -\infty, \; \text{and} \; \bar{F}(x) \sim M_{F^r_{max}}(b_2) \bar{F}^r_{max}(x), \; x \to \infty,$$

where $b_1 = \min(\alpha_1 + \beta_1, \alpha_2 + \beta_2)$, $b_2 = \min(\alpha_1 - \beta_1, \alpha_2 - \beta_2)$, and $F^l_{max}(x)$, $F^r_{max}(x)$ are the distribution functions of the GH distribution whose parameters α_i, β_i attain the value b_1 resp. b_2. The assertions remain valid if one or both factors are VG distributed instead.

If $\alpha_1 + \beta_1 = \alpha_2 + \beta_2$ or $\alpha_1 - \beta_1 = \alpha_2 - \beta_2$, the left resp. right tail behaviour cannot be precisely specified, and one only has the weaker result $F \in \mathcal{L}_{b_1,b_2}$.

Since the convolution tails are asymptotically equivalent to the tail of one factor distribution, multiplied by a constant, approximate quantile values of the convolution for probabilities close to zero or one can be computed with help of Lemma 2(b). It can be shown that the latter also applies under the weaker assumption $F \in \mathcal{L}_{b_1,b_2}$. Therefore, we obtain that the asymptotic behaviour of the quantile function of a convolution of GH distributions with $\alpha_1 + \beta_1 = \alpha_2 + \beta_2$ is given by $F^{-1}(u) \sim \frac{\log(u)}{\alpha_1+\beta_1}$, $u \downarrow 0$, and similarly, $F^{-1}(u) \sim -\frac{\log(1-u)}{\alpha_1-\beta_1}$ for $u \uparrow 1$ if $\alpha_1 - \beta_1 = \alpha_2 - \beta_2$.

A corresponding result for the regularly varying tails of the t-distributions can be obtained by applying [7, Theorem 1.1 and the Theorem on p. 54] which yields

Corollary 2 *Let F_1, F_2 be the distribution functions of $t(\lambda_1, \delta_1, \mu_1)$ and $t(\lambda_2, \delta_2, \mu_2)$ with corresponding densities f_1, f_2, then*

$$\lim_{|x|\to\infty} \frac{(f_1 * f_2)(x)}{f_1(x) + f_2(x)} = 1 \; \text{and} \; \lim_{x\to-\infty} \frac{(F_1 * F_2)(x)}{F_1(x) + F_2(x)} = \lim_{x\to\infty} \frac{\overline{(F_1 * F_2)}(x)}{\bar{F}_1(x) + \bar{F}_2(x)} = 1.$$

If $\lambda_1 < \lambda_2$, then with the above notations we have $f_1(x) = o\big(f_2(x)\big)$ as $|x| \to \infty$ and $F_1(x) = o\big(F_2(x)\big), x \to -\infty$, as well as $\bar{F}_1(x) = o\big(\bar{F}_2(x)\big), x \to \infty$, consequently

$$\lim_{|x|\to\infty} \frac{(f_1 * f_2)(x)}{f_2(x)} = 1 \;\text{and}\; \lim_{x\to-\infty} \frac{(F_1 * F_2)(x)}{F_2(x)} = \lim_{x\to\infty} \frac{\overline{(F_1 * F_2)}(x)}{\bar{F}_2(x)} = 1$$

(see also [7, Theorem 2.1]). Hence, also in this case the tail behaviour of the convolution and the asymptotic behaviour of the convolution density is determined by the factor with the heavier tails. Approximate quantile values can then, similarly as before, be calculated using Lemma 2(a).

4 Multivariate Normal Mean-Variance Mixtures and GH Distributions

Let us first fix some notation which will be used throughout the rest of the paper: The vectors $u = (u_1, \dots, u_d)^\top$ and $x = (x_1, \dots, x_d)^\top$ are elements of $\mathbb{R}^d$, the superscript $^\top$ stands for the transpose of a vector or matrix. $\langle u, x\rangle = u^\top x = \sum_{i=1}^d u_i x_i$ denotes the scalar product of the vectors u, x and $\|u\| = (u_1^2 + \cdots + u_d^2)^{1/2}$ the Euclidean norm of u. If A is a real-valued $d \times d$-square matrix, then $\det(A)$ denotes the determinant of A. The $d \times d$-identity matrix is labeled I_d. In contrast to u and x, the letters y, s and t are reserved for univariate real variables, that is, we assume $y, s, t \in \mathbb{R}$ or $\mathbb{R}_+$. To properly distinguish between the real number zero and the zero vector, we write $0 \in \mathbb{R}$ and $\mathbf{0} := (0, \dots, 0)^\top \in \mathbb{R}^d$. Note that here and in the following $d \geq 2$ indicates the dimension, whereas n is usually used as running index for all kinds of sequences. In particular, the notation $N_d(\mu, \Delta)$ will be used for the d-dimensional normal distribution with mean vector μ and covariance matrix Δ.

With these preliminaries, we can formulate the multivariate version of Definition 1 as follows:

Definition 4 An $\mathbb{R}^d$-valued random variable X is said to have a *multivariate normal mean-variance mixture distribution* if

$$X \stackrel{d}{=} \mu + Z\beta + \sqrt{Z} A W,$$

where $\mu, \beta \in \mathbb{R}^d$, A is a real-valued $d \times d$-matrix such that $\Delta := AA^\top$ is positive definite, W is a standard normally distributed random vector ($W \sim N_d(\mathbf{0}, I_d)$) and $Z \sim G$ is a real-valued, non-negative random variable independent of W. Equivalently, a probability measure F on $(\mathbb{R}^d, \mathscr{B}^d)$ is said to be a multivariate normal mean-variance mixture if

$$F(\mathrm{d}x) = \int_{\mathbb{R}_+} N_d(\mu + y\beta, y\Delta)(\mathrm{d}x)\, G(\mathrm{d}y),$$

where the mixing distribution G is a probability measure on $(\mathbb{R}_+, \mathscr{B}_+)$. We shall use the short hand notation $F = N_d(\mu + y\beta, y\Delta) \circ G$.

Remark 1 Note that one can further assume w.l.o.g. $|\det(A)| = \det(\Delta) = 1$, since a (positive) multiplicative constant can always be included within the variable Z. Further observe that the use of a single univariate mixing variable Z causes dependencies between all entries of X, as we shall see in Sect. 5.

The straightforward generalization of Lemma 1 in Sect. 2 is

Lemma 3 *Let $\mathbb{G}$ be a class of probability distributions on $(\mathbb{R}_+, \mathscr{B}_+)$ and suppose $G, G_1, G_2 \in \mathbb{G}$.*

(a) If G possesses a moment generating function $M_G(y)$ on some open interval (a, b) with $a < 0 < b$, then $F = N_d(\mu + y\beta, y\Delta) \circ G$ also possesses a moment

generating function $M_F(u) = e^{\langle u,\mu\rangle} M_G\big(\frac{\langle u,\Delta u\rangle}{2} + \langle u,\beta\rangle\big)$ *that is defined for all* $u \in \mathbb{R}^d$ *with* $a < \frac{\langle u,\Delta u\rangle}{2} + \langle u,\beta\rangle < b$.

(b) If $G = G_1 * G_2 \in \mathbb{G}$, *then* $\big(N_d(\mu_1 + y\beta, y\Delta) \circ G_1\big) * \big(N_d(\mu_2 + y\beta, y\Delta) \circ G_2\big) = N_d(\mu_1 + \mu_2 + y\beta, y\Delta) \circ G$.

(c) If $(\mu_n)_{n\ge1}$ *and* $(\beta_n)_{n\ge1}$ *are convergent sequences of real vectors with finite limits* $\mu, \beta \in \mathbb{R}^d$ *(that is,* $\|\mu\|, \|\beta\| < \infty$*), and* $(G_n)_{n\ge1}$ *is a sequence of mixing distributions with* $G_n \xrightarrow{w} G$, *then* $N_d(\mu_n + y\beta_n, y\Delta) \circ G_n \xrightarrow{w} N_d(\mu + y\beta, y\Delta) \circ G$.

For further reference, we also briefly highlight the relationship between multivariate normal mean-variance mixtures and elliptical distributions. From a financial point of view, the latter are of some interest because they have the nice property that within this class the Value-at-Risk (VaR) is a coherent risk measure in the sense of [3] (this has been shown in [22, Theorem 1], see also [33, Theorem 8.28]).

Definition 5 An $\mathbb{R}^d$-valued random vector X has an *elliptical distribution* if there exists a function $\psi : \mathbb{R}_+ \to \mathbb{R}$, a symmetric, positive semidefinite $d \times d$-matrix Σ and some $\mu \in \mathbb{R}^d$ such that the characteristic function $\phi_X(u) = E[e^{i\langle u,X\rangle}]$ of X admits the representation

$$\phi_X(u) = e^{i\langle u,\mu\rangle}\psi\big(\langle u, \Sigma u\rangle\big) \qquad \forall\, u \in \mathbb{R}^d.$$

The elliptical distribution $\mathcal{L}(X)$ then is denoted by $E_d\big(\mu, \Sigma, \psi(t)\big)$.

It can be shown that if an elliptical distribution has a density f, then it must necessarily be of the form

$$f(x) = \frac{1}{\sqrt{\det(\Sigma)}}\, h\big(\langle x-\mu, \Sigma^{-1}(x-\mu)\rangle\big)$$

for some measurable function $h : \mathbb{R} \to \mathbb{R}_+$. The level sets of such a density obviously are the ellipsoids $\{x \in \mathbb{R}^d \mid \langle x-\mu, \Sigma^{-1}(x-\mu)\rangle = \bar{c}\}, \bar{c} > 0$, which explains where the name of this class of distributions stems from. The following characterization of elliptically distributed random vectors is provided in [33, Proposition 6.27]:

Proposition 4 $X \sim E_d\big(\mu, \Sigma, \psi(t)\big)$ *if and only if*

$$X \stackrel{d}{=} \mu + RAS$$

where R *is an* $\mathbb{R}_+$*-valued random variable,* S *is a random vector which is independent of* R *and uniformly distributed on the unit sphere* $\mathcal{S} := \{\xi \in \mathbb{R}^d \mid \|\xi\| = 1\}$, *and* A *is a* $d \times d$*-matrix fulfilling* $AA^\top = \Sigma$.

Let us note here for further reference that the multivariate normal distribution obviously is elliptic ($N_d(\mu, \Delta) = E_d\big(\mu, \Delta, e^{-\frac{t}{2}}\big)$), and if $X \sim N_d(\mu, \Delta)$, then for the random variable R in the corresponding representation of Proposition 4 it holds that $R \stackrel{d}{=} \sqrt{Y}$ with $Y \sim \chi^2_d$, see [33, Example 6.23, p. 199].

The connection between elliptical distributions and multivariate normal mean-variance mixtures is given in

Corollary 3 *A normal mean-variance mixture $F = N_d(\mu + y\beta, y\Delta) \circ G$ is an elliptical distribution if and only if $\beta = \mathbf{0}$, that is, if and only if it is a normal variance mixture.*

Proof The characteristic function of F can be shown to have the form $\phi_F(u) = e^{i\langle u,\mu\rangle} \mathscr{L}_G\big(\frac{\langle u,\Delta u\rangle}{2} - i\langle u, \beta\rangle\big)$ which evidently has the representation $e^{i\langle u,\mu\rangle}\psi\big(\langle u, \Sigma u\rangle\big)$ required by Definition 5 with $\Sigma = \Delta$ and $\psi(t) = \mathscr{L}_G\big(\frac{t}{2}\big)$ if and only if $\beta = \mathbf{0}$. □

Now we leave the general theory and turn our attention to the multivariate GH distributions. These have already been introduced as a natural generalization of the univariate case at the end of the seminal paper [5] and were investigated further in [8, 9]. They are defined as normal mean-variance mixtures with GIG mixing distributions in the following way:

$$GH_d(\lambda, \alpha, \beta, \delta, \mu, \Delta) := N_d(\mu + y\Delta\beta, y\Delta) \circ GIG\big(\lambda, \delta, \sqrt{\alpha^2 - \langle\beta, \Delta\beta\rangle}\big), \quad (27)$$

where it is usually assumed without loss of generality (see p. 15) that $\det(\Delta) = 1$ which we shall also do in the following if not stated otherwise. Due to the parameter restrictions of GIG distributions, the other GH parameters have to fulfil the constraints

$$\lambda \in \mathbb{R},\quad \alpha, \delta \in \mathbb{R}_+,\quad \beta, \mu \in \mathbb{R}^d \quad \text{and} \quad \begin{array}{ll} \delta \geq 0,\ 0 \leq \sqrt{\langle\beta, \Delta\beta\rangle} < \alpha, & \text{if } \lambda > 0, \\ \delta > 0,\ 0 \leq \sqrt{\langle\beta, \Delta\beta\rangle} < \alpha, & \text{if } \lambda = 0, \\ \delta > 0,\ 0 \leq \sqrt{\langle\beta, \Delta\beta\rangle} \leq \alpha, & \text{if } \lambda < 0. \end{array}$$

The meaning and influence of the parameters is essentially the same as in the univariate case (see p. 7). Again, parametrizations with $\delta = 0$, $\alpha = 0$ or $\sqrt{\langle\beta, \Delta\beta\rangle} = \alpha$ have to be understood as limiting cases.

Note that the above definition of multivariate GH distributions as normal mean-variance mixtures of the form $N_d(\mu + y\Delta\beta, y\Delta) \circ G$ is of course equivalent to the representation $N_d(\mu + y\tilde{\beta}, y\Delta) \circ G$ used above because the $d \times d$-matrix Δ is always regular by assumption. The modification of the mean term just simplifies some formulas as we shall see below. For notational consistency with Sect. 2, the term $GH_d(\lambda, \alpha, \beta, \delta, \mu, \Delta)$ will be reserved for multivariate GH distributions with $\beta, \mu \in \mathbb{R}^d$, whereas $GH(\lambda, \alpha, \beta, \delta, \mu)$ denotes a univariate GH distribution with $\beta, \mu \in \mathbb{R}$ as before.

If $\delta > 0$ and $\sqrt{\langle\beta, \Delta\beta\rangle} < \alpha$, the density of $GH_d(\lambda, \alpha, \beta, \delta, \mu, \Delta)$ is given by

$$\begin{aligned} d_{GH_d(\lambda,\alpha,\beta,\delta,\mu,\Delta)}(x) &= \int_0^\infty d_{N_d(\mu+y\Delta\beta,y\Delta)}(x)\, d_{GIG\left(\lambda,\delta,\sqrt{\alpha^2-\langle\beta,\Delta\beta\rangle}\right)}(y)\, dy \\ &= \frac{(\alpha^2 - \langle\beta, \Delta\beta\rangle)^{\frac{\lambda}{2}}}{(2\pi)^{\frac{d}{2}}\alpha^{\lambda-\frac{d}{2}}\delta^\lambda K_\lambda(\delta\sqrt{\alpha^2 - \langle\beta, \Delta\beta\rangle})} \big(\langle x - \mu, \Delta^{-1}(x - \mu)\rangle + \delta^2\big)^{(\lambda-\frac{d}{2})/2} \\ &\quad \times K_{\lambda-\frac{d}{2}}\big(\alpha\sqrt{\langle x - \mu, \Delta^{-1}(x - \mu)\rangle + \delta^2}\big)\, e^{\langle\beta, x-\mu\rangle}. \end{aligned} \quad (28)$$

Remark 2 If the $d \times d$-matrix Δ is replaced by a matrix $\bar{\Delta}$ of the same dimensions with $\det(\bar{\Delta}) \neq 1$, then the normal density $d_{N_d(\mu+y\bar{\Delta}\beta, y\bar{\Delta})}(x)$ has an additional factor $\det(\bar{\Delta})^{-1/2}$ which will be incorporated in the norming constant of $d_{GH_d(\lambda,\alpha,\beta,\delta,\mu,\bar{\Delta})}(x)$. Suppose $\bar{\Delta} = c^{1/d}\Delta$ for some $c > 0$, then $\det(\bar{\Delta}) = c$, and if we also replace $\lambda, \alpha, \beta, \delta, \mu$ by the barred parameters

$$\bar{\lambda} := \lambda, \quad \bar{\alpha} := c^{\frac{1}{2d}}\alpha, \quad \bar{\beta} := \beta, \quad \bar{\delta} := c^{-\frac{1}{2d}}\delta, \quad \bar{\mu} := \mu,$$

then it is easily seen from (28) that the densities of $GH_d(\lambda, \alpha, \beta, \delta, \mu, \Delta)$ and $GH_d(\bar{\lambda}, \bar{\alpha}, \bar{\beta}, \bar{\delta}, \bar{\mu}, \bar{\Delta})$ and thus both distributions coincide. Note that these considerations also remain true for all subsequently defined limit distributions. This again shows that the assumption $\det(\Delta) = 1$ is not an essential restriction. The barred parameters will be used later at some points in Sect. 5 to indicate that $\det(\bar{\Delta}) = 1$ is not assumed there.

If multivariate GH distributions would have been defined as a mixture of the form $N_d(\mu + y\beta, y\Delta) \circ GIG\big(\lambda, \delta, \sqrt{\alpha^2 - \langle\beta, \Delta\beta\rangle}\big)$ (see the remark on the previous page), then the last factor of the density (28) would be $e^{\langle\Delta^{-1}\beta, x-\mu\rangle}$ instead of $e^{\langle\beta, x-\mu\rangle}$, and $\bar{\beta}$ would have to be defined by $\bar{\beta} = c^{1/d}\beta$.

With the special choice $\lambda = -\frac{1}{2}$, one obtains the multivariate normal inverse Gaussian distribution $NIG_d(\alpha, \beta, \delta, \mu, \Delta)$ possessing the density

$$\begin{aligned} d_{NIG_d(\alpha,\beta,\delta,\mu,\Delta)}(x) = \sqrt{\frac{2}{\pi}}\, \frac{\delta\alpha^{\frac{d+1}{2}} e^{\delta\sqrt{\alpha^2-\langle\beta,\Delta\beta\rangle}}}{(2\pi)^{\frac{d}{2}}} \big(\langle x-\mu, \Delta^{-1}(x-\mu)\rangle + \delta^2\big)^{-\frac{d+1}{4}} \\ \times K_{\frac{d+1}{2}}\big(\alpha\sqrt{\langle x-\mu, \Delta^{-1}(x-\mu)\rangle + \delta^2}\big)\, e^{\langle\beta, x-\mu\rangle}. \end{aligned} \tag{29}$$

Let us briefly mention possible weak limits of multivariate GH distributions here. If $\lambda > 0$ and $\delta \to 0$, then by Eqs. (27), (3), and Lemma 3(c) we have convergence to a multivariate Variance-Gamma distribution

$$GH_d(\lambda, \alpha, \beta, \delta, \mu, \Delta) \xrightarrow{w} N_d(\mu + y\Delta\beta, y\Delta) \circ G\big(\lambda, \tfrac{\alpha^2-\langle\beta,\Delta\beta\rangle}{2}\big) = VG_d(\lambda, \alpha, \beta, \mu, \Delta)$$

which has the density

$$\begin{aligned} d_{VG_d(\lambda,\alpha,\beta,\mu,\Delta)}(x) = \frac{(\alpha^2 - \langle\beta, \Delta\beta\rangle)^{\lambda}}{(2\pi)^{\frac{d}{2}}\alpha^{\lambda-\frac{d}{2}}2^{\lambda-1}\Gamma(\lambda)} \big(\langle x-\mu, \Delta^{-1}(x-\mu)\rangle\big)^{(\lambda-\frac{d}{2})/2} \\ \times K_{\lambda-\frac{d}{2}}\big(\alpha\sqrt{\langle x-\mu, \Delta^{-1}(x-\mu)\rangle}\big)\, e^{\langle\beta, x-\mu\rangle}. \end{aligned} \tag{30}$$

For $\lambda < 0$ and $\alpha \to 0$ as well as $\beta \to \mathbf{0}$, we arrive at the multivariate scaled and shifted t-distribution with $f = -2\lambda$ degrees of freedom:

$$GH_d(\lambda, \alpha, \beta, \delta, \mu, \Delta) \xrightarrow{w} N_d(\mu, y\Delta) \circ iG(\lambda, \tfrac{\delta^2}{2}) = t_d(\lambda, \delta, \mu, \Delta).$$

It has the density

$$d_{t_d(\lambda,\delta,\mu,\Delta)}(x) = \frac{\Gamma(-\lambda + \frac{d}{2})}{(\delta^2\pi)^{\frac{d}{2}}\,\Gamma(-\lambda)}\left(1 + \frac{\langle x-\mu, \Delta^{-1}(x-\mu)\rangle}{\delta^2}\right)^{\lambda-\frac{d}{2}}. \tag{31}$$

If $\lambda < 0$, but $\langle \beta, \Delta\beta\rangle \to \alpha^2$, then we have weak convergence to the normal mean-variance mixture

$$GH_d(\lambda, \alpha, \beta, \delta, \mu, \Delta) \xrightarrow{w} N_d(\mu + y\Delta\beta, y\Delta) \circ iG(\lambda, \tfrac{\delta^2}{2})$$

possessing the density

$$\begin{aligned} d_{GH_d(\lambda,\sqrt{\langle\beta,\Delta\beta\rangle},\beta,\delta,\mu,\Delta)}(x) &= \frac{2^{\lambda+1-\frac{d}{2}}\,\delta^{-2\lambda}}{\pi^{\frac{d}{2}}\Gamma(-\lambda)\alpha^{\lambda-\frac{d}{2}}}\left(\langle x-\mu, \Delta^{-1}(x-\mu)\rangle + \delta^2\right)^{(\lambda-\frac{d}{2})/2} \\ &\quad\times K_{\lambda-\frac{d}{2}}\left(\alpha\sqrt{\langle x-\mu, \Delta^{-1}(x-\mu)\rangle + \delta^2}\right) e^{\langle\beta, x-\mu\rangle}, \end{aligned} \tag{32}$$

where $\alpha = \sqrt{\langle\beta, \Delta\beta\rangle}$.

The most important properties of multivariate GH distributions are summarized in the following theorem which goes back to [8, Theorem 1], see also [9, p. 49f]. It shows that this distribution class is closed under forming marginals, conditioning and affine transformations.

Theorem 1 *Suppose $X \sim GH_d(\lambda, \alpha, \beta, \delta, \mu, \Delta)$. Let $(X_1, X_2)^\top$ be a partition of X where X_1 has the dimension r and X_2 the dimension $k = d - r$, and let $(\beta_1, \beta_2)^\top$ and $(\mu_1, \mu_2)^\top$ be similar partitions of β and μ. Furthermore, let*

$$\Delta = \begin{pmatrix} \Delta_{11} & \Delta_{12} \\ \Delta_{21} & \Delta_{22} \end{pmatrix}$$

be a partition of Δ such that Δ_{11} is an $r \times r$-matrix. Then the following holds:

(a) $X_1 \sim GH_r(\lambda^*, \alpha^*, \beta^*, \delta^*, \mu^*, \Delta^*)$ with starred parameters given by $\lambda^* = \lambda$, $\alpha^* = \det(\Delta_{11})^{-\frac{1}{2r}}\sqrt{\alpha^2 - \langle\beta_2, (\Delta_{22} - \Delta_{21}\Delta_{11}^{-1}\Delta_{12})\beta_2\rangle}$, $\beta^* = \beta_1 + \Delta_{11}^{-1}\Delta_{12}\beta_2$, $\delta^* = \det(\Delta_{11})^{\frac{1}{2r}}\delta$, $\mu^* = \mu_1$, and $\Delta^* = \det(\Delta_{11})^{-\frac{1}{r}}\Delta_{11}$.

(b) The conditional distribution of X_2 given $X_1 = x_1$ is $GH_k(\tilde\lambda, \tilde\alpha, \tilde\beta, \tilde\delta, \tilde\mu, \tilde\Delta)$ with tilded parameters $\tilde\lambda = \lambda - \frac{r}{2}$, $\tilde\alpha = \det(\Delta_{11})^{\frac{1}{2k}}\alpha$, $\tilde\beta = \beta_2$, $\tilde\delta = \det(\Delta_{11})^{-\frac{1}{2k}} \times \sqrt{\delta^2 + \langle x_1 - \mu_1, \Delta_{11}^{-1}(x_1 - \mu_1)\rangle}$, $\tilde\mu = \mu_2 + \Delta_{21}\Delta_{11}^{-1}(x_1 - \mu_1)$, and $\tilde\Delta = \det(\Delta_{11})^{\frac{1}{k}} \times (\Delta_{22} - \Delta_{21}\Delta_{11}^{-1}\Delta_{12})$.

(c) Suppose $Y = BX + b$ *where* B *is a regular* $d \times d$*-matrix and* $b \in \mathbb{R}^d$*, then* $Y \sim GH_d(\hat{\lambda}, \hat{\alpha}, \hat{\beta}, \hat{\delta}, \hat{\mu}, \hat{\Delta})$ *where* $\hat{\lambda} = \lambda$*,* $\hat{\alpha} = |\det(B)|^{-\frac{1}{d}}\alpha$*,* $\hat{\beta} = (B^{-1})^\top \beta$*,* $\hat{\delta} = |\det(B)|^{\frac{1}{d}}\delta$*,* $\hat{\mu} = B\mu + b$*, and* $\hat{\Delta} = |\det(B)|^{-\frac{2}{d}} B\Delta B^\top$*.*

Remark 3 An important fact we want to stress here is that the above theorem remains also valid for all multivariate GH limit distributions considered before. Thus, one can in particular conclude from part (b) that the limiting subclass of VG distributions itself is, in contrast to the t limit distributions, not closed under conditioning. This holds because the parameter $\tilde{\delta}$ of the conditional distribution in general is greater than zero, and the parameter $\tilde{\lambda} = \lambda - \frac{r}{2}$ may become negative if the subdimension r is sufficiently large.

Moreover, all margins of $t_d(\lambda, \delta, \mu, \Delta)$ are again t distributed $t_r(\lambda, \delta^*, \mu^*, \Delta^*)$ because if the joint distribution has the parameters $\alpha = 0$ and $\beta = \mathbf{0}$, part (a) of the theorem implies that $\alpha^* = 0$ and $\beta^* = \mathbf{0}$ for every marginal distribution. Similarly, all margins of $VG_d(\lambda, \alpha, \beta, \mu, \Delta)$ are again VG distributions because if $\delta = 0$, then also $\delta^* = 0$. In addition it can be shown that all margins of $GH_d\big(\lambda, \sqrt{\langle\beta, \Delta\beta\rangle}, \beta, \delta, \mu, \Delta\big)$-distributions are of the same limiting type as their joint distribution, too.

Let us finally take a closer look at the moments of multivariate GH distributions. By Definition 4, every random variable X possessing a multivariate normal mean-variance mixture distribution admits the stochastic representation $X \stackrel{d}{=} \mu + Z\beta + \sqrt{Z}AW$ with independent random variables Z and $W \sim N_d(\mathbf{0}, I_d)$. The standardization of W and its independence from Z imply that

$$\begin{aligned} E(X) &= \mu + E(Z)\beta \\ \mathrm{Cov}(X) &= E\big[(X - E(X))(X - E(X))^\top\big] = E(Z)\Delta + \mathrm{Var}(Z)\beta\beta^\top, \end{aligned} \tag{33}$$

with $\Delta = AA^\top$, provided that $E(|Z|), \mathrm{Var}(Z) < \infty$. If $X \sim GH_d(\lambda, \alpha, \beta, \delta, \mu, \Delta)$, then by (27) $X \stackrel{d}{=} \mu + Z\Delta\beta + \sqrt{Z}AW$ and $Z \sim GIG\big(\lambda, \delta, \sqrt{\alpha^2 - \langle\beta, \Delta\beta\rangle}\big)$. Using Proposition 1 and Eq. (7), one obtains explicit expressions for $E(Z)$ and $\mathrm{Var}(Z)$ which can be inserted into the general equations above to finally obtain

$$\begin{aligned} E[GH_d(\lambda, \alpha, \beta, \delta, \mu, \Delta)] &= \mu + \frac{\delta^2}{\zeta}\frac{K_{\lambda+1}(\zeta)}{K_\lambda(\zeta)}\beta, \\ \mathrm{Cov}[GH_d(\lambda, \alpha, \beta, \delta, \mu, \Delta)] &= \frac{\delta^2}{\zeta}\frac{K_{\lambda+1}(\zeta)}{K_\lambda(\zeta)}\Delta + \frac{\delta^4}{\zeta^2}\left(\frac{K_{\lambda+2}(\zeta)}{K_\lambda(\zeta)} - \frac{K^2_{\lambda+1}(\zeta)}{K^2_\lambda(\zeta)}\right)\beta\beta^\top \end{aligned} \tag{34}$$

with $\zeta = \delta\sqrt{\alpha^2 - \langle\beta, \Delta\beta\rangle}$. In case of the Variance-Gamma limits we have

$$E[VG_d(\lambda, \alpha, \beta, \mu, \Delta)] = \mu + \frac{2\lambda}{\alpha^2 - \langle\beta, \Delta\beta\rangle}\beta,$$
$$\text{Cov}[VG_d(\lambda, \alpha, \beta, \mu, \Delta)] = \frac{2\lambda}{\alpha^2 - \langle\beta, \Delta\beta\rangle}\Delta + \frac{4\lambda}{(\alpha^2 - \langle\beta, \Delta\beta\rangle)^2}\beta\beta^\top. \tag{35}$$

Observe that by Lemma 3 both multivariate GH and VG distributions possess moment generating functions and hence finite moments of arbitrary order because the mixing GIG and Gamma distributions do have this property. This is no longer true for the limit distributions with $\lambda < 0$ because the corresponding inverse Gamma mixing distributions only have finite moments up to order $r < -\lambda$. By Theorem 1(a), the marginal distributions of $t_d(\lambda, \delta, \mu, \Delta)$ are given by $t(\lambda, \sqrt{\Delta_{ii}}\delta, \mu_i)$, $1 \le i \le d$ (recall that $\alpha = 0$ and $\beta = \mathbf{0}$ in this case), and from their tail behaviour (see p. 10) one can easily conclude that mean vector and covariance matrix of the t limit distributions are well defined and finite only if $\lambda < -\frac{1}{2}$ resp. $\lambda < -1$. If these constraints are fulfilled, then

$$E[t_d(\lambda, \delta, \mu, \Delta)] = \mu \quad \text{and} \quad \text{Cov}[t_d(\lambda, \delta, \mu, \Delta)] = \frac{\delta^2}{-2\lambda - 2}\Delta. \tag{36}$$

In the other limiting case where $\langle\beta, \Delta\beta\rangle = \alpha^2 > 0$, Eqs. (33) state that necessary and sufficient conditions for the existence of a mean vector and covariance matrix of the limit distributions are that the inverse Gamma mixing distributions have finite means and variances which holds true if and only if $\lambda < -1$ and $\lambda < -2$, respectively. If λ is appropriately small, then

$$E\big[GH_d\big(\lambda, \sqrt{\langle\beta, \Delta\beta\rangle}, \beta, \delta, \mu, \Delta\big)\big] = \mu + \frac{\delta^2}{-2\lambda - 2}\beta,$$
$$\text{Cov}\big[GH_d\big(\lambda, \sqrt{\langle\beta, \Delta\beta\rangle}, \beta, \delta, \mu, \Delta\big)\big] = \frac{\delta^2}{-2\lambda - 2}\Delta + \frac{\delta^4}{4(\lambda+1)^2(-\lambda-2)}\beta\beta^\top. \tag{37}$$

5 On the Dependence Structure of Multivariate GH Distributions

Correlation is probably the most established dependence measure due to its simplicity and its predominant role within the normal world where it characterizes dependencies almost completely. This follows from the fact that the components W_i, $1 \le i \le d$, of a standard normally distributed random vector $W \sim N_d(\mathbf{0}, I_d)$ are independent from each other (the joint density is just the product of the marginal ones in this case) and the stochastic representation

$$X \sim N_d(\mu, \Delta) \iff X \stackrel{d}{=} \mu + AW \text{ where } W \sim N_d(\mathbf{0}, I_d) \text{ and } AA^\top = \Delta.$$

Since X in distribution is nothing but a linear transform of a random vector W with independent (normally distributed) entries, the components of X can, roughly speaking, exhibit at most linear dependencies, and exactly these are specified and quantified by the pairwise correlations. However, things completely change if we depart from normality and consider normal variance mixtures instead. Suppose

$$X \sim N_d(\mu, y\Delta) \circ G, \quad \text{that is,} \quad X \stackrel{d}{=} \mu + \sqrt{Z}AW$$

where $\mathcal{L}(Z) = G$, $W \sim N_d(\mathbf{0}, I_d)$ and $AA^\top = \Delta$ according to Definition 4. As we already remarked on p. 15, the mixing variable Z causes dependencies between the components of X, but these are typically not captured by correlation as the following lemma shows. It is a slightly more general version of [33, Lemma 6.5] which we adopt here since—in our opinion—the result is as simple as illustrative.

Lemma 4 *Suppose that* $X \stackrel{d}{=} \mu + \sqrt{Z}AW$ *has a normal variance mixture distribution where* $E(Z) < \infty$, *and* $\Delta = AA^\top$ *is a* $d \times d$*-diagonal matrix such that* $\mathrm{Cov}(X_i, X_j) = 0$, $1 \le i, j \le d$, $i \ne j$, *by* (33). *Then the* X_i, $1 \le i \le d$, *are independent if and only if* Z *is almost surely constant, that is, if and only if* X *is multivariate normally distributed.*

Proof Because Δ is diagonal (and positive definite by Definition 4), we can assume without loss of generality that also the matrix A is diagonal and $A_{ii} = \sqrt{\Delta_{ii}}$, $1 \le i \le d$. The independence of Z and W and Jensen's inequality then imply

$$\begin{aligned} E\left(\prod_{i=1}^d |X_i - \mu_i|\right) &= E\left((\sqrt{Z})^d \prod_{i=1}^d |\sqrt{\Delta_{ii}}\, W_i|\right) = E\big((\sqrt{Z})^d\big) \prod_{i=1}^d E\big(|\sqrt{\Delta_{ii}}\, W_i|\big) \\ &\ge E\big(\sqrt{Z}\big)^d \prod_{i=1}^d E\big(|\sqrt{\Delta_{ii}}\, W_i|\big) = \prod_{i=1}^d E\big(|X_i - \mu_i|\big). \end{aligned}$$

Since the function $f(x) = x^d$ is strictly convex on $\mathbb{R}_+$ for $d \ge 2$, equality throughout holds if and only if Z is constant almost surely. □

Remark 4 The above result can even be extended: If $X \stackrel{d}{=} \mu + Z\beta + \sqrt{Z}AW$ has a normal mean-variance mixture distribution with $0 < \mathrm{Var}(Z) < \infty$, $\Delta = AA^\top$ is a $d \times d$-diagonal matrix and $\mathrm{Cov}(X_i, X_j) = 0$ for some $1 \le i \ne j \le d$, then X_i and X_j are not independent either. This can be seen as follows: Since Δ is diagonal and $\mathrm{Var}(Z) > 0$, by (33) $\mathrm{Cov}(X_i, X_j) = 0$ implies that $(\beta\beta^\top)_{ij} = 0$. This means, either $\beta_i = 0$ or $\beta_j = 0$ (or both, but then we would be within the setting of Lemma 4 again). Suppose $\beta_i \ne 0$ and $\beta_j = 0$, then we calculate similarly as above

$$\begin{aligned}
E\big((X_i-\mu_i)\,|X_j-\mu_j|\big) &= E\big((\beta_i Z+\sqrt{Z}\sqrt{\Delta_{ii}}W_i)\,|\sqrt{Z}\sqrt{\Delta_{jj}}W_j|\big)\\
&= E\big((\beta_i Z^{\frac{3}{2}}+Z\sqrt{\Delta_{ii}}W_i)\big)E\big(|\sqrt{\Delta_{jj}}W_j|\big) = \beta_i\,E\big(Z^{\frac{3}{2}}\big)\,E\big(|\sqrt{\Delta_{jj}}W_j|\big)\\
&> \beta_i\,E(Z)^{\frac{3}{2}}\,E\big(|\sqrt{\Delta_{jj}}W_j|\big) = E(\beta_i Z)\,E(Z)^{\frac{1}{2}}\,E\big(|\sqrt{\Delta_{jj}}W_j|\big)\\
&> E(\beta_i Z)\,E\big(Z^{\frac{1}{2}}\big)E\big(|\sqrt{\Delta_{jj}}W_j|\big) = E(X_i-\mu_i)\,E\big(|X_j-\mu_j|\big),
\end{aligned}$$

and the inequalities are strict because $f(x)=x^{\frac{3}{2}}$ and $g(x)=\sqrt{x}$ are strictly convex resp. concave and $\mathcal{L}(Z)$ is non-degenerate by assumption.

Thus, in general zero correlation within multivariate normal mean-variance mixture models must not be interpreted as independence. In particular, the components X_i of a generalized hyperbolic distributed random vector $X \sim GH_d(\lambda, \alpha, \beta, \delta, \mu, \Delta)$ can never be independent because Theorem 1(b) states that the conditional distribution $\mathcal{L}\big(X_i \mid (X_1,\ldots,X_{i-1},X_{i+1},\ldots,X_d)^\top = \bar{x}\big) = GH(\tilde\lambda, \tilde\alpha, \tilde\beta, \tilde\delta, \tilde\mu)$ always depends on the vector $\bar{x}$ (at least the parameter $\tilde\delta$ does so) for every $1 \le i \le d$. Moreover, it should be observed that for generalized hyperbolic distributed random variables the maximal attainable absolute correlation is usually strictly smaller than one: the Cauchy–Schwarz inequality states that $|\mathrm{Corr}(X_1,X_2)|=1$ can occur if and only if $X_2 = aX_1+b$ almost surely for some $a,b\in\mathbb{R}$ and $a\neq 0$, but if $X_1 \sim GH(\lambda_1,\alpha_1,\beta_1,\delta_1,\mu_1)$ and $X_2 \sim GH(\lambda_2,\alpha_2,\beta_2,\delta_2,\mu_2)$, the required linear relationship imposes some conditions on the GH parameters. Recall that $aX_1+b \sim GH(\lambda, \frac{\alpha}{|a|}, \frac{\beta}{a}, \delta|a|, a\mu+b)$ by Theorem 1(c). Thus, using the scale- and location-invariant parameters $\zeta_i = \delta_i(\alpha_i^2-\beta_i^2)^{\frac{1}{2}}$ and $\rho_i = \frac{\beta_i}{\alpha_i}$, $i=1,2$, we conclude that $X_2 = aX_1+b$ can hold only if $\zeta_1=\zeta_2$, $|\rho_1|=|\rho_2|$, and $\lambda_1=\lambda_2$.

Having seen that correlation is in general not the tool to precisely describe and measure dependencies in multivariate models, one may ask if there exists a more powerful notion for this purpose. The answer is provided by

Definition 6 A *d-dimensional copula* $C : [0,1]^d \to [0,1]$ is a distribution function on $[0,1]^d$ with standard uniform marginal distributions.

Clearly, the k-dimensional margins of a copula C are also copulas for every $2 \le k < d$. The central role of copulas in the study of multivariate distributions is highlighted by the following fundamental result which goes back to [42]. It not only shows that copulas are inherent in every multivariate distribution, but also that the latter can be constructed by plugging the desired marginal distributions into a suitably chosen copula. A short and elegant proof of Sklar's Theorem which is based on the distributional transform can be found in [37].

Theorem 2 (Sklar's Theorem) *Let F be a d-dimensional distribution function with margins $F_1,\ldots,F_d$. Then there exists a copula* $C : [0,1]^d \to [0,1]$ *such that for all* $x=(x_1,\ldots,x_d)^\top \in [-\infty,\infty]^d$

$$F(x_1,\ldots,x_d) = C\big(F_1(x_1),\ldots,F_d(x_d)\big). \tag{38}$$

If $F_1, \ldots, F_d$ are all continuous, then C is unique, otherwise C is uniquely determined on $F_1(\mathbb{R}) \times \cdots \times F_d(\mathbb{R})$ where $F_i(\mathbb{R})$ denotes the range of F_i.
Conversely, if $C : [0,1]^d \to [0,1]$ is a copula and $F_1, \ldots, F_d$ are univariate distribution functions, then the function $F(x)$ defined by (38) *is a multivariate distribution function with margins $F_1, \ldots, F_d$.*

If all marginal distribution functions F_i of F are continuous and their generalized inverses F_i^{-1} are defined by $F_i^{-1}(u_i) := \inf\{y \mid F_i(y) \geq u_i\}$ (with the usual convention $\inf \emptyset = \infty$), then $F_i\big(F_i^{-1}(u_i)\big) = u_i$. Thus it immediately follows from (38) by inserting $x_i = F_i^{-1}(u_i)$, $u_i \in [0,1]$, $1 \leq i \leq d$, that in this case the unique copula $C_F(u)$ contained in F is given by

$$C_F(u_1, \ldots, u_d) = F\big(F_1^{-1}(u_1), \ldots, F_d^{-1}(u_d)\big). \tag{39}$$

The computation of this so-called implied copula $C_F(u)$ is in general numerically demanding if the distribution function $F(x)$ is not known explicitly. Suppose for example that only the density $f(x)$ of F can be expressed in closed form, then already the determination of a single value $F(x_0)$ requires to evaluate a d-dimensional integral which especially for higher dimensions d can hardly be done sufficiently precise in reasonable time. The same problem arises if one tries to calculate the implied copula from the moment generating function $M_F(u)$ via Fourier inversion (see [35, Theorem 1]). But for multivariate normal mean-variance mixtures it is sometimes possible to significantly reduce the numerical complexity: Suppose that $F = N_d(\mu + y\beta, y\Delta) \circ G$ with known margins F_i possessing Lebesgue densities f_i as above, and let O be an orthogonal $d \times d$-matrix such that $O\Delta O^\top$ is diagonal, then

$$\begin{aligned} C_F(u_1, \ldots, u_d) &= F\big(F_1^{-1}(u_1), \ldots, F_d^{-1}(u_d)\big) \\ &= \int_{-\infty}^{F_d^{-1}(u_d)} \cdots \int_{-\infty}^{F_1^{-1}(u_1)} \int_0^\infty d_{N_d(\mu+y\beta, y\Delta)}(x_1, \ldots, x_d)\, G(\mathrm{d}y)\, \mathrm{d}x_1 \ldots \mathrm{d}x_d \qquad (40) \\ &= \int_0^\infty \prod_{i=1}^d \Phi_{((O(\mu+y\beta))_i, (O\Delta O^\top)_{ii})}\Big(\big(O(F_1^{-1}(u_1), \ldots, F_d^{-1}(u_d))^\top\big)_i\Big)\, G(\mathrm{d}y), \end{aligned}$$

where $\Phi_{(\mu,\sigma^2)}$ denotes the (univariate) distribution function of $N(\mu, \sigma^2)$. The last expression can be evaluated much easier on a computer since it only requires the calculation of one-dimensional integrals (possibly more than one because the values $F_i^{-1}(u_i)$ of the marginal quantile functions may only be obtained by integrating the corresponding densities $f_i(x_i)$ numerically).

If in addition to the marginal distributions F_i also F itself possesses a Lebesgue density $f(x)$, a further simplification can be achieved by using the (implied) copula density $c_F(u)$ which is defined by

$$c_F(u_1, \ldots, u_d) := \frac{\partial C_F(u_1, \ldots, u_d)}{\partial u_1 \ldots \partial u_d} = \frac{f\big(F_1^{-1}(u_1), \ldots, F_d^{-1}(u_d)\big)}{f_1\big(F_1^{-1}(u_1)\big) \cdots f_d\big(F_d^{-1}(u_d)\big)}, \tag{41}$$

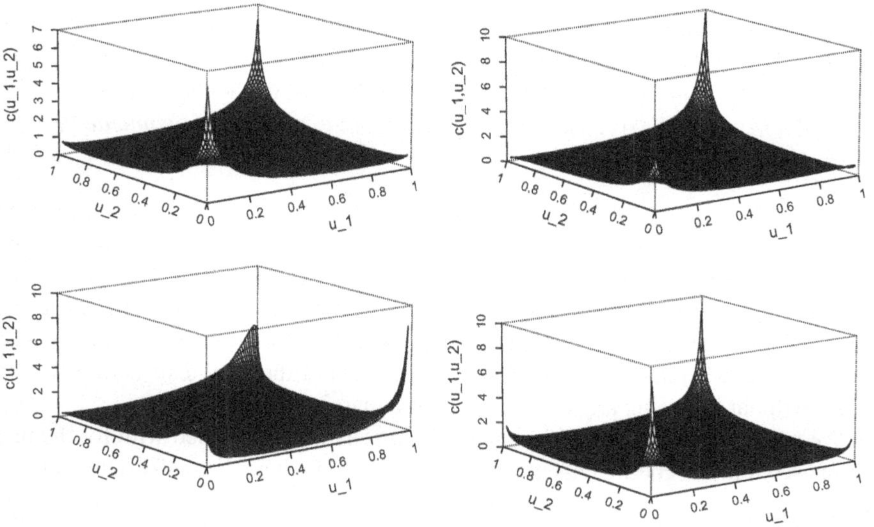

Fig. 1 Densities of implied copulas of bivariate GH distributions and their limits. The underlying distributions are as follows: *top left* Symmetric $NIG_2(10, \mathbf{0}, 0.2, \mathbf{0}, \bar{\Delta})$, *top right* Skewed $NIG_2(10, \left(\begin{smallmatrix}4\\1\end{smallmatrix}\right), 0.2, \mathbf{0}, \bar{\Delta})$, *bottom left* Skewed $NIG_2(4, \left(\begin{smallmatrix}3\\-2\end{smallmatrix}\right), 0.2, \mathbf{0}, \bar{\Delta})$, *bottom right* $t_2(-2, 2, \mathbf{0}, \bar{\Delta})$. For all distributions $\bar{\Delta} = \left(\begin{smallmatrix}1 & \rho\\ \rho & 1\end{smallmatrix}\right)$ with $\rho = 0.3$

where the last equation immediately follows from (39). Combining (41) and Theorem 1(a) allows to calculate the copula densities $c_{GH_d(\lambda,\alpha,\beta,\delta,\mu,\Delta)}(u)$ of all multivariate GH distributions including the aforementioned limits. Some results for the bivariate case are visualized in Fig. 1. Note that the choice of $\rho = 0.3$ implies $\det(\bar{\Delta}) = 1 - \rho^2 < 1$, so the parameters of the t- and NIG distributions are the barred ones $(\bar{\lambda}, \bar{\alpha}, \bar{\beta}, \bar{\delta}, \bar{\mu})$ defined in the remark on page 18. If $\bar{\beta} = \beta = \mathbf{0}$, then by Eqs. (34)–(36) $\bar{\Delta}$ equals the correlation matrix of the related distribution.

Apart from being inherent in every multivariate distribution, the importance of copulas relies on the fact that they encode the dependencies between the margins F_i of F. Many popular dependence measures like, for example, Kendall's tau, Spearman's rho, or the Gini coefficient can be expressed and calculated solely in terms of the associated copulas (see [33, Proposition 7.32 and Corollary 7.35], and [34, Corollary 5.1.14]). Thus the assertion of Sklar's Theorem might alternatively be stated in the following way: Every multivariate distribution can be split up into two parts, the marginal distributions and the dependence structure. The next proposition shows that copulas and hence all dependence measures that can be derived from them are invariant under strictly increasing transformations of the margins. A proof can be found in, e.g., [33, Proposition 7.7].

Proposition 5 *Suppose that* $(X_1, \dots, X_d)^\top$ *is a random vector with joint distribution function* F, *continuous margins* F_i, $1 \le i \le d$, *and implied copula* C_F *given by* (39). *Let* $T_1, \dots, T_d$ *be strictly increasing functions and* G *be the joint distribution function of* $\big(T_1(X_1), \dots, T_d(X_d)\big)^\top$. *Then the implied copulas of* F *and* G *coincide, that is,* $C_F = C_G$.

From the above proposition it especially follows that the correlation of two random variables does not depend on the inherent copula of their joint distribution alone because correlation is invariant under (strictly) increasing linear transformations only, but not under arbitrary increasing mappings. Correlation is also linked to the marginal distributions since it requires them to possess finite second moments to be well defined, whereas by Sklar's Theorem a copula of the joint distribution always exists without imposing any conditions on the margins.

We now turn to the dependence measure we shall be concerned with for the rest of this section, the coefficients of tail dependence, which are formally defined by

Definition 7 Let F be the joint distribution function of the bivariate random vector $(X_1, X_2)^\top$ and F_1, F_2 be the marginal distribution functions of X_1 and X_2, then the *(strong) coefficient of upper tail dependence* of F resp. X_1 and X_2 is

$$\lambda_u := \lambda_u(F) = \lambda_u(X_1, X_2) = \lim_{q \uparrow 1} P\big(X_2 > F_2^{-1}(q) \,|\, X_1 > F_1^{-1}(q)\big),$$

provided a limit $\lambda_u \in [0, 1]$ exists. If $0 < \lambda_u \le 1$, then F resp. X_1 and X_2 are said to be *upper tail dependent*; if $\lambda_u = 0$, they are called *upper tail independent* or asymptotically independent in the upper tail. Similarly, the *(strong) coefficient of lower tail dependence* is

$$\lambda_l := \lambda_l(F) = \lambda_l(X_1, X_2) = \lim_{q \downarrow 0} P\big(X_2 \le F_2^{-1}(q) \,|\, X_1 \le F_1^{-1}(q)\big),$$

again provided a limit $\lambda_l \in [0, 1]$ exists. If $\lambda_u = \lambda_l = 0$, then F resp. X_1 and X_2 are *tail independent*.

If the distribution functions F_1 and F_2 are not continuous and strictly increasing, F_1^{-1} and F_2^{-1} in the previous definition again have to be understood as generalized inverses as defined on p. 24.

The larger (or smaller) q, the more rare is the event $\{X_i > F_i^{-1}(q)\}$ (respectively $\{X_i \le F_i^{-1}(q)\}$). Thus the coefficients of tail dependence are nothing but the limits of the conditional probabilities that the second random variable takes extremal values given the first one also does so. In other words, they may be regarded as the probabilities of joint extremal outcomes of X_1 and X_2. This concept also is of some importance in finance: Suppose, for example, that X_1 and X_2 represent two risky assets. If their joint distribution is lower tail dependent, the possibility that both of them suffer severe losses at the same time cannot be neglected. In portfolio credit risk models, X_1 and X_2 may be the state variables of two different firms or credit instruments, and the coefficient of lower tail dependence can then be interpreted as the probability of a joint default. Tail dependence is a copula property, which is illustrated by the subsequent

Proposition 6 *Let $(X_1, X_2)^\top$ be a bivariate random vector with joint distribution function F, continuous margins F_1, F_2, and implied copula C_F as defined in* (39). *Then the following holds:*

(a) The coefficients of lower and upper tail dependence can be calculated by

$$\lambda_l = \lim_{q \downarrow 0} \frac{C_F(q, q)}{q} \quad \text{and} \quad \lambda_u = \lim_{q \uparrow 1} \frac{1 - 2q + C_F(q, q)}{1 - q}.$$

(b) If in addition F_1, F_2 are strictly increasing, λ_l and λ_u can be obtained by

$$\lambda_l = \lim_{q \downarrow 0} P\big(X_2 \le F_2^{-1}(q) \,|\, X_1 = F_1^{-1}(q)\big) + \lim_{q \downarrow 0} P\big(X_1 \le F_1^{-1}(q) \,|\, X_2 = F_2^{-1}(q)\big),$$
$$\lambda_u = \lim_{q \uparrow 1} P\big(X_2 > F_2^{-1}(q) \,|\, X_1 = F_1^{-1}(q)\big) + \lim_{q \uparrow 1} P\big(X_1 > F_1^{-1}(q) \,|\, X_2 = F_2^{-1}(q)\big).$$

The assertion of part (a) of the proposition can be found in many textbooks on copulas and dependence, and part (b) essentially follows from the ideas of [33, pp. 233 and 249]. A detailed proof can be found in [27, Proposition 2.22].

With the help of these preliminaries, we are now able to give a complete answer to the question which members of the multivariate GH family show tail dependence and which do not. To our knowledge, only symmetric GH distributions have been considered in this regard in the literature so far. By Eq. (27) and Corollary 3, every multivariate GH distribution with parameter $\beta = \mathbf{0}$ belongs to the class of elliptical distributions, thus the tail independence of $GH_d(\lambda, \alpha, \mathbf{0}, \delta, \mu, \Delta)$ (apart from the t limit case with $\alpha = 0$) can be deduced from the more general result below of [28, Theorem 4.3]. It uses the representation $X \stackrel{d}{=} \mu + RAS$ of an elliptically distributed random vector X which was introduced in Proposition 4.

Theorem 3 *Let $X \stackrel{d}{=} \mu + RAS \sim E_d\big(\mu, \Sigma, \psi(t)\big)$ be an elliptically distributed random vector with $\Sigma_{ii} > 0$, $1 \le i \le d$, and $|\rho_{ij}| := |\Sigma_{ij}/\sqrt{\Sigma_{ii}\Sigma_{jj}}| < 1$ for all $i \ne j$. Then the following statements are equivalent:*

(a) The distribution function F_R of R is regularly varying with exponent $p < 0$, that is, $F_R \in \mathscr{R}_p$ (see Definition 3).
(b) $(X_i, X_j)^\top$ is tail dependent for all $i \ne j$.

Moreover, if $F_R \in \mathscr{R}_p$ with $p < 0$, then for all $i \ne j$

$$\lambda_u(X_i, X_j) = \lambda_l(X_i, X_j) = \frac{\int_{(\pi/2 - \arcsin(\rho_{ij}))/2}^{\pi/2} \cos^{|p|}(t)\, dt}{\int_0^{\pi/2} \cos^{|p|}(t)\, dt}.$$

If $X \sim N_d(\mu, y\Delta) \circ G$ has a normal variance mixture distribution which is elliptical by Corollary 3, then X admits the stochastic representations $\mu + \sqrt{Z}AW \stackrel{d}{=}$

$X\mu + RAS$ where the vector μ and the $d \times d$-matrix A on the left and right hand side coincide. This equation suggests that the tail behaviour of the distribution F_R of R is mainly influenced by the distribution G of Z and vice versa. Indeed, one can show that F_R is regularly varying with exponent $2p < 0$ ($F_R \in \mathscr{R}_{2p}$) if and only if $G \in \mathscr{R}_p$ (see [33, pp. 199 and 575f]).

Suppose now $X \sim GH_d(\lambda, \alpha, \mathbf{0}, \delta, \mu, \Delta)$ (excluding the t limiting case for a moment), then by Eq. (1) the density of the corresponding mixing distribution $GIG(\lambda, \delta, \alpha)$ has a right tail of the form described in Eq. (24) with constants $a_2 = \lambda - 1$, $b_2 = \frac{\alpha^2}{2}$, and $c_2 = \frac{(\alpha/\delta)^\lambda}{2K_\lambda(\delta\alpha)}$. (In case of the VG limit, the density of the mixing Gamma distribution $G\big(\lambda, \frac{\alpha^2}{2}\big)$ also has a right tail of the form (24) with the same constants a_2 and b_2, but $c_2 = \frac{(\alpha^2/2)^\lambda}{\Gamma(\lambda)}$.) By Proposition 2 and Definition 2, the distribution functions of $GIG(\lambda, \delta, \alpha)$ and $G\big(\lambda, \frac{\alpha^2}{2}\big)$ both have an exponential right tail with rate b_2. In view of Definition 3 and the subsequent remark, distribution functions with exponential right tails can be regarded as regularly varying with exponent $-\infty$. Consequently, for the distribution function F_R of R in the representation $X \stackrel{d}{=} \mu + RAS$ we have $F_R \in \mathscr{R}_{-\infty}$ as well. Theorem 3 thus yields

$$\lambda_u(X_i, X_j) = \lambda_l(X_i, X_j) = \lim_{p \to -\infty} \frac{\int_{(\pi/2 - \arcsin(\rho_{ij}))/2}^{\pi/2} \cos^{|p|}(t)\,\mathrm{d}t}{\int_0^{\pi/2} \cos^{|p|}(t)\,\mathrm{d}t} = 0,$$

showing the tail independence of all symmetric $GH_d(\lambda, \alpha, \mathbf{0}, \delta, \mu, \Delta)$-distributions with parameter $\alpha > 0$. An alternative way to obtain this result is to show that the *weak* tail dependence coefficient of $GH_d(\lambda, \alpha, \mathbf{0}, \delta, \mu, \Delta)$ is always strictly smaller than 1, which was done in [38].

In the t limiting case, however, we have $X \stackrel{d}{=} \mu + \sqrt{Z} AW \sim t_d(\lambda, \delta, \mu, \Delta)$ with $Z \sim iG\big(\lambda, \frac{\delta^2}{2}\big)$, and from Eq. (4) it is easily seen that the density $d_{iG(\lambda, \delta^2/2)}$ is regularly varying with exponent $\lambda - 1$. Hence $G = F_Z \in \mathscr{R}_\lambda$ and thus, as pointed out above, $F_R \in \mathscr{R}_{2\lambda}$, so we conclude from Theorem 3 that $\lambda_u(X_i, X_j) = \lambda_l(X_i, X_j) > 0$ for all t distributions $t_d(\lambda, \delta, \mu, \Delta)$. The coefficients are quantified more accurately in Theorem 4 below.

This main result of the present section shows that the dependence behaviour can change dramatically if we move from symmetric to skewed GH distributions with parameter $\beta \neq \mathbf{0}$: in addition to tail independence also complete dependence can occur, that is, both of the coefficients λ_l and λ_u may be equal to one.

Theorem 4 *Let* $X \sim GH_2(\lambda, \alpha, \beta, \delta, \mu, \Delta)$ *and define* $\rho := \frac{\Delta_{12}}{\sqrt{\Delta_{11}\Delta_{22}}}$ *as well as* $\bar{\beta}_i := \sqrt{\Delta_{ii}}\beta_i$ *for* $i = 1, 2$. *Then the following holds:*

(a) If $0 \leq \sqrt{\langle \beta, \Delta\beta \rangle} < \alpha$, *then the GH distribution (including possible VG limits) is tail independent if* $-1 < \rho \leq 0$. *If* $0 < \rho < 1$, *then*

$$\lambda_l(X_1, X_2) = \lambda_u(X_1, X_2) = \begin{cases} 0, & c_*, c_*^{-1} > \rho, \\ 1, & \min(c_*, c_*^{-1}) < \rho, \end{cases}$$

where $c_* := \frac{\sqrt{\alpha^2-(1-\rho^2)\bar\beta_2^2}+\bar\beta_1+\rho\bar\beta_2}{\sqrt{\alpha^2-(1-\rho^2)\bar\beta_1^2}+\bar\beta_2+\rho\bar\beta_1}$.

(b) If $\lambda < 0$ *and* $\alpha = 0$, *then* $X \sim t_2(\lambda, \delta, \mu, \Delta)$ *and*

$$\lambda_u(X_1, X_2) = \lambda_l(X_1, X_2) = 2F_{t(\lambda-\frac{1}{2},\sqrt{-2\lambda+1},0)}\left(-\sqrt{\frac{(-2\lambda+1)(1-\rho)}{1+\rho}}\right),$$

where $F_{t(\lambda-\frac{1}{2},\sqrt{-2\lambda+1},0)}$ *is the distribution function of the univariate Student's t-distribution* $t\big(\lambda-\frac{1}{2}, \sqrt{-2\lambda+1}, 0\big)$ *with* $f = -2\lambda+1$ *degrees of freedom.*

(c) Let $\lambda < 0$ *and* $0 < \sqrt{\langle\beta, \Delta\beta\rangle} = \alpha$. *If* $(\bar\beta_1+\rho\bar\beta_2)(\bar\beta_2+\rho\bar\beta_1) < 0$, *then*

$$\lambda_u(X_1, X_2) = \lambda_l(X_1, X_2) = \begin{cases} 0, & \rho < 0, \\ 1, & \rho > 0. \end{cases}$$

If $(\bar\beta_1+\rho\bar\beta_2)(\bar\beta_2+\rho\bar\beta_1) > 0$, *then*

$$\lambda_u(X_1, X_2) = \lambda_l(X_1, X_2) = \begin{cases} 0, & c_*, c_*^{-1} > \rho, \\ 1, & \min(c_*, c_*^{-1}) < \rho, \end{cases} \quad \textit{where } c_* := \frac{\bar\beta_1+\rho\bar\beta_2}{\bar\beta_2+\rho\bar\beta_1}.$$

Proof Propositions 6 and 5 state that tail dependence is a copula property and therefore invariant under strictly increasing transformations of X_1 and X_2. But if $X \sim GH_2(\lambda, \alpha, \beta, \delta, \mu, \Delta)$, the linear transformation $Y = \begin{pmatrix} 1/\sqrt{\Delta_{11}} & 0 \\ 0 & 1/\sqrt{\Delta_{22}} \end{pmatrix}(X-\mu)$ obviously is strictly increasing in each component, and Theorem 1(c) implies that $Y \sim GH_2(\bar\lambda, \bar\alpha, \bar\beta, \bar\delta, \mathbf{0}, \bar\Delta)$ with $\bar\lambda = \lambda, \bar\alpha = \alpha, \bar\beta = \begin{pmatrix} \sqrt{\Delta_{11}} & 0 \\ 0 & \sqrt{\Delta_{22}} \end{pmatrix}\beta, \bar\delta = \delta, \bar\Delta = \begin{pmatrix} 1 & \rho \\ \rho & 1 \end{pmatrix}$ and $\rho := \Delta_{12}/\sqrt{\Delta_{11}\Delta_{22}}$. Note that we here use the barred parameters defined in Remark 2 because in general $\det(\bar\Delta) = 1-\rho^2 < 1$. As already pointed out in Remarks 2 and 3 on pages 18 and 20, these considerations remain also valid for all GH limit distributions. Hence we can and will always assume $X \sim GH_2(\lambda, \alpha, \bar\beta, \delta, \mathbf{0}, \bar\Delta)$ in the following. The fact that Δ is supposed to be positive definite with $\det(\Delta) = 1$ by definition implies the inequality $0 < \frac{1}{\Delta_{11}\Delta_{22}} = \frac{\Delta_{11}\Delta_{22}-\Delta_{12}^2}{\Delta_{11}\Delta_{22}} = 1-\rho^2$, thus $|\rho| < 1$.

(a) If $X \sim GH_2(\lambda, \alpha, \bar\beta, \delta, \mathbf{0}, \bar\Delta)$ and $0 \le \sqrt{\langle\beta, \Delta\beta\rangle} < \alpha$, then by Theorem 1(a) the marginal distributions are $X_1 \sim GH\big(\lambda, (\alpha^2-(1-\rho^2)\bar\beta_2^2)^{1/2}, \bar\beta_1+\rho\bar\beta_2, \delta, 0\big)$ and $X_2 \sim GH\big(\lambda, (\alpha^2-(1-\rho^2)\bar\beta_1^2)^{1/2}, \bar\beta_2+\rho\bar\beta_1, \delta, 0\big)$. To simplify notations we set $\hat\alpha_1 := (\alpha^2-(1-\rho^2)\bar\beta_2^2)^{1/2}$, $\hat\beta_1 := \bar\beta_1+\rho\bar\beta_2$, and $\hat\alpha_2 := (\alpha^2-(1-\rho^2)\bar\beta_1^2)^{1/2}$, $\hat\beta_2 := \bar\beta_2+\rho\bar\beta_1$, then we obtain $\hat\alpha_1^2-\hat\beta_1^2 = \hat\alpha_2^2-\hat\beta_2^2 = \alpha^2-\langle\beta, \Delta\beta\rangle > 0$. Thus the densities of $\mathcal{L}(X_1)$ and $\mathcal{L}(X_2)$ both have tails of the form (24) (see also the remark thereafter), and Proposition 2 implies that the corresponding distribution functions F_1 and F_2 fulfill the assumptions of Lemma 2(b) with $b_1 = \hat\alpha_i+\hat\beta_i$ and $b_2 = \hat\alpha_i-\hat\beta_i$, $i = 1, 2$. From this we conclude that $F_1^{-1}(q) \sim c_l F_2^{-1}(q)$ for $q \downarrow 0$ as well as $F_1^{-1}(q) \sim c_u F_2^{-1}(q)$ for $q \uparrow 1$ where $c_l := \frac{\hat\alpha_2+\hat\beta_2}{\hat\alpha_1+\hat\beta_1} > 0$ and $c_u := \frac{\hat\alpha_2-\hat\beta_2}{\hat\alpha_1-\hat\beta_1} > 0$. Note that $c_l c_u = \frac{\hat\alpha_2^2-\hat\beta_2^2}{\hat\alpha_1^2-\hat\beta_1^2} = 1$ and thus $c_u = c_l^{-1}$. All this also holds in the VG limit case with $\delta = 0$ because Theorem 1(a) still applies there, and the univariate VG marginal densities have tails of the form (24), too (see p. 9).

Theorem 1(b) states that the conditional distribution of X_i given $X_j = x_j$ (where here and in the following $i, j \in \{1, 2\}$ as well as $i \neq j$) is given by $P(X_i \,|\, X_j = x_j) = GH\big(\lambda - \frac{1}{2}, \alpha(1-\rho^2)^{-1/2}, \bar{\beta}_i, \sqrt{\delta^2 + x_j^2}\sqrt{1-\rho^2}, \rho x_j\big)$, and part (c) of the same theorem then yields

$$
\begin{aligned}
P\Big(\frac{X_i - \rho x_j}{\sqrt{\delta^2 + x_j^2}\sqrt{1-\rho^2}} \,\Big|\, X_j = x_j\Big) \\
&= GH\Big(\lambda - \tfrac{1}{2}, \alpha\sqrt{\delta^2 + x_j^2}, \bar{\beta}_i\sqrt{\delta^2 + x_j^2}\sqrt{1-\rho^2}, 1, 0\Big) \\
&=: GH^*_{i|j}\big(\lambda - \tfrac{1}{2}, \alpha, \bar{\beta}_i, \delta, \rho, x_j\big).
\end{aligned}
$$

Again, this also remains true in the VG limit case (see Remark 3 on p. 20). Let $F^q_{i|j}$ denote the distribution function of $GH^*_{i|j}\big(\lambda - \frac{1}{2}, \alpha, \bar{\beta}_i, \delta, \rho, F_j^{-1}(q)\big)$ and set

$$
h_{i|j}(q) := (1-\rho^2)^{-\frac{1}{2}} \frac{F_i^{-1}(q) - \rho F_j^{-1}(q)}{\sqrt{\delta^2 + \big(F_j^{-1}(q)\big)^2}} \qquad \text{for } q \in (0,1),
$$

then we have

$$
\begin{aligned}
\lim_{q\downarrow 0} P\big(X_i \le F_i^{-1}(q) \,\big|\, X_j = F_j^{-1}(q)\big) &= \lim_{q\downarrow 0} F^q_{i|j}\big(h_{i|j}(q)\big), \\
\lim_{q\uparrow 1} P\big(X_i > F_i^{-1}(q) \,\big|\, X_j = F_j^{-1}(q)\big) &= \lim_{q\uparrow 1} 1 - F^q_{i|j}\big(h_{i|j}(q)\big).
\end{aligned}
$$

Moreover, if $\alpha > |\beta| \ge 0$, then $GH(\lambda, r\alpha, r\beta, \delta, \mu) \xrightarrow{w} \epsilon_\mu$ for $r \to \infty$ because

$$
\begin{aligned}
&\lim_{r\to\infty} \phi_{GH(\lambda, r\alpha, r\beta, \delta, \mu)}(u) = \\
&\quad = \lim_{r\to\infty} e^{iu\mu} \left(\frac{(r\alpha)^2 - (r\beta)^2}{(r\alpha)^2 - (r\beta + iu)^2}\right)^{\frac{\lambda}{2}} \frac{K_\lambda\big(\delta\sqrt{(r\alpha)^2 - (r\beta + iu)^2}\big)}{K_\lambda\big(\delta\sqrt{(r\alpha)^2 - (r\beta)^2}\big)} \\
&\quad = \lim_{r\to\infty} e^{iu\mu} \left(\frac{\alpha^2 - \beta^2}{\alpha^2 - (\beta + \frac{iu}{r})^2}\right)^{\frac{\lambda}{2}} \frac{K_\lambda\big(r\delta\sqrt{\alpha^2 - (\beta + \frac{iu}{r})^2}\big)}{K_\lambda\big(r\delta\sqrt{\alpha^2 - \beta^2}\big)} = e^{iu\mu}
\end{aligned}
$$

which implies that $GH^*_{i|j}\big(\lambda - \frac{1}{2}, \alpha, \bar{\beta}_i, \delta, \rho, F_j^{-1}(q)\big)$ converges weakly to the degenerate distribution ϵ_0 if $q \downarrow 0$ or $q \uparrow 1$. From the asymptotic relations of the quantile functions $F_1^{-1}(q)$ and $F_2^{-1}(q)$ we further obtain

$$
\lim_{q\downarrow 0} h_{i|j}(q) = (1-\rho^2)^{-\frac{1}{2}}\big(\rho - c_l^{j-i}\big), \quad \text{and} \quad \lim_{q\uparrow 1} h_{i|j}(q) = (1-\rho^2)^{-\frac{1}{2}}\big(c_l^{i-j} - \rho\big)
$$

(remember $c_u = c_l^{-1}$), consequently

$$\lim_{q\downarrow 0} P\big(X_i \le F_i^{-1}(q)\,\big|\,X_j = F_j^{-1}(q)\big) = F_{\varepsilon_0}\left(\frac{\rho - c_l^{j-i}}{\sqrt{1-\rho^2}}\right) = \begin{cases} 0, & c_l^{j-i} > \rho, \\ 1, & c_l^{j-i} < \rho, \end{cases}$$

as well as

$$\lim_{q\uparrow 1} P\big(X_i > F_i^{-1}(q)\,\big|\,X_j = F_j^{-1}(q)\big) = 1 - F_{\varepsilon_0}\left(\frac{c_l^{i-j} - \rho}{\sqrt{1-\rho^2}}\right) = \begin{cases} 0, & c_l^{i-j} > \rho, \\ 1, & c_l^{i-j} < \rho, \end{cases}$$

and Proposition 6(b) finally implies that $\lambda_l(X_1, X_2) = \lambda_u(X_1, X_2) = 0$ if and only if $c_l, c_l^{-1} > \rho$. Since $c_l > 0$, the conditions are trivially met if $\rho \le 0$. If $0 < \rho < 1$, then at most one of the quantities c_l and c_l^{-1} can be smaller than ρ (note that the convergence to a well-defined limit cannot be assured if $c_l^{j-i} = \rho > 0$, therefore we exclude these possibilities in our considerations). This completes the proof of (a).

(c) Because Theorem 1(a) still applies if $X \sim GH_2(\lambda, \alpha, \bar{\beta}, \delta, \mathbf{0}, \bar{\Delta})$, $\lambda < 0$, and $0 < \sqrt{\langle \beta, \Delta\beta \rangle} = \alpha$, we have, using the notations from above, $X_i \sim GH(\lambda, \hat{\alpha}_i, \hat{\beta}_i, \delta, 0)$, $i = 1, 2$. However, in this case $\hat{\alpha}_i^2 - \hat{\beta}_i^2 = \alpha^2 - \sqrt{\langle \beta, \Delta\beta \rangle} = 0$, hence both marginal distributions are univariate GH limit distributions with $\lambda < 0$ and $\hat{\alpha}_i = |\hat{\beta}_i|$. If $\hat{\beta}_i > 0$, we conclude from Eqs. (20), (21), and Proposition 2 that the tail behaviour of the distribution function is given by $F_i(y) \sim c_{i1}|y|^{\lambda-1}e^{-2\hat{\alpha}_i|y|}$ for $y \to -\infty$ and $1 - F_i(y) \sim c_{i2}|y|^{\lambda}$ as $y \to \infty$ where

$$c_{i1} = \frac{2^{\lambda-1}}{\hat{\alpha}_i^{\lambda+1}\delta^{2\lambda}\Gamma(|\lambda|)} \qquad \text{and} \qquad c_{i2} = \frac{2^{\lambda}}{|\lambda|\hat{\alpha}_i^{\lambda}\delta^{2\lambda}\Gamma(|\lambda|)}.$$

Lemma 2 now states that $F_i^{-1}(q) \sim \frac{\log(q)}{2\hat{\alpha}_i}$ for $q \downarrow 0$ and $F_i^{-1}(q) \sim \big(\frac{c_{i2}}{1-q}\big)^{\frac{1}{|\lambda|}}$ for $q \uparrow 1$. If $\hat{\beta}_i < 0$, then we analogously obtain $F_i^{-1}(q) \sim -\big(\frac{c_{i2}}{q}\big)^{\frac{1}{|\lambda|}}$ as $q \downarrow 0$ and $F_i^{-1}(q) \sim -\frac{\log(1-q)}{2\hat{\alpha}_i}$ as $q \uparrow 1$. Because the case $\hat{\beta}_i = \bar{\beta}_i + \rho\bar{\beta}_j = 0$ is ruled out by assumption, the equality $0 = \hat{\alpha}_i^2 - \hat{\beta}_i^2 = \alpha^2 - (1-\rho^2)\bar{\beta}_j^2 - (\bar{\beta}_i + \rho\bar{\beta}_j)^2$ implies that $\alpha > \sqrt{1-\rho^2}|\bar{\beta}_i|$. Thus, we can proceed along the same lines as in the proof of part (a) and get

$$\lim_{q\downarrow 0} P\big(X_i \le F_i^{-1}(q)\,\big|\,X_j = F_j^{-1}(q)\big) = F_{\epsilon_0}\Big(\lim_{q\downarrow 0} h_{i|j}(q)\Big),$$
$$\lim_{q\uparrow 1} P\big(X_i > F_i^{-1}(q)\,\big|\,X_j = F_j^{-1}(q)\big) = 1 - F_{\epsilon_0}\Big(\lim_{q\uparrow 1} h_{i|j}(q)\Big)$$

if we again exclude the cases where $h_{i|j}(q) \to 0$ for the same reasons as above.

Suppose $\hat{\beta}_1, \hat{\beta}_2 > 0$, then $F_1^{-1}(q) \sim c_l F_2^{-1}(q)$ with $c_l = \frac{\hat{\alpha}_2}{\hat{\alpha}_1} = \frac{\hat{\beta}_2}{\hat{\beta}_1} > 0$ as $q \downarrow 0$ and $F_1^{-1}(q) \sim c_u F_2^{-1}(q)$ with $c_u = \big(\frac{c_{12}}{c_{22}}\big)^{1/|\lambda|} = \big(\frac{\hat{\alpha}_2^{\lambda}}{\hat{\alpha}_1^{\lambda}}\big)^{1/|\lambda|} = \frac{\hat{\beta}_1}{\hat{\beta}_2} = c_l^{-1}$ for $q \uparrow 1$. Consequently, we again have

$$\lim_{q\downarrow 0} h_{i|j}(q) = (1-\rho^2)^{-\frac{1}{2}}\left(\rho - c_l^{j-i}\right), \qquad \lim_{q\uparrow 1} h_{i|j}(q) = (1-\rho^2)^{-\frac{1}{2}}\left(c_l^{i-j} - \rho\right)$$

and conclude, analogously as before, that $\lambda_l(X_1, X_2) = \lambda_u(X_1, X_2) = 0$ if and only if $c_l, c_l^{-1} > \rho$. If $\hat{\beta}_1, \hat{\beta}_2 < 0$, the tail behaviour of the quantile functions is just exchanged ($c_l \rightsquigarrow c_l^{-1}$ and $c_u = c_l^{-1} \rightsquigarrow c_l$), hence the assertion remains also valid in this case.

Finally, let $\hat{\beta}_1 > 0$ and $\hat{\beta}_2 < 0$, then $F_1^{-1}(q) \sim \frac{\log(q)}{2\hat{\alpha}_1}$ and $F_2^{-1}(q) \sim -\left(\frac{c_{22}}{q}\right)^{\frac{1}{|\lambda|}}$ as $q \downarrow 0$, thus $\lim_{q\downarrow 0} \frac{F_1^{-1}(q)}{F_2^{-1}(q)} = 0$ and

$$\lim_{q\downarrow 0} h_{i|j}(q) = \begin{cases} (1-\rho^2)^{-\frac{1}{2}}\rho, & i-j=-1, \\ -\infty, & i-j=1, \end{cases}$$

hence $\lambda_l(X_1, X_2) = 0$ if and only if $\rho < 0$. Further $F_1^{-1}(q) \sim \left(\frac{c_{12}}{1-q}\right)^{\frac{1}{|\lambda|}}$ and $F_2^{-1}(q) \sim -\frac{\log(1-q)}{2\hat{\alpha}_1}$ for $q \uparrow 1$, consequently $\lim_{q\uparrow 1} \frac{F_2^{-1}(q)}{F_1^{-1}(q)} = 0$ and

$$\lim_{q\uparrow 1} h_{i|j}(q) = \begin{cases} -(1-\rho^2)^{-\frac{1}{2}}\rho, & i-j=1, \\ \infty, & i-j=-1, \end{cases}$$

which implies that also $\lambda_u(X_1, X_2) = 0$ if and only if $\rho < 0$. Trivially, all conclusions remain true if $\hat{\beta}_1 < 0$ and $\hat{\beta}_2 > 0$.

(b) The proof of this part goes back to [22], see also [33, p. 250]. If $\lambda < 0$ and $\alpha = 0$, we can assume $X \sim GH_2(\lambda, 0, \mathbf{0}, \delta, \mathbf{0}, \bar{\Delta}) = t_2(\lambda, \delta, \mathbf{0}, \bar{\Delta})$, and the marginal distributions are given by $\mathcal{L}(X_1) = \mathcal{L}(X_2) = GH(\lambda, 0, 0, \delta, 0) = t(\lambda, \delta, 0)$ according to Theorem 1(a), hence we have $F_1^{-1}(q) = F_2^{-1}(q)$ for all $q \in (0,1)$ in this case. By Theorem 1(b), the conditional distributions also coincide, that is, $P(X_2 \,|\, X_1 = x) = P(X_1 \,|\, X_2 = x) = t\big(\lambda, \sqrt{\delta^2 + x^2}\sqrt{1-\rho^2}, \rho x\big)$, and part (c) of the same theorem implies

$$P\left(\frac{\sqrt{-2\lambda+1}}{\sqrt{1-\rho^2}}\,\frac{X_2 - \rho x}{\sqrt{\delta^2+x^2}} \,\middle|\, X_1 = x\right) = P\left(\frac{\sqrt{-2\lambda+1}}{\sqrt{1-\rho^2}}\,\frac{X_1 - \rho x}{\sqrt{\delta^2+x^2}} \,\middle|\, X_2 = x\right)$$
$$= t\big(\lambda - \tfrac{1}{2}, \sqrt{-2\lambda+1}, 0\big).$$

(Note that, in principle, the additional scaling factor $\sqrt{-2\lambda+1}$ is not necessary, but leads to the relation $\delta^2 = -2\lambda + 1 = -2(\lambda - \frac{1}{2})$ of the parameters of the conditional distribution which therewith becomes a classical Student's t-distributiom with $f = -2\lambda + 1$ degrees of freedom.) If we set

$$h(q) := \frac{\sqrt{-2\lambda+1}}{\sqrt{1-\rho^2}} \frac{F_2^{-1}(q) - \rho F_1^{-1}(q)}{\sqrt{\delta^2 + (F_1^{-1}(q))^2}} \quad \text{for } q \in (0,1),$$

we get, using that $F_1^{-1}(q) = F_2^{-1}(q)$,

$$\lim_{q\downarrow 0} h(q) = -\frac{\sqrt{-2\lambda+1}\,(1-\rho)}{\sqrt{1-\rho^2}} = -\sqrt{\frac{(-2\lambda+1)(1-\rho)}{1+\rho}} = -\lim_{q\uparrow 1} h(q),$$

consequently

$$\lim_{q\downarrow 0} P\big(X_2 \le F_2^{-1}(q)\,\big|\,X_1 = F_1^{-1}(q)\big) = \lim_{q\downarrow 0} P\big(X_1 \le F_1^{-1}(q)\,\big|\,X_2 = F_2^{-1}(q)\big)$$
$$= \lim_{q\downarrow 0} F_{t(\lambda-\frac{1}{2},\sqrt{-2\lambda+1},0)}\big(h(q)\big) = F_{t(\lambda-\frac{1}{2},\sqrt{-2\lambda+1},0)}\left(-\sqrt{\frac{(-2\lambda+1)(1-\rho)}{1+\rho}}\right)$$

and

$$\lim_{q\uparrow 1} P\big(X_2 > F_2^{-1}(q)\,\big|\,X_1 = F_1^{-1}(q)\big) = \lim_{q\uparrow 1} P\big(X_1 > F_1^{-1}(q)\,\big|\,X_2 = F_2^{-1}(q)\big)$$
$$= \lim_{q\uparrow 1} 1 - F_{t(\lambda-\frac{1}{2},\sqrt{-2\lambda+1},0)}\big(h(q)\big)$$
$$= 1 - F_{t(\lambda-\frac{1}{2},\sqrt{-2\lambda+1},0)}\left(\sqrt{\frac{(-2\lambda+1)(1-\rho)}{1+\rho}}\right).$$

The symmetry relation $F_{t(\lambda-1/2,\sqrt{-2\lambda+1},0)}(-x) = 1 - F_{t(\lambda-1/2,\sqrt{-2\lambda+1},0)}(x)$ and Proposition 6(b) now yield the desired result. □

The conditions $c_* > \rho$ and $c_*^{-1} > \rho$ in Theorem 4(a) are trivially fulfilled if $\bar{\beta}_1 = \bar{\beta}_2$, because then $c_* = c_*^{-1} = 1$. This, in particular, includes the case $\beta = \mathbf{0}$ which provides an alternative proof for the tail independence of symmetric GH distributions (apart from the t limit case). In general, however, it might seem to be a little bit cumbersome to check these conditions. The following corollary provides a simpler criterion for tail independence of GH distributions.

Corollary 4 *Suppose that* $X \sim GH_2(\lambda, \alpha, \beta, \delta, \mu, \Delta)$ *and* $\rho := \frac{\Delta_{12}}{\sqrt{\Delta_{11}\Delta_{22}}} > 0$. *Then we have* $\lambda_l(X_1, X_2) = \lambda_u(X_1, X_2) = 0$ *if either* $\sqrt{\langle\beta, \Delta\beta\rangle} < \alpha$ *and* $\beta_1\beta_2 \ge 0$, *or* $0 < \sqrt{\langle\beta, \Delta\beta\rangle} = \alpha$ *and* $\beta_1\beta_2 > 0$.

Proof According to Theorem 4(a) and (c), we just have to show that the conditions $\beta_1\beta_2 \ge 0$ resp. > 0 imply $c_*, c_*^{-1} > \rho$. Assume $\sqrt{\langle\beta, \Delta\beta\rangle} < \alpha$ first. If both $\beta_1, \beta_2 \ge 0$, then so are $\bar{\beta}_1 = \sqrt{\Delta_{11}}\beta_1$ and $\bar{\beta}_2 = \sqrt{\Delta_{22}}\beta_2$. Since $\rho > 0$, we see from the inequality $0 < \alpha^2 - \langle\beta, \Delta\beta\rangle = \alpha^2 - \bar{\beta}_1^2 - 2\rho\bar{\beta}_1\bar{\beta}_2 - \bar{\beta}_2^2$ that $\bar{\beta}_i < \alpha$, $i = 1, 2$. Therewith we obtain

$$c_* = \frac{\sqrt{\alpha^2 - (1-\rho^2)\bar{\beta}_2^2} + \bar{\beta}_1 + \rho\bar{\beta}_2}{\sqrt{\alpha^2 - (1-\rho^2)\bar{\beta}_1^2} + \bar{\beta}_2 + \rho\bar{\beta}_1} > \frac{\sqrt{\alpha^2 - (1-\rho^2)\alpha^2} + \rho\bar{\beta}_1 + \rho\bar{\beta}_2}{\alpha + \bar{\beta}_1 + \bar{\beta}_2} = \rho,$$

and an analogous estimate shows that also $c_*^{-1} > \rho$. If $\beta_1 \leq 0$ and $\beta_2 \leq 0$, we use the fact that c_*^{-1} may alternatively be represented by $c_*^{-1} = \frac{\sqrt{\alpha^2-(1-\rho^2)\bar{\beta}_2^2}-\bar{\beta}_1-\rho\bar{\beta}_2}{\sqrt{\alpha^2-(1-\rho^2)\bar{\beta}_1^2}-\bar{\beta}_2-\rho\bar{\beta}_1}$ and similarly conclude that $c_*, c_*^{-1} > \rho$.

Now, let $0 < \sqrt{\langle\beta, \Delta\beta\rangle} = \alpha$ and note that the condition $\beta_1\beta_2 > 0$ implies $(\bar{\beta}_1 + \rho\bar{\beta}_2)(\bar{\beta}_2 + \rho\bar{\beta}_1) > 0$. If both $\beta_1, \beta_2 > 0$, then $c_* = \frac{\bar{\beta}_1+\rho\bar{\beta}_2}{\bar{\beta}_2+\rho\bar{\beta}_1} > \frac{\rho\bar{\beta}_1+\rho\bar{\beta}_2}{\bar{\beta}_2+\bar{\beta}_1} = \rho$, and $c_*^{-1} > \rho$ follows analogously. If $\beta_1, \beta_2 < 0$, the same result is obtained by using the representation $c_* = \frac{-\bar{\beta}_1-\rho\bar{\beta}_2}{-\bar{\beta}_2-\rho\bar{\beta}_1}$. □

An immediate consequence of the preceding corollary is that complete dependence ($\lambda_l(X_1, X_2) = \lambda_u(X_1, X_2) = 1$) within bivariate GH distributions can only occur if the parameters β_1 and β_2 have opposite signs, and one might conjecture that the conditions $c_*, c_*^{-1} > \rho$ are also always fulfilled in these cases such that a two-dimensional GH distribution would be tail independent for almost any choice of parameters. However, this is not true, and it is fairly easy to construct counterexamples: Take $\alpha = 4$, $\bar{\beta}_1 = 3$, $\bar{\beta}_2 = -2$, and $\rho = 0.3$, then $\alpha^2 - \langle\beta, \Delta\beta\rangle = \alpha^2 - \bar{\beta}_1^2 - 2\rho\bar{\beta}_1\bar{\beta}_2 - \bar{\beta}_2^2 = 6.6$ and

$$c_*^{-1} = \frac{\sqrt{\alpha^2 - (1-\rho^2)\bar{\beta}_1^2} + \bar{\beta}_2 + \rho\bar{\beta}_1}{\sqrt{\alpha^2 - (1-\rho^2)\bar{\beta}_2^2} + \bar{\beta}_1 + \rho\bar{\beta}_2} \approx 0.286 < \rho.$$

The corresponding copula density is shown in Fig. 1. In view of Theorem 4, the densities displayed there represent all possible tail dependencies of GH distributions: $NIG_2(10, \mathbf{0}, 0.2, \mathbf{0}, \bar{\Delta})$ and $NIG_2(10, \binom{4}{1}, 0.2, \mathbf{0}, \bar{\Delta})$ are tail independent, $NIG_2(4, \binom{3}{-2}, 0.2, \mathbf{0}, \bar{\Delta})$ is completely dependent, and $t_2(-2, 2, \mathbf{0}, \bar{\Delta})$ lies in between.

The fact that for GH distributions the coefficients of tail dependence can only take the most extreme values 0 and 1 may surely be surprising at first glance, but this phenomenon can also be observed in other distribution classes (making it possibly less astonishing). For example, in [4] a similar behaviour for the upper tail dependence coefficient $\lambda_u(X_1, X_2)$ of a skewed grouped t distribution is found. An alternative derivation and discussion of their results can also be found in [24].

6 Some Further Remarks and Developments

In the present paper, we have restricted ourselves to a thorough discussion of the "classical" uni- and multivariate GH distributions as introduced in [5]. In recent years, however, several possible extensions and generalizations have been suggested in the literature. Therefore, we conclude with a short overview over some of the latter and explain how they emerge from resp. fit into the present context.

Multivariate affine GH models (MAGH) As shown above, the dependence structure of multivariate GH distributions is fairly strict in some sense since it neither allows independent components nor non-trivial values of the tail dependence coefficients. A possible way to relax these restrictions is to consider affine mappings of random vectors with independent GH distributed components: If $Y \stackrel{d}{=} AX + \mu$, where $\mu \in \mathbb{R}^d$, A is a lower triangular $d \times d$-matrix, and $X = (X_1, \ldots, X_d)^\top$ with independent $X_i \sim GH(\lambda_i, \alpha_i, \beta_i, 1, 0)$, $1 \le i \le d$, then Y is said to have a multivariate affine GH distribution. Dependent on the choice of A, $\mathcal{L}(Y)$ can either possess independent margins or show upper and lower tail dependence. A thorough discussion of this model is provided in [39].

Generalized GH distributions (GGH) This extension of the multivariate symmetric GH family is introduced in [25]. It is obtained by allowing for more general radial components R in the elliptical representation. If $X \sim GH_d(\lambda, \alpha, \mathbf{0}, \delta, \mu, \Delta)$, then by Definition 4, Corollary 3, Proposition 4 and the remark thereafter one may represent X by $X \stackrel{d}{=} \mu + \sqrt{Z}\sqrt{Y}AS$ where A is a $d \times d$-matrix fulfilling $AA^\top = \Delta$, the random variables Z, Y, S are independent and $Z \sim GIG(\lambda, \delta, \alpha)$, $Y \sim \chi^2_d$, and S is uniformly distributed on the d-dimensional unit sphere $\mathscr{S}$.

The distribution of $\sqrt{Z}$ from the above representation is a special case of an extended GIG distribution (EGIG) which is defined in [25] as follows: Suppose that $U \sim GIG(\lambda, \delta, \gamma)$, then $U^{\frac{1}{\nu}} \sim EGIG(\nu, \lambda, \delta, \gamma)$ where $\nu > 0$, and the corresponding density is given by

$$d_{EGIG(\nu,\lambda,\delta,\gamma)}(x) = \left(\frac{\gamma}{\delta}\right)^{\lambda} \frac{\nu}{2K_\lambda(\delta\gamma)} x^{\nu\lambda-1} e^{-\frac{1}{2}(\delta^2 x^{-\nu} + \gamma^2 x^{\nu})} \mathbb{1}_{(0,\infty)}(x). \tag{42}$$

Moreover, a generalized Gamma distribution $GG(\nu, a, b)$ with $\nu \in \mathbb{R} \setminus \{0\}$ and $a, b > 0$ is defined by having the density

$$d_{GG(\nu,a,b)}(x) = \frac{|\nu| a^{|\nu| b}}{\Gamma(b)} x^{\nu b-1} e^{-a^{|\nu|} x^{\nu}} \mathbb{1}_{(0,\infty)}(x). \tag{43}$$

These distributions also emerge as limits of extended GIG distributions if $\delta \to 0$; one has $EGIG(\nu, \lambda, 0, \gamma) = GG\big(\nu, \lambda, (\gamma^2/2)^{\frac{1}{\nu}}\big)$. Further, if $Y \sim \chi^2_d$, then $\mathcal{L}\big(\sqrt{Y}\big) = GG(2, 2^{-\frac{1}{2}}, \frac{d}{2})$, and $\sqrt{\frac{Y}{2}} \sim GG\big(2, 1, \frac{d}{2}\big)$. Summing up, a multivariate symmetric GH distributed random vector $X \sim GH_d(\lambda, \alpha, \mathbf{0}, \delta, \mu, \Delta)$ admits the stochastic representation $X \stackrel{d}{=} \mu + UV\bar{A}S$ with independent $U \sim EGIG(2, \lambda, \delta, \alpha)$, $V \sim GG\big(2, 1, \frac{d}{2}\big)$, $\bar{A} = \sqrt{2}A$, and A and S as above. The multivariate generalized GH distributions (GGH) can now be defined as the law of a d-dimensional random vector X admitting the representation $X \stackrel{d}{=} \mu + \bar{U}\bar{V}\bar{A}S$ with independent $\bar{U} \sim EGIG(|2\nu|, \lambda, \delta, \alpha)$, $\bar{V} \sim GG\big(2\nu, 1, \frac{d}{2\nu}\big)$, a random vector S that is uniformly distributed on the unit sphere

$\mathscr{S}$, and a $d \times d$-matrix $\bar{A}$ satisfying $\bar{A}\bar{A}^\top = \bar{\Delta}$ (see [25, Theorem 5]). We then write $X \sim GGH_d(\lambda, \alpha, \delta, \mu, \bar{\Delta}, \nu)$. Apart from the multivariate symmetric GH distributions ($\nu = 1$), this class also contains multivariate generalized t-distributions as well as generalized multivariate VG distributions as special cases for appropriate parameter choices. By construction and Proposition 4, all these distributions are elliptic, so their tail dependence coefficients can be derived with help of Theorem 3 along the same lines as pointed out on p. 27f.

Remark 5 We changed the order and notation of the GGH parameters here to make them fit better into the present context and to more clearly explain how this class emerges as a quite natural extension of multivariate symmetric GH distributions. In [25], GGH distributions are denoted by $GGH_d(\mu, \Sigma, \beta, a, b, p)$, which is related to our notation as follows:

$$\mu = \mu, \quad \Sigma = \bar{\Delta}, \quad \beta = \nu, \quad a = \delta, \quad b = \alpha, \quad p = \lambda.$$

GH factor models and copulas We finally want to remark that the dependence structure of factor models for credit portfolios which have already been mentioned on p. 12 significantly differs from that of multivariate GH distributions discussed above. Recall that the state variables X_i in general factor models are given by

$$X_i := \sqrt{\rho}M + \sqrt{1-\rho}Z_i, \qquad 0 \le \rho < 1, \quad i = 1, \ldots, N, \tag{44}$$

where $M, Z_1, \ldots, Z_N$ are assumed to be independent and, in addition, the Z_i are identically distributed (hence so are the X_i). The corresponding distribution functions are denoted by F_M, F_Z, F_X and are usually supposed to be continuous and strictly increasing on $\mathbb{R}$. If M and the Z_i are standard normally distributed ($M, Z_i \sim N(0, 1)$), then also the joint distribution of the X_i is a multivariate normal distribution with the associated implied copula. However, if we assume the factors M and Z_i to follow a GH distribution ($M \sim GH(\lambda_M, \alpha_M, \beta_M, \delta_M, \mu_M)$, $Z_i \sim GH(\lambda_Z, \alpha_Z, \beta_Z, \delta_Z, \mu_Z)$ for all $1 \le i \le N$), then the distribution of the random vector $X = (X_1, \ldots, X_N)^\top$ is not a multivariate generalized hyperbolic one. This can easily be deduced from the fact that the X_i in general are not GH distributed due the lack of stability under convolutions of the GH class, whereas a multivariate GH distribution must always have univariate GH margins according to Theorem 1(a). Consequently, the implied copula C_{G_X} of the distribution G_X of the vector X also differs from the implied copula of a multivariate GH distribution. The factor copula C_{G_X} can be calculated by

$$\begin{aligned} C_{G_X}(u_1, \ldots, u_N) &= G_X\big(F_X^{-1}(u_1), \ldots, F_X^{-1}(u_N)\big) \\ &= E\big[P\big(X_1 \le F_X^{-1}(u_1), \ldots, X_N \le F_X^{-1}(u_N) \,|\, M\big)\big] \\ &= \int_{\mathbb{R}} \prod_{i=1}^{N} F_Z\left(\frac{F_X^{-1}(u_i) - \sqrt{\rho}y}{\sqrt{1-\rho}}\right) F_M(\mathrm{d}y) \end{aligned} \tag{45}$$

and admits tail dependence $(\lambda_u(X_i, X_j), \lambda_l(X_i, X_j) > 0, \quad 1 \leq i \neq j \leq N)$ if and only if the M is heavy tailed, that is, $F_M \in \mathscr{R}_p$ for some $-\infty < p < 0$ (see Definition 3). This has been shown in [32]. Hence, we can conclude that factor models with GH distributions can show tail dependence if and only if $F_M = t(\lambda, \delta, \mu)$ or $F_M = GH(\lambda, \alpha, \pm\alpha, \delta, \mu)$.

References

1. Aas, K., Haff, I.H.: The generalized hyperbolic skew Student's t-distribution. J. Fin. Econ. **4**, 275–309 (2006)
2. Abramowitz, M., Stegun, I.A.: Handbook of Mathematical Functions, 5th edn. Dover Publications, New York (1968)
3. Artzner, P., Delbaen, F., Eber, J.-M., Heath, D.: Coherent measures of risk. Math. Fin. **9**(3), 203–228 (1999)
4. Banachewicz, K., van der Vaart, A.: Tail dependence of skewed grouped t-distributions. Stat. Prob. Lett. **78**, 2388–2399 (2008)
5. Barndorff-Nielsen, O.E.: Exponentially decreasing distributions for the logarithm of particle size. Proc. Roy. Soc. London A **353**, 401–419 (1977)
6. Barndorff-Nielsen, O.E.: Processes of normal inverse Gaussian type. Fin. Stoch. **2**, 41–68 (1998)
7. Bingham, N.H., Goldie, C.M., Omey, E.: Regularly varying probability densities. Publ. Inst. Math. Beograd **80**(94), 47–57 (2006)
8. Blæsild, P.: The two-dimensional hyperbolic distribution and related distributions, with an application to Johannsen's bean data. Biometrika **68**, 251–263 (1981)
9. Blæsild, P., Jensen, J.L.: Multivariate distributions of hyperbolic type. In: Taillie, C., Patil, G.P., Baldessari, B. (eds.) Statistical Distributions in Scientific Work 4, pp. 45–66. Reidel, Dordrecht (1981)
10. Daul, S., De Giorgi, E., Lindskog, F., McNeil, A.: Using the grouped t-copula. Risk 73–76 (2003)
11. Eberlein, E.: Application of generalized hyperbolic Lévy motions to finance. In: Barndorff-Nielsen, O.E., Mikosch, T., Resnick, S. (eds.) Lévy Processes: Theory and Applications, pp. 319–336. Birkhäuser, Boston (2001)
12. Eberlein, E., Frey, R., v. Hammerstein, E.A.: Advanced credit portfolio modeling and CDO pricing. In: Jäger, W., Krebs, H.-J. (eds.) Mathematics - Key Technology for the Future: Joint Projects between Universities and Industry 2004–2007, pp. 253–280. Springer, Berlin (2008)
13. Eberlein, E., v. Hammerstein, E.A.: Generalized hyperbolic and inverse Gaussian distributions: limiting cases and approximation of processes. In: Dalang, R.C., Dozzi, M., Russo, F. (eds.) Seminar on Stochastic Analysis, Random Fields and Applications IV, Progress in Probability Vol. 58, pp. 221–264, Birkhäuser, Basel (2004)
14. Eberlein, E., Keller, U.: Hyperbolic distributions in finance. Bernoulli **1**, 281–299 (1995)
15. Eberlein, E., Kluge, W.: Exact pricing formulae for caps and swaptions in a Lévy term structure model. J. Comp. Fin. **9**, 99–125 (2006)
16. Eberlein, E., Kluge, W.: Calibration of Lévy term structure models. In: Fu, M.C., Jarrow, R.A., Yen, J.-Y., Elliott, R.J. (eds.) Advances in Mathematical Finance, pp. 147–172. Birkhäuser, Boston (2007)
17. Eberlein, E., Koval, N.: A cross-currency Lévy market model. Quant. Fin. **6**, 465–480 (2006)
18. Eberlein, E., Özkan, F.: Time consistency of Lévy models. Quant. Fin. **3**, 40–50 (2003)
19. Eberlein, E., Prause, K.: The generalized hyperbolic model: financial derivatives and risk measures. In: Geman, H., Madan, D.B., Pliska, S., Vorst, T. (eds.) Mathematical Finance—Bachelier Congress 2000, pp. 245–267. Springer, Berlin (2002)

20. Eberlein, E., Raible, S.: Term structure models driven by general Lévy processes. Math. Fin. **9**, 31–54 (1999)
21. Embrechts, P., Goldie, C.M.: On closure and factorization properties of subexponential distributions. J. Austr. Math. Soc. A **29**, 243–256 (1980)
22. Embrechts, P., McNeil, A.J., Straumann, D.: Correlation and dependence in risk management: properties and pitfalls. In: Dempster, M.A.H. (ed.) Risk Management: Value at Risk and Beyond, pp. 176–223. Cambridge University Press, Cambridge (2002)
23. Entringer, R.C.: Functions and inverses of asymptotic functions. Am. Math. Monthly **74**(9), 1095–1097 (1967)
24. Fung, T., Seneta, E.: Tail dependence for two skew t distributions. Stat. Prob. Letters **80**, 784–791 (2010)
25. Fung, T., Seneta, E.: Extending the multivariate generalised t and generalised VG distributions. J. Mult. Ana. **101**, 154–164 (2010)
26. Good, I.J.: On the population frequencies of species and the estimation of population parameters. Biometrika **40**, 237–264 (1953)
27. v. Hammerstein, E.A.: Generalized hyperbolic distributions: Theory and applications to CDO pricing, Ph.D. Thesis, University of Freiburg (2011). http://www.freidok.uni-freiburg.de/volltexte/7974/
28. Hult, H., Lindskog, F.: Multivariate extremes, aggregation and dependence in elliptical distributions. Adv. Appl. Prob. **34**, 587–608 (2002)
29. Jørgensen, B.: Statistical Properties of the Generalized Inverse Gaussian distribution. Lecture Notes in Statistics, vol. 9. Springer, New York (1982)
30. Madan, D.B., Carr, P.P., Chang, E.C.: The variance gamma process and option pricing. Europ. Fin. Rev. **2**, 79–105 (1998)
31. Madan, D.B., Seneta, E.: The variance gamma (V.G.) model for share market returns. J. Bus. **63**, 511–524 (1990)
32. Malevergne, Y., Sornette, D.: How to account for extreme co-movements between individual stocks and the market. J. Risk **6**, 71–116 (2004)
33. McNeil, A.J., Frey, R., Embrechts, P.: Quantitative Risk Management, Revised edn. Princeton University Press, Princeton (2015)
34. Nelsen, R.B.: An Introduction to Copulas, 2nd edn. Springer Series in Statistics, Springer, New York (2006)
35. Papapantoleon, A.: Computation of copulas by Fourier methods. In: Glau, K., Scherer, M., Zagst, R. (eds.) Innovations in Quantitative Risk Management, pp. 347–354. Springer, Heidelberg (2015)
36. Pillichshammer, F., Leobacher, G.: A method for approximate inversion of the hyperbolic CDF. Computing **69**(4), 291–303 (2002)
37. Rüschendorf, L.: On the distributional transform, Sklar's theorem, and the empirical copula process. J. Stat. Plan. Inf. **139**(11), 3921–3927 (2009)
38. Schlueter, S., Fischer, M.: The weak tail dependence coefficient of the elliptical generalized hyperbolic distribution. Extremes **15**(2), 159–174 (2012)
39. Schmidt, R., Hrycej, T., Stützle, E.: Multivariate distribution models with generalized hyperbolic margins. Comp. Stat. Data Ana. **50**, 2065–2096 (2006)
40. Sichel, H.: Statistical valuation of diamondiferous deposits. J. South. Afric. Inst. Min. Metal. **73**, 235–243 (1973)
41. Sichel, H.: On a distribution representing sentence-length in written prose. J. Roy. Stat. Soc. A. **137**, 25–34 (1974)
42. Sklar, A.: Fonctions de répartition à n dimensions et leurs marges. Publ. Inst. Stat. Univ. Paris **8**, 229–231 (1959)
43. Watson, G.N.: A Treatise on the Theory of Bessel Functions (reprinted 2nd edn.). Cambridge University Press, Cambridge (1952)

Gamma Kernels and *BSS/LSS* Processes

Ole E. Barndorff-Nielsen

Abstract This paper reviews the roles of gamma type kernels in the theory and modelling for Brownian and Lévy semistationary processes. Applications to financial econometrics and the physics of turbulence are pointed out.

Keywords Ambit stochastics · Semistationary · Volatility · Green's function · Stochastic calculus · Identification

1 Introduction

The use of gamma kernels in modelling Brownian and Lévy semistationary (*BSS* and *LSS*) processes, at first introduced as a simple convenient choice, has turned out to be of a more significant nature than first envisaged. This paper reviews the roles the kernels have had in the study of these and related types of processes and their applications.[1]

BSS and *LSS* processes are prominent examples of the types of continuous time stationary processes on $\mathbb{R}$ studied in Ambit Stochastics, a concept introduced in [14]. Two main areas of applications of such processes are financial econometrics and the physics of turbulence, cf. for instance [21] respectively [43].

In its full generality Ambit Stochastics is a framework for modelling tempo-spatial dynamic fields. A main point of Ambit Stochastics is that it specifically incorporates terms modelling stochastic volatility. This is true in particular of *BSS* and *LSS* processes.

[1]Proofs and technical details are, in most cases, not presented here. For these and related literature we refer to the papers cited.

O.E. Barndorff-Nielsen (✉)
Department of Mathematics and CREATES (Funded by the Danish National Research Foundation Under Grant DNRF78), Thiele Centre, University of Aarhus, Aarhus, Denmark
e-mail: oebn@math.au.dk

© Springer International Publishing Switzerland 2016

J. Kallsen and A. Papapantoleon (eds.), *Advanced Modelling in Mathematical Finance*, Springer Proceedings in Mathematics & Statistics 189,
DOI 10.1007/978-3-319-45875-5_2

The papers [8, 11, 49] review recent developments in the theory and applications of Ambit Stochastics.

Section 2 recalls the definitions of *BSS* and *LSS* processes and presents some instances of the role of the gamma kernels, including an illustration of the modelling capability. Section 3 points out that the gamma kernel has an interpretation as a Green's function corresponding to a certain fractional differential operator.

The asymptotic behaviour of the autocorrelation functions of *BSS* and *LSS* processes is of crucial importance for their applications, not least in regard to the modelling of turbulence, and this is reviewed in Sect. 4 under the gamma kernel assumption.

An outstanding issue is the establishment of an Ito type stochastic calculus for *BSS* and *LSS* processes; the point here is that these types of processes are not in general semimartingales. An important step in this direction has been a detailed study of the path properties of such processes, as discussed in Sect. 5.

The questions of what can be deduced about the ingredients of a *BSS* or *LSS* process based on knowledge of its law and/or from high frequency observations of its sample path form the topic of Sect. 6. This involves both purely theoretical reasoning and central questions of inference.

2 *BSS* and *LSS* Processes

The concept of Brownian semistationary processes, or *BSS* processes, was introduced in [16], cf. also [14, 15]. Such a process is of the form

$$Y_t = \mu + \int_{-\infty}^{t} g(t-s)\sigma_s \mathrm{d}B_s + \int_{-\infty}^{t} q(t-s)a_s \mathrm{d}s \tag{1}$$

where B is Brownian motion, σ and a are stochastic processes and g and q are deterministic kernels with $g(t) = h(t) = 0$ for $t \leq 0$. The process Y is stationary provided σ and a are stationary, as we shall henceforth assume. The intended role of the processes σ and a is to model volatility, or intermittency as it is called in turbulence.

For simplicity we assume from now on that σ and a are independent of the Brownian motion B. We note however that a very general treatment of stochastic integration theory for Ambit Stochastics is given in [28].

The specification (1) is a particular case of the general concept of *LSS* (Lévy semistationary) processes defined as

$$Y_t = \mu + \int_{-\infty}^{t} g(t-s)\sigma_s \mathrm{d}L_s + \int_{-\infty}^{t} q(t-s)a_s \mathrm{d}s, \tag{2}$$

where L denotes an arbitrary Lévy process on $\mathbb{R}$. This concept was introduced in [7] and has been further studied for instance in [21, 48, 49, 54], and references given there. One of the roles of processes of type *LSS* is to model volatility/intermittency. General conditions for existence of the stochastic integrals in (1) and (2) are given in [18].

We refer to

$$G_t = \int_{-\infty}^{t} g\,(t-s)\,\mathrm{d}B_s \tag{3}$$

as the Gaussian base process.

The case where $a = \sigma^2$, i.e.

$$Y_t = \mu + \int_{-\infty}^{t} g(t-s)\sigma_s \mathrm{d}B_s + \int_{-\infty}^{t} q(t-s)\sigma_s^2 \mathrm{d}s, \tag{4}$$

can be seen as a stationary process analogue of the so-called *BNS* model, discussed extensively in financial econometrics, see for instance [22].

In this paper we discuss cases where g, and possibly also q, is of the gamma type

$$g\,(t;\nu,\lambda) = \frac{\lambda^\nu}{\Gamma(\nu)} t^{\nu-1} e^{-\lambda t}. \tag{5}$$

The form of the gamma kernel means that small and large lag behaviour of Y can be controlled separately. The more general form $g\,(t) = t^{\nu-1} f\,(t)$ with f continuous and slowly varying at 0 offers the same type of control, and many of the asymptotic results in the literature on *BSS/LSS* processes are derived under this latter assumption. However, the gamma kernel has some very particular properties of key relevance.

We note that with g as the gamma kernel the restriction $\nu > \frac{1}{2}$ is needed for the stochastic integral in (1) to be well defined and that (3) constitutes a semimartingale only if $\nu = 1$ or $\nu > \frac{3}{2}$. For $\nu \in \left(\frac{1}{2}, \frac{3}{2}\right)$ the process is Hölder continuous with index less than $\nu - \frac{1}{2}$. These aspects and some of their consequences are discussed in [9, 10, 16, 19] and will be touched upon later in the paper.

As models for the timewise development of the main component of the velocity vector in a homogeneous turbulent field stochastic processes of *BSS* type have been extensively studied probabilistically and compared to empirical and simulated data, see [14, 15, 34, 43]. Of special interest in the context of turbulence are the cases where the roughness parameter ν of the gamma kernel satisfies either $\frac{1}{2} < \nu < 1$ or $1 < \nu < \frac{3}{2}$. Near 0 the gamma kernel behaves quite differently depending on whether $\frac{1}{2} < \nu < 1$ or $1 < \nu < \frac{3}{2}$, tending respectively to 0 and ∞. The dynamics of the process (1) is significantly different in the two cases and this has strong consequences with respect to path behaviour and to inference on volatility/intermittency, see Sects. 5 and 6.

Considering the setting (4) we note that to accomodate the manifest observed skewness in the distribution of velocity differences in turbulence, cf. [3], the conditional mean $\mathrm{E}\,\{Y_{t+u} - Y_t|\sigma\}$ should be of the same order as $\mathrm{V}\,\{Y_{t+u} - Y_t|\sigma\}^{\frac{1}{2}}$ for small values of u. This can be achieved by having

$$q^2(u) \sim \int_0^u g^2(s)\,\mathrm{d}s \tag{6}$$

for $u \downarrow 0$. With g as the gamma kernel a natural way of obtaining this is to take $q(t) = g\left(t; \nu - \frac{1}{2}, \lambda\right)$. For $\nu = \frac{5}{6}$ the resulting process may, for suitable choice of the volatility/intermittency process σ^2, be considered as a temporal stochastic model for fully developed turbulence; cf. the following Sect. 4.

It is natural to extend the concept of *LSS* processes to the specification

$$Y_t = \mu + \int_{-\infty}^t g(t-s)\sigma_s \mathrm{d}L_s^T + \int_{-\infty}^t q(t-s)a_s \mathrm{d}s \tag{7}$$

where T denotes a time change and $L_t^T = L(T(t))$. Here T and σ represent the two different aspects of the volatility: intensity and amplitude. For a discussion of time change in stochastic processes and its role in modelling volatility/intermittency see [17].

Note 1 *Convolution* Let

$$Y_t = \int_{-\infty}^t g(t-s)\sigma_s \mathrm{d}L_s^T$$

and let h be a shift kernel. Then under mild conditions (cf. the Fubini Theorem presented in [2]) we have

$$\int_{-\infty}^t h(t-s)Y_s \mathrm{d}s = \int_{-\infty}^t h * g(t-s)\sigma_s \mathrm{d}L_s^T$$

where $h * g$ is the convolution of h and g. The resulting process is again of *LSS* type, the left hand side constituting a natural operation on *LSS* processes while convolution of kernels, as on the right hand, is a useful way to flexible modelling. As a concrete example, in [43] the convolution of two gamma kernels is used successfully as description of the spectral density function in well developed turbulence.

The paper [48] discusses the case where

$$Y_t = \int_{-\infty}^t g(t-s; \nu, \lambda)\mathrm{d}L_s \tag{8}$$

in great detail. Necessary and sufficient conditions for the existence of Y are given in terms of ν and the Lévy measure v of L, under the assumption that σ is predictable, strongly stationary and square integrable and satisfies $\mathrm{E}\left\{\sigma_0^{2(1-\nu)}\right\} < \infty$. Provided the Lévy measure has log moment, existence is guaranteed for all $\nu > \frac{1}{2}$ while for $0 < \nu \leq \frac{1}{2}$ an additional condition on the Lévy measure is needed.

Note 2 *Selfdecomposability* A striking example of the special character of the gamma kernel is the fact that whenever the process (8) is well-defined the one-dimensional law of Y is self-decomposable even if the driving Lévy process L does not have that property. In view of how the class of selfdecomposable distributions is defined this is both remarkable and difficult to explain. (The proof given in [48] is by direct analytical derivation and does not throw light on the probabilistic aspect.)

On the other hand, as shown in [13], the process Y as such is selfdecomposable if and only if L is selfdecomposable.

We conclude this Section by an illustration of the flexibility of modelling using gamma kernels. Here, as is often convenient, the volatility/intermittency process σ^2 is taken to be of *LSS* form.

Example 1 BSS process with GH marginals A stationary *BSS* processes with generalised hyperbolic (GH) marginals was used in [7] in connection with a study on modelling electricity spot prices by Lévy semistationary processes. As an illustration of the versatility of *BSS/LSS* processes we here represent the proof of the existence of such a GH related process.

We recall that the GH laws are analytically very tractable and have been found to fit empirical distributions in a wide range of applications (cf. for instance [3, 30, 31, 34]. Also, in a recent extensive development of the Kolmogorov-Obukhov statistical theory of turbulence, Björn Birnir [23–25] has formulated a stochastic version of the Navier-Stokes equations under which the velocity differences follow GH distributions and this theoretical study is backed by a detailed empirical and simulation based analysis showing excellent agreement to the GH form.

The existence of stationary *BSS* processes having generalised hyperbolic marginals is established on the basis of the form (4) by suitable choice of g and q as gamma kernels and by taking σ^2 as a particular *LSS* process, specifically as a generalised inverse Gaussian Ornstein-Uhlenbeck process (*GIG*-OU process).

Note first that, whatever g, q and σ^2, the conditional law of Y_t given σ is normal:

$$Y_t|\sigma \overset{law}{=} N\left(\mu + \beta \int_{-\infty}^{t} q(t-s)\,\sigma_s^2 \mathrm{d}s, \int_{-\infty}^{t} g^2(t-s)\,\sigma_s^2 \mathrm{d}s\right).$$

Now suppose that σ^2 follows an *LSS* process given by

$$\sigma_t^2 = \int_{-\infty}^{t} h(t-s)\,\mathrm{d}L_s \tag{9}$$

where L is a subordinator. Then, by the stochastic Fubini theorem we find

$$\begin{aligned}\int_{-\infty}^{t} q(t-s)\,\sigma_s^2 \mathrm{d}s &= \int_{-\infty}^{t}\int_{u}^{t} q(t-s)\,h(s-u)\,\mathrm{d}s\mathrm{d}L_u\\ &= \int_{-\infty}^{t} k(t-u)\,\mathrm{d}L_u\end{aligned}$$

where $k = q * h$, the convolution of q and h. Similarly,

$$\int_{-\infty}^{t} g^2(t-s)\,\sigma_s^2 \mathrm{d}s = \int_{-\infty}^{t} m(t-u)\,\mathrm{d}L_u$$

with $m = g^2 * h$.

Next, for $\frac{1}{2} < \nu < 1$ define g by

$$g(t) = \left(\lambda \frac{\Gamma(2\nu-1)}{\Gamma(\nu)^2}\right)^{-\frac{1}{2}} g\left(t;\nu,\frac{\lambda}{2}\right).$$

Then we have

$$g^2(t) = g(t;2\nu-1,\lambda).$$

Hence, if

$$h(t) = g(t;2(1-\nu),\lambda)$$

and if, moreover,

$$q(t) = g(t;2\nu-1,\lambda)$$

we obtain

$$k(t) = m(t) = e^{-\lambda t}.$$

In other words,

$$Y_t|\sigma \overset{law}{=} N\left(\mu + \beta\vartheta_t^2, \vartheta_t^2\right)$$

where

$$\vartheta_t^2 = \int_{-\infty}^{t} e^{-\lambda(t-u)}\mathrm{d}L_u. \tag{10}$$

It follows that if the subordinator L is such that ϑ_t^2 has the generalised inverse Gaussian law $GIG(\delta,\gamma)$ then the law of Y_t is the generalised hyperbolic $GH(\alpha,\beta,\mu,\delta)$ (where $\alpha = \sqrt{\beta^2+\gamma^2}$). The existence of such a subordinator follows from a theorem of Jurek and Verwaat, see [38], according to which a random variable X is representable in law on the form

$$X \overset{law}{=} \int_0^\infty e^{-\lambda t}\mathrm{d}L_t^T \tag{11}$$

if and only if the distribution of X is selfdecomposable; and selfdeconposability of GIG has been established in [33].

3 Gamma Kernel as Green's Function

For any $\gamma \in (0, 1)$ and $n \in \mathbb{N}$ the Caputo fractional derivative $D^{n,\gamma}$ is, in its basic form, defined by

$$D^{n,\gamma} f(x) = \Gamma(1-\gamma)^{-1} \int_c^x (x-\xi)^{-\gamma} f^{(n)}(\xi)\,\mathrm{d}\xi$$

where f denotes any function on the interval $[c, \infty)$ which is n times differentiable there and such that $f^{(n)}$ is absolutely continuous on $[c, \infty)$. This concept was introduced by [27] and has since been much generalised and extensively applied in a great variety of scientific and technical areas. For a comprehensive exposition of this and other concepts of fractional differentiation, see [42], cf. also [1, 44, 45].

For functions f on $\mathbb{R}$ let M_λ with $\lambda \geq 0$ be the operator $M_\lambda f(x) = e^{\lambda x} f(x)$ and, for $0 < \gamma < 1$ and $c \in \mathbb{R}$, define the operator $\mathbb{D}_\lambda^{n,\gamma}$ by

$$\mathbb{D}_\lambda^{n,\gamma} f(x) = M_\lambda^{-1} D D^{n,\gamma} M_\lambda f(x)$$

where D indicates ordinary differentiation and $D^{n,\gamma}$ is the Caputo fractional derivative.[2]

Now, suppose that $1 < \nu < \frac{3}{2}$ and consider the equation

$$\mathbb{D}_\lambda^{1,\nu-1} f(x) = \phi(x) \tag{12}$$

where ϕ is assumed known. We seek the solution f to this equation, stipulating that $f(c)$ should be equal to 0, and it turns out to be

$$f(x) = \Gamma(\nu)^{-1} \int_c^x (x-\xi)^{\nu-1} e^{-\lambda(x-\xi)} \phi(\xi)\mathrm{d}\xi. \tag{13}$$

In other words,

$$g(x) = g(x; \nu, \lambda) = \Gamma(\nu)^{-1} x^{\nu-1} e^{-\lambda x} \tag{14}$$

is the Green's function corresponding to the operator $\mathbb{D}_\lambda^{1,\nu-1}$ when $1 < \nu < \frac{3}{2}$.

The verification is by direct calculation. With f given by (13) we find

$$(M_\lambda f)(x) = \Gamma(\nu)^{-1} \int_c^x (x-\xi)^{\nu-1} e^{\lambda\xi} \phi(\xi)\mathrm{d}\xi$$

---

[2]The differentiation term $DD^{n,\gamma}$ may be viewed as a special case of the more general definition

$$D^{m,n,\gamma} = D^m D^{n,\gamma}$$

where m, like n, is a nonnegative integer and $0 < \gamma < 1$. Then $D^{m,0,\gamma}$ equals the Riemann-Liouville fractional derivative while $D^{0,n,\gamma}$ is the Caputo fractional derivative.

so

$$(M_\lambda f)'(x) = \Gamma(\nu-1)^{-1}\int_c^x (x-\xi)^{\nu-2}e^{\lambda\xi}\phi(\xi)\mathrm{d}\xi$$

and hence, for any $\gamma \in (0,1)$,

$$\begin{aligned}
\mathbb{D}_\lambda^{1,\gamma} f(x) &= \Gamma(\nu-1)^{-1}\Gamma(1-\gamma)^{-1}e^{-\lambda x}D\int_c^x (x-\xi)^{-\gamma}\int_c^\xi (\xi-\eta)^{\nu-2}e^{\lambda\eta}\phi(\eta)\mathrm{d}\eta\mathrm{d}\xi \\
&= \Gamma(\nu-1)^{-1}\Gamma(1-\gamma)^{-1}e^{-\lambda x}D\int_c^x e^{\lambda\eta}\phi(\eta)\int_\eta^x (x-\xi)^{-\gamma}(\xi-\eta)^{\nu-2}\mathrm{d}\xi\mathrm{d}\eta \\
&= \Gamma(\nu-1)^{-1}\Gamma(1-\gamma)^{-1}e^{-\lambda x}D\int_c^x (x-\eta)^{-\gamma+\nu-1}e^{\lambda\eta}\phi(\eta)\mathrm{d}\eta \\
&\quad \times \int_0^1 (1-w)^{(1-\gamma)-1}w^{(\nu-1)-1}\mathrm{d}w \\
&= \Gamma(\nu-\gamma)^{-1}e^{-\lambda x}D\int_c^x (x-\eta)^{-\gamma+\nu-1}e^{\lambda\eta}\phi(\eta)\mathrm{d}\eta.
\end{aligned}$$

Consequently, for $\gamma = \nu - 1$ we have

$$\mathbb{D}_\lambda^{1,\nu-1} f(x) = e^{-\lambda x}D\int_c^x e^{\lambda\eta}\phi(\eta)\mathrm{d}\eta = \phi(x).$$

On the other hand, in case $\nu \in \left(\frac{1}{2},1\right)$ the relevant equation is

$$\mathbb{D}_\lambda^{0,\nu} f(x) = \phi(x)$$

and the solution is again of the form (13). In fact,

$$\begin{aligned}
\mathbb{D}_\lambda^{0,\gamma} f(x) &= \Gamma(\nu)^{-1}\Gamma(1-\gamma)^{-1}e^{-\lambda x}D\int_c^x (x-\xi)^{-\gamma}\int_c^\xi (\xi-\eta)^{\nu-1}e^{\lambda\eta}\phi(\eta)\mathrm{d}\eta\mathrm{d}\xi \\
&= \Gamma(\nu)^{-1}\Gamma(1-\gamma)^{-1}e^{-\lambda x}D\int_c^x e^{\lambda\eta}\phi(\eta)\mathrm{d}\eta\int_\eta^x (x-\xi)^{-\gamma}(\xi-\eta)^{\nu-1}\mathrm{d}\xi \\
&= \Gamma(\nu)^{-1}\Gamma(1-\gamma)^{-1}e^{-\lambda x}D\int_c^x (x-\eta)^{-\gamma+\nu}\, e^{\lambda\eta}\phi(\eta)\mathrm{d}\eta \\
&\quad \times \int_0^1 (1-w)^{(1-\gamma)-1}w^{(\nu-1)-1}\mathrm{d}w \\
&= \Gamma(\nu)^{-1}\Gamma(1-\gamma)^{-1}e^{-\lambda x}D\int_c^x (x-\eta)^{-\gamma+\nu}\, e^{\lambda\eta}\phi(\eta)\mathrm{d}\eta \\
&\quad \times \int_0^1 (1-w)^{(1-\gamma)-1}w^{(\nu-1)-1}\mathrm{d}w \\
&= \Gamma(1-\gamma+\nu)^{-1}e^{-\lambda x}D\int_c^x (x-\eta)^{-\gamma+\nu}\, e^{\lambda\eta}\phi(\eta)\mathrm{d}\eta
\end{aligned}$$

and with $\gamma = \nu$ we have

$$\mathbb{D}_{\lambda}^{0,\nu} f(x) = \phi(x) .$$

Thus, in both cases, $\frac{1}{2} < \nu < 1$ and $1 < \nu < \frac{3}{2}$, the gamma kernel (14) occurs as the Green's function. In the former case the differential operator is of Riemann-Liouville type and in the latter of Caputo type.

This suggests, in particular, that for suitable choice of g and q as gamma kernels there may exist an extension of the definition of Caputo derivatives (corresponding to taking the limit $c \to -\infty$) such that the process (4) may be viewed as the solution to a stochastic differential equation of the form $\mathbb{D}Y_t = \sigma_t \dot{B} + \beta\sigma_t^2$.

Note 3 Introducing the operator I_{λ}^{ν} by

$$I_{\lambda}^{\nu}\phi = M_{\lambda}^{-1} D^{0,\nu-1} M_{\lambda}\phi$$

we may reexpress formula (13) as

$$f = I_{\lambda}^{\nu}\phi$$

and the calculation above shows that

$$\mathbb{D}_{\lambda}^{1,\nu-1} I_{\lambda}^{\nu} = I \tag{15}$$

where I denotes the identity operator. Thus $\mathbb{D}_{\lambda}^{1,\nu-1}$ is the left inverse of I_{λ}^{ν}.

The operator I_{λ}^{ν} also has a right inverse (where, again, $1 < \nu < \frac{3}{2}$). To determine that, let

$$J_{\lambda}^{1,\gamma} = M_{\lambda}^{-1} D^{1,\gamma} M_{\lambda}$$

(where $\gamma \in (0, 1)$). Then

$$\begin{aligned}
I_{\lambda}^{\nu} J_{\lambda}^{1,\gamma} f(x) &= M_{\lambda}^{-1} D^{0,\nu-1} M_{\lambda} \Gamma(\gamma)^{-1} e^{-\lambda x} \int_c^x (x-\xi)^{-\gamma} \left(e^{\lambda\xi} f(\xi)\right)' \mathrm{d}\xi \\
&= \Gamma(\gamma)^{-1} e^{-\lambda x} D^{0,\nu-1} \int_c^x (x-\xi)^{-\gamma} \left(e^{\lambda\xi} f(\xi)\right)' \mathrm{d}\xi \\
&= \frac{1}{\Gamma(\gamma)\Gamma(\nu-1)} e^{-\lambda x} \int_c^x (x-\xi)^{1-\nu} \int_c^{\xi} (\xi-\eta)^{-\gamma} \left(e^{\lambda\eta} f(\eta)\right)' \mathrm{d}\eta \mathrm{d}\xi \\
&= \frac{1}{\Gamma(\gamma)\Gamma(\nu-1)} e^{-\lambda x} \int_c^x \left(e^{\lambda\eta} f(\eta)\right)' \mathrm{d}\eta \int_{\eta}^x (x-\xi)^{1-\nu} (\xi-\eta)^{-\gamma} \mathrm{d}\xi \\
&= \frac{1}{\Gamma(\gamma)\Gamma(\nu-1)} e^{-\lambda x} \int_c^x (x-\eta)^{2-\nu-\gamma} \left(e^{\lambda\eta} f(\eta)\right)' \mathrm{d}\eta \\
&\quad \times \int_0^1 (1-w)^{2-\nu-1} w^{1-\gamma-1} \mathrm{d}w \\
&= \frac{B(2-\nu, 1-\gamma)}{\Gamma(\gamma)\Gamma(1-\nu)} e^{-\lambda x} \int_c^x (x-\eta)^{2-\nu-\gamma} \left(e^{\lambda\eta} f(\eta)\right)' \mathrm{d}\eta.
\end{aligned}$$

So, for $\gamma = 2 - \nu$ we have

$$I_\lambda^\nu J_\lambda^{1,\gamma} f(x) = f(x),$$

i.e. $J_\lambda^{1,2-\nu}$ is the right inverse of I_λ^ν.

Remark 1 In view of the link to fractional differentiation established in the present Section it is pertinent briefly to refer to the broad range of studies of the relevance of (multi)fractional calculus to turbulence modelling existing in the literature. Some links between that literature and the ambit modelling approach are given in [52].

Another related line of study is that of space-time fractional diffusion equations and the possibility of interpreting the associated Green's functions as probability densities, see [46, 55].

4 Autocorrelation

The autocorrelation function r of the Gaussian base process (3) is

$$r(u) = \frac{\int_0^\infty g(u+s)\, g(s)\, \mathrm{d}s}{\int_0^\infty g^2(s)\, \mathrm{d}s} \tag{16}$$

and

$$\mathrm{E}\left\{(G_{t+u} - G_t)^2\right\} = \int_0^\infty \psi_u(v)\, \mathrm{d}v = 2\int_0^\infty g^2(s)\, \mathrm{d}s \bar{r}(u) \tag{17}$$

where

$$\bar{r}(u) = 1 - r(u) \tag{18}$$

is the complementary autocorrelation function of G.

When g is the gamma kernel (5) the autocorrelation function is expressible in terms of the type K Bessel functions. Specifically

$$r(u) = 2^{-\nu+\frac{3}{2}} \Gamma\left(\nu - \frac{1}{2}\right)^{-1} \bar{K}_{\nu-\frac{1}{2}}(\lambda u) \tag{19}$$

where, for all real ν, $\bar{K}$ is defined as $\bar{K}_\nu(x) = x^\nu K_\nu(x)$. This function (19) equals the Whittle-Matérn autocorrelation function which is widely used in geostatistics and other areas of spatial statistics as the autocorrelation between two points a distance u apart in d-dimensional Euclidean space, see [37].

The asymptotic behaviour of r as u tends to 0 is of special interest and this is given by

$$\bar{r}(u) \sim \begin{cases} 2^{-2\nu+1}\dfrac{\Gamma\left(\frac{3}{2}-\nu\right)}{\Gamma\left(\frac{1}{2}+\nu\right)}u^{2\nu-1} + O(u^2) & \text{for } \frac{1}{2} < \nu < \frac{3}{2} \\ \dfrac{1}{2}u|\ln\dfrac{u}{2}| + O(u^3|\ln(u)|) & \text{for } \nu = \frac{3}{2} \\ \dfrac{1}{2}\dfrac{\Gamma(\nu-\frac{5}{2})}{\Gamma\left(\nu-\frac{1}{2}\right)}u^2 + O(u^3|\ln(u)|) & \text{for } \frac{3}{2} < \nu \end{cases} \tag{20}$$

In a paper from 1948 [40] von Karmann discussed the behaviour of the double correlation functions in three dimensional homogeneous and isotropic turbulence. These functions are defined by

$$\phi(r) = \frac{\overline{u(x_1, x_2, x_3)\, u(x_1 + r, x_2, x_3)}}{\overline{u^2}} \tag{21}$$

and

$$\psi(r) = \frac{\overline{u(x_1, x_2, x_3)\, u(x_1, x_2 + r, x_3)}}{\overline{u^2}} \tag{22}$$

where u denotes the main component of the three-dimensional velocity vector (i.e. the component in the mean wind direction) and the overbar indicates mean value. Due to the continuity equation for incompressible fluids the functions ψ and ϕ are related by

$$\psi(r) = \phi(r) + \frac{r}{2}\phi'(r), \tag{23}$$

see [41], cf. also Sect. 6.2.1 of [32]. Von Karmann sets up a series of physically based assumptions concerning this type of turbulence and supplementing these assumptions with some speculative reasoning he arrived at the following proposal for the functional form of ϕ

$$\phi(r) = \frac{2^{2/3}}{\Gamma(1/3)} r^{1/3} K_{1/3}(r). \tag{24}$$

A main point in von Karmann's argument was that the spectral density corresponding to this functional form interpolates smoothly between behaving as a fourth power near the origin and decaying at exponential rate $-5/3$ for large frequencies.[3] Both of these traits correspond to well documented empirical behaviour, and the $5/3$ rate is the spectral counterpart to Kolmogorov's $2/3$ law. In the same paper von Karmann compared this form, or rather that of the transversal correlation function ψ, to wind tunnel data obtained at California Institute of Technology and found a fair agreement between the observations and ψ, as determined from (23).

[3]The $-5/3$ behaviour is very manifest in the socalled inertial range but, as documented by later, extensive and accurate measurements, for the largest frequencies (the dissipation range) the spectral density decreases at a much faster rate. The total behaviour of the spectral density is accurately described by a formula due to Skharofsky, see for instance Fig. 5 in [35].

We note that if ϕ has the form (19) then ψ as determined from (23) is given by

$$\psi(r) = 2^{-\nu+\frac{3}{2}}\Gamma\left(\nu - \frac{1}{2}\right)^{-1}\left(\bar{K}_{\nu-\frac{1}{2}}(r) + \frac{1}{2}r\bar{K}'_{\nu-\frac{1}{2}}(r)\right). \tag{25}$$

It follows from elementary properties of the Bessel functions K that we have the simple relation

$$\bar{K}'_{\nu}(x) = -x\bar{K}_{\nu-1}(x). \tag{26}$$

Hence

$$\begin{aligned}\psi(r) &= 2^{-\nu+\frac{3}{2}}\Gamma\left(\nu - \frac{1}{2}\right)^{-1}\left(\bar{K}_{\nu-\frac{1}{2}}(r) - \frac{1}{2}r^2\bar{K}_{\nu-\frac{3}{2}}(r)\right)\\ &= 2^{-\nu+\frac{3}{2}}\Gamma\left(\nu - \frac{1}{2}\right)^{-1}\left(\bar{K}_{\nu-\frac{1}{2}}(r) - \frac{1}{2}r^{\nu+\frac{1}{2}}K_{\nu-\frac{3}{2}}(r)\right)\\ &= 2^{-\nu+\frac{3}{2}}\Gamma\left(\nu - \frac{1}{2}\right)^{-1}\left(\bar{K}_{\nu-\frac{1}{2}}(r) - \frac{1}{2}r^{\nu+\frac{1}{2}}K_{\frac{3}{2}-\nu}(r)\right)\end{aligned}$$

i.e.

$$\psi(r) = 2^{-\nu+\frac{3}{2}}\Gamma\left(\nu - \frac{1}{2}\right)^{-1}\left(\bar{K}_{\nu-\frac{1}{2}}(r) - \frac{1}{2}r^{2\nu-1}\bar{K}_{\frac{3}{2}-\nu}(r)\right). \tag{27}$$

One sees that $r(u) = \bar{\phi}(u)$ where $\bar{\phi}(u) = 1 - \phi(u)$; thus the asymptotic behaviour of $\bar{\phi}(u)$ as $u \to 0$ is determined by (20). As regards the asymptotic properties of the complementary autocorrelation function $\bar{\psi}(u) = 1 - \psi(u)$ of the transversal velocities it follows immediately from the Table that, for $\frac{1}{2} < \nu < \frac{3}{2}$ and $u \to 0$, the leading terms of the expansions of $\bar{\phi}(u)$ and $\bar{\psi}(u)$ are both of order $u^{2\nu-1}$.

Formula (24) is a special case, obtained for $\nu = \frac{5}{6}$, of the general form of autocorrelation function (19). von Karmann's derivation was not based on any specified probability structure or model. The general form (19), obtained by a Fourier inversion, was proposed as correlation function by [53] (Russian Edition 1959). As mentioned above, that form is also known as the Whittle-Matérn correlation function. Note that the von Karmann-Tatarski specification refers to spatial correlations whereas that of (19) concerns timewise correlation. However, the Taylor Frozen Field Hypothesis[4] provides a direct physical link between the two results.

Note 4 Expressed in terms of $\phi(r)$ itself, rather than the spectrum of ϕ, the basis for von Karmann's proposal was that $\bar{\phi}(r)$ provides a good fit to the behaviour of the second order structure function both over the inertial subrange (the 2/3 law) and at large lags. It follows from the Table (20) that the 2/3 behaviour in fact extends all the way down to 0. Turning this observation around, the indication is that, by the

[4]The Hypothesis states that spatial and temporal increments of the main component of the velocity vector are equivalent in law up to a proportional change of time. Cf. for instance [43].

nature of the gamma kernel, the asymptotic behaviour of the third order structure function near 0, which is linear, extends to the inertial subrange.

Note 5 *Moving average processes with bi-gamma kernel* The results in (20) may also be used for describing the small scale behaviour of the following moving average process with a gamma type kernel.

Consider the stationary Gaussian moving average process given by

$$Y_t = \int_{-\infty}^{\infty} g(t-s;\nu,\lambda,\mu,\kappa)\,\mathrm{d}B_s \tag{28}$$

where B is Brownian motion on $\mathbb{R}$ and

$$g(t;\nu,\lambda,\mu,\kappa) = \begin{cases} t^{\nu-1}e^{-\lambda t} & \text{for } t>0; \\ |t|^{\mu-1}\,e^{-\kappa|t|} & \text{for } t<0. \end{cases}$$

We refer to this as the bi-gamma kernel. In case $\mu=\nu$ and $\kappa=\lambda$ the kernel is symmetric around 0 and may be written as $g(|t|;\nu,\lambda)$. The process Y is well defined provided both ν and μ are greater than $\frac{1}{2}$, and it may be rewritten as

$$Y_t = \int_{-\infty}^{t} (t-s)^{\nu-1}\,e^{-\lambda(t-s)}\mathrm{d}B_s + \int_{t}^{\infty} (s-t)^{\mu-1}\,e^{-\kappa(s-t)}\mathrm{d}B_s.$$

Here

$$\begin{aligned} \mathrm{E}\{Y_0Y_u\} &= \int_0^{\infty} g(s;\nu,\lambda)\,g(u+s;\nu,\lambda)\,\mathrm{d}s \\ &\quad + \int_0^{u} g(s;\mu,\kappa)\,g(u-s;\nu,\lambda)\,\mathrm{d}s \\ &\quad + \int_u^{\infty} g(s;\mu,\kappa)\,g(s-u;\mu,\kappa)\,\mathrm{d}s. \end{aligned}$$

By Formula 3.383.1 in [36] we find

$$\begin{aligned} \int_0^{u} g(s;\mu,\kappa)\,g(u-s;\nu,\lambda)\,\mathrm{d}s &= \int_0^{u} s^{\mu-1}e^{-\kappa s}\,(u-s)^{\nu-1}\,e^{-\lambda(u-s)}\mathrm{d}s \\ &= e^{-\lambda u}\int_0^{u} s^{\mu-1}\,(u-s)^{\nu-1}\,e^{-(\lambda+\kappa)s}\mathrm{d}s \\ &= e^{-\lambda u}B(\nu,\mu)\,{}_1F_1(\mu;\nu+\mu;\lambda+\kappa)\,u^{\lambda+\kappa-1} \end{aligned}$$

where ${}_1F_1$ is the general hypergeometric function, and we have ([36], Formula 9.14.1)

$${}_1F_1(\mu;\nu+\mu;(\lambda+\kappa)) = \sum_{k=0}^{\infty} \frac{(\mu)_k}{(\nu+\mu)_k}\frac{(\lambda+\kappa)^k}{k!}.$$

All in all this implies that for $u \to 0$ the second order structure function of (28) is of the form

$$S_2(u) = cu^{2\nu-1} + c'u^{2\mu-1} - c''u^{\nu+\mu-1}$$

where c, c' and c'' are positive constants. Thus, in particular, if $\frac{1}{2} < \nu < \frac{3}{2}$ and $\frac{1}{2} < \kappa < \frac{3}{2}$ then

$$S_2(u) \sim \begin{cases} cu^{2\nu-1} & \text{for } \nu < \mu; \\ \left(c + c' - c''\right) u^{2\nu-1} & \text{for } \nu = \mu; \\ c'u^{2\mu-1} & \text{for } \nu > \mu. \end{cases} \tag{29}$$

5 Pathwise Behaviour

The fine structure of *BSS* and *LSS* processes is discussed in [47, 49].

In [47] the authors establish a connection between the path behaviour of the *BSS* process

$$Y_t = \int_{-\infty}^{t} (t-s)^{\nu-1} e^{-\lambda(t-s)} \mathrm{d}B_s \tag{30}$$

and that of the fractional Ornstein-Uhlenbeck process Y^H with index H, that is

$$Y_t^H = \int_{-\infty}^{t} e^{-\lambda(t-s)} \mathrm{d}B_s^H, \tag{31}$$

under the assumption that $H = \nu - \frac{1}{2}$ and $\nu \in \left(\frac{1}{2}, 1\right) \cup \left(1, \frac{3}{2}\right)$ so that we are in the nonsemimartingale case. While both of these are stationary Gaussian processes, the former is in many situations more realistic in regard to applications, particularly for $\nu = 5/6$ which corresponds to von Karmann's spectral density function. This is the case in particular for turbulence modelling. The key difference lies in the tail behaviour of the increments for large lags.

The following result is established in [47].

Theorem 1 *For all $t > 0$*

$$Y_t = Y_t^H - D_t \tag{32}$$

where $D \in C^1([0, \infty))$. A concrete representation of the process D is available:

$$D_t = \int_{-\infty}^{t} \int_{s}^{t} \left(e^{-\lambda(t-u)} - e^{-\lambda(t-s)}\right) \frac{\partial}{\partial u} K^H(u, s) \, \mathrm{d}u \mathrm{d}B_s$$

where K^H denotes the kernel for the fractional Brownian motion, i.e.

$$K^H(t, s) = (t-s)_+^{H-\frac{1}{2}} - (-s)_+^{H-\frac{1}{2}}.$$

Similarly Y may be represented as

$$Y_t = B_t^H - V_t$$

where V is an absolutely continuous process.

The stochastic analysis for volatility modulated Lévy-driven Volterra processes, developed in [4, 5, 13], is an important tool in the derivation of these results. Together these three papers constitute a foundation for an Ito calculus for BSS *and* LSS *processes.*

6 Recovery and Inference

Once a *BSS* or *LSS* model has been formulated the question arises as to what can be learned about the components of the model, either in law or pathwise. In discussing this we will, unless otherwise mentioned, assume that the skewness terms are absent.

Thus let

$$Y_t = \int_{-\infty}^{t} g(t-s)\sigma_s \mathrm{d}L_s^T \tag{33}$$

where L^T denotes a Lévy process L time changed by a chronometer T (i.e. a cadlag increasing process such that $T(t) \to -\infty$ for $t \to -\infty$ and $T(t) \to \infty$ for $t \to \infty$). The question has several aspects: (i) In case the process has been observed continuously over an interval, can any of the model components be exactly determined (ii) Under what conditions does the law of Y uniquely determine the kernel g or the laws of σ or L or T (iii) If the data available consists of high frequency observations, what inference procedures for assessing some or all of the components might be available; in particular, what can be said about the volatility/intermittency process σ^2.

Below we exemplify these aspects. For additional results, proofs and references see [50, 51].

(i) The following example presents a case where concrete pathwise recovery of L^T over the interval $(-\infty, t)$ is possible if Y has been observed continuously over the same interval

Example 2 Suppose that g is the gamma kernel and that $\sigma \equiv 1$ and $T = t$, that is

$$Y_t = \int_{-\infty}^{t} g(t-s;\nu,\lambda)\,\mathrm{d}L_s.$$

By the stochastic Fubini theorem (cf. [2]) we find for $\nu \in \left(\frac{1}{2}, 1\right)$ and letting

$$Z_t = \int_{-\infty}^{t} g(t-u;1-\nu,\lambda)\,Y_u \mathrm{d}u$$

that

$$\begin{aligned} Z_t &= \int_0^t \int_{-\infty}^u g(t-u;1-\nu,\lambda)\, g(u-s;\nu,\lambda)\, \mathrm{d}L_s \mathrm{d}u \\ &= \int_{-\infty}^t \int_s^t g(t-u;1-\nu,\lambda)\, g(u-s;\nu,\lambda)\, \mathrm{d}u \mathrm{d}L_s \\ &= \int_{-\infty}^t \int_0^{t-s} g(w;1-\nu,\lambda)\, g(t-s-w;\nu,\lambda)\, \mathrm{d}w \mathrm{d}L_s \\ &= \int_{-\infty}^t g(t-s;1,\lambda)\, \mathrm{d}L_s^T = c\int_{-\infty}^t e^{-\lambda(t-s)} \mathrm{d}L_s \end{aligned}$$

for a constant c. Hence

$$\begin{aligned} Z_t^+ &= \int_0^t Z_s \mathrm{d}s = \int_0^t \int_{-\infty}^s e^{-\lambda(s-u)} \mathrm{d}L_u \mathrm{d}s \\ &= \int_{-\infty}^t \int_u^t e^{-\lambda(s-u)} \mathrm{d}s \mathrm{d}L_u \\ &= \lambda^{-1} \int_{-\infty}^t \left(1-e^{-\lambda(t-u)}\right) \mathrm{d}L_u = \lambda^{-1}(L_t - Z_t) \end{aligned}$$

or

$$L = \lambda Z^+ + Z.$$

It is noteworthy here that L is explicitly recoverable in spite of the fact that with $\nu \in \left(\frac{1}{2}, 1\right)$ the kernel $g(t;\nu,\lambda)$ tends to ∞ for $t \to 0$.

Note 6 Under a minor regularity condition on the time change T, the same argument goes through, giving that $L^T = \lambda Z^+ + Z$.

Note 7 For general moving average processes driven by Lévy noise

$$Y_t = \int_{-\infty}^{\infty} g(t-s)\, L(\mathrm{d}s)$$

recovery of L from complete knowledge of the realisation of Y on $\mathbb{R}$ is, subject to regularity restrictions on g and L, possible in terms of linear limit operations. This applies in particular for the gamma kernel and in that case $\{L_s : s \le t\}$ is recoverable from $\{Y_s : s \in \mathbb{R}\}$, see [50].

Consider the class $\mathcal{G}$ of kernels g such that g is integrable with non-vanishing Fourier transform. This is the case in particular for the gamma kernel who's Fourier transform is

$$\hat{g}(\zeta;\nu,\lambda) = \frac{\lambda^\nu}{\Gamma(\nu)}(\lambda - i\zeta)^{-\nu}.$$

It is shown in [51] that if $g \in \mathcal{G}$ and

$$Y_t = \int_{-\infty}^{t} g(t-s)\mathrm{d}L_s^T,$$

with T a subordinator independent of L, then the law of T is completely determined by the laws of L and Y. This is thus, in particular, the case when g is of the gamma type.

(ii) The paper [39], cf. also [26], introduces a powerful nonparametric procedure for estimation of the kernel function for *BSS* processes

$$Y_t = \int_{-\infty}^{t} g\,(t-s)\,\sigma_s \mathrm{d}B_s$$

with kernel function g in $\mathscr{G}$. This is done through determining g from the autocorrelation function

$$r\,(u) = \int_{0}^{\infty} g\,(s+|u|)\,g\,(s)\,\mathrm{d}s.$$

Numerically the gamma kernel is used as a test case. (Earlier approaches to the estimation of the kernel are presented in [26].)

(iii) A key question of inference for *BSS* and *LSS* models is how to assess the inherent volatility σ^2. More specifically, the wish will typically be to draw accurate inference on the accumulated volatility

$$\sigma_t^{2+} = \int_0^t \sigma_s^2 \mathrm{d}s.$$

The realised quadratic variation is a natural initial tool to this end and for *BSS* processes Y, as given by (1), that will under mild conditions yield a consistent estimator of σ_t^{2+} provided Y is a semimartingale, i.e. it will hold that

$$[Y_\delta]_t \xrightarrow{p} \sigma_t^{2+} \quad \text{as} \quad \delta \to 0.$$

However, suppose that

$$Y_t = \int_{-\infty}^{t} g\,(t-s;\, \nu, \lambda)\,\sigma_s \mathrm{d}B_s$$

with $\nu \in \left(\frac{1}{2}, 1\right) \cup (1, \frac{3}{2}]$. Then Y is not a semimartingale and the realised quadratic variation $[Y_\delta]_t$ converges to ∞ for $\nu \in \left(\frac{1}{2}, 1\right)$ and to 0 for $\nu \in (1, \frac{3}{2}]$. The rate of these convergences is determined by ν. For instance, when $\nu \in \left(\frac{1}{2}, 1\right)$ we have

$$c\delta^{2(1-\nu)}\,[Y_\delta]_t \xrightarrow{p} \sigma_t^{2+} \quad \text{as} \quad \delta \to 0$$

where $c = \lambda 2^{2(1-\nu)} \Gamma\,(2\nu - 1)\,/\,\Gamma\,(\nu)^2$.

Obtaining estimates of ν as a way to inference on σ_t^{2+}, in particular through stable central limit theorems, requires advanced reasoning. The papers [9, 10, 29] develop the theory of multipower variations for this and related purposes. In particular, the latter two papers discuss the use of COF (change-of-frequency) statistics. Similar points for *LSS* processes are discussed in [49].

So far we have, for simplicity of discussion, assumed that the skewness terms in the *BSS/LSS* processes are 0. When this is not the case it is still possible, under certain conditions, to establish stable central limit theorems, as shown in the above-mentioned papers. However, consider the case.where Y is of the form (4) and g and q are given respectively as $g\,(t;\,\nu,\lambda)$ and $g\left(t;\,\nu-\frac{1}{2},\lambda\right)$ for $\nu\in\left(\frac{1}{2},1\right)$, cf. (6). Then the skewness term contains information on the volatility/intermittency process. In fact, it can be shown that then

$$\delta^{2(1-\nu)}\,[Y_\delta]_t \xrightarrow{p} c\int_0^t \sigma_s^2 \mathrm{d}s + c'\int_0^t \sigma_s^4 \mathrm{d}s$$

for certain constants c and c' and where $[Y_\delta]_t$ is the quadratic variation of Y over the interval $(0,t)$ at lag δ.

Recently a powerful procedure for simulation of *BSS* processes is presented in [20], under the assumption that the kernel function is regularly varying at 0, as is the case for the gamma kernel with $\nu\in\left(\frac{1}{2},1\right)$, and the method is applied successfully for inference on ν (referred to as the roughness parameter). Approximation of *LSS* processes by Fourier methods is discussed in [21].

As long as the interest is solely in regard to relative volatility, inference on ν can be avoided by using the realised relative volatility, defined as $[Y_\delta]_t\,/\,[Y_\delta]_T$ for $0<t<T$. This yields

$$[Y_\delta]_t\,/\,[Y_\delta]_T \xrightarrow{p} \sigma_t^{2+}/\sigma_T^{2+} \quad \text{as} \quad \delta\to 0.$$

Associated confidence intervals based on a stable central limit theorem have been developed in [12].

Acknowledgements I am grateful to the reviewers of the paper and to the editors for their helpful comments and suggestions regarding the original version of my manuscript and to Orimar Sauri for a careful checking of the revised version.

References

1. Adda, F.B., Cresson, J.: About non-differentiable functions. J. Math. Anal. Appl. **263**, 721–737 (2005)
2. Barndorff-Nielsen, O.E., Basse-O'Connor, A.: Quasi Ornstein-Uhlenbeck processes. Bernoulli **17**, 916–921 (2009)
3. Barndorff-Nielsen, O.E., Blæsild, P., Schmiegel, J.: A parsimonious and universal description of turbulent velocity increments. Eur. Phys. J. B **41**, 345–363 (2004)

4. Barndorff-Nielsen, O.E., Benth, F.E., Szozda, B.: Stochastic integration for $\mathscr{V}\mathscr{M}\mathscr{V}\mathscr{L}$ processes via white noise analysis. Infinite Dimensional Anal., Q. Probab. Relat. Top. **17**, 1450011 (2014)
5. Barndorff-Nielsen, O.E., Benth, F.E., Pedersen, J., Veraart, A.: On stochastic integration theory for volatility modulated Lévy-driven Volterra processes. Stoch. Proc. Appl. **124**, 812–847 (2013)
6. Barndorff-Nielsen, O.E., Benth, F.E., Veraart, A.: Ambit processes and stochastic partial differential equations. In: DiNunno, G., Øksendal, B. (eds.) Advanced Mathematical Methods for Finance, pp. 35–74. Springer, Berlin (2011)
7. Barndorff-Nielsen, O.E., Benth, F.E., Veraart, A.: Modelling electricity spot prices by Lévy semistationary processes. Bernoulli **19**, 803–845 (2013)
8. Barndorff-Nielsen, O.E., Benth, F.E., Veraart, A.: Recent advances in ambit stochastics with a view towards tempo-spatial stochastic volatility/intermittency. Banach Centre Publ. **104**, 25–60 (2015)
9. Barndorff-Nielsen, O.E., Corcuera, J.M., Podolskij, M.: Multipower variation for Brownian semistationary processes. Bernoulli **17**, 1159–1194 (2011)
10. Barndorff-Nielsen, O.E., Corcuera, J.M., Podolskij, M.: Limit theorems for functionals of higher order differences of Brownian semistationary processes. In: Shiryaev, A.N., Varadhan, S.R.S., Presman, E. (eds.) Prokhorov and Contemporary Probability Theory, pp. 69–96. Springer, Berlin (2013)
11. Barndorff-Nielsen, O.E., Hedevang, E., Schmiegel, J., Szozda, B.: Some recent developments in Ambit Stochastics with particular reference to the Statistical Theory of Turbulence. In: Benth, F.E., Nunno, G. (eds.) Stochastics of Environmental and Financial Economics, pp. 3–25. Springer, Heidelberg (2015)
12. Barndorff-Nielsen, O.E., Pakkanen, M., Schmiegel, J.: Assessing relative volatility/intermittency/energy dissipation. Elec. J. Stat. **8**, 1996–2021 (2013)
13. Barndorff-Nielsen, O.E., Sauri, O., Szozda, B.: Selfdecomposable fields. J. Theor. Probab. **28**, 1–35 (2015)
14. Barndorff-Nielsen, O.E. and Schmiegel, J.: Ambit processes; with applications to turbulence and tumour growth. In Benth, F.E., Nunno, G.D., Linstrøm, T., Øksendal, B., Zhang, T. (eds.) Stochastic Analysis and Applications: The Abel Symposium 2005, pp. 93–124. Springer, Heidelberg (2007)
15. Barndorff-Nielsen, O.E., Schmiegel, J.: A stochastic differential equation framework for the timewise dynamics of turbulent velocities. Theory Prob. Appl. **52**, 372–388 (2008)
16. Barndorff-Nielsen, O.E. Schmiegel, J.: Brownian semistationary processes and volatility/intermittency. In Albrecher, H., Rungaldier, W., Schachermeyer, W.: (eds.) Advanced Financial Modelling. Radon Series Comp. Appl. Math. **8**. pp. 1–26. W. de Gruyter, Berlin (2009)
17. Barndorff-Nielsen, O.E., Shiryaev, A.N.: Change of Time and Change of Measure, 2nd edn. World Scientific, Singapore (2015)
18. Basse-O'Connor, A., Graversen, S.-E., Pedersen, J.: Stochastic integration on the real line. Theory Probab. Appl. **58**, 193–215 (2014)
19. Basse-O'Connor, A., Pedersen, J.: Lévy driven moving averages and semimartingales. Stoc. Process. Their Appl. **119**(9), 2970–2991 (2009)
20. Bennedsen, M., Lunde, A., Pakkanen, M.: Hybrid scheme for Brownian semistationary processes. CREATES Research Paper 2015–43. (Submitted.)
21. Benth, F.E., Eyjolffson, H., Veraart, A.E.D.: Approximating Lévy semistationary processes via Fourier methods in the context of power markets. SIAM J. Finan. Math. **5**, 71–98 (2014)
22. Benth, F.E., Ortiz-Latorre, S.: A change of measure preserving the affine structure in the Barndorff-Nielsen and Shephard model for commodity markets. Int. J. Theor. Appl. Financ. **18**, 1550038 40pp (2015)
23. Birnir, B.: The Kolmogorov-Obukhov statistical theory of turbulence. J. Nonlinear Sci. **0938–8974**, 1–32 (2013)
24. Birnir, B.: The Kolmogorov-Obukhov Theory of Turbulence. Springer, Heidelberg (2013)
25. Birnir, B.: The Kolmogorov Obukhov-She-Leveque scaling in Turbulence. Com. Pure Appl. Anal. **13**, 1737–1757 (2014)

26. Brockwell, P.J., Ferrazzano, V., Klüppelberg, C.: High-frequency sampling and kernel estimation for continuous-time moving average processes. J. Time Ser. Anal. **33**, 152–160 (2012)
27. Caputo, M.: Linear models of dissipation whose Q is almost frequency independent. II.Geophys. J. Roy. Astronom. Soc. **13**, 529–539 (1967)
28. Chong, C., Klüppelberg, C.: Integrability conditions for space-time stochastic integrals: Theory and applications. Bernoulli **21**, 2190–2216
29. Corcuera, J.M., Hedevang, E., Pakkanen, M.S., Podolskij, M.: Asymptotic theory for Brownian semi-stationary processes with application to turbulence. Stoch. Proc. Their Appl. **123**, 2552–2574 (2013)
30. Eberlein, E., Hammerstein, E.A.V.: Generalized Hyperbolic and Inverse Gaussian distributions. Birkhäuser, In Seminar on Stochastic Analysis, Random Fields and Applications IV Basel (2004)
31. Eberlein, E.: Generalised hyperbolic models. In: Cont, R. (ed.) Encyclopedia of Quantitative Finance. Wiley, London (2010)
32. Frisch, U.: Turbulence. Cambridge University Press, The Legacy of A.N. Kolmogorov (1995)
33. Halgreen, C.: Self-decomposability of the generalised inverse Gaussian and hyperbolic distributions. Z. Wahrsch. Verw. Gebiete. **47**, 13–17 (1979)
34. Hedevang, E., Schmiegel, J.: A causal continuous time stochastic model for the turbulent energy dissipation in a helium jet flow. J. Turb. **14**, 1–26 (2013)
35. Hedevang, E., Schmiegel, J.: A Lé vy based approach to random vector fields: With a view towards turbulence. J. Nonlinear Sci. Numer. Simul. **15**, 411–436 (2014)
36. Gradshteyn, I.S., Ryzhik, I.M.: Table of Integrals, Series and Products. (Fifth Ed.). Academic Press, London (1996)
37. Guttorp, P., Gneiting, T.: On the Whittle-Matérn correlation family. NrCSE Technical report Series NrCSE-TS No. 080 (2005)
38. Jurek, Z.J., Vervaat, W.: An integral representation for self-decomposable Banach space valued random variables. Z. Wahrsch. Verw. Gebiete. **62**, 247–262 (1983)
39. Kanaya, S., Lunde, A., Sauri, O.: Nonparametric estimation of kernel functions of Brownian semi-stationary processes. (To appear.) (2015)
40. von Karman, T.: Progress in the statistical theory of turbulence. J. Marine Res. **7**, 252–264 (1948)
41. von Karman, T., Howarth, L.: On the statistical theory of isotropic turbulence. Proc. Roy. Soc. Lond. A **164**, 192–215 (1938)
42. Kilbas, A.A., Srivastava, H.M., Trujillo, J.J.: Theory and Applications of Fractional Differential Equations. North-Holland, Amsterdam (2006)
43. Márquez, J.U., Schmiegel, J.: Modelling turbulent time series by BSS-processes. In: Podolskij M., Stelzer R., Thorbj ørnsen, S., Veraart, A. (eds.) The Fascination of Probability, Statistics and their Applications; in Honour of Ole E. Barndorff-Nielsen. Springer, Heidelberg (To appear) (2015)
44. Li, C., Qian, D., Chen, Y.: On Riemann-Liouville and Caputo derivatives. Discrete Dyn. Nat. Soc. **2011**, Article ID 562494 (2011)
45. Mainardi, F.: Fractional Calculus and Wawes in Linear Visocelasticity. Imperial College Press, London (2010)
46. Mainardi, F, Pagnini, G., Gorenflo, R.: Probability distributions as solutions to fractional diffusion equations. Mini-proceedings: 2nd MaPhySto Conference on Lévy Processes: Theory and Applications, pp. 197–205 (2002)
47. Sauri, O.: Brownian semistationary processes and related processes. In Lévy semistationary models with applications in energy markets. PhD thesis. Department of economics and business economics, Aarhus University, pp. 2–27 (2015)
48. Pedersen, J., Sauri, O.: On Lévy semistationary processes with a gamma kernel. In Ramses, H., Mena, R.H., Pardo, J.C., Rivero, V., Bravo, G. (eds.) XI Symposium of Probability and Stochastic Processes CIMAT, Mexico, November 18–22, 2013. Springer, Heidelberg. Progress in Probability **69**, 217–239 (2015)

49. Podolskij, M.: Ambit Fields: survey and new challenges. In Ramses, H., Mena, R.H., Pardo, J.C., Rivero, V., Bravo, G.: XI Symposium of Probability and Stochastic Processes CIMAT, Mexico, November 18–22, 2013. Springer, Heidelberg. Progress in Probability **69**, 241–279 (2014)
50. Sauri, O.: Invertibility of infinitely divisible continuous-time moving average processes (2015). arXiv:1505.00196
51. Sauri, O., Veraart, A.E.D.: On the class of distributions of subordinated Lévy processes and bases. (Submitted.) (2015)
52. Schmiegel, J., Barndorff-Nielsen, O.E., Eggers, H.C.: A class of spatio-temporal and causal stochastic processes with application to multiscaling and multifractality. S. Afr. J. Sci. **101**, 513–519 (2005)
53. Tatarski, V.I.: Wave Propagation in a Turbulent Medium. McGraw-Hill, New York (1961)
54. Veraart, A.E., Veraart, L.A.: Modelling electricity day-ahead prices by multivariate Lévy semi-stationary processes. In: Benth, F.E., Kholodnyi, V.A., Laurence, P. (eds.) Quantitative Energy Finance, pp. 157–188. Springer, New York (2014)
55. Wolpert, R.L., Taqqu, M.S.: Fractional Ornstein-Uhlenbeck Lévy processes and the Telekom process: Upstairs and downstairs. Sig. Process. **85**, 1523–1545 (2005)

Explicit Computations for Some Markov Modulated Counting Processes

Michel Mandjes and Peter Spreij

Abstract In this paper we present elementary computations for some *Markov modulated* counting processes, also called counting processes with *regime switching*. Regime switching has become an increasingly popular concept in many branches of science. In finance, for instance, one could identify the background process with the 'state of the economy', to which asset prices react, or as an identification of the varying default rate of an obligor. The key feature of the counting processes in this paper is that their intensity processes are functions of a finite state Markov chain. This kind of processes can be used to model default events of some companies. Many quantities of interest in this paper, like conditional characteristic functions, can all be derived from conditional probabilities, which can, in principle, be *analytically* computed. We will also study limit results for models with rapid switching, which occur when inflating the intensity matrix of the Markov chain by a factor tending to infinity. The paper is largely expository in nature, with a didactic flavor.

Keywords Counting process · Markov chain · Markov modulated process · Regime switching

AMS subject classification: 60G44 · 60G55 · 60J27

M. Mandjes · P. Spreij (✉)
Korteweg-de Vries Institute for Mathematics, Universiteit van Amsterdam,
PO Box 94248, 1090 GE Amsterdam, The Netherlands
e-mail: spreij@uva.nl

M. Mandjes
e-mail: m.r.h.mandjes@uva.nl

P. Spreij
IMAPP, Radboud Universiteit, Heyendaalseweg 135, 6525AJ Nijmegen, The Netherlands

© Springer International Publishing Switzerland 2016

J. Kallsen and A. Papapantoleon (eds.), *Advanced Modelling in Mathematical Finance*,
Springer Proceedings in Mathematics & Statistics 189,
DOI 10.1007/978-3-319-45875-5_3

1 Introduction

In this paper we present some elementary computations concerning some *Markov modulated* (MM) counting processes, denoted N, also called counting processes with *regime switching*. Such processes fall into the class of *hybrid models* [29] and are in fact Hidden Markov processes [10]. Although in the present paper we restrict ourselves to certain counting processes, it is worth mentioning that owing to its various attractive features, regime switching has become an increasingly popular concept in many branches of science. In a broad spectrum of application domains it offers a natural framework for modeling situations in which the stochastic process under study reacts to an autonomously evolving environment. In finance, for instance, one could identify the background process with the 'state of the economy', to which asset prices react, or as an identification of the varying default rate of an obligor. In operations research, in particular in wireless networks, the concept can be used to model the channel conditions that vary in time, and to which users react. In the literature in the latter field there is a sizeable body of work on Markov-modulated queues, see e.g. [2, Chap. XI] and [27], while Markov modulation has been intensively used in insurance and risk theory as well [3]. In the economics literature, the use of regime switching dates back to at least the late 1980s [16]. Various specific models have been considered since then, see for instance [1, 11, 12]. For other direct applications of models with regime switching in finance (hedging of claims, interest rate models, credit risk, application to pension funds) we refer to [8, 22, 23, 30, 31] for recent results.

The key feature of the counting processes, commonly denoted N, in this paper is that their intensity processes are of the form $\lambda_t = \lambda(X_t, N_t)$, where X is a finite state Markov chain whose jumps with probability one never coincide with the jumps of the counting process. For mathematical convenience we assume without loss of generality that X takes its values in the set of d-dimensional basis vectors.

This kind of processes can be used to model default events of some companies. We restrict our treatment to models where the intensity is of a special form, leading to the MM one point process which can be used to model the default event of a single company, its extension to the situation of defaults of various companies and an MM Poisson process, which can be used to model defaults for a large pool of obligors whose individual intensities of default are all the same and small.

The intensities $\lambda_t = \lambda(X_t, N_t)$ that we use will be affine in X_t, i.e. $\lambda_t = \lambda^\top X_t f(N_t)$ for some $\lambda \in \mathbb{R}^d$ and some function f. It is possible to show that the joint process (X, N) is Markov, in fact it is an affine process after a state transformation. This means that for many quantities of interest, like conditional characteristic functions, one can in principle use the full technical apparatus that has become available for affine process, see [9]. However, as these quantities can all be derived from conditional probabilities (our processes are finite, or at most countably, valued), using these techniques is like making a detour since the conditional probabilities can be derived by more straightforward methods. Moreover these conditional probabilities give a *direct* insight into the probabilistic structure of the process and can in principle be *analytically* computed. Therefore, we circumvent the theory of affine processes and focus on direct computation of all conditional probabilities of interest.

We will also study limit results for models with rapid switching, which occur when inflating the intensity matrix of the Markov chain by a factor tending to infinity. Rapid switching between (macro) economic states is unrealistic, but in models for the profit and loss of trading positions, especially in high frequency trading, rapid switching may take place, see [15]. We will see that the limit processes have intensities that are expectations under the invariant distribution of the chain. This is similar to what happens in the context of Markov modulated Ornstein-Uhlenbeck processes [18], see also [19], whereas comparable results under scaling in the operations research literature can be found in [5, 6].

The paper is largely expository in nature, with a didactic flavor. We do not claim novelty of all results below. Rather we emphasize the uniform approach that we follow, using martingale methods, that may also lead to alternative proofs of known results, e.g. those concerning transition probabilities by using 'ε-arguments' as in [27]. The organization of the paper is as follows. In Sect. 2 we study Markov modulated model for the total number of defaults when there are n obligors. As a primer, in Sect. 2.1 we extensively study the Markov modulated model for a single obligor, in particular its distributional properties. Then we switch to the more general situation of Sect. 2.2, where our approach is inspired by the easier case of the previous section. All results are basically obtained by exploiting the Markovian nature of the joint process (X, N). Section 3 gives a few results for the Markov modulated Poisson process. Conditional probabilities of future values of the counting processes, when only its own past can be observed (and not the underlying Markov chain) can be computed using filtering theory, which is the topic of Sect. 4. In Sect. 5 we obtain the limit results for processes where the Markov chain is rapidly switching.

2 The MM Model for Multiple Obligors

We assume throughout that a probability space $(\Omega, \mathscr{F}, \mathbb{P})$ is given. Suppose we have n obligors with default times τ^i for obligor i, $i = 0, \dots, n$. Let $Y_t^i = \mathbf{1}_{\{\tau^i \leq t\}}$, $t \in [0, \infty)$. Here we encounter the canonical set-up for the intensity based approach in credit risk modelling, see [13, Chap. 12] or [4, Chap. 6] for further details on probabilistic aspects. We postulate for each $i \in \{1, \dots, n\}$

$$\mathrm{d}Y_t^i = \lambda_t(1 - Y_t^i)\,\mathrm{d}t + \mathrm{d}m_t^i, \tag{1}$$

for λ_t a nonnegative process to be specified, but which is the same for each obligor i Here each m^i is a martingale w.r.t. to the filtration, call it $\mathbb{F}^i$, generated by Y^i and the process λ. We impose that the τ_i are conditionally independent given λ. Hence, simultaneous defaults occur with probability zero, as the τ^i have a continuous distribution. By the conditional independence assumption, the m^i are also martingales w.r.t. $\mathbb{F} = \vee_{i=1}^{n} \mathbb{F}^i$. The process λ is assumed to be predictable w.r.t. $\mathbb{F}$. In all what follows in this section we take $N_t = \sum_{i=1}^{n} Y_t^i$.

2.1 The MM One Point Process

For a better understanding of what follows, we single out the special case $n = 1$ and we write τ instead of τ^1. There is some advantage in starting with a simpler case that allows for more explicit formulas, is more transparent, and that at the same time can serve as a warming up for the more general setting.

2.1.1 The General One Point Process with Intensity

Let us consider the basic case, the random variable τ has an exponential distribution with parameter λ, and $Y_t = \mathbf{1}_{\{\tau \le t\}}$, $t \in [0, \infty)$. Then Y has semimartingale decomposition

$$dY_t = \lambda(1 - Y_t)\, dt + dm_t, \tag{2}$$

where $\lambda > 0$ and m a martingale w.r.t. the filtration generated by the process Y. As a matter of fact, the distributional property of τ is equivalent to the decomposition of Y in (2). Clearly Y_t is a Bernoulli random variable, so $y(t) := \mathbb{E}Y_t = \mathbb{P}(Y_t = 1) = \mathbb{P}(\tau \le t)$. Alternatively, taking expectations, we get the ODE

$$\dot{y}(t) = \lambda(1 - y(t)),$$

which is, with $y(0) = 0$, indeed solved by

$$y(t) = 1 - \exp(-\lambda t).$$

Let Λ^τ be the compensator of Y, then

$$\Lambda_t^\tau = \int_0^t \lambda(1 - Y_s)\, ds = \int_0^t \lambda \mathbf{1}_{\{s<\tau\}}\, ds = \int_0^{t\wedge\tau} \lambda\, ds = \lambda(\tau \wedge t).$$

Note that Y can be considered as N^τ, the at τ stopped Poisson process with intensity λ. The compensator Λ of N stopped at τ indeed yields Λ^τ.

As a first generalization we change the above setup in the sense that we postulate

$$dY_t = \lambda_t(1 - Y_t)\, dt + dm_t, \tag{3}$$

where λ is a nonnegative locally integrable Borel function, also known as the (time varying) hazard rate. As above one can show that

$$y(t) = 1 - \exp\left(-\int_0^t \lambda_s \, \mathrm{d}s\right).$$

In a next generalization we suppose that λ becomes a random process defined on an auxiliary probability space $(\Omega', \mathscr{F}', \mathbb{P}')$. We can look at the product probability space $(\Omega \times \Omega', \mathscr{F} \otimes \mathscr{F}', \mathbb{P} \otimes \mathbb{P}')$ and redefine in the obvious way Y, τ and λ on this product space. The filtration we will use consists of the σ-algebras $\mathscr{F}_t^Y \otimes \mathscr{F}_t^\lambda$.

It is assumed that λ is predictable and a.s. locally integrable w.r.t. Lebesgue measure. For a given trajectory $\lambda_t = \lambda_t(\omega')$ we define Y on $(\Omega, \mathscr{F}, \mathbb{P})$ as in (3). With $\mathscr{F}^\lambda$ the σ-algebra generated by the full process λ, we have that

$$\mathbb{E}[Y_t | \mathscr{F}^\lambda] = 1 - \exp\left(-\int_0^t \lambda_s \, \mathrm{d}s\right),$$

and hence

$$y(t) = \mathbb{E} Y_t = 1 - \mathbb{E} \exp\left(-\int_0^t \lambda_s \, \mathrm{d}s\right).$$

Alternatively, one can construct the point process Y as follows. Let $(\Omega, \mathscr{F}, \mathbb{Q})$ be a probability space on which is defined a standard Poisson process Y and *independently* of Y the nonnegative predictable process λ. Put $L_t = \mathscr{E}(\mu)_t$, the Doléans exponential of the $\mathbb{Q}$-local martingale μ given by $\mu_t = \int_0^t (\lambda_s \mathbf{1}_{\{Y_{s-}=0\}} - 1) \, \mathrm{d}(Y_s - s)$. Note that $L_0 = 1$. Let τ_k be the consecutive jump times of Y, $\tau_0 = 0$. Note that the differences $\tau_k - \tau_{k-1}$ have a standard exponential distribution under $\mathbb{Q}$. The assertion of the following lemma is a variation on Eq. (4.23) in [4].

Lemma 1 *The density process L allows the following explicit expression,*

$$L_t = (\lambda_{\tau_1})^{Y_t} \exp\left(t - \int_0^{\tau_1 \wedge t} \lambda_s \, \mathrm{d}s\right) \mathbf{1}_{\{Y_t \le 1\}}.$$

If λ is a bounded process, L is a martingale, hence $\mathbb{E} L_t = L_0 = 1$.

Proof By construction, L is a local martingale. For bounded λ we have $\mathbb{E} \int_0^t L_s^2 \, \mathrm{d}s \le C \exp(2t)$ for some constant C, which yields L a square integrable martingale. The given expression for L_t can be verified by an elementary, but slightly tedious computation.

Under the assumption that L is a martingale (guaranteed for bounded λ), by Girsanov's theorem, see [7, Chap. VI, T3 and T4], we can define for every $T > 0$ a probability $\mathbb{P}$ on $(\Omega, \mathscr{F}_T)$ such that

$$m_t := Y_t - t - \langle Y, \mu \rangle_t = Y_t - \int_0^t \lambda_s \mathbf{1}_{\{Y_{s-}=0\}} \, \mathrm{d}s$$

is a local martingale under $\mathbb{P}$. Note that $\mathbb{P}(Y_T > 1) = \mathbb{E}_{\mathbb{Q}} \mathbf{1}_{\{Y_T > 1\}} L_T = 0$. Hence, under $\mathbb{P}$ we have $\mathbf{1}_{\{Y_s=0\}} = 1 - Y_s$ and the expression for m_t coincides with (3) for $t \le T$.

Note that L cannot be uniformly integrable, since $L_\infty = 0$, which follows from $L_{\tau_2} = 0$. Hence it is not automatic that one can define a probability $\mathbb{P}$ on $(\Omega, \mathscr{F})$ such that m is a martingale on $[0, \infty)$. Note that the laws under $\mathbb{P}$ and $\mathbb{Q}$ of λ are the same.

2.1.2 The One Point Process with MM Intensity

In this section we consider (3), where we specify λ_t as a function of a finite state Markov chain X_t, i.e. $\lambda_t = \lambda(X_t)$. We see that, trivial cases excluded, unlike the constant hazard rate λ in (2), we now have a rate that assumes different values according to the states of the Markov chain. We thus have a rate that is subject to *regime switching*, one also says that we have a *Markov modulated* rate. In order to pose a precise mathematical model, we make some conventions. Let d be the size of the state space of the Markov chain X. Then w.l.o.g. we may assume that X takes its values in the set $\{e_1, \dots, e_d\}$ of d-dimensional standard basis vectors. This implies that any function of X_t can be written as a linear map of X_t, in particular $\lambda(X_t) = \lambda^\top X_t$, where on the right hand side λ is a vector in $\mathbb{R}^d_+$.

Let Q be the transition matrix of X, for which we use the convention that Q_{ij} for $i \neq j$ is the intensity of a transition from state j to state i. As a consequence the *column sums* of Q are equal to zero. We then have

$$\mathrm{d}X_t = QX_t\,\mathrm{d}t + \mathrm{d}M^X_t,$$

where M^X a martingale with values in $\mathbb{R}^d$. We also assume that Q is irreducible and we denote by π the vector representing the invariant distribution.

Furthermore it will be throughout assumed that Y and X have no simultaneous jumps, hence the quadratic variation process $[X, Y]$ ($[X, Y]_t = \sum_{s\leq t} \Delta X_s \Delta Y_s$) is identically zero.

For the single obligor case, we pose the following model with regime switching,

$$\mathrm{d}Y_t = \lambda^\top X_t(1 - Y_t)\,\mathrm{d}t + \mathrm{d}m_t,$$

where $\lambda \in \mathbb{R}^d_+$.

One way of constructing this model is by realizing it on a product space with $\lambda_t = \lambda^\top X_t$ as in Sect. 2.1.1. Alternatively, one can realize Y as standard Poisson process and independently of it, X as a Markov chain on the auxiliary space under $\mathbb{Q}$. By independence, one has $[X, Y] = 0$ under $\mathbb{Q}$ and as these brackets remain the same under an absolutely continuous change of measure using the $\mathbb{Q}$-martingale μ of the previous section, we are then guaranteed to have $[X, Y] = 0$ under $\mathbb{P}$ as well. In this case it is possible to have $\mathbb{P}$ defined on $(\Omega, \mathscr{F})$ for $\mathscr{F} = \mathscr{F}_\infty$, where we use the filtration generated by Y and X. As a side remark we note that $\mathbb{P}$ will not be absolutely continuous w.r.t. $\mathbb{Q}$ on $\mathscr{F}_\infty$.

In all what follows in this paper we adopt the following *Convention: we will use the generic notation M for a martingale, possibly even of varying dimensions, whose precise form is not important.*

An important role will be played by the matrices $Q_{k\lambda} := Q - k\,\mathrm{diag}(\lambda)$ for $k \geq 0$. Here $\mathrm{diag}(\lambda)$ is the diagonal matrix with ii-element equal to λ_i. Here is a, possibly known, stability result for the matrix Q_λ (we take $k = 1$, but a similar result is obviously true for all positive k).

Lemma 2 *Let $\lambda_i > 0$ for all i. Then the matrix Q_λ is invertible and* $\exp(Q_\lambda t) \to 0$ *for $t \to \infty$.*

Proof That Q_λ is invertible, can be seen as follows. Write

$$Q_\lambda = -(I - Q\mathrm{diag}(\lambda)^{-1})\mathrm{diag}(\lambda)$$

and note that $Q\mathrm{diag}(\lambda)^{-1}$ is also the intensity matrix of a Markov chain, as its off-diagonal elements are positive and $\mathbf{1}^\top Q\mathrm{diag}(\lambda)^{-1} = 0$. Therefore $I - Q\mathrm{diag}(\lambda)^{-1}$ is invertible, and so is Q_λ.

In proving the limit result, we give a probabilistic argument.[1] Consider the augmented matrix

$$Q^a_\lambda = \begin{pmatrix} 0 & -\mathbf{1}^\top Q_\lambda \\ 0 & Q_\lambda \end{pmatrix},$$

which is the transition matrix of a Markov chain taking values in $\{e^a_0, \ldots, e^a_d\}$, labelled as the standard basis vectors of $\mathbb{R}^{d+1}$. Clearly, 0 is an absorbing state, and the only one. Hence whatever initial state $x^a(0)$, we have that $\exp(Q^a_\lambda t)x^a(0) \to e^a_0$ for $t \to \infty$. Computing the exponential and taking $x^a(0) \neq e^a_0$, we find

$$\exp(Q^a_\lambda t)x^a = \begin{pmatrix} 1 & \mathbf{1}^\top(I - \exp(Q_\lambda t)) \\ 0 & \exp(Q_\lambda t) \end{pmatrix} x^a(0) = \begin{pmatrix} \mathbf{1}^\top(I - \exp(Q_\lambda t))x(0) \\ \exp(Q_\lambda t)x(0) \end{pmatrix}.$$

Hence $\exp(Q_\lambda t) \to 0$.

In a next section, see Remark 2, we shall see how to compute $\mathbb{P}(Y_t = 1)$. It turns out to be the case that

$$\mathbb{P}(Y_t = 1) = 1 - \mathbf{1}^\top \exp(Q_\lambda t)x(0).$$

We conclude in view of Lemma 2 that $\mathbb{P}(Y_t = 1) \to 1$ for $t \to \infty$. Hence, with probability one, the obligor eventually defaults, as expected.

---

[1] This argument has been provided by Koen de Turck, University of Ghent.

2.2 The MM Model for Multiple Obligors

In Sect. 2.1.2 we have seen results for default processes in the situation of a single obligor. In the present section we generalize those results, at the cost of considerably more complexity, to the situation of multiple obligors.

2.2.1 Multiple Obligors with Time-Varying Intensity

Recall (1). Let's first look at the constant intensity case, $\lambda_t = \lambda > 0$. Then $N_t = \sum_{i=1}^n Y_t^i$ satisfies

$$\mathrm{d}N_t = \lambda(n - N_t)\,\mathrm{d}t + \mathrm{d}m_t, \tag{4}$$

where $m = \sum_{i=1}^n m^i$. By the independence of the default times, m is a martingale w.r.t. $\mathbb{F}$ and N_t has the $\mathrm{Bin}(n, 1 - \exp(-\lambda t))$ distribution. Moreover, given N_u, $u \le s$, $N_t - N_s$ has for $t > s$ the $\mathrm{Bin}(n - N_s, 1 - \exp(-\lambda(t - s)))$ distribution. This model has long ago been used in software reliability going back to [21], with various refinements, like in a Bayesian set up the parameters n and λ being random, see [25, 26] or with time varying but deterministic intensity function $\lambda(t)$, see [14].

Next we look at the case of time varying, possibly random, λ. By the assumed conditional independence of the τ^i given λ we have, similar to the constant λ case, that N_t, conditional on the process λ, has a $\mathrm{Bin}(n, 1 - \exp(-\Lambda_t))$ distribution with $\Lambda_t = \int_0^t \lambda_s\,\mathrm{d}s$.

Let $p^k(t) = \mathbb{P}(N_t = k|\mathscr{F}^\lambda)$, put

$$p(t) = \begin{pmatrix} p^0(t) \\ \vdots \\ p^n(t) \end{pmatrix}$$

and

$$A = \begin{pmatrix} -n & 0 & \cdots & \cdots\cdots & 0 \\ n & -(n-1) & 0 & \cdots\cdots & 0 \\ 0 & n-1 & -(n-2) & 0 \;\cdots & 0 \\ \vdots & \ddots & \ddots & \ddots\;\ddots & \vdots \\ \vdots & & \ddots & \ddots\; -1 & 0 \\ 0 & \cdots & \cdots & 0 \;\; 1 & 0 \end{pmatrix}. \tag{5}$$

Then we have for $p(t)$ the system of differential equations

$$\dot{p}(t) = \lambda_t A p(t),$$

which has solution (here we use that λ is real-valued)

$$p(t) = \exp(\Lambda_t A)e_0,$$

where $\Lambda_t = \int_0^t \lambda_s \, \mathrm{d}s$ and e_0 is the first standard basis vector of $\mathbb{R}^{n+1}$. For the vector whose elements are the unconditional probabilities $\mathbb{P}(N_t = k)$ one has to take the expectation and it depends on the specification of λ whether this results in analytic expressions. We will see that this happens in case of a Markov modulated rate process.

2.2.2 The MM Case

We assume to have a finite state Markov process as in Sect. 2.1.2 and let $\lambda_t = \lambda^\top X_{t-}$. For N_t one now has its submartingale decomposition

$$\mathrm{d}N_t = \lambda^\top X_t(n - N_t)\,\mathrm{d}t + \mathrm{d}m_t.$$

This is the model of Sect. 2.1.2 extended to more obligors. The default rate for each obligor has become random ($\lambda^\top X_t$), but is taken the same for all of them.

Let $\nu_t^k = \mathbf{1}_{\{N_t = k\}}$, $k = 0, \ldots, n$. For notational convenience we set $\nu_t^{-1} = 0$. It follows that $\Delta\nu_t^k = 1$ iff N_t jumps from $k-1$ to k at t, and $\Delta\nu_t^k = -1$ iff N_t jumps from k to $k+1$. This can be summarized by

$$\mathrm{d}\nu_t^k = (\nu_{t-}^{k-1} - \nu_{t-}^k)\,\mathrm{d}N_t.$$

In vector form this becomes

$$\mathrm{d}\nu_t = (J - I)\nu_{t-}\,\mathrm{d}N_t, \tag{6}$$

where

$$J = \begin{pmatrix} 0 & & & & \\ 1 & 0 & & & \\ 0 & 1 & \ddots & & \\ \vdots & \ddots & \ddots & \ddots & 0 \\ 0 & \cdots & 0 & 1 & 0 \end{pmatrix}.$$

Using the dynamics for N, we get

$$\begin{aligned} \mathrm{d}\nu_t^k &= (\nu_{t-}^{k-1} - \nu_{t-}^k)(\lambda^\top X_{t-}(n - N_t)\,\mathrm{d}t + \mathrm{d}m_t) \\ &= \lambda^\top X_t((n-k+1)\nu_t^{k-1} - (n-k)\nu_t^k)\,\mathrm{d}t + \mathrm{d}M_t. \end{aligned}$$

Letting $\nu_t = \begin{pmatrix} \nu_t^0 \\ \vdots \\ \nu_t^n \end{pmatrix}$, we get from the above display

$$\mathrm{d}\nu_t = \lambda^\top X_t A \nu_t \,\mathrm{d}t + \mathrm{d}M_t, \tag{7}$$

where A is as in (5). This equation for ν is a main ingredient in the next result.

Proposition 1 *Let $\zeta_t = \nu_t \otimes X_t$. The process ζ is Markov with transition matrix $\mathbf{Q}$, where $\mathbf{Q} = (A \otimes \mathrm{diag}(\lambda) + I \otimes Q)$. It follows that $\mathbb{E}[\zeta_t | \mathscr{F}_s] = \exp(\mathbf{Q}(t-s))\zeta_s$.*

Proof We will use Eq. (7) together with the dynamics of X. Using the product rule and the fact that N and X do not jump at the same time and summarizing again all martingale terms again as M, we get (recall the multiplication rule $(A \otimes B)(C \otimes D) = (AC) \otimes (BD)$)

$$\begin{aligned}
\mathrm{d}(\nu_t \otimes X_t) &= \big((A\nu_t\lambda^\top X_t) \otimes X_t + \nu_t \otimes (QX_t)\big)\,\mathrm{d}t + \mathrm{d}M_t \\
&= \big((A\nu_t) \otimes (X_t\lambda^\top X_t) + \nu_t \otimes (QX_t)\big)\,\mathrm{d}t + \mathrm{d}M_t \\
&= ((A\nu_t) \otimes (\mathrm{diag}(\lambda)X_t) + I\nu_t \otimes (QX_t))\,\mathrm{d}t + \mathrm{d}M_t \\
&= (A \otimes \mathrm{diag}(\lambda) + I \otimes Q)(\nu_t \otimes X_t)\,\mathrm{d}t + \mathrm{d}M_t \\
&= \mathbf{Q}(\nu_t \otimes X_t)\,\mathrm{d}t + \mathrm{d}M_t.
\end{aligned}$$

Note that ζ_t by construction consists of the indicators of the values of the joint process (ν, X). Hence the equation $\mathrm{d}\zeta_t = \mathbf{Q}\zeta_t\,\mathrm{d}t + \mathrm{d}M_t$ reveals, cf. Lemma 1.1 in Appendix B of [10], that ζ (and hence (ν, X)) is Markov.

An explicit computation shows

$$\mathbf{Q} = \begin{pmatrix}
Q_{n\lambda} & 0 & \cdots & \cdots & \cdots & 0 \\
n\,\mathrm{diag}(\lambda) & Q_{(n-1)\lambda} & 0 & \cdots & \cdots & 0 \\
0 & (n-1)\,\mathrm{diag}(\lambda) & Q_{(n-2)\lambda} & 0 & \cdots & 0 \\
\vdots & \ddots & \ddots & \ddots & \ddots & \vdots \\
\vdots & & \ddots & \ddots & Q_\lambda & 0 \\
0 & \cdots & \cdots & 0 & \mathrm{diag}(\lambda) & Q
\end{pmatrix}, \tag{8}$$

where for $k \in \mathbb{N}$ we have $Q_{k\lambda} = Q - k\,\mathrm{diag}(\lambda)$.

Remark 1 The original dynamic equations for X_t and N_t can be retrieved from Proposition 1. Realizing the relations $X_t = (\mathbf{1}^\top \otimes I)\zeta_t$ and $(\mathbf{1}^\top \otimes I)\mathbf{Q} = \mathbf{1}^\top \otimes Q$, and $\mathbf{1}^\top A = 0$, we obtain from Proposition 1

$$\begin{aligned}
\mathrm{d}X_t &= (\mathbf{1}^\top \otimes I)\,(\mathbf{Q}(\nu_t \otimes X_t))\,\mathrm{d}t + \mathrm{d}M_t \\
&= (\mathbf{1}^\top \otimes Q)(\nu_t \otimes X_t)\,\mathrm{d}t + \mathrm{d}M_t \\
&= QX_t\,\mathrm{d}t + \mathrm{d}M_t.
\end{aligned}$$

Similarly, we get from $\nu_t = (I \otimes \mathbf{1}^\top)\zeta_t$,

$$
\begin{aligned}
\mathrm{d}\nu_t &= (I \otimes \mathbf{1}^\top)\,(\mathbf{Q}(\nu_t \otimes X_t))\,\mathrm{d}t + \mathrm{d}M_t \\
&= (A \otimes \lambda^\top)(\nu_t \otimes X_t)\,\mathrm{d}t + \mathrm{d}M_t \\
&= A\nu_t\lambda^\top X_t\,\mathrm{d}t + \mathrm{d}M_t.
\end{aligned}
$$

Using $\begin{pmatrix}0 & 1 & \cdots & n\end{pmatrix} A\nu_t = \begin{pmatrix}n & \cdots & 1 & 0\end{pmatrix} \nu_t = n - N_t$, we get from the last display the decomposition $\mathrm{d}N_t = (n - N_t)\lambda^\top X_t\,\mathrm{d}t + \mathrm{d}m_t$ back.

Letting $\pi(t) = \mathbb{E}\zeta_t$, we obtain from Proposition 1 the ODE

$$
\dot{\pi}(t) = \mathbf{Q}\pi(t), \tag{9}
$$

with the initial condition $\pi(0) = e_0 \otimes x(0)$, where e_0 has 1 as its first element, all other elements being zero. We will give a rather explicit expression for $\pi(t) = \exp(\mathbf{Q}t)\pi(0)$, for which we need some additional results.

The differential equation for π is the following type of forward equation,

$$
\dot{F} = \mathbf{Q}F.
$$

Here F can be any matrix valued function of appropriate dimensions. We will block-diagonalize the matrix $\mathbf{Q}$. The transformation that is needed for that is given by the matrix V whose ij-block $(i, j = 0, \ldots, n)$ is

$$
V_{ij} = \binom{n-j}{n-i}(-1)^{i-j} I.
$$

Note that $V_{ij} = 0$ for $i < j$, V is block lower-triangular. The inverse matrix is also block lower-triangular with blocks

$$
V_{ij}^{-1} = \binom{n-j}{n-i} I.
$$

One may check by direct computation that indeed $VV^{-1} = I$. It is straightforward to verify that $\mathbf{Q}^V := V^{-1}\mathbf{Q}V$ is block-diagonal with ith block $(i = 0, \ldots, n)$ equal to

$$
\mathbf{Q}_i^V = Q_{(n-i)\lambda}.
$$

Putting $G = V^{-1}F$ we obtain

$$
\dot{G} = \mathbf{Q}^V G,
$$

whose solution satisfying $G(0) = I$ is block diagonal with ith block $G_i(t) = \exp(Q_{(n-i)\lambda}t)$. We thus obtain the following lemma.

Lemma 3 *The solution to the forward ODE $\dot{F} = \mathbf{Q}F$ with initial condition $F(0)$ is given by $F(t) = \exp(\mathbf{Q}t)F(0)$, where*

$$\exp(\mathbf{Q}t) = V \begin{pmatrix} \exp(Q_{n\lambda}t) & & \\ & \ddots & \\ & & \exp(Qt) \end{pmatrix} V^{-1}.$$

If $F(t) = \exp(\mathbf{Q}t)$, its blocks $F_{ij}(t)$ can be explicitly computed. One has $F_{ij}(t) = 0$ if $i < j$, and for $i \geq j$ it holds that

$$F_{ij}(t) = \binom{n-j}{n-i} \sum_{k=j}^{i} (-1)^{i-k} \binom{i-j}{i-k} \exp(Q_{(n-k)\lambda}t).$$

Proof We use the block triangular structure of V and V^{-1} together with the block diagonal structure of $\mathbf{Q}^V$ to compute

$$\begin{aligned} F_{ij}(t) &= \sum_{k=j}^{i} V_{ik} \exp(Q_{(n-k)\lambda}t) V_{kj} \\ &= \sum_{k=j}^{i} \binom{n-k}{n-i} (-1)^{i-k} \exp(Q_{(n-k)\lambda}t) \binom{n-j}{n-k} \\ &= \binom{n-j}{n-i} \sum_{k=j}^{i} (-1)^{i-k} \binom{i-j}{i-k} \exp(Q_{(n-k)\lambda}t), \end{aligned}$$

as stated.

Proposition 2 *The solution $\pi(t)$ to the system (9) of ODEs under the initial condition $\pi(0) = e_0 \otimes x(0)$ has components $\pi^i(t) \in \mathbb{R}^d$ given by*

$$\pi^i(t) = \binom{n}{i} \sum_{k=0}^{i} (-1)^{i-k} \binom{i}{k} \exp(Q_{(n-k)\lambda}t) x(0). \tag{10}$$

Proof We use Lemma 3 and recall the specific form of the initial condition $\pi(0)$. We have to compute $\exp(\mathbf{Q}t)\pi(0)$ and obtain from Lemma 3 with $j = 0$ for $\pi^i(t) = F_{i0}(t)$

$$\begin{aligned} \pi^i(t) &= \binom{n}{n-i} \sum_{k=0}^{i} (-1)^{i-k} \binom{i}{i-k} \exp(Q_{(n-k)\lambda}t) x(0) \\ &= \binom{n}{i} \sum_{k=0}^{i} (-1)^{i-k} \binom{i}{k} \exp(Q_{(n-k)\lambda}t) x(0). \end{aligned}$$

Remark 2 Let us look at a special case, $n = 1$. Then we can write $N_t = Y_t$ and it is sufficient to compute

$$\pi^1(t) = \mathbb{E}(Y_t X_t) = (\exp(Qt) - \exp(Q_\lambda t))\,x(0). \tag{11}$$

As a consequence we are able to compute $\mathbb{P}(Y_t = 1) = \mathbf{1}^\top \mathbb{E}(Y_t X_t)$,

$$\mathbb{P}(Y_t = 1) = 1 - \mathbf{1}^\top \exp(Q_\lambda t) x(0),$$

since $\mathbf{1}^\top \exp(Q_t) = \mathbf{1}^\top$. As $\exp(Qt) \to \pi \mathbf{1}^\top$, we conclude in view of Lemma 2 from (11) that $\pi^1(t) \to \pi$ for $t \to \infty$. This result should be obvious, as Y_t eventually becomes 1 and X_t converges in distribution to its invariant law.

For the case $n > 1$ the expressions for $\pi^i(t)$ are a bit complicated, but their asymptotic values for $t \to \infty$, are as expected, $\pi^i(t) \to 0$ for $i < n$, whereas $\pi^n(t) \to \pi$. This again follows from Lemma 2.

Proposition 2 has the following corollary.

Corollary 1 *Let $\phi(t, u) = \mathbb{E} \exp(\mathrm{i} u N_t) X_t$. It holds that*

$$\phi(t, u) = \sum_{k=0}^{n} \binom{n}{k} \exp(\mathrm{i}uk)(1 - \exp(\mathrm{i}u))^{n-k} \exp(Q_{(n-k)\lambda} t) x(0).$$

Proof We shall use the elementary identity

$$\sum_{k=j}^{n} \beta^k \binom{n}{k}\binom{k}{j} = \binom{n}{j} \beta^j (1 + \beta)^{n-j}$$

for $\beta = -e^{-\mathrm{i}u}$ in the last step in the chain of equalities below. From Proposition 2 we obtain

$$\begin{aligned}
\mathbb{E} \exp(\mathrm{i}uN_t) X_t &= \sum_{k=0}^{n} e^{\mathrm{i}uk} \pi^k(t) \\
&= \sum_{k=0}^{n} e^{\mathrm{i}uk} \binom{n}{k} \sum_{j=0}^{k} (-1)^{k-j} \binom{k}{j} \exp(Q_{(n-j)\lambda} t) x(0) \\
&= \sum_{j=0}^{n} \sum_{k=j}^{n} (-e^{\mathrm{i}u})^k \binom{n}{k}\binom{k}{j} (-1)^j \exp(Q_{(n-j)\lambda} t) x(0) \\
&= \sum_{j=0}^{n} \binom{n}{j} e^{\mathrm{i}ju} (1 - e^{\mathrm{i}u})^{n-j} \exp(Q_{(n-j)\lambda} t) x(0).
\end{aligned}$$

Remark 3 Alternatively, one can compute a moment generating function $\psi(t, v) = \mathbb{E} \exp(-vN_t) X_t$ for $v \geq 0$. Let B have a binomial distribution with parameters n and $p = 1 - \exp(-v)$. Then we have for $\psi(t, v)$ the compact expression $\psi(t, v) = \mathbb{E} \exp((Q - B\mathrm{diag}(\lambda))t) x(0) = \mathbb{E} \exp(Q_{\lambda B} t) x(0)$.

Remark 4 There appears to be no simpler representation for $\phi(t,u)$. We note that this function also satisfies the PDE

$$\dot{\phi}(t,u) = (Q + n(e^{iu} - 1)\mathrm{diag}(\lambda))\phi(t,u) + \mathrm{i}(e^{iu} - 1)\mathrm{diag}(\lambda)\frac{\partial \phi(t,u)}{\partial u}. \quad (12)$$

Just by computing the partial derivatives, one verifies that this equation holds. Alternatively, one can apply the Itô formula to $\exp(\mathrm{i}uN_t)X_t$ followed by taking expectations.

2.2.3 Conditional Probabilities

The vehicle we use is the process ζ, recall $\zeta_t = \nu_t \otimes X_t$. Our aim is to find expressions for $\zeta_{t|s} = \mathbb{E}[\zeta_t|\mathscr{F}_s]$ for $t > s$, from which one can deduce the conditional probabilities $\mathbb{E}[\nu_t|\mathscr{F}_s]$ and $\mathbb{E}[N_t|\mathscr{F}_s]$. By the Markov property, Proposition 1, we have $\mathbb{E}[\zeta_t|\mathscr{F}_s] = \exp(\mathbf{Q}(t-s))\zeta_s$. Let $\zeta_{t|s} = \mathbb{E}[\zeta_t|\mathscr{F}_s]$ and $\zeta^k_{t|s} = \mathbb{E}[\mathbf{1}_{\{N_t=k\}}X_t|\mathscr{F}_s]$. We aim at a more explicit representation of the conditional probabilities $\zeta^k_{t|s}$ for $k \geq 0$. Note that $\zeta^k_{t|s} = (e_k^\top \otimes I)\zeta_{t|s}$. Hence $\zeta^k_{t|s} = (e_k^\top \otimes I)\exp(\mathbf{Q}(t-s))\zeta_s$. Using Lemma 3, we have

$$\zeta^k_{t|s} = (e_k^\top \otimes I)V \begin{pmatrix} \exp(Q_{n\lambda}(t-s)) & & \\ & \ddots & \\ & & \exp(Q(t-s)) \end{pmatrix} V^{-1}\zeta_s.$$

By matrix computations as before this leads to the following result.

Proposition 3 *It holds that*

$$\zeta^k_{t|s} = \sum_{j=0}^{k} \binom{n-j}{k-j} \sum_{i=0}^{k} (-1)^{k-i} \binom{k-j}{k-i} \exp(Q_{(n-i)\lambda}(t-s))\zeta^j_s.$$

Note that in the formula of this proposition, only one of the ζ^j_s is different from zero and then equal to X_s. Effectively, the sum over j thus reduces to one term only. The conditional probabilities $\nu^k_{t|s} = \mathbb{P}(N_t = k|\mathscr{F}_s)$ can now simply be computed as $\mathbf{1}^\top \zeta^k_{t|s}$. Note that these still depend on X_s, and one has the explicit expression

$$\mathbb{E}[\nu^k_t|\mathscr{F}_s] = \sum_{j=0}^{n} \binom{n-j}{n-k} \sum_{i=j}^{k} (-1)^{k-i} \binom{k-j}{k-i} \mathbf{1}^\top \exp(Q_{(n-i)\lambda}(t-s))X_s\nu^j_s.$$

Remark 5 Consider the special case $n = 1$ and let $Z_t = Y_tX_t$, Y_t as in Sect. 2.1.2. This amounts to taking $k = n = 1$ in Proposition 3 and one gets for $Z_{t|s} = \mathbb{E}[Z_t|\mathscr{F}^Y_s]$ the simpler expression

$$Z_{t|s} = \exp(Q_\lambda(t-s))Z_s + \big(\exp(Q(t-s)) - \exp(Q_\lambda(t-s))\big)X_s. \qquad (13)$$

The next purpose is to compute $\mathbb{E}[e^{\mathrm{i}uN_t}X_t|\mathscr{F}_s]$ and from that one $\mathbb{E}[e^{\mathrm{i}uN_t}|\mathscr{F}_s] = \mathbf{1}^\top\mathbb{E}[e^{\mathrm{i}uN_t}X_t|\mathscr{F}_s]$.

Proposition 4 *The following hold.*

$$\mathbb{E}[e^{\mathrm{i}uN_t}X_t|\mathscr{F}_s] = \sum_{k=0}^{n}\sum_{j=k}^{n}\binom{n-k}{j-k}(1-e^{\mathrm{i}u})^{n-j}e^{\mathrm{i}uj}\exp(Q_{(n-j)\lambda}(t-s))\zeta_s^k,$$

$$\mathbb{E}[e^{\mathrm{i}uN_t}|\mathscr{F}_s] = \sum_{k=0}^{n}\sum_{j=k}^{n}\binom{n-k}{j-k}(1-e^{\mathrm{i}u})^{n-j}e^{\mathrm{i}uj}\mathbf{1}^\top\exp(Q_{(n-j)\lambda}(t-s))\zeta_s^k. \qquad (14)$$

Proof We start from the identity $e^{\mathrm{i}uN_t}X_t = \mathbf{F}\zeta_t$, with $\mathbf{F} = e(u)\otimes I$, where $e(u) = \big(1\ e^{\mathrm{i}u}\ \cdots\ e^{n\mathrm{i}u}\big)$. Hence we have

$$\mathbb{E}[e^{\mathrm{i}uN_t}X_t|\mathscr{F}_s] = (e(u)\otimes I)\exp(\mathbf{Q}(t-s))\zeta_s.$$

This can be put into the asserted more explicit representation, involving the matrices $Q_{k\lambda}$ by application of Proposition 3. The second assertion is a trivial consequence.

It is conceivable that only N is observed, and not the background process X. In such a case one is only able to compute conditional expectation of quantities as above conditioned on $\mathscr{F}_s^N$ instead of $\mathscr{F}_s$, see Sect. 4.1 for results.

3 The Markov Modulated Poisson Process

In this section we study MM Poisson processes. These have an intensity process $\lambda_t = \lambda^\top X_t$, using the same notation as before. In terms of defaultable obligors, such processes occur as limits of the total number of defaults N_t as in Sect. 2.2 where $n\to\infty$ and the vector λ is scaled to become λ/n, as we shall see later. So we can use this to approximate the total number of defaults in a market with a large number of obligors, where each of them has small default rate.

3.1 The Model

The point of departure is to postulate the dynamics of the counting process N as

$$\mathrm{d}N_t = \lambda^\top X_t\,\mathrm{d}t + \mathrm{d}m_t.$$

We follow the same approach as before. So we use that conditionally on $\mathscr{F}^X$ we have that N_t has a Poisson(Λ_t) distribution with $\Lambda_t = \int_0^t \lambda^\top X_s \,\mathrm{d}s$. It follows that

$$\mathbb{E}[\mathbf{1}_{\{N_t=k\}}X_t|\mathscr{F}^X] = \frac{1}{k!}\Lambda_t^k \exp(-\Lambda_t)X_t =: p^k(t)X_t,$$

and

$$\frac{\mathrm{d}}{\mathrm{d}t}p^k(t) = p^{k-1}(t) - p^k(t)\lambda^\top X_t.$$

Then we obtain

$$\mathrm{d}\mathbb{E}[\mathbf{1}_{\{N_t=k\}}X_t|\mathscr{F}^X] = \big(p^{k-1}(t) - p^k(t)\big)\mathrm{diag}(\lambda)X_t\,\mathrm{d}t + p^k(t)(QX_t\,\mathrm{d}t + \mathrm{d}M_t),$$

and with $\pi^k(t) = \mathbb{E}(p^k(t)X_t)$ we find

$$\dot{\pi}^k(t) = \mathrm{diag}(\lambda)\pi^{k-1}(t) + (Q - \mathrm{diag}(\lambda))\pi^k(t).$$

For $k = 0$, one immediately finds the solution $\pi^0(t) = \exp(Q_\lambda t)x(0)$. For $k > 0$ there seems to be no simply expression in terms of exponential of Q and $Q_{k\lambda}$ as in Proposition 2, not even for $k = 1$, although one has

$$\pi^1(t) = \int_0^t \exp(-Q_\lambda(t-s))\mathrm{diag}(\lambda)\exp(Q_\lambda s)\,\mathrm{d}s\,x(0).$$

However, it is possible to get a formula for the vector

$$\Pi^n(t) = \begin{pmatrix} \pi^0(t) \\ \vdots \\ \pi^n(t) \end{pmatrix},$$

since it satisfies the ODE

$$\dot{\Pi}^n(t) = \mathbf{Q}_n\Pi^n(t),$$

where $\mathbf{Q}_n \in \mathbb{R}^{(n+1)d\times(n+1)d}$ is given by

$$\mathbf{Q}_n = \begin{pmatrix} Q-\mathrm{diag}(\lambda) & 0 & \cdots & \cdots & 0 \\ \mathrm{diag}(\lambda) & Q-\mathrm{diag}(\lambda) & 0 & & 0 \\ 0 & \mathrm{diag}(\lambda) & \ddots & \ddots & \vdots \\ \vdots & & \ddots & Q-\mathrm{diag}(\lambda) & 0 \\ 0 & \cdots & 0 & \mathrm{diag}(\lambda) & Q-\mathrm{diag}(\lambda) \end{pmatrix}.$$

Together with the initial conditions $\pi^k(0) = \delta_{k0}x(0)$, one obtains

$$\Pi^n(t) = \exp(\mathbf{Q}_n t)(e_0^n \otimes x(0)),$$

where e_0^n is the first basis vector of $\mathbb{R}^{n+1}$. An elementary expression for $\exp(\mathbf{Q}_n t)$ is not available due to the fact that $Q - \text{diag}(\lambda)$ and $\text{diag}(\lambda)$ do not commute. Besides, $\mathbf{Q}_n$ is block lower triangular with identical blocks on the main diagonal and therefore cannot be block diagonalized.

However, in the present case there is a nice expression for the characteristic function $\phi(t, u) = \mathbb{E}\exp(\mathrm{i}uN_t)X_t$, unlike the situation of Corollary 1. To determine $\phi(t, u)$, we apply the Itô formula (note that $[N, X] = 0$) and obtain

$$\mathrm{d}\exp(\mathrm{i}uN_t)X_t = (e^{\mathrm{i}u} - 1)e^{\mathrm{i}uN_{t-}}X_{t-}\mathrm{d}N_t + e^{\mathrm{i}uN_{t-}}\mathrm{d}X_t, \tag{15}$$

which yields after taking expectations and using the dynamics of X and N

$$\dot{\phi}(t, u) = ((e^{\mathrm{i}u} - 1)\text{diag}(\lambda) + Q)\phi(t, u).$$

Hence

$$\phi(t, u) = \exp\big(((e^{\mathrm{i}u} - 1)\text{diag}(\lambda) + Q)t\big)x(0).$$

Contrary to the $\pi^k(t)$ of Proposition 2 we thus found a *simple* formula for $\phi(t, u)$. This formula is in line with [2, Proposition 1.6] for Markovian arrival processes.

Remark 6 It is possible to obtain the above results as limits from results in Sect. 2.2.2, by replacing there λ by λ/n and letting $n \to \infty$.

If we look at the moment generating functions $\psi(t, v) = \mathbb{E}\exp(-vN_t)X_t$, we have $\psi(t, v) = \exp\big((Q - (1 - e^{-v})\text{diag}(\lambda))t\big)x(0)$. Replace in Remark 3 the parameter λ with λ/n and let $n \to \infty$ and write B_n instead of B. Then we have $\psi_n(t, v) = \mathbb{E}\exp\big((Q - \text{diag}(\lambda)B_n/n)t\big)x(0)$. As $B_n/n \to 1 - e^{-v}$ a.s., we obtain $\exp\big((Q - \text{diag}(\lambda)B_n/n)t\big) \to \exp\big((Q - \text{diag}(\lambda)(1 - e^{-v}))t\big)$ a.s. Since the exponentials are bounded, we also have convergence of the expectations by dominated convergence. Replacing $-v$ with $\mathrm{i}u$ gives the characteristic function.

3.2 Conditional Probabilities

Mimicking the approach of Sect. 2.2.2, we consider again the $\nu_t^k = \mathbf{1}_{\{N_t = k\}}$. Let

$$\bar{\nu}_t^n = \begin{pmatrix} \nu_t^0 \\ \vdots \\ \nu_t^n \end{pmatrix}.$$

Then $\bar{\nu}^n$ still satisfies Eq. (6). Combining this with the dynamics of N, we obtain the semimartingale decomposition

$$\mathrm{d}\bar{\upsilon}_t^n = \lambda^\top X_t(J-I)\bar{\upsilon}_t^n\,\mathrm{d}t + \mathrm{d}M_t.$$

Letting $\bar{\zeta}_t^n = \bar{\upsilon}_t^n \otimes X_t$, then we can derive, similar to the approach of Sect. 2.2.2,

$$\mathrm{d}\bar{\zeta}_t^n = \mathbf{Q}_n\bar{\zeta}_t^n\,\mathrm{d}t + \mathrm{d}M_t.$$

This is for each n a finite dimensional system, which can be extended to an infinite dimensional system for ζ_t. The resulting infinite coefficient matrix will be lower triangular again,

$$\mathrm{d}\zeta_t = \mathbf{Q}_\infty\zeta_t\,\mathrm{d}t + \mathrm{d}M_t,$$

where $\mathbf{Q}_\infty = I_\infty \otimes Q_\lambda - J_\infty \otimes \mathrm{diag}(\lambda)$ with I_∞ the infinite dimensional identity matrix and J_∞ the infinite dimensional counterpart of the earlier encountered matrix J. It follows that for the vector of conditional probabilities we have

$$\mathbb{E}[\zeta_t|\mathscr{F}_s] = \exp(\mathbf{Q}_\infty(t-s))\bar{\zeta}_s.$$

This looks like an infinite dimensional expression, but $\mathbb{E}[\mathbf{1}_{\{N_t=n\}}X_t|\mathscr{F}_s]$ can be computed from $\mathbb{E}[\bar{\zeta}_t^n|\mathscr{F}_s] = \exp(\mathbf{Q}_n(t-s))\bar{\zeta}_s^n$, which effectively reduces the infinite dimensional system to a finite dimensional one. One can now also compute, with $\ell_n^\top = \begin{pmatrix}0 \cdots 0 \; 1\end{pmatrix} \in \mathbb{R}^{1\times(n+1)}$,

$$\mathbb{P}(N_t = n, X_t = e_j|\mathscr{F}_s) = (\ell_n^\top \otimes e_j^\top)\exp(\mathbf{Q}_n(t-s))\bar{\zeta}_s^n.$$

3.3 Conditional Characteristic Function

Our aim is to find an expression for $\phi_{t|s} := \mathbb{E}[\exp(\mathrm{i}uN_t)X_t|\mathscr{F}_s]$. Since we deal in the present section with the MM Poisson process N, the bivariate process (X, N), unlike its counterpart in Sect. 2, is an instance of a Markov additive process [2], and $\mathbb{E}[\exp(\mathrm{i}u(N_t - N_s))X_t|\mathscr{F}_s]$ will only depend on X_s. We first follow the forward approach.

Proposition 5 *It holds that*

$$\phi_{t|s} = \exp\big(((e^{\mathrm{i}u}-1)\mathrm{diag}(\lambda) + Q)(t-s)\big)\, e^{\mathrm{i}uN_s}X_s. \tag{16}$$

Proof Starting point is Eq. (15). We use the dynamics of N and X to get the semi-martingale decomposition

$$\begin{aligned}\mathrm{d}\exp(\mathrm{i}uN_t)X_t &= (e^{\mathrm{i}u}-1)e^{\mathrm{i}uN_t}\mathrm{diag}(\lambda)X_t\,\mathrm{d}t + e^{\mathrm{i}uN_t}QX_t\,\mathrm{d}t + \mathrm{d}M_t \\ &= ((e^{\mathrm{i}u}-1)\mathrm{diag}(\lambda) + Q)e^{\mathrm{i}uN_t}X_t\,\mathrm{d}t + \mathrm{d}M_t.\end{aligned}$$

Let $t \geq s$. We obtain (differentials w.r.t. t)

$$d\phi_{t|s} = ((e^{iu} - 1)\mathrm{diag}(\lambda) + Q)\phi_{t|s}\,dt,$$

which has the desired solution.

Next we outline the backward approach. Observe first that $\phi_{t|s}$ is a martingale in the s-parameter and that due to the fact that (N, X) is Markov, we can write for some function Φ, $\phi_{t|s} = \Phi(t-s, N_s)X_s$. We identify Φ as follows, using the Itô formula w.r.t. s. We obtain

$$\begin{aligned} d\phi_{t|s} &= \left(-\dot{\Phi}(t-s, N_s)\,ds + (\Phi(t-s, N_{s-}+1) - \Phi(t-s, N_{s-}))dN_s\right) X_{s-} \\ &\quad + \Phi(t-s, N_{s-})\,dX_s \\ &= \left(-\dot{\Phi}(t-s, N_s) + (\Phi(t-s, N_s+1) - \Phi(t-s, N_s))\mathrm{diag}(\lambda)\right) X_s\,ds \\ &\quad + \Phi(t-s, N_s)QX_s\,ds + dM_s. \end{aligned}$$

The above mentioned martingale property leads to the system of ODEs ($n \geq 0$)

$$\dot{\Phi}(t, n) = \Phi(t, n+1)\mathrm{diag}(\lambda) + \Phi(t, n)\,(Q - \mathrm{diag}(\lambda))\,. \tag{17}$$

We have the initial conditions $\Phi(0, n) = \exp(iun)$. To know $\Phi(t, n)$ it seems necessary to know $\Phi(t, n+1)$, which suggest that the ODEs are difficult to solve constructively. Instead, we pose a solution, we will verify that

$$\Phi(t, n) = \exp\left(((e^{iu} - 1)\mathrm{diag}(\lambda) + Q)t\right) e^{iun}.$$

Differentiation of the given expression for $\Phi(t, n)$ gives

$$\dot{\Phi}(t, n) = \Phi(t, n)((e^{iu} - 1)\mathrm{diag}(\lambda) + Q).$$

Note that $\Phi(t, n+1) = \Phi(t, n)e^{iu}$. Insertion of this into the ODE gives

$$\dot{\Phi}(t, n) = \Phi(t, n)(e^{iu}\mathrm{diag}(\lambda) + (Q - \mathrm{diag}(\lambda))),$$

which coincides with (17).

4 Filtering

Let N be a counting process with predictable intensity process λ. In many cases it is conceivable that λ is an unobserved process and expressions in terms of λ are not always useful. Let $\hat{\lambda}_t = \mathbb{E}[\lambda_t | \mathscr{F}_t^N]$. Then the semimartingale decomposition of N w.r.t. the filtration $\mathbb{F}^N$ is given by

$$dN_t = \hat{\lambda}_t\,dt + d\hat{m}_t,$$

where $\hat{m}$ is a (local) martingale w.r.t. $\mathbb{F}^N$. The general filter of the Markov chain X, $\hat{X}_t = \mathbb{E}[X_t|\mathscr{F}_t^N]$ satisfies the following well known formula (see [7], originating from [28]) with Q as in Sect. 2.1.2

$$\mathrm{d}\hat{X}_t = Q\hat{X}_t\,\mathrm{d}t + \hat{\lambda}_{t-}^{+}(\widehat{X\lambda}_{t-} - \hat{X}_{t-}\hat{\lambda}_{t-})(\mathrm{d}N_t - \hat{\lambda}_t\,\mathrm{d}t),$$

where $\widehat{X\lambda}_t = \mathbb{E}[X_t\lambda_t|\mathscr{F}_t^N]$ and where we use the notation $x^+ = \mathbf{1}_{x\neq 0}/x$ for a real number x. For any of the previously met models for the counting process N we have a predictable intensity process of the form $\lambda_t = \lambda^\top X_{t-}f(N_{t-})$, where f depends on the specific model at hand. It follows that $\hat{\lambda}_t = \lambda^\top\hat{X}_{t-}f(N_{t-})$. In all cases we consider it happens that $f(N_t)$ remains zero after it has reached zero, and hence N stops jumping as soon as $f(N_t) = 0$. Since $\lambda^\top X_t > 0$, with the convention $\frac{0}{0} = 0$ the above filter equation reduces to

$$\mathrm{d}\hat{X}_t = Q\hat{X}_t\,\mathrm{d}t + \frac{1}{\lambda^\top\hat{X}_{t-}}(\mathrm{diag}(\lambda)\hat{X}_{t-} - \hat{X}_{t-}\lambda^\top\hat{X}_{t-})(\mathrm{d}N_t - \hat{\lambda}_t\,\mathrm{d}t). \tag{18}$$

For the specific models we have encountered we give in the next sections more results on $\hat{X}$.

4.1 Filtering for the MM Multiple Point Process

The notation of this section is as in Sect. 2.2.2 and subsequent sections. Let $\hat{\zeta}_t = \mathbb{E}[\zeta_t|\mathscr{F}_t^N]$. Then $\hat{\zeta}_t = \nu_t \otimes \hat{X}_t$, where $\hat{X}_t = \mathbb{E}[X_t|\mathscr{F}_t^N]$. For $\hat{X}_t$ we have from (18),

$$\mathrm{d}\hat{X}_t = Q\hat{X}_t\,\mathrm{d}t + \frac{1}{\lambda^\top\hat{X}_{t-}}\left(\mathrm{diag}(\lambda)\hat{X}_{t-} - \hat{X}_{t-}\hat{X}_{t-}^\top\lambda\right)(\mathrm{d}N_t - (n - N_t)\lambda^\top\hat{X}_t\,\mathrm{d}t).$$

At the jump times τ_k ($k = 1, \ldots, n$) (these are the order statistics of the original default times τ^i) of N we thus have

$$X_{\tau_k} = \frac{1}{\lambda^\top\hat{X}_{\tau_k-}}\mathrm{diag}(\lambda)\hat{X}_{\tau_k-}.$$

Between the jump times, $\hat{X}$ evolves according to the ODE

$$\frac{\mathrm{d}\hat{X}_t}{\mathrm{d}t} = Q\hat{X}_t - (n - N_t)(\mathrm{diag}(\lambda)\hat{X}_{t-} - \hat{X}_{t-}\hat{X}_{t-}^\top\lambda),$$

which is also valid after the last jump of N. It follows that for $t \geq \tau_n$ we have $\hat{X}_t = \exp(Q(t - \tau_n))\hat{X}_{\tau_n}$.

Below we need $[\nu,\hat{X}]^{\otimes}_t=\sum_{s\le t}\Delta\nu_s\otimes\Delta\hat{X}_s$. Using the equations for ν and $\hat{X}$, we find

$$\mathrm{d}[\nu,\hat{X}]^{\otimes}_t=\frac{1}{\lambda^\top\hat{X}_{t-}}((J-I)\otimes(\mathrm{diag}(\lambda)-\lambda^\top\hat{X}_{t-}I))\hat{\zeta}_{t-}\mathrm{d}N_t.$$

For $\hat{\zeta}_t$ we have, using the product formula for tensors,

$$\mathrm{d}\hat{\zeta}_t=\mathrm{d}\nu_t\otimes\hat{X}_{t-}+\nu_{t-}\otimes\mathrm{d}\hat{X}_t+\mathrm{d}[\nu,\hat{X}]^{\otimes}_t.$$

This yields after some tedious computations the following semimartingale decomposition for $\hat{\zeta}$

$$\begin{aligned}\mathrm{d}\hat{\zeta}_t&=\big(I\otimes Q+(n-N_t)(J-I)\otimes\mathrm{diag}(\lambda)\big)\hat{\zeta}_t\,\mathrm{d}t\\&\quad+\frac{1}{\lambda^\top\hat{X}_{t-}}\big(J\otimes\mathrm{diag}(\lambda)-\lambda^\top\hat{X}_{t-}I\otimes I\big)\hat{\zeta}_{t-}\,\mathrm{d}\hat{m}_t\\&=\mathbf{Q}\hat{\zeta}_t\,\mathrm{d}t+\frac{1}{\lambda^\top\hat{X}_{t-}}\big(J\otimes\mathrm{diag}(\lambda)-\lambda^\top\hat{X}_{t-}I\otimes I\big)\hat{\zeta}_{t-}\,\mathrm{d}\hat{m}_t,\end{aligned}$$

where $\mathrm{d}\hat{m}_t=\mathrm{d}N_t-(n-N_t)\lambda^\top\hat{X}_t\,\mathrm{d}t$ and $\mathbf{Q}$ as in Sect. 2.2.2.

Here are two applications. One can now compute

$$\mathbb{P}(N_t=k|\mathscr{F}^N_s)=\mathbf{1}^\top\mathbb{E}[\zeta^k_{t|s}|\mathscr{F}^N_s]=\mathbf{1}^\top\hat{\zeta}^k_{t|s},$$

for which we can use $\hat{\zeta}_{t|s}=\exp(\mathbf{Q}(t-s))\hat{\zeta}_s$. Formula (14) yields for the conditional characteristic function of N_t given its own past until time $s<t$ the explicit expression

$$\mathbb{E}[e^{\mathrm{i}uN_t}|\mathscr{F}^N_s]=\sum_{k=0}^{n}\sum_{j=k}^{n}\binom{n-k}{j-k}(1-e^{\mathrm{i}u})^{n-j}e^{\mathrm{i}uj}\mathbf{1}^\top\exp(Q_{(n-j)\lambda}(t-s))\hat{X}_s\nu^k_s.$$

In case $n=1$ the above formulas simplify considerably. Here are a few examples, where we use the notation of Sect. 2.1.2. Suppose that only Y is observed. Let $\mathscr{F}^Y_t=\sigma(Y_s,0\le s\le t)$. With $Z_t:=Y_tX_t$ we want to compute $\hat{Z}_{t|s}:=\mathbb{E}[Z_t|\mathscr{F}^Y_s]$ for $t\ge s$. Let $\hat{X}_t=\mathbb{E}[X_t|\mathscr{F}^Y_t]$, then obviously, $\hat{Z}_{t|s}=\hat{X}_{t|s}Y_s$. Moreover, one has from (13)

$$\hat{Z}_{t|s}=\exp(Q(t-s))\hat{X}_s-\exp(Q_\lambda(t-s))\hat{X}_s(1-Y_s).$$

As a consequence we have for $\hat{Y}_{t|s}=\mathbf{1}^\top\hat{Z}_{t|s}$

$$\hat{Y}_{t|s}=1-\mathbf{1}^\top\exp(Q_\lambda(t-s))\hat{X}_s(1-Y_s).$$

4.2 Filtering for the MM Poisson Process

The filter equations now take the familiar form

$$\mathrm{d}\hat{X}_t = Q\hat{X}_t\,\mathrm{d}t + \frac{1}{\lambda^\top \hat{X}_{t-}}\left(\mathrm{diag}(\lambda)\hat{X}_{t-} - \hat{X}_{t-}\hat{X}_{t-}^\top\lambda\right)(\mathrm{d}N_t - \lambda^\top\hat{X}_t\,\mathrm{d}t).$$

For $\bar{\nu}_t$ we have the infinite dimensional analogue of (6). This leads for $\hat{\zeta}_t = \bar{\nu}_t \otimes \hat{X}_t$ as in a Sect. 4.1 to

$$\mathrm{d}\hat{\zeta}_t = \mathbf{Q}_\infty\hat{\zeta}_t\,\mathrm{d}t + \frac{1}{\lambda^\top\hat{X}_{t-}}(J_\infty \otimes \mathrm{diag}(\lambda) - \lambda^\top\hat{X}_{t-}I_\infty \otimes I_\infty)\hat{\zeta}_{t-}\,(\mathrm{d}N_t - \lambda^\top\hat{X}_t\,\mathrm{d}t).$$

Note that this system is infinite dimensional, but for each n we also have for $\hat{\hat{\zeta}}_t^n = \mathbb{E}[\hat{\zeta}_t^n|\mathscr{F}_t^N]$ the truncated finite dimensional system

$$\mathrm{d}\hat{\hat{\zeta}}_t^n = \mathbf{Q}_n\hat{\hat{\zeta}}_t^n\,\mathrm{d}t + \frac{1}{\lambda^\top\hat{X}_{t-}}(J \otimes \mathrm{diag}(\lambda) - \lambda^\top\hat{X}_{t-}I \otimes I)\hat{\hat{\zeta}}_{t-}^n\,(\mathrm{d}N_t - \lambda^\top\hat{X}_t\,\mathrm{d}t).$$

For the conditional characteristic function $\mathbb{E}[\exp(\mathrm{i}uN_t)X_t|\mathscr{F}_s^N]$ we have

$$\mathbb{E}[\exp(\mathrm{i}uN_t)X_t|\mathscr{F}_s^N] = \exp\left(((e^{\mathrm{i}u}-1)\mathrm{diag}(\lambda)+Q)(t-s)\right)e^{\mathrm{i}uN_s}\hat{X}_s,$$

whereas $\psi_t = e^{\mathrm{i}uN_t}\hat{X}_t$ satisfies the equation ($\mathrm{d}\hat{m}_t = \mathrm{d}N_t - \lambda^\top\hat{X}_t\,\mathrm{d}t$)

$$\mathrm{d}\psi_t = (\frac{e^{\mathrm{i}u}}{\lambda^\top\hat{X}_{t-}}\mathrm{diag}(\lambda) - I)\psi_{t-}\mathrm{d}\hat{m}_t + \left(Q + (e^{\mathrm{i}u}-1)\mathrm{diag}(\lambda)\right)\psi_t\,\mathrm{d}t.$$

5 Rapid Switching

In this section we present some auxiliary results that we shall use in obtaining limits for the various default processes when the Markov chain evolves under a rapid switching regime, i.e. the transition matrix Q will be replaced with αQ, where $\alpha > 0$ tends to infinity. In the first two results and their proofs we use the notation $C(M)$ for the matrix of cofactors of a square matrix M. Throughout this section we write λ_∞ for $\lambda^\top\pi$.

Lemma 4 *Let Q have a unique invariant vector π. Then*

$$C(Q) = q\pi\mathbf{1}^\top,$$

where the constant q can be computed as $\det(\hat{Q})$, *where $\hat{Q}$ is obtained from Q by replacing its last row with* $\mathbf{1}^\top$.

Proof Note first that π can be obtained as the solution to $\hat{Q}\pi = e_d$, where e_d is the last basis vector of $\mathbb{R}^d$. By Cramer's rule π can be expressed using the cofactors of $\hat{Q}$. In particular, $\pi_d = \hat{C}_{dd}/\det(\hat{Q})$, where $\hat{C}$ is the cofactor matrix of $\hat{Q}$. But $\hat{C}_{dd} = C_{dd}$, so $\pi_d = C_{dd}/\det(\hat{Q})$.

Write $C = C(Q)$ and recall that $CQ = \det(Q)$ and hence zero. It follows that every row of C is a left eigenvector of Q. Since Q has rank $d-1$ by its assumed irreducibility, every row of C is a multiple of $\mathbf{1}^\top$. Hence $C = \alpha\mathbf{1}^\top$, for some $\alpha \in \mathbb{R}^{d\times 1}$. By similar reasoning, $C = \pi\beta$ for some $\beta \in \mathbb{R}^{1\times d}$. We conclude that $C = q\pi\mathbf{1}^\top$ for some real constant q. Use now $C_{dd} = q\pi_d$ and the above expression for π_d to arrive at $q = \det(\hat{Q})$.

Proposition 6 *Let Q have a unique invariant vector π and let all λ_i be positive. Then* $(\alpha Q - \mathrm{diag}(\lambda))^{-1} \to -\frac{\pi\mathbf{1}^\top}{\lambda_\infty}$ *for* $\alpha \to \infty$.

Proof We have seen in Sect. 2.1.2 that $Q - \mathrm{diag}(\lambda)$ is invertible if all $\lambda_i > 0$ and so the same is true for $\alpha Q - \mathrm{diag}(\lambda)$. Both $\det(\alpha Q - \mathrm{diag}(\lambda))$ and the cofactor matrix of $\alpha Q - \mathrm{diag}(\lambda)$ are polynomials in α and we compute the leading term. The determinant is computed by summing products of elements of $\alpha Q - \mathrm{diag}(\lambda)$, from each row and each column one. The α^d term in this determinant has coefficient $\det(Q)$, which is zero. Consider the term with α^{d-1}. It is seen to be equal to $-\sum_{i=1}^d \lambda_i C(\alpha Q - \mathrm{diag}(\lambda))_{ii} = -\alpha^{d-1}\sum_{i=1}^d \lambda_i C(Q - \mathrm{diag}(\lambda/\alpha))_{ii}$. For the cofactor matrix itself a similar procedure applies. We get $C(\alpha Q - \mathrm{diag}(\lambda)) = \alpha^{d-1}C(Q - \mathrm{diag}(\lambda)/\alpha)$ and it results from Lemma 4 that for $\alpha \to \infty$

$$\frac{C(\alpha Q - \mathrm{diag}(\lambda))}{\det(\alpha Q - \mathrm{diag}(\lambda))} \to \frac{C(Q)}{-\sum_{i=1}^d \lambda_i C(Q)_{ii}} = -\frac{q\pi\mathbf{1}^\top}{q\sum_{i=1}^n \lambda_i\pi_i} = -\frac{\pi\mathbf{1}^\top}{\lambda_\infty}.$$

Proposition 7 *For $\alpha \to \infty$ it holds that*

$$\exp\big((\alpha Q - \mathrm{diag}(\lambda))t\big) \to \exp(-\lambda_\infty t)\pi\mathbf{1}^\top.$$

Proof For any analytic function $f : \mathbb{C} \to \mathbb{C}$, $f(z) = \sum_{k=0}^\infty a_k z^k$, one defines $f(M) := \sum_{k=0}^\infty a_k M^k$ for $M \in \mathbb{C}^{d\times d}$ (assuming that the power series converges on the spectrum of M). It then holds (see also Higham [17, Definition 1.11], where this is taken as a definition of $f(M)$) that

$$f(M) = \frac{1}{2\pi \mathrm{i}}\oint_\Gamma (zI - M)^{-1} f(z)\,\mathrm{d}z,$$

where Γ is a closed contour such that all eigenvalues of M are inside it. Take $M = \alpha Q - \mathrm{diag}(\lambda)$. It follows from Proposition 6, note that also λ_∞ lies inside Γ as it is a convex combination of the λ_i, that $(zI - \alpha Q + \mathrm{diag}(\lambda))^{-1} \to \frac{1}{z+\lambda_\infty}\pi\mathbf{1}^\top$. Hence

$$f(\alpha Q - \mathrm{diag}(\lambda)) \to \pi \mathbf{1}^\top f(-\lambda_\infty).$$

Apply this to $f(z) = \exp(tz)$.

5.1 Rapid Switching for the MM Multiple Point Process

Suppose we scale the Q matrix with $\alpha \geq 0$, and we let X^α be Markov with transition matrix αQ. Many (random) variables below will be indexed by α as well. Here is a way to get accelerated dynamics for N_t^α (previously denoted N_t).

Suppose that one takes the original Markov chain X and replaces the dynamics of N with one in which X is accelerated,

$$N_t^\alpha = \int_0^t (n - N_s^\alpha)\lambda^\top X_{\alpha s}\,\mathrm{d}s + m_t. \tag{19}$$

Indeed the process X^α defined by $X_t^\alpha = X_{\alpha t}$ has intensity matrix αQ, and its invariant measure is π again. Recall that, conditionally on $\mathscr{F}^X$, N_t^α has a Bin$(n, 1 - \exp(-\int_0^t \lambda^\top X_{\alpha s}\,\mathrm{d}s))$ distribution.

The ergodic property of X gives $\int_0^t X_{\alpha s}\,\mathrm{d}s = \frac{1}{\alpha}\int_0^{\alpha t} X_s\,\mathrm{d}s \to \pi t$ a.s. and hence by dominated convergence for the expectations we have that the limit distribution of N_t^α for $\alpha \to \infty$ is Bin$(n, 1 - \exp(-\lambda_\infty t))$. One immediately sees that the default times $\tau^{\alpha,k}$ convergence in distribution to τ^k that are independent and have an exponential distribution with parameter λ_∞. Keeping this in mind, the other results in this section are easily understandable.

We recall the content of Proposition 7. Replacing λ with $k\lambda$ for $k \geq 0$ (zero included) yields

$$\exp\big((\alpha Q - k\mathrm{diag}(\lambda))t\big) \to \exp(-k\lambda_\infty t)\pi \mathbf{1}^\top. \tag{20}$$

To express the dependence of the matrix $\mathbf{Q}$ given by (8) on α in the present section, we write $\mathbf{Q}^\alpha$ (so $\mathbf{Q}^\alpha = A \otimes \mathrm{diag}(\lambda) + I \otimes \alpha Q$) and $F^\alpha(t)$ instead of $F(t)$ as given in Lemma 3.

Lemma 5 *The solution F^α to the equation $\dot{F} = \mathbf{Q}^\alpha F$, has for $\alpha \to \infty$ limit F^∞ given by its blocks*

$$F_{ij}^\infty(t) = f_{ij}^\infty(t)\pi \mathbf{1}^\top,$$

where the $f_{ij}^\infty(t)$ are the binomial probabilities on $n - i$ 'successes' of a Bin$(n - j, \exp(-\lambda_\infty t))$ distribution,

$$f_{ij}^\infty(t) = \binom{n-j}{n-i} \exp(-(n-i)\lambda_\infty t)(1 - \exp(-\lambda_\infty t))^{i-j}.$$

Proof We depart from Lemma 3 and the expression for $F^\alpha_{ij}(t)$ given there when we replace Q with αQ. Taking limits for $\alpha \to \infty$ yields

$$
\begin{aligned}
F^\infty_{ij}(t) &= \binom{n-j}{n-i}\sum_{k=j}^{i}(-1)^{i-k}\binom{i-j}{i-k}\exp(-(n-k)\lambda_\infty t)\pi\mathbf{1}^\top \\
&= \binom{n-j}{n-i}(-1)^{i-j}\exp(-(n-j)\lambda_\infty t)\sum_{l=0}^{i-j}\binom{i-j}{l}(-\exp(\lambda_\infty t))^l\pi\mathbf{1}^\top \\
&= \binom{n-j}{n-i}\exp(-(n-i)\lambda_\infty t)(1-\exp(-\lambda_\infty t))^{i-j}\pi\mathbf{1}^\top,
\end{aligned}
$$

from which the assertion follows.

Remark 7 One can also use this proposition to show that N^α_t in the limit has the $\mathrm{Bin}(n, 1-\exp(-\lambda_\infty t))$ distribution. Indeed, since $\nu^i_0 = \delta_{i0}$, we get $\mathbb{P}(N^\alpha_t = i, X_t = e_j) \to F^\infty_{i0}(t) = f^\infty_{i0}(t)\pi$ and hence $\mathbb{P}(N^\alpha_t = i) \to f^\infty_{i0}(t)$.

For conditional probabilities one has the following result.

Corollary 2 *Let N be a process like in Eq. (4), with λ replaced with λ_∞. For $\alpha \to \infty$ one has in the limit $\zeta^i_{t|s} = 0$ for $i < N_s$ and for $i \geq N_s$*

$$
\zeta^i_{t|s} = \binom{n-N_s}{n-i}\exp(-(n-i)\lambda_\infty(t-s))(1-\exp(-\lambda_\infty(t-s)))^{i-N_s}\pi.
$$

It follows that, conditional on $\mathcal{F}_s$, $N_t - N_s$ has a $\mathrm{Bin}(n-N_s, 1-\exp(-\lambda_\infty(t-s)))$ distribution. In fact, one has weak convergence of the N^α to N.

Proof We compute in the limit $\zeta^i_{t|s} = \mathbb{E}[\nu^i_t X_t|\mathcal{F}_s]$ and obtain from Lemma 5

$$
\begin{aligned}
\zeta^i_{t|s} &= \sum_{j=0}^{n} F^\infty_{ij}(t-s)\zeta^j_s \\
&= \sum_{j=0}^{n} f^\infty_{ij}(t-s)\nu^j_s\pi \\
&= \sum_{j=0}^{i}\binom{n-j}{n-i}\exp(-(n-i)\lambda_\infty(t-s))(1-\exp(-\lambda_\infty(t-s)))^{i-j}\nu^j_s\pi \\
&= \binom{n-N_s}{n-i}\exp(-(n-i)\lambda_\infty(t-s))(1-\exp(-\lambda_\infty(t-s)))^{i-N_s}\pi,
\end{aligned}
$$

from which the first assertion follows.

Weak convergence can be proved in many ways. Let us first look at the case of one obligor, $n = 1$. The integral in Eq. (19) is, with $\tau^\alpha = \tau^{1,\alpha}$ equal to

$$\frac{1}{\alpha}\int_0^{\alpha(\tau^\alpha\wedge t)} \lambda^\top X_u\,\mathrm{d}u.$$

Replacing the upper limit of the integral by t, this almost surely converges to $\lambda_\infty t$ for $\alpha \to \infty$. In fact this convergence is a.s. uniform. Having already established the convergence in distribution of the τ^α, and by switching to an auxiliary space on which the τ^α a.s. converge to τ^∞, we get

$$\frac{1}{\alpha}\int_0^{\alpha(\tau^\alpha\wedge t)} \lambda^\top X_u\,\mathrm{d}u \to \lambda_\infty(\tau^\infty \wedge t).$$

This is sufficient, see [24] or [20, Sect. 8.3d] to conclude the weak convergence result for the case $n = 1$.

For the general case, one first notices that the process N^α is a sum of MM one point processes that are conditionally independent given $\mathscr{F}^X$ and become independent in the limit. Combine this with the result for $n = 1$. Alternatively, one could apply the results in [20, Sect. 7.3d] again, although the computations will now be more involved.

5.2 *Rapid Switching for the MM Poisson Process*

As before we replace Q with αQ and let $\alpha \to \infty$ and denote N^α the corresponding counting process. We apply Proposition 7 to the matrix exponential $\exp\big(((e^{iu}-1)\mathrm{diag}(\lambda) + \alpha Q)(t-s)\big)$, and we find that the limit for $\alpha \to \infty$ equals $\exp((e^{iu}-1)\lambda_\infty(t-s))\pi\mathbf{1}^\top$. Hence, by virtue of (16), we obtain $\mathbb{E}[\exp(iuN_t^\alpha)X_t|\mathscr{F}_s] \to \exp((e^{iu}-1)\lambda_\infty(t-s))\pi$ for the limit of the conditional characteristic function. This is just one of the many ways that eventually lead to the conclusion that for $\alpha \to \infty$ the process N^α converge weakly to an ordinary Poisson process with constant intensity λ_∞. In [24] one can find the stronger result that the variational distance between the MM law of N_t^α, $t \in [0, T]$ and the limit law is of order α^{-1}.

References

1. Ang, A., Bekaert, G.: Regime switches in interest rates. J. Bus. Econ. Stat. **20**, 163–182 (2002)
2. Asmussen, S.: Applied Probability and Queues, 2nd edn. Springer, New York (2003)
3. Asmussen, S., Albrecher, H.: Ruin Probabilities, 2nd edn. World Scientific, New Jersey (2010)
4. Bielecki, T.R., Rutkowski, M.: Credit Risk: Modelling, Valuation and Hedging. Springer, Berlin (2002)
5. Blom, J., de Turck, K., Mandjes, M.: Analysis of Markov-modulated infinite-server queues in the central-limit regime. Probab. Eng. Inf. Sci. **29**, 433–459 (2015)
6. Blom, J., Kella, O., Mandjes, M., Thorsdottir, H.: Markov-modulated infinite-server queues with general service times. Queueing Syst. **76**, 403–424 (2014)
7. Brémaud, P.: Point Processes and Queues. Springer, New York-Berlin (1980)

8. Chen, A., Delong, L.: Optimal Investment for a Defined-Contribution Pension Scheme under a Regime Switching Model, ASTIN Bulletin, 1–23 March 2015
9. Duffie, D., Filipović, D., Schachermayer, W.: Affine processes and applications in finance. Ann. Appl. Probab. **13**, 984–1053 (2003)
10. Elliott, R.J., Aggoun, L., Moore, J.B.: Hidden Markov Models. Springer, New York (1995)
11. Elliott, R., Mamon, R.: An interest rate model with a Markovian mean-reverting level. Quant. Financ. **2**, 454–458 (2002)
12. Elliott, R., Siu, T.K.: On Markov-modulated exponential-affine bond price formulae. Appl. Math. Financ. **16**, 1–15 (2009)
13. Filipović, D.: Term-structure Models. Springer, Berlin (2009)
14. Goel, A.L., Okumoto, K.: Bayesian software prediction models, vol. 1: An imperfect debugging model for reliability and other quantitative measures of software systems. Rome Air Develop. Center, NY, RADC-TR-78-155 (1978)
15. Di Graziano, G.: Optimal Trading Stops and Algorithmic Trading. SSRN (2014). http://ssrn.com/abstract=2381830
16. Hamilton, J.: A new approach to the economic analysis of nonstationary time series and the business cycle. Econometrica **57**, 357–384 (1989)
17. Higham, N.J.: Functions of Matrices: Theory and Computation. SIAM, Philadelphia (2008)
18. Huang, G., Mandjes, M., Spreij, P.: Weak convergence of Markov-modulated diffusion processes with rapid switching. Statist. Probab. Lett. **86**, 74–79 (2014)
19. Huang, Gang, Jansen, Marijn, Mandjes, Michel, Spreij, Peter, De Turck, Koen: Markov-modulated Ornstein-Uhlenbeck processes. Adv. Appl. Probab. **48**(1), 235–254 (2015)
20. Jacod, J., Shiryaev, A.N.: Limit theorems for stochastic processes, 2nd edn. Grundlehren der Mathematischen Wissenschaften 288. Springer, Berlin (2003)
21. Jelinski, Z., Moranda, P.B.: Software reliability research. In: Freiberger, W. (ed.) Statistical Computer Performance Evaluation, pp. 465–484. Academic, NewYork (1972)
22. Jiang, Z., Pistorius, M.R.: On perpetual American put valuation and first passage in a regime-switching model with jumps. Financ. Stochast. **12**, 331–355 (2008)
23. Jiang, Z., Pistorius, M.: Optimal dividend distribution under Markov regime switching. Financ. Stochast. **16**, 449–476 (2012)
24. Kabanov, Y.M., Liptser, R.S., Shiryaev, A.N.: Weak and strong convergence of distributions of counting processes. Teor. Veroyatnost. i Primenen. **28**, 288–319 (1983)
25. Koch, G., Spreij, P.J.C.: Software reliability as an application of martingale and filtering theory. IEEE Trans. Reliab. **R-32**, 342–345 (1983)
26. Littlewood, B.: Theories of software reliability: how good are they and how can they be improved? IEEE Trans. Softw. Eng. **SE-6**, 489–500 (1980)
27. Neuts, M.: Matrix-Geometric Solutions in Stochastic Models: An Algorithmic Approach. Courier Dover Publications, New York (1981)
28. van Schuppen, J.H.: Filtering, prediction and smoothing for counting process observations, a martingale approach. SIAM J. Appl. Math. **32**, 552–570 (1977)
29. Yin, G., Zhu, C.: Hybrid Switching Diffusions: Properties and Applications. Springer, New York (2010)
30. Yin, G.: Asymptotic expansions of option price under regime-switching diffusions with a fast-varying switching process. Asymptotic Anal. **65**, 203–222 (2009)
31. Zhou, X.Y., Yin, G.: Markowitz mean-variance portfolio selection with regime switching: A continuous-time model. SIAM J. Control Optim. **42**, 1466–1482 (2003)

Part II
Statistics and Risk

Introducing Distances Between Commodity Markets: The Case of the US and UK Natural Gas

Hélyette Geman and Bo Liu

Abstract The goal of the paper is twofold: first to present the energy markets in 2015 after the revolution of shale oil and shale gas; secondly investigate whether the two major gas markets, the US and the UK, are integrated at a time when natural gas is a much preferred source of electricity compared to coal, in particular in feeding the 'peakers' providing electricity that complements solar, wind and other intermittent renewables. Introducing the novel concept of *distances* between two commodity markets through quantities accounting for fundamental financial economic indicators, our conclusion is negative, illuminating the importance of transportation costs yet in the case of natural gas.

Keywords Natural gas · Forward curve · Metrics in commodity markets

1 Introduction

The energy world is experiencing dramatic changes, with a decline of prices across the board—except possible for some renewables—, in a context of a very weak world economy growth and over supply of production. The Russian giant Gazprom saw its revenues in the first semester of 2015 decline by 13 % year on year, because of a production that was the lowest since the company was incorporated in its current form after the disintegration of the former Soviet Union and the decline of the Russian ruble with respect to the dollar—remembering that commodities are denom-

H. Geman (✉)
Commodity Finance Centre, Birkbeck-University of London, London, UK
e-mail: hgeman@hotmail.com

H. Geman
Johns Hopkins University, Baltimore, USA

B. Liu
Department of Applied Mathematics and Statistics, Johns Hopkins University, Baltimore, USA

© Springer International Publishing Switzerland 2016

J. Kallsen and A. Papapantoleon (eds.), *Advanced Modelling in Mathematical Finance*, Springer Proceedings in Mathematics & Statistics 189,
DOI 10.1007/978-3-319-45875-5_4

inated in dollars. The company has a large capacity available, but both domestic and international demands have collapsed.

Even though the coal output increased by 37 % in the period 2004–2014, investments in coal mining projects are abandoned by 'ethical' funds, followed by the big banks, like Goldman Sachs that withdrew from the financing of the extension of the mining complex in the Galileo basin in Australia, the biggest coal extraction project in the world.

The three fossil fuels, coal, crude oil and natural gas represent in 2015 over 80 % of the world energy supply: 31 % for natural gas, 29 % for oil and 21 % for coal. They are used in a very large spectrum of the economic activity, including heating and power generation, steel making and transportation. With increasing emissions standards, as expressed for instance by the COP 21 in Paris in December 2015, natural gas and low-carbon fuels will gradually replace oil and coal in the energy mix. The three fossil fuels are forecast to each account for 25 % of the global energy demand by 2040.

The biggest story over the past decade in the energy world is the shale revolution, thanks to one of the most innovative technologies of the 21st century: horizontal drilling and hydraulic fracturing, or fracking, methods used to unlock natural gas and oil from shale rocks previously seen as inaccessible. The US shale boom had profound economic and political implications, enhancing US energy independence and reshaping global energy balance, in addition to boosting local economies in the basins of extraction (Fig. 1).

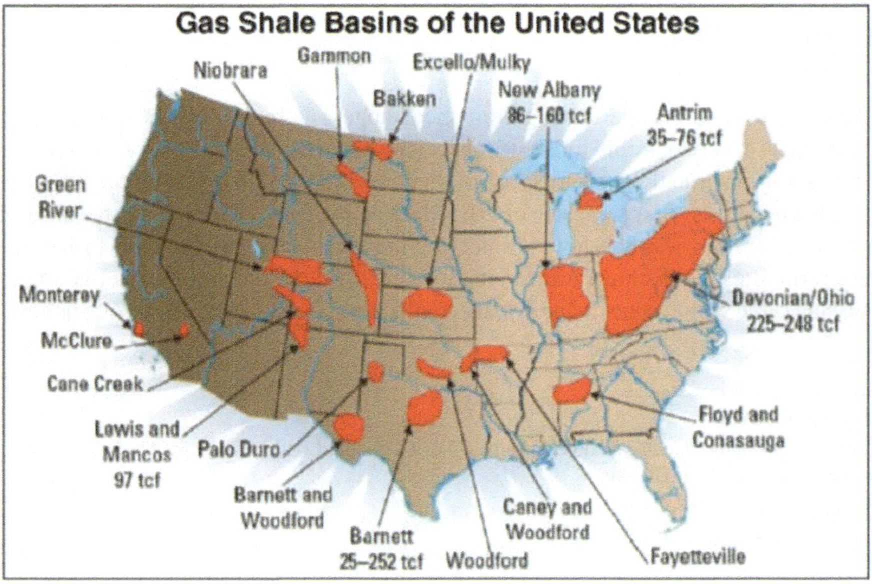

Fig. 1 Shale gas basins

The first US shale oil well was drilled in South Texas's Eagle Ford field in September 2008, at which time US crude oil production was only 4.7 million barrels per day. From 2009 to 2011, oil prices recovered from $40 to around $100 per barrel, amid post-financial crisis recovery. At the same time, the US shale industry was booming. New drilling rigs were blanketing shale fields around the country, from Texas to North Dakota and Pennsylvania. US oil production has been steadily increasing since then and reached a peak of 9.6 million barrels a day in April 2015. The Organization of Oil Producing countries (OPEC), now exports half of the oil it did at the time of the first oil shale well. The displaced oil has gone elsewhere in the world, creating a global oil glut and triggering the price collapse.

Besides crude oil, the US gas industry is the biggest beneficiary of the shale revolution. There has been a 72 % increase in the natural gas production between 2007 and 2015. The share of shale in the US production has jumped from less than 2 % in 2001 to 40 % in 2014 Q1 and is projected to grow to 53 % by 2040. In 2011, the International Energy Agency (IEA) based in Paris had issued an ominous special report suggesting that natural gas could play a more prominent role in the global energy mix in the near term with surging post-crisis demand recovery, ample supply and expansion of liquid natural gas (LNG) trade. US gas production has been so vibrant that some old importing LNG terminals were converted into liquefaction entities. Several gas export projects got federal approvals since 2011. Cheniere's Sabine Pass terminal in Louisiana should start exporting by the end of 2015, the first US gas export project in 46 years!

Besides the large supply, world weak demand has contributed to the oil price crash in the recent period. The opening of two export terminals in July 2014 in Libya was one of the elements that triggered the collapse. Both the US WTI benchmark and the UK/international benchmark Brent tumbled from roughly $110 in June 2014 to less than $50 by January 2015, as depicted in the Fig. 2.

They momentarily rebounded above $60 in spring 2015 before further plunging to a six-year low in August 2015—driving the Bloomberg Commodity index to its weakest value since 2000—and a ten year low in December 2015. In December 2015, OPEC decided again not to cut its production, led by the number one oil exporter, Saudi Arabia, which decided to protect its market share from US shale oil producers and increased output from Iran ramping up production after the end of economic sanctions against the country and the refurbishment of oil producing facilities. Countries in South America and Africa, which have a higher cost of production like Venezuela and Nigeria, are obviously suffering greatly from the current situation.

While the number of oil rigs fell by 64 % from a highest number of 1609 between October 2014 and October 2015, the Permian shale oil field in Texas continued to increase its production, thanks to its unique combination of a large quantity of available oil, highest quality of shale rocks and advanced shipping and storage infrastructure.

Fig. 2 Crude oil prices from 1986 to 2015

2 Trading Strategies in Natural Gas Markets

Natural gas, a fossil fuel, has long been prized for its ability to burn clearly and provides relatively high levels of energy. In 1821, the first well specifically designed to recover natural gas was dug in the US. The pure form of natural gas consists of a mixture of hydrocarbon gases, mainly methane, a molecule consisting of *one* carbon atom and four hydrogen atoms. Natural gas and oil are generally found together in deposits beneath the earth. A main difference between natural gas and crude oil is the fact that natural gas requires very specific types of transmission and storage facility, respectively pipelines and saline caverns or former gas fields; and spot trading of natural gas is much less feasible than in the case of crude oil.

The International Energy Agency (based in Paris) estimates that for the first time since 1982, the US will have produced in 2015 more natural gas than Russia. The graph below depicts the trajectory of the Henry Hub natural gas index (on which all Futures and options traded on the NYMEX are written), over the period—the US market was the first deregulated gas market in the world and the Henry Hub index has been observed for decades. It went from 2 \$ per Million British thermal units (MMBtus) in 1996, reached an absolute maximum of \$15 in 2006, a local maximum (like most commodities except for gold) in July 2008 at \$13.5, to fluctuate today around \$3 or below. The US gas is the unique commodity whose price today has returned to its level of 1998 (Fig. 3).

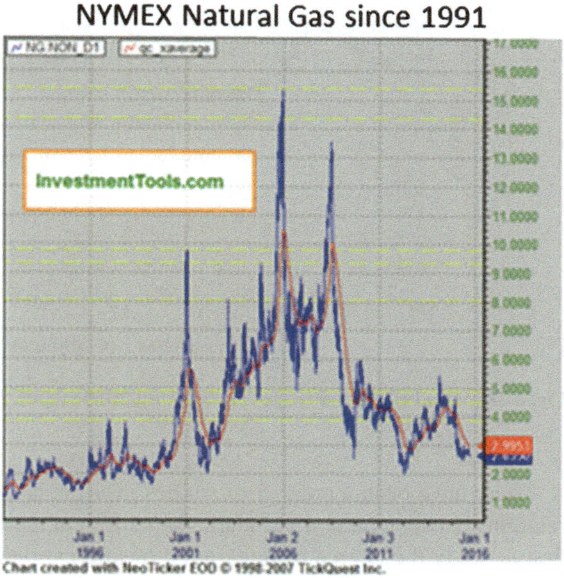

Fig. 3 Price Trajectory of the Henry Hub index—1991–2015

So far, Europe has been relying on Russia for about 30 % of its natural gas supplies. Gas from the existing Nord Stream pipeline first arrives in Germany, is then fed into a European gas network that serves consumers in Germany and other countries that include France, the UK and the Netherlands. The project of a South Stream pipeline, that would have included the Italian energy giant ENI together with Gazprom and delivered gas to the EU through Bulgaria, was opposed by Washington and Brussels after the economic sanctions decided on Russia in July 2014 and renewed in June 2015, following the annexation by Moscow of the Crimean peninsula.

A new project, called Nord Stream 2, is defended by Germany for its own needs and European needs. In Europe (as well as in Japan, an economy where natural gas is a crucial source of energy), the long—term contracts, where the price of natural gas was indexed to oil prices according to a linear formula, are disappearing under the combined effect of lower oil prices—making the formulas less profitable to natural gas producers, and the pressure of increased production worldwide, with LNG tankers touring the world. At the same time, the large development in Europe of renewable energy has made flexible gas-fired turbines necessary to provide electricity at times when the renewable but intermittent source of energy, solar or wind, is not available.

In Australia, Coal Seam Gas (CSG) is another prospective source of energy. New projects in East Coast Australia will increase LNG capacity by 150 % and by 2016, Australia should be a major player, likely to relieve pressure on Asian prices, Japan in particular. Figure 4 shows the very large difference between the price of natural gas in the US (represented by the HH index) and in Japan (represented by the Japan Korea Marker).

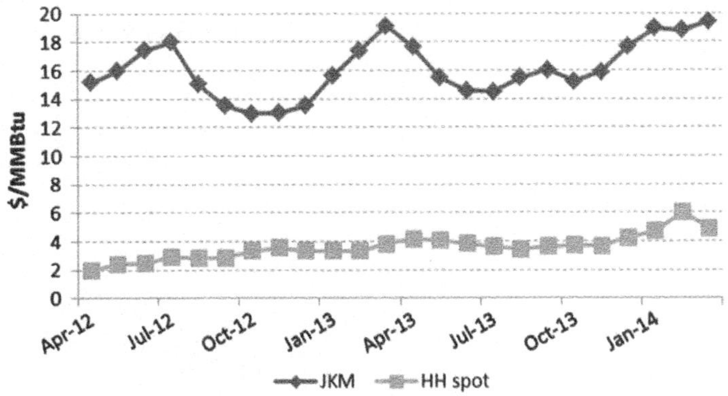

Fig. 4 JKM and HH monthly spot prices, April 2010–March 2014

There have been a number of studies over the last decade investigating the price relations between natural gas markets in different regions of the world. [12] were the first to study the degree of integration of international gas markets. They conclude that the North American markets were not integrated with European or Japanese ones during the period 1993–2004. [11] focus on European spot gas markets for the period 2000–2005. They have shown UK and Continental European markets had converged via the Interconnector; but markets within European continent had not. [10] finds an increasing convergence of transatlantic natural gas spot prices from 1999 to 2008. The author argues that LNG was the key driver of this convergence. [1] study a few European spot gas markets during the period following deregulation, particularly three major hubs: NBP in UK, Zeebrugge in Belgium and TTF in the Netherlands. Their results indicate a highly integrated region with respect to natural gas, with probably crude oil as the intermediary. [3] also conclude that the co-movement between gas prices in US and UK are mediated through crude oil prices. Their results show that the relationship between the two prices is not stable over time—the prices may be related, but the cointegrating relationship could change over time.

3 Natural Gas Forward Markets

As said before, spot trading is hardly feasible, except if one has access to LNG tankers that may tour the world to stop at points of greater profitability. Hence, gas traders need strategies based on liquid Futures contracts, which has only been the case for the US Henry Hub index for many decades. And the main trading strategy for many decades was to play the changes in the seasonal spread Winter/Summer, this one increasing from a prior value to a much higher one in the case of a very cold winter or a problem in the supply of natural gas. A famous example of this strategy is provided by the example of the Amaranth hedge fund which had taken in March 2005 a very large position in winter long/summer short Futures. It made a profit of $1.25 billion by the end of August: the hurricanes Rita and Katrina destroyed oil platforms in the Gulf of Mexico, making natural gas the *substitution* commodity and sending January Futures prices to very high values, as well as the Futures spreads. A year later, the same position—which was vastly increased in size from the previous year—lost $6.25 billion because hurricanes did not occur and there was a large amount of gas in inventory, a property the star trader did not pay attention to. Note that in the oil markets of 2015, any news about a larger inventory in major countries sends prices down further, another evidence of the crucial role of inventories.

With the arrival of shale gas and the collapse of prices, the seasonality of the US natural gas forward curve became less pronounced (as shown in the graph below in 2013) because of the large amount of natural gas available all year long, and the strategies based on seasonality less profitable; hence the necessity of turning to other types of strategies (Fig. 5).

The second gas market to be deregulated was the United Kingdom, and the liquidity of Futures contracts written on the National Balancing Point (NBP) index has been quite reasonable 60 months out, as depicted in Fig. 6. Note that, beyond the seasonality, one clearly sees an increasing shape in the two forward curves, indicating that in 2010, market participants had the view that spot prices were too low and had to increase, both in the US and in the UK. In order to play a locational strategy between two markets of the same commodity (or two different commodities)—like betting on their convergence or divergence—it is important to be able to evaluate the relative evolution of the two forward curves. In this order, we propose to use in this section the notion of *distance* between the HH and the NBP forward curves.

In order to model commodity forward curves dynamics, [2] introduced a new state variable representing the average of liquid forward prices. This quantity, denoted $\overline{F}$, and capturing the entirety of the forward curve was meant to replace the previously widely used spot price [7, 9] when managing a portfolio of Futures positions. In agreement with the geometric Brownian motion-based reference models, $\overline{F}$ was

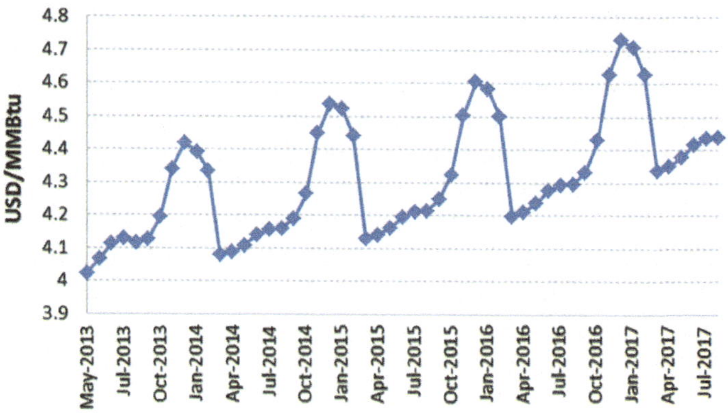

Fig. 5 Henry Hub natural gas forward *curve* in April 2013

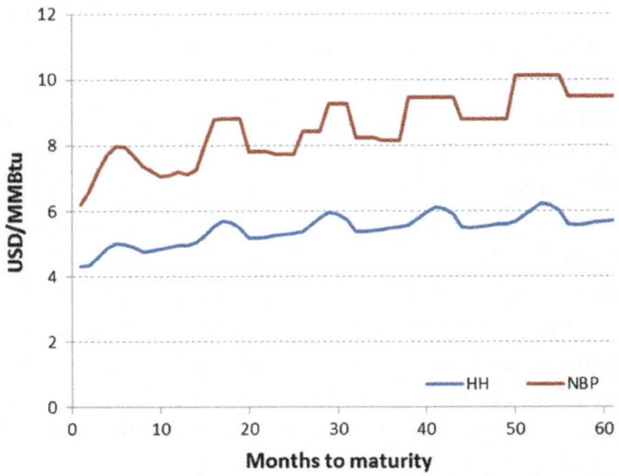

Fig. 6 HH and NBP forward *curves* on 8/13/2010

defined as the geometric average (quite close to the arithmetic average) of all liquid forward contract prices:

$$\overline{F}(t) = \sqrt[N]{\prod_{T=1}^{N} F(t, T)}.$$

where N is a multiple of 12 in the case of a seasonal commodity.

Compared to the spot price, $\overline{F}$ has several advantages. Firstly, it is the 'center of gravity' of the forward curve, a more robust quantity able to capture all the information available at time t in the liquid forward markets. Secondly, this average is meant to avoid all the problems related to the instability (natural gas or electricity) or opacity (agricultural goods) of the commodity spot price. Note that in electricity for instance, the spot forward relationship does not even hold since the convenience yield of Keynes (1936) loses its economic interpretation as electricity is generally not storable (as of 2015).

Thirdly, $\overline{F}$ absorbs the seasonality if we use a multiple of 12 maturities when calculating the average. Lastly, it is the quantity (up to the discount factors) that underlies a commodity swap spanning the period up to the maturity of the swap; and commodity swaps are experiencing a gigantic liquidity in energy markets. In the same way, $\overline{F}$ reflects the anticipation of the average of spot prices at stake in the Asian options traded on a number of commodity exchanges.

Next, Borovkova and Geman define the seasonal premia $s(M)$ for $M = 1, \ldots, 12$ as the set of long-term average premia (expressed in %) in Futures expiring in the calendar month M over the average forward price F and assume these premia to be deterministic.

Lastly, they introduce a stochastic cost of carry net of seasonal premium $g(t, T)$ defined by the adjustment of the famous spot forward relationship in terms of $\overline{F}(t)$, the seasonal premia and the cost of carry, namely

$$F(t, T) = \overline{F}(t) \exp\left[s(T) + (T - t)g(t, T)\right]$$

where the seasonal premium s has the merit to depend only on the specific maturity month and is independent of the length of the period.

Remarks

(i) Even in absence of seasonality, the so-called 'normal backwardation', namely Futures prices decreasing with the maturity property has been invalidated in commodity markets as of 2006, where different shapes have been observed for the forward curve, depending on the market conditions in terms of spot prices, long term views and available inventories.
(ii) But the Samuelson effect—namely the decrease of the volatility with the maturity of the Future contract—continued(s) to be empirically observed in the forward curves. Indeed, it seems reasonable that the effects of positive or negative news arrival has more effect on the volatility of short dated maturities than long term ones given the reduced length of the adjustment period.

[2] build the graph of the volatilities of the functions $g(t, T)$ and exhibit for them the same Samuelson effect, as illustrated below by UK Futures contracts (Fig. 7).

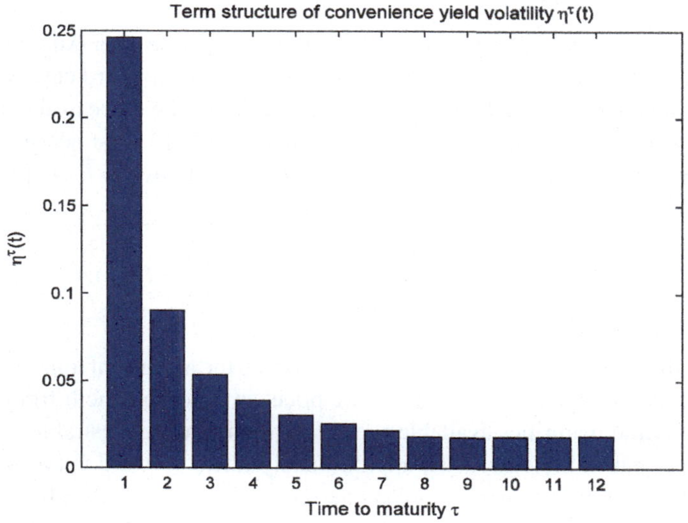

Fig. 7 Volatilities of natural gas stochastic convenience yields

Introducing Distances between Commodity Forward Curves

[5, 8] introduced the novel concept of 'distance' between forward curves. In fact, several definitions of distances are appropriate in order to make accurate statements and relate to economic fundamentals discussed in the literature on commodities. Working [13] established in a seminal paper the use of the spread of the forward curve (long term Future price minus shorted minus the short dated one) as an indication for low inventory when the spread is negative (backwardation). In order to investigate whether there has been a march towards integration of the US and UK natural gas markets, we propose below measures of distances which incorporate characteristics of the forward curve (and spot price, its limit point).

We propose three definitions of *distances* between two commodity markets and will apply them to the ones of NBP and HH indexes Let $F_1(t, i)$ and $F_2(t, i)$ be the forward prices of two commodities with maturity i months at date t; $S_1(t)$ and $S_2(t)$ be spot prices at date t. We first define the $\overline{F}(t)$ distance

$$D_1(t) = |\overline{F}_1(t) - \overline{F}_2(t)|.$$

It is the absolute value of the difference between the two curves averages. In the period 2010–2014, the NBP forward curve was always higher than the HH forward curve. Hence, the absolute value disappears and D_1 is approximately the arithmetic average of the distances between each pair of forward prices of the same maturity—a very natural distance to introduce. The second distance is the simple spot distance

$$D_2(t) = |S_1(t) - S_2(t)|.$$

where again, the absolute value can be deleted for the period after Aug 2010.

In order to account for possible differences in slopes of forward curves, we wish to add a 'slope term' in the first distance. We first define the slope as the difference between the average of the 12 furthermost points and the 12 most nearby points on the forward curve.[1] It is expressed in dollars, hence is additive to $\overline{F}$ or spot prices:

$$\text{dollar slope} = \sqrt[12]{\prod_{T=50}^{61} F(t, T)} - \sqrt[12]{\prod_{T=2}^{13} F(t, T)}.$$

We define the 'dollar slope' as the difference between the average of long-term Future prices and the average of short term Future prices in order to benefit from the large number of liquid maturities available as well as remove any undesirable seasonality effect. Note that in the case of an approximately linear forward curve, this dollar slope coincides with the famous *spread* analyzed in the commodities literature. The spread introduced by Working in 1949 as a representation of inventory was later used by [4] who analyzed 26 commodity markets where the inventory numbers were not available. [6] studied a database of US oil and gas prices and inventories, and

[1] The first nearby is avoided as it is very volatile.

validated Working's conjecture by directly exhibiting a high correlation between spreads and inventories. Our definition is meant to delete the seasonality effect in the distance and use a mix of contracts to avoid an excessive influence of a maturity of low liquidity.

We introduce now a third distance that mixes both the average forward prices as well as the shapes of the forward curves, namely

$$D_3(t) = \left|\overline{F}_1(t) - \overline{F}_2(t)\right| + \frac{|dollar\ slope_1 - dollar\ slope_2|}{2}.$$

Note that

(a) In the case of two parallel forward curves, the distance D_3 is identical to D_1 since the second term is zero
(b) In the case of approximately linear and parallel forward curves, both increasing for instance, the distance D_3 is reduced to D_1 again (Fig. 8)

(b') In the case of approximately linear forward curves with one increasing and the other decreasing, the first term in D_3 is zero and the value of D_3 is the same as the one obtained in (b), a satisfactory result since the forward curves are in both cases not "integrated", but for different reasons. The division by 2 of the dollar slope term in D_3 was necessary to recover these same values of D_3 in (b) and (b'), a property we view as desirable.

The Case of HH and NBP: Are Distances Declining Over the Period 2010–2014?

The complete 61 months of NBP forward curve data are not available before August 2010. In order to be able to calculate D_3, and to keep consistency of the approach across the time horizon for D_1, we focus the analysis of distances on the period with full NBP data, namely August 13, 2010 to April 10, 2014.

In Fig. 9 we plot all distances using weekly data in the usual unit of \$/Million Btu. Compared to the brisk spot prices moves, the distances seem more stable. The augmented Dickey-Fuller tests show that they are all stationary, D_1 and D_2 at 5 %, D_3 at 10 % significance level.

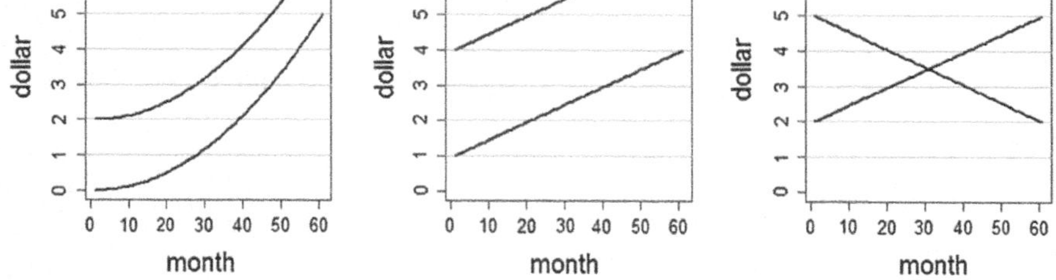

Fig. 8 Hypothetical forward *curves*

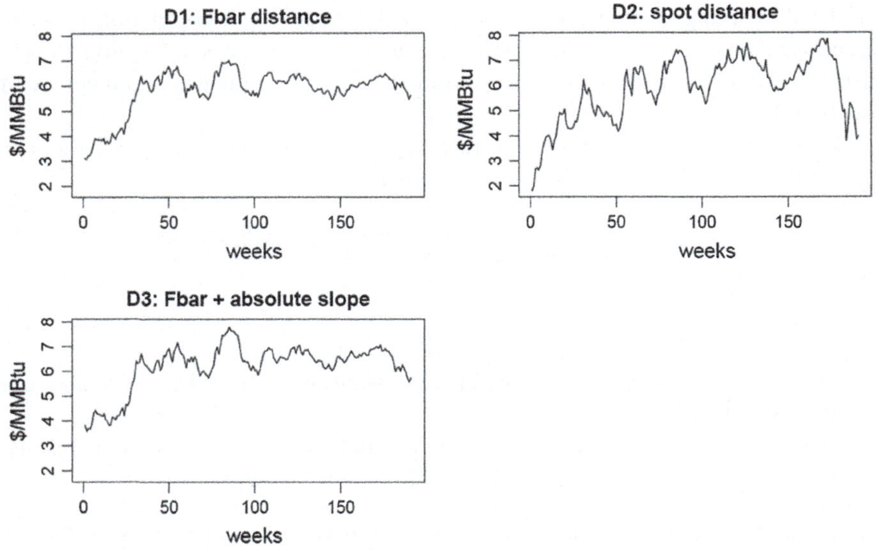

Fig. 9 Distances between NBP and HH, August 2010–April 2014

Analyzing a database covering the period January 2005 to April 2014 and using various econometric tests, [5] conclude that no convergence yet has occurred between the US and UK natural gas markets; this is visually confirmed by the Fig. 10.

Fig. 10 NBP and HH forward *curve* on Aug 10, 2015

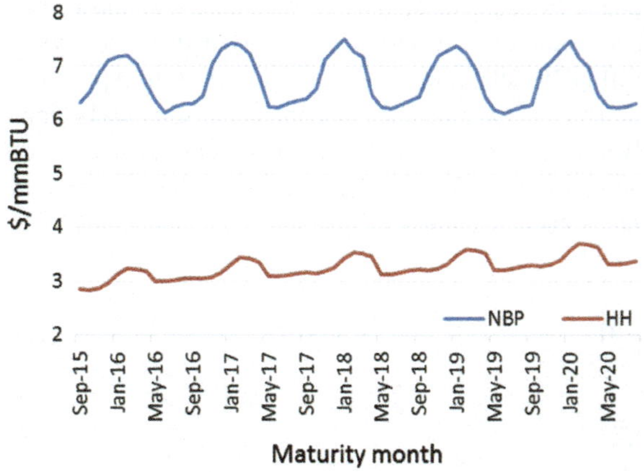

4 Conclusion

We have presented in this paper the outlook of energy markets in 2015, with a focus on natural gas, in particular investigated whether the march toward integration of world natural gas markets had already started between the US and the UK—the only developed gas markets for which full forward curves have been observed for a number of years. To this end, we analyze a database of spot and forward curves of the Henry Hub and National Balancing Point Indexes over the period January 2005 to April 2014. In order to go beyond the standard cointegration analysis used in the existing literature on the subject, we benefit from the information embedded in the forward markets by using the novel concept of *distance* between two commodity markets. In fact, we define three distances—two of them meant to strongly capture the signals provided by the forward curves. Using these different perspectives, our answer is negative as of April 2014.

This has an importance at a time when new fleets of LNG tankers are cruising the oceans in search for discrepancies in natural gas prices.

References

1. Asche, F., Misund, B., Sikveland, M.: The relationship between spot and contract gas prices in Europe. Energy Econ. **38**, 212–217 (2013)
2. Borovkova, S., Geman, H.: Seasonal and stochastic effects in commodity forward curves. Rev. Deriv. Res. **9**(2), 167–186 (2006)
3. Brown,S.P., Yucel, M.K.: Market arbitrage: European and North American natural gas prices. Energy J. **30**(Special Issue), 167–186 (2009)
4. Fama, E.F., French, K.R.: Commodity futures prices: Some evidence on forecast power, premiums, and the theory of storage. J. Bus. 55–73 (1987)
5. Geman, H., Liu, B.: Are natural gas markets marching to integration? Evidence from the HH and NBP indexes. J. Energy Markets. **8**(2), 47–65 (2015)
6. Geman, H., Ohana, S.: Forward curves, scarcity and price volatility in oil and natural gas markets. Energy Econ. **31**(4), 576–585 (2009)
7. Geman, H., Vasicek, O.: Forwards and futures on non-storable commodities: the case of electricity. Risk, 12–27 Aug 2001
8. Geman, H., Vergel, P.: Live cattle as a new frontier in commodity markets. J. Agric. Sustain. **7**(1), (2015)
9. Gibson, R., Schwartz, E.S.: Stochastic convenience yield and the pricing of oil contingent claims. J. Financ. **45**(3), 959–976 (1990)
10. Neumann, A.: Linking natural gas markets-is LNG doing its job? Energy J **30**(Special Issue), 187–200 (2009)
11. Neumann, A., Siliverstovs, B., Hirschhausen, C.V.: Convergence of European spot market prices for natural gas? A real-time analysis of market integration using the Kalman Filter. Appl. Econ. Lett. **13**(11), 727–732 (2006)
12. Siliverstovs, B., L'Hégaret, G., Neumann, A., von Hirschhausen, C.: International market integration for natural gas? A cointegration analysis of prices in Europe, North America and Japan. Energy Econ. **27**(4), 603–615 (2005)
13. Working, H.: The theory of price of storage. Am. Econ. Rev. 1254–1262 (1949)

Three Non-Gaussian Models of Dependence in Returns

Dilip B. Madan

Abstract Three particular models of dependence in asset returns with non-Gaussian marginals are investigated on daily return data for sector exchange traded funds. The first model is a full rank Gaussian copula (*FGC*). The second models returns as a linear mixture of independent Lévy processes (*LML*). The third correlates Gaussian components in a variance gamma representation (*VGC*). On a number of occasions all three models are comparable. More generally, in some by sectors, we get a superior performance from the *LML* model followed by *VGC* and *FGC* as measured by the proportion of portfolios with higher p-values. There are occasions when the *VGC* and *FGC* dominate. The concept of local correlation is introduced to help discriminate between the models and it is observed that the *LML* models display higher levels of local correlation especially in the tails when compared with either the *VGC* or *FGC* models.

Keywords Variance gamma model · Exchange traded funds · Independent components analysis · Local correlation

A number of applications in financial modelling call for the description of the joint law of asset returns over some horizon of interest. For many of these applications it is well recognized that the marginal return distribution of each asset return taken individually is not Gaussian (Jondeau et al. [15], Menn et al. [28], McNeil et al. [27], Frey and Embrechts [5]). Hence the focus on non-Gaussian multivariate return distributions. Applications include the design of optimal portfolios where the interest is in the physical multivariate return distribution, or the pricing of options on a basket of stocks for which the relevant return distribution is risk neutral. The marginal distributions reflect varying degrees of skewness and excess kurtosis, features that may be inherited and even exaggerated in portfolios. We investigate and report on the comparative performance of three tractable multivariate models for asset returns that have recently appeared in the literature. The particular feature of tractability for

D.B. Madan (✉)
Robert H. Smith School of Business, University of Maryland,
College Park, MD 20742, USA
e-mail dbm@rhsmith.umd.edu

© Springer International Publishing Switzerland 2016

J. Kallsen and A. Papapantoleon (eds.), *Advanced Modelling in Mathematical Finance*, Springer Proceedings in Mathematics & Statistics 189,
DOI 10.1007/978-3-319-45875-5_5

the three chosen models is the ability to estimate the models in dimensions as high as 50, by reduction to a suitable sequence of univariate estimation problems. There are a variety of multivariate elliptical distributions like the multivariate *t-distribution* (Kotz and Nadarajah [16]), or the multivariate variance gamma (Madan and Seneta [23], Schoutens and Cariboni [31] that impose a uniform tail structure across the different assets that we do not study here. Some of the methods developed here could however be applied to these and other models, provided one has access to the distribution of portfolio returns and the pairwise joint densities.

The first of our three models is a full rank Gaussian copula (FGC) that has been proposed and studied by Malevergne and Sornette [26] and is discussed in Bouchaud and Potters [4] along with a variety of other non-Gaussian models. In this model each asset return is a nonlinear transform of a set of standard normal variates that are correlated with a correlation matrix C of possibly full rank. The second model follows the idea implicit in a multivariate normal model where all the variables are linear transformations of independent Gaussian variates. We now consider a linear mixture of independent but non-Gaussian variates, that like the Gaussian variable, are infinitely divisible and associated with the unit time distribution of a Lévy process. The model was implemented for portfolio design in asset allocation by Madan and Yen [24] using independent components analysis (ICA et al. [13]) to identify the independent variables. It was also used by Madan [19] in an equilibrium asset pricing model. We denote this model *LML* for Lévy mixture. Models of this type have appeared in risk neutral studies and we cite for example Ballotta and Bonfiglioli [1] and Itkin and Lipton [14], our use here leverages the use of independent components analysis as applied to time series data with a view to maximizing non-Gaussianity. The third model writes the marginals as following the variance gamma law (Madan and Seneta [23], Madan et al. [22]). The marginals are gamma time changed Brownian motion at unit time and we correlate the Brownian motions. The model was proposed by Eberlein and Madan [8] and employed by Madan [20] in a study pricing options on a basket of stocks. We term this model *VGC* for correlated variance gamma.

In order to investigate models of dependence it is helpful to consider data where there is some presumption of the presence of dependencies. Though this is expected of stock returns in general as they presumably share exposure to common macro movements of the economies in which they trade, one would expect such dependencies to be even greater for sector specific exchange traded funds (*ETF's*) that constitute diversified portfolios of similar collections of stocks. Additionally we have daily data on the market values of these funds, thereby providing us with a fertile environment in which to test our models of multivariate dependence.

With these considerations in mind the three models are estimated on a number of *ETF* returns partitioned by economic sector, as well as one set that selects a single *ETF* from each of nine sectors. The question then arises as to how one may evaluate model performance on this data. For many applications one is interested in the return on portfolios and so we ask how well the models explain the univariate distribution of returns on arbitrary portfolios. For this evaluation we construct a thousand arbitrary randomly generated long short portfolio returns on the unit sphere of dimension matching the number of *ETF's*. We construct both the actual portfolio return in our data and the distribution of this return as predicted by each of our three estimated models. We then construct the p-value on a chi-square test for whether the actual return comes from each of the three models in turn. Under the null hypothesis that the data constitute independent draws from the given distribution the number of observations in each prespecified interval has asymptoticaly a normal distribution with a resulting chi-square distribution for the test statistic. A rule of thumb is to use $2n^{2/5}$ intervals for a sample of size n and we employ 20 regular intervals for a sample of 700. Finally we graph the proportion of portfolios with a p-value greater than x for a range of x values. A model with a higher proportion of high p-values for each candidate probability level does a better job in explaining the univariate laws of arbitrary portfolios and is therefore a superior model for the data set in question.

Different models appear to dominate on different occasions. For example, within sector one gets a better performance from *LML* while across sectors FGC and *VGC* dominate LML. These observations lead us to enquire deeper into the structure of dependence in the different models. We follow the ideas of Longin and Solnik [18] related to extreme correlation and localize further. For this purpose, we develop the concept of local correlation. Later in the paper we comment on the relationship between local correlation as we define it and the formulations of Burtschell, Gregory and Laurent [6] and Langnau [17]. We observe that for the *LML* model there is greater correlation in the tails of the distribution than in the center while for *FGC* and *VGC* correlation drops of in the tails. As the local correlation turns out to be a real valued function defined on a subdomain for the values of the two variables, we call this function the correlation signature of the model and we present the correlation signatures for our three models as they are estimated for the energy sector and a cross sector grouping. A richer understanding of correlations is being called for in recent research and we note Embrechts [10] in this regard.

The outline of the rest of the paper is as follows. Section 1 presents the details of the three models being contrasted. Section 2 outlines the estimation procedures. In Sect. 3 we present our procedure for investigating the "goodness-of-fit". Section 4 introduces the concept of local correlation and the correlation signature of a model. Section 5 describes the data and the portfolio partitions employed. Results on performance evaluation are presented in Sect. 6 including the correlation signatures. Section 7 concludes.

1 The Models Studied

We present the three models studied, *FGC, LML,* and *VGC* in three subsections. There are numerous other alternatives to modeling non-Gaussian dependence and we refer the reader to Bouchaud and Potters [4] for other possibilities. Our attention to these three models is motivated partly by the property that these models may be estimated quite easily in dimensions up to 50 or more by essentially a sequential application of univariate methods. We have also used these three models separately in recent papers and we cite Eberlein and Madan [8, 9] and Madan [21]. Among non-Gaussian models we also have the class of stochastic volatility models, however their estimation in high dimensions via EM algorithms is quite involved especially when one begins to consider the structure of dependencies between the volatilities as well as the returns.

1.1 The Model FGC

From one perspective it is uninformative to compute correlations of non-Gaussian variates as the result does not lead us to any ability at writing down the joint probability law. We merely have correlation estimates and plenty of them if the dimension is high, but there we stop. However, if the data are transformed to standard normal variates first, before the correlation is computed then the computed correlations may be used to write down a candidate joint multivariate normal law for these transformed Gaussian variates. The original data is then modeled as a non-linear transformation of correlated Gaussian variates. This formulation results in a specific joint multivariate probability element.

Let $X = (X_1, \ldots, X_N)$ be a vector of dimension N with continuous marginal distributions for each X_i given by

$$P\left(X_i \leq x\right) = F_i(x).$$

One may transform the marginal laws to standard normal variates by

$$Z_i = \Phi^{-1}(F_i(X_i)),$$

where Φ is the standard normal distribution function.

By construction Z_i is a standard normal variate and one may recover X_i as

$$X_i = F_i^{-1}\left(\Phi(Z_i)\right). \tag{1}$$

It is supposed that the vector $Z = (Z_1, \cdots, Z_N)$ is standard multivariate normal with correlation matrix C. The joint probably density of X may be expressed in terms of the multivariate normal density for Z by a simple change of variable. In our application we shall take the marginal distributions F_i to come from the variance gamma class of distributions.

1.2 The Model LML

The Lévy mixture model postulates that

$$X = AY, \tag{2}$$

for a mixing matrix A and an N-dimensional vector Y, with each variable Y_i being independent of $(Y_j,\ j \neq i)$. We further suppose that each Y_j has a variance gamma distribution.

Given characteristic functions

$$\phi_j(u) = E\left[\exp\left(iuY_j\right)\right]$$

the joint characteristic function of X, with u now a vector is easily derived to be

$$\phi_X(u) = \prod_{j=1}^{N} \phi_j\left((A'u)_j\right).$$

1.3 The Model VGC

The marginal distributions are here postulated to be in the centered variance gamma class with

$$X_i = \theta_i\left(g_i - 1\right) + \sigma_i\sqrt{g_i}Z_i, \tag{3}$$

where Z_i are standard normal variates and the g_i's are a sequence of independent gamma variates with unit mean and variance ν_i. The uncentered variance gamma process (Madan and Seneta [23], Madan [22]) at unit time is a Brownian motion with drift θ_i and volatility σ_i time changed by a gamma process with unit mean rate and variance rate ν_i. The unit time random variable may equivalently be specified as in Eq. (3). The methods of this section could extended to other specifications for marginal distributions that are Lévy processes written as time changed Brownian motions. This includes the Meixner process and the *CGMY* model as shown in Madan and Yor [25] as well as the Normal inverse Gaussian model of Barndorff-Nielsen [3] among other possibilities.

In the *VGC* specification we now further suppose that Z is multivariate normal with correlation matrix C. The joint probability density and characteristic functions are not available in closed form as one has to integrate out a large number of independent gamma densities which appear as products of square roots that do not separate out in either the density or the characteristic function. The joint law, however, is easily simulated from a multivariate normal simulation coupled with drawings from gamma densities.

1.4 Comparative Remarks on the Three Models

The model *FGC* creates dependence by taking the nonlinear transform (1) of correlated Gaussian variates. On the other hand in *VGC* the transformation is linear as seen in equation (3) but both the intercept and slope are stochastic but simultaneously generated, for each asset, by a single gamma variate. The gamma variate is different for different assets. Hence, in this case we have a stochastic linear transformation of correlated Gaussian variates. In the model *LML* Gaussian variates do not appear at all, as we now take a multivariate linear transform (2) of independent non-Gaussian variates. However to the extent the components are modeled as time changes Brownian motions independent Gaussian variates do appear in the components. The three models create dependence in apparently quite different ways. We employ the variance gamma model for our univariate model here, but one could easily extend to the case of the generalized hyperbolic distribution (Eberlein [7]) or its numerous special cases.

2 Estimation Procedures

We suppose we have data $X_t = (X_{1t}, \cdots, X_{Nt})$ for $t = 1, \cdots T$ independent draws from an unknown distribution. The data is supposed to be centered with a zero sample mean. For *FGC* one first estimates the marginal distribution functions on the univariate data. We employ distributions in the variance gamma class for this purpose (Madan and Seneta [23], Madan [22]). This gives us a matrix of marginal *VG* parameters

$$\sigma_i, \nu_i, \theta_i, i = 1, \cdots, N.$$

We then form the univariate data

$$Z_{it} = \Phi^{-1}(F_{VG}(X_{it}; \sigma_i, \nu_i, \theta_i)),$$

where F_{VG} is the distribution function of a *VG* random variable with the specified parameters. We then estimate the correlation matrix C by

$$C_{jk} = \frac{1}{T}\sum_{t=1}^{T} Z_{jt}Z_{kt}.$$

For the purpose of generating observations from this model we first simulate say 10,000 readings from a multivariate normal density $Z_s = (Z_{1s}, \cdots Z_{Ns})$, $s = 1, \ldots, 10{,}000$, with this correlation matrix and then generate simulated readings on the variables by

$$X_{js} = F_{VG}^{-1}\left(\Phi\left(Z_{js}\right), \sigma_j, \nu_j, \theta_j\right).$$

Such simulated draws from the estimated model are used in our subsequent analysis of the model's goodness of fit.

For the *LML* model we first identify the mixing matrix following Madan and Yen [24] and employ independent components analysis for this purpose. The hypothesis of independent components analysis is precisely the statement that one is observing a linear mixture of independent variates and this procedure first performs a principal components analysis (PCA) to generate a set of unit variance orthogonal random variables constructed as linear combinations of the original observed variables. It is then observed that an equivalent PCA is obtained on multiplication by any rotation matrix. The procedure is based on recognizing that a mixing of non-Gaussian signals induces a convergence to Gaussianity and hence the path back to the original signals amounts to maximizing a metric of non-Gaussianity. Such a criterion is employed to search over the class of rotation matrices to construct the matrix A that is the product of the matrix delivering the PCA followed by the non-Gaussianity maximizing rotation matrix. The specific criterion used is the maximization of the expected logarithm of the hyperbolic cosine (Hyvärinen [12]). Once the matrix A has been identified, one obtains data on the independent components on premultiplication of the observed data matrix by the inverse of A. We further postulate that these independent components are variance gamma random variables. We then estimate the parameters of the variance gamma model on the data for these components for obtaining a full estimated specification for the joint law under the *LML* model.

Again, for the purpose of simulating say 10,000 draws from this estimated dependence model we adopt the following procedure. First we generate an $N \times 1$ vector of independent variance gamma random variables 10,000 times, from the N estimated variance gamma laws. We then multiply each $N \times 1$ vector by the $N \times N$ matrix A to sample a draw of N observations from this probability law. The result is a matrix of N by 10,000 readings from the *LML* dependence model.

For our third model, *VGC* we employ the same *VG* marginal laws estimated in *FGC* and then infer the correlations between the Gaussian variates from the observed matrix of covariances between observed returns. This procedure inflates Gaussian component correlations relative to observed correlations by a factor that just depends on the estimated marginal laws. This inflation factor is explicitly described in Eberlein and Madan [8]. On occasion these inflation factors can lead to an estimated correlation matrix with some entries above unity. In this case we construct the closest correlation to our symmetric matrix using the procedures of Qi and Sun [30]. We thus generate 10,000 readings from this law by generating correlated Gaussian random variables and independent gamma variates to form a reading on an N vector in line with Eq. (3). The result is a N by 10,000 matrix of draws from the *VGC* law.

3 Investigating Goodness of Fit

We have estimated three joint laws on asset returns in dimensions ranging from 3 to 7. It is of interest to enquire into the quality of the estimated models, or their ability to describe the data. We do not have available in closed form the relevant joint densities and hence we cannot compute likelihoods and the models are not nested in any case. We also do not have estimates of asymptotic distributions of parameter estimates or likelihoods and hence cannot employ the procedures of non-nested tests either. We consider a performance based evaluation as opposed to testing whether the data comes from the proposed model. In fact these are a tractable class of models available to us and the data may well not come from any of them as the modeling of multidimensional financial return data is a fairly complex exercise. We enquire instead into how well these models of dependence explain the univariate laws of randomly chosen linear mixtures. A multivariate model that explains well the univariate law of all linear mixtures is clearly a good candidate model for the joint law.

With such a performance evaluation in mind, we randomly generate 1000 linear combinations with coefficients located on the unit sphere of N dimensional space. For each linear combination we construct readings on returns for this linear combination of asset returns in our data. Next, for each of our three models with 10,000 simulated paths we construct three sets of 10,000 simulated readings for the same linear combination. The simulated readings are employed to construct the expected number of observations in 20 equally spaced cells covering the interquantile range from 5 to 95 %. We then count the observed number of readings in each of these cells for each of a thousand linear combinations. Given the counts on observed and expected number of observations in each cell we construct a chi-square test p-value. We thus have three sets of a thousand p-values. Finally we graph against x, a candidate p-value between zero and unity, the proportion of portfolios with an observed p-value above x. There are three such graphs for each of our three models. A model whose graph dominates that for another model clearly has a higher proportion of portfolios with high p-values than the dominated model and hence provides us with a superior explanation of the univariate laws of arbitrary linear combinations. The dominating model therefore constitutes a better candidate model of dependence for this data.

4 Local Correlation

With a view towards taking a deeper look at how dependence is modeled in an arbitrary joint density we consider the formulation of local correlation in the neighbourhood of an arbitrary point in space. We note that Burtschell, Gregory and Laurent [6] have considered a form of local correlation in the context of credit modeling where the correlation coefficient depends on the level of the systematic factor in an otherwise one factor Gaussian copula construction of dependence. Langnau [17] introduces deterministic time dependent correlations between driving Brownian motions to construct a form of correlation that is local in time. We work instead with the joint density of two random variables, focussing attention on the possibility of correlations varying when sampling is censored to different interval ranges for the two variables being studied.

To partially motivate what we are after we consider the following economic scenario. Suppose that in normal times stocks move around with a low volatility and some low level of correlation. As a boom develops and money starts moving from fixed income markets towards the stock market, arbitrage traders trying to maintain price relativities introduce greater correlation as volatilities also rise. A similar pattern occurs as we go into a bear market. On occasion one may see some sectors expand at the expense of another and here we may have negative correlations occuring between a pair of stocks.

By way of a theoretical model underlying such scenarios we note that any density in two dimensional space may be expressed as a mixture of bivariate normal densities and we consider 5 such densities. They have the following means, variances and correlations for our two variates respectively. The values for the means for X are, $\mu_X = (0.03, 0.03, 0, -0.03, -0.03)$ and for Y we have $\mu_Y = (0.03, -0.03, 0, 0.03, -0.03)$. The standard deviations in the five cases for X are $\sigma_X = (0.3, 0.2, 0.1, 0.2, 0.3)$ while for Y they are $\sigma_Y = (0.4, 0.3, 0.15, 0.3, 0.4)$, where we have induced high volatility in up and down markets, a lower volatility when we have a movement from one asset to the other and a still lower volatility for normal times. We take the five correlations to be $\rho_{XY} = (0.8, -0.5, 0, -0.5, 0.8)$, with the bull and bear markets being highly positively correlated, the movement between markets being negatively correlated and there being no correlation in normal times. We suppose the total sample comes from a mixture of these five joint densities in the proportions $w = (0.1, 0.05, 0.7, 0.05, 0.1)$. One may explicitly write the joint density as

$$q(x, y) = \sum_{i=1}^{5} w_i b\left(\frac{x - \mu_X(i)}{\sigma_X(i)}, \frac{y - \mu_Y(i)}{\sigma_Y(i)}; \rho_{XY}(i)\right),$$

where $b(x, y; \rho)$ is the bivariate normal density with correlation coefficient ρ. We present in Fig. 1 a sample of size 10,000 from such a density.

One may compute sensored correlations for data in the four corners for X, Y being above or below plus or minus one percent. In this sample these correlations are 0.6220, −0.3201, −0.2731 and 0.6038 respectively. In principle therefore one may have correlation varying as we move around in the spatial domain.

For an analysis of such situations we consider an arbitrary joint density for two random variables $q(x, y)$. We now define

$$h(x, y) = -2 \ln q(x, y).$$

Consider expanding the function h to second order around the point (a, b) to obtain

$$\begin{aligned} h(x, y) \approx{} & h(a, b) + h_a(x - a) + h_b(y - b) \\ & + \frac{1}{2} h_{aa}(x - a)^2 + \frac{1}{2} h_{bb}(y - b)^2 \\ & + h_{ab}(x - a)(y - b). \end{aligned}$$

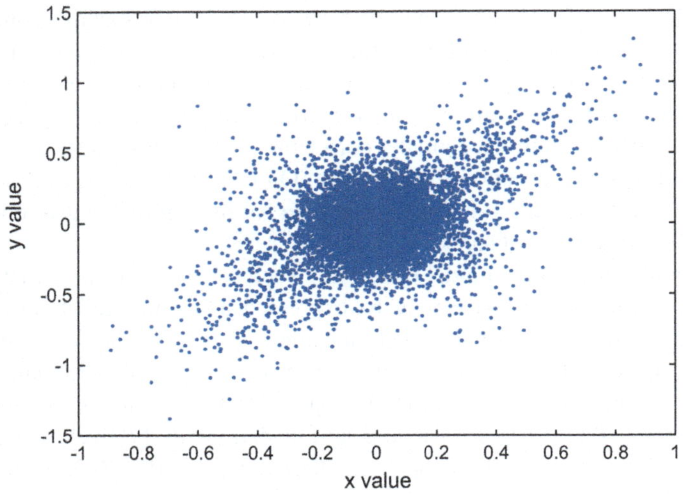

Fig. 1 Sample of locally correlated data

A bivariate normal distribution has such an exact quadratic expression for the log likelihood where we identify

$$\Sigma^{-1} = \begin{pmatrix} h_{aa} & h_{ab} \\ h_{ba} & h_{bb} \end{pmatrix}.$$

In this case we would have

$$\Sigma = \frac{1}{h_{aa}h_{bb} - h_{ab}^2} \begin{pmatrix} h_{bb} & -h_{ab} \\ -h_{ba} & h_{aa} \end{pmatrix}$$

and the correlation would be

$$\rho_{ab} = \frac{-h_{ab}}{\sqrt{h_{aa}h_{bb}}}. \tag{4}$$

We might consider defining this value generally as the local correlation. It is then a question as to whether this value is between -1 and 1. For this we require the square to be less than one or

$$h_{ab}^2 \leq h_{aa}h_{bb}. \tag{5}$$

This is precisely the condition for the negative of the log likelihood to be a convex function. We must have $h_{aa}, h_{bb} > 0$ along with (5). For many models with unimodal joint densities we have the concavity of the density near the mode and hence in this region the negative of the log likelihood is convex. The proposed definition for local correlation will yield magnitudes dominated by unity in absolute value in this region. There is therefore quite generally a local domain in which one may investigate the shape of local correlation. We call the map in this local domain the correlation signature of the model. Additionally there are models with universally log concave densities and for these models the local correlation is universally well defined. This is an important class of densities, much studied in its own right by Barlow and Proschan [2], and Prèkopa [29].

We now investigate the nature of this local correlation surface for our three models *FGC, VGC* and finally *LML* in separate subsections.

4.1 Local Correlation for FGC

Here the joint law is that of nonlinear transforms of correlated Gaussians. The joint density is therefore

$$f(x, y) = b(g(x), h(y))g'(x)h'(y),$$

where $b(z_1, z_2)$ is again the bivariate normal density with correlation coefficient ρ. The functions g, h are the two nonlinear transforms, where the transformed variates are taken to be distributed with a bivariate normal density. The negative of twice the log density is

$$\begin{aligned}\Psi(x, y) &= -2\ln(f(x, y)) \\ &= -2\ln b(g(x), h(y)) - 2\ln g'(x) - 2\ln h'(y).\end{aligned}$$

Define $\widetilde{b}(z_1, z_2) = -2\ln b(z_1, z_2)$. It follows that

$$\Psi_{xy} = \widetilde{b}_{xy} g'(x)h'(y).$$

We have $\widetilde{b}_{xy} = \rho$ the assumed constant correlation coefficient of the bivariate normal and g', h' are high in the center and low in the tails by virtue of being derivatives of distribution functions. Hence this model gives higher correlation in the neck of the joint density and correspondingly lower correlation in the tails.

4.2 Local Correlation for VGC

For the *VGC* structure we have

$$\begin{aligned}x &= \theta_x(g_x - 1) + \sigma_x\sqrt{g_x}Z_x \\ y &= \theta_y(g_y - 1) + \sigma_y\sqrt{g_y}Z_y.\end{aligned}$$

The joint density is now

$$E\left[\frac{1}{\sigma_x\sqrt{g_x}}b\left(\frac{x-\theta(g_x-1)}{\sigma_x\sqrt{g_x}},\frac{y-\theta(g_y-1)}{\sigma_y\sqrt{g_y}}\right)\frac{1}{\sigma_y\sqrt{g_y}}\right],$$

where E denotes the expectation operator with respect to integrating out the gamma variates g_x, g_y.

The critical function now is

$$\Psi(x,y)=-2\ln E\left[\frac{1}{\sigma_x\sqrt{g_x}}b\left(\frac{x-\theta(g_x-1)}{\sigma_x\sqrt{g_x}},\frac{y-\theta(g_y-1)}{\sigma_y\sqrt{g_y}}\right)\frac{1}{\sigma_y\sqrt{g_y}}\right]$$

and we may consider in its place the lower bound given by expectation of the logarithm. This yields the function

$$\widetilde{\Psi}(x,y)=-2E\left[\ln\left(\frac{1}{\sigma_x\sqrt{g_x}}b\left(\frac{x-\theta(g_x-1)}{\sigma_x\sqrt{g_x}},\frac{y-\theta(g_y-1)}{\sigma_y\sqrt{g_y}}\right)\frac{1}{\sigma_y\sqrt{g_y}}\right)\right]$$

and this generates a possibly flat correlation the expected cross partial is a scaling of ρ.

4.3 *Local Correlation for LML*

For the *LML* model we have the joint density

$$f(x,y)=g(ax+by)h(cx+dy)\kappa,$$

where κ is a normalization constant.

In this case we get

$$\Psi(x,y)=-2\ln g(ax+by)-2\ln h(cx+dy)-2\ln\kappa.$$

If we compute the cross partial derivatives on defining $\widetilde{g}, \widetilde{h}$ to be $-2\ln g$, $-2\ln h$ respectively we get

$$\begin{aligned}\Psi_x&=\widetilde{g}'a+\widetilde{h}'c,\\ \Psi_{xy}&=\widetilde{g}''ab+\widetilde{h}''cd,\\ \Psi_{xx}&=\widetilde{g}''a^2+\widetilde{h}''c^2,\\ \Psi_{yy}&=\widetilde{g}''b^2+\widetilde{h}''d^2.\end{aligned}$$

For our convexity condition we require that

$$\Psi_{xx}\Psi_{yy} \geq \Psi_{xy}^2$$

or equivalently that

$$\left(\widetilde{g}''a^2 + \widetilde{h}''c^2\right)\left(\widetilde{g}''b^2 + \widetilde{h}''d^2\right) \geq \left(\widetilde{g}''ab + \widetilde{h}''cd\right)^2$$

and this yields the condition

$$\widetilde{g}''\,\widetilde{h}''\left(a^2d^2 + b^2c^2 - 2abcd\right) \geq 0$$

or

$$\widetilde{g}''\,\widetilde{h}''\,(ad - bc)^2 \geq 0$$

and hence we just need that

$$\widetilde{g}''\,\widetilde{h}'' \geq 0.$$

This condition is satisfied if the marginal laws are themselves log convex. We note that the second derivatives peak up in the tails and the center and are small in the middle. This model can therefore lead to tail and central correlations with flat correlations in the middle.

5 The Data Employed

We obtained data on the time series of Exchange Traded Funds (henceforth *ETF*) that follow various sectors of the US economy. As we are also interested in risk neutral laws we focused attention on funds that also have options trading on the *ETF* market

Table 1 ETF groupings by sector

Sectors	ETF Tickers
Consumer discretionary	xly,rth,xrt,itb,xhb
Energy	xle,iye,ieo,oih,xop
Financials	xlf,iyf,iai,kbe,kre,rkh,kce
Health and Pharmaceuticals	xlv,bbh,pph
Industrials and Technology	xli,iyt,iyw,xlk
Internet, Networking, Semiconductor, Software	hhh, bdh, igw, smh, swh
Materials, Real estate, Telecommunications	xlb, iyr, iyz, tth
Natural resources	ige, gdx, slx, xme
Utilities	idm, xlu, uth
Cross sectors	xly,xlp,xle,xlf,xlv,xli,xlk,xlb,xlu

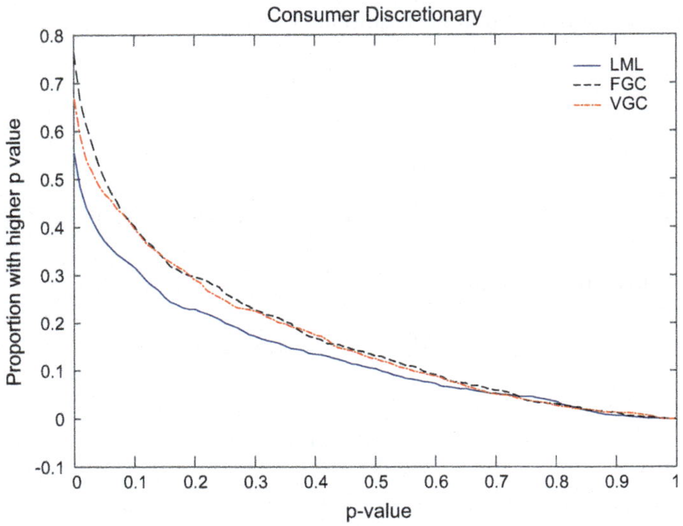

Fig. 2 Portfolio Proportions with given probability values in the Consumer Discretionary sector

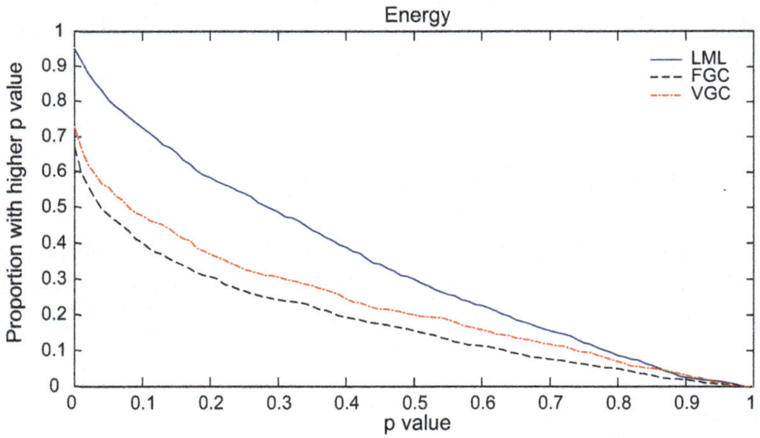

Fig. 3 Portfolio Proportions with given probability values in the Energy sector

and for which we had a time series exceeding 700 days of daily data ending on July 21 2009. There are nine industry groups and the *ETF's* in the group are displayed in Table 1.

For each of these nine groups we present nine graphs with three curves each, one for *FGC* in black, another for *LML* in blue and the third for *VGC* in red. Each curve displays the proportion of a thousand random linear combinations with a *p*-value

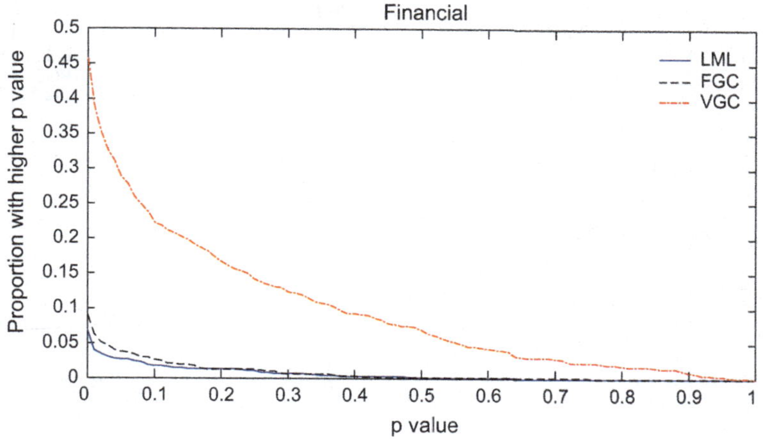

Fig. 4 Portfolio Proportions with given probability values in the Financial sector

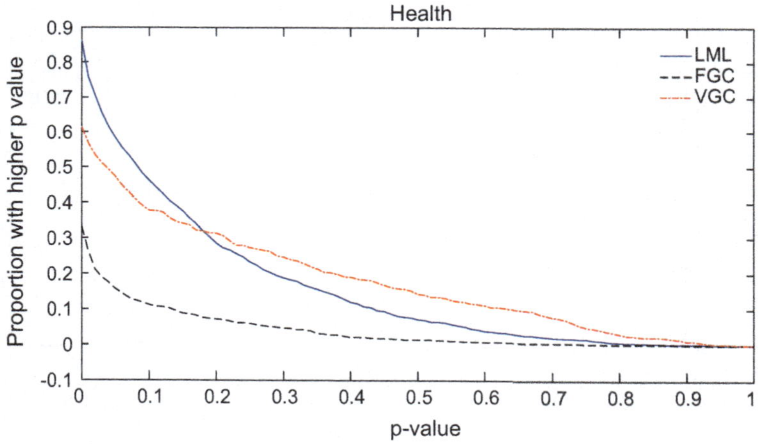

Fig. 5 Portfolio Proportions with given probability values in the Health related sector

for a chi-square test on the univariate law for this linear combination exceeding the candidate value given by the x axis. Figure 2 displays the result for the Consumer Discretionary sector where *FGC* and *VGC* perform equally well and dominate LML. In Fig. 3 we have the results for the Energy sector where *LML* dominates, followed by *VGC* and the *FGC*. For the Financial sector, Fig. 4, that saw a lot of movement in this period, *VGC* dominates by far the other two models. In the Health related sector Fig. 5 *LML* and *VGC* criss cross and dominate *FGC*. For the industrial sector, Fig. 6 all three models are equivalent. The technology sector Fig. 7 like Energy has *LML*

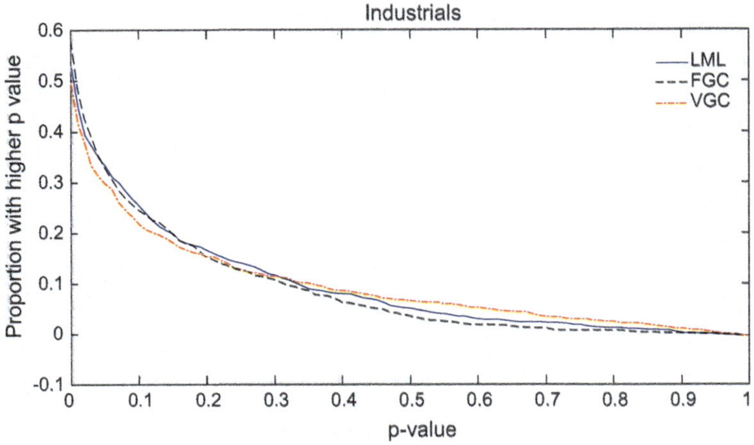

Fig. 6 Portfolio Proportions with given probability values in the Industrial sector

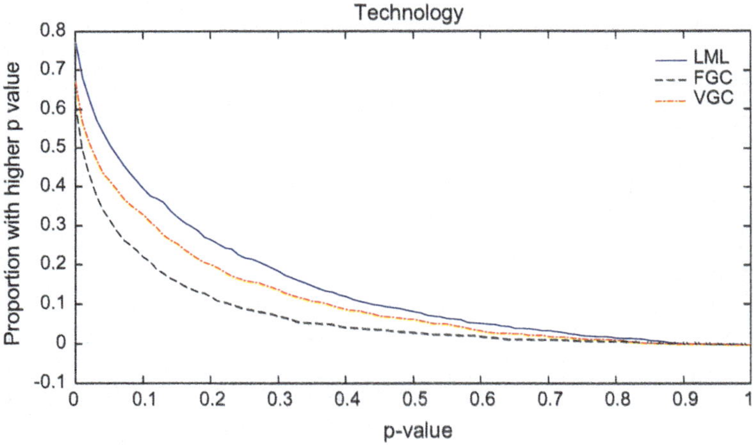

Fig. 7 Portfolio Proportions with given probability values in the Technology sector

dominating followed by *VGC* and *FGC*. All three models are equivalent for Natural Resources Fig. 8. Telecom, Fig. 9 sees the order *VGC* followed by *LML* and *FGC*. Finally the Utility sector Fig. 10 has *LML* followed by *VGC* and *FGC*.

We observe from focusing in some cases around the 10 % point that in five of the nine groups we have *LML* dominating *VGC* that dominates *FGC*. In a further two cases all three models are equivalent. In one case, Financials, VGC dominates the other two by far. There is some broad preference for *LML* followed by *VGC* and then *FGC*.

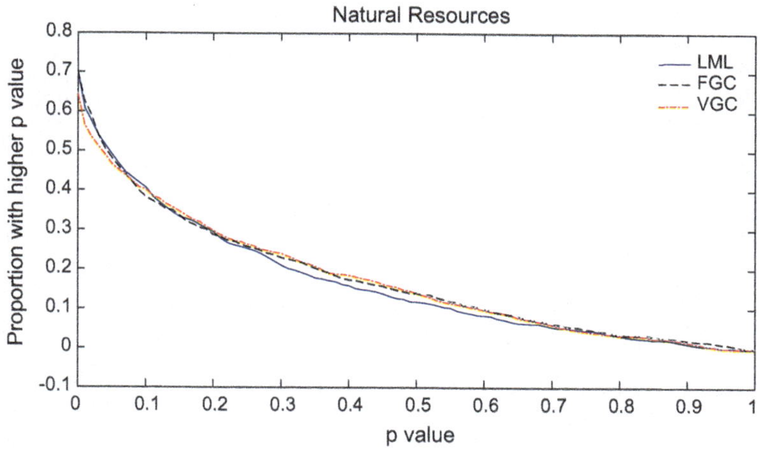

Fig. 8 Portfolio Proportions with given probability values in the Natural Resource related sector

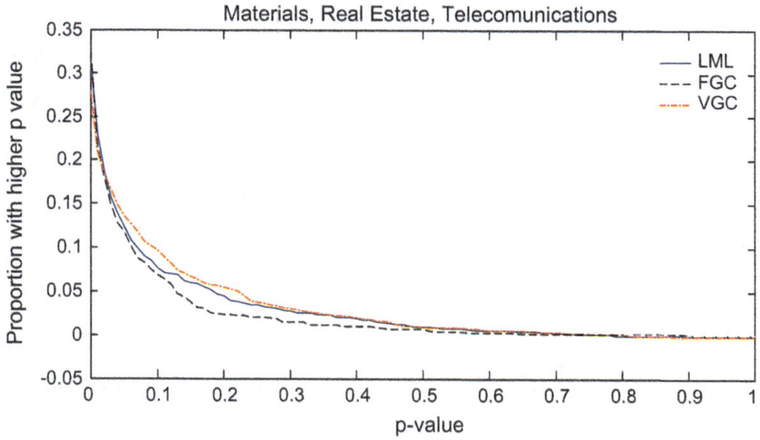

Fig. 9 Portfolio Proportions with given probability values in the Telecom sector

We next consider the cross sector group Fig. 11 with one *ETF* from each of the nine sectors. In this grouping we have a clear domination by VGC over FGC and *LML* that are somewhat equivalent.

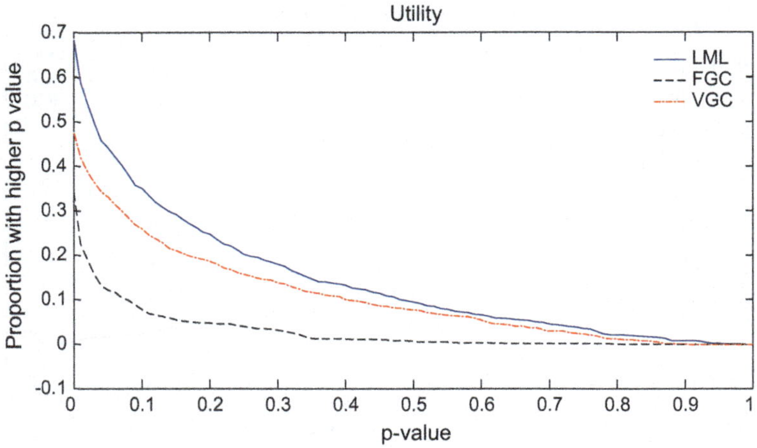

Fig. 10 Portfolio Proportions with given probability values in the Utility sector

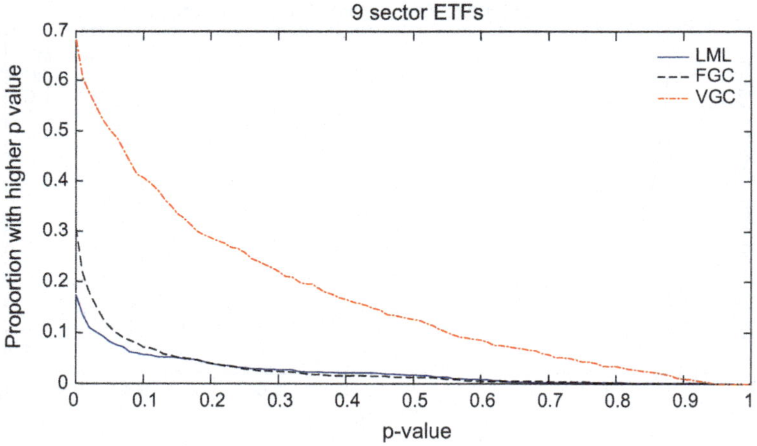

Fig. 11 Portfolio Proportions with given probability values in the Cross Sector group

6 Model Correlation Signatures

We first present the details on how the correlation signatures are constructed for each model. In each case we extract the joint density for a pair of returns in the set of returns jointly modeled. We then evaluate the local correlation numerically by evaluating the appropriate derivatives of the negative of the log likelihood. We describe in separate subsections the procedure for constructing the joint density for a pair of variables.

6.1 FGC

For the FGC model the joint density is obtained as follows. Let the marginal distribution functions be $F(x)$, $G(y)$. We then have that

$$z_1 = \Phi^{-1}(F(x))$$
$$z_2 = \Phi^{-1}(G(y))$$

are distributed bivariate normal with correlation ρ. The density of $z = (z_1, z_2)$ is

$$b(z_1, z_2).$$

It follows that the density of x, y is

$$q(x, y) = b(\Phi^{-1}(F(x)), \Phi^{-1}(G(y)))\frac{f(x)}{\phi(z_1)}\frac{g(y)}{\phi(z_2)}.$$

Once we have q we may apply our local correlation surface construction to extract the correlation signature of this model.

We now wish to incorporate scaling to unit variance. We may do this via the marginals as

$$X = \frac{x}{\sigma}$$

and

$$F_X(a) = P(X \le a) = P\left(\frac{x}{\sigma} \le a\right) = F_x(\sigma a).$$

6.2 LML

We wish to construct the correlation signatures of our models estimated on one of the more volatile sectors, and we take by way of an example, the energy sector. There are five *ETF's* for which the joint law was estimated and these are

$$xle, iye, ieo, oih, xop.$$

We consider xle, iye and ieo, xop. We have modeled daily returns as

$$x_i = \phi_i' y$$
$$x_j = \phi_j' y.$$

and we have the joint characteristic functions but we shall compute the correlation signatures for standardized variates. The variance of x is given by

$$\sigma_i^2 = \sum_k \phi_{ik}^2 \left(\sigma_k^2 + \theta_k^2 \nu_k\right).$$

The standardized vector is

$$X_i = \frac{x_i}{\sigma_i} = \frac{\phi_i'}{\sigma_i} y = \Phi_i' y$$

and it is centered of unit variance by construction.

We easily obtain from the joint characteristic function the joint characteristic function of any 2 variates out of the full set modeled. We may invert this joint characteristic function using two dimensional Fourier inversion for the joint density as described for example in Hurd and Zhou [11].

6.3 VGC

For the *VGC* model the construction is

$$X = \theta_x(g_x - 1) + \sigma_x\sqrt{g_x} Z_x,$$
$$Y = \theta_y(g_y - 1) + \sigma_y\sqrt{g_y} Z_y.$$

Hence given the density of $Z = (Z_x, Z_y)$ we may write

$$q(x,y) = E\left[b\left(\frac{x - \theta_x(g_x - 1)}{\sigma_x\sqrt{g_x}}, \frac{y - \theta_y(g_y - 1)}{\sigma_y\sqrt{g_y}} \right) \frac{1}{\sigma_x\sqrt{g_x}} \frac{1}{\sigma_y\sqrt{g_y}} \right]$$
$$= \int_0^\infty \int_0^\infty b\left(\frac{x - \theta_x(g_x - 1)}{\sigma_x\sqrt{g_x}}, \frac{y - \theta_y(g_y - 1)}{\sigma_y\sqrt{g_y}} \right) p_x(g_x) p_y(g_y) dg_x dg_y,$$

where p_x, p_y are the gamma densities for the two gamma time changes.

We may write the joint density more explicitly as

$$q(x,y) = \int_0^\infty \int_0^\infty b\left(\frac{x - \theta_x(g_x - 1)}{\sigma_x\sqrt{g_x}}, \frac{y - \theta_y(g_y - 1)}{\sigma_y\sqrt{g_y}} \right) \times$$
$$\frac{1}{\sigma_x \nu_x^{\frac{1}{\nu_x}} \Gamma\left(\frac{1}{\nu_x}\right)} g_x^{\frac{1}{\nu_x} - \frac{3}{2}} e^{-\frac{g_x}{\nu_x}} \frac{1}{\sigma_y \nu_y^{\frac{1}{\nu_y}} \Gamma\left(\frac{1}{\nu_y}\right)} g_y^{\frac{1}{\nu_y} - \frac{3}{2}} e^{-\frac{g_y}{\nu_y}} dg_x dg_y.$$

We make the change of variable to

$$w_x = \frac{g_x}{\nu_x}$$
$$w_y = \frac{g_y}{\nu_y}$$

to get

$$q(x, y) = \int_0^\infty \int_0^\infty b\left(\frac{x - \theta_x(\nu_x w_x - 1)}{\sigma_x\sqrt{\nu_x w_x}}, \frac{y - \theta_y(\nu_y w_y - 1)}{\sigma_y\sqrt{\nu_y w_y}}\right) \times \frac{1}{\sigma_x\sqrt{\nu_x w_x}\Gamma\left(\frac{1}{\nu_x}\right)} w_x^{\frac{1}{\nu_x}-1} e^{-w_x} \frac{1}{\sigma_y\sqrt{\nu_y w_y}\Gamma\left(\frac{1}{\nu_y}\right)} w_y^{\frac{1}{\nu_y}-1} e^{-w_y} dw_x dw_y.$$

We evaluate this as a double sum using Gauss-Laguerre quadrature for the construction of the joint density in two dimensions. This joint density computed on grid is then input into the program that constructs the local correlation surface by computing the required derivatives. We explicitly evaluate the joint density as

$$q(x, y) = \sum_{ij} p_i p_j b\left(\frac{x - \theta_x(\nu_x w_i - 1)}{\sigma_x\sqrt{\nu_x w_i}}, \frac{y - \theta_y(\nu_y w_j - 1)}{\sigma_y\sqrt{\nu_y w_j}}\right) \times \frac{1}{\sigma_x\sqrt{\nu_x w_i}\Gamma\left(\frac{1}{\nu_x}\right)} w_i^{\frac{1}{\nu_x}-1} \frac{1}{\sigma_y\sqrt{\nu_y w_j}\Gamma\left(\frac{1}{\nu_y}\right)} w_j^{\frac{1}{\nu_y}-1},$$

where p_i are the Laguerre weights and w_i are the points.

6.4 Correlation Signature Results for Energy and the Cross Sector Group

We present in Table 2 the correlation signatures for two pairs of stocks from the Energy sector, ieo,xop and xle,iye and three pairs of stocks from the cross sector group xly,xlp, xli,xlk and xly,xli. We observe that the local correlations in *LML* tend to be substantially higher and particularly so in the tails. The local correlations are computed at the center of the distributions and 10 % age points up and down from this level. Economically one experiences higher correlations in times of crisis that are viewed as rare tail events. To the extent local correlation captures such spatially contingent notions of correlation these results support the economic relevance of such models.

7 Conclusion

Three models of dependence in asset returns with non-Gaussian marginals are investigated on *ETF* daily return data. The first is a full rank Gaussian copula also studied and proposed in Malvergne and Sornette (2005) termed *FGC*. The second is a linear

Table 2 Correlation signatures of models for selected stock pairings

Correlation signatures for LML, FGC and VGC respectively									
ieo	xop			xop			xop		
	0.663066	0.702658	0.725284	0.151467	0.15983	0.179276	0.206153	0.211492	0.2177
	0.639887	0.694015	0.735871	0.160283	0.168552	0.188268	0.212761	0.21722	0.222705
	0.589141	0.664609	0.73184	0.179742	0.188198	0.209177	0.21984	0.223406	0.228671
xle	iye			iye			iye		
	0.709821	0.64253	0.543917	0.073226	0.080281	0.092234	0.184961	0.203322	0.188705
	0.725103	0.689161	0.627705	0.076677	0.083968	0.096331	0.162205	0.176877	0.163207
	0.712397	0.721196	0.707472	0.083994	0.091854	0.105202	0.162894	0.177886	0.164916
xly	xlp			xlp			xlp		
	0.870662	0.824027	0.789156	0.518616	0.539077	0.591593	0.586448	0.593867	0.598697
	0.80326	0.791869	0.800942	0.539412	0.554894	0.601744	0.591685	0.599928	0.604753
	0.737392	0.775675	0.828535	0.590685	0.599728	0.641581	0.599142	0.604244	0.608796
xli	xlk			xlk			xlk		
	0.88267	0.823123	0.768254	0.692328	0.705675	0.746557	0.686074	0.698671	0.704153
	0.831342	0.79638	0.779932	0.724125	0.727674	0.758185	0.681454	0.696105	0.704073
	0.790777	0.788804	0.809007	0.795676	0.785003	0.803508	0.688898	0.69427	0.702349
xly	xli			xli			xli		
	0.881328	0.833928	0.799507	0.649106	0.675192	0.735665	0.719777	0.713182	0.714502
	0.828022	0.800518	0.792809	0.66263	0.680397	0.729319	0.726997	0.725125	0.722426
	0.780602	0.784001	0.806905	0.702033	0.71057	0.749299	0.72911	0.730034	0.728464

mixture of independent Lévy processes as proposed in Madan and Yen [24] and studied in [19] termed *LML*. The third correlates Gaussian components in a variance gamma representation of the marginals as proposed in Eberlein and Madan [8] termed *VGC*. All three models are easily estimated in fairly high dimensions as most of the work is done at a univariate level. The models are evaluated on the basis of their ability to explain the univariate laws of randomly generated portfolios.

It is observed that on a number of occasions all three models are at a comparable level of performance. In some cases we get a superior performance from the *LML* model followed by *VGC* and *FGC*. There are occasions when the *VGC* and *FGC* dominate. The three models are tractable in different ways with the *LML* model yielding closed form characteristic functions.

With a view to exploring more deeply the different forms of dependence modeling the concept of local correlation is introduced. It is shown that the *LML* model displays higher levels of local correlation than that obtained in the *FGC* and *VGC* models.

References

1. Ballotta, L., Bonfiglioli, E.: Multivariate asset models using Lévy processes and applications. Eur. J. Financ. (2014). http://dx.doi.org/10.1080/1351847X.2013.870917
2. Barlow, R., Proschan, F.: Statistical Theory of Reliability and Life Testing. Holt Rinehart and Winston, New York (1981)
3. Barndorff-Nielsen, O.E.: Processes of normal inverse Gaussian type. Financ. Stochast. **2**, 41–68 (1998)
4. Bouchaud, J.-P., Potters, M.: Theory of Financial Risk and Derivative Pricing. Cambridge University Press, Cambridge (2003)
5. Boyarchenko, S.I., Levendorskii, S.Z.: Non-Gausian Merton-Black-Scholes Theory. World Scientific, Singapore (2002)
6. Burtschell, X., Gregory, J., Laurent, J-P.: Beyond the Gaussian Copula: Stochastic and Local Correlation, Working paper, ISFA Actuarial School, University Claude Bernard of Lyon (2005)
7. Eberlein, E.: Application of generalized hyperbolic Lévy motions to Finance. In: Barndorff-Nielsen, O., Mikosch, T., Resnick, S. (eds.) Lévy Processes: Theory and Applications. Birkhäuser, Boston (2001)
8. Eberlein, E., Madan, D.B.: On correlating Lévy processes. J. Risk. (2010)
9. Eberlein, E., Madan, D.B.: Unlimited liabilities, reserve capital requirements and the taxpayer put option. Quant. Financ. **12**, 709–724 (2010b)
10. Embrechts, P.: Linear correlations and EVT: properties and caveats. J. Finan. Econ. **7**, 30–39 (2009)
11. Hurd, T., Zhou, Z.: A fourier transform method for spread option pricing. SIAM J. Finan. Math. **1**, 142–157 (2009)
12. Hyvärinen, A.: Fast and robust fixed-point algorithms for independent component analysis IEEE Trans. Neural Netw. **10**, 626–634 (1999)
13. Hyvärinen, A., Karhunen, J., Oja, E.: Independent Component Analysis. John Wiley & Sons Inc., New York (2001)
14. Itkin, A., Lipton, A.: Efficient solutions of structural default models with correlated jumps and mutual obligations. Int. J. Comput. Math. **92**, 2380–2405 (2015)
15. Jondeau, E., Poon, S., Rockinger, M.: Financial Modeling Under Non Gaussian Distributions. Springer, London (2007)
16. Kotz, S., Nadarajah, S.: Mutlivariate t Distributions and Their Applications. Cambridge University Press, Cambridge (2004)

17. Langnau, A.: Introduction to Local Correlation Modelling (2009). arXiv:0909.3441v3
18. Longin, F., Solnik, B.: Extreme correlation of international equity markets. J. Financ. **56**, 649–676 (2001)
19. Madan, D.B.: Equilibrium asset pricing with non-Gaussian returns and exponential utility. Quant. Financ. **6**, 455–463 (2006)
20. Madan, D.B.: Pricing and hedging basket options to prespecified levels of acceptability. Quant. Financ. (2009)
21. Madan, D.B.: Variance swap portfolio theory. In: Chiarella, Carl, Novikov, Alexander (eds.) Contemporary Quantitative Finance: Essays in Honor of Eckhard Platen, pp. 183–194. Springer, Berlin (2010)
22. Madan, D., Carr, P., Chang, E.: The variance gamma process and option pricing. Eur. Financ. Rev. **2**, 79–105 (1998)
23. Madan, D., Seneta, E.: The Variance Gamma (VG) model for share market returns. J. Bus. **63**, 511–524 (1990)
24. Madan, D.B., Yen, J.-Y.J.: Asset allocation with multivariate non-Gaussian returns. In: Birge, J.R., Linetsky, V (eds.) Handbooks in Operations Research and Management Science: Financial Engineering, vol. 15, pp. 949–969. North-Holland, The Netherlands (2008)
25. Madan, D.B., Yor, M.: Representing the CGMY and Meixner Lévy processes as time changed Brownian motions. J. Comput. Financ. **12**, 27–47 (2008)
26. Malevergne, Y., Sornette, D.: High-order moments and cumulants of multivariate Weibull asset return distributions: Analytical theory and empirical tests: II. Financ. Lett. **3**, 54–63 (2005)
27. McNeil, A.J., Frey, R., Embrechts, P.: Quantitative Risk Management. Princeton University Press, Princeton (2005)
28. Menn, C., Fabozzi, F.J., Rachev, S.T.: Fat Tailed and Skewed Asset Return Distributions: Implications for Risk Management, Portfolio Selection and Option Pricing. John Wiley and Sons, New York
29. Prèkopa, A.: On logarithmic concave measures and functions. Act. Sci. Math. (Szeged) **34**, 335–343 (1973)
30. Qi, Houduo, Sun, D.: A quadratically convergent Newton method for computing the nearest correlation matrix. SIAM J. Matrix Anal. Appl. **28**, 360–385 (2006)
31. Schoutens, W., Carriboni, J.: Lévy Processes in Credit Risk. John Wiley and Sons, New York (2009)

Estimation of Correlation Between Latent Processes

Akitoshi Kimura and Nakahiro Yoshida

Dedicated to Professor Ernst Eberlein on the occasion of his 70th birthday

Abstract This paper discusses estimation of correlation between hidden semimartingales. We show the consistency and the asymptotic mixed normality of the proposed correlation estimator in a high frequency setting. As an example, estimation of covariance between intensity processes of doubly stochastic point processes will be mentioned.

Keywords High frequency data · Latent correlation · Asymptotic mixed normality

1 Introduction

Epps [15] showed that the sample correlation between two stock prices has downward bias when the sampling frequency increases. In explanation, it is recognized that non-synchronicity and microstructure cause this phenomenon for high frequency financial

This work was in part supported by Japan Society for the Promotion of Science Grants-in-Aid for Scientific Research Nos. 24340015 (Scientific Research), Nos. 24650148 and 26540011 (Challenging Exploratory Research); CREST Japan Science and Technology Agency; NS Solutions Corporation; and by a Cooperative Research Program of the Institute of Statistical Mathematics.

A. Kimura (✉) · N. Yoshida
Graduate School of Mathematical Sciences, University of Tokyo, 3-8-1 Komaba, Meguro-ku, Tokyo 153-8914, Japan
e-mail: kimura@ms.u-tokyo.ac.jp

N. Yoshida
e-mail: narahiro@ms.u-tokyo.ac.jp

A. Kimura · N. Yoshida
Japan Science and Technology Agency, CREST, Tokyo, Japan

© Springer International Publishing Switzerland 2016
J. Kallsen and A. Papapantoleon (eds.), *Advanced Modelling in Mathematical Finance*, Springer Proceedings in Mathematics & Statistics 189,
DOI 10.1007/978-3-319-45875-5_6

data. Non-synchronous covariance estimation schemes and denoising techniques have been developed in recent years. Among many others, see Malliavin and Mancino [28, 29], Hayashi and Yoshida [17–19], Voev and Lunde [38], Griffin and Oomen [16], Mykland [31], Zhou [41], Zhang et al. [39], Zhang [40], Podolskij and Vetter [36], Jacod et al. [23], Christensen et al. [9], Bibinger [5, 6], Ogihara and Yoshida [34], Koike [24–26], Ogihara [32, 33] for developments of this field. The microstructure is often expressed as noise added to the latent efficient prices. This modeling was successful, at least for developing theory, though the reality of noise is not so clearly explained. The lead-lag phenomenon is also an issue related to the non-synchronicity, see e.g., de Jong and Nijman [11], Hoffmann et al. [21] and Abergel and Huth [1].

Jump processes have been successfully applied to analyses of financial data; see [12–14] and others for pioneering works by Ernst Eberlein for applications of Lévy processes and related distributions. Recently, along with the developments in measurement and storage technologies, the timing of sampling is becoming more and more precise. In order to model non-synchronicity and microstructure as well as lead-lags, modeling of ultra high frequency phenomena is going toward use of jump processes: Hewlett [20], Large [27], Bowsher [8], Bacry et al. [4], Cont et al. [10], Abergel and Jedidi [2, 3], Smith et al. [37], Muni Toke and Pomponio [30], and Ogihara and Yoshida [35]. In this stream, the aim of this paper is to present a simple estimator of correlation between two latent processes indirectly observed high frequently. An application is correlation estimation of intensity processes of two observable counting processes. We will prove asymptotic mixed normality in the situation where the processes are observed in finite time horizon and the trajectory of each underlying process is asymptotically estimable, for example where the market is active and the number of transactions becomes large.

2 Model

We consider a stochastic basis $\mathscr{B} = (\Omega, \mathscr{F}, \mathbb{F}, P)$, $\mathbb{F} = (\mathscr{F}_t)_{t\in[0,1]}$.

On $\mathscr{B}$, let $\mathbb{X} = (X^1, X^2)$ be an $\mathbb{R}^2$-valued Itô process given by

$$\mathbb{X}_t = \mathbb{X}_0 + \int_0^t \mathbb{X}_s^0 ds + \int_0^t \mathbb{X}_s^1 dw_s \quad (t \in [0, 1]), \tag{1}$$

where w is an r-dimensional $\mathbb{F}$-Wiener process, $\mathbb{X}_0$ is an $\mathscr{F}_0$-measurable random variable, $\mathbb{X}^0$ is a two-dimensional $\mathbb{F}$-adapted process, $\mathbb{X}^1$ is an $\mathbb{R}^2 \otimes \mathbb{R}^{\mathrm{r}}$-valued $\mathbb{F}$ -adapted process specified more precisely later, cf. Condition [A] below.

Let a_n be a positive number depending on $n \in \mathbb{N}$. On $\mathscr{B}$, consider a two-dimensional measurable process $\mathbb{Y}^n = (Y^{n,1}, Y^{n,2})$ having a decomposition

$$\mathbb{Y}_t^n = \mathbb{Y}_0^n + \int_0^t a_n \mathbb{X}_s ds + \mathbb{M}_t^n,$$

where $\mathbb{M}^n = (M^{n,\alpha})_{\alpha=1,2}$ is a two-dimensional measurable process with $\mathbb{M}_0^n = 0$. We will put additional conditions on $\mathbb{M}^n$ in Condition [B$^\flat$], [B], [B$'$] and [B$^\sharp$] of Sect. 3

Example 1 Suppose that each X^α is $\mathbb{R}_+ = [0, \infty)$-valued and $Y^{n,\alpha}$ is a counting process with intensity process $a_n X^\alpha$. This model describes the high frequency counting data of the orders or transactions in the active market, for example.

Example 2 Let $t_j = j/n$, $\mathbb{X} = w$ and $\mathbb{M}_t^n = \sum_{j=1}^n \{(w_{t_j \wedge t} - w_{t_{j-1}\wedge t})^{\otimes 2} - (t_j \wedge t - t_{j-1} \wedge t)I_r\}$. The increment $\mathbb{M}_{t_j}^n - \mathbb{M}_{t_{j-1}}^n$ is dependent on $\mathscr{F}_{t_j}^{\mathbb{X}} = \sigma\{\mathbb{X}_s;\ s \in [0, t_j]\}$, and it has bias under the conditional probability $P[\,\cdot\,|\mathscr{F}_{t_j}^{\mathbb{X}}]$.

Suppose that $\mathbb{X}$ is unobservable but we observe $\mathbb{Y}^n$ instead. In this article, we will discuss estimation of the covariation $\langle X^1, X^2\rangle$ from the data of $\mathbb{Y}$. The idea is very simple. First we make a filter to estimate the state of X^α and then estimate the covariation between X^1 and X^2 from the filtered data. We will discuss asymptotic properties of our estimator when $a_n \to \infty$ as $n \to \infty$. Our method applies to an estimation problem with ultra high frequency data in finite time horizon. From another point of view, we can say the bracket of the drift $\mathbb{X}_t$ is estimated from the observations $\mathbb{Y}_{t_j}^n/a_n$ that are contaminated by the small noises $\mathbb{M}_t^n/a_n$.

Let $I_j = [t_{j-1}, t_j)$ for a sampling design $\Pi = (t_j)_{j=0,\dots,b_n}$ with $0 = t_0 < t_1 < \cdots < t_{b_n} = 1$, $b_n \in \mathbb{N}$, and let $h_j = t_j - t_{j-1}$. Numbers t_j and h_j are depending on n. Write $\Delta_j V = V_{t_j} - V_{t_{j-1}}$ for a process V. The matrix transpose will be denoted by $\star$. For a matrix A, $A^{\otimes} = A \otimes A = AA^\star$ and the (i, j)-element is denoted by $A^{\otimes(i,j)}$. We assume $\lim_{n\to\infty} b_n = \infty$.

We define a covariance estimator between intensity processes associated with sampling design Π by

$$S_n^{\alpha\beta} = \sum_{j=2}^{b_n} \left(\frac{\Delta_j \mathbb{Y}}{a_n h_j} - \frac{\Delta_{j-1}\mathbb{Y}}{a_n h_{j-1}}\right)^{\otimes(\alpha,\beta)} \qquad (\alpha, \beta = 1, 2).$$

This estimator depends on the scaling parameter a_n. We also define a correlation estimator of intensity processes associated with sampling design Π by

$$C_n^{12} = \frac{S_n^{12}}{\sqrt{S_n^{11} S_n^{22}}}.$$

That is,

$$C_n^{12} = \sum_{j=2}^{b_n} \left(\frac{\Delta_j \mathbb{Y}}{h_j} - \frac{\Delta_{j-1}\mathbb{Y}}{h_{j-1}}\right)^{\otimes(1,2)} \times \left[\sum_{j=2}^{b_n} \left(\frac{\Delta_j \mathbb{Y}}{h_j} - \frac{\Delta_{j-1}\mathbb{Y}}{h_{j-1}}\right)^{\otimes(1,1)} \sum_{j=2}^{b_n} \left(\frac{\Delta_j \mathbb{Y}}{h_j} - \frac{\Delta_{j-1}\mathbb{Y}}{h_{j-1}}\right)^{\otimes(2,2)}\right]^{-1/2}.$$

This estimator does not need the value of the scaling parameter a_n.

The aim of this paper is to study consistency and asymptotic mixed normality of these estimators. The organization of the paper is as follows. In the next Sect. 3, we state assumptions and the results of this paper. In Sect. 4, we prove the theorems.

3 Results

Hereafter we will assume $t_j = j/b_n$ for simplicity. In particular, $h_j = 1/b_n =: \delta_n$ independently of j, and

$$S_n^{\alpha\beta} = \frac{1}{(a_n\delta_n)^2}\sum_{j=2}^{b_n}\left(\Delta_j\mathbb{Y} - \Delta_{j-1}\mathbb{Y}\right)^{\otimes(\alpha,\beta)}$$
$$= \frac{1}{(a_n\delta_n)^2}\sum_{j=2}^{b_n}\left(\mathbb{Y}_{t_j} - 2\mathbb{Y}_{t_{j-1}} + \mathbb{Y}_{t_{j-2}}\right)^{\otimes(\alpha,\beta)}$$

for $\alpha, \beta = 1, 2$. The equi-spaced case will be considered but some extensions to irregular sampling cases are possible. We will often use h_j when emphasizing the length of I_j rather than the uniform length δ_n.

In this paper, we will consider the following condition.

[A] Process $\mathbb{X}$ admits the representation (1) for an $\mathbb{R}^2$-valued $\mathscr{F}_0$-measurable random variable $\mathbb{X}_0$ and coefficients $\mathbb{X}^\kappa$ ($\kappa = 0, 1$) such that $\mathbb{X}^0$ is a càdlàg $\mathbb{F}$-adapted process and that $\mathbb{X}^1$ has a representation

$$\mathbb{X}_t^1 = \mathbb{X}_0^1 + \int_0^t \mathbb{X}_s^{10}ds + \int_0^t \sum_{\kappa'=1}^{r'} \mathbb{X}_s^{1\kappa'}d\tilde{w}_s^{\kappa'} \quad (t \in [0,1]),$$

where $\mathbb{X}_0^1$ is an $\mathbb{R}^2 \otimes \mathbb{R}^r$-valued $\mathscr{F}_0$-measurable random variable, $\tilde{w} = (\tilde{w}^1, \dots, \tilde{w}^{r'})$ is an r'-dimensional $\mathbb{F}$-Wiener process (not necessarily independent of w), and $\mathbb{X}_t^{1\kappa'}$ ($\kappa' = 0, 1, \dots, r'$) are $\mathbb{R}^2 \otimes \mathbb{R}^r$-valued càdlàg $\mathbb{F}$-adapted processes.

We will consider the following conditions [B$^\flat$], [B], [B$'$] and [B$^\sharp$] sorted according to the rate of b_n.

[B$^\flat$]

(i) $\lim_{n\to\infty} b_n^2/a_n = 0$.
(ii) $\sum_{j=1}^{b_n} |\Delta_j \mathbb{M}^n|^2 = O_p(a_n)$ as $n \to \infty$.

[B] $\mathbb{M}^n = (M^{n,\alpha})_{\alpha=1,2}$ is a two-dimensional $\mathbb{F}$-local martingale with $\mathbb{M}_0^n = 0$ and such that

(i) $\lim_{n\to\infty} b_n^{5/2}/a_n = 0$.

(ii) $\sum_{j=1}^{b_n} |\Delta_j \mathbb{M}^n|^2 = O_p(a_n)$ as $n \to \infty$ and $\sup_{t\in[0,1]} |\Delta \mathbb{M}^n_t| \le c\, a_n^{1/2}$ for a constant c independent of n, where $\Delta \mathbb{M}^n_t = \mathbb{M}^n_t - \mathbb{M}^n_{t-}$.

(iii) The absolutely continuous (with respect to the Lebesgue measure a.s.) mapping $[0,1] \ni t \mapsto \langle \mathbb{M}^n, w\rangle_t \in \mathbb{R}^2 \otimes \mathbb{R}^r$ satisfies $\sup_{t\in[0,1]} |d\langle \mathbb{M}^n, w\rangle_t / dt| = O_p(b_n)$ as $n \to \infty$.

Here, $\langle \mathbb{M}^n, w\rangle_t$ is the $2 \times r$ matrix of angle brackets $\langle \mathbb{M}^{n,\alpha}, w^k\rangle_t$ for $\mathbb{M}^n = (\mathbb{M}^n)_{\alpha=1,2}$ and $w = (w^k)_{k=1,\dots,r}$.

[B′] $\mathbb{M}^n = (M^{n,\alpha})_{\alpha=1,2}$ is a two-dimensional $\mathbb{F}$-local martingale with $\mathbb{M}^n_0 = 0$ satisfying [B] (i), (iii) and

(ii′) $E\left[\sum_{j=1}^{b_n} |\mathbb{M}^n_{t_j} - \mathbb{M}^n_{t_{j-1}}|^2\right] = O(a_n)$ as $n \to \infty$.

[B$^\sharp$]

(i) $\lim_{n\to\infty} b_n^3/a_n = 0$.

(ii) $\sum_{j=1}^{b_n} |\Delta_j \mathbb{M}^n|^2 = O_p(a_n)$ as $n \to \infty$.

Remark 1 [B$^\flat$] is for consistency. [B], [B′] and [B$^\sharp$] are for asymptotic mixed normality. These conditions do not follow from each other.

Remark 2 For [B′] (ii′), it suffices to show that $\sup_{t\in[0,1-\delta_n]} E\left[|\mathbb{M}^n_{t+\delta_n} - \mathbb{M}^n_t|^2\right] = O(a_n\delta_n)$. We note that $a_n\delta_n = a_n/b_n = (a_n/b_n^{5/2}) \times b_n^{3/2} \to \infty$ when $b_n^{5/2}/a_n \to 0$.

Remark 3 In practice, $\langle \mathbb{M}^n\rangle_t$ can be of order a_n. Then [B] (iii) requires fairly fast asymptotic orthogonality between $\mathbb{M}^n$ and w. This condition is satisfied for example when $\mathbb{Y}^n$ are counting processes.

Let $x\tilde{\otimes}y = ((x_iy_j + x_jy_i)/2) \in \mathbb{R}^r \otimes \mathbb{R}^r$ for $x = (x_i)$, $y = (y_i) \in \mathbb{R}^r$, and let $x^{\tilde{\otimes}(\alpha,\beta)} = x^\alpha \tilde{\otimes} x^\beta$ for $x = (x_i^\alpha) \in \mathbb{R}^2 \otimes \mathbb{R}^r$. We write $x \cdot y = \sum_{i=1}^r x_i y_i$ for $x = (x_i)$, $y = (y_i) \in \mathbb{R}^r$, and $x \cdot y = \sum_{i,j=1}^r x_{i,j} y_{i,j}$ for $x = (x_{i,j})$, $y = (y_{i,j}) \in \mathbb{R}^r \otimes \mathbb{R}^r$.

Let $S_n = (S_n^{12}, S_n^{11}, S_n^{22})^\star$,

$$U^{\alpha\beta} = \frac{2}{3}\langle X^{\alpha,c}, X^{\beta,c}\rangle_1 = \frac{2}{3}\int_0^1 X_t^{\alpha 1} \cdot X_t^{\beta 1} dt \quad (\alpha, \beta = 1, 2)$$

and let $U = (U^{12}, U^{11}, U^{22})^\star$, where $X^{\alpha,c}$ is the continuous part of X^α and $X_t^{\alpha 1}$ is the α-th row of $\mathbb{X}^1_t$. Let

$$\Gamma = (\gamma^{pq})_{p,q=(1,2),(1,1),(2,2)}, \quad \gamma^{pq} = \int_0^1 (\mathbb{X}^1_s)^{\tilde{\otimes} p} \cdot (\mathbb{X}^1_s)^{\tilde{\otimes} q}\, ds.$$

For example,

$$(\mathbb{X}^1_s)^{\tilde{\otimes} p} = \left(\frac{X_{i,s}^{\alpha 1} X_{j,s}^{\beta 1} + X_{j,s}^{\alpha 1} X_{i,s}^{\beta 1}}{2}\right)_{i,j=1,\dots,r} \qquad (p = (\alpha, \beta))$$

and

$$\gamma^{(\alpha_1,\beta_1),(\alpha_2,\beta_2)} = \int_0^1 \sum_{i,j=1}^{r} \frac{X_{i,s}^{\alpha_1 1}X_{j,s}^{\beta_1 1} + X_{j,s}^{\alpha_1 1}X_{i,s}^{\beta_1 1}}{2} \frac{X_{i,s}^{\alpha_2 1}X_{j,s}^{\beta_2 1} + X_{j,s}^{\alpha_2 1}X_{i,s}^{\beta_2 1}}{2} ds,$$

where $X_{i,s}^{\alpha 1}$ is the (α, i)-element of $\mathbb{X}_s^1$.

Theorem 1 *(a) $S_n \to^p U$ as $n \to \infty$ under $[A]$ and $[B^\flat]$, where p denotes the convergence in probability.*
(b) Under $[A]$ and any one of $[B]$, $[B']$ and $[B^\sharp]$,

$$b_n^{1/2}\left(S_n - U\right) \to^{d_s} \Gamma^{\frac{1}{2}}\zeta$$

as $n \to \infty$, where ζ is an $\mathbb{R}^3$-valued standard normal variable independent of $\mathscr{F}$ and d_s denotes the $\mathscr{F}$-stable convergence.

Proof of Theorem 1 is given in Sect. 4.

Let $R = U^{12}/\sqrt{U^{11}U^{22}}$. For the correlation estimator, we have

Theorem 2 *Suppose that $U^{11}U^{22} \neq 0$ a.s. Then*

(a) $C_n^{12} \to^p R$ as $n \to \infty$ under $[A]$ and $[B^\flat]$.
(b) Under $[A]$ and any one of $[B]$, $[B']$ and $[B^\sharp]$,

$$b_n^{1/2}\left(C_n^{12} - R\right) \to^{d_s} \left(\frac{1}{\sqrt{U^{11}U^{22}}}, \frac{-U^{12}}{2\sqrt{(U^{11})^3U^{22}}}, \frac{-U^{12}}{2\sqrt{U^{11}(U^{22})^3}}\right) \cdot \Gamma^{\frac{1}{2}}\zeta$$

as $n \to \infty$, where ζ is an $\mathbb{R}^3$-valued standard normal variable independent of $\mathscr{F}$.

Proof These results are easy consequences of Theorem 1. Apply the so-called Delta-method for stable convergence to the second assertion. □

Example 3 (Cox process) Let $\mathbb{F}' = (\mathscr{F}_t')_{t\in[0,1]}$ be a filtration on a probability space $(\Omega', \mathscr{F}', P')$. Suppose that $\mathbb{X}$ is a two-dimensional nonnegative Itô process given by (1) for an r-dimensioal $\mathbb{F}'$-Wiener process w and an $\mathscr{F}_0'$-measurable random variable $\mathbb{X}_0$, and two-dimensional $\mathbb{F}'$-predictable processes $\mathbb{X}_t^\kappa$ $(\kappa = 0, 1)$. Let μ^α $(\alpha = 1, 2)$ be independent Poisson random measures on $\mathbb{R}_+ \times \mathbb{R}_+$ defined on $(\Omega'', \mathscr{F}'', P'')$ with intensity measure $\nu(dt, dx) = dtdx$. Let $Y_t^{n,\alpha} = \int_0^t \int 1_{[0,a_nX_s^\alpha]}(x)\mu^\alpha(ds, dx)$. Let $\Omega = \Omega' \times \Omega''$, $\mathscr{F} = \mathscr{F}' \times \mathscr{F}''$, $P = P' \times P''$, and $\mathbb{F} = (\mathscr{F}_t)_{t\in[0,1]}$ with $\mathscr{F}_t = \cap_{u>t}(\mathscr{F}_u' \times \mathscr{F}_u'')$, where $\mathscr{F}_t''$ is the filtration on Ω'' generated by $\mu^\alpha([0, s] \times B)$ $(s \leq t,\ B \in \mathbb{B}(\mathbb{R}_+),\ \alpha \in \{1, 2\})$. Then, $\mathbb{X}$-conditionally, $Y^{n,\alpha}$ is a time-inhomogeneous Poisson process with intensity function $a_nX_t^\alpha$. The process w naturally extended to Ω is an r-dimensional $\mathbb{F}$-Wiener process, $\mathbb{M}_t^{n,\alpha} = \int_0^t \int 1_{[0,a_nX_s^\alpha]}(x)(\mu^\alpha - \nu)(ds, dx)$ and $\langle \mathbb{M}^n, w\rangle = 0$, that is, [B] (iii) holds. Condition [B] (ii) is easily verified by localization.

4 Proof

By a localization argument, we may assume that there exists a constant K,

$$\sup_{t\in[0,1]}\left(|\mathbb{X}_{t-}|+|\mathbb{X}^0_{t-}|+|\mathbb{X}^1_{t-}|+|\mathbb{X}^{10}_{t-}|+\sum_{\kappa'=1}^{r'}|\mathbb{X}^{1\kappa'}_{t-}|\right)\leq K.$$

We write

$$\tilde{V}_j=\frac{\Delta_j V}{a_n\delta_n}$$

for a stochastic process V. Define $\chi_t=(\chi^\alpha_t)_{\alpha=1,2}$ by

$$\chi^\alpha_t=\int_0^t a_n X^\alpha_s ds.$$

Let $T_n=(T^{12}_n,T^{11}_n,T^{22}_n)^\star$ for

$$T^{\alpha\beta}_n=\frac{1}{(a_n\delta_n)^2}\sum_{j=2}^{b_n}\left(\Delta_j\chi-\Delta_{j-1}\chi\right)^{\otimes(\alpha,\beta)}$$

for $\alpha,\beta=1,2$. By definition, $M^{n,\alpha}=Y^{n,\alpha}-Y^{n,\alpha}_0-\chi^\alpha$ is an $\mathbb{F}$-martinagle for each $\alpha=1,2$.

Lemma 1 *(a)* $S_n-T_n\to^p 0$ *if* $[B^\flat]$ *holds.*
(b) $b_n^{1/2}(S_n-T_n)\to^p 0$ *if any one of* $[B]$, $[B']$ *and* $[B^\sharp]$ *holds.*

Proof Let

$$s^{\alpha\beta}_j=\frac{1}{(a_n\delta_n)^2}\left(\Delta_j\mathbb{Y}^n-\Delta_{j-1}\mathbb{Y}^n\right)^{\otimes(\alpha,\beta)}=\left(\tilde{\mathbb{Y}}^n_j-\tilde{\mathbb{Y}}^n_{j-1}\right)^{\otimes(\alpha,\beta)}$$

and

$$t^{\alpha\beta}_j=\frac{1}{(a_n\delta_n)^2}\left(\Delta_j\chi-\Delta_{j-1}\chi\right)^{\otimes(\alpha,\beta)}=\left(\tilde{\chi}_j-\tilde{\chi}_{j-1}\right)^{\otimes(\alpha,\beta)}.$$

By definition,

$$\begin{aligned}s^{\alpha\beta}_j-t^{\alpha\beta}_j&=(\tilde{\mathbb{M}}^n_j-\tilde{\mathbb{M}}^n_{j-1})^{\otimes(\alpha,\beta)}+(\tilde{M}^\alpha_j-\tilde{M}^\alpha_{j-1})(\tilde{\chi}^\beta_j-\tilde{\chi}^\beta_{j-1})\\&\quad+(\tilde{\chi}^\alpha_j-\tilde{\chi}^\alpha_{j-1})(\tilde{M}^\beta_j-\tilde{M}^\beta_{j-1}).\end{aligned}\quad(2)$$

From (2),

$$|S_n - T_n| \le \sum_{j=2}^{b_n} |\tilde{\mathbb{M}}_j^n - \tilde{\mathbb{M}}_{j-1}^n|^2 + 2\left\{\sum_{j=2}^{b_n} |\tilde{\mathbb{M}}_j^n - \tilde{\mathbb{M}}_{j-1}^n|^2\right\}^{1/2}\left\{\sum_{j=2}^{b_n} |\tilde{\chi}_j - \tilde{\chi}_{j-1}|^2\right\}^{1/2}.$$

Since $\sum_{j=1}^{b_n} |\Delta_j \mathbb{M}^n|^2 = O_p(a_n)$ in any case, we have

$$\sum_{j=2}^{b_n} |\tilde{\mathbb{M}}_j^n - \tilde{\mathbb{M}}_{j-1}^n|^2 \le 4\sum_{j=1}^{b_n} |\tilde{\mathbb{M}}_j^n|^2 = O_p(b_n^2/a_n). \tag{3}$$

It is easy to see that

$$E[|\tilde{\chi}_j - \tilde{\chi}_{j-1}|^{2m}] \le \frac{C_{m,K}}{b_n^m} \tag{4}$$

for $m \in \mathbb{N}$. In particular, $\sum_{j=1}^{b_n} |\tilde{\chi}_j - \tilde{\chi}_{j-1}|^2 = O_p(1)$. Thus, we have $S_n - T_n \to^p 0$ under [B$^\flat$], as well as $b_n^{1/2}(S_n - T_n) \to^p 0$ under [B$^\sharp$].

Suppose that [B] is satisfied. We will show (b) by estimating the terms on the right-hand side of (2). We have $\tilde{\chi}_j - \tilde{\chi}_{j-1} = \Phi_j + \Phi'_j$, where

$$\Phi_j = \mathbb{X}_{t_{j-2}}^1 \delta_n^{-1} \int_{t_{j-2}}^{t_{j-1}} \int_s^{s+\delta_n} dw_r\, ds$$

and

$$\Phi'_j = \delta_n^{-1} \int_{t_{j-2}}^{t_{j-1}} \left\{\int_s^{s+\delta_n} (\mathbb{X}_r^1 - \mathbb{X}_{t_{j-2}}^1) dw_r + \int_s^{s+\delta_n} \mathbb{X}_r^0\, dr\right\} ds.$$

It is easy to see that $\sum_{j=2}^{b_n} |\Phi'_j|^2 = O_p(\delta_n)$. Therefore, by (3),

$$\begin{aligned} b_n^{1/2}\left|\sum_{j=2}^{b_n} (\tilde{\mathbb{M}}_j^n - \tilde{\mathbb{M}}_{j-1}^n) \otimes \Phi'_j\right| &\le b_n^{1/2}\left\{\sum_{j=2}^{b_n} |\tilde{\mathbb{M}}_j^n - \tilde{\mathbb{M}}_{j-1}^n|^2\right\}^{1/2}\left\{\sum_{j=2}^{b_n} |\Phi'_j|^2\right\}^{1/2} \\ &= O_p(\sqrt{b_n^2/a_n}) = o_p(1). \end{aligned}$$

Due to [B] (iii), for any sequence of positive numbers K_n tending to ∞, we have $\lim_{n\to\infty} P[\rho_n < 1] = 0$ for

$$\rho_n = \inf\left\{t;\ b_n^{-1} \sup_{r\in[0,t]} \left|\frac{d\langle \mathbb{M}^n, w\rangle}{dr}(r)\right| \ge K_n\right\} \wedge 1.$$

By the continuity of $\langle \mathbb{M}^n, w\rangle$, $b_n^{-1}\sup_{r\in[0,1]}\left|\frac{d\langle \mathbb{M}^n, w^{\rho_n}\rangle}{dr}(r)\right| \le K_n$, where $w_t^\tau = w_{(t\wedge\tau)}$ for a stopping time τ. Let

$$\tau_n = \rho_n \wedge \sigma^n(k_n) \wedge \inf\left\{t;\ \sup_{\substack{u,v\in[0,t]\\|u-v|\le\delta_n}} |w_v - w_u| \ge \delta_n^{1/3}\right\}$$

$$\wedge \inf\left\{t;\ a_n^{-1}\sum_{j=1}^{b_n}\left|\mathbb{M}^n_{t_j\wedge t} - \mathbb{M}^n_{t_{j-1}\wedge t}\right|^2 \ge \delta_n^{-1/6}\right\},$$

where $(\sigma^n(k))_{k\in\mathbb{N}}$ is a localizing sequence for the locally bounded process $\mathbb{M}^n$ and $k_n \in \mathbb{N}$. Since $\lim_{k\to\infty} P[\sigma^n(k) = 1] = 1$ for each $n \in \mathbb{N}$, there exists a sequence $(k_n)_{n\in\mathbb{N}}$ such that $\lim_{n\to\infty} P[\sigma^n(k_n) = 1] = 1$. Obviously,

$$P[\tau_n < 1] \to 0 \quad (n \to \infty). \tag{5}$$

We shall write Φ_j below for Φ_j defined for the stopped w^{τ_n} in place of w, similarly $\mathbb{M}^n$ and $\tilde{\mathbb{M}}^n$ for their stopped versions respectively. By Doob's inequality, [B] (ii) holds for the stopped $\mathbb{M}^n$. For $j' \in \{j, j-1\}$,

$$\begin{aligned}
\left|E\left[\tilde{\mathbb{M}}^n_{j'} \otimes \Phi_j \middle| \mathscr{F}_{t_{j-2}}\right]\right| &\le C\,(a_n\delta_n)^{-1}|\mathbb{X}^1_{t_{j-2}}|\,\delta_n^{-1} \\
&\quad \left|\int_{t_{j-2}}^{t_{j-1}} E\left[\int_{t_{j-2}}^{t_j} 1_{(t_{j'-1},t_{j'}]\cap(s,s+\delta_n]}\, d\langle \mathbb{M}^n, w^{\tau_n}\rangle_r \middle| \mathscr{F}_{t_{j-2}}\right] ds\right| \\
&\le C\,|\mathbb{X}^1_{t_{j-2}}|\, E\left[\sup_{r\in[0,1]}\left|\frac{d\langle \mathbb{M}^n, w^{\rho_n}\rangle}{dr}(r)\right| \middle| \mathscr{F}_{t_{j-2}}\right] a_n^{-1} \\
&\le C_K b_n a_n^{-1} K_n.
\end{aligned}$$

Therefore

$$b_n^{1/2}\sum_{j=2}^{b_n}\left|E\left[(\tilde{\mathbb{M}}^n_j - \tilde{\mathbb{M}}^n_{j-1}) \otimes \Phi_j \middle| \mathscr{F}_{t_{j-2}}\right]\right| \to^p 0 \tag{6}$$

for $K_n = (b_n^{5/2}/a_n)^{-1/2}$. We have

$$\begin{aligned}
(b_n^{1/2})^2\sum_{j=2}^{b_n} E\left[|\tilde{\mathbb{M}}^n_{j'}|^2|\Phi_j|^2\right] &\le b_n(a_n\delta_n)^{-2}E\left[\sum_{j=2}^{b_n}|\Delta_{j'}\mathbb{M}^n|^2\right](K\delta_n^{1/3})^2 \\
&= O\left(b_n(a_n\delta_n)^{-2}(a_n\delta_n^{-1/6} + a_n)\delta_n^{2/3}\right) \\
&= O(b_n^{5/2}/a_n) \;=\; o(1).
\end{aligned} \tag{7}$$

From (6), (7) and (5), we obtain $b_n^{1/2}(S_n - T_n) \to^p 0$, by dividing the sum $S_n - T_n$ into sums for even j's and odd j's.

Under [B′], for the square-integrable martingales $\mathbb{M}^n$, we can obtain (6) and (7) with τ_n made of ρ_n and the stopping time for increments of w. □

Obviously,

$$\begin{aligned}
&\frac{1}{h_j}\int_{I_j}\left(\int_{t_{j-2}}^{t} V_s\,dw_s\right)dt - \frac{1}{h_{j-1}}\int_{I_{j-1}}\left(\int_{t_{j-2}}^{t} V_s\,dw_s\right)dt \\
&= \frac{1}{h_j}\int_{I_j}\left(\int_{t_{j-2}}^{t_{j-1}} V_s\,dw_s\right)dt + \frac{1}{h_j}\int_{I_j}\left(\int_{t_{j-1}}^{t} V_s\,dw_s\right)dt - \frac{1}{h_{j-1}}\int_{I_{j-1}}\left(\int_{t_{j-2}}^{t} V_s\,dw_s\right)dt \\
&= \int_{I_{j-1}} H_j(s)V_s\,dw_s + \int_{I_j} K_j(s)V_s\,dw_s
\end{aligned}$$

for a suitably integrable process V_s, where

$$H_j(s) = 1 - \frac{t_{j-1}-s}{h_{j-1}} \quad \text{and} \quad K_j(s) = \frac{t_j - s}{h_j}.$$

We write $\mathbb{X}^\kappa = (X^{\alpha\kappa})_{\alpha=1,2}$ for $\kappa = 0, 1$ with $X^{\alpha\kappa}$ taking values in $\mathbb{R}^r$, and $\mathbb{X}^{1\kappa'} = (X^{\alpha 1\kappa'})_{\alpha=1,2;\kappa'=1,\dots,r'}$ with $X^{\alpha 1\kappa'}$ taking values in $\mathbb{R}^r$. Let

$$\begin{aligned}
A_j^\alpha &= X_{t_{j-2}}^{\alpha 1}\int_{I_{j-1}} H_j(s)\,dw_s + X_{t_{j-2}}^{\alpha 1}\int_{I_j} K_j(s)\,dw_s, \\
B_j^\alpha &= \frac{1}{h_j}\int_{I_j}\int_{t_{j-2}}^{t}\left(X_s^{\alpha 1} - X_{t_{j-2}}^{\alpha 1}\right)dw_s\,dt - \frac{1}{h_{j-1}}\int_{I_{j-1}}\int_{t_{j-2}}^{t}\left(X_s^{\alpha 1} - X_{t_{j-2}}^{\alpha 1}\right)dw_s\,dt, \\
C_j^\alpha &= \frac{1}{h_j}\int_{I_j}\int_{t_{j-2}}^{t} X_{t_{j-2}}^{\alpha 0}ds\,dt - \frac{1}{h_{j-1}}\int_{I_{j-1}}\int_{t_{j-2}}^{t} X_{t_{j-2}}^{\alpha 0}ds\,dt \text{ and} \\
D_j^\alpha &= \frac{1}{h_j}\int_{I_j}\int_{t_{j-2}}^{t}\left(X_s^{\alpha 0} - X_{t_{j-2}}^{\alpha 0}\right)ds\,dt - \frac{1}{h_{j-1}}\int_{I_{j-1}}\int_{t_{j-2}}^{t}\left(X_s^{\alpha 0} - X_{t_{j-2}}^{\alpha 0}\right)ds\,dt.
\end{aligned}$$

Let

$$\hat{\iota}_j^\alpha = \frac{\Delta_j \chi^\alpha}{a_n h_j} - \frac{\Delta_{j-1}\chi^\alpha}{a_n h_{j-1}} = \frac{1}{a_n\delta_n}\left(\Delta_j\chi^\alpha - \Delta_{j-1}\chi^\alpha\right).$$

Then $\hat{\iota}_j^\alpha = A_j^\alpha + B_j^\alpha + C_j^\alpha + D_j^\alpha$. Note that $T_n^{\alpha\beta} = \sum_{j=2}^{b_n}\hat{\iota}_j^\alpha\hat{\iota}_j^\beta$. In what follows, $\rho = 0$ for the proof of the consistency and $\rho = 1/2$ for the proof of the mixed normality of the estimators.

Lemma 2 $b_n^\rho \sum_{j=2}^{b_n} F_j G_j \to^p 0$ *for all pairs* $(F_j, G_j) \in \{A_j^\alpha, B_j^\alpha, C_j^\alpha, D_j^\alpha; \alpha = 1, 2\}^2 \backslash \{(A_j^\alpha, A_j^\beta); \alpha, \beta = 1, 2\}$.

Proof Let $\mathcal{Q}_n = b_n^\rho \sum_{j=2}^{b_n} F_j G_j$. For $F_j = A_j^\alpha$ and $G_j = C_j^\beta$, $E[\mathcal{Q}_n^2] = O(b_n^{1+2\rho-3}) = o(1)$. For $F_j = A_j^\alpha$ and $G_j = D_j^\beta$, $E[|\mathcal{Q}_n|] = O(b_n^{1+\rho-3/2})o(1) = o(1)$ because $\mathbb{X}^0$ is càdlàg; see Billingsley [7] Lemma 1 of Sect. 12. For $F_j = A_j^\alpha$ and $G_j = B_j^\beta$, the principal part consists of $\mathcal{Q}_n$ with $F_j = A_j^\alpha$ and $G_j = \dot{B}_j^\beta$ with $\dot{B}_j^\beta$ taking the form of

$$\dot{B}_j^\beta = \frac{1}{h_m} \int_{I_m} \int_{t_{j-2}}^{t} \int_{t_{j-2}}^{s} X_u^{\beta 1 \kappa'} d\tilde{w}_u^{\kappa'} \, dw_s \, dt$$

for $m = j-1, j$. Since $X^{\beta 1 \kappa'}$ is càdlàg , considering L^1-estimate of the gap, we can replace this functional by

$$\dot{B}_j^\beta = \frac{X_{t_{j-2}}^{\beta 1 \kappa'}}{h_m} \int_{I_m} \int_{t_{j-2}}^{t} \int_{t_{j-2}}^{s} d\tilde{w}_u^{\kappa'} \, dw_s \, dt.$$

Then we can find orthogonality between terms in $\mathcal{Q}_n$ to obtain $E[\mathcal{Q}_n^2] = O(b_n^{1+2\rho-3}) = o(1)$. It is easy to show that the L^1-norm vanishes in the limit for other pairs. □

Let $Q_n^{1,\alpha\beta} = \sum_{j=2}^{b_n} A_j^\alpha A_j^\beta$. Then Itô's formula and simple calculus yield $Q_n^{1,\alpha\beta} = Q_n^{2,\alpha\beta} + Q_n^{3,\alpha\beta}$, where

$$Q_n^{2,\alpha\beta} = \sum_{j=2}^{b_n} \left\{ X_{t_{j-2}}^{\alpha 1} \cdot X_{t_{j-2}}^{\beta 1} \frac{h_{j-1}}{3} + X_{t_{j-2}}^{\alpha 1} \cdot X_{t_{j-2}}^{\beta 1} \frac{h_j}{3} \right\}$$

and

$$Q_n^{3,\alpha\beta} = \sum_{j=2}^{b_n} 2(\mathbb{X}_{t_{j-2}}^1)^{\tilde{\otimes}(\alpha,\beta)} \cdot \bigg(\int_{I_{j-1}} \int_{t_{j-2}}^{t} H_j(s)\, dw_s \otimes H_j(t)\, dw_t$$
$$+ \int_{I_j} \int_{t_{j-1}}^{t} K_j(s)\, dw_s \otimes K_j(t)\, dw_t + \int_{I_{j-1}} H_j(s)\, dw_s \otimes \int_{I_j} K_j(s)\, dw_s \bigg).$$

Lemma 3 $b_n^\rho (Q_n^{2,\alpha\beta} - U^{\alpha\beta}) \to^p 0$.

Proof It is sufficient to prove that

$$\sum_{j=1}^{b_n} b_n^\rho \int_{I_j} \{(\mathbb{X}_s^1)^{\otimes(\alpha,\beta)} - (\mathbb{X}_{t_{j-1}}^1)^{\otimes(\alpha,\beta)}\} ds \to^p 0.$$

By definition,

$$(\mathbb{X}_s^1)^{\otimes(\alpha,\beta)} - (\mathbb{X}_{t_{j-1}}^1)^{\otimes(\alpha,\beta)} = (\mathbb{X}_s^1 - \mathbb{X}_{t_{j-1}}^1)^{\otimes(\alpha,\beta)} + ((\mathbb{X}_s^1 - \mathbb{X}_{t_{j-1}}^1)\mathbb{X}_{t_{j-1}}^{1\ \star})^{(\alpha,\beta)}$$
$$+(\mathbb{X}_{t_{j-1}}^1(\mathbb{X}_s^1 - \mathbb{X}_{t_{j-1}}^1)^\star)^{(\alpha,\beta)}.$$

Obviously,

$$\left|\sum_{j=1}^{b_n} E\left[b_n^\rho \int_{I_j} (\mathbb{X}_s^1 - \mathbb{X}_{t_{j-1}}^1)^{\otimes(\alpha,\beta)}\, ds \middle| \mathscr{F}_{t_{j-1}}\right]\right| = O_p(b_n^{1+\rho-2}).$$

By assumption [A] on $\mathbb{X}^1$ and the property of stochastic integral,

$$\left|\sum_{j=1}^{b_n} E\left[b_n^\rho \int_{I_j} (X_s^{\alpha 1} - X_{t_{j-1}}^{\alpha 1})(X_{t_{j-1}}^{\beta 1})^\star\, ds | \mathscr{F}_{t_{j-1}}\right]\right|$$
$$= \left|\sum_{j=1}^{b_n} b_n^\rho \int_{I_j} E\left[\left(\int_{t_{j-1}}^s X_u^{\alpha 10}\, du\right)\middle| \mathscr{F}_{t_{j-1}}\right](X_{t_{j-1}}^{\beta 1})^\star\, ds\right| = O_p(b_n^{1+\rho-2}).$$

Therefore, we have

$$\sum_{j=1}^{b_n} E\left[b_n^\rho \int_{I_j} \left((\mathbb{X}_s^1)^{\otimes(\alpha,\beta)} - (\mathbb{X}_{t_{j-1}}^1)^{\otimes(\alpha,\beta)}\right) ds \middle| \mathscr{F}_{t_{j-1}}\right] \to^p 0. \tag{8}$$

From the continuity of $\mathbb{X}^1$, it holds that

$$E\left[\left|\sum_{j=1}^{b_n} E\left[\left(b_n^\rho \int_{I_j} (\mathbb{X}_s^1)^{\otimes(\alpha,\beta)} - (\mathbb{X}_{t_{j-1}}^1)^{\otimes(\alpha,\beta)}\, ds\right)^2 \middle| \mathscr{F}_{t_{j-1}}\right]\right|\right] = O(b_n^{1+2\rho-2})o(1) = o(1).$$

Therefore, we have

$$\sum_{j=1}^{b_n} E\left[\left(b_n^\rho \int_{I_j} (\mathbb{X}_s^1)^{\otimes(\alpha,\beta)} - (\mathbb{X}_{t_{j-1}}^1)^{\otimes(\alpha,\beta)}\, ds\right)^2 \middle| \mathscr{F}_{t_{j-1}}\right] \to^p 0. \tag{9}$$

By (8) and (9), we conclude the proof. □

By recombination, we obtain $Q_n^{3,\alpha\beta} = V_n^{\alpha\beta} + o_p(b_n^{-1/2})$ with

$$V_n^{\alpha\beta} = \sum_{j=2}^{b_n-1} \int_{I_j} \Big\{ 2(\mathbb{X}^1_{t_{j-1}})^{\tilde{\otimes}(\alpha,\beta)} \cdot \int_{t_{j-1}}^{t} H_{j+1}(s)\, dw_s H_{j+1}(t) \otimes dw_t$$

$$+2(\mathbb{X}^1_{t_{j-2}})^{\tilde{\otimes}(\alpha,\beta)} \cdot \int_{t_{j-1}}^{t} K_j(s)\, dw_s K_j(t) \otimes dw_t$$

$$+2(\mathbb{X}^1_{t_{j-2}})^{\tilde{\otimes}(\alpha,\beta)} \cdot \int_{I_{j-1}} H_j(s)\, dw_s K_j(t) \otimes dw_t \Big\}.$$

Lemma 4

$$b_n^{\frac{1}{2}} (V_n^{12}, V_n^{11}, V_n^{22})^\star \to^{d_s} \Gamma^{1/2}\zeta.$$

Proof Write $v_j^{\alpha\beta} = \int_{I_j} L_j^{\alpha\beta} dw_t$, where

$$L_j^{\alpha\beta}(t)u = 2(\mathbb{X}_{t_{j-1}})^{\tilde{\otimes}(\alpha,\beta)} \cdot \int_{t_{j-1}}^{t} H_{j+1}(s)\, dw_s H_{j+1}(t) \otimes u$$

$$+2(\mathbb{X}^1_{t_{j-2}})^{\tilde{\otimes}(\alpha,\beta)} \cdot \int_{t_{j-1}}^{t} K_j(s)\, dw_s K_j(t) \otimes u$$

$$+2(\mathbb{X}^1_{t_{j-2}})^{\tilde{\otimes}(\alpha,\beta)} \cdot \int_{I_{j-1}} H_j(s)\, dw_s K_j(t) \otimes u$$

for $u \in \mathbb{R}^r$. Then, $V_n^{\alpha\beta} = \sum_{j=2}^{b_n-1} v_j^{\alpha\beta}$. Set $f_t^{\alpha\beta} = 2(\mathbb{X}_t^1)^{\tilde{\otimes}(\alpha,\beta)}$, $V_n = (V_n^{12}, V_n^{11}, V_n^{22})^\star$, $v_j = (v_j^{12}, v_j^{11}, v_j^{22})^\star$ and let A, B take values in $\{12, 11, 22\}$.

Let $T \in [0, 1]$. Note that $v_j^{\alpha\beta}$ is $\mathscr{F}_{t_j}$-measurable. Let $[b_nT]' = \min\{[b_nT], (b_n - 1)\}$. We have

$$\sum_{j=2}^{[b_nT]'} E[b_n^{1/2} v_j^{\alpha\beta} | \mathscr{F}_{t_{j-1}}] = 0.$$

By estimating the L^2-norm of

$$R_n = \sum_{j=2}^{[b_nT]'} b_n \int_{I_j} f_{t_{j-2}}^A \cdot f_{t_{j-2}}^B \Big\{ \Big| \int_{I_{j-1}} H_j(s)\, dw_s \Big|^2 - \int_{I_{j-1}} H_j(s)^2\, ds \Big\} K_j(t)^2\, dt,$$

it is easy to see that $R_n = o_p(1)$ as $n \to \infty$. Now it holds that

$$
\begin{aligned}
&\sum_{j=2}^{[b_nT]'} E[(b_n^{1/2}v_j)^{\otimes(A,B)}|\mathscr{F}_{t_{j-1}}] \\
&= \sum_{j=2}^{[b_nT]'} b_n \int_{I_j} \Big\{ f_{t_{j-1}}^A \cdot f_{t_{j-1}}^B \int_{t_{j-1}}^t H_{j+1}(s)^2\, ds H_{j+1}(t)^2 \\
&\quad + f_{t_{j-1}}^A \cdot f_{t_{j-2}}^B \int_{t_{j-1}}^t H_{j+1}(s)K_j(s)\, ds H_{j+1}(t)K_j(t) \\
&\quad + f_{t_{j-2}}^A \cdot f_{t_{j-1}}^B \int_{t_{j-1}}^t K_j(s)H_{j+1}(s)\, ds K_j(t)H_{j+1}(t) \\
&\quad + f_{t_{j-2}}^A \cdot f_{t_{j-2}}^B \int_{t_{j-1}}^t K_j(s)^2\, ds K_j(t)^2 \\
&\quad + f_{t_{j-2}}^A \cdot f_{t_{j-2}}^B \int_{I_{j-1}} H_j(s)^2\, ds K_j(t)^2 \Big\}\, dt + R_n \\
&= \sum_{j=2}^{[b_nT]'} b_n \Big\{ f_{t_{j-1}}^A \cdot f_{t_{j-1}}^B \frac{h_j^2}{18} + f_{t_{j-1}}^A \cdot f_{t_{j-2}}^B \frac{h_j^2}{72} \\
&\quad + f_{t_{j-2}}^A \cdot f_{t_{j-1}}^B \frac{h_j^2}{72} + f_{t_{j-2}}^A \cdot f_{t_{j-2}}^B \frac{h_j^2}{18} + f_{t_{j-2}}^A \cdot f_{t_{j-2}}^B \frac{h_{j-1}h_j}{9} \Big\} + o_p(1) \\
&\to^p \int_0^T \frac{1}{4} f_s^A \cdot f_s^B\, ds.
\end{aligned}
$$

Similarly,

$$
\begin{aligned}
\sum_{j=2}^{[b_nT]'} E\Big[b_n^{1/2}v_j^{\alpha\beta}\Delta_j w\Big|\mathscr{F}_{t_{j-1}}\Big] &= \sum_{j=2}^{[b_nT]'} E\Big[b_n^{1/2}\int_{I_j} L_j^{\alpha\beta}(t)\, dt\Big|\mathscr{F}_{t_{j-1}}\Big] \\
&= \sum_{j=2}^{[b_nT]'} b_n^{1/2} \int_{I_j} f_{t_{j-2}}^{\alpha\beta} \cdot \int_{I_{j-1}} H_j(s)\, dw_s K_j(t)\, dt = o_p(1),
\end{aligned}
$$

which is the asymptotic orthogonality between $\sum_{j=2}^{[b_nT]'} b_n^{1/2}v_j^{\alpha\beta}$ and w. For the Lindeberg type condition,

$$
\sum_{j=2}^{[b_nT]'} E\Big[\big|b_n^{1/2}v_j\big|^2 1_{\{|b_n^{1/2}v_j|>\epsilon\}}\Big|\mathscr{F}_{t_{j-1}}\Big] \le \sum_{j=2}^{b_n} E\Big[\frac{1}{\epsilon^2}\big|b_n^{1/2}v_j\big|^4\Big|\mathscr{F}_{t_{j-1}}\Big] = O_p(b_n^{-1}).
$$

For an $\mathbb{F}$- bounded martingale N which is orthogonal to w, i.e. $\langle N, w\rangle = 0$,

$$\sum_{j=2}^{[b_nT]'} E\left[b_n^{1/2} v_j^{\alpha\beta} \Delta_j N \middle| \mathscr{F}_{t_{j-1}}\right] = \sum_{j=2}^{[b_nT]'} E\left[b_n^{1/2} \int_{I_j} L_j^{\alpha\beta}(t)\, dw_t \int_{I_j} dN_t \middle| \mathscr{F}_{t_{j-1}}\right] = 0.$$

After all that, by Theorem 3.1 of [22], we conclude that $b_n^{1/2} V_n \to^{d_s} \Gamma^{1/2}\zeta$. □

Proof of Theorem 1
If $b_n^{5/2}/a_n \to 0$, *by Lemmas* 1, 2, 3 *and* 4, *we have*

$$b_n^{1/2}\Big(S_n - U\Big) \to^{d_s} \Gamma^{1/2}\zeta.$$

From Lemmas 1, 2 *and* 3 *and the above, it holds that* if $b_n^2/a_n \to 0$, *then* $S_n \to^p U$ *as* $n \to \infty$. □

References

1. Abergel, F., Huth, N.: High frequency lead/lag relationships. Empirical Facts (2012)
2. Abergel, F., Jedidi, A.: A mathematical approach to order book modeling. Int. J. Theor. Appl. Financ. **16**(05), 1350,025 (2013)
3. Abergel, F., Jedidi, A.: Long-time behavior of a Hawkes process-based limit order book. SIAM J. Finan. Math. **6**(1), 1026–1043 (2015)
4. Bacry, E., Delattre, S., Hoffmann, M., Muzy, J.F.: Modelling microstructure noise with mutually exciting point processes. Quant. Financ. **13**(1), 65–77 (2013)
5. Bibinger, M.: Efficient covariance estimation for asynchronous noisy high-frequency data. Scand. J. Stat. **38**(1), 23–45 (2011)
6. Bibinger, M.: An estimator for the quadratic covariation of asynchronously observed Itô processes with noise: asymptotic distribution theory. Stochast. Process. Appl. **122**(6), 2411–2453 (2012)
7. Billingsley, P.: Convergence of Probability Measures. John Wiley & Sons (2013)
8. Bowsher, C.G.: Modelling security market events in continuous time: intensity based, multivariate point process models. J. Econometrics **141**(2), 876–912 (2007)
9. Christensen, K., Kinnebrock, S., Podolskij, M.: Pre-averaging estimators of the ex-post covariance matrix in noisy diffusion models with non-synchronous data. J. Econometrics **159**(1), 116–133 (2010)
10. Cont, R., Stoikov, S., Talreja, R.: A stochastic model for order book dynamics. Oper. Rese. **58**(3), 549–563 (2010)
11. De Jong, F., Nijman, T.: High frequency analysis of lead-lag relationships between financial markets. J. Empir. Financ. **4**(2), 259–277 (1997)
12. Eberlein, E.: Application of generalized hyperbolic Lévy motions to finance. In: Lévy processes, pp. 319–336. Springer (2001)
13. Eberlein, E., Keller, U.: Hyperbolic distributions in finance. Bernoulli **1**, 281–299 (1995)
14. Eberlein, E., Prause, K.: The generalized hyperbolic model: financial derivatives and risk measures. In: Mathematical Finance—Bachelier Congress 2000, pp. 245–267. Springer (2002)
15. Epps, T.W.: Comovements in stock prices in the very short run. J. Am. Stat. Assoc. **74**(366a), 291–298 (1979)

16. Griffin, J.E., Oomen, R.C.: Covariance measurement in the presence of non-synchronous trading and market microstructure noise. J. Econometrics **160**(1), 58–68 (2011)
17. Hayashi, T., Yoshida, N.: On covariance estimation of non-synchronously observed diffusion processes. Bernoulli **11**(2), 359–379 (2005)
18. Hayashi, T., Yoshida, N.: Asymptotic normality of a covariance estimator for nonsynchronously observed diffusion processes. Ann. Inst. Stat. Math. **60**(2), 367–406 (2008)
19. Hayashi, T., Yoshida, N.: Nonsynchronous covariation process and limit theorems. Stochast. Process. Appl. **121**(10), 2416–2454 (2011)
20. Hewlett, P.: Clustering of order arrivals, price impact and trade path optimisation. In: Workshop on Financial Modeling with Jump processes, Ecole Polytechnique, pp. 6–8 (2006)
21. Hoffmann, M., Rosenbaum, M., Yoshida, N.: Estimation of the lead-lag parameter from nonsynchronous data. Bernoulli **19**(2), 426–461 (2013)
22. Jacod, J.: On continuous conditional Gaussian martingales and stable convergence in law. In: Séminaire de Probabilités XXXI, pp. 232–246. Springer (1997)
23. Jacod, J., Li, Y., Mykland, P.A., Podolskij, M., Vetter, M.: Microstructure noise in the continuous case: the pre-averaging approach. Stochast. Process. Appl. **119**(7), 2249–2276 (2009)
24. Koike, Y.: Estimation of integrated covariances in the simultaneous presence of nonsynchronicity, microstructure noise and jumps. Econometric Theory pp. 1–79 (2013)
25. Koike, Y.: An estimator for the cumulative co-volatility of asynchronously observed semimartingales with jumps. Scand. J. Stat. **41**(2), 460–481 (2014)
26. Koike, Y.: Limit theorems for the pre-averaged Hayashi-Yoshida estimator with random sampling. Stochast. Process. Appl. **124**(8), 2699–2753 (2014)
27. Large, J.: Measuring the resiliency of an electronic limit order book. J. Finan. Markets. **10**(1), 1–25 (2007)
28. Malliavin, P., Mancino, M.E.: Fourier series method for measurement of multivariate volatilities. Financ. Stochast. **6**(1), 49–61 (2002)
29. Malliavin, P., Mancino, M.E., et al.: A Fourier transform method for nonparametric estimation of multivariate volatility. Ann. Stat. **37**(4), 1983–2010 (2009)
30. Muni Toke, I., Pomponio, F.: Modelling trades-through in a limited order book using Hawkes processes. Economics Discussion Paper (2011-32) (2011)
31. Mykland, P.A.: A Gaussian calculus for inference from high frequency data. Ann. Financ. **8**(2–3), 235–258 (2012)
32. Ogihara, T.: Parametric inference for nonsynchronously observed diffusion processes in the presence of market microstructure noise. arXiv:1412.8173 (2014)
33. Ogihara, T.: Local asymptotic mixed normality property for nonsynchronously observed diffusion processes. Bernoulli **21**(4), 2024–2072 (2015)
34. Ogihara, T., Yoshida, N.: Quasi-likelihood analysis for stochastic regression models with nonsynchronous observations. arXiv:1212.4911 (2012)
35. Ogihara, T., Yoshida, N.: Quasi likelihood analysis for point process regression models. Preprint (2015)
36. Podolskij, M., Vetter, M., Sommer, M.: Estimation of volatility functionals in the simultaneous presence of microstructure noise and jumps. Available at SSRN 950344 (2007)
37. Smith, E., Farmer, J.D., Gillemot, L.S., Krishnamurthy, S., et al.: Statistical theory of the continuous double auction. Quant. Financ. **3**(6), 481–514 (2003)
38. Voev, V., Lunde, A.: Integrated covariance estimation using high-frequency data in the presence of noise. J. Finan. Econometrics **5**(1), 68–104 (2007)
39. Zhang, L., Mykland, P.A., Aït-Sahalia, Y.: A tale of two time scales. J. Am. Stat. Assoc. (2012)
40. Zhang, L., et al.: Efficient estimation of stochastic volatility using noisy observations: A multiscale approach. Bernoulli **12**(6), 1019–1043 (2006)
41. Zhou, B.: High-frequency data and volatility in foreign-exchange rates. J. Bus. Econ. Stat. **14**(1), 45–52 (1996)

Hunting for Black Swans in the European Banking Sector Using Extreme Value Analysis

Jan Beirlant, Wim Schoutens, Jan De Spiegeleer, Tom Reynkens and Klaus Herrmann

Abstract In financial risk management, a Black Swan refers to an event that is deemed improbable yet has massive consequences. In this communication we propose a way to investigate if the recent financial crisis was a Black Swan event for a given bank based on weekly closing prices and derived log-returns. More specifically, using techniques from extreme value methodology we estimate the tail behavior of the negative log-returns over two specific horizons:

- Pre-crisis: from January 1, 1994 until August 7, 2007 (often referred to as the official starting date of the credit crunch crisis);
- Post-crisis: from August 8, 2007 until September 23, 2014 (the cut-off date of our study).

We illustrate this approach with Barclays and Credit Suisse data, and argue that Barclays can be considered as having experienced a Black Swan and Credit Suisse not. We then link the differences in tail risk behavior between these banks with capitalization and leverage indicators. We emphasize the statistical methods for modeling univariate extremes linked with graphical support.

Keywords Extreme value methods · Extreme value index · Scale estimator · Weekly returns

AMS 2000 subject classifications 62G32

1 Introduction

Clearly, the recent financial crisis that started in 2007 can be used as a motivating example necessitating the use of extreme value analysis (EVA) in financial statistics. Bollerslev and Todorov [5] studied the effect of the crisis for the S&P500 considering

J. Beirlant (✉) · W. Schoutens · J. De Spiegeleer · T. Reynkens · K. Herrmann
Department of Mathematics, LStat and LRisk, KU Leuven, Leuven, Belgium
e-mail: jan.beirlant@kuleuven.be

© Springer International Publishing Switzerland 2016

J. Kallsen and A. Papapantoleon (eds.), *Advanced Modelling in Mathematical Finance*, Springer Proceedings in Mathematics & Statistics 189,
DOI 10.1007/978-3-319-45875-5_7

the contrast between the statistical probability measure and the risk neutral measure. Here, we study the tail behavior of the negative log-returns of the weekly closing prices of listed stocks. Using techniques from extreme value methodology we propose to analyze the tail behavior of a bank over two specific horizons:

- Pre-Crisis: from January 1, 1994 until August 7, 2007 (often referred to as the official starting date of the credit crunch crisis);
- Post-Crisis: from August 8, 2007 until September 23, 2014 (the cut-off date of our study).

More specifically, we will investigate how one could decide if the recent financial crisis was a Black Swan event for a given bank based on statistical differences between both sets of return data. We illustrate this approach using data from Barclays and Credit Suisse, two major European banks. Of course one should also connect such a statistical finding with economic indicators of a bank, whether it experienced a Black Swan event from an EVA perspective or not.

We restrict ourselves here to *weekly* return data. Indeed, financial return series may suffer from serial dependence such as volatility clustering, which violates the classical assumption of independence. Such serial dependence is at least much weaker in weekly returns. Using results from Hsing [15] our statistical tests, however, will take serial dependence into account. In Fig. 1 the negative weekly returns are plotted against time for the selected banks. The vertical scales are identical allowing to appreciate the impact of the crisis on the weekly losses for the different banks. We also add a vertical line indicating August 7, 2007.

EVA is designed for estimating extreme quantities of a statistical variable, such as Value at Risk, which has become a popular risk measure. The models underlying EVA contain scale and shape parameters, and the statistical methods on which estimation of the scale and shape has been built, offer tools that can be used for general statistical inference such as the definition of appropriate tail models for a distribution at hand. GARCH models constitute a popular approach in analyzing financial time series which exhibit volatility clustering. Here, however, we follow the approach outlined in Sun and Zhou [19], using the results from Hsing [15] that Hill's [14] estimator is still consistent for certain types of dependent data, such as GARCH processes.

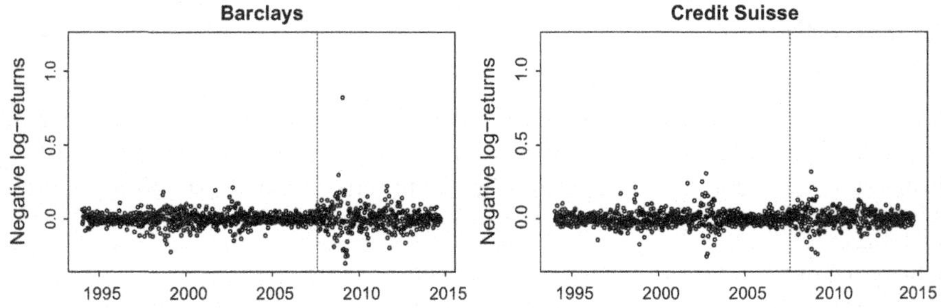

Fig. 1 Negative weekly log-returns for Barclays and Credit Suisse

Moreover, for a bank which was badly hit by the crisis, the fitted shape parameter can lead to a near integrated-GARCH situation, which entails an inappropriate GARCH fit and unreliable estimates of the GARCH innovations.

In this paper we look for indicators for truly significant changes in the log-returns through statistical tests for changes in scale and shape parameters, and by calculating the return period of the largest post-crisis loss, in view of the data before the crisis. We emphasize the use of graphical methods that support decision making. In the next section we recall the most important facts from EVA, and review the graphical and estimation methods along the above specifications. Next, we propose estimators for the scale parameter in case of Pareto-type distributions and provide some new asymptotic results. In Sect. 4 we go into the problem of threshold selection when performing inference on the shape and scale parameters. In particular we stress the use of bias reduction techniques which helps to come around the problem of choosing a particular threshold when performing statistical inferences on the parameters. In the final section we make the link with economical indicators.

2 A Recollection from Univariate Extreme Value Methodology

2.1 *Max-Domain of Attraction*

We briefly recollect some facts from EVA. Recent books that have appeared on the subject provide more details: Embrechts et al. [12], Coles [7], Reiss and Thomas [17], Beirlant et al. [2], Castillo et al. [6], de Haan and Ferreira [10], and Resnick [18]. Beirlant et al. [3] give an overview of EVA and apply it in a financial risk context. EVA is based on the limit result for normalized partial maxima of i.i.d. random variables $X_1, \ldots, X_n$. Let $X_{1,n} \leq X_{2,n} \leq \ldots \leq X_{n,n}$ denote the ordered observations and hence $X_{n,n} = \max\{X_1, \ldots, X_n\}$. The limit theorem is then formulated as follows: if there exist normalizing constants $a_n > 0$ and b_n such that for all x,

$$\lim_{n\to\infty} P\left((X_{n,n} - b_n)/a_n \leq x\right) = G(x), \tag{1}$$

for some non-degenerate distribution function G, then G is necessarily of extreme value type; that is, up to an affine change of variables, one has

$$G(x) = G_\xi(x) := \exp\left\{-(1+\xi x)^{-1/\xi}\right\} \text{ if } x > -1/\xi \tag{2}$$

for some real value ξ. The parameter ξ is termed *the extreme value index* (EVI), which is of prime interest in EVA. When $\xi = 0$, $G_0(x)$ is to be read as $\exp\{-\exp(-x)\}$. If (1) holds, we say that F, which is underlying the data $X_1, X_2, \ldots$, is in *the max-domain of attraction* (MDA) of G_ξ. The limiting distribution functions in (1) are then max-stable. They are indeed the unique max-stable laws.

The EVI ξ governs the behavior of the right-tail of F. The Fréchet domain of attraction ($\xi > 0$) contains heavy-tailed distributions like the Pareto and the Student t-distributions, i.e., tails of a negative polynomial type and infinite right endpoint. Short-tailed distributions, with a finite right endpoint like the beta distributions, belong to the Weibull MDA with $\xi < 0$. Finally the Gumbel MDA corresponding to $\xi = 0$ contains a great variety of distributions with an exponentially decreasing tail, such as the exponential, the normal and the gamma distributions, but not necessarily with an infinite right endpoint.

In order to characterize the MDAs in a mathematically correct way, there are now two possibilities: model descriptions through the distribution function $F(x) = P(X \leq x)$ (*probability view*) or through the quantile function Q, defined as the inverse function of F (*quantile view*).

Firstly, one can describe the MDAs through the stochastic behavior of the so-called *peaks over threshold* (POTs) $X - t$ given that $X > t$. Pickands' [16] theorem states that X is in the MDA of G_ξ if and only if for some sequence $\sigma_t > 0$ the conditional distribution of the scaled excesses as $t \to Q(1)$ converges to *the generalized Pareto distribution* (GPD) P_ξ

$$P((X - t)/\sigma_t \leq x | X > t) \to P_\xi(x) = 1 - (1 + \xi x)^{-1/\xi} \qquad (3)$$

with $1 + \xi x > 0$ and $x > 0$. Remark that in case $\xi = 0$ the GPD is nothing else than the exponential distribution with distribution function $1 - \exp(-x)$ for $x > 0$. From this, one chooses an appropriate threshold t and hopes for a reasonable rate of convergence in (3). Fitting the GPD with survival function $\left(1 + \frac{\xi}{\sigma} x\right)^{-1/\xi}$ to the excesses $X_i - t$ for those data X_i for which $X_i > t$, one estimates the shape parameter ξ and the scale σ for instance by maximum likelihood. In practice t can be chosen as one of the largest data, e.g. the $(k+1)$th largest data point $X_{n-k,n}$, for some $1 < k < n$.

Secondly, through the work of de Haan [8, 9] the MDA characterization was constructed on the basis of the regular varying behavior of the *tail function U*, which is associated with the quantile function Q by $U(x) := Q(1 - \frac{1}{x})$. The MDAs can indeed be characterized by the *extended regular variation property* specifying the difference between high quantiles corresponding to tail proportions that differ by $100x\%$:

$$F \in MDA(\xi) \iff \lim_{u \to \infty} \frac{U(ux) - U(u)}{a(u)} = (x^\xi - 1)/\xi \qquad (4)$$

for every real valued x and some positive function a, and where the expression on the right equals $\log x$ for $\xi = 0$.

In the specific case of the Fréchet MDA with $\xi > 0$, the extended regular variation property (4) corresponds to regular variation of U with index $\xi > 0$:

$$F \in MDA(\xi > 0) \iff U(x) = x^\xi \ell(x), \qquad (5)$$

where ℓ is a slowly varying function defined by $\lim_{u\to\infty} \frac{\ell(ux)}{\ell(u)} = 1$, for all $x > 0$. Then, condition (3) specifies the regular variation of the right tail function $\bar{F} := 1 - F$ with index $1/\xi$. The elements of this MDA are termed Pareto-type distributions. Remark that the regular variation of $\bar{F}$ is equivalent to stating that as $t \to \infty$

$$P(X/t > x|X > t) \to x^{-1/\xi}, \quad x > 1, \tag{6}$$

which then forms a simplified POT approach in comparison with (3).

Almost all authors consider the following subclass of Pareto-type distributions, which was first introduced in Hall [13]:

$$\bar{F}(x) = Ax^{-1/\xi}\left(1 + bx^{-\beta}(1 + o(1))\right), \tag{7}$$

$$U(x) = A^{\xi}x^{\xi}\left(1 + \xi b A^{-\xi\beta}x^{-\xi\beta}(1 + o(1))\right), \quad \text{as } x \to \infty, \tag{8}$$

where $A > 0$ is then the scale parameter, while $\beta > 0$ and b are the second-order shape and scale parameters. This extra assumption then allows to derive specific approximations for the bias and variance of the estimators, and to derive bias reduced estimators as discussed below.

2.2 Estimation when $\xi > 0$

Assumption (5) can be graphically verified using log-log plots, i.e. Pareto QQ-plots,

$$\left(\log((n + 1)/i), \log X_{n-i+1,n}\right), \quad i = 1, \dots, n, \tag{9}$$

which, for some k, should then be ultimately linear for a set of largest values $X_{n-k,n} \le X_{n-k+1,n} \le \dots \le X_{n,n}$. The classical Hill [14] estimator $H_{k,n}$ of $\xi > 0$

$$H_{k,n} = \frac{1}{k}\sum_{j=1}^{k} \log X_{n-j+1,n} - \log X_{n-k,n}, \tag{10}$$

can be motivated as an estimator of the slope of the least squares regression line based on the final k points in the log-log plot and passing through an appropriately chosen anchor point $\left(\log((n + 1)/(k + 1)), \log X_{n-k,n} \dots\right)$, see Beirlant et al. [1]. It can also be derived as a maximum likelihood estimator of ξ using the simple Pareto model in the right hand side of (6) based on the relative excesses $X_{n-j+1,n}/X_{n-k,n}$, $j = 1, \dots, k$, over the random threshold $X_{n-k,n}$.

Hsing [15] derived the asymptotic distribution of $H_{k,n}$ for weakly dependent series. Under (7), as $k, n \to \infty$ and $k/n \to 0$, this leads to

$$\sqrt{k}\left(H_{k,n} - \xi - \frac{B(n/k)}{1 + \xi\beta}\right) \to_d \mathcal{N}\left(0, \xi^2(1 + \chi + \omega - 2\psi)\right) \tag{11}$$

as $n \to \infty$ and $k/n \to 0$, where $B(n/k) = -\xi\beta b A^{-\xi\beta}(k/n)^{\xi\beta}$, and χ, ω, ψ are parameters of serial dependence, being 0 in case of independence. Under the condition $\sqrt{k}B(n/k) \to \lambda$ as $k, n \to \infty$ and $k/n \to 0$ we then obtain

$$\sqrt{k}\left(H_{k,n} - \xi\right) \to_d \mathcal{N}\left(\frac{\lambda}{1+\xi\beta}, \xi^2(1+\chi+\omega-2\psi)\right).$$

Estimators $\hat{\chi}, \hat{\omega}, \hat{\psi}$ are given in (3.6) in Hsing [15]. Furthermore, Sun and Zhou [19] showed that a GARCH(1,1) dependence structure fits to the approach of Hsing [15].

From (11) it follows that $H_{k,n}$ can have high bias for a large range of k values. This bias originates from the fact that the estimators are based on (6) replacing the limit by an equality, which is inaccurate for too large values of k. Theoretically, this k-region is represented by $\sqrt{k}B(n/k) \to \lambda > 0$ as $k, n \to \infty$, $k/n \to 0$. Accommodation of bias has been considered recently in a number of papers in case of i.i.d. data. Bias reduced estimators typically exhibit plots which are more horizontal as a function of k. In case the tail under consideration is a composition of two different Pareto components, the corresponding levels of the estimates are better visible. In that sense such estimators are useful as a diagnostic tool in order to interpret Hill plots $(k, H_{k,n})$ and plots of other tail estimators. For instance, choosing a value of k as large as possible, with the original and bias reduced version of the estimator approximately equal, leads to an estimate with a smaller bias and a variance as small as possible. Along the probability view, bias reduction can be obtained by replacing the Pareto fit in (6) by an *extended Pareto distribution* (EPD) with distribution function

$$G_{\xi,\kappa,\beta}(y) = 1 - \left(y(1+\kappa(1-y^{-\beta}))\right)^{-1/\xi}, \quad y > 1,$$

to the relative excesses $X_{n-j+1,n}/X_{n-k,n}$, $j = 1, \ldots, k$ using maximum likelihood (see Beirlant et al. [4]). This EPD approximation follows when approximating the left hand side of (6) under (7) with $\kappa = \kappa_t = \xi b t^{-\beta}$.

In Fig. 2, we gather the log-log plots (Pareto QQ-plots). The Hill plots with the original $H_{k,n}$ and using the EPD approximation with (ξ, κ) estimated by maximum likelihood per k are shown in Fig. 3. We set $\rho = -\beta\xi$ equal to 1 and hence $\hat{\beta} = -\rho/\hat{\xi} = 1/\hat{\xi}$. Because of the different sample sizes for pre- and post-crisis data we plot the estimates against the ratio k/n. The log-log plots show that there is barely any difference in slope between the pre- and post-crisis data. The plots for the shape estimators seem to confirm this. For Barclays, k/n values around 0.1 seem to be suitable along the abovementioned guideline, since for $k/n \leq 0.1$ for both periods, the Hill and EPD estimates remain rather close, in contrast to the larger k values. In the case of Credit Suisse the ultimate top portion of the log-log plot appears to be concave leading to decreasing Hill estimates as k decreases, meeting the bias reduced estimator only at the smallest k values, say up to $k/n \approx 0.02$. For both banks, the Hill estimates for the pre- and post-crisis are close in the suitable region for k/n. However, in case of Barclays, the Pareto QQ-plot of the post-crisis data lies higher than the one of the pre-crisis data, indicating a change in scale since it follows from

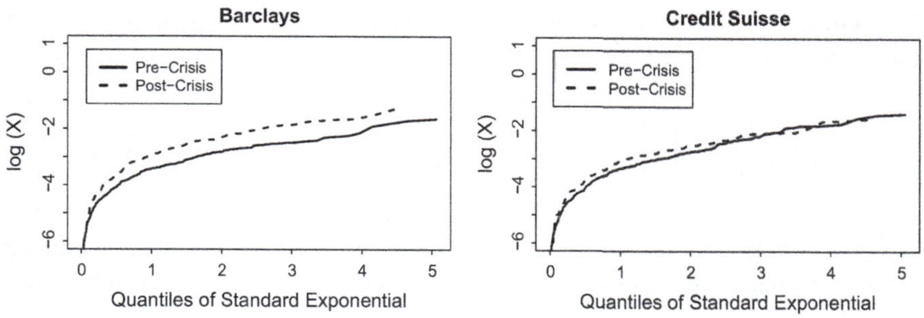

Fig. 2 Log-log plots for the pre- (*solid line*) and post-crisis (*dashed line*) negative log-returns for Barclays and Credit Suisse

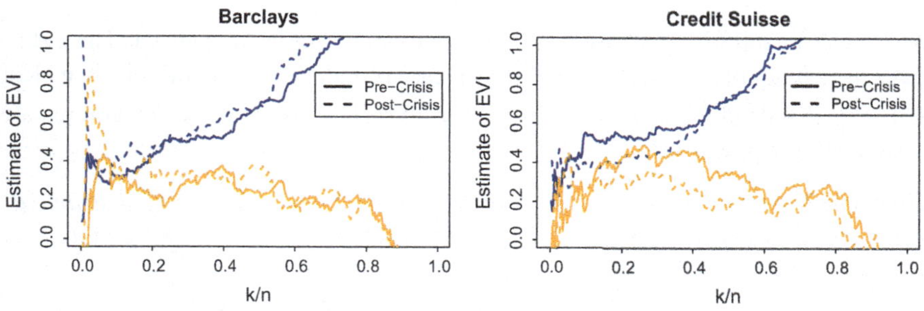

Fig. 3 Hill (*blue*) and EPD (*orange*) estimates as a function of k/n for the pre- (*solid line*) and post-crisis (*dashed line*) negative log-returns for Barclays and Credit Suisse

(8) that the log-log plot is an approximation of the graph $(\log \frac{n+1}{i}, \log U(\frac{n+1}{i})) = (\log \frac{n+1}{i}, \xi \log A + \xi \log \frac{n+1}{i})$ for $i = 1, \ldots, n$. In the Credit-Suisse case, both Pareto QQ-plots are close and hence no change in scale can be deduced.

3 Estimating the Scale Parameter

Following the suggestion made in Einmahl et al. [11] one can also inspect for changes in the scale parameter A introduced in (7)–(8). An initial estimator for A is given by

$$\hat{A}_{k,n} = \frac{k+1}{n+1} X_{n-k,n}^{1/H_{k,n}}. \tag{12}$$

The following theorem provides an asymptotic normality result for this estimator which is valid for dependent data.

Theorem 1 *Under the conditions of Theorem 3.3 in Hsing [15] and under* (7), *when* $k, n \to \infty$, $k/n \to 0$ *and* $\sqrt{k}B(n/k) \to \lambda$, *we have that*

$$\frac{\sqrt{k}\xi}{\log U(n/k)}\left(\frac{\hat{A}_{k,n}}{A}-1\right) \to_d \mathcal{N}\left(\frac{-\lambda}{1+\xi\beta}, 1+\chi+\omega-2\psi\right).$$

Theorem 1 shows that the scale estimator can have large bias. Using $\hat{\xi}_{k,n}$ and $\hat{\kappa}_{k,n}$, the EPD estimators of ξ and κ, we get the following bias reduced estimator of A:

$$\hat{A}^{EP}_{k,n} = \frac{k+1}{n+1} X^{1/\hat{\xi}_{k,n}}_{n-k,n}\left(1-\frac{\hat{\kappa}_{k,n}}{\hat{\xi}_{k,n}}\right). \tag{13}$$

We provide an intuitive derivation for both scale estimators in Appendix 1. Theorem 2 gives the asymptotic distribution of $\hat{A}^{EP}_{k,n}$ in case of independent data. It is then clear that $\hat{A}^{EP}_{k,n}$ is indeed a bias reduced estimator of A. The proofs of both theorems are postponed to Appendix 2.

Theorem 2 *Assuming* $X_1, \ldots, X_n$ *are independent and identically distributed following* (7), *when* $k, n \to \infty$, $k/n \to 0$ *and* $\sqrt{k}B(n/k) \to \lambda$, *we have that*

$$\frac{\sqrt{k}\xi}{\log U(n/k)}\left(\frac{A}{\hat{A}^{EP}_{k,n}}-1\right) \to_d \mathcal{N}\left(0, \left(\frac{1+\xi\beta}{\xi\beta}\right)^2\right).$$

In Fig. 4, $\log \hat{A}_{k,n}$ and $\log \hat{A}^{EP}_{k,n}$ are plotted for the two selected banks with the pre- and post-crisis series. We can again select suitable regions based on the closeness of the scale estimator and the bias reduced version. We then choose $k/n \approx 0.1$ for Barclays and $k/n \approx 0.02$ for Credit Suisse. We see that there is some difference in scale estimates between the pre- and post-crisis data for Barclays while much less for Credit Suisse (for these values of k/n).

4 Testing for Black Swans

We define a Black Swan as a highly improbable event with large consequences. Therefore, as a first indicator we consider the probability of obtaining a loss at least as big as the largest loss post-crisis, *in view of the data information before the crisis*. We express this in terms of the corresponding return period. Secondly, we test for significant differences in scale and shape parameters between pre- and post-crisis periods. While it is difficult to define a Black Swan through a minimal return period and/or a maximal p-value level, we will argue that the financial crisis can be considered as a Black Swan in the Barclays case, while it is not in the Credit Suisse case.

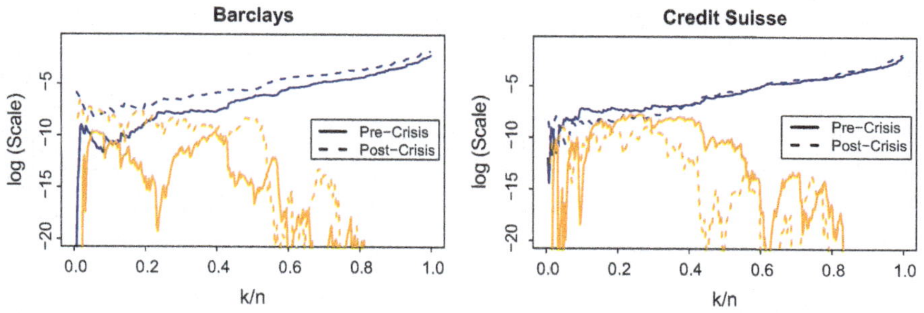

Fig. 4 Scale estimates $\hat{A}_{k,n}$ (*blue*) and bias reduced scale estimates $\hat{A}_{k,n}^{EP}$ (*orange*), in log-scale, as a function of k/n for the pre- (*solid line*) and post-crisis (*dashed line*) negative log-returns for Barclays and Credit Suisse

4.1 Return Periods of Worst Negative Log-Returns

Here, and in the sequel, we denote the number of pre-crisis, respectively post-crisis, negative log-returns by n_1, respectively n_2, and the ordered pre-crisis, respectively post-crisis, negative log-returns by $x_{1,n_1}, \ldots, x_{n_1,n_1}$, respectively $y_{1,n_2}, \ldots, y_{n_2,n_2}$. We also use the superscripts (X) and (Y) to indicate the pre-crisis, respectively, post-crisis data. The return period can now be denoted by $r_{\max} = 1/P(X > y_{n_2,n_2})$. Then, applying the Weissman [20] estimator following from the approximation (6),

$$\hat{r}_{\max,k} = 1/\hat{P}_k(X > y_{n_2,n_2}) = \frac{n_1+1}{k+1}\left(\frac{y_{n_2,n_2}}{x_{n_1-k,n_1}}\right)^{1/H_{k,n_1}^{(X)}}, \quad k = 1, \ldots, n_1.$$

In a similar way as in the proof of Theorem 1 in Appendix 2 (see also Theorem 4.4.7 in de Haan and Ferreira [10]), one can show that, treating y_{n_2,n_2} as a fixed number,

$$\frac{\sqrt{k}}{\sqrt{1+\log^2\left(\frac{k}{n_1}r_{\max}\right)}}\left(\log r_{\max} - \log \hat{r}_{\max,k}\right)$$

is asymptotically normal with asymptotic variance $1 + \chi + \omega - 2\psi$. Hence an approximate 95 % asymptotic lower confidence bound for $\log r_{\max}$ is given by

$$\log \hat{r}_{\max,k} - \frac{1.645}{\sqrt{k}}\sqrt{1+\log^2\left(\frac{k}{n_1}\hat{r}_{\max,k}\right)}\sqrt{1+\hat{\chi}+\hat{\omega}-2\hat{\psi}}. \tag{14}$$

As described in Beirlant et al. [4], a bias reduced version for return periods can be constructed by replacing the simple Pareto distribution by the EPD in the right hand side of (6):

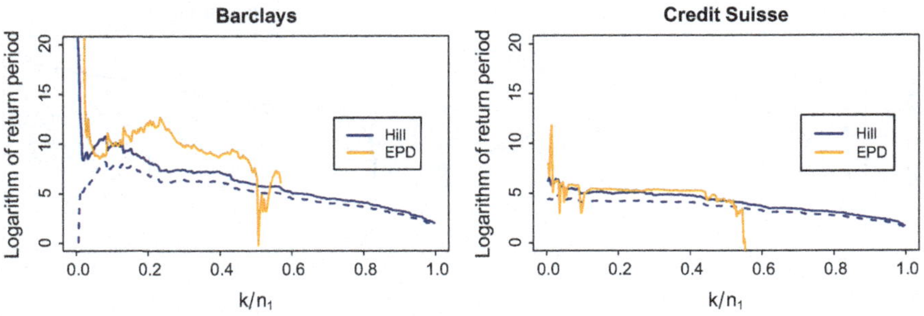

Fig. 5 Estimates of the return periods for obtaining a weekly loss as big as the largest loss post-crisis in view of the data information before the crisis: $\log \hat{r}_{\max,k}$ (*blue*) and $\log \hat{r}^{EP}_{\max,k}$ (*orange*), as a function of k/n_1, for Barclays and Credit Suisse. Approximate 95 % asymptotic lower confidence bounds for $\log \hat{r}_{\max}$ are shown by the *dashed lines*

$$\hat{r}^{EP}_{\max,k} = \frac{n_1+1}{k+1}\left(1 - G_{\hat{\xi}_{k,n_1},\hat{\kappa}_{k,n_1},\hat{\beta}_{k,n_1}}\left(\frac{y_{n_2,n_2}}{x_{n_1-k,n_1}}\right)\right)^{-1}, \quad k = 1,\ldots,n_1.$$

In Fig. 5 we plot $\log \hat{r}_{\max,k}$ and $\log \hat{r}^{EP}_{\max,k}$ as a function of k/n_1 for the two selected banks, jointly with the lower bounds (14) (dashed lines). Choosing $k/n_1 \approx 0.1$ where the different estimators coincide, we obtain for Barclays a return period $e^{10} \approx 22{,}000$ weeks. This return period corresponds to $2 \times 423 = 846$ years using an equal frequency for negative and positive log-returns. This is in sharp contrast with the corresponding return period $e^6 \approx 400$ weeks or $2 \times 7.7 = 15.4$ years for Credit Suisse.

4.2 Testing for Differences in Shape or Scale

We now want to test more formally if there is a significant difference in at least the shape or the scale parameter. We consider the $\alpha = 5\,\%$ significance level.

In order to test $H_0^{(\xi)}: \xi^{(X)} \geq \xi^{(Y)}$ versus $H_1^{(\xi)}: \xi^{(X)} < \xi^{(Y)}$ we can use the test statistic

$$T^{(\xi)}_{k_1,k_2,n_1,n_2} = \frac{H^{(Y)}_{k_2,n_2} - H^{(X)}_{k_1,n_1}}{\sqrt{\frac{(H^{(Y)}_{k_2,n_2})^2\,(1+\hat{\chi}_2+\hat{\omega}_2-2\hat{\psi}_2)}{k_2} + \frac{(H^{(X)}_{k_1,n_1})^2\,(1+\hat{\chi}_1+\hat{\omega}_1-2\hat{\psi}_1)}{k_1}}}$$

with k_1 and k_2 appropriately selected number of extremes for pre- and post-crisis data, and $\hat{\chi}_1, \hat{\omega}_1, \hat{\psi}_1$ and $\hat{\chi}_2, \hat{\omega}_2, \hat{\psi}_2$ are the corresponding estimates for χ, ω, ψ for the pre- and post-crisis period respectively. Under equality of the tail indices the asymptotic distribution of $T^{(\xi)}_{k_1,k_2,n_1,n_2}$ is then standard normal for small values of k_1, k_2 such that $\sqrt{k_1}B^{(X)}(n_1/k_1) \to 0$ and $\sqrt{k_2}B^{(Y)}(n_2/k_2) \to 0$.

Similarly, to test $H_0^{(A)}\colon A^{(X)} \geq A^{(Y)}$ versus $H_1^{(A)}\colon A^{(X)} < A^{(Y)}$ we use

$$T^{(A)}_{k_1,k_2,n_1,n_2} = \frac{\log \hat{A}^{(Y)}_{k_2,n_2} - \log \hat{A}^{(X)}_{k_1,n_1}}{\sqrt{\frac{\log^2(n_2/k_2)\,(1+\hat{\chi}_2+\hat{\omega}_2-2\hat{\psi}_2)}{k_2} + \frac{\log^2(n_1/k_1)\,(1+\hat{\chi}_1+\hat{\omega}_1-2\hat{\psi}_1)}{k_1}}}$$

which also follows asymptotically a standard normal distribution under equality in $H_0^{(A)}$.

As shown in Appendix 3, the two tests are not independent. We therefore have to be prudent drawing conclusions. The joint test combines information of the two separate tests and uses the following hypotheses: $H_0\colon H_0^{(\xi)} \cap H_0^{(A)}$ versus $H_1\colon H_1^{(\xi)} \cup H_1^{(A)}$. From (16) in Appendix 2 it follows that the determinant of the covariance matrix is asymptotically 0 and hence a bivariate Hotelling T^2 cannot be performed. It is critical to control the probability for a type I error of the joint test, hence asking even more statistical evidence before concluding a Black Swan event. Using the Bonferroni correction we obtain that the probability for a type I error for the joint test is smaller than $\alpha = 5\,\%$ when using the $\alpha/2 = 2.5\,\%$ significance level for each test separately.

In Figs. 6 and 7 we plot the p-values of the two asymptotic tests for equality of shape and scale against k/n under equality of the ratios $k_1/n_1 = k_2/n_2$. The red lines

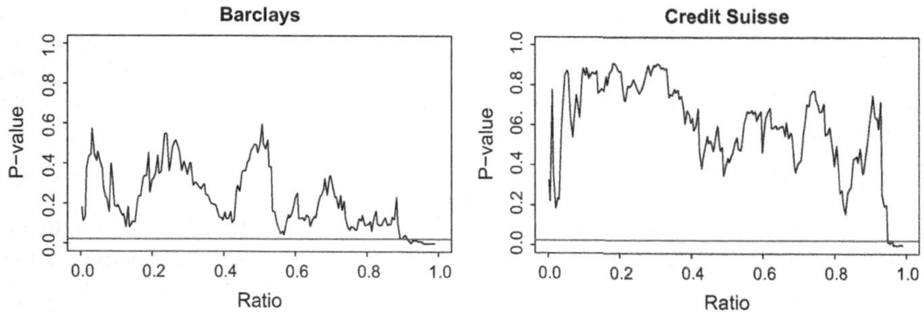

Fig. 6 p-values for testing differences in shape using $T^{(\xi)}_{k_1,k_2,n_1,n_2}$ as a function of the ratio $k_1/n_1 = k_2/n_2$ for pre- and post-crisis negative log-returns for Barclays and Credit Suisse

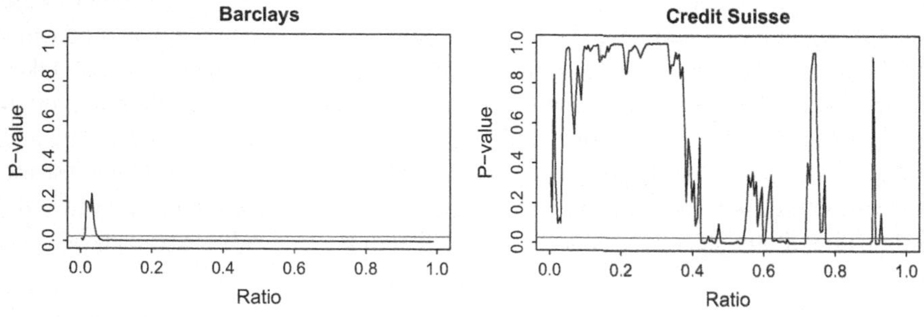

Fig. 7 p-values for testing differences in scale using $T^{(A)}_{k_1,k_2,n_1,n_2}$ as a function of the ratio $k_1/n_1 = k_2/n_2$ for pre- and post-crisis negative log-returns for Barclays and Credit Suisse

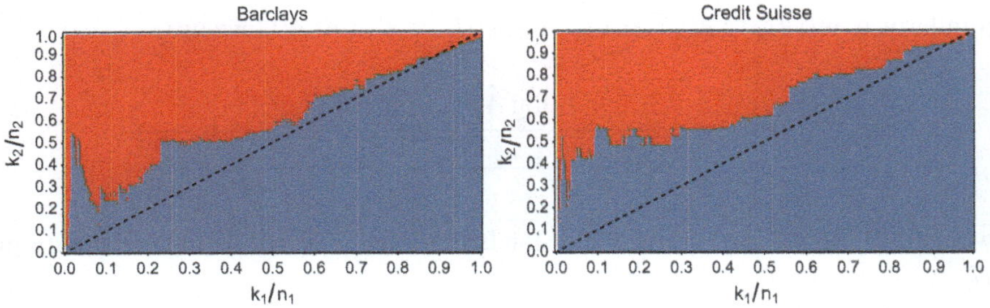

Fig. 8 Indicator function for the event "p-value for the test using $T^{(\xi)}_{k_1,k_2,n_1,n_2}$ is below $\alpha/2 = 2.5\,\%$" for all possible choices of k_1 and k_2 for pre- and post-crisis negative log-returns for Barclays and Credit Suisse

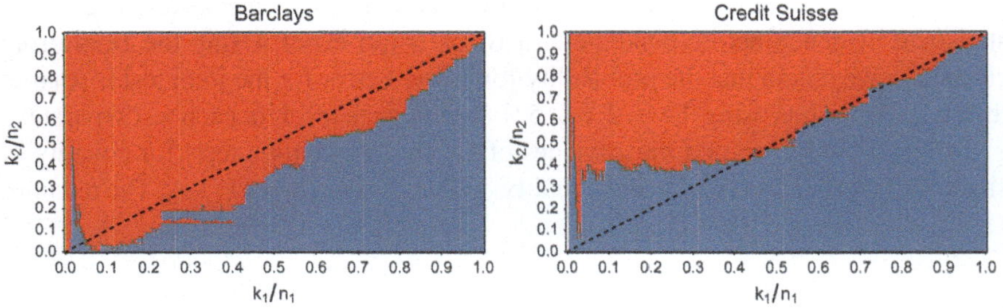

Fig. 9 Indicator function for the event "p-value for the test using $T^{(A)}_{k_1,k_2,n_1,n_2}$ is below $\alpha/2 = 2.5\,\%$" for all possible choices of k_1 and k_2 for pre- and post-crisis negative log-returns for Barclays and Credit Suisse

show the 2.5 % significance level. From the discussion following Figs. 3 and 4, using ratios around 0.1 for Barclays and around 0.02 for Credit Suisse corresponding to the the lowest bias. The shape parameters do not show significant differences. The scale parameters show significant results for Barclays except for $k/n \leq 0.05$, whereas for Credit Suisse the scale parameters show strongly non-significant results for k/n smaller than 0.4. We now consider the p-values for testing scale and shape differences for all possible choices of k_1, k_2 for Barclays and Credit Suisse. The 3-dimensional plots showing the p-values can be found in Appendix 4 where a red plane indicates the 2.5 % significance level. Here, we consider the indicator function which takes value 1 when the p-value is below 2.5 % and 0 otherwise. This function is plotted in Figs. 8 and 9 where (light) blue and red correspond to 0 and 1, respectively, and the black dashed line indicates $k_1/n_1 = k_2/n_2$. In case of Barclays, the test for the scale difference is non-significant only for large values of k_1, while in case of Credit Suisse non-significance also appears for small values of k_1 and k_2 together.

5 Relating Statistical Conclusions with Economic Indicators

Above we provided statistical indicators for a Black Swan event in financial return data linked with the recent financial crisis, measuring the probability for the experienced losses in view of the a priori return data, and by testing for significant differences in the scale parameters of the Pareto tail before and after the crisis. For Barclays the return period for the experienced loss as a result of the financial crisis is extremely large and we find a significant difference in the scale parameters before and after the crisis, and so we label Barclays as having experienced a Black Swan event during the recent crisis *in view of the pre-crisis return data only*. In contrast,

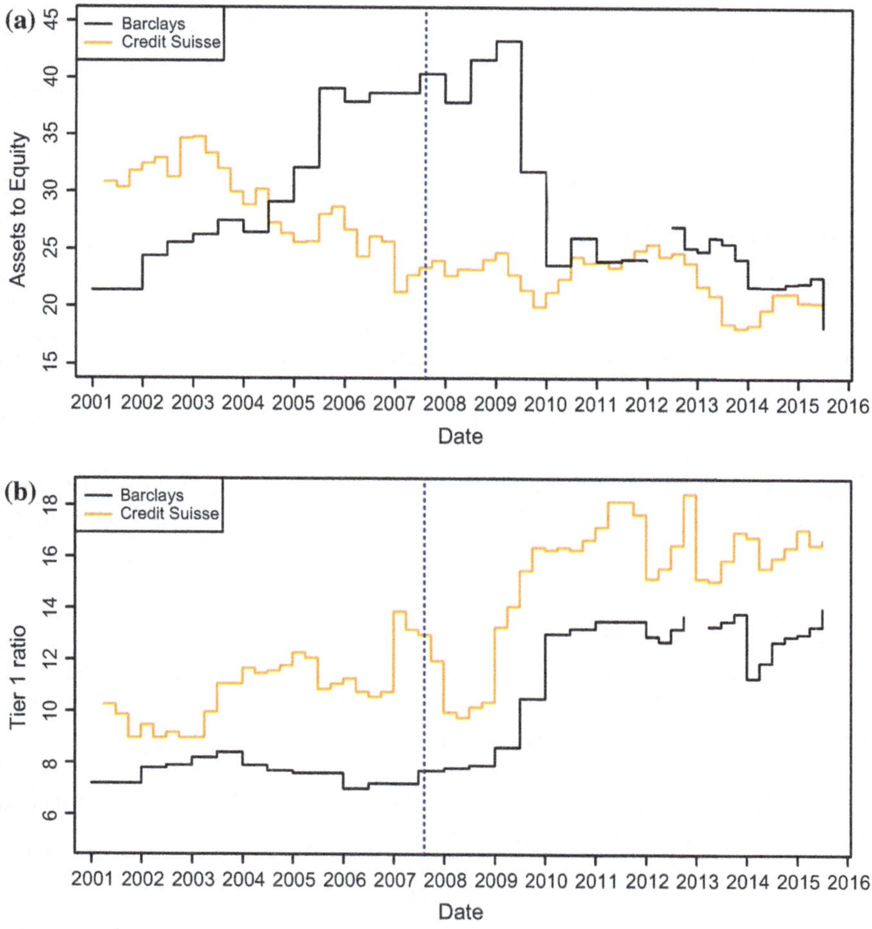

Fig. 10 Assets to equity ratio (**a**) and Tier 1 ratio (in %) (**b**) for Barclays (*black*) and Credit Suisse (*orange*)

for Credit Suisse the statistical significance is not met and the return period is more than 50 times smaller than in the Barclays case.

Of course one should be able to explain the vulnerability of a bank to such a financial crisis in terms of its economic parameters. At the time of the financial crisis, Barclays was a bank with an outspoken amount of leverage. Barclays' ratio of the assets to the equity base was almost twice as large compared to the leverage of Credit Suisse (Fig. 10a). Credit Suisse had indeed much less assets for every dollar of equity. This made Credit Suisse less susceptible to a shock in the financial system.

When studying the Tier 1 ratio of both banks, the same conclusion holds (Fig. 10b). This ratio relates the Tier 1 capital to the risk-weighted assets of a financial institution. Here, Barclays stands out again as a more vulnerable bank compared to Credit Suisse. Because of its vulnerability, Barclays witnessed a true Black Swan event, whereas this was not the case for Credit Suisse. This provides some explanation for the statistical conclusions obtained above.

Acknowledgements The research of Klaus Herrmann is supported by IAP Research Network P7/06 of the Belgian State (Belgian Science Policy), and the project GOA/12/014 of the KU Leuven Research Fund.

Appendix 1: Derivation of the Scale Estimators $\hat{A}_{k,n}$ and $\hat{A}^{EP}_{k,n}$

Starting from the Hall model (7) and ignoring the second order terms yields the approximation

$$\bar{F}(x) \sim Ax^{-1/\xi}, \quad \text{as } x \to \infty. \tag{15}$$

Alternatively, for intermediate order statistics $X_{n-k,n}$, the tail probability $\bar{F}(X_{n-k,n})$ can be estimated by the empirical probability $k/n \approx (k+1)/(n+1)$, leading to the defining equation

$$\hat{A}_{k,n} X_{n-k,n}^{-1/H_{k,n}} = \frac{k+1}{n+1},$$

where ξ in (15) is estimated by the Hill estimator $H_{k,n}$. This immediately gives (12). In order to reduce the bias in estimating the scale parameter, ξ first needs to be estimated by the EPD estimator $\hat{\xi}_{k,n}$ to lift up the bias caused by the estimation of ξ. The other source of bias originates from ignoring the second order terms when approximating A. Following a similar reasoning as before, now taking the second order terms into account, the defining equation is

$$\hat{A} X_{n-k,n}^{-1/\hat{\xi}_{k,n}} \left(1 + bX_{n-k,n}^{-\beta}(1+o(1))\right) = \frac{k+1}{n+1}.$$

Since $\kappa = \kappa_t = \xi b t^{-\beta}(1+o(1))$, we can estimate $bX_{n-k,n}^{-\beta}(1+o(1))$ by $\hat{\kappa}_{k,n}/\hat{\xi}_{k,n}$ with $\hat{\kappa}_{k,n}$ the EPD estimator for κ at the threshold $t = X_{n-k,n}$. In order to obtain numerically stable results, we can use that $(1+\kappa_t/\xi)^{-1} \sim 1 - \kappa_t/\xi$ since $\kappa_t \to 0$ as $t \to \infty$, which leads to the bias reduced scale estimator in (13).

Appendix 2: Proofs for Sect. 3

Proof (Proof of Theorem 1)
Remark that

$$\sqrt{k}\left(\log \hat{A}_{k,n} - \log A\right) = T_{k,n}^{(1)} + T_{k,n}^{(2)},$$

with

$$T_{k,n}^{(1)} = \sqrt{k}\left(\frac{\log X_{n-k,n}}{H_{k,n}} - \frac{\log U(n/k)}{\xi}\right)$$
$$T_{k,n}^{(2)} = \sqrt{k}\left(\frac{\log U(n/k)}{\xi} + \log\left(\frac{k+1}{n+1}\right) - \log A\right).$$

First, as $U(x) = A^{\xi}x^{\xi}(1 + \xi b A^{-\xi\beta}x^{-\xi\beta}(1 + o(1))$ when $x \to \infty$,

$$T_{k,n}^{(2)} = -\sqrt{k}B(n/k)\frac{1}{\xi\beta}(1 + o(n/k)),$$

as $n/k \to \infty$. Next, with $\tilde{H}_{k,n} := \frac{1}{k}\sum_{j=1}^{k}(\log X_{n-j+1,n} - \log U(n/k))$ and $\mathbb{E}(\tilde{H}_{k,n}) = \xi + B(n/k)/(1+\xi\beta)$, see Hsing [15], we get

$$\begin{aligned}
T_{k,n}^{(1)} &= -\frac{\log U(n/k)}{H_{k,n}\xi}\sqrt{k}\left(H_{k,n} - \xi\right) + \frac{\sqrt{k}}{H_{k,n}}(\log X_{n-k,n} - \log U(n/k))\\
&= -\frac{\log U(n/k)}{H_{k,n}\xi}\sqrt{k}\left(\tilde{H}_{k,n} - \mathbb{E}(\tilde{H}_{k,n})\right)\\
&\quad + \frac{1}{H_{k,n}}\left(\frac{\log U(n/k)}{\xi} + 1\right)\sqrt{k}(\log X_{n-k,n} - \log U(n/k))\\
&\quad - \frac{\log U(n/k)}{H_{k,n}\xi}\frac{\sqrt{k}B(n/k)}{1+\xi\beta}.
\end{aligned}$$

Hence,

$$\begin{aligned}
\sqrt{k}\left(\log \hat{A}_{k,n} - \log A\right) &= -\frac{\log U(n/k)}{H_{k,n}\xi}\sqrt{k}\left(\tilde{H}_{k,n} - \mathbb{E}(\tilde{H}_{k,n})\right)\\
&\quad + \frac{1}{H_{k,n}}\left(\frac{\log U(n/k)}{\xi} + 1\right)\sqrt{k}(\log X_{n-k,n} - \log U(n/k))\\
&\quad - \frac{1}{\xi}\left(\frac{\log U(n/k)}{H_{k,n}(1+\beta\xi)} + \frac{1}{\beta}\right)\sqrt{k}B(n/k).
\end{aligned}$$

Using the fact that $\log U(n/k)/\log(n/k) \to \xi$ as $n/k \to \infty$, the result now follows from Lemma 2.1 and Corollary 3.4 in Hsing [15]. □

Proof (Proof of Theorem 2)

Using the approach from Beirlant et al. [4], we have with $\kappa_n = \kappa(X_{n-k,n})$

$$
\begin{aligned}
k^{-1/2} Z_{k,n} &:= \frac{n}{k}\bar{F}(X_{n-k,n}) - 1 \\
&= \frac{A}{(k/n) X_{n-k,n}^{1/\xi}} \left(1 + \frac{\kappa_n}{\xi}(1 + o_p(1))\right) - 1 \\
&= \frac{A}{\hat{A}_{k,n}^{EP}} X_{n-k,n}^{1/\hat{\xi}_{k,n} - 1/\xi} \left(1 + \frac{\kappa_n}{\xi}(1 + o_p(1))\right)\left(1 - \frac{\hat{\kappa}_{k,n}}{\hat{\xi}_{k,n}}\right) - 1 \\
&= \frac{A}{\hat{A}_{k,n}^{EP}} \left(1 - \frac{1}{\xi \hat{\xi}_{k,n}} (\hat{\xi}_{k,n} - \xi) \log X_{n-k,n} (1 + o_p(1))\right) \\
&\quad \times \left(1 + \{\frac{\kappa_n}{\xi} - \frac{\hat{\kappa}_{k,n}}{\hat{\xi}_{k,n}}\}(1 + o_p(1))\right) - 1,
\end{aligned}
$$

from which it follows, using $\hat{\kappa}_{k,n} - \kappa_n = O_p(k^{-1/2})$ and $\hat{\xi}_{k,n} - \xi = O_p(k^{-1/2})$ from Theorem 3.1 in Beirlant et al. [4],

$$
\begin{aligned}
&\frac{A}{\hat{A}_{k,n}^{EP}} - 1 \\
&= \frac{k^{-1/2} Z_{k,n} + 1}{\left(1 - \frac{\hat{\xi}_{k,n} - \xi}{\xi \hat{\xi}_{k,n}} \log X_{n-k,n}(1 + o_p(1))\right)\left(1 + \{\frac{\kappa_n}{\xi} - \frac{\hat{\kappa}_{k,n}}{\hat{\xi}_{k,n}}\}(1 + o_p(1))\right)} - 1 \\
&= (k^{-1/2} Z_{k,n} + 1)\left(1 + \frac{1}{\xi \hat{\xi}_{k,n}} (\hat{\xi}_{k,n} - \xi) \log X_{n-k,n}(1 + o_p(1))\right) \\
&\quad \times \left(1 - \{\frac{\kappa_n}{\xi} - \frac{\hat{\kappa}_{k,n}}{\hat{\xi}_{k,n}}\}(1 + o_p(1))\right) - 1.
\end{aligned}
$$

This implies that $\sqrt{k}(A/\hat{A}_{k,n}^{EP} - 1)$ has the same limit distribution as

$$
\frac{\log U(n/k)}{\xi^2} \sqrt{k}(\hat{\xi}_{k,n} - \xi) - \xi^{-1} \sqrt{k}(\hat{\kappa}_{k,n} - \kappa_n) + Z_{k,n}.
$$

From Theorem 3.1 in Beirlant et al. [4] it follows that this stochastic sum is asymptotically unbiased when $\sqrt{k} B(n/k) \to \lambda$, while the asymptotic variance follows from the variance of $\hat{\xi}_{k,n}$ which has the asymptotic dominating coefficient $\log U(n/k)/\xi^2$ in this asymptotic representation. □

Appendix 3: The Dependence Between Tests on Scale and Shape

We now derive the asymptotic covariance matrix of

$$\left(\frac{\xi\sqrt{k}}{\log U\left(\frac{n}{k}\right)}(\log \hat{A}_{k,n} - \log A),\ \sqrt{k}\left(\frac{H_{k,n}}{\xi} - 1\right)\right).$$

From

$$\frac{\log X_{n-k,n}}{H_{k,n}} - \frac{\log U\left(\frac{n}{k}\right)}{\xi} = \frac{1}{\xi}\left(\log X_{n-k,n} - \log U\left(\frac{n}{k}\right)\right) - \frac{\log X_{n-k,n}}{\xi H_{k,n}}(H_{k,n} - \xi),$$

we have using the notation from the proof of Theorem 1 that

$$\begin{aligned}\frac{\xi T^{(1)}_{k,n}}{\log U\left(\frac{n}{k}\right)} &= \sqrt{k}\left(\frac{\log X_{n-k,n} - \log U\left(\frac{n}{k}\right)}{\log U\left(\frac{n}{k}\right)}\right) - \sqrt{k}\left(\frac{H_{k,n} - \xi}{H_{k,n}}\right)\frac{\log X_{n-k,n}}{\log U(n/k)} \\ &\sim_p \sqrt{k}\left(\frac{\log X_{n-k,n} - \log U\left(\frac{n}{k}\right)}{\log U\left(\frac{n}{k}\right)}\right) - \sqrt{k}\frac{H_{k,n} - \xi}{\xi}.\end{aligned}$$

We hence have concerning the asymptotic covariance

$$\begin{aligned}Acov\left(\frac{\xi T^{(1)}_{k,n}}{\log U\left(\frac{n}{k}\right)}, \sqrt{k}\frac{H_{k,n} - \xi}{\xi}\right) &= Acov\left(\sqrt{k}\frac{\log X_{n-k,n} - \log U\left(\frac{n}{k}\right)}{\log U\left(\frac{n}{k}\right)}, \sqrt{k}\frac{H_{k,n} - \xi}{\xi}\right) \\ &\quad - Avar\left(\sqrt{k}\frac{H_{k,n} - \xi}{\xi}\right).\end{aligned}$$

From (11) we know that the asymptotic variance in this expression is asymptotically equal to $1 + \chi + \omega - 2\psi$.

$$\begin{aligned}Acov\left(\frac{\xi T^{(1)}_{k,n}}{\log U\left(\frac{n}{k}\right)}, \sqrt{k}\frac{H_{k,n} - \xi}{\xi}\right) &= \frac{k}{\xi \log U\left(\frac{n}{k}\right)} Acov\left(\log X_{n-k,n} - \log U\left(\frac{n}{k}\right), H_{k,n} - \xi\right) \\ &\quad - (1 + \chi + \omega - 2\psi).\end{aligned}$$

Following Hsing [15], approximating $H_{k,n}$ by $H^+_{k,n} - \left(\log X_{n-k,n} - \log U\left(\frac{n}{k}\right)\right)$ with $H^+_{k,n} = \frac{1}{k}\sum_{j=1}^{k} \max\{\log X_{n-j+1,n} - \log U\left(\frac{n}{k}\right), 0\}$, we find

$$k\,Acov\left(\log X_{n-k,n} - \log U\left(\frac{n}{k}\right), H_{k,n} - \xi\right) \approx k\,Acov\left(\log X_{n-k,n} - \log U\left(\frac{n}{k}\right), H^+_{k,n} - \xi\right)$$
$$- k\,Avar\left(\log X_{n-k,n} - \log U\left(\frac{n}{k}\right)\right).$$

From Corollary 3.4 in Hsing [15] it then follows that

$$k\,Acov\left(\log X_{n-k,n} - \log U\left(\frac{n}{k}\right), H^+_{k,n} - \xi\right) = \xi^2(1+\psi),$$
$$k\,Avar\left(\log X_{n-k,n} - \log U\left(\frac{n}{k}\right)\right) = \xi^2(1+\omega),$$

which results in

$$Acov\left(\frac{\xi T^{(1)}_{k,n}}{\log U\left(\frac{n}{k}\right)}, \sqrt{k}\frac{H_{k,n}-\xi}{\xi}\right) = \frac{\xi}{\log U\left(\frac{n}{k}\right)}(\psi-\omega) - (1+\chi+\omega-2\psi).$$

Since $T^{(2)}_{k,n}$ is deterministic it does not play a role in the calculation of the covariance matrix. We then get

$$Acov\left(\frac{\xi\sqrt{k}}{\log U\left(\frac{n}{k}\right)}(\log \hat{A}_{k,n} - \log A), \sqrt{k}\frac{H_{k,n}-\xi}{\xi}\right) = -(1+\chi+\omega-2\psi) + \frac{\xi}{\log U\left(\frac{n}{k}\right)}(\psi-\omega).$$

Using the obtained expression for the asymptotic variance of both components (see Theorem 1 and (11)) and the fact that $\log U\left(\frac{n}{k}\right)/\log(n/k) \to \xi$ as $n/k \to \infty$ gives the asymptotic covariance matrix of $\left(\frac{\xi\sqrt{k}}{\log U\left(\frac{n}{k}\right)}(\log \hat{A}_{k,n} - \log A), \sqrt{k}\left(\frac{H_{k,n}}{\xi} - 1\right)\right)$:

$$\begin{pmatrix} 1+\chi+\omega-2\psi & -(1+\chi+\omega-2\psi) + \frac{\psi-\omega}{\log\left(\frac{n}{k}\right)} \\ -(1+\chi+\omega-2\psi) + \frac{\psi-\omega}{\log\left(\frac{n}{k}\right)} & 1+\chi+\omega-2\psi \end{pmatrix}$$
$$= (1+\chi+\omega-\psi)\begin{pmatrix} 1 & -1 \\ -1 & 1 \end{pmatrix} + \frac{\psi-\omega}{\log\left(\frac{n}{k}\right)}\begin{pmatrix} 0 & 1 \\ 1 & 0 \end{pmatrix}. \tag{16}$$

Appendix 4: 3D Plots of p-values for Tests

Figs. 11 and 12

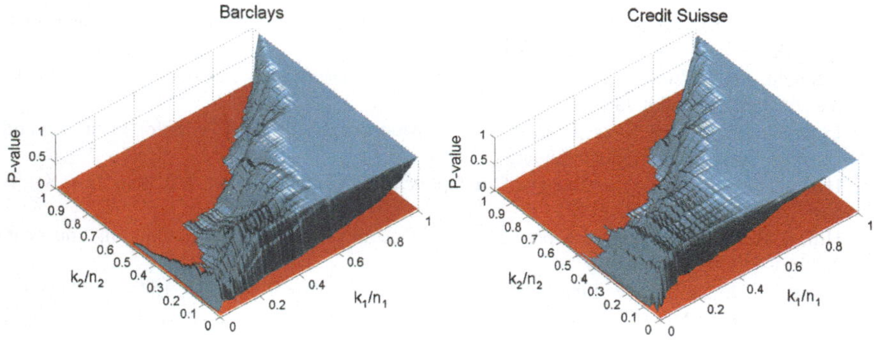

Fig. 11 p-values for testing differences in shape using $T^{(\xi)}_{k_1,k_2,n_1,n_2}$ for all possible choices of k_1 and k_2 for pre- and post-crisis negative log-returns for Barclays and Credit Suisse

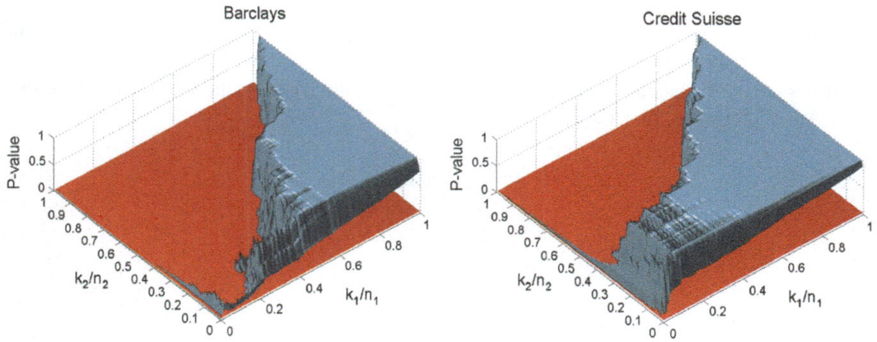

Fig. 12 p-values for testing differences in scale using $T^{(A)}_{k_1,k_2,n_1,n_2}$ for all possible choices of k_1 and k_2 for pre- and post-crisis negative log-returns for Barclays and Credit Suisse

References

1. Beirlant, J., Vynckier, P., Teugels, J.L.: Tail index estimation, Pareto quantile plots and regression diagnostics. J. Am. Stat. Assoc. **91**, 1659–1667 (1996)
2. Beirlant, J., Goegebeur, Y., Teugels, J., Segers, J.: Statistics of Extremes: Theory and Applications. Wiley, Chichester (2004)
3. Beirlant, J., Schoutens, W., Segers, J.: Mandelbrot's extremism. Wilmott Magazine, pp. 97–103 (2005)
4. Beirlant, J., Joossens, E., Segers, J.: Second-order refined peaks-over-threshold modelling for heavy-tailed distributions. J. Stat. Plann. Infer. **139**, 2800–2815 (2009)
5. Bollerslev, T., Todorov, V.: Tails, fears and risk premia. J. Financ. **66**, 2165–2211 (2011)
6. Castillo, E., Hadi, A., Balakrishnan, N., Sarabia, J.: Extreme Value and Related Models with Applications in Engineering and Science. Wiley, Hoboken, NJ (2005)
7. Coles, S.: An Introduction to Statistical Modelling of Extreme Calues. Springer Series in Statistics, Springer, London (2001)
8. de Haan, L.: On regular variation and its applications to the weak convergence of sample extremes. Mathematical Centre Tract 32, Amsterdam (1970)
9. de Haan, L.: Slow variation and characterization of domains of attraction. In: de Oliveira, T. (ed.) Statistical Extremes and Applications, D, pp. 31–48. Reidel, Dordrecht (1984)

10. de Haan, L., Ferreira, A.: Extreme Value Theory: An Introduction. Springer Science and Business Media, LLC, New York, NY (2006)
11. Einmahl, J.H.J., Zhou, C., de, Haan, L.: Statistics of heteroscedastic extremes. J. R. Stat. Soc. Ser. B. Stat. Methodol **78**(1), 31–51 (2016)
12. Embrechts, P., Klüppelberg, C., Mikosch, T.: Modelling Extremal Events for Insurance and Finance. Springer, Berlin, Heidelberg (1997)
13. Hall, P.: On some simple estimates of an exponent of regular variation. J. R. Stat. Soc. Ser. B Stat. Methodol **44**, 37–42 (1982)
14. Hill, B.M.: A simple general approach to inference about the tail of a distribution. Ann. Stat. **3**, 1163–1174 (1975)
15. Hsing, T.: On tail index estimation using dependent data. Ann. Statist. **19**, 1547–1569 (1991)
16. Pickands III, J.: Statistical inference using extreme order statistics. Ann. Statist. **3**, 119–131 (1975)
17. Reiss, R.D., Thomas, M.: Statistical Analysis of Extreme Values, with Applications to Insurance, Finance, Hydrology and Other Fields, 2nd edn. Birkhaüser, Basel (2001)
18. Resnick, S.I.: Heavy-Tail Phenomena: Probabilistic and Statistical Modeling. Springer, New York, NY (2007)
19. Sun, P., Zhou, C.: Diagnosing the distribution of GARCH innovations. J. Empir. Financ. **29**, 287–303 (2014)
20. Weissman, I.: Estimation of parameters and large quantiles based on the *k* largest observations. J. Am. Stat. Assoc. **73**, 812–815 (1978)

Collateralized Borrowing and Default Risk

Eva Lütkebohmert and Yajun Xiao

Abstract We study how margin requirements in the collateralized borrowing affect banks' risk exposure. In a model where a firm's asset value and margin requirement follow correlated geometric Brownian motions, we derive analytic expressions for firm's default probability and debt value. Our results show that variations in margin requirements, reflecting funding liquidity shocks in the short-term collateralized lending market, can lead to a significant increase in firms' default risks, in particular for those firms heavily relying on short-term collateralized borrowing. Moreover, our results imply that reducing margin in liquidity crises can be very effective to restore market lending confidence.

Keywords Collateralized borrowing · Funding liquidity · Margin requirements · Structural credit risk models

1 Introduction

The collateralized short-term lending market has been growing rapidly before the outbreak of the 2008 financial crisis. [8] document that asset-backed commercial papers (ABCP) outstanding in the United States had grown up to 1.1 trillion USD by the end of 2006, dominating the amount of unsecured (non-asset-backed) commercial papers (provided to firms with high-quality debt ratings) outstanding. While the repo market, operating mostly over the counter and under collateral, is short on official statistics [11] argue that its volume is estimated at roughly 12 trillion USD. Given

E. Lütkebohmert (✉)
Department of Quantitative Finance, Institute for Economic Research,
University of Freiburg, Platz der Alten Synagoge, 79098 Freiburg
im Breisgau, Germany
e-mail: eva.luetkebohmert@finance.uni-freiburg.de

Y. Xiao
Business School, University of Technology Sydney, Broadway, PO Box 123,
Sydney, NSW 2007, Australia
e-mail: Yajun.Xiao@uts.edu.au

© Springer International Publishing Switzerland 2016
J. Kallsen and A. Papapantoleon (eds.), *Advanced Modelling in Mathematical Finance*,
Springer Proceedings in Mathematics & Statistics 189,
DOI 10.1007/978-3-319-45875-5_8

the importance and the size of the collateralized lending market, it seems highly necessary to account for the risks inherent in collateralized borrowing within firms' internal risk management.

An important indicator of funding liquidity in the collateralized lending market is the margin requirement. Margin, also called haircut, is a percentage cut from the value of assets that are used as collateral for borrowing. For example, when a financial institution pledges assets worth 100 dollars as collateral but can only borrow 80 dollars, the margin rate is $\frac{100-80}{100} = 20\,\%$, meaning that $20\,\%$ is sliced off from the assets' value. When margins increase significantly, financial institutions may not be able to raise enough funds any more as they cannot provide the required collateral for borrowing. Effectively, it means that the short-term creditors decide not to roll over maturing debt, or in other words, a run on short-term debt occurs. Today, it is well understood that funding liquidity dry-ups in the short-term lending market have played a central role in the most recent crisis in which we have witnessed soaring credit spreads accompanying runs on short-term financial instruments. Examples are the run on ABCP starting in August 2007, on repo in September 2007, and on money market mutual funds in September 2008.

In this paper, we provide a framework to model defaults caused by scarcity of collateral in the short-term lending market and study its implications on default risk for those financial institutions relying on short-term collateralized borrowings.

We build a structural model for a financial institution in which we model both the firm's asset value and the required margin for collateralized borrowing as two correlated geometric Brownian motions. More precisely, we consider a financial institution financed by a mixture of long- and short-term debt. Both types of debt are collateralized by the firm's assets. Short-term debt needs to be rolled over periodically while long-term debt is locked until maturity. The firm defaults due to insolvency when its asset value deteriorates and falls below an exogenous threshold. This corresponds to the classical Black-Cox type default (see [3]). Besides, a negative shock in the collateralized lending market, modelled by variations in the margin process, can increase margin requirements and hence decrease collateral value. When margins are so high that the available collateral is not sufficient to support borrowing the outstanding amount of short-term debt, the latter cannot be rolled over. Hence, the short-term creditors run on debt, which triggers an illiquidity default. In this way, our model allows for two different default scenarios.

We obtain semi-analytic solutions for both firm's default probability and debt value in terms of Bessel functions. Although in this paper, the implementation is conducted under Monte Carlo simulations, the semi closed-form solutions provide an accurate and numerically efficient alternative to compute firm's default probability and thus are particularly interesting for practical applications of the suggested model for firms' internal risk management procedure. Our numerical results show that firm's default probabilities tend to increase when margins increase. The increase is more pronounced for those firms heavily relying on short-term financing or with high rollover frequency. When lending conditions deteriorate and margin exceeds the moderate level of $25\,\%$, the probability of a default due to tightened margin requirements increases dramatically and can eventually dominate the probability of an insolvency default for firms relying to a large extent on short-term financing. We also find that not only financing structure matters but also rollover frequency has a strong influence on default risk. In the limiting case when short-term debt is rolled over on a daily basis, the default probability is extremely sensitive to changes in margin requirements. In that situation, an illiquidity default can easily be triggered when lending conditions tighten. Such defaults can also be understood as debt runs as discussed in [1, 8, 17]. Hence, our model is especially suitable for financial institutions relying on periodically rolled over collateralized borrowing as, for instance, commercial paper conduits or, more general, banks in the shadow banking system.[1] Additionally, we show that variations in margin requirements over short time periods are important to explain firms' default probabilities. Default probabilities increase in both margin volatility and correlation between firm's asset value and margin process.

The remainder of the paper is structured as follows. Section 2 presents the model setup and discusses the default mechanism. The semi closed-form expressions for default probability and debt value are derived in Sect. 3. The numerical results on the effects of funding liquidity on a firm's default probability are presented in Sect. 4 while Sect. 5 concludes. The Appendix provides details on the derivation of the debt value.

2 Model

In this section, we will first describe the firm's asset and liability structure and then discuss the default mechanism in our setting which depends not only on the performance of the firm's asset value but also on the level of the margin requirement.

2.1 *Firm Assets*

Consider a firm which finances its risky assets by debt and equity. The firm's asset value $(V_t)_{t\geq 0}$ is assumed to follow a geometric Brownian motion under the risk-neutral measure

$$\frac{dV_t}{V_t} = r_f dt + \sigma dW_t^1, \tag{1}$$

[1] Former Federal Reserve Chair Ben Bernanke provided a definition in April 2012 at the 2012 Federal Reserve Bank of Atlanta Financial Markets Conference: "Shadow banking, as usually defined, comprises a diverse set of institutions and markets that, collectively, carry out traditional banking functions–but do so outside, or in ways only loosely linked to, the traditional system of regulated depository institutions. Examples of important components of the shadow banking system include securitization vehicles, asset-backed commercial paper (ABCP) conduits, money market mutual funds, markets for repurchase agreements (repos), investment banks, and mortgage companies".

where r_f is the constant expected return on the firm's asset value, which equals the risk-free rate, σ is the asset volatility and $(W_t^1)_{t\geq 0}$ is a standard Brownian motion representing shocks to the firm's asset value.

2.2 Debt Structure

The firm finances its risky assets by a mixture of short- and long-term debt. Long-term debt with maturity T has principal L and coupon C_L while short-term debt with principal S and coupon C_S has to be periodically renewed. We assume all debt to be collateralized. Therefore, the firm needs to pledge its assets as collateral. To simplify the analysis, we assume that the firm uses the entire firm asset value as collateral to borrow. However, the firm cannot borrow the full value but only a fraction of the collateral. The cut on the collateral is called margin or haircut, denoted by m_t at time t, and provides a measure for funding liquidity in the short-term debt market. Hence, at each instant in time t, the firm can at most borrow

$$(1 - m_t)V_t.$$

Thus, the firm can only renew maturing short-term debt S when $(1 - m_t)V_t$ is larger than S. Otherwise, short-term creditors withdraw funds and the firm defaults to a run on short-term debt. In contrast to an insolvency default, such a failure is caused by tightened lending conditions or scarcity of collateral. The latest financial crisis exhibited that the former played a critical role in liquidity dry-ups in the short-term lending market. In the following we model $n_t = 1 - m_t$ as a second geometric Brownian motion

$$\frac{dn_t}{n_t} = \eta dW_t^2, \tag{2}$$

where η is the constant volatility parameter, and $(W_t^2)_{t\geq 0}$ is a standard Brownian motion with $\mathrm{Cov}(W_t^1, W_t^2) = \rho t$. The coefficient ρ representing correlation between firm fundamental and one minus margin is assumed to be constant and positive such that asset value and margin are negatively correlated. This is motivated by recent work of [2] who show that leverage is procyclical and thus, margin as the reciprocal of leverage is countercyclical versus the firm's asset value. Note that $m_t = 1 - n_t$ becomes negative when n_t is larger than 1. In such a case, however, defaults will be driven by insolvency risk, i.e., the firm defaults when its asset value V_t drops below some exogenous insolvency threshold, and the margin constraint is not binding as we will see in the next subsection.

2.3 Default Timing

When liquidity dries up in the short-term lending market, the margin will be so high that the firm is unable to maintain the continuously renewed short-term debt profile and thereby the firm defaults at time

$$\tau_m = \inf\{t > 0 | n_t V_t \leq e^{\lambda_m t} S\},$$

where λ_m is a constant. Such a default can be regarded as a bank run because the short-term creditors are unwilling to lend. Besides the firm can default because of insolvency as in the classical structural credit risk models at the first hitting time

$$\tau_i = \inf\{t > 0 | V_t \leq e^{\lambda_i t} B\},$$

where λ_i is a constant and $e^{\lambda_i t} B$ is the exogenous insolvency threshold at time t. The default time τ is hence given by

$$\tau = \min(\tau_m, \tau_i).$$

The parameters λ_m and λ_i are exogenously given and we set

$$\lambda_i = r_f - \frac{1}{2}\sigma^2 \quad \text{and} \quad \lambda_m = r_f - \frac{1}{2}\sigma^2 - \frac{1}{2}\eta^2. \tag{3}$$

This particular choice of λ_i and λ_m has two advantages. First, it makes $n_t V_t e^{-\lambda_m t}$ and $V_t e^{-\lambda_i t}$ driftless[2] and hence achieves an analytic distribution of the default time τ and the joint distribution of times τ_m and τ_i. Secondly, it reserves a reasonable economic meaning that collateral value and short-term debt value (firm value and debt value) will have the same expected growth rate. This implies a constant leverage ratio of short-term debt to collateral value and total debt to firm asset value in the steady state (see [19]). Further [19] shows that the choices of the parameters λ_i and λ_m has little effect on default correlation for maturities less than 5 years and the impact is still rather small for maturities longer than 5 years. The choice of these parameters, however, does affect the default probability and debt evaluation, in particular, for firms with low asset quality or high margin requirements. In practical application, the parameters should ideally be calibrated to the firm's short-term debt and total liability profile such that they reflect the corresponding growth rates of the latter.

From the definition of the default times due to insolvency τ_i and illiquidity τ_m, it is clear that when n_t is larger than one implying a negative margin m_t, we have $\tau_i < \tau_m$ whenever $S > B$ which usually is the case as B should reflect the firm's total liabilities while S only refers to the principal of short-term debt. Thus, negative margin m_t

[2]This transformation facilitates the analysis and will make the drift parameter redundant in the numerical experiments. However, with drifts present, the semi close-form solutions are still achievable and it causes no loss of efficiency in the simulations.

actually does not imply any problems in our model as it refers to situations where funding liquidity is unproblematic and defaults are solely due to insolvency. Hence, in the simulations one could even set one as an absorbing level for the process n_t.

3 Default Probability and Debt Value

In this section, we derive analytic expressions for the firm's default probability and for the joint density of first passage times τ_m and τ_i following the derivations in [15]. Based on these, we then provide semi-analytic formulas for the firm's debt and equity value. Our approach relies on the methodology introduces in [15, 19], but is also related to work by [4, 16] who use similar techniques to study correlated defaults in the context of pricing of credit derivatives.

3.1 *Analytic Representation of Default Probability*

We first reformulate the problem that the geometric Brownian motion V_t, resp. $n_t V_t$, breaches the threshold $e^{\lambda_i t} B$, resp. $e^{\lambda_m t} S$, to the problem that a two-dimensional driftless Brownian motion hits zero. With the choice of λ_i and λ_m specified in equation (3), the processes

$$
\begin{aligned}
X_t^1 &:= \ln V_t - \ln e^{\lambda_i t} B &&= \ln\left(\frac{V_0}{B}\right) + (r_f - \frac{1}{2}\sigma^2 - \lambda_i)t + \sigma W_t^1 = \ln\left(\frac{V_0}{B}\right) + \sigma W_t^1,\\
X_t^2 &:= \ln n_t V_t - \ln e^{\lambda_m t} S &&= \ln\left(\frac{n_0 V_0}{S}\right) + (r_f - \frac{1}{2}\sigma^2 - \frac{1}{2}\eta^2 - \lambda_m)t + \sigma W_t^1 + \eta W_t^2\\
& &&= \ln\left(\frac{n_0 V_0}{S}\right) + \sigma W_t^1 + \eta W_t^2.
\end{aligned}
$$

turn into driftless Brownian motions. Observe that the first hitting time τ_i that V_t hits $e^{\lambda_i t} B$ is equivalent to the one that X^1 reaches zero (vertical axis), and the first time τ_m that $n_t V_t$ hits $e^{\lambda_m t} S$ is equivalent to the one that X^2 reaches zero (horizontal axis). We can rewrite X in SDE form as

$$
\begin{pmatrix} dX_t^1 \\ dX_t^2 \end{pmatrix} = \Omega \begin{pmatrix} dZ_t^1 \\ dZ_t^2 \end{pmatrix}
$$

with the initial condition

$$
\begin{pmatrix} X_0^1 \\ X_0^2 \end{pmatrix} = \begin{pmatrix} \ln V_0/B \\ \ln n_0 V_0/S \end{pmatrix},
$$

where $Z = (Z^1, Z^2)'$ are uncorrelated standard Brownian motions, and the volatility matrix equals

$$\Omega = \begin{pmatrix} \sigma\sqrt{1-\rho_x^2} & \sigma\rho_x \\ 0 & \sqrt{\sigma^2 + 2\rho\sigma\eta + \eta^2} \end{pmatrix} = \begin{pmatrix} \sigma_1\sqrt{1-\rho_x^2} & \sigma_1\rho_x \\ 0 & \sigma_2 \end{pmatrix}.$$

Here we use $\sigma_1 = \sigma$ and $\sigma_2 = \sqrt{\sigma^2 + 2\rho\sigma\eta + \eta^2}$ for notational convenience in the sequel and we denote the covariance between X^1 and X^2 by

$$\rho_x = \frac{\sigma + \rho\eta}{\sqrt{\sigma^2 + 2\rho\sigma\eta + \eta^2}}.$$

In our application, $0 < \rho < 1$ such that asset value and margin are negatively correlated, and volatilities $\sigma, \eta > 0$, which implies that $\rho_x > 0$ and Ω is invertible. Thus, we are able to define the process $Z = \Omega^{-1}X$ where

$$\Omega^{-1} = \begin{pmatrix} \frac{1}{\sigma_1\sqrt{1-\rho_x^2}} & -\frac{\rho_x}{\sigma_2\sqrt{1-\rho_x^2}} \\ 0 & \frac{1}{\sigma_2} \end{pmatrix}.$$

The transformation $T : \mathbb{R}^2 \to \mathbb{R}^2$ defined by $T(x) = \Omega^{-1}x$ keeps the horizontal axis unchanged but rotates the vertical axis to the line $z_2 = -z_1\sqrt{1-\rho_x^2}/\rho_x$ such that the angle between the horizontal axis ($z_2 = 0$) and the line $z_2 = -z_1\sqrt{1-\rho_x^2}/\rho_x$ is

$$\alpha = \pi + \tan^{-1}\left(-\frac{\sqrt{1-\rho_x^2}}{\rho_x}\right)$$

with $0 < \alpha < \pi$. The transformation gives that τ^m is the first time that Z_t^2 crosses the horizontal axis and τ^i is the first time that Z_t^1 crosses the line $z_2 = z_1 \tan\alpha$. Following [15] we observe that the process $Z_t = \Omega^{-1}X$ initially stays in the wedge between $z_2 = 0$ and $z_2 = -z_1\sqrt{1-\rho_x^2}/\rho_x$ which can be expressed in polar coordinates as

$$\mathscr{C}_\alpha = \{(r\cos\theta, r\sin\theta) : r > 0, 0 < \theta < \alpha\} \subset \mathbb{R}^2.$$

The default time τ is then the exit time when the process Z_t leaves the wedge C_α. In particular, Z_τ lives on the boundary of this wedge which we denote by

$$\partial\mathscr{C}_\alpha = \{(r\cos\theta, r\sin\theta) : r \geq 0, \theta \in \{0, \alpha\}\} \subset \mathbb{R}^2.$$

Using these notations and denoting by I_ν the modified Bessel function of the first kind of order ν, we obtain the following result which is based on corresponding derivations in [15].

Proposition 1 (Compare [15], Eqs. (1.4), (3.2), and (3.3)) *The survival probability of $\tau > t$ is given by*[3]

$$P(\tau > t) = \frac{2r_0}{\sqrt{2\pi t}} e^{-r_0^2/4t} \sum_{n \text{ odd}} \frac{1}{n} \sin\frac{n\pi\theta_0}{\alpha} \left[I_{(\nu_n-1)/2}(r_0^2/4t) + I_{(\nu_n+1)/2}(r_0^2/4t) \right], \tag{4}$$

where $\nu_n = n\pi/\alpha$ and the density of default time is given by

$$P(\tau \in dt) = -\tfrac{\partial}{\partial t} P(\tau > t). \tag{5}$$

Additionally, the joint density $P(\tau_m \in ds, \tau_i \in dt)$ of the first passage times (τ^m, τ^i) is given by

$$P_{s<t}(\tau_m \in ds, \tau_i \in dt) = \frac{dsdt\pi \sin\alpha}{2\alpha^2\sqrt{s(t - s\cos^2\alpha)(t-s)}} \exp\{-\frac{r_0^2}{2s} \frac{t - s\cos 2\alpha}{(t-s) + (t - s\cos 2\alpha)}\} \cdot \sum_{n=1}^{\infty} n \sin\left(\frac{n\pi(\alpha - \theta_0)}{\alpha}\right) I_{n\pi/2\alpha}\left(\frac{r_0^2}{2s} \frac{t-s}{(t-s) + (t - s\cos 2\alpha)}\right), \tag{6}$$

for $s < t$, and for $s > t$ by

$$P_{t<s}(\tau_m \in ds, \tau_i \in dt) = \frac{dsdt\pi \sin\alpha}{2\alpha^2\sqrt{t(s - t\cos^2\alpha)(s-t)}} \exp\{-\frac{r_0^2}{2t} \frac{s - t\cos 2\alpha}{(s-t) + (s - t\cos 2\alpha)}\} \cdot \sum_{n=1}^{\infty} n \sin\left(\frac{n\pi\theta_0}{\alpha}\right) I_{n\pi/2\alpha}\left(\frac{r_0^2}{2t} \frac{s-t}{(s-t) + (s - t\cos 2\alpha)}\right). \tag{7}$$

Although semi-analytic, the numerical computation of these expressions is rather difficult. Firstly, the singularity in the joint density causes problems. More precisely, it is mentioned in [15] that the joint density $P_{t<s}(\tau_m \in ds, \tau_i \in dt)$, resp. $P_{s<t}(\tau_m \in ds, \tau_i \in dt)$, tends to infinity as $|s - t| \to 0$ in case $X_0^1/\sigma_1 < \rho_x X_0^2/\sigma_2$ which implies that the initial angle θ_0 lies in the second quadrant $\pi/2 < \theta_0 < \pi$. Secondly, the truncation error can be challenging as well in evaluating the series of Bessel functions. For instance, it is hard to determine the critical number N up to which the sum over odd n in equation (4) needs to be computed in order to obtain a good approximation when we evaluate (5) numerically for small t. Finally, the modified Bessel functions of the first kind $I_\nu(r_0^2/4t)$ explode for small t. In such circumstances Monte Carlo simulation provides an attractive alternative, either to obtain outright approximations or to provide simple checks on the accuracy of the numerical methods. A different approach to calculate these quantities has been suggested in a recent paper by [14]. The authors derive analytic solutions for the Laplace transform of the first passage times by solving a non-homogeneous modified Helmholtz equation in an infinite wedge using finite Fourier transform. The resulting solution for the Laplace transforms can then be numerically inverted and the authors show that the

[3] This result can already be found in [19].

implementation is very accurate and much more efficient in terms of computation time when compared to Monte Carlo simulation. The following proposition states the corresponding result in our setting. The proof is a straightforward application of the method in [14].

Proposition 2 (Compare [14], Theorem 2) *The Laplace transform of the first passage times* τ_i, τ_m *is given by*

$$\begin{aligned} L(x_1, x_2) &= \mathbb{E}^{(x_1,x_2)}\left[e^{-p_1\tau_i - p_2\tau_m}\right] = \mathbb{E}\left[e^{-p_1\tau_i - p_2\tau_m} | X(0) = (x_1, x_2)\right] \\ &= \sum_{n=1}^{\infty} \sqrt{\frac{2}{\alpha}} \sin(v_n\theta) U_n(r) + exp(-D_1x_1 - D_2x_2), \end{aligned} \tag{8}$$

where $\alpha = \pi + \tan^{-1}\left(-\frac{\sqrt{1-\rho_x^2}}{\rho_x}\right)$, $v_n = n\pi/\alpha$, $D_1 = \sqrt{2p_1}/\sigma_1$, $D_2 = \sqrt{2p_2}/\sigma_2$, *and*

$$\begin{aligned} U_n(r) = \frac{1}{2} A(c, D_1, D_2) \int_{\varphi=0}^{\alpha} \sqrt{\frac{2}{\alpha}} \sin(v_n\varphi) \Bigg[K_{v_n}(ar) \int_{l=0}^{r} e^{-G(\varphi)l} l I_{v_n}(al) dl \\ + I_{v_n}(ar) \int_{l=r}^{\infty} e^{-G(\varphi)l} l K_{v_n}(al) dl \Bigg] d\varphi, \end{aligned}$$

with $c = p_1 + p_2$, $A(c, D_1, D_2) = \sigma_1^2 D_1 + 2\rho_x\sigma_1\sigma_2 D_1 D_2 + \sigma_2^2 D_2^2 - 2c$, $a = a(c) = \sqrt{2c}$, *and*

$$G(\varphi) := D_1\sigma_1 \sin(\alpha - \varphi) + D_2\sigma \sin\varphi.$$

Here I_v *and* K_v *denote the modified Bessel functions of the first and the second kind, resp., of order* v*. Moreover, the Laplace transform of the default time* $\tau = \min\{\tau_i, \tau_m\}$ *is given by*

$$\mathbb{E}^{(x_1,x_2)}\left[e^{-p\tau}\right] = \mathbb{E}^{x_1}\left[e^{-p\tau_i}\right] + \mathbb{E}^{x_2}\left[e^{-p\tau_m}\right] - \mathbb{E}^{(x_1,x_2)}\left[e^{-p\max\{\tau_i,\tau_m\}}\right],$$

where

$$\mathbb{E}^{(x_1,x_2)}\left[e^{-p\max\{\tau_i,\tau_m\}}\right] = \frac{2}{\pi} \int_{v=0}^{\infty} \frac{K_{iv}(\sqrt{2p}r)}{\sinh(\alpha v)} [\cosh(\beta_1 v)\sinh((\alpha - \theta)v) + \cosh(\beta_2 v)\sinh(\theta v)] dv$$

with functions $\beta_j(c) = \arccos(H_j/a(c)) = -i \log\left(\frac{H_j}{a(c)} + i\sqrt{1 - \frac{H_j^2}{a(c)^2}}\right)$ *for* $j = 1, 2$ *and where* $H_1 = G(0) = D_1\sigma_1 \sin\alpha$ *and* $H_2 = G(\alpha) = D_2\sigma_2 \sin\alpha$.

The Laplace transforms can be numerically inverted using the double Laplace inversion formula provided in [5]. This method to compute default probabilities is shown to be 120–180 times faster in terms of computation time than Monte Carlo simulation with Richardson extrapolation (compare [14], Table 3). Although the numerical results in this paper are based on Monte Carlo simulation, the availability of the above analytic expressions for default probabilities in our model setting is thus very important for practical applications.

3.2 Calculation of Debt and Firm Value

In this subsection we will derive semi-analytic expressions for debt and firm value based on the results of the previous subsection. We will first calculate the value of short-term debt. Therefore, denote the asset value at default time τ by $\bar{V}_\tau$ with $\bar{V}_\tau = e^{\lambda_i \tau_i} B$ if $\tau = \tau_i$, and $\bar{V}_\tau = e^{\lambda_m \tau_m} S$ if $\tau = \tau_m$. Recall that short-term debt needs to be periodically renewed at debt maturities. Thus, when short-term debt principal is due at some maturity T_S, the firm needs to pay off existing short-term creditors and issues new short-term debt with the same principal S and coupon C_S. The firm is exposed to funding liquidity problems if the margin requirement at short-term debt rollover dates significantly increases such that the firm is unable to borrow enough short-term funds to ensure financing of its asset holdings. Assuming equal seniority of short- and long-term debt, the total value at time 0 of periodically rolled over short-term debt until time T is given then by

$$D_S(V, B) = \mathbb{E}\left[\int_0^{T \wedge \tau} C_S e^{-r_f s} ds\right] + \mathbb{E}\left[e^{-r_f T} \chi_{\{T < \tau\}} S\right] + \frac{RS}{S+L} \mathbb{E}\left[e^{-r_f \tau} \bar{V}_\tau \chi_{\{\tau \leq T\}}\right],$$

where R is the recovery rate in case of a default. This is the value of all short-term debt that needs to be issued until time T to finance the firm's risky assets. The first term is the coupon payment before default, the second term is the principal payment when there is no default prior to time T, and the third term is the recovered value in case of default. The value of the long-term debt at time 0 can be computed similarly as

$$D_L(V, B) = \mathbb{E}\left[\int_0^{T \wedge \tau} C_L e^{-r_f s} ds\right] + \mathbb{E}\left[e^{-r_f T} \chi_{\{T < \tau\}} L\right] + \frac{RL}{S+L} \mathbb{E}\left[e^{-r_f \tau} \bar{V}_\tau \chi_{\{\tau \leq T\}}\right].$$

It seems that short-term debt is equivalent to long-term debt, which is true to certain degree since short-term debt can be considered as long-term financing through rollover.[4] However, the periodic rollover exposes the firm to funding liquidity risk which is represented by the default time $\tau = \min\{\tau_m, \tau_i\}$. The impact of firm's financing structure and the dynamics of the margin process on debt value will be investigated in the next section. The value of total debt at time 0 is given by

---

[4]Rolling over short-term debt is common practice as firms can reclassify short-term debt as long-term according to the Statement of Financial Accounting Standards.

$$D(V, B) = D_S(V, B) + D_L(V, B).$$

A semi-analytic representation of the total debt value is provided in Proposition 3 below.

Given the values of periodically rolled over short-term debt $D_S(V, B)$ and long-term debt $D_L(V, B)$, the debt yield y_S, resp. y_L, at time 0 computed as the equivalent return on debt conditional on it being held to maturity without default, is determined by solving

$$D_j(V, B) = \frac{C_j}{y_j}(1 - e^{-y_j T}) + je^{-y_j T} \tag{9}$$

for $j \in \{L, S\}$. The difference between debt yield and the risk-free rate gives the credit spread on the firm's short-, resp. long-term debt. The average yield y is calculated by setting $D_S + D_L$ for the debt value D_j, the sum $S + L$ for principal j, and $C_S + C_L$ for the coupon C_j in the above formula. The average firm credit spread is then given by $y - r_f$.

We now turn to the total firm value that is the firm's asset value plus the tax benefit and minus the bankruptcy costs. Assume that the tax rate on the interests is ι. Then the total firm value at time 0 is given by

$$v(V, B) = V_0 + \mathbb{E}\left[\int_0^{T\wedge\tau} \iota e^{-r_f s}(C_S + C_L)ds\right] - (1 - R)\mathbb{E}\left[e^{-r_f \tau} V_\tau \chi_{\{\tau\le T\}}\right].$$

The equity value is the total firm value net the total debt value

$$E(V, B) = v(V, B) - D(V, B).$$

The evaluation of debt and equity is summarized in the following proposition.

Proposition 3 *The value of the debt at time* 0 *is*

$$\begin{aligned}
&D(V, B)\\
&= \tfrac{C_S+C_L}{r_f}\left(1 - e^{-r_f T} P(\tau > T)\right) - \tfrac{C_S+C_L}{r_f}\int_0^T e^{-r_f s} P(\tau \in ds) + (S + L)e^{-r_f} P(\tau > T)\\
&+ R\Bigg(\int_0^T\int_0^u e^{-r_f v}\bar{V}_v P_{v<u}(\tau_m \in dv, \tau_i \in du) + \int_0^T\int_u^T e^{-r_f u}\bar{V}_u P_{u<v}(\tau_m \in dv, \tau_i \in du)\\
&\quad + \int_T^\infty\int_0^T e^{-r_f v}\bar{V}_v P_{v<u}(\tau_m \in dv, \tau_i \in du) + \int_0^T\int_T^\infty e^{-r_f u}\bar{V}_u P_{u<v}(\tau_m \in dv, \tau_i \in du)\Bigg)
\end{aligned} \tag{10}$$

and the total firm value is

$$
\begin{aligned}
&v(V,B)\\
&= V_0\frac{\iota(C_S+C_L)}{r_f}\left(1-e^{-r_fT}P(\tau>T)\right)-\frac{\iota(C_S+C_L)}{r_fT}\int_0^T e^{-r_fs}P(\tau\in ds)\\
&-(1-R)\Bigg(\int_0^T\int_0^u e^{-r_fv}\bar{V}_vP_{v<u}(\tau_m\in dv,\tau_i\in du)+\int_0^T\int_u^T e^{-r_fu}\bar{V}_uP_{u<v}(\tau_m\in dv,\tau_i\in du)\\
&\quad+\int_T^\infty\int_0^T e^{-r_fv}\bar{V}_vP_{v<u}(\tau_m\in dv,\tau_i\in du)+\int_0^T\int_T^\infty e^{-r_fu}\bar{V}_uP_{u<v}(\tau_m\in dv,\tau_i\in du)\Bigg),
\end{aligned}
\tag{11}
$$

where the value of $\bar{V}$ *takes*

$$
\begin{aligned}
\bar{V}_v &= e^{\lambda_m v}S, \quad \lambda_m = r_f-\frac{1}{2}\sigma^2-\frac{1}{2}\eta^2,\\
\bar{V}_u &= e^{\lambda_i u}B, \quad \lambda_i = r_f-\frac{1}{2}\sigma^2.
\end{aligned}
$$

The analytic form for the survival probability $P(\tau>T)$, *the default density* $P(\tau\in ds)$, *and the joint default density* $P(\tau_m\in dv,\tau_i\in du)$ *is given in Proposition 1.*

The proof can be found in the Appendix. It is based on the expressions for survival probability and default time density stated in Proposition 1 which follow from [15].

4 Numerical Results

4.1 Model Parameters

We calibrate our model to parameters used in the literature on structural credit risk models. As in [13], we set the risk-free rate equal to 8 %, the historical average of treasury rates between 1973 and 1998. For the sensitivity analysis we choose an initial asset value equal to $V_0 = 100$ monetary units. The volatility of the firm's assets is set to $\sigma = 25\,\%$ as in [18]. We choose the recovery rate $R = 60\,\%$ following [6], who finds that bonds have default recovery rates of around 60 % across nine different aggregate states. [9] argue that the financial firms tend to have shorter debt maturities as they rely heavily on repo transactions with maturities from one day to three months and commercial papers with maturities of less than 9 months. In our empirical analysis, we therefore assume the maturity of long-term debt to be $T = 1$ year and short-term debt is rolled over on a quarterly basis, i.e., every three months short-term debt matures and new short-term debt has to be issued which then matures three months later and so on. For comparison, we also consider a daily rollover frequency. This choice is motivated by the fact that our analysis is mainly focusing on short-term effects of negative shocks in the lending market, reflected by increasing margin requirements, on firms' default risk. In other words, our model is suitable for stress testing the effects of liquidity risk over short time periods.

Long-term debt principal is set to 40 monetary units with a continuously paid coupon of 3.8 monetary units. Short-term debt principal is 20 monetary units with coupon of 1.8 monetary units. This implies that the coupon rate on long-term debt is larger than that on short-term debt. Throughout our analysis, we keep the total debt outstanding fixed as 60 monetary units and the coupon rate to every type of debt is constant. As in [7], we set the default threshold $B = 61.40$ monetary units which yields an appropriate credit spread as a benchmark.

Market data on margins unfortunately is not publicly available. However, there are financial institutions collecting margin data (see e.g. [10]), and there are reports from banks revealing margin requirements for some asset classes in certain periods.[5] The Term Asset-Backed Securities Loan Facility (TALF) and the Public-Private Investment Program (PPIP), announced in early 2009, provide bond lending at exactly 50 % margin. The latter is an intermediate level between the 5 % margin required at the peak of the leverage bubble and the 70–90 % margin demanded during the crisis in 2008. Since then, the asset market enjoyed a sound rebound in prices and the bond market saw a solid drop in spread. Therefore, we choose an initial margin of $m_0 = 10\,\%$, resp. $n_0 = 90\,\%$, a level considered in a booming market. We set the volatility of the process n_t to $\eta = 0.25$ so that the unconditional mean of n_t in a year is still close to 95 %. The firm fundamental and its margin requirement are affected by the common market factor. The correlation between the two driving processes $(V_t)_{t\geq 0}$ and $(n_t)_{t\geq 0}$ is assumed to be $\rho = 50\,\%$ such that asset value and margin are negatively correlated. The sensitivity of our results with respect to the choice of the volatility and correlation parameters will be discussed in the next subsection. The calibrated model parameters will produce 250 basis points credit spread on aggregate debt calculated according to equation (9) with principal $L + S$ and coupon $C_L + C_S$ and maturity in 1 year. We do not investigate the evaluation of debt and equity in our numerical analysis. The debt value, however, is used in order to determine an appropriate default barrier. Table 1 summarizes these baseline parameters of our numerical analysis.

4.2 Default Probability

Margin requirements measure a firm's capability to raise funding through collateralized borrowing. A large negative shock in the lending market causes significant increases in margins, which means creditors are reluctant to lend and run on debt. Highly leveraged firms are thus especially vulnerable w.r.t. runs on collateralized debt. By comparing the total default probability in Panel A with that in Panel C of Fig. 1, we immediately see that default risk is substantial for banks relying on

---

[5] See, e.g., "International banking and financial market developments", BIS Quarterly Review December 2011, or "The role of margin requirements and haircuts in procyclicality", CGFS Papers, No 36.

Table 1 Baseline parameters

Firm characteristic	
Initial firm fundamental	$V_0 = 100$
Volatility of firm fundamental	$\sigma = 25\,\%$
Bankruptcy recovery rate	$R = 60\,\%$
Insolvency threshold	$B = 61.40$
Debt structure	
Long-term debt principal	$L = 40$
Long-term debt coupon	$C_L = 3.8$
Maturity of long-term debt	$T = 1$ year
Short-term debt principal	$S = 20$
Short-term debt coupon	$C_S = 1.8$
Short-term debt rollover frequency	$\Delta t = 3$ months
Margin	
Initial margin	$m_0 = 10\,\%$
Volatility of margin	$\eta = 25\,\%$
Correlation parameter	$\rho = 0.5$
Macro variables	
Risk-free interest rate	$r_f = 8\,\%$

periodically rolled over funding.[6] The default probability in Panel C shoots up dramatically, more than 10 % in one year, when the ratio of short-term debt to total debt increases form 1/3 to 1/2 and when the initial margin level is larger than 25 %. Panel B shows that the default probability due to funding liquidity risk for firms with one-third short-term debt is almost negligible even at higher margin levels. In stark contrast, the default probability due to funding liquidity risk picks up quickly when margin climbs over 25 % in Panel D. It completely dominates the default probability due to insolvency if the margin is larger than 35 %. The default probability caused by illiquidity is about 20 % in one year for an initial margin level of 50 %. This is not very surprising as the initial collateral value is $100 \times 0.5 = 50$ whereas short-term debt principal is 30. Thus, firms will very likely fail to roll over maturing short-term debt and hence default when margin requirements tighten. Moreover, the results in Fig. 1 indicate that the rollover frequency matters. The higher the rollover frequency of short-term debt is, the higher is also the firm's default probability as the firm is exposed to higher rollover risk. Especially in the limiting case when debt is rolled over on a daily basis, the default probability caused by illiquidity completely dom-

[6]Be aware that we use $n_0 = 1 - m_0$ in our simulations not m_0 directly.

inates the one due to insolvency when the initial margin is larger than 35 %. This leads to a dramatic increase in the total default probability. A similar phenomenon has been found in [12].

Figure 1 further reveals a stylized fact observed in the crisis. Firms, taking huge leverage in boom times in which funding liquidity seemingly can be neglected, were severely hit by the liquidity dry-ups in the collateralized short-term debt market. For the same reason, in the sequel analysis, we assume a mild level of margin or a short-term debt ratio higher than the one in the base model in the interest of comparison. We investigate how default changes against short-term debt ratio, correlation and margin volatility.

Our results show that the default probability is highly sensitive to the firm's debt financing structure. Panel A of Fig. 2 indicates that even for a mild level of initial margin, $m_0 = 25\,\%$, the total default probability dramatically increases from about 6.5 % a year to roughly 10 % a year for both rollover frequencies when the short-term debt ratio increases to 0.6. Panel B reveals that this increase is mainly caused by a significant rise in the default probability due to funding liquidity. Additionally,

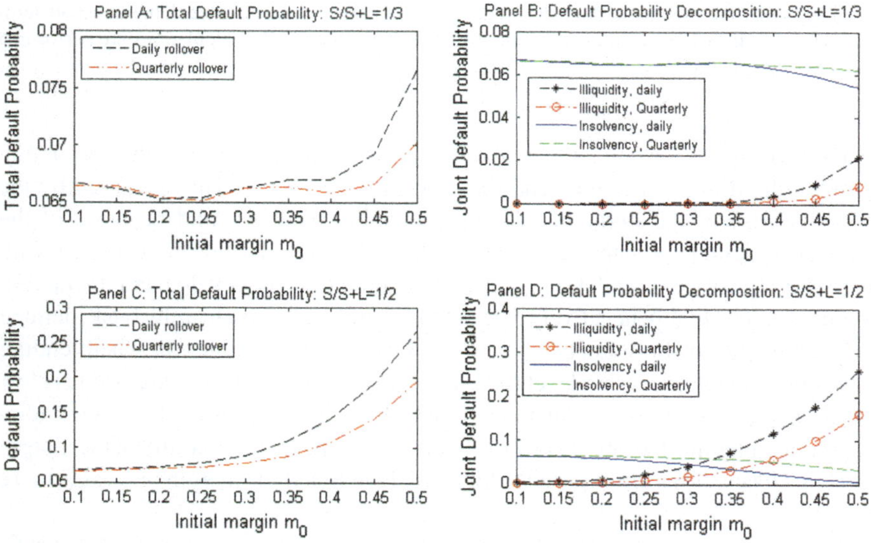

Fig. 1 Dependence of Default Probability on Initial Margin.
Panel A shows the firm's total default probability while *Panel B* illustrates its individual components separately for short-term debt ratio equal to 1/3, the default probability due to insolvency $P(\tau_i \leq T, \tau_i < \tau_m)$, and the default probability due to illiquidity $P(\tau_m \leq T, \tau_m < \tau_i)$, for daily and quarterly rollover frequency. *Panel C* and *Panel D* plot the same probailities for short-term debt ratio equal to 1/2. *Panel A* and *B* use the baseline parameters listed in Table 1 to produce a credit spread of 250 bps on aggregate debt. *Panel C* and *D* use the same parameters apart from the short-term debt ratio which is set to 1/2 here. Calculations are performed under the risk-neutral measure

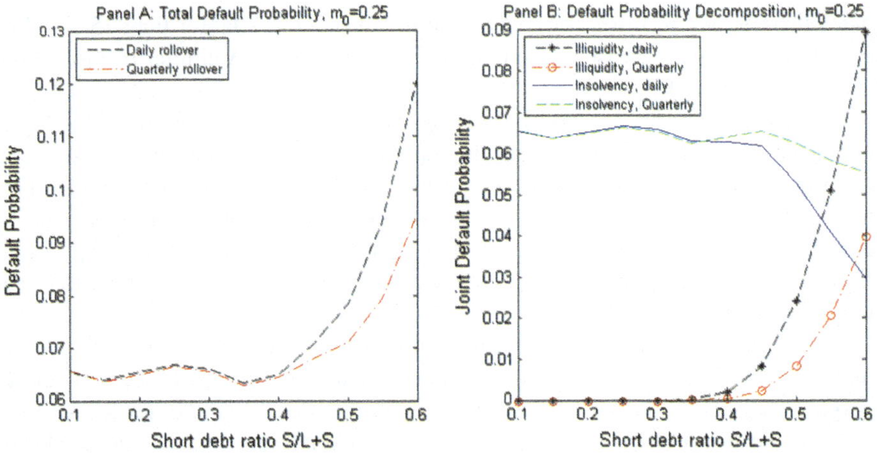

Fig. 2 Dependence of Default Probability on Short-term Debt Ratio.
Panel A shows the firm's total default probability while *Panel B* illustrates its individual components separately, the insolvency default probability $P(\tau_i \leq T, \tau_i < \tau_m)$, and the illiquidity default probability $P(\tau_m \leq T, \tau_m < \tau_i)$, for daily and quarterly rollover frequency. The baseline parameters listed in Table 1 are used to produce the credit spread of 250 bps on aggregate debt with short-term debt ratio equalt to $1/3$ except that the intial margin is fixed at 25 %. Calculations are performed under the risk-neutral measure

Panel A of Fig. 3 shows that the total default probability increases in the correlation between the firm's asset value and the process $n_t = 1 - m_t$. This is caused by the steep increase of the illiquidity default probability and the mild decrease of the insolvency default probability shown in Panel B. The intuition here is that with increasing (positive) correlation coefficient ρ, the firm's asset value and the margin process become highly negatively correlated. Therefore, even a healthy bank, defined by good quality of firm assets, is exposed to significant default risk when lending conditions tighten as margin level tends to be high when the firm's asset value is low which makes rolling over collateralized short-term debt very difficult. Furthermore, Fig. 4 illustrates that the default probability is increasing in the volatility of the margin process. Hence, firms are exposed to higher default risk if the margin process is more volatile.

Thus, our results show that tightened funding conditions can increase a firm's default risk greatly even if it holds high quality assets. Tightened lending conditions can arise from increasing margin requirements, higher fluctuations in the margin process, or increasing negative correlation between firm fundamental value and margin process. Firms heavily relying on collateralized short-term funding will then be exposed to significant default risk. Reports have shown that during the European debt

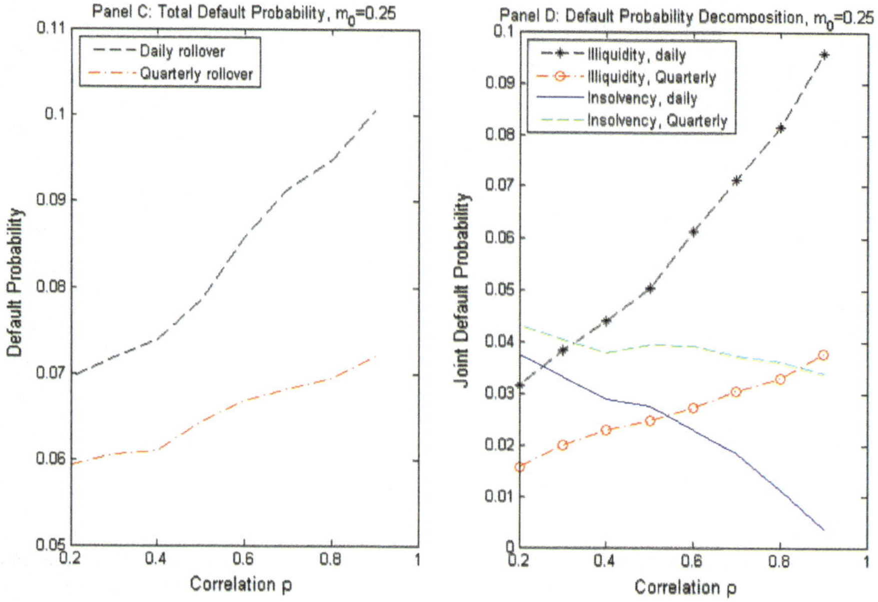

Fig. 3 Dependence of Default Probability on Correlation.
Panel A shows the firm's total default probability while *Panel B* illustrates its individual components separately, the default probability due to insolvency $P(\tau_i \leq T, \tau_i < \tau_m)$, and the default probability due to illiquidity $P(\tau_m \leq T, \tau_m < \tau_i)$, for daily and quarterly rollover frequency. The baseline parameters listed in Table 1 are used to produce the credit spread of 250 bps on aggregate debt with short-term debt ratio equalt to 1/3 except that the inital margin is 25 %. Calculations are performed under the risk-neutral measure

crisis the market margin on mortgage-backed, asset-backed, and structural securities was well above 50 %.[7] The lenders' confidence collapsed and the private lending activity basically stopped before ECB stepped up to rescue. In boom times, the funding liquidity risk could be negligible with very loose margin requirement. However, when the market switches into a regime of distress with significantly high margin, our results show that the default probability of a firm can dramatically increase even if its fundamental performs well. Thus, our model can be implemented for distress testing of firm's exposure to rollover risk arising from collateralized borrowing.

[7]There is no authoritative data on the use of haircuts/initial margins in the repo market in either Europe or the US. Table 1 in the research report published by Committee on the Global Financial System Study Group shows margin data in bilateral interviews in various financial centers with various market users, including banks, prime brokers, custodians, asset managers, pension funds and hedge funds. For reference see http://www.bis.org/publ/cgfs36.pdf.

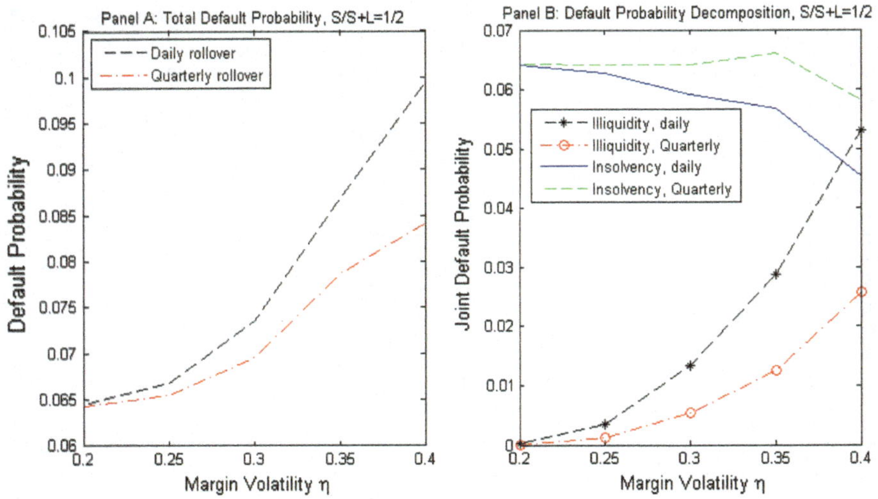

Fig. 4 Dependence of Default Probability on Margin Volatility.
Panel A shows the firm's total default probability while *Panel B* illustrates its individual components separately, the insolvency default probability $P(\tau_i \leq T, \tau_i < \tau_m)$, and the illiquidity default probability $P(\tau_m \leq T, \tau_m < \tau_i)$, for daily and quarterly rollover frequency. The baseline parameters listed in Table 1 are used to produce the credit spread of 250 bps on aggregate debt except that the short-term debt ratio is set to $1/2$ here. Calculations are performed under the risk-neutral measure

5 Conclusion

We propose a structural credit risk model incorporating funding liquidity risk represented by variations in margin requirements in the collateralized short-term lending market. By modelling the firm's asset value and margin processes as two correlated geometric Brownian motions, we account for two different types of default: The firm defaults due to insolvency when its asset value hits an exogenous default threshold while the firm defaults due to illiquidity when the margin process hits another barrier reflecting that the firm's collateral value is insufficient for rolling over short-term debt. We transform the default timing into a first passage time of two correlated Brownian motions and derive explicit expressions for both default probabilities and debt values in terms of Bessel functions and their integrals. These semi-analytic representations are very useful for practical applications as numerical methods based on such expressions are several times faster than Monte Carlo simulations. Our results show that fluctuations in margin requirements can significantly expose a firm to rollover risk, especially for firms heavily relying on short-term financing. Thus, funding liquidity risk should be taken into account in firm's internal risk management as well as in debt pricing.

Acknowledgements This work was jointly supported by the German Research Foundation through the project *Modelling of market, credit and liquidity risks in fixed-income-markets* and the UTS Business School Research Funds. The financial support is gratefully acknowledged by the authors.

Appendix

Proof of Proposition 3

We first calculate the value of short-term debt D_S,

$$\begin{aligned} & D_S(V,B) \\ & = \mathbb{E}\left[\int_0^{T\wedge\tau} C_S e^{-r_f s} ds\right] + \mathbb{E}\left[e^{-r_f T}\chi_{\{T<\tau\}}S\right] + \tfrac{S}{S+L}\mathbb{E}\left[e^{-r_f\tau}\bar{V}_\tau\chi_{\{\tau\le T\}}\right] \qquad \text{(A.1)} \\ & = T_1 + T_2 + T_3. \end{aligned}$$

The first term is

$$\begin{aligned} T_1 &= \mathbb{E}\left[\int_0^{T\wedge\tau} C_k e^{-r_f s} ds\right] \\ &= \tfrac{C_S}{r_f}\left(1 - e^{-r_f(T\wedge\tau)}\right) \\ &= \tfrac{C_S}{r_f} - \tfrac{C_k}{r_f}\mathbb{E}\left[e^{-r_f T}\chi_{\{\tau>T\}}\right] - \tfrac{C_S}{r_f}\mathbb{E}\left[e^{-r_f\tau}\chi_{\{\tau\le T\}}\right] \\ &= \tfrac{C_S}{r_f}\left(1 - e^{-r_f T}P(\tau>T)\right) - \tfrac{C_S}{r_f}\int_0^T e^{-r_f s}P(\tau\in ds), \end{aligned}$$

where the survival probability $P(\tau > T)$ and the default density function are given in (4) and (5). The second term can be computed as

$$T_2 = \mathbb{E}\left[e^{-r_f T}\chi_{\{T<\tau\}}S\right] = Se^{-r_f T}P(\tau > T).$$

The last term is

$$\begin{aligned} T_3 &= \tfrac{RS}{S+L}\mathbb{E}\left[e^{-r_f\tau}\bar{V}_\tau\chi_{\{\tau\le T\}}\right] \\ &= \tfrac{RS}{S+L}\mathbb{E}\left[e^{-r_f\tau_m}\bar{V}_{\tau_m}\chi_{\{\tau_m<\tau_i<T\}} + e^{-r_f\tau_i}\bar{V}_{\tau_i}\chi_{\{\tau_i<\tau_m<T\}}\right. \\ &\qquad \left. + e^{-r_f\tau_m}\bar{V}_{\tau_m}\chi_{\{\tau_m<T<\tau_i\}} + e^{-r_f\tau_i}\bar{V}_{\tau_i}\chi_{\{\tau_i<T<\tau_m\}}\right] \\ &= \tfrac{RS}{S+L}\left(\int_0^T\int_0^u e^{-r_f v}\bar{V}_v P_{v<u}(\tau_m\in dv, \tau_i\in du) + \int_0^T\int_u^T e^{-r_f u}\bar{V}_u P_{u<v}(\tau_m\in dv, \tau_i\in du)\right. \\ &\qquad \left. + \int_T^\infty\int_0^T e^{-r_f v}\bar{V}_v P_{v<u}(\tau_m\in dv, \tau_i\in du) + \int_0^T\int_T^\infty e^{-r_f u}\bar{V}_u P_{u<v}(\tau_m\in dv, \tau_i\in du)\right), \end{aligned} \qquad \text{(A.2)}$$

where the joint default density is given in (6) and (7). For the default time $v < u$, we know

$$\bar{V}_v = e^{\lambda_m v}S,$$

and for $u < v$

$$\bar{V}_u = e^{\lambda_i u}B.$$

The long-term debt value will be the same by replacing principal and coupon. The total firm value is the unlevered firm value plus tax shields minus bankruptcy costs

$$
\begin{aligned}
& v(V, B) \\
&= V_0 + \mathbb{E}\left[\int_0^{T\wedge\tau} \iota(C_S + C_L)e^{-r_f s} ds\right] - (1-R)\mathbb{E}\left[e^{-r_f \tau}\bar{V}_\tau \chi_{\{\tau\le T\}}\right] \\
&= V_0 \tfrac{\iota(C_S+C_L)}{r_f}\left(1 - e^{-r_f T} P(\tau > T)\right) - \tfrac{\iota(C_S+C_L)}{r_f}\int_0^T e^{-r_f s} P(\tau \in ds) \\
&\quad - (1-R)\Bigg(\int_0^T\int_0^u e^{-r_f v}\bar{V}_v P_{v<u}(\tau_m \in dv, \tau_i \in du) \\
&\qquad + \int_0^T\int_u^T e^{-r_f u}\bar{V}_u P_{u<v}(\tau_m \in dv, \tau_i \in du) \\
&\qquad + \int_T^\infty\int_0^T e^{-r_f v}\bar{V}_v P_{v<u}(\tau_m \in dv, \tau_i \in du) \\
&\qquad + \int_0^T\int_T^\infty e^{-r_f u}\bar{V}_u P_{u<v}(\tau_m \in dv, \tau_i \in du)\Bigg).
\end{aligned}
\tag{A.3}
$$

Finally, equity value is calculated as total firm value net the debt value

$$E(V, B) = v(V, B) - D(V, B). \tag{A.4}$$

□

References

1. Acharya, V.V., Schnabl, P., Suarez, G.A.: Securitization without risk transfer. J. Financ. Econ. **107**(3), 515–536 (2013)
2. Adrian, T., Shin, H.S.: Procyclical leverage and value-at-risk. Rev. Finan. Stud. **27**(2), 373–403 (2014)
3. Black, F., Cox, J.: Valuing corporate securities: Some effects of bond indenture provisions. J. Financ. **31**, 351–367 (1976)
4. Blanchet-Scalliet, C., Patras, F.: Structural counterparty risk valuation for credit default swaps. In: Brigo, D., Bielecki, T., Patras, F. (eds.) Credit Risk Frontiers: Subprime Crisis, Pricing and Hedging, CVA, MBS, Ratings, and Liquidity. Wiley (2011)
5. Cai, N., Kou, S., Liu, Z.: A two-sided Laplace inversion algorithm with computable error bounds and its applications in financial engineering. Adv. Appl. Probab. **46**, 766–789 (2014)
6. Chen, H.: Macroeconomic conditions and the puzzles of credit spreads and capital structure. J. Financ. **65**(6), 2171–2212 (2010)
7. Chen, L., Collin-Dufresne, P., Goldstein, R.: On the relation between the credit spread puzzle and the equity premium puzzle. Rev. Finan. Stud. **22**(9), 3367–3409 (2009)
8. Covitz, D., Liang, N., Suarez, G.A.: The evolution of a financial crisis: collapse of the asset-backed commercial paper market.The. J. Financ. **68**(3), 815–848 (2013)
9. Custódio, C., Ferreira, M.A., Laureano, L.: Why are US firms using more short-term debt? J. Financ. Econ. **108**(1), 182–212 (2013)

10. Geanakoplos, J.: Solving the present crisis and managing the leverage cycle. Fed. Reserve Bank N.Y. Econ. Policy Rev. 101–131 (2010)
11. Gorton, G., Metrick, A.: Securitized banking and the run on repo. J. Financ. Econ. **104**(3), 425–451 (2012)
12. He, Z., Xiong, W.: Rollover risk and credit risk. J. Financ. **67**(2), 391–430 (2012)
13. Huang, J., Huang, M.: How much of the corporate-treasury yield spread is due to credit risk? Rev. Asset Pricing Stud. **2**, 153–202 (2003)
14. Kou, S., Zhong, H.: First passage times of two-dimensional Brownian motion. Working paper (2015)
15. Metzler, A.: On the first passage problem for correlated Brownian motion. Stat. Probab. Lett. **80**(5), 277–284 (2010)
16. Patras, F.: A reflection principle for correlated defaults. Stochast. Process. Appl. **116**, 690–698 (2006)
17. Schroth, E., Suarez, G.A., Taylor, L.A.: Dynamic debt runs and financial fragility: Evidence from the 2007 ABCP crisis. J. Financ. Econ. **112**(2), 164–189 (2014)
18. Zhang, B., Zhou, H., Zhu, H.: Explaining credit default swap spreads with the equity volatility and jump risks of individual firms. Rev. Finan. Stud. **22**(12), 5099–5131 (2009)
19. Zhou, C.: An analysis of default correlations and multiple defaults. Rev. Finan. Stud. **14**(2), 555–576 (2001)

Model Uncertainty in a Holistic Perspective

Gerhard Stahl

Abstract This paper focuses on model uncertainty within a holistic perspective. The latter is characterized by a consistent approach to risk measurement by combining stochastic, economic, operational and regulatory elements. This paper is a plea to account for model uncertainties on the level of consequences and not at the level of risk factors. This has important implications for validation, auditing and is of use testing of internal models. In line with risk management approaches, uncertainties have to managed. The starting point for this process is the identification and measurement of uncertainties. To achieve this goal further specific criteria for validity and resilience, are introduced in this paper. Examples from real world internal models highlight the practical relevance of the introduced concepts. A concluding section summarizes the main insights.

Keywords Solvency II · Internal models · Standard formula · Model validation · Model uncertainty

1 Introduction

Looking back on the development of the financial industry in the 20-th century, in particular two components of technical progress, namely the invention of personal computers and advances in financial engineering put their stamp on. Their combined application changed the face of financial markets by dealing in real time with derivatives and other financial instruments.

As a consequence thereof, modern risk management was integrated in the *value creation chain* of financial institutions, see Wilson for an overview [1]. It turned out to be questionable, whether the determination of regulatory capital for trading portfolios, with a significant amount of derivatives and *diversification* effects due to international activities, by means of so-called (regulatory) standardized methods was

G. Stahl (✉)
Talanx AG, 30659 Hanover, Germany
e-mail: Gerhard.Stahl@talanx.com

© Springer International Publishing Switzerland 2016

J. Kallsen and A. Papapantoleon (eds.), *Advanced Modelling in Mathematical Finance*, Springer Proceedings in Mathematics & Statistics 189,
DOI 10.1007/978-3-319-45875-5_9

still adequate *and* economically efficient. An important lesson of those times was that the complexity of derivative portfolios can in general not be captured by so-called standardized methods. From a regulatory perspective, the solution was seen in the application of so-called *internal models*, by which the adequate and yet economic efficient capital cushion is determined by firm-specific stochastic models. Furthermore, in addition to this quantitative component, further qualitative requirements fostered good risk management practices. It is remarkable that Egdeworth suggested already in 1888, see [2], to base banking regulation on stochastics, however it took more then 100 years until this approach came true.

The application of internal models in the financial industry for regulatory purposes started in the mid-nineties, see for example the book of Pada-Schioppa [3], and internal models are going to be implemented for insurances under Solvency II, see [4].

In the aftermath of the financial crisis in 2007 market participants and their stakeholders pointed however critically on the *uncertainty* related to such model based approaches. To tackle with *model uncertainty* in a *holistic* manner is the motivation of this paper.

The post-crisis perception of model uncertainty of internal models shows their Janus-faced ambivalence—either applied for pricing or for risk management purposes. The society reacted with regulations which are based on *improbable events*, in order to increase the *safety* of the financial system. Approaches to root decisions by rational calculations based on extreme improbable events are not only bearing the risk to fail their goal, but even destroys the rationality of the methods and procedures underlying an internal model, see Luhmann's book [5], p. 4.

This paradox or dilemma of risk-based regulation is a main motivation for this paper. This paper provides a closer look at the interplay of model uncertainty, available background knowledge (including data and sample sizes) and the regulatory required safety cushion of a risk management system. Starting point to grasp and to analyze this interplay is the definition of a *model in a wide sense*. This definition of a holistic model is crucial, because only this perspective ensures to capture all components in an integrated manner. This definition is in line with the general understanding of an internal model under Solvency II and fits perfectly to the ISO-norm of risk management.

The model uncertainty is split-up into *epistemic, aleatory and imprecision* components. This structure is embedded in the recent contributions of Aven to risk management. From this work the insights to deal with very different aspects of risk management systems either on the level of *risk factors* or *consequences* are applied to the assessment of model uncertainties.

The recent research contributions of Rüschendorf, Vanduffel, Puccetti, Embrechts and Bernard, see [6–8] amoung others are applied to real-life internal models and considered in the perspective of the quantification of uncertainties as given in Dubois. The reader may get an impression on the complexity of real-life internal models by inspecting SCOR's model documentation, see [9].

Furthermore the recent work of Cambou and Fillipovic on regulatory framework in Switzerland is interpreted within Aven's framework. Last but not least applications of model uncertainties under Solvency II are given.

The D-VaR Approach of RiskMetrics In October 1994 JP Morgan launched its 1st edition of RiskMetrics, a model based stochastic approach to control and to report risks related to investment portfolios, including derivatives, and to allocate economic capital. The latter is used as a buffer against potential future losses. JP Morgan's initiative also raised a new dimension: *transparency*. It was the very first time that a financial institution shared openly with the public their internal documents and made business knowledge, which was expensive to be developed, available to third parties. This strategy of transparency was surprisingly successful, because at one hand it sent out a world-wide signal of competence to the markets about JP Morgan's expertise in issues of risk management. Hence, JP Morgan was acknowledged as the leader in questions of risk management. On the other hand regulators took RiskMetrics as a blue print to formulate the *market risk amendment* in 1996, see [10] which came in law under the capital adequacy directive in 1997. The latter offered banks the option to choose internal models or the standard formula for the calculation of regulatory capital. This approach was rolled out to other risk categories for banks (credit risk and operational risk) under Basel II and is nowadays going to be applied for insurance companies under Solvency II.

From the early days on, regulators not only feared to be assessed by arbitrage by financial institutions which apply internal models but furthermore questioned the *precision* of such models. In other words, *model uncertainty* was judged highly and played an important role from the very beginning of the application of internal models for regulatory purposes. For that reason a safety factor was applied in order to tame model uncertainty. Based on a contribution by Stahl [11], the *safety factor* of 3 was applied to internal models used by banks to capture market risk. This factor is still—even in the aftermath of the financial crisis—in force, see [12].

The following two examples illustrate the last two paragraphs. We start with JP Morgan's *Delta-Normal-Method*.

Example 1 Let us consider a portfolio whose value may depend on n risk factors, say $\mathbf{R} = (R^1, \ldots, R^d)$. The D-VaR (Delta-VaR) approach is characterized by a first order Taylor expansion of the valuation function. Hence the approximation for the change in value of the portfolio V due to fluctuations of risk factors over a fixed period of time is:

$$\triangle V = V(\mathbf{R} + \triangle\mathbf{R}) - V(\mathbf{R}) \approx \sum_{i=1}^{d} \frac{\partial V}{\partial R^i} \triangle R^i = \mathbf{D}\triangle\mathbf{R}, \tag{1}$$

where $\mathbf{D}$ denotes the vector of partial derivatives

$$\mathbf{D} = (\partial V/\partial R^1, \ldots, \partial V/\partial R^d).$$

Important examples for the risk factors R^i, commonly used in practice, are the log prices of basic instruments such as stocks.

Stochastic models come into play in order to capture the uncertainty, how the value of a portfolio may change over a certain holding period. Recall that the prices of basic instruments at time $t+1$ are uncertain at t, hence ΔV is a random variable. Consider a Black Scholes framework, i.e. it is assumed that the log price changes $\Delta \mathbf{R} = \Delta \ln \mathbf{P}$ follow a multivariate Gaussian distribution of dimension d, with mean vector μ and covariance matrix Σ:

$$\Delta \mathbf{R} \sim N_d(\mu, \Sigma). \tag{2}$$

By definition of the multivariate Gaussian distribution, a linear combination of $\Delta \mathbf{R}$ with exposure $\Lambda_t = (\lambda_1, \ldots, \lambda_d)$ follows a univariate normal distribution. Hence, the distribution of the change of a linearized portfolio, denoted by $\mathcal{D}(\Delta V)$, is given by the following univariate normal distribution:

$$\mathcal{D}(\Delta V) \approx N(\mathbf{D}\, diag(\Lambda_t)\, \mu,\ \ \mathbf{D}^T\, diag(\Lambda_t)\, \Sigma\, diag(\Lambda_t)\mathbf{D}). \tag{3}$$

In practice the forecast horizon for market risk is rather short say one or ten trading days. This allows a further simplification by setting $\mu = 0$.

$$\mathcal{D}(\Delta V) \approx N(0, \sigma^2), \tag{4}$$

where $\sigma^2 = \mathbf{D}^T\, diag(\Lambda_t)\, \Sigma\, diag(\Lambda_t)\mathbf{D}$.

The D-VaR approach—and its generalizations—are elaborated at length in [13]. The simplicity of the D-VaR approach represents a tradeoff between mathematical accuracy and operational feasibility of the underpinning business processes of an internal model as well as the mathematical complexity. Its main theoretical drawbacks are: possible poor accuracy for non-linear portfolios, positions which are delta neutral call for extra caution. Its main advantages are: high transparency, speedy calculation, rules of thumb for determining marginal VaR, applicable also for intra-day VaR etc. If a *backtesting* for one trading year (i.e. 250 observations) shows no evidence for a re-specification, the pros of the D-VaR approach rule out the cons, because the advantages support and contribute to the use of the model by the company and the regulators.

However the approximations (1), (2) and (4) raised the question of *model uncertainty*, hence a safety factor was looked for. The following example shows how a safety factor was determined in order to capture model uncertainty. The factor found empirical support in the aftermath of the turmoil after September 11, see [14]. Note that Roy [15], was the first in the middle of the last century who applied the *Chebycheff inequality* in the context of risk management.

Example 2 Let F denote a symmetrical distribution function with expectation zero and variance σ^2, then the Chebycheff inequality yields:

$$F^{-1}(1 - \frac{1}{k^2}) \leq \frac{1}{2k^2}\sigma \tag{5}$$

Compared with the Gaussian (G) case, (4) the following inequality yields at the level of significance of $\alpha = 99\,\%$:

$$\frac{VaR_F}{VaR_G} \leq 3.03 \tag{6}$$

It was emphasized by Stahl in [11] that the crude nature of the Chebycheff inequality gives sufficient reason that **all kinds of uncertainties** are captured.

Reflecting the examples above, model uncertainties may be evaluated at two different levels: it is reasonable to start with (1) and (2) at the levels of risk factors or inputs—this is the perspective of **modelers**—, or the perspective of a **decision maker**, i.e. a user of the model is taken. In this view uncertainties are considered at the level of consequences. Aven [16], formalizes these two perspectives by means of the triple

$$(A, C, U), \tag{7}$$

where A denotes objects like events, risk factors or input models, whereas C is related to the consequences, i.e. the level where the economic consequences from decisions are measured. U denotes the uncertainty in this system, which can be related to A, C and both as well.

2 Basic Definitions—Framework

This section provides basic definitions, laying the ground for the concept of a *holistic* model and the associated *model uncertainty*.

First, we reference to the ISO-Norm framework on risk management, see [17], for the definition of the term *risk*, because it is grounded on *uncertainty*. This framework fits smoothly to the concept of an *internal model in wide sense* which is introduced in this section. This representation goes beyond a purely stochastic model by recognizing the full context. This approach captures and reflects the regulatory context given by the (three Pillars) structure of Solvency II and related requirements of rating agencies with respect to a *holistic* and *firm-wide approach*, see [18–20].

2.1 Holistic Models

Definition 1 Risk is defined as an effect of uncertainty on objectives, where an effect is a deviation from the expected—positive or negative. Objectives can have different aspects and can apply at different levels (strategic, organization-wide, project, product, process).

Aven generalized and improved the ISO-Norm framework on risk management with respect to a clarification of the terminology, see Appendix B in [16] and the formal setting for the assessment of uncertainty. However the general concept of the ISO-Norm is valid and followed here.
The following definition taken from the ISO-Norm framework introduces the term uncertainty.

Definition 2 Uncertainty is the state, even partial, of deficiency of information related to understanding or knowledge of an event, its consequence or likelihood.

This paper differentiates three components of uncertainty: *epistemic, aleatory and imprecision*, where the latter refers to numerical imprecision. These components are captured by an extension of (7)

$$(A, C, U, P, K), \tag{8}$$

where K denotes the *background knowledge* and P a stochastic model for the sources of uncertainty, see Aven [16] for details. This split of uncertainty into three components is also considered in contributions in Diebold et al. [21], as well as in a recent publication by the British Insurers [22].

Definition 3 The functional relation defined by

$$F(X_{t+h} \mid \circ) := F(X_{t+h} \mid \mathfrak{I}_t, Z_{H|t}, \mathbb{R}, \mathbb{O}) \tag{9}$$

denotes a **model in a wide sense**, where F is a forecast distribution, F, with forecast horizon h of the variable of interest, X. F is understood as the output of the model. The definitions of the conditioning sets are given in the remarks.

The following remarks highlight some of the interpretations, which tackle the meaning of a model in wide sense.

Remark 1
1. The relational reference (9) takes the information sets $\mathfrak{I}_t, Z_{H|t}, \mathbb{R}, \mathbb{O}$ as sources of risk. It is the explicit incorporation of **all** sources of uncertainty in an holistic approach, which is new.
2. $\mathfrak{I}_t$ denotes ordinary data sets which are produced and updated continuously. It is further split into:
$$\mathfrak{I}_t := (\mathfrak{D}_t, \Lambda_t), \tag{10}$$
where $\mathfrak{D}_t$ are data related to risk factors and Λ_t denotes an exposure vector.
In order to calibrate e.g. an Economic Scenario Generator, time series of prices of financial instruments are an example for $\mathfrak{I}_t$. In addition to such empirical data, experts judgements might be used, e.g. for operational risks. In any case $\mathfrak{I}_t$ denotes input data for sources of risk represented by particular models.
3. The information set $Z_{H|t}$ denotes some background knowledge, given at time t for a time horizon $H \gg h$, H put to 5 years; i.e. $H = 5 \times h$. On one hand, $Z_{H|t}$ is used during the modelling process (e.g. variable selection), and on the

other hand it reflects forward looking information, expertise (e.g. strategic issues, especially targets). Compared with the information set $\mathfrak{I}_t$, $Z_{H|t}$ is not the output of a physical production process but merely gathering external and internal opinions, expectations etc.

4. With $\mathbb{R}$ regulatory requirements are denoted, whose compliance is a prerequisite. (EIOPA frameworks, as well as requirements from rating agencies and investors). External stakeholders may impact the model (9) significantly. Important examples: forecast horizon h, level of significance, granularity of risk categories, requirements on the organization and business processes. With the latter the requirements determine explicitly the costs and also influence operational risk.
5. $\mathbb{O}$ denotes the organizational set-up of the undertaking. For insurance groups with their separation of business lines this goes hand in hand with complex segregation of duties and responsibilities. Furthermore group-wide separated business processes are a standard.

2.2 Components of Uncertainty

The following example elaborates on these components.

Example 3 Let $\mathcal{F}$ denote a whole firm, an insurance say. Its total risk is assumed to be derived from a model according to (9) and measured by some risk measure $\rho(F(X_{t+h} \mid \circ))$.

How good (9) approximates reality of $\mathcal{F}$ is known as **epistemic** uncertainty, see [23]. Typically epistemic uncertainties are captured by subjective probabilities, interval probabilities etc. Assumed, (9) is specified, its parameters estimated, by means of the information sets $\mathfrak{I}_t$ and $Z_{H|t}$, the related business processes fulfilling all requirements imposed by $\mathbb{R}$ and the operational constraints stemming from $\mathbb{O}$. The result is:

$$\widehat{F}(X_{t+h} \mid \mathfrak{I}_t, Z_{H|t}, \mathbb{R}, \mathbb{O}); \tag{11}$$

a specific parameterized version of (9). Typically the estimated model is prone to **aleatory** uncertainty, which may be measured by standard statistical techniques. Often a final numerical step is necessary, in order to calculate $\rho(\widehat{F}(X_{t+h} \mid \circ))$.

In practice, this calculation is realized by a Monte-Carlo-Simulation, $\widehat{F}^*(X_{t+h} \mid \circ)$ drawn from $\widehat{F}(X_{t+h} \mid \circ)$. Epistemic uncertainty, denoted by U_e is given by

$$U_e = \mid \rho(\mathcal{F}) - \rho(F(X_{t+h} \mid \circ)) \mid . \tag{12}$$

Aleatory uncertainty, denoted by U_s is given by

$$U_s = \mid \rho(F(X_{t+h} \mid \circ)) - \rho(\widehat{F}(X_{t+h} \mid \circ)) \mid . \tag{13}$$

Uncertainty related to imprecision, denoted by U_i is given by

$$U_i = \mid \rho(\widehat{F}(X_{t+h} \mid \circ)) - \rho(\widehat{F}^*(X_{t+h} \mid \circ)) \mid . \quad (14)$$

To determine the total uncertainty,

$$U_T = U_e + U_s + U_i \quad (15)$$

further assumptions about their dependencies are crucial. Often, only two extremes are considered: total dependence or independence, see e.g. [24].

The next subsection recalls well known approaches to quantify U_e.

2.3 Measurement of Uncertainty

There are various established approaches to determine uncertainties in (12)–(14):

- probability theory; including frequentistic and bayesian approaches,
- imprecise probabilities,
- possibility theory and
- evidence theory,

see Dubois [25], and Baudrit and Dubois [26] for a general reference to these concepts. For applications in the context of risk management see: the books of Aven [16] and [23] as a reference for the general framework as well as the ISO standard [27], which is devoted to the question: How to express uncertainties in measurements? The book of Oberkampf and Roy addresses applications in the field of validation, see [28], and last but least the book of Cruz et al. which apply these techniques in the area of operational risk, see [29].
Within the possibility theory the definition of a *possibility function* denoted by π is key:

Definition 4 A possibility distribution π is a mapping from a set Ω to the unit interval such that $\pi(\omega) = 1$ for at least one element $\omega \in \Omega$. For $A \subset \Omega$ the possibility function $\Pi(A)$ and the related necessity functions $(N(A))$ are derived by:

$$\Pi(A) = \sup_{\omega \in A} \pi(\omega) \quad (16)$$

$$N(A) = 1 - \Pi(A^c) = \inf_{\omega \in A^c} (1 - \pi(\omega)). \quad (17)$$

For a given $\alpha \in [0, 1]$, the α-cuts, A_α are defined by

$$A_\alpha = \{\omega \in \Omega | \pi(\omega) \geq \alpha\}. \quad (18)$$

These α-**cuts** are nested, i.e.: $\alpha > \beta$ yields $A_\alpha \subseteq A_\beta$

The next definition links possibility and necessity functions to *stochastic models* represented by a family of distributions $\mathcal{P}$ on Ω.

Definition 5 Let $\mathcal{P}$ denote a set of probability distributions on Ω. Lower and upper probability bounds are defined as follows:

$$P_\star(A) = \inf_{P \in \mathcal{P}} P(A) \tag{19}$$

$$P^\star(A) = \sup_{P \in \mathcal{P}} P(A). \tag{20}$$

A possibility degree can be viewed as an upper bound of probability degree. Let $\mathcal{P}_\pi = \{P|A \subseteq \Omega : N(A) \leq P(A) \leq \Pi(A)\}$ be the set of probability distributions encoded by a possibility function π. This representation is coherent since upper and lower probabilities induced by $\mathcal{P}_\pi$ are just N and Π.

3 Management of Model Uncertainty

In a landmarking paper from the Office of the Comptroller of the Currency (OCC) on model validation, see [30], the importance of the *management of model uncertainty* is highlighted. It is outlined there in detail that the outcome of the *validation process is a quantitative or qualitative assessment of model uncertainty*, where a model consists of three components: an *information* input component—which refers to $\mathfrak{I}_t$ in (9)–, a processing component and a reporting component, both refer to $\mathbb{O}$ in (9). The definition of a model in a wide sense proposed in this paper is in so far more general, as $\mathbb{R}$ and $Z_{H|t}$ are considered explicitely.

If OCC's view on validation is accepted, the risk management principles and processes as laid down by the ISO norm may be followed in order to *identify and analyze* model uncertainty. Furthermore the use test is seen as part of the model validation, because senior management assess model uncertainties in so far as their decisions take uncertainties into account. By means of the validation process (together with internal audit process) model uncertainty is monitored and reviewed. The outcomes of these processes are gathered and communicated by means of validation (and audit) reports.

In [30] it is emphasized (at page 1) that model uncertainty is related to potential *indirect costs* such as possible adverse *consequences* by relying on incorrect models or misusing of correct ones. This aspect of focusing on the consequences shows the link to the use test and the role to be played by senior management. The OCC considers in respect to the management of model uncertainty both perspectives: that of the level of the consequences, i.e. the component C in (A, C, U, P, K)—here senior management plays the important role with respect to assessment or evaluation—and the level of events, i.e. the component A in (A, C, U, P, K), where the modelers and validators identify and analyze model uncertainty.

Example 4 **Management of Model Uncertainty in Practice** For insurance undertakings which are going to opt for internal models under Solvency II it is regulatory required, to employ a validation process. The outcome of this process showed up a significant amount of model uncertainty both at the level of **valuation models** and **risk models** as well. Whereas it is common knowledge that the functional form, F as well as X_{t+h}, $\mathfrak{I}_t$, $Z_{H|t}$ and the organizational set-up $\mathbb{O}$ in (9) contribute at the model uncertainty, it is a new insight that also the regulatory framework $\mathbb{R}$ **contributes** to model uncertainty. On one hand $\mathbb{R}$ contributes to U_e via operational risk, because the regulation of internal models under Solvency II is very prescriptive and extensive. On the other hand Solvency II anchored the required level of capital at 99.5 %. Hence U_s is of significant amount due to the large quantile.
Talanx Group applies the OCC principles by catching model uncertainties at the level of consequences in putting aside capital for U_e, U_s and U_i, where the lion part of U_T stems from U_s in determining an SCR for $\alpha = 99.5\,\%$. Hence U_T is a combination of statistical analysis, insights from validation and expert judgement and estimated as roughly 6.3 % of the Group's own funds. This is roughly an amount of one billion Euro.

Whereas the topic of model uncertainty in the context of internal models is reflected widely by various stakeholders, the fact that the *Standard Formula* (SF) under Solvency II is also prone to model uncertainty is worth to note. The assumptions underlying the standard formula are summarized by EIOPA in [31]. Practical experience has shown that the SF is *not conservative* compared with the option to apply an internal model. The following example represents the SF analogue to (9):

Example 5 **Model uncertainty for the Standard Formula** With

$$\widehat{G}(X_{t+h} \mid \mathfrak{I}_{\mathbb{R}}, Z_{H|\mathbb{R}}, \mathbb{R}_{SF}), \tag{21}$$

the anlogue of (9) for the SF is denoted. Compared to an internal model, the conditioning information sets $\mathfrak{D}_{\mathbb{R}}$ and $Z_{H|\mathbb{R}}$ are static, i.e. are not adapted over time. How $Z_{H|\mathbb{R}}$, the forward looking perspective was incorporated in the SF is unknown for the user.
The main drawback of the SF however is that it is impossible for the user to determine the uncertainty

$$U_T(\widehat{G}(X_{t+h} \mid \mathfrak{I}_{\mathbb{R}}, Z_{H|\mathbb{R}}, \mathbb{R}_{SF}).$$

Hence an important requirement of the ISO-norm: to determine the uncertainty of a risk measure can not be realized. Also OCC's main outcome of validation - the model uncertainty can not be determined without an *additional model*.
Furthermore $(\mathfrak{D}_{\mathbb{R}}, \Lambda_t) = \mathfrak{I}_{\mathbb{R}} \subset \mathfrak{I}_t = (\mathfrak{D}_t, \Lambda_t)$ and $\mathbb{R}_{SF}$ is less challenging then $\mathbb{R}$, the regulatory framework for an internal model. Experience with the SF has shown that:

- $U_e(\widehat{G}) > U_e(F)$
- $U_s(\widehat{G}) > U_s(F)$
- $U_i(\widehat{G}) < U_i(F)$

In total we have:

$$U_T(\widehat{G}) > U_T(F)$$

Note that SF employs the **cybernetic principle** of a feedback loop, which is the backbone of risk management, only in a limited way, because latest information $\mathfrak{D}_{\mathbb{R}}$ from markets can not be taken into account. Taking capital market risk as an example $\mathfrak{D}_{\mathbb{R}}$ ends in 2009 in the SF. This implies that economic issues related to unbalanced national households are not captured. This will cause challenges for future ORSA exercises based on the SF. Furthermore, the SF violates **economic principles** such as: arbitrage freeness, diversification and economy of scales. In the light that SF is not conservative, internal models are the superior approach.

In the light of these examples it is important to formulate criteria in order to estimate model uncertainty. In this respect the contributions of Aven and his co-authors as well as the contributions of Oberkampf are important. The following sub-section refers to Aven.

3.1 Realiability, Validity and Model Uncertainty

From the preceding sections we conclude that model uncertainty is unavoidable. For that reason users and stakeholders look for criteria of *reliability* and *validity* of internal models, which increase transparency and trustiness of reported figures. Within Aven's framework, see [32] and [16],

$$(A, C, U, P, K)$$

the authors formulate the following criteria. The criteria for reliability, $R1-R3$, are considered first:

R1 The degree to which the risk analysis methods produce the same results at reruns of these methods.
R2 The degree to which the risk analysis produces identical results when conducted by different analysis teams, but using the same methods and data.
R3 The degree to which the risk analysis produces identical results when conducted by different analysis teams with the same analysis scope and objectives, but no restrictions on methods and data.

These realibility criteria deserve some remarks:

Remark 2 1. Requirement $R1$ seems trivial, however given the challenge caused by the complexity of $\mathbb{O}$ and $\mathbb{R}$ it takes several years of hard work to ensure this indispensable criterion for real-life internal models. The requirement of the technical repeatability of a risk analysis is of highly practical importance. Given the complexity of an internal model formalized by (9), $R1$ can only be assured, if the business precesses underpinning the risk analysis are sufficiently mature. Operational risks from weaknesses in business processes leading to non repeatable results are considered as unacceptable. Criterion $R1$ sets a minimum requirement on the interplay of the stochastic model based on $\mathfrak{I}_t$ and its **operational implementation,** $\mathbb{O}$ in (9). Hence, $R1$ formulates a minimum standard in respect to operational risk of running an internal model.

2. Criterion $R2$ refers to the fact that internal models are **socio-technical** systems. Human factors come into play e.g. in the model building process and in the calibration of the model. Aspects related to human factors are captured by $\mathbb{O}$ in (9) and through K in (8). $R2$ refers to the impact of expert judgements on the components (12)–(14) of U_T.
The relevance of criterion $R2$ relates implicitly to the sample size n. For moderate or small sample sizes U_e might of significant amount and in addition is definitely important for U_s. Hence, criterion $R2$ refers **modeler's** contribution to (13) and (14); his influence decreases as the sample size increases.
3. Criterion $R3$ offers a high degree of freedom with respect to data and methods this assumes implicitly the regulation $\mathbb{R}$ component does not impose restrictions on the components $F, \mathfrak{I}_t$ and $Z_{H|t}$. In regulated industries this might be too optimistic. $R3$ considers

$$G(X_{t+h} \mid \circ) := G(X_{t+h} \mid \tilde{\mathfrak{I}}_t, \tilde{Z}_{H|t}, \mathbb{R}, \mathbb{O}) \tag{22}$$

as an alternative to

$$F(X_{t+h} \mid \circ) := F(X_{t+h} \mid \mathfrak{I}_t, Z_{H|t}, \mathbb{R}, \mathbb{O}),$$

as introduced in (9). Obviously all components of U_T are touched by (22), however the focus will in general lay on the **epistemic** uncertainty

$$U_e = \rho(\mathcal{F}) - \rho(G(X_{t+h} \mid \circ)). \tag{23}$$

As in the case of $R2$, $R3$ refers to K and $\mathbb{O}$, if however the sample size is large, $R3$ is of lower importance. In this case $U_T G(X_{t+h} \mid \circ)$ is expected to be close to $U_T(F(X_{t+h} \mid \circ)$. If the information content in $\tilde{\mathfrak{I}}_t$ and $\tilde{Z}_{H|t}$ is close to that in $\mathfrak{I}_t$ and $Z_{H|t}$ then U_s should be close for the two approaches.

Now the criteria for validity are considered:

V1 The degree to which the produced risk numbers are accurate compared to the underlying true risk.

V2 The degree to which the assigned probabilities adequately describe the assessor's uncertainties of the unknown quantities considered.
V3 The degree to which the epistemic uncertainty assessments are complete.
V4 The degree to which the analysis addresses the right quantities.

The criteria for validity $V1$–$V4$ are more common, compared to $R1-R3$. This is mainly due to the fact that reliability refers to aspects related to $\mathbb{O}$, which are in general only of practical interest and go beyond pure scientific modelling. However, it is important to note that the criteria $V1$–$V4$ are closely related to the framework as laid down in the ISO-norm. In particular these criteria require to include all components of U_T into consideration. The following example considers aspects of $V4$:

Example 6 Under Solvency II the regulatory required level of significance is $\alpha = 99.5\,\%$. For many risk categories the sample size n however is smaller then 100. This implies that

$$F_n^{-1}(0.995) > x_{(n)}, \tag{24}$$

where $x_{(1)}, \ldots, x_{(n)}$ denotes the ordered sample. From (24) follows that only a lower bound for $F_n^{-1}(0.995)$ is **observed**. Hence, only by means of a model the information given by the sample can be extrapolated. Each extrapolation is exposed yet to model uncertainties. From this perspective it is questionable,that the Solvency II level of significance is the **right quantity** in the sense of $V4$, because it depends very much on U_e and increases—compared to lower quantiles—U_s. This is true for both, **internal models** and the **standard formula** as well.

This example highlights the regulatory dilemma in analogy to the quotes of Luhmann in the introduction. The following example, taken from [25], shows that the set of probability distributions related to the Chebychev inequality may be related to possibility functions.

Example 7 Again the Chebychev inequality is given by

$$P(X \in [\mu - k\sigma, \mu + k\sigma]) \geq 1 - \frac{1}{k^2}. \tag{25}$$

Hence the intervals $[\mu - k\sigma, \mu + k\sigma]$ can be seen as α-cuts of an associated possibility function π, with

$$\pi([\infty, \mu - k\sigma]) = \pi([\mu + k\sigma, \infty]) = \frac{1}{k^2}.$$

This possibility distribution defines a family $\mathcal{P}_\pi$ such that $\mathcal{P}^{\mu,\sigma} \subseteq \mathcal{P}_\pi$ containing all distributions with known mean μ and standard deviation σ.

Remark 3 As pointed out in Dubois [25], the knowledge about X is *rather weakly* informative, if it is expressed by (25). Barrieu and Scandolo [33], take this view on (25) to criticize how Stahl derived the safety factor in [11]. In [33] they apply more advanced methods in order to derive sharper bounds. However their critique neglects some important parts of the context of internal models as applied in practice by the financial industry, which is captured by the model in the wide sense. According to the Bank of International Settlements factor three will be applied also in the revised market risk framework, see [34], p. 95.

1. In [33] an internal model is simplified to

$$F(X_{t+h} \mid \circ) := F(X_{t+h} \mid \mathfrak{I}_t), \tag{26}$$

   hence the context given by $Z_{H|t}$, $\mathbb{R}$, $\mathbb{O}$ is omitted. This neglects important aspects of $R1-R3$ and $V1-V4$ and as a consequence underestimates U_T.
2. As a consequence, epistemic uncertainty and imprecision, U_e and U_i are not considered in [33], whereas both are addressed in [11].
3. Furthermore and this aspect is even more important, factor three relates to a 10-day time horizon. In practice, the time horizon for allocation economic capital is however 250 trading days. Hence due to

$$\rho(\widehat{F}(X_{t+250} \mid \circ)) \geq 3\,\rho(\widehat{F}(X_{t+10} \mid \circ)), \tag{27}$$

   factor three is economically of minor importance and only but a regulatory side condition that does not come directly into play in the allocation of capital. Hence the application of factor three is just a regulatory minimum requirement, not more. It is far from being conservative on a one-year time horizon.

3.2 Vulnerability, Resilience and Robustness

The concepts of reliability and validity cover important topics related to the Standardized Formula and internal models as well under Solvency II. However one intuitive and very appealing concept is still missing yet: *robustness*.

Definition 6 In [16], Aven introduced his concept of *vulnerability*, which is an antonym to robustness, by **conditional** consequences:

$$(C, U \mid A). \tag{28}$$

Hence, vulnerability is a two dimensional combination of consequences and uncertainties, given the occurrence of an event A. Most *stress tests* may be represented along the concept based on (28). Closely related to vulnerability is the concept of *resilience*:

Definition 7 In [16], Aven introduced his concept of *resilience* by **conditional** consequences:

$$(C, U \mid \textit{any } A \textit{ including new types of } A). \tag{29}$$

Aven emphasizes that for the last definition some boundary, B, for A is important to be introduced in order to become meaningful. Obviously, (29) is related to *unknown uncertainties* as discussed in Diebold et al. [21]. In the field of insurance so-called *emerging* risks are a real-life example, see [19]. The interpretation of Avens concepts of vulnerability and resilience must be given in the context of the criteria of realiability and validity.
If $\nu(X)$ denotes the *value* of portfolio X, then for an adverse a (28) will often yield:

$$\mid \nu(X \mid A) \mid \rightarrow \infty. \tag{30}$$

However

$$P(A) \rightarrow 0. \tag{31}$$

is difficult to assess exactly, because here model uncertainties will come into play. The following example considers two situations of practical relevance:

Example 8 This exemplifies practical implications of (30) and (31):

1. In combination (30) and (31) show that stress tests are a tool to consider the effect of extreme events (30) on the value of a portfolio. This will surely have an impact on the **capitalization** of a firm and hence stresses *capital adequacy ratios*. However all requirements of $V1-V4$ seem in general not fulfilled.
2. Hence stress tests in the sense outlined here do not contribute in a confirmatory way to the plausibility of the **risk** derived from an internal model. A realistic example for the illustrated situation (31) might, if the implications of the Lehman default in 2007 should be incorporated in an internal model by means of redeliberation of the Economic Scenario Generator. To relate the observed credit spread widening to a specific level of significance is merely an expert setting then an empirically founded probability.

In statistics the concept of robustness is well established, see e.g. Huber and Ronchetti [35]. Cont et al. applied these concepts in the context of risk management in stringent way, see [36]. Heyde et al. required robustness as an important axiom for risk measures, see [12]. Stahl et al. emphasized that the Wasserstein metric is the canonical one, to describe continuity concepts in the framework of risk management, see [37]. A very comprehensive overview on risk measures also considering aspects of Knightian uncertainty is provided by Föllmer and Weber in [38]. Because risk management is interested in *valuation* and calculation of *risk* it is in general not possible to robustify both.

Example 9 In the light of criterion $R1$ and $R3$ it seems natural to require that small perturbations should not change the results too much. Essentially this requires a continuous **influence function**, and where the derivative of the influence function—denoted by IF, see [35] for a definition,—of IF exists, it should be appropriately bounded:

$$\lambda^* = \sup_{x \neq y} \frac{\mid IF(y; T, F) - IF(y; T, F) \mid}{\mid y - x \mid}. \tag{32}$$

The robustness requirement (32) known as *local-shift-sensitivity* is key in respect to fulfill $R1$ and $R3$.

4 Uncertainty Under Solvency II

In this section, examples from practice for epistemic and stochastic uncertainties are considered. The first subsection considers uncertainties at the level of consequences, C in (8). In particular the impact and interplay of sample size and epistemic uncertainty is analyzed. The second subsection bridges the gap of current research on *VaR-bounds* and model uncertainty as considered in the last section.

4.1 Uncertainty at the Level of Consequences

If model uncertainty at the level of consequences is to be considered, it is reasonable to look at the forecast distribution of a whole insurance undertaking. Experience from *validation* of internal models over the years has shown that $\mathcal{P}$ in (19) and (20) can be chosen by

$$\mathcal{P} := \{\mathbf{G}, \mathbf{GEV}, \mathbf{B}\}, \tag{33}$$

which consists of three reasonable parametric models: the Gaussian (G), the Generalized Extreme Value distribution (GEV) and the Burr XII (B); their definitions are given in the Appendix. The determination of U_e will in this section be based on $\mathcal{P}$. Furthermore for each element in $\mathcal{P}$ the Maximum-Likelihood-Estimator (MLE), $\rho(X)_n$, and a non-parametric estimator (NPE), $\rho(X)_{np}$, defined by:

$$\rho(X)_{np} := \int x dF_n - F_n^{-1}(\alpha)$$

for the risk measure $\rho(X)$

$$\rho(X) = \mathbb{E}(X) - \mathbb{V}aR_\alpha(X)$$

are compared by their efficiency for the sample sizes $n = 50$, 100 and 400 by means of a Monte-Carlo simulation. For internal models under Solvency II, $\rho(X)$ is used in order to determine the solvency capital required.

Example 10 **Uncertainty in** $\rho(X)$ The following table summarizes the results. The columns in the following table show also the ratio between the standard error and the $\mathbb{V}aR_\alpha(X)$ for $\alpha = 99.5\,\%$ and $\alpha = 80\,\%$ for the considered candidates, models and sample sizes.

As outlined in Jaschke and Stahl [39], the **calibration test** under Solvency II may be applied to determine the 99.5 % quantile by **rescaling** 80 % one. In the light of the examples above such an approach would reduce both, epistemic and aleatory uncertainty. Furthermore SCR figures anchored on the 80 % level could be easily challenged by senior management, e.g. within the Own Risk and Solvency Assessment. From the examples given in the Tables 1, 2, 3 and 4 it is also evident that uncertainty increases with tail weights, hence conservative approaches are penalized twice: first at the level of determination risk and then as a consequence at the level

Table 1 This table depicts the relative estimation error $\frac{2\sigma}{q_{0.005}}$ for the parametric and non-parametric estimates for $\rho_p(X)$ resp. $\rho_{np}(X)$ for the sample sizes $n = 50$, 100 and 400 for $\alpha = 0.995$. As expected the uncertainty increases with the tail weight and decreases with the amount of prior knowledge. The latter is expressed by the parametric versus non-parametric MLE

$\rho_p(X)$, $\rho_{np}(X)$	n = 50	50	n = 100	100	n = 400	400
G	18	42	13	29	6	15
GEV	34	110	24	78	12	39
B	61	91	43	65	22	32

Table 2 At the 99.5 % level of significance $G \in \mathcal{P}$ **underestimates the risk. The result is based on the Monte-Carlo simulation (MC) as a yardstick**

	MC	G	GEV	BURR
SCR	100	82	105	103

Table 3 This table depicts the relative estimation error $\frac{2\sigma}{q_{0.005}}$ for the parametric and non-parametric estimates for $\rho(X)_p$ resp. $\rho_{np}(X)$ for the sample sizes $n = 50$, 100 and 400 for $\alpha = 0.8$. In this case the uncertainty does not monotonically move neither with tail weight and nor with prior knowledge. The latter is expressed by the parametric versus non-parametric MLE. **The closeness of the parametric and non-parametric MLE indicate that the 80 % quantile suits much better in the sense of V4**

$\rho(X)_p$, $\rho_{np}(X)$	n = 50	50	n = 100	100	n = 400	400
G	3	7	4	5	2	2
GEV	8	10	6	7	3	4
B	6	7	4	5	2	3

Table 4 At the 80 % level of significance **all members of** $\mathcal{P}$ **give a reasonable approximation**

	MC	G	GEV	BURR
SCR	1	108	108	98

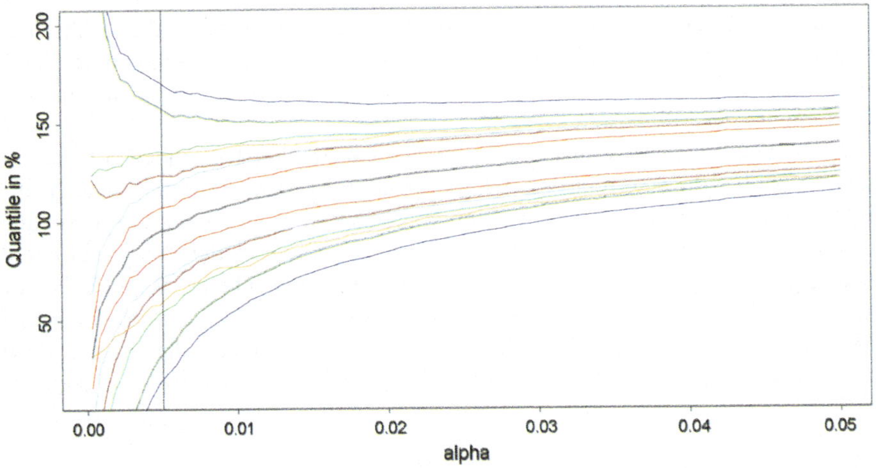

Fig. 1 Quantiles in percent for different standard errors, n = 100. *Black line* Distribution based on the Monte-Carlo simulation, *Red line* Parametric normal model, *Dark Red line* Non-parametric normal model, *Light Blue line* Parametric GEV model, *Blue line* Non-parametric GEV model, *Green line* Parametric BURR model, *Dark Green line* Non-parametric BURR model, *Orange line* 95 % Confidence band

of uncertainty. Note that this **adverse** effect is a consequence of the too large 99.5 % quantile. The different amount of uncertainty in non-parametric versus parametric MLE's is an example for that. In Fig. 1 it is highlighted that $\alpha = 0$ model uncertainty is considered near a singularity. Hence it is not surprising that unexpected effects are to be observed. This again sheds some light on the paradox mentioned by Luhmann.

In addition, Fig. 1 depicts for $n = 100$ and $0 \leq \alpha \leq 0.05$ the standard error for the considered candidate models.

4.2 Uncertainty Induced by the Copula

In a series of papers [6–8] laid the foundations to estimate the epistemic uncertainty caused by the *aggregation* of risk categories. In this subsection the risk factors **R** are defined on the d-dimensional Euclidian space, $\mathbb{R}^d$. For the random vector

$$\mathbf{R} = (R_1, \ldots, R_d) \tag{34}$$

the associated marginal distributions, F_i

$$R_i \sim F_i \tag{35}$$

are assumed to be *known*, i.e. $U_e(R_i) = 0$, whereas about the copula of $\mathbf{R}$ different levels of a priori information are considered. In [6] the *Frechet* class, denoted by

$$\mathcal{F}_d(R_1, \dots, R_d), \tag{36}$$

of all copulas with given marginal distributions are considered in order to derive *upper* and *lower* bounds for the $\mathbb{V}aR$ of the sum

$$S = \sum_{i=1}^{d} R_i. \tag{37}$$

In the notation of (19) and (20) this reads as

$$\overline{\mathbb{V}aR} = P^*(A) = \sup_{C \in \mathcal{F}} P(A) \tag{38}$$

$$\underline{\mathbb{V}aR} = P_*(A) = \inf_{C \in \mathcal{F}} P(A). \tag{39}$$

According to [6] the difference (38)–(39)

$$DU = \overline{\mathbb{V}aR} - \underline{\mathbb{V}aR} = \sup_{C \in \mathcal{F}} P(A) - \inf_{C \in \mathcal{F}} P(A) = P^*(A) - P_*(A) \tag{40}$$

is denoted as the *dependence uncertainty* spread (DU). The class of probability distributions related to Frechet class $\mathcal{F}_d$ will be denoted by $\mathcal{P}^{\mathcal{F}}$. By calculating DU-spreads (40), again the perspective moves back to the level of consequences, C. Hence, it is worthwhile to compare heuristically the impact of either using $\mathcal{P}^{\mu,\sigma}$ or $\mathcal{P}^{\mathcal{F}}$ on the amount of $U_e(C)$. With respect to $\mathcal{P}^{\mu,\sigma}$ only the correlation of $\mathbf{R}$ and the expectations $\mathbb{E}(R_i)$ are implicitly assumed to be known. Whereas with respect to $\mathcal{P}^{\mathcal{F}}$ the marginal distributions R_i are assumed to be completely known and only but their copula is unknown. Given the richness of $\mathcal{F}_d$ it is to be expected that for both approaches U_e measured by DU-spreads will be very large. For that reason [7] and [8] introduced further restrictions on $\mathcal{P}^{\mathcal{F}}$ in order to sharpen (40).

The following Theorem is taken from [8], Theorem 3.3 highlights their approach.

Theorem 1 *Let* $q \in (0, 1)$; $R_i \sim F_i (i = 1, \dots, d)$ *and* $S = \sum_{i=1}^{d} R_i$ *satisfy* $var(S) \leq s^2$. *Then we have:*

$$a := \max(\mu - s\sqrt{\frac{1-\alpha}{\alpha}}, A) \leq m \leq \underline{\mathbb{V}aR}_\alpha(S) \leq \overline{\mathbb{V}aR}_\alpha(S) \leq M \leq b := \min(\mu + s\sqrt{\frac{1-\alpha}{\alpha}}, B) \tag{41}$$

In particular if $s^2 \geq \alpha(A - \mu)^2 + (1 - \alpha)(B - \mu)^2$ *then* $a = A$ *and* $b = B$ *and the unconditional bounds are not improved by the presence of the constraint on variance.*

In Remark 3.3, in [8], the authors relate their results to so-called *Cantelli* bounds, see also [33], by considering the $\mathcal{P}^{\mu,\sigma}$ approach within Aven's level of events, A, for calculating upper VaR bounds for (37) yielding the following estimation:

$$P(S \leq \overline{VaR}_{\alpha}(S)) = \sup_{P\ in \mathcal{P}^{\mu,\sigma}} P(A_{\alpha}) \leq \mu + \sigma\sqrt{\frac{\alpha}{1-\alpha}}, \tag{42}$$

where A_{α} denotes an α-cut as defined in (18).

Remark 4 1. For $\alpha = 99.5\,\%$—the required significance level under Solvency II—the value for $\sqrt{\frac{1-\alpha}{\alpha}}$ in (41) and (42) is about 14, which is quite high.
2. Note, that $\sqrt{\frac{1-\alpha}{\alpha}}$ has an essential singularity at $\alpha = 1$. Hence U_T will be non-robust in this region. Basing capital requirements on such high quantiles induces model risk, i.e. large values for U_T. This is cumbersome for all stakeholders.

Example 11 The following representative example shows an insurance expose to six risk categories $(R_1, \ldots, R_6)$. Based on the rearrangement algorithm, see [6] and [7] the upper and lower VaR bound were calculated.

The approaches considered so far applied universal concepts to estimate uncertainty in aggregation—only but high-level information based on I_t is considered. In order to improve the a-priori information [7] used *structural* information - also available by I_t. In this context, the vector **R** is now interpreted as representing risk categories which are typically encountered in insurance undertakings; e.g.: natural catastrophes, investment risk and operational risk, say. Assume on the other hand that the vector **R** represents the organizational structure of and insurance group, then for these variables a hierarchical structure is reasonable to be assumed. This approach leads to a significant improvement of the DU-spreads.
As outlined for a large insurance group the uncertainty with respect to aggregation may be considered from two perspectives: the group may be considered on one hand as a *portfolio of entities* owned by the group:

$$G = \sum_{i=1}^{e} E_i, \tag{43}$$

where G stands for the group and E_i for entity $i, i = 1, \ldots, e$, where each E_i is exposed to risk categories **R** (Fig. 2):

$$E_i = \sum_{i=1}^{e} \beta_i R_i \tag{44}$$

and hence

$$G = \sum_{i=1}^{e} \sum_{j=1}^{d} \beta_i^j R_i. \tag{45}$$

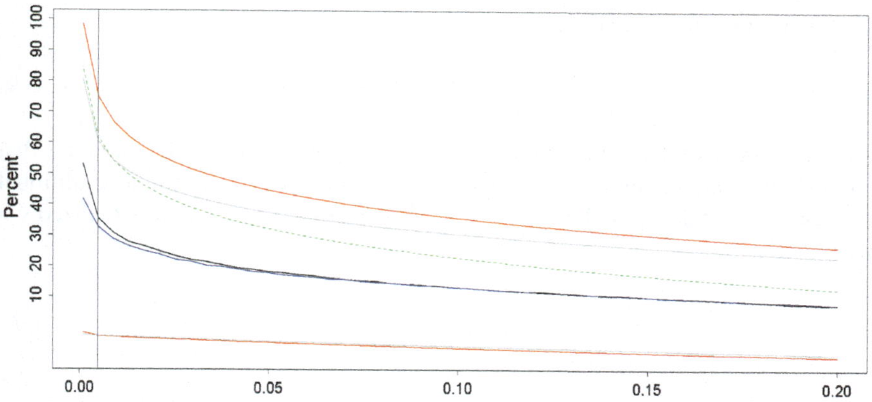

Fig. 2 Upper and lower VaR bounds for $(R_1, \ldots, R_6)$. Overview of value-at-risk bounds for different information sets of dependence structures on the risks. *Red* No information (only marginals R_i). *Grey* Additional information of stochastic independence between the categories market risk and natural catastrophes. *Green* Comonotonic dependence. *Black* Dependence given by the Monte-Carlo simulation. *Blue* Approximation of the empirical dependence structure through the Bernstein-copula approach, see Pfeifer et al. [40]

Furthermore aggregation with respect to risk categories can be considered:

$$R_i^G = \sum_{i=1}^{e} \beta_i^j R_i. \tag{46}$$

and then

$$G = \sum_{i=1}^{d} R_i^G \tag{47}$$

where R_i^G denotes the aggregate of risk category R_i on group level.

4.3 *Regulatory Implications*

In a recent paper by Cambou and Filipovic [41], the approach followed by FINMA—the SWISS regulatory authority—to tackle model uncertainty is represented. For a given list of r prescribed scenarios s_i for $i = 1, \ldots, r$, where for scenario i an auxiliary probability weight c_i is given by the regulator. The associated loss $l_i, i = 1, \ldots, r$ has to be determined by the insurance company:

$$l_i = \mathbb{E}(L \mid s_i) - \mathbb{E}(L). \tag{48}$$

Note that for insurance undertakings

$$\rho(L) = \mathbb{E}(L) - \mathbb{V}aR_{\alpha}(L) \tag{49}$$

is used in order to determine the required capital, i.e. L denotes the variables that captures the undertaking's annual losses. The stressed loss distribution conditional on scenario i is then set to be $F_L(x - l_i)$. The scenarion aggregation is realized via mixing:

$$F_L^{SST}(x) = p_0 F_L(x) + \sum_{i=1}^{r} p_i F_L(x - l_i), \tag{50}$$

is determined, where $p_0 = 1 - \sum_{i=1}^{r} p_i$ and SST is a shorthand natation for the Swiss Solvency Test—the swissian regulatory regime.

Remark 5 1. The regulatory intervention in (50) is applied at the level of consequences, C, which can be interpreted as a *capital add-on*. Note that the stochastic model F remains unchanged.

2. The mixture (50) may be interpreted in different ways:
   - The sum $\sum_{i=1}^{r} p_i F_L(x - z_i)$ may be interpreted as *epistemic uncertainties*, because the p_i are *subjective*
   - the associated weights p_i may be interpreted as an expert judgement
   - the mixture may interpreted as a *linear pooling* of expert knowledge
   - the mixture may be interpreted within a *Bayesian framework* which combines $r + 1$ models according to a multinomial prior function, see [42]
   - the mixture may be related to Huber's gross error model, see [35]. For a mixture distribution $H = (1 - \alpha)F + \alpha G$, the distance $d(F, H)$ may be interpreted as U_e.

3. The practice of internal models has shown, that the coefficient of variation cv are in general small. Hence the expectation is the dominating parameter of the forecast distribution. Hence it is very reasonable to base regulatory interventions on this parameter.
4. The approach of Swiss Solvency Test is close to the theory of Schmeidler and Gilboa, see [43] where the *utility* of state x is valuated by:

$$V(x) = c \sum_{i=1}^{n} p_i u_i(x) + (1 - c) \min\{u_i(x) \mid i = 1, \ldots, n\}. \tag{51}$$

   if the outcome of the adverse scenarios are just applied to the own funds the factor $1 - c$ may be interpreted as an uncertainty which do not allow to determine the probabilities p_i. However the *consequences* of the states of nature are incorporated.

5 Conclusions

1. In practice the interplay of capital and risk are often confounded by misinterpreted stress test concepts. In the light of (28), stress tests should solely be applied to valuation of $\nu(X)$. The focus on own funds would allow stakeholders to take *point-in-time* aspects into account. Either pro-cyclical or anti-cyclical. The pros and cons of such an approach is seen in discussions about the market consistent embedded value, MCEV.
2. Own funds (OF) resp. $E(X)$ have a greater impact on the capital adequacy ratio (CAR)
$$\frac{OF}{SCR} \tag{52}$$
then $\sigma(X)$ or $\mathbb{V}aR(X)$. Again this speaks for $\nu(X)$.
3. Risk models (9) should be robustified as outlined in [37]. The risk model should be treated in the *through-the-cycle* spirit of the standard formula. Then a multiplier might be applied.
4. The superiority of internal models over the standard formula can be seen by comparing their *value of information*, VoI:
$$VoI(G(X_{t+h} \mid \tilde{\mathfrak{J}}_t, \tilde{Z}_{H|t}, \mathbb{R}, \mathbb{O}) \leq VoI(F(X_{t+h} \mid \mathfrak{J}_t, Z_{H|t}, \mathbb{R}, \mathbb{O}) \tag{53}$$
This is a re-formulation of the saying that the application of internal models swaps on the other hand information to stakeholders by increasing transparency. With an internal model U_T in the approximation of $\rho(\mathcal{F})$ can be estimated, but this can be hardly done for the SF. The *idiosyncratic* contribution of a firm $\mathcal{F}$ in $\rho_{SF}(\mathcal{F})$ is unknown.
5. Model uncertainties should be captured by multipliers at the level of consequences, C.
6. Local-shift-sensitivity is a silent feature to guarantee that U_i is under control.
7. Often an (A, C, P) approach is taken for granted, though the situation at hand requires (A, C, U). In this context, the relevance of $V4$ is underestimated.
8. For auditing, validating or approving internal models the costs differ very much, whether (A, C, P) or (A, C, U) is assumed or appropriate. Assuming (A, C, P) though it is (A, C, U) faces the risk to **increase** model uncertainties. The (A, C, U) based approach will merely focus on **capital** and act on the level of consequences.
9. With respect to $V4$ and (24), the regulatory capital requirement should satisfy $\rho(X) \leq x_{(n)}$, else $\rho(X)$ depends too much on expert judgements. In the case that the sample size is to small, a regulatory multiplier could be applied in order to ensure a sufficient capital cushion. To apply a 80 % level together with a multiplier is preferable compared to a 99.5 % level, because in the latter case the amount of epistemic uncertainty is not specified. This approach should be followed both, for internal and external purposes.
10. The focus of regulatory interventions should be on the level of consequences, C.

11. Though both, decision makers of a firm and regulators might be considered as users, the latter emphasize more the perspective of a modeler, which considers the whole process that constitutes a model in a wide sense, i.e. the stochastic model including all underpinning business processes. This is a significant change compared to (6) which is based on the level of the consequences. In [3] this change of the regulatory perspective is described.

Appendix

Gaussian Distribution

Non parametric estimator

For the non parametric case we use the estimator

$$\epsilon_{np} = \hat{\sigma} * \sqrt{\frac{\alpha * (1-\alpha)}{n * f^2(\Phi^{-1}(\alpha))}} * \Phi^{-1}\left(1 - \frac{\vartheta}{2}\right), \tag{54}$$

with $\hat{\sigma}$ the estimated standard deviation, α the quantile for which the standard error is calculated, n the sample size, f the density of the standard normal distribution, Φ^{-1} the quantile function of the standard normal distribution and ϑ the confidence level.

Parametric estimator

For the parametric case we use the estimator

$$\epsilon_p = \frac{1}{\sqrt{n}} * \Phi^{-1}\left(1 - \frac{\vartheta}{2}\right) * \hat{\sigma} * \sqrt{1 + \frac{(\Phi^{-1}(\alpha))^2}{2}}, \tag{55}$$

with the same notation as before.

GEV Distribution

The density of the GEV distribution is given by

$$f(x;\mu,\sigma,\xi) = \frac{1}{\sigma}\left[1 + \xi\left(\frac{x-\mu}{\sigma}\right)\right]^{(-1/\xi)-1} \exp\left\{-\left[1 + \xi\left(\frac{x-\mu}{\sigma}\right)\right]^{-1/\xi}\right\}$$

for $1 + \xi(x-\mu)/\sigma > 0$, where $\mu \in \mathbb{R}$ is the location parameter, $\sigma > 0$ is the scale parameter and $\xi \in \mathbb{R}$ denotes the shape parameter.

Non parametric estimator

For the non parametric case we use the estimator

$$\epsilon_{np} = \sqrt{\frac{\alpha * (1-\alpha)}{n * \bar{f}^2(\bar{F}^{-1}(\alpha))}} * \Phi^{-1}\left(1 - \frac{\vartheta}{2}\right), \tag{56}$$

with the same notations as in (54) and $\bar{f} = f(x; \hat{\mu}, \hat{\sigma}, \hat{\xi}) = f(x; -8530.60, 739.99, -0.116)$ the density of the estimated GEV distribution $\bar{F}$. The parameters were estimated with the Log-likelihood function using the method of Nelder and Mead to determine the maximum of the function.

Parametric estimator

For the parametric case we use the estimator

$$\epsilon_p = \sqrt{\frac{1}{n} * \bar{a}^{-1} * (I(\hat{\Theta}))^{-1} * \bar{a}} * \Phi^{-1}\left(1 - \frac{\vartheta}{2}\right), \tag{57}$$

with

$$\bar{a} = \begin{pmatrix} 1 \\ \frac{1}{\xi} * ((-\log(\alpha))^{-\hat{\xi}} - 1) \\ \hat{\sigma} * \left[-\frac{1}{\hat{\xi}^2} * ((-\log(\alpha))^{-\hat{\xi}} - 1) - \log(-\log(\alpha)) * (-\log(\alpha))^{-\hat{\xi}}\right] \end{pmatrix}, \tag{58}$$

where $I(\hat{\Theta})$ denotes the observed Fisher information matrix. For different sample sizes we get the following results for the 0.05 % quantile and 95 % confidence level:

BURR Distribution

The density of the BURR distribution is given by

$$f(x; a, b, q) = \frac{aq(x - c)^{a-1}}{b^a \left[1 + \left(\frac{x-c}{b}\right)^a\right]^{1+q}}, \quad \text{for } x > c.$$

In addition to the shape parameters $a > 0$ and $q > 0$ and the scaling parameter $b > 0$ we introduce a location parameter $c \in \mathbb{R}$ to relax the property that the density lives on the positive half line.

Non parametric estimator

For the non parametric case we use the estimator

$$\epsilon_{np} = \sqrt{\frac{\alpha * (1 - \alpha)}{n * \tilde{f}^2(\tilde{F}^{-1}(\alpha))}} * \Phi^{-1}\left(1 - \frac{\vartheta}{2}\right), \tag{59}$$

with the same notations as in (54) and $\tilde{f} = f(x; \hat{a}, \hat{b}, \hat{q}) = f(x; 14.24, 6912.24, 1.28)$ the density of the estimated BURR distribution $\tilde{F}$. The parameters were estimated with the Log-likelihood function using the method of Nelder and Mead to determine the maximum of the function. The parameter c was estimated with 1.33*min(data).

Parametric Estimator

For the parametric case we use the estimator

$$\epsilon_p = \sqrt{\frac{1}{n} * \bar{E}^{-1} * (I(\hat{\Theta}))^{-1} * \bar{E}} * \Phi^{-1}\left(1 - \frac{\vartheta}{2}\right), \tag{60}$$

with

$$\bar{E} = \begin{pmatrix} -\frac{\hat{b}}{\hat{a}^2} * \log(\beta) * \beta^{\frac{1}{\hat{a}}} \\ \beta^{\frac{1}{\hat{a}}} \\ \frac{\hat{b}}{\hat{a}} * \frac{\log(1-\alpha)}{\hat{q}^2} * (1+\beta) * \beta^{\frac{1}{\hat{a}}-1} \end{pmatrix} \tag{61}$$

and $\beta = (1-\alpha)^{-\frac{1}{\hat{q}}} - 1$ and the same notations as before.

References

1. Wilson, T.: Value and Capital Management. Wiley, New York (2015)
2. Edgeworth, F.Y.: The mathematical theory of banking. J. Roy. Stat. Soc. **51**, 113–127.1 (1888)
3. Padoa-Schioppa, T.: Regulating Finance (2004)
4. Bennemann, C., Oehlenberg, L., Stahl, G.: Handbuch Solvency II. Schäffer Poeschel (2011)
5. Luhmann, N.: Soziologie des Risikos, Walter de Gruyter, Berlin (2003)
6. Embrechts, P., Puccetti, G., Rüschendorf, L., Wang, R., Beleraji, A.: An Academic Response on Basel 3.5, 2013, Preprint, 45, 45,
7. Puccetti, G., Rüschendorf, L., Small, D., Vanduffel, S.: Reduction of Value-at-Risk bounds by independence and variance information. Scand. Actuarial J **9** (2016). Forthcoming
8. Vanduffel, S.: Value-at-risk bounds with variance constraints, Preprint, 6, 8 (2015)
9. SCOR Switzerland AG: From Principle-Based Risk Management to Solvency Requirements, Analytical Framework for the Swiss Solvency Test, Second Edn. (2008)
10. BIS, Amendment to the Basel Accord to Incorporate Market Risk. BIS, Basel (1996)
11. Stahl, G.: Three Cheers, 1997 May, Risk Magazine
12. Heyde, C.C., Kou, S.G., Peng, X.H.: What Is a Good Risk Measure: Bringing the Gaps between Data, Coherent Risk Measures, and Insurance Risk Measures, Preprint (2006)
13. Morgan, J.P.: Riskmetrics—technical document
14. Jaschke, S., Stahl, G., Stehle, S.: Value-at-Risk forecasts under scrutiny—the German experience. Quant. Financ. **7**(6), 621–636 (2006)
15. Roy, A.D.: Safety First and the Holding of Assets. Econometrica **20**, 1351–1360 (1952)
16. Aven, T.: Quantitative Risk Assessment, Cambridge (2011)
17. International Standard: Risk management—Principles and guidelines, First Edn. (2009)
18. BaFin, Minimum Requirements for Risk Management in Insurance Undertakings, 2009, Circular 3/2009,
19. Standard & Poor's: Enterprise Risk Management For Financial Institutions: Rating Criteria and Best Practices (2005)
20. Standard & Poor's: Insurance criteria: refining the focus of insurer enterprise risk management criteria (2006)
21. Diebold, F.X., Doherty, N.A., Herring, R.J.: The Konwn, the Unknown and the Unknowable in Financial Risk Management. Princeton University Press (2010)
22. Association of British Insurers, Non-modeled Risks—A guide to more complete catastrophe risk assessment for (re)insurers (2014)

23. Aven, T., Baraldi, P., Flage, R., Zino, E.: Uncertainty in Risk Assessment, Wiley (2014)
24. Pedroni, N., Zio, E., Ferrario, E., Pasanisi, A., Couplet, M.: Propagation of aleatory and epistemic uncertainties in the model for the design of a flood protection dike. Oper. Res. Lett. **32**, 399–408 (2012)
25. Dubois, D.: Possibility theory and statistical reasoning. Comput. Stat. Data Anal. **51**, 47–69 (2006)
26. Baudrit, C., Dubois, D.: Practical representations of incomplete probabilistic knowledge. Compuat. Stat. Data Anal. **51**, 86–108 (2006)
27. International Standard: Guide to the Expression of Uncertainty in Measurement, First Edn. (1995)
28. Oberkampf, W., Roy, C.: Validation and Verfication in Scientific Computing. Cambridge University Press (1010)
29. Cruz, M., Peters, G., Shevchenko, P.: Fundamental Aspects of Operational Risk and Insurance Analytics. Wiley, New York (2015)
30. Comptroller of the Currency: Model Validation, 2011, OCC Bulletin 2011–12,
31. EIOPA, The underlying assumptions in the standard formulafor the Solvency Capital Requirement calculation, 2014, EIOPA-14-322,
32. Aven, T., B.: Heide, Realibility and validity of risk analysis, Eng. Safety. Syst. **94**, 1491–1498 (2009)
33. Barrieu, P., Scandolo, G.: Assessing Financial Model Risk. Electron. J. Oper. Res. (2013)
34. Bank for International Settlements: Fundamental review of the trading book: A revised market risk framework. BIS October 2014 (2014)
35. Huber, P.J., Ronchetti, E.: Robust Statistics, Wiley (2009)
36. Cont, R., Dequest, R., Scandolo, G.: Robustness and sensitivity analysis of risk measurement procedures, 2007, Financial Engineering Report No. 2007-06
37. Stahl, G., Kiesel, R., Zheng, J., Rühlicke, R.: Conceptionalizing robustness in risk managment. Preprint (2012)
38. Föllmer, H., Weber, S.: The Axiomatic Approach to Risk Measures for Capital Determination. Preprint (2015)
39. Jaschke, S., Stahl, G., Internal models in Solvency II, 2007 March, Life and Pensions
40. Pfeifer, D., Strassburger, D., Philipps, J.: Modelling and simulation of dependence structures in nonlife insurance with Bernstein copulas. Preprint (2009)
41. Cambou, M., Filipovic, D.: Model Uncertainty and Scenario Aggregation. Preprint (2015)
42. Clemen, R.T., Winkler, R.L.: Combining probability distributions from experts in risk analysis. J. Risk Anal. **2**, 187–203 (1999)
43. Gilboa, I., Schmeidler, D.: Maximum expected utility with a non-unique prior. J. Math. Econ. **22**, 173–188 (1989)

Part III
Derivative Pricing, Hedging, and Optimisation

Option Pricing in Affine Generalized Merton Models

Christian Bayer and John Schoenmakers

Abstract In this article we consider affine generalizations of the Merton jump diffusion model Merton (J Finan Econ 3:125–144, 1976 [8]) and the respective pricing of European options. On the one hand, the Brownian motion part in the Merton model may be generalized to a log-Heston model, and on the other hand, the jump part may be generalized to an affine process with possibly state dependent jumps. While the characteristic function of the log-Heston component is known in closed form, the characteristic function of the second component may be unknown explicitly. For the latter component we propose an approximation procedure based on the method introduced in Belomestny et al. (J Funct Anal 257(4):1222–1250, 2009 [1]). We conclude with some numerical examples.

Keywords Affine jump models · Characteristic function approximations · Fourier option pricing

1 Introduction

The Merton jump diffusion model [8] can be considered one of the first asset models beyond Black-Scholes that may produce non-flat implied volatility surfaces. On the other hand, European options within this model can be priced quasi-analytically by means of an infinite series of Black-Scholes type expressions. From a mathematical point of view, the logarithm of the Merton model is the sum of a compound Poisson process and an independent Brownian motion, and as such can be seen as the sum of two independent degenerate affine processes. The goal of this article is to enlarge the flexibility of the Merton model by generalizing the Brownian motion to a continuous Heston model and replacing the compound Poisson process by another, independent,

C. Bayer · J. Schoenmakers (✉)
Berlin, Germany
e-mail: schoenma@wias-berlin.de

C. Bayer
e-mail: christian.bayer@wias-berlin.de

© Springer International Publishing Switzerland 2016

J. Kallsen and A. Papapantoleon (eds.), *Advanced Modelling in Mathematical Finance*, Springer Proceedings in Mathematics & Statistics 189,
DOI 10.1007/978-3-319-45875-5_10

affine model that may incorporate both stochastic volatility and jumps. In financial modeling affine processes have become very popular the last decades, both due to their flexibility and their analytical tractability. The theoretical analysis of affine processes is developed in the seminal papers [4, 5]. Once the characteristic functions of the affine ingredients of our new generalized Merton model are known, we may price European options by the meanwhile standard Carr-Madan Fourier based method [2]. For a variety of affine models, such as the Heston model and several stochastic volatility models with state independent jumps, the characteristic function is explicitly known. However if, for instance, in an affine jump model the jump intensity depends on the present state, a closed form expression for the characteristic function is not known to the best of our knowledge. Yet, such models make sense in certain applications such as crisis modeling. For example, one may wish to model an increased intensity of downward jumps in regimes of increased volatility. In order to cope with such kind of processes numerically, we recap and apply the general series expansion representation for the characteristic function of an affine process developed in [1] and present some numerical examples.

2 Merton Jump Diffusion Models

Merton [8] introduced and studied stock price models of the form

$$S_t = S_0 e^{rt+Y_t},$$

where Y is the sum of a Brownian motion with drift and an independent compound Poisson process,

$$Y_t = \gamma t + \sigma W_t + J_t, \tag{1}$$

and r is a constant, continuously compounded risk-free rate. In (1) J may be represented as

$$J_t = \sum_{l=1}^{N_t} U_l,$$

where $U_1, U_2, \ldots$ are i.i.d. real valued random variables and N_t denotes a Poisson process with parameter λ. The extended characteristic function of Y_t is given by,

$$\begin{aligned}\Phi_t(z) &= \mathbb{E}\left[e^{\mathrm{i}zY_t}\right] = e^{\mathrm{i}z\gamma t}\mathbb{E}\left[e^{\mathrm{i}z\sigma W_t}\right]\mathbb{E}\left[e^{\mathrm{i}zJ_t}\right] \\ &= \exp\left[\mathrm{i}z\gamma t - \frac{z^2\sigma^2}{2}t + \lambda t\int\left(e^{\mathrm{i}zu} - 1\right)\mu(du)\right],\end{aligned} \tag{2}$$

for a certain jump probability measure μ on $\mathcal{B}(\mathbb{R})$ due to the distribution of U_1.

We henceforth assume that the model is given under a risk-neutral pricing measure. Due to no-arbitrage arguments, we must have that $S_t e^{-rt}$ is a martingale under this measure. This implies that

$$S_0 = \mathbb{E}\left[S_t e^{-rt}\right] = S_0 \mathbb{E}\left[\exp\left(Y_t\right)\right] = S_0 \Phi_t(-\mathrm{i}), \text{ hence } \Phi_t(-\mathrm{i}) = 1. \tag{3}$$

By (2) we then get

$$\gamma = -\frac{\sigma^2}{2} - \lambda \int \left(e^u - 1\right) \mu(du). \tag{4}$$

As an example, with $\lambda = 0$ (no jumps), $\gamma = -\frac{\sigma^2}{2}$ and we retrieve the risk neutral Black-Scholes model. Merton particularly studied the case where U is normally distributed and derived a representation for a call (or put) option in terms of an infinite series of Black-Scholes expressions. In this paper we are interested in generalizations of (1) of the form

$$Y_t = \gamma t + \sigma W_t + X_t^1, \tag{5}$$

or even,

$$Y_t = \gamma t + H_t + X_t^1, \tag{6}$$

where H is the first component of a log-Heston type model with $H_0 = 0$, whereas X_t^1 is the first component of some generally multidimensional affine (eventually jump) process X, independent of W and H respectively, with $X_0^1 = 0$. In particular, the characteristic function of X^1 is possibly not known in closed form. We note that γ later might be time-dependent, i.e., $\gamma = \gamma(t)$.

At this stage, the separation between X^1 and W (in (5)) and H (in (6)), respectively, seems somewhat artificial. As we shall see in the subsequent sections, we will use an asymptotic approximation for the characteristic function of X^1, The exact characteristic function for W and H, respectively, improves the overall accuracy of the approximation, especially regarding the tail behavior.

3 Recap of Affine Processes and Approximate Characteristic Functions

We consider an affine process X in the state space $\mathfrak{X} \subset \mathbb{R}^d$, $d \in \mathbb{N}_+$, with generator given by

$$\begin{aligned} Af(x) = &\frac{1}{2} \sum_{i,j=1}^{d} a^{ij}(x) \frac{\partial^2 f}{\partial x^i \partial x^j} + \sum_{i=1}^{d} b^i(x) \frac{\partial f}{\partial x^i} \\ &+ \int_{\mathbb{R}^d} \left[f(x+z) - f(x) - z^\top \frac{\partial f}{\partial x} \right] v(x, dz), \end{aligned} \tag{7}$$

where a^{ij} and b^i are suitably defined *affine* functions in x on $\mathbb{R}^d$, and

$$v(x, dz) =: v^0(dz) + x^\top v^1(dz)$$

with v^0 and v_i^1, $i = 1, \ldots, d$, being suitably defined locally finite measures on $\mathcal{B}(\mathbb{R}^d \setminus \{0\})$ with finite first moment. Alternatively, the dynamics of X are described by the Itô-Lévy SDE:

$$dX_t = b(X_t)dt + \sigma(X_t)dW(t) + \int_{\mathbb{R}^d} z\widetilde{N}(X_{t-}, dt, dz), \quad X_0 = x, \tag{8}$$

where W is a Wiener process in $\mathbb{R}^m$ and the function $\sigma : \mathbb{R}^d \to \mathbb{R}^d \times \mathbb{R}^m$ satisfies

$$\sum_{k=1}^m \sigma_{ik}(x)\sigma_{jk}(x) = a^{ij}(x).$$

Further, in (8)

$$\widetilde{N}(x, dt, dz) := \widetilde{N}(x, dt, dz, \omega) := N(x, dt, dz, \omega) - v(x, dz)dt,$$

is a compensated Poisson point process on $\mathbb{R}_+ \times \mathbb{R}^d$, such that

$$\mathbb{P}[N(x, (0, t], B) = k] = \exp(-tv(x, B))\frac{t^k v^k(x, B)}{k!}, \quad k = 0, 1, 2, \ldots$$

for bounded $B \in \mathcal{B}(\mathbb{R}^d \setminus \{0\})$. It is assumed that the coefficients in (8) (and so in (7)) satisfy sufficient conditions such that (8) has a unique weak solution X, and that X is an affine process with generator (7). For details regarding these assumptions, in particular the admissibility conditions that are to be fulfilled, we refer to [1], [4], see also [5].

The characteristic function of $X_t^{0;x}$, with initial value $X_0^{0;x} = x \in \mathbb{R}^d$, is denoted by,

$$\widehat{p}(t, x, u) := \mathbb{E}\left[e^{\mathrm{i}u^\top X_t^{0;x}}\right], \quad x \in \mathfrak{X}, \ u \in \mathbb{R}^d, \ t \geq 0. \tag{9}$$

For a variety of affine processes the characteristic function is explicitly known. However, in general the characteristic function of an affine process involves the solution of a multi-dimensional generalized Riccati equation that may not be solved explicitly. In particular, for affine jump processes with state dependent jump part a closed form expression for the characteristic function generally does not exist. In this section we recall the approach by Belomestny, Kampen, and Schoenmakers [1], who developed in general a series expansion for the log-characteristic function in terms of the ingredients of the generator of the affine process under consideration. By truncating this expansion one may obtain an approximation of the characteristic function that may subsequently be used for approximate option pricing.

Henceforth, $x \in \mathfrak{X}$ is fixed. It is assumed that the characteristic function (9) satisfies:

Assumption **HE:** There exists a non-increasing function $R : (0, \infty) \ni r \to R(r) \in (0, \infty]$, such that for any $u \in \mathbb{R}^d$, the function $[0, \infty) \ni s \to \widehat{p}(s, x, u) \in \mathbb{C}$ has a holomorphic extension to the region

$$G_u := \{z \in \mathbb{C} : |z| < R(\|u\|)\} \cup \{z \in \mathbb{C} : \operatorname{Re} z \geq 0 \text{ and } |\operatorname{Im} z| < R(\|u\|)\}$$

(cf. Proposition 3.7, 3.8, and Theorem 4.1 and Corollary 4.2–4.4 in [1]).

Under Assumption **HE**, Theorem 3.4 in [1] is particularly fulfilled for each u. Moreover, by taking in [1], Theorem 3.4–(ii),

$$\eta_u = \eta(\|u\|) := \frac{\pi}{2R(\|u\|)}, \tag{10}$$

we arrive at the log-series representation [1]–(5.12) for the characteristic function (w.r.t. the principal branch of the logarithm),

$$\begin{aligned} \ln \widehat{p}(t, x, u) = \ln \left(\sum_{r \geq 0} h_{r,0}(u; \eta_u)(1 - e^{-\eta_u t})^r \right) + \mathrm{i}u^\top x \\ + x^\top \frac{\sum_{r \geq 1} h_r(u; \eta_u)(1 - e^{-\eta_u t})^r}{\sum_{r \geq 0} h_{r,0}(u; \eta_u)(1 - e^{-\eta_u t})^r}, \quad u \in \mathbb{R}^d, \ t \geq 0, \end{aligned} \tag{11}$$

where the coefficients $h_{r,0}(u; \eta_u) \in \mathbb{C}$ and $h_r(u; \eta_u) = \left[h_{r,e_1}(u; \eta_u), \ldots, h_{r,e_d}(u; \eta_u)\right] \in \mathbb{C}^d$ with $e_i := \left(\delta_{ij}\right)_{j=1,\ldots,d}$, can be computed algebraically from the coefficients of the affine generator A in a way that is described below, see Eq. (15).

Alternatively, in [1] a direct expansion of the form

$$\widehat{p}(t, x, u) = e^{\mathrm{i}u^\top x} \sum_{r=0}^{\infty} q_r(x, u; \eta_u)(1 - e^{-\eta_u t})^r, \ u \in \mathbb{R}^d, \ t \geq 0, \tag{12}$$

is derived with

$$q_r(x, u; \eta_u) = \sum_{|\gamma| \leq r} h_{r,\gamma}(u; \eta_u) x^\gamma,$$

and the $h_{r,\gamma}$ are computed by the recursion (15) as described below.

Remark 1 Because of Assumption **HE**, if Theorem 3.4–(i) applies for some u, it applies for any u' with $\|u'\| \leq \|u\|$. As a consequence, one may take in (11) any $\eta_u = \eta(\|u''\|)$ with $\|u''\| \geq \|u\|$.

In order to outline the construction of the expansion (11), let us denote

$$f_u(x) := e^{\mathrm{i}u^\top x}, \quad z \in \mathbb{R}^d. \tag{13}$$

Then for each multi-index $\beta \in \mathbb{N}_0^d$ we may compute algebraically

$$\mathfrak{b}_\beta(x,u) := \mathrm{i}^{-|\beta|}\partial_{u^\beta}\frac{Af_u(x)}{f_u(x)} =: \mathfrak{b}_\beta^0(u) + \sum_{\kappa,\,|\kappa|=1} \mathfrak{b}_{\beta,\kappa}^1(u)\,x^\kappa \tag{14}$$

(in multi-index notation), provided that for the jump part in the generator (7),

$$\frac{1}{f_u(x)}\int_{\mathbb{R}^d}\left(f_u(x+z) - f_u(x) - z^\top\frac{\partial f_u}{\partial x}\right)\nu(x,dz)$$
$$= \int_{\mathbb{R}^d}\left(e^{\mathrm{i}u^\top z} - 1 - \mathrm{i}u^\top z\right)\nu^0(dz) + x^\top\int_{\mathbb{R}^d}\left(e^{\mathrm{i}u^\top z} - 1 - \mathrm{i}u^\top z\right)\nu^1(dz)$$

is explicitly known. That is, the cumulant generating functions of ν^0 and ν_i^1, $i = 1, \ldots, d$, are explicitly known. We note that the expression $Af_u(x)/f_u(x)$ in (14) is termed the *symbol* of the operator A. As such the $\mathfrak{b}_\beta$ in (14) are, modulo some integer power of the imaginary unit, derivatives of the symbol of A.

Let us next consider a fixed $u \in \mathbb{R}^d$ and $\eta_u > 0$. Then for each multi-index γ and integer $r \geq 0$ we are going to construct $h_{r,\gamma} = h_{r,\gamma}(u;\eta_u)$ as follows. For $|\gamma| > r$ we set $h_{r,\gamma} \equiv 0$ and for $0 \leq r \leq |\gamma|$, the $h_{r,\gamma}$ are determined by the following recursion. As initialization we take $h_{0,0} \equiv 1$, and for $0 \leq r < |\gamma|$ we have (cf. [1]–(4.6)),

$$(r+1)h_{r+1,\gamma} = \sum_{|\beta|\leq r-|\gamma|}\eta_u^{-1}\binom{\gamma+\beta}{\beta}h_{r,\gamma+\beta}\mathfrak{b}_\beta^0 \tag{15}$$
$$+\sum_{|\kappa|=1,\,\kappa\leq\gamma}\ \sum_{|\beta|\leq r+1-|\gamma|}\eta_u^{-1}\binom{\gamma-\kappa+\beta}{\beta}h_{r,\gamma-\kappa+\beta}\mathfrak{b}_{\beta,\kappa}^1 + rh_{r,\gamma},$$

where $|\gamma| \leq r+1$, and empty sums are defined to be zero. We next set

$$h_r(u;\eta_u) := \left[h_{r,e_i}(u;\eta_u)\right]_{i=1,\ldots,d}.$$

In view of Theorem 4.1 in [1] suitable choices of η_u are

$$\eta_u \gtrsim 1 + \|u\|^2 \text{ in case of pure affine diffusions,}$$
$$\eta_u \gtrsim e^{\zeta\|u\|},\ \zeta > 0, \text{ for affine jump processes with thinly tailed large jumps.}$$

In practice the best choice of η_u can be determined in view of the particular problem under consideration. Generally, on the one hand, η_u should be large enough to guarantee convergence of the series (11), but not too large in order to keep fast speed of convergence.

As a natural approximation to (11) and (12) we consider for $K = 1, 2, \ldots,$

$$\ln \widehat{p}_K(t,x,u) = \ln\left(\sum_{r=0}^{K} h_{r,0}(u;\eta_u)(1-e^{-\eta_u t})^r\right) + \mathrm{i}u^\top x \tag{16}$$

$$+x^\top \frac{\sum_{r=1}^{K} h_r(u;\eta_u)\,(1-e^{-\eta_u t})^r}{\sum_{r=0}^{K} h_{r,0}(u;\eta_u)(1-e^{-\eta_u t})^r},\ u\in\mathbb{R}^d,\ t\geq 0,$$

and the "ground" expansion based approximation

$$\widehat{p}(t,x,u) = e^{\mathrm{i}u^\top x}\sum_{r=0}^{K} q_r(x,u;\eta_u)(1-e^{-\eta_u t})^r,\ u\in\mathbb{R}^d,\ t\geq 0, \tag{17}$$

respectively.

Remark 2 In connection with approximations (16) and (17) it seems natural to estimate R using Cauchy's criterion, and η_u according to (10). That is, we could take

$$\eta_u \approx \frac{\pi}{2}\sqrt[K]{\frac{\left|A^K f_u(x)\right|}{K!}},$$

where the sequence $g_r(x,u) := A^r f_u(x)/f_u(x)$ can be obtained from the recursion

$$g_{r+1,\gamma} = \sum_{|\beta|\leq r-|\gamma|}\binom{\gamma+\beta}{\beta} g_{r,\gamma+\beta}\mathfrak{b}^0_\beta \tag{18}$$

$$+\sum_{|\kappa|=1,\,\kappa\leq\gamma}\ \sum_{|\beta|\leq r+1-|\gamma|}\binom{\gamma-\kappa+\beta}{\beta} g_{r,\gamma-\kappa+\beta}\mathfrak{b}^1_{\beta,\kappa},$$

with $g_{0,0}=1$ (cf [1]–(4.6)).

4 Generalized Merton Models

We now consider generalized Merton models of the form (5) and (6). For the characteristic function of (5) we have,

$$\Phi_t(z) = e^{\mathrm{i}z\gamma t}\mathbb{E}e^{\mathrm{i}z\sigma W_t}\mathbb{E}e^{\mathrm{i}zX_t^{0;\left(0,x^2,\ldots,x^d\right);1}}$$

$$= \exp\left[\mathrm{i}z\gamma t - \frac{z^2\sigma^2}{2}t\right]\widehat{p}\left(t,\left(0,x^2,\ldots,x^d\right),(z,0,\ldots,0)\right), \tag{19}$$

where $X_t^{\cdot;\cdot;1}$ denotes the first component of $X_t^{\cdot;\cdot}$ cf. (2). Firstly, the martingale condition (3) can now be formulated as

$$\gamma = \gamma(t) = -\frac{\sigma^2}{2} - t^{-1} \ln \widehat{p}(t, \left(0, x^2, \ldots, x^d\right), (-\mathrm{i}, 0, ..., 0)), \tag{20}$$

that is, γ may in principle depend on time t. More generally, the characteristic function of (6) takes the form,

$$\Phi_t(z) = e^{\mathrm{i}z\gamma(t)t} \widehat{p}_H(t, z) \widehat{p}(t, \left(0, x^2, \ldots, x^d\right), (z, 0, \ldots, 0)), \tag{21}$$

with $\widehat{p}_H(t, z) := \mathbb{E}\left[\exp(\mathrm{i}zH_t)\right]$, and

$$\gamma(t) = -t^{-1} \ln \widehat{p}_H(t, -\mathrm{i}) - t^{-1} \ln \widehat{p}(t, \left(0, x^2, ..., x^d\right), (-\mathrm{i}, 0, ..., 0)). \tag{22}$$

In a situation where $\widehat{p}$ in (19) and (21), respectively, is unknown in closed form, we propose to replace it with an approximation $\widehat{p}_K$ due to (16) for some level K large enough. It is convenient to choose X_t^1 and H such that $\exp\left(X_\cdot^1\right)$ and $\exp\left(H_\cdot\right)$ are martingales, respectively. Since $X_0^1 = H_0 = 0$, we then have $\gamma = 0$ in (22).

Before considering affine processes with really unknown characteristic function, in the next section we recall the known characteristics of a log-Heston type model.

4.1 Heston Model

Let us consider for $X = (X^1, X^2)$ a log-Heston type model with dynamics

$$\begin{aligned} dX^1 &= -\frac{1}{2}\alpha^2 X^2 dt + \alpha\sqrt{X^2} dW, \quad X^1(0) = 0, \\ dX^2 &= \kappa\left(\theta - X^2\right) dt + \sigma\sqrt{X^2}\left(\rho dW + \sqrt{1-\rho^2} d\overline{W}\right), \quad X^2(0) = \theta, \end{aligned} \tag{23}$$

for some $\alpha, \sigma, \kappa, \theta > 0$, and $-1 \leq \rho \leq 1$. Note that the initial value of X^2 is taken to be the expectation of the long-run stationary distribution of X^2. The characteristic function X^1 due to (23) is known as follows (we take Lord and Kahl's representation [7], using the principal branch of the square root and logarithm[1]):

$$\ln \widehat{p}(t, \theta, z) := \ln \widehat{p}(t, (0, \theta), (z, 0)) = A(z; t) + B(z; t)\theta, \text{ with} \tag{24}$$

$$\begin{aligned} A(z; t) &:= \frac{\theta\kappa}{\sigma^2}\left((a - d)t - 2\ln\frac{e^{-dt} - g}{1 - g}\right), \\ B(z; t) &:= \frac{a + d}{\sigma^2}\frac{1 - e^{dt}}{1 - ge^{dt}} \text{ with} \end{aligned} \tag{25}$$

[1]Roger Lord confirmed to J.S. a typo in the published version and so we refer to the preprint version.

$$a := \kappa - \mathrm{i}z\alpha\sigma\rho,\ d := \sqrt{a^2 + \alpha^2\sigma^2\left(\mathrm{i}z + z^2\right)},\ g := \frac{a+d}{a-d},$$

while abusing notation in (24) slightly. By construction, $\exp\left(X_t^1\right)$ is a martingale and so it holds that $\ln \widehat{p}(t, \theta, -\mathrm{i}) = 0$. This can be easily seen from the Heston dynamics (23) and also by taking $z = -\mathrm{i}$ in (25), where we then have that $a = \kappa - z\alpha\sigma\rho \in \mathbb{R}$, so $d = |a|$. Thus $|g| = \infty$ if $a > 0$ and $|g| = 0$ if $a < 0$ and for both cases we get that $A(-\mathrm{i}; t) \equiv B(-\mathrm{i}; t) \equiv 0$. As a consequence we have $\gamma = -\sigma^2/2$ in (20).

4.2 Heston Model with State Dependent Jumps

We now consider a generalized Heston model with state dependent jumps in the first component, henceforth termed the HSDJ model, of the following form:

$$\begin{aligned} dX^1 &= -\lambda_0 a_0 dt - \left(\lambda_1 a_1 + \frac{1}{2}\alpha^2\right) X^2 dt + \alpha\sqrt{X^2}dW \\ &\quad + \int_{\mathbb{R}} y\left(N(X_-^2, dt, dy) - \lambda_0\mu_0(y)dydt - X^2\lambda_1\mu_1(y)dydt\right), \\ dX^2 &= \kappa\left(\theta - X^2\right)dt + \sigma\sqrt{X^2}\left(\rho dW + \sqrt{1-\rho^2}d\overline{W}\right) \end{aligned} \tag{26}$$

with $X^1(0) = 0,\ X^2(0) = \theta$ and with t suppressed in X_{t-} (cf. (8)). In this model $N(w, dt, dy)$ is for each $w > 0$ a Poisson point process on $\mathbb{R}_+ \times \mathbb{R}$ and μ_0 and μ_1 are considered to be probability densities of jumps that arrive at rate $\lambda_0 > 0$ and $w\lambda_1 > 0$, respectively. Further in (26), a_0 and a_1 are non-negative constants given by

$$a_0 = \int \left(e^y - y - 1\right)\mu_0(y)dy \text{ and } a_1 = \int \left(e^y - y - 1\right)\mu_1(y)dy, \tag{27}$$

hence in particular it is assumed that the measures associated with μ_0 and μ_1 have exponential moments. In the HSDJ model the density μ_0 may have support $\mathbb{R}$, for example Gaussian, while the density μ_1 may be concentrated on $(-\infty, 0)$ for example. In this way λ_0 and μ_0 are responsible for the "normal" random jumps in (26), while λ_1 and μ_1 are responsible for downward jumps which, due to the (state) dependence on X^2, arrive with increasing intensity as the volatility X^2 increases. As such the model covers a stylized empirical fact observed for several underlying quantities, such as assets, indices, or interest rates. Since μ_0 and μ_1 are assumed to be probability densities, the dynamics of X^1 in (26) may also be written as

$$dX^1 = \left(-\lambda_0 \left(\mathfrak{m}_0 + a_0\right) - \left(\frac{1}{2}\alpha^2 + \lambda_1 \left(\mathfrak{m}_1 + a_1\right)\right) X^2\right) dt$$
$$+ \alpha\sqrt{X^2}dW + \int_{\mathbb{R}} yN(X^2_-, dt, dy), \tag{28}$$

with

$$\mathfrak{m}_0 := \int_{\mathbb{R}} y\mu_0(y)dy \quad \text{and} \quad \mathfrak{m}_1 := \int_{\mathbb{R}} y\mu_1(y)dy. \tag{29}$$

One can show rigorously that $e^{X_t^1}$ is a martingale with $\mathbb{E}\left[e^{X_t^1}\right] = 1$, and so we may take in (20) $\gamma = -\sigma^2/2$ again (see for example [3]).

In Appendix A we spell out the generator, cf. (7), and its corresponding symbol derivatives (14) corresponding to the HSDJ model (26).

Example 3 In the case where $\lambda_1 = 0$, the characteristic function $\widehat{p}_{\lambda_0,\mu_0}$ of X^1 is simply given by (see (28), (27) and (29))

$$\begin{aligned}\ln \widehat{p}_{\lambda_0,\mu_0}(t,\theta,z) &= \ln \widehat{p}(t,\theta,z) - t\lambda_0 \left(a_0 + \mathfrak{m}_0\right) \mathrm{i}z + t\lambda_0\psi_0(z) \\ &= \ln \widehat{p}(t,\theta,z) - t\lambda_0\psi_0(-\mathrm{i})\mathrm{i}z + t\lambda_0\psi_0(z),\end{aligned}$$

where

$$\psi_0(z) := \int \left(e^{\mathrm{i}yz} - 1\right) \mu_0(y)dy = \int e^{\mathrm{i}yz}\mu_0(y)dy - 1$$

follows from the characteristic function of the jump measure and $\widehat{p}(t,\theta,z)$ is given by (24). Note that we have $\ln \widehat{p}_{\lambda_0,\mu_0}(t,\theta,-\mathrm{i}) = 0$ again indeed. For example if the jumps are $\mathcal{N}(c,\nu^2)$ distributed we have the well known expression

$$\psi_0(z) = e^{\mathrm{i}cz - \frac{1}{2}\nu^2 z^2} - 1,$$

hence

$$\ln \widehat{p}_{\lambda_0,c,\nu^2}(t,\theta,z) := \ln \widehat{p}(t,\theta,z) + t\lambda_0 \left(e^{\mathrm{i}cz - \frac{1}{2}\nu^2 z^2} - \mathrm{i}ze^{c + \frac{1}{2}\nu^2} + \mathrm{i}z - 1\right).$$

5 Numerical Examples

In this section we will price European options by a Fourier based method due to Carr-Madan [2]. Let the stock price at maturity T be given as

$$S_T = S_0 e^{rT + Y_T},$$

where $\exp[Y_.]$ is a martingale with $Y_0 = 0$. If the characteristic function

$$\Phi_T(z) := \mathbb{E}\left[e^{\mathrm{i}zY_T}\right]$$

is known, then the price of a European call option with strike K at time $t = 0$ is given by

$$C(K) = (S_0 - Ke^{-rT})^+ + \frac{S_0}{2\pi}\int_{-\infty}^{\infty} \frac{1 - \Phi_T(z - \mathrm{i})}{z(z - \mathrm{i})} e^{-\mathrm{i}z \ln \frac{Ke^{-rT}}{S_0}} dz \tag{30}$$

(Carr-Madan's formula). For more general Fourier valuation formulas, see [6]. In general, the decay of the integrand in (30) is of order $O(|z|^{-2})$ as $|z| \to \infty$, hence relatively slow. We therefore use a kind of variance reduction for integrals using the formula

$$\mathcal{BS}(S_0, T, r, \sigma_B) = (S_0 - Ke^{-rT})^+ + \frac{S_0}{2\pi}\int_{-\infty}^{\infty} \frac{1 - \Phi_T^{\mathcal{BS}}(z - \mathrm{i})}{z(z - \mathrm{i})} e^{-\mathrm{i}z \ln \frac{Ke^{-rT}}{S_0}} dz, \tag{31}$$

where $\mathcal{BS}$ is the well-known Black-Scholes formula based on the risk-neutral Black-Scholes model

$$S_t^{\mathcal{B}} := S_0 e^{rT - \sigma_B^2 T/2 + \sigma_B W_T}, \text{ with}$$
$$\Phi_T^{\mathcal{BS}}(z) := \mathbb{E}\left[e^{\mathrm{i}z\left(-\sigma_B^2 T/2 + \sigma_B W_T\right)}\right] = e^{-(z^2 + \mathrm{i}z)\sigma_B^2 T/2},$$

for a suitable but in principle arbitrary $\sigma_B > 0$. Next, subtracting (30) and (31) gives the variance reduced formula

$$C(K) = \mathcal{BS}(S_0, T, r, \sigma_B) + \frac{S_0}{2\pi}\int_{-\infty}^{\infty} \frac{\Phi_T^{\mathcal{BS}}(z - \mathrm{i}) - \Phi_T(z - \mathrm{i})}{z(z - \mathrm{i})} e^{-\mathrm{i}z \ln \frac{Ke^{-rT}}{S_0}} dz, \tag{32}$$

where the integrand decays at a rate $|z^{-2}| \max[|\Phi_T^{\mathcal{BS}}(z - \mathrm{i})|, |\Phi_T(z - \mathrm{i})|]$ which is typically (much) faster than in (30), provided that $\Phi_T(z - \mathrm{i})$ tends to zero as $|z| \to \infty$.

5.1 Product of Heston Models

We first consider a model where the stock price S_t is obtained as the product of two independent Heston factors, i.e., (6) with X_t^1 another Heston model. Clearly, in this case a closed form expression for the characteristic function of $\ln S_t$ exists, and therefore the asymptotic expansion presented in this paper is not needed for pricing. This allows us to easily compute accurate reference prices, and thus assess the numerical accuracy of prices obtained from the expansion of the characteristic function. All calculations were done using Mathematica. Using its symbolic capabilities, we have implemented the recursion (15) in full generality.

Table 1 Parameters of the Heston+Heston-model. v denotes the initial variance in both components

	H_t	X_t^1
α	1.0	1.0
κ	1.5	1.5
σ	0.6	0.3
θ	0.04	0.0225
ρ	−0.2	−0.3
v	0.04	0.0225

The Heston parameters for the components H_t and X_t^1 are presented in Table 1. Additionally, we choose $S_0 = 10$ and $r = 0.05$ for option pricing. Based on these parameters, we compute the asymptotic expansion $\widehat{p}_K$ of the characteristic function using (12) with $K = 8$, i.e., including the first *nine* terms in the expansion.

In Fig. 1, we compare the exact and the approximate characteristic functions of the (normalized) logarithm of the stock prices—i.e., with $S_0 = 1$ for convenience. We can clearly see that the approximation deteriorates when $|u|$ becomes large, but then both the exact and the approximate characteristic functions tend to 0. Moreover,

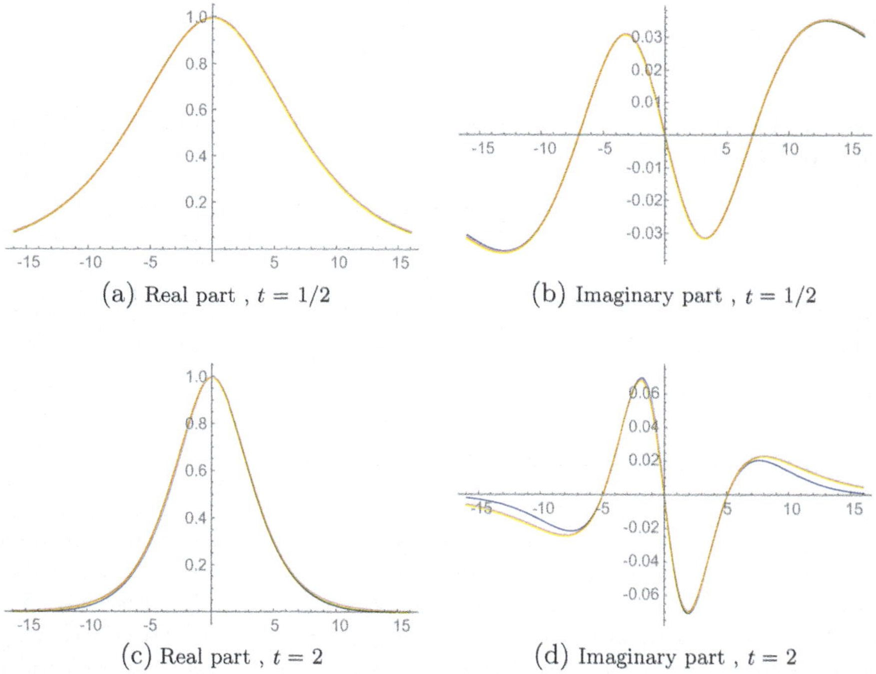

Fig. 1 Exact (*blue*) and approximate (*orange*) characteristic functions of the logarithm of the normalized stock price in the generalized Merton model with two Heston factors evaluated at time $t = 1/2$ and $t = 2$ (years)

Table 2 Price of ATM call option with maturity $T = 1$ computed using domain of integration $[-L, L]$ for both the exact characteristic function and the approximate formula, together with the relative error for using the approximate formula—w.r.t. the most accurate price obtained from the exact formula

L	2	4	8	16	32	64
Exact	0.8350	0.9621	1.1105	1.1832	1.1884	1.884
Approx.	0.8353	0.9626	1.1111	1.1842	1.1896	1.1896
(Rel. error)	0.2981	0.1912	0.0665	0.0054	0.0010	0.0010

the approximation formula is more accurate for small t (cf. (11), (16) and (12), (17, respectively).

When we come to option pricing, we plug the approximate formula for the characteristic function into the Fourier pricing formula (32). For the implementation, we clearly need to replace the infinite domain of integration by a finite one, i.e., we use (32) integrating from $-L$ to L, $L \in \mathbb{R}_+$. This cut-off is potentially critical for our approximation procedure, as large integration domains (and, hence, large $|u|$) may correspond to large errors of the approximate formula. Fortunately, Table 2 indicates that this effect does not materialize.

Remark 4 At this stage, we would like to highlight once more the heuristic choice of η proposed in Remark 2. Without a good choice of η, it is very easy to run into situations, where the approximation error is already too large for the needed domain of integration.

Let us consider option prices and the corresponding errors for maturities from 1/2 to 5 years and for strike prices between 7 (deep in) and 13 (deep out of) the money. Figure 2 shows that errors remain small ($\leq 2\,\%$ ATM) for maturities up to 2 years. For (deep) OTM options, it seems to be more reasonable to look at absolute instead of relative errors, which give a similar impression.

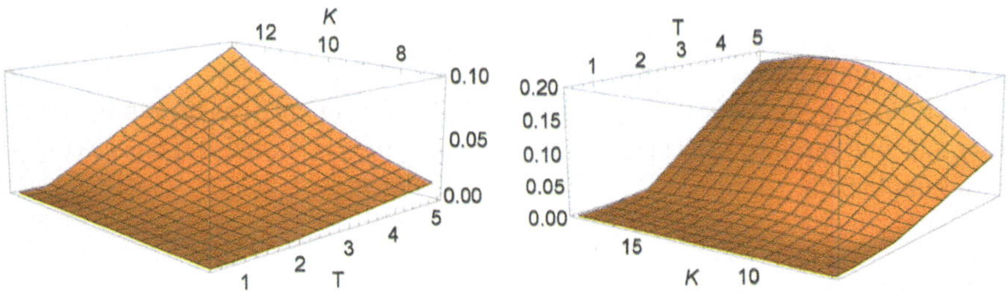

(a) Relative error. (b) Absolute error.

Fig. 2 Relative and absolute errors of European call option prices

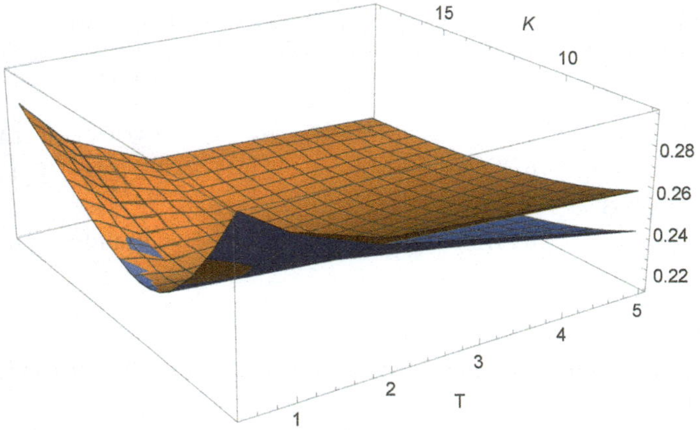

Fig. 3 Implied volatility of the generalized Merton model based on two Heston factors based on exact (*blue*) and approximate (*orange*) characteristic functions

Finally, the implied volatility in this model is plotted in Fig. 3. Considerable deviations between the exact and the approximate formula are only observed for higher maturities.

5.2 Generalized Merton Model with State-Dependent Jumps

Let us consider a generalized Merton model of the form (6) where X^1 is an affine jump process with state-dependent jump-intensity in the sense of (26). The parameters corresponding to the diffusive parts of both H and X^1 are chosen as in Table 1. Regarding the jump part of X^1, we set $\lambda_0 = 0$, $\mu_0 = 0$, thereby turning off the jumps with constant, i.e., not state dependent, intensities. The jump parameters of X^1 are chosen according to Table 3.

This means that jumps in the log-price have exponentially distributed magnitude and negative sign. The mean jump of the log-price is around 0.22, i.e., in case of a downward jump ("crisis"), the stock loses about 20 % of its value on average. The intensity λ_1 seems excessively high, but recall that this intensity is multiplied by the instantaneous variance of the Heston component, which is started at 0.04 see (Fig. 4) for a sample path.

Table 3 Jump parameters of X^1

	X_t^1
λ_1	10
$\mu_1(y)$	$\mathbf{1}_{y<0} p e^{py}$
p	4.48

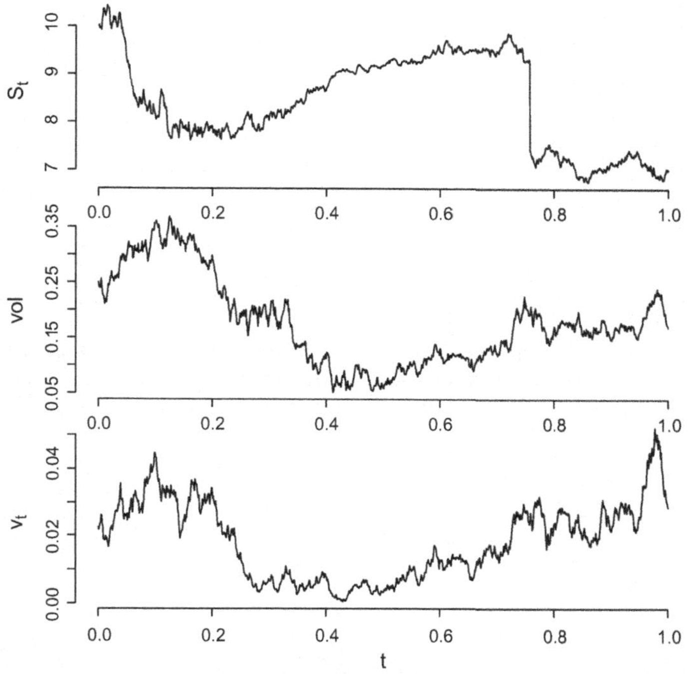

Fig. 4 Sample path of S_t in the generalized Merton model with state-dependent jumps (*first panel*), volatility (more precisely, the square root of the sum of both variance components) of S_t (*second panel*), and of the variance component of the second Heston factor. A jump occurs shortly after time 0.75

By (33), and (34) below, we obtain

$$\psi_0(\xi) = 0, \qquad \mathfrak{m}_0 + a_0 = \psi_0(-\mathrm{i}) = 0,$$
$$\psi_1(\xi) = \int_{-\infty}^{0} \left(e^{\mathrm{i}\xi y} - 1\right) p e^{py} dy = -\frac{\mathrm{i}\xi}{p + \mathrm{i}\xi}, \qquad \mathfrak{m}_1 + a_1 = \psi_1(-\mathrm{i}) = -\frac{1}{p+1}.$$

Figure 5 shows the approximate characteristic function including jumps at time $t = 1/2$, compared with the exact characteristic function. As expected, the jumps lead to a considerable change in the characteristic function. We compare the characteristic function to another numerical approximation based on Monte Carlo simulation. Both approximations lead to very close results especially in the real part. The results are less close for the imaginary part, but notice that the graphical representation exaggerates the differences as the scale is much smaller in the second plot (from -0.1 to 0.1 instead of 0 to 1).

These changes in the distribution have the expected changes in the option prices. In particular, the implied volatilities become larger, and also the smile becomes much more pronounced, comparing Fig. 6 with Fig. 3.

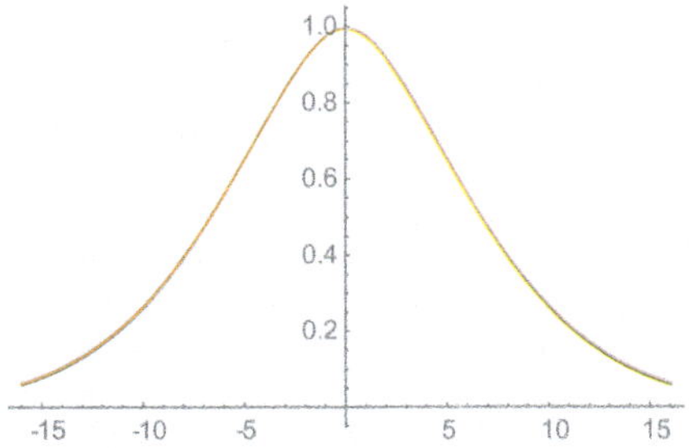

(a) Real part, $t = 1/2$

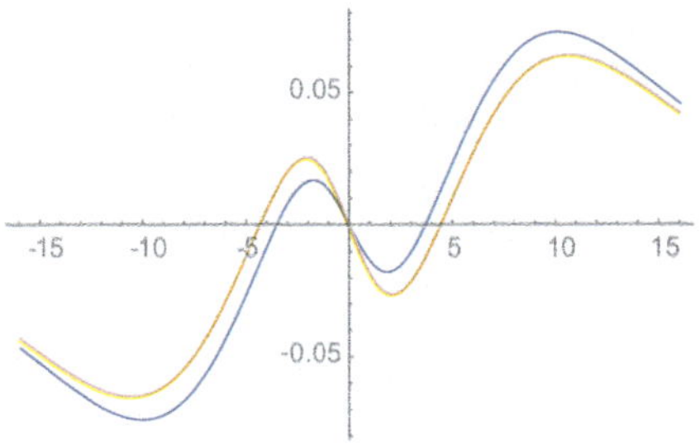

(b) Imaginary part, $t = 1/2$

Fig. 5 Approximate characteristic function (*orange*) of the logarithm of the normalized stock price in the generalized Merton model with one Heston factor and one Heston factor with jumps evaluated at time $t = 1/2$ (year). Comparison with the characteristic function computed by a Monte Carlo simulation (*blue*)

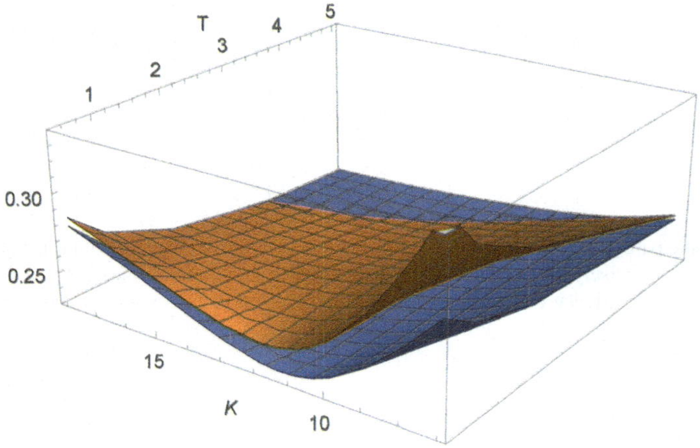

Fig. 6 Implied volatility of the generalized Merton model with one Heston factor and one Heston factor with jumps (*orange*), compared with the implied volatilities computed with the exact characteristic function in Fig. 3

Table 4 Option prices for maturity $T = 1/2$ for various strike prices in the Heston model plus jumps. We compare prices obtained by the asymptotic expansion of the characteristic function with prices obtained by Monte Carlo simulation

K	7	8	9	10	11	12	13
Monte Carlo	3.2719	2.3688	1.5511	0.8888	0.4427	0.2006	0.0884
Asym. formula	3.2279	2.3276	1.5144	0.8583	0.4217	0.1880	0.0818
Rel. error	0.0134	0.0174	0.0237	0.0343	0.0476	0.0627	0.0744
$\frac{\text{MC stat. error}}{\text{Ref. price}}$	0.0018	0.0023	0.0031	0.0044	0.0067	0.0106	0.0166

Finally, let us directly compare the price for some European call options with reference prices obtained by Monte Carlo simulation, see Table 4. Once again, we used $S_0 = 10$ and $r = 0.05$. The Monte Carlo prices are based on 100,000 trajectories with 1000 time-steps each, the statistical error, i.e., the standard deviation divided by the square root of the number of samples, is considerable smaller than the observed difference.

Unfortunately, the results of Table 4 are not as convincing as the accuracy of the approximation in the pure diffusion case suggested, compare Table 2 and Fig. 2. We suspect a combination of slow decay of the characteristic function, sub-optimal choice of the damping parameter η and higher truncation error of the asymptotic characteristic function, see the conclusions below for some further comments.

6 Conclusions

From the examples we conclude that for times being not too large the approximation procedure based on [1] performs rather well. More specifically, if no jumps are in the play the procedure works very good, but with incorporated (state dependent) jumps the accuracy is somewhat worse. In order to resolve this issue one could investigate different directions. One reason for lower accuracy may be a diminished effect of the Black-Scholes ingredients in the Fourier pricing formula (32) in the presence of state dependent jumps. This in turn might require a larger integration range where that approximation gets worse at the upper and lower end, respectively. As a way out, it looks natural to replace the role of the Black-Scholes ingredients in (32) by an affine model with state independent jumps for which the characteristic function is known, leading to a representation of the form

$$C^{\mathrm{appr}}(K) = (S_0 - Ke^{-rT})^+ + \frac{S_0}{2\pi}\int_{-\infty}^{\infty} \frac{1-\Phi_T^{\mathrm{known}}(z-\mathrm{i})}{z(z-\mathrm{i})} e^{-\mathrm{i}z\ln\frac{Ke^{-rT}}{S_0}}\,dz$$

$$+ \frac{S_0}{2\pi}\int_{-\infty}^{\infty} \frac{\Phi_T^{\mathrm{known}}(z-\mathrm{i}) - \Phi_T^{\mathrm{appr}}(z-\mathrm{i})}{z(z-\mathrm{i})} e^{-\mathrm{i}z\ln\frac{Ke^{-rT}}{S_0}}\,dz =: I_{\mathrm{known}} + I_{\mathrm{appr}}.$$

The integral I_{known} can be computed with any desired accuracy while for the integral I_{appr} a relatively small integration range may be sufficient.

Other reasons for the decreased accuracy in Sect. 5.2 for instance, may be a too small η chosen due to Remark 2, or not enough iterations. However, we leave all these investigations for further research, since this article is considered merely a first guide on numerical implementation of the method in [1].

Appendix: Generator and $\mathfrak{b}_\beta$ for the HSDJ Model

By conferring (7), (8), and (26), we have in fact

$$v(x, dz) = v^0(dz) + x^\top v^1(dz) = \lambda_0\mu_0(z_1)\delta_0(z_2)dz_1dz_2 + x_2\lambda_1\mu_1(z_1)\delta_0(z_2)dz_1dz_2$$

with δ_0 being the Dirac delta function, that is the (singular) density of the Dirac probability measure $\mathbb{R}$ concentrated in $\{0\}$. Thus, the generator of the HSDJ model is given by

$$\begin{aligned} Af(x_1, x_2) &= \left(-\lambda_0 a_0 - \left(\frac{1}{2}\alpha^2 + \lambda_1 a_1\right) x_2\right) \partial_{x_1} f + \kappa(\theta - x_2)\,\partial_{x_2} f \\ &+ \frac{1}{2}\alpha^2 x_2 \partial_{x_1x_1} f + \alpha\sigma\rho x_2 \partial_{x_1x_2} f + \frac{1}{2}\sigma^2 x_2 \partial_{x_2x_2} f \\ &+ \int_{\mathbb{R}} \left[f(x_1 + z_1, x_2) - f(x_1, x_2) - z_1 \partial_{x_1} f\right] (\lambda_0\mu_0(z_1)dz_1 + x_2\lambda_1\mu_1(z_1)dz_1)\,. \end{aligned}$$

Since we are dealing with jump probability densities rather than infinite jump measures, as in the case of infinite activity processes, the generator may be written as

$$\begin{aligned} Af(x_1, x_2) &= \left(-\lambda_0(\mathfrak{m}_0 + a_0) - \left(\frac{1}{2}\alpha^2 + \lambda_1(\mathfrak{m}_1 + a_1)\right) x_2\right) \partial_{x_1} f \\ &+ \kappa(\theta - x_2)\,\partial_{x_2} f + \frac{1}{2}\alpha^2 x_2 \partial_{x_1x_1} f + \alpha\sigma\rho x_2 \partial_{x_1x_2} f + \frac{1}{2}\sigma^2 x_2 \partial_{x_2x_2} f \\ &+ \lambda_0 \int_{\mathbb{R}} \left[f(x_1 + y, x_2) - f(x_1, x_2)\right] \mu_0(y)dy \\ &+ x_2\lambda_1 \int_{\mathbb{R}} \left[f(x_1 + y, x_2) - f(x_1, x_2)\right] \mu_1(y)dy, \end{aligned}$$

using (29).

With $f_u(x) = e^{\mathrm{i}u^\top x}$ we so obtain,

$$\begin{aligned} \frac{Af_u(x)}{f_u(x)} &= \left(-\lambda_0(\mathfrak{m}_0 + a_0) - \left(\frac{1}{2}\alpha^2 + \lambda_1(\mathfrak{m}_1 + a_1)\right) x_2\right) \mathrm{i}u_1 \\ &+ \kappa(\theta - x_2)\,\mathrm{i}u_2 - \frac{1}{2}\alpha^2 x_2 u_1^2 - \alpha\sigma\rho x_2 u_1 u_2 - \frac{1}{2}\sigma^2 x_2 u_2^2 \\ &+ \lambda_0\psi_0(u_1) + x_2\lambda_1\psi_1(u_1) \end{aligned}$$

with

$$\psi_i(\xi) := \int_{\mathbb{R}} \left(e^{\mathrm{i}\xi y} - 1\right) \mu_i(y)dy, \quad i = 0, 1. \tag{33}$$

Note that we have

$$\mathfrak{m}_i + a_i = \psi_i(-\mathrm{i}), \quad i = 0, 1. \tag{34}$$

The first order derivatives w.r.t. u are,

$$\begin{aligned}\partial_{u_1}\frac{Af_u(x)}{f_u(x)} &= -\lambda_0\left(\mathfrak{m}_0+a_0\right)\mathrm{i}-\left(\frac{1}{2}\alpha^2+\lambda_1\left(\mathfrak{m}_1+a_1\right)\right)\mathrm{i}x_2\\ &\quad-\alpha^2x_2u_1-\alpha\sigma\rho x_2u_2+\lambda_0\partial_{u_1}\psi_0(u_1)+x_2\lambda_1\partial_{u_1}\psi_1(u_1)\\ \partial_{u_2}\frac{Af_u(x)}{f_u(x)} &= \kappa\left(\theta-x_2\right)\mathrm{i}-\alpha\sigma\rho x_2u_1-\sigma^2x_2u_2.\end{aligned}$$

For the second order derivatives we have

$$\begin{aligned}\partial_{u_1u_1}\frac{Af_u(x)}{f_u(x)} &= -\alpha^2x_2+\lambda_0\partial_{u_1u_1}\psi_0(u_1)+x_2\lambda_1\partial_{u_1u_1}\psi_1(u_1)\\ \partial_{u_1u_2}\frac{Af_u(x)}{f_u(x)} &= -\alpha\sigma\rho x_2,\quad \partial_{u_2u_2}\frac{Af_u(x)}{f_u(x)}=-\sigma^2x_2,\end{aligned}$$

and for multi-indices β with $|\beta|\geq 3$, i.e. the higher order ones,

$$\partial_{u^\beta}\frac{Af_u(x)}{f_u(x)}=\begin{cases}\lambda_0\partial_{u_1^{|\beta|}}\psi_0(u_1)+x_2\lambda_1\partial_{u_1^{|\beta|}}\psi_1(u_1) \text{ for } \beta=(|\beta|,0),\\ 0 \text{ if } \beta\neq(|\beta|,0).\end{cases}\tag{35}$$

Hence the ingredients (14) of the recursion (15) are in multi-index notation as follows.
$|\beta|=0$:

$$\begin{aligned}&\mathfrak{b}_0(x,u)=-\lambda_0\left(\mathfrak{m}_0+a_0\right)\mathrm{i}u_1+\kappa\theta\mathrm{i}u_2+\lambda_0\psi_0(u_1)\\ &+x_2\left(\lambda_1\psi_1(u_1)-\left(\frac{1}{2}\alpha^2+\lambda_1\left(\mathfrak{m}_1+a_1\right)\right)\mathrm{i}u_1-\kappa\mathrm{i}u_2-\frac{1}{2}\alpha^2u_1^2-\alpha\sigma\rho u_1u_2-\frac{1}{2}\sigma^2u_2^2\right),\end{aligned}$$

whence

$$\begin{aligned}\mathfrak{b}_0^0(u) &= -\lambda_0\left(\mathfrak{m}_0+a_0\right)\mathrm{i}u_1+\kappa\theta\mathrm{i}u_2+\lambda_0\psi_0(u_1),\\ \mathfrak{b}^1_{0,e_1}(u) &= 0,\quad \mathfrak{b}^1_{0,e_2}(u)=\lambda_1\psi_1(u_1)-\left(\frac{1}{2}\alpha^2+\lambda_1\left(\mathfrak{m}_1+a_1\right)\right)\mathrm{i}u_1\\ &\quad-\kappa\mathrm{i}u_2-\frac{1}{2}\alpha^2u_1^2-\alpha\sigma\rho u_1u_2-\frac{1}{2}\sigma^2u_2^2.\end{aligned}$$

For $|\beta|=1$, (14) yields

$$\begin{aligned}\mathfrak{b}_{(1,0)}(x,u) &= -\lambda_0\left(\mathfrak{m}_0+a_0\right)-\lambda_0\partial_{u_1}\psi_0(u_1)\mathrm{i}-\left(\frac{1}{2}\alpha^2+\lambda_1\left(\mathfrak{m}_1+a_1\right)\right)x_2\\ &\quad+\alpha^2x_2u_1\mathrm{i}+\alpha\sigma\rho x_2u_2\mathrm{i}-x_2\lambda_1\partial_{u_1}\psi_1(u_1)\mathrm{i}\\ \mathfrak{b}_{(0,1)}(x,u) &= \kappa\left(\theta-x_2\right)+\alpha\sigma\rho x_2u_1\mathrm{i}+\sigma^2x_2u_2\mathrm{i},\end{aligned}$$

whence

$$\mathfrak{b}^0_{(1,0)}(u) = -\lambda_0 (\mathfrak{m}_0 + a_0) - \lambda_0 \partial_{u_1} \psi_0(u_1)\mathfrak{i}, \ \mathfrak{b}^1_{(1,0),e_1}(u) = 0,$$
$$\mathfrak{b}^1_{(1,0),e_2}(u) = -\left(\frac{1}{2}\alpha^2 + \lambda_1 (\mathfrak{m}_1 + a_1)\right) + \alpha^2 u_1 \mathfrak{i} + \alpha\sigma\rho u_2 \mathfrak{i} - \lambda_1 \partial_{u_1} \psi_1(u_1)\mathfrak{i}$$

and

$$\mathfrak{b}^0_{(0,1)}(u) = \kappa\theta, \ \mathfrak{b}^1_{(0,1),e_1}(u) = 0,$$
$$\mathfrak{b}^1_{(0,1),e_2}(u) = -\kappa + \alpha\sigma\rho u_1 \mathfrak{i} + \sigma^2 u_2 \mathfrak{i}.$$

Next, for $|\beta| = 2$, (14) yields

$$\mathfrak{b}_{(2,0)}(x,u) = \alpha^2 x_2 - \lambda_0 \partial_{u_1 u_1} \psi_0(u_1) - x_2 \lambda_1 \partial_{u_1 u_1} \psi_1(u_1),$$
$$\mathfrak{b}_{(1,1)}(x,u) = \alpha\sigma\rho x_2,$$
$$\mathfrak{b}_{(0,2)}(x,u) = \sigma^2 x_2,$$

whence

$$\mathfrak{b}^0_{(2,0)}(u) = -\lambda_0 \partial_{u_1 u_1} \psi_0(u_1), \ \mathfrak{b}^1_{(2,0),e_1}(u) = 0,$$
$$\mathfrak{b}^1_{(2,0),e_2}(u) = \alpha^2 - \lambda_1 \partial_{u_1 u_1} \psi_1(u_1),$$
$$\mathfrak{b}^0_{(1,1)}(u) = \mathfrak{b}^1_{(1,1),e_1}(u) = 0, \quad \mathfrak{b}^1_{(1,1),e_2}(u) = \alpha\sigma\rho,$$
$$\mathfrak{b}^0_{(0,2)}(u) = \mathfrak{b}^1_{(0,2),e_1}(u) = 0, \quad \mathfrak{b}^1_{(0,2),e_2}(u) = \sigma^2.$$

For multi-indices β with $|\beta| \geq 3$ we get

$$\mathfrak{b}_\beta(x,u) = \begin{cases} \lambda_0 \mathfrak{i}^{-|\beta|} \partial_{u_1^{|\beta|}} \psi_0(u_1) + x_2 \lambda_1 \mathfrak{i}^{-|\beta|} \partial_{u_1^{|\beta|}} \psi_1(u_1) \text{ for } \beta = (|\beta|, 0), \\ 0 \text{ if } \beta \neq (|\beta|, 0), \end{cases}$$

whence

$$\mathfrak{b}^0_\beta(u) = \begin{cases} \lambda_0 \mathfrak{i}^{-|\beta|} \partial_{u_1^{|\beta|}} \psi_0(u_1) \text{ for } \beta = (|\beta|, 0), \\ 0 \text{ if } \beta \neq (|\beta|, 0), \end{cases}$$

and

$$\mathfrak{b}^1_{\beta,e_1}(u) = 0,$$
$$\mathfrak{b}^1_{\beta,e_2}(u) = \begin{cases} \lambda_1 \mathfrak{i}^{-|\beta|} \partial_{u_1^{|\beta|}} \psi_1(u_1) \text{ for } \beta = (|\beta|, 0), \\ 0 \text{ if } \beta \neq (|\beta|, 0). \end{cases}$$

References

1. Belomestny, D., Kampen, J., Schoenmakers, J.: Holomorphic transforms with application to affine processes. J. Funct. Anal. **257**(4), 1222–1250 (2009)
2. Carr, P., Madan, D.: Option valuation using the fast Fourier transform. J. Comput. Financ. **2**, 61–74 (1999)
3. Cont, R., Tankov, P.: Financial modelling with jump processes. Chapman & Hall/CRC Financial Mathematics Series. Chapman & Hall/CRC, Boca Raton, FL (2004)
4. Duffie, D., Filipović, D., Schachermayer, W.: Affine processes and applications in finance. Ann. Appl. Probab. **13**(3), 984–1053 (2003)
5. Duffie, D., Pan, J., Singleton, K.: Transform analysis and asset pricing for affine jump-diffusions. Econometrica **68**(6), 1343–1376 (2000)
6. Eberlein, E., Glau, K., Papapantoleon, A.: Analysis of Fourier transform valuation formulas and applications. Appl. Math. Financ. **17**(3), 211–240 (2010)
7. Lord, R., Kahl, C.: Complex logarithms in Heston-like models. Math. Fin. **20**(4), 671–694 (2010)
8. Merton, R.C.: Option pricing when underlying stock returns are discontinuous. J. Financ. Econ. **3**, 125–144 (1976)

Approximate Pricing of Call Options on the Quadratic Variation in Lévy Models

Giso Jahncke and Jan Kallsen

Abstract In this note we consider approximate pricing of volatility options on an underlying which follows an exponential Lévy process. More specifically, we study call options on the realized variance. The key idea of our approach is to interpret the compensated quadratic variation of the Lévy process as a perturbed Brownian motion. The approximation involves even cumulants of the Lévy process and option price sensitivities (greeks) in the limiting Bachelier model. We illustrate numerically that our formulas work well if the cumulants of the Lévy process are not too large.

Keywords Realized variance · Option pricing · Lévy processes · Approximation

MSC subject classification (2010): 91G20 · 60G51

1 Introduction

The valuation of options written on the realized variance

$$\sum_{n=1}^{N} \log(S_{t_n}/S_{t_{n-1}})^2 = \sum_{n=1}^{N}(X_{t_n} - X_{t_{n-1}})^2, \quad 0 = t_0 < \ldots < t_N = T \tag{1}$$

of a stock $S = \exp(X)$ has received some attention in the Mathematical Finance literature. For better mathematical tractability, most work focuses on using the quadratic variation $[X, X]$ of the logarithmic price X as a continuous time approximation for (1). This is justified by the well known fact that

G. Jahncke · J. Kallsen (✉)
Mathematisches Seminar, Christian-Albrechts-Universität Zu Kiel,
Westring 383, 24098 Kiel, Germany
e-mail: kallsen@math.uni-kiel.de

G. Jahncke
e-mail: jahncke@math.uni-kiel.de

© Springer International Publishing Switzerland 2016
J. Kallsen and A. Papapantoleon (eds.), *Advanced Modelling in Mathematical Finance*,
Springer Proceedings in Mathematics & Statistics 189,
DOI 10.1007/978-3-319-45875-5_11

$$\sum_{n=1}^{N}(X_{t_n}-X_{t_{n-1}})^2 \to [X,X]_T \text{ as } \sup_{n=1,\dots,N}|t_n-t_{n-1}|\to 0,$$

cf. e.g. [13, I.4.7]. This approximation works quite well for daily fixings and longer maturities, as is confirmed by [4, 21]. We refer to [15, 23] for results on discretely sampled realized variance.

The pricing of options written on the quadratic variation $[X,X]$ is widely studied in the literature. If the underlying stock process is modelled as a continuous semimartingale, [8, 11] provide model-free valuation approaches based on a replicating portfolio of European options. [3] deals with the Barndorff-Nielsen-Shephard [1] model without leverage, whereas [21, 22] focuses on a Heston-type model [12] with jumps in stock price and volatility. In [7] the pricing problem is discussed for exponential Lévy models, whereas [14] deals with affine stochastic volatility models. Recent studies on pricing options on discretely sampled variance in models with jumps and stochastic volatility include [15, 23].

In this note we focus on European call options written on the quadratic variation $[X,X]$ of a Lévy process representing a logarithmic stock price. For such options semi-explicit formulas are known, using a suitable integral transform of the payoff, cf. e.g. [14, Lemma 6.3] using Laplace transform. However, these formulas depend on the characteristic function of $[X,X]$, which is not known in closed form for most Lévy processes. The key idea of this paper is to apply the perturbation approach of [9] to the present setup. More specifically, we connect the compensated quadratic variation process via a curve in the space of Lévy processes with a Brownian motion. This leads to a formal series resembling an Edgeworth expansion. We obtain approximate pricing formulas involving cumulants of the Lévy process X and sensitivities of call option prices in a Bachelier model. For numerical illustration we consider a Merton jump-diffusion model with Gaussian jumps, showing that our approximation works quite well if the even cumulants of X are not too large.

The paper is organized as follows. We introduce the mathematical setup in Sect. 2 and state our main results in Sect. 3. Subsequently, we provide some numerical illustration. Section 5 contains proofs.

2 Mathematical Setup

2.1 Market Model

We consider a market with two traded assets, a bond and a non-dividend paying stock. The price process B of the bond is given by $B_t = e^{rt}$ for a deterministic interest rate $r \geq 0$. In what follows, we work with discounted quantities, using B as numéraire. We assume absence of arbitrage and model directly under a martingale measure P. Relative to this measure, the discounted price process S of the stock is assumed to be of the form

$$S_t = S_0 e^{X_t}, \quad t \in \mathbb{R}_+ \tag{2}$$

with $S_0 > 0$ and some real-valued Lévy process X satisfying $E(e^{X_1}) = 1$. Moreover, we make the following

Assumption 2.1 1. $E\left(e^{2R[X,X]_1}\right) < \infty$ for some $R > 0$,
2. there exists $\gamma > 0$ such that $\liminf_{r\to 0} r^{\gamma-2} \int_{-r}^{r} x^4 \nu^X(dx) > 0$, where ν^X denotes the Lévy measure of X.

The second assumption is slightly stronger than to assume that the Lévy process has infinitely many jumps on time intervals of positive length.

2.2 *Option Payoff Function*

We focus on pricing a call option on the quadratic variation of X with strike $K > 0$ and maturity $T > 0$, which has payoff

$$f([X, X]_T) := ([X, X]_T - K)^+.$$

We write $f([X, X]_T) = g(L_T)$ with $g(x) := (x - \tilde{K})^+$,

$$L_t := [X, X]_t - E([X, X]_t)\,,\ t \in \mathbb{R}^+,$$

and $\tilde{K} := K - E([X, X]_T)$. Note that the compensated quadratic variation L is a Lévy process as well with $E(L_1) = 0$ und $\mathrm{Var}\,(L_1) =: \tilde{\sigma}^2 > 0$. Since L is of finite variation, it does not have a Brownian motion part. Its Lévy measure, on the other hand, is given by

$$\nu^L(B) = \int 1_B(x^2)\nu^X(dx), \quad B \in \mathcal{B}.$$

The goal of this paper is to come up with a numerical approximation of $E(f([X, X]_T))$ $= E(g(L_T))$.

2.3 *Perturbation Approach*

Our goal is to obtain an explicit approximate formula for $E(g(L_T))$. The idea is to view the compensated quadratic variation L as a perturbed Brownian motion. To this end, we connect the process L through a curve in the space of Lévy processes with a Brownian motion. This is inspired by a parallel approach in [9], where L represents the logarithmic stock price X itself rather than its compensated quadratic variation. For $\lambda \in (0, 1]$ we define the process L^λ via

$$L_t^\lambda := \lambda L_{\frac{t}{\lambda^2}}, \quad t \in \mathbb{R}^+. \tag{3}$$

Observe that L^λ is again a Lévy process, satisfying $E\left(L_t^\lambda\right) = 0$ and $\mathrm{Var}\left(L_t^\lambda\right) = t\tilde{\sigma}^2$. Definition (3) does not make sense for $\lambda = 0$ but we obtain Brownian motion in the limit:

Lemma 2.2 *For $\lambda \to 0$ the family of Lévy processes $(L^\lambda)_{\lambda\in(0,1]}$ converges in law with respect to the Skorokhod topology (cf. [13] for details) to a Brownian motion with the same drift and volatility, i.e.*

$$L^\lambda \xrightarrow{\mathcal{D}} \tilde{\sigma} W \text{ as } \lambda \to 0,$$

where W denotes standard Brownian motion.

We denote the limiting process by $L_t^0 := \tilde{\sigma} W_t$, $t \in \mathbb{R}^+$.

2.4 nth Order Approximation

We define a function $q : [0, 1] \to \mathbb{R}^+$ by

$$q(\lambda) := E\left(g(L_T^\lambda)\right), \ \lambda \in [0, 1]. \tag{4}$$

Then

- $q(1) = E(g(L_T))$ is our original option price of interest,
- $q(0) = E(g(\tilde{\sigma} W_T))$ is the price of a European call option in a Bachelier model with volatility $\tilde{\sigma}$,
- $q(\lambda)$ corresponds to an interpolation between the upper two cases.

In order for (5) below to make sense we remark that:

Lemma 2.3 *Function q is infinitely often differentiable on* [0, 1].

Lemma 2.3 implies that the nth order Taylor polynomial $q_n(\lambda) := \sum_{k=0}^n \frac{q^{(k)}(0)}{k!}\lambda^k$ is formally defined for any $n \in \mathbb{N}$. The idea now is to approximate $q(1)$ by the Taylor polynomial of q in 0:

Definition 2.4 We call

$$q_n(1) := q(0) + \sum_{k=1}^{n} \frac{q^{(k)}(0)}{k!} \tag{5}$$

the *nth order approximation* to $q(1)$.

The goal in this note is to compute $q_n(1)$ explicitly. In this context two questions naturally come to mind. Firstly, one may wonder whether the Taylor series converges for $n \to \infty$ to the true value $q(1)$. We chose not to study this desirable property in detail because it seems to require even stronger moment conditions on $[X, X]_1$, which are satisfied in hardly any model of practical relevance. Alternatively, one may come up with remainder terms $o(\lambda^n)$ similarly as in [17]. We do not pursue this direction either because the artificial parameter λ of interest is not small in our case; it equals 1 by construction. Explicit estimates of the remainder term, however, are of use only if they can be computed more easily than the unknown quantity $q(1)$ itself. We leave this discussion for future research.

3 Computation of the Approximation

In order to compute the nth order approximation, we consider two ingredients.

3.1 Cumulants of a Lévy Process

Definition 3.1 The *Laplace exponent* $\psi_Y : U \to \mathbb{C}$ of a Lévy process Y is the unique continuous logarithm of its Laplace transform that vanishes in 0. By *Laplace transform* we refer to

$$E(e^{zY_t}) = \exp(\psi_Y(z)t)$$

for $t \geq 0$ and $z \in U := \{a + ib \in \mathbb{C} : E(e^{aY_1}) < \infty\}$. We denote the Laplace exponent of L^λ by ψ^λ.

Definition 3.2 For $n \in \mathbb{N}$, the *nth cumulant* of a Lévy process Y is defined as

$$\kappa_n^Y := \psi_Y^{(n)}(0),$$

where ψ_Y denotes the Laplace exponent of Y.

Note that $\kappa_n^L = \kappa_n^{[X,X]}$ for $n > 1$. Recall that $\kappa_n^Y = \int x^n \nu^Y(dx)$, $n > 2$ for the Lévy measure ν^Y of Y. This also holds for $n = 2$ if Y has no Brownian motion part. Since the jumps of $[X, X]$ and X are related via $\Delta[X, X]_t = \Delta X_t^2$, the cumulants of $[X, X]$ are obtained from the cumulants of X as $\kappa_n^{[X,X]} = \kappa_{2n}^X$ for $n \geq 1$.

3.2 Bachelier Greeks

The option with payoff $g(L_T^0)$ is in fact a call option with strike $\tilde{K}$ and maturity T in a corresponding Bachelier model, where the discounted stock price moves according to $\tilde{S} := \tilde{\sigma} W$ with standard Brownian motion W. Its initial fair price is given by $C(0)$, where function $C : \mathbb{R}^+ \to \mathbb{R}^+$ is defined as

$$C(s) := \int g(s + \sqrt{\tilde{\sigma}^2 T}x)\varphi(x)dx,$$

and $\varphi(x) := (2\pi)^{-1/2}\exp(-x^2/2)$ denotes the probability density function (pdf) of the standard normal distribution. A simple calculation shows that

$$C(s) = \sqrt{\tilde{\sigma}^2 T}\alpha\left(\frac{s - \tilde{K}}{\sqrt{\tilde{\sigma}^2 T}}\right) \tag{6}$$

with $\tilde{\sigma}^2 = \text{Var}(L_1) = \int x^2 \nu^L(dx) = \kappa_4^X$,

$$\alpha(x) := x\Phi(x) + \varphi(x) \tag{7}$$

and $\Phi(x) := \int_{-\infty}^x \varphi(y)dy$ denoting the cumulative distribution function (cdf) of the standard normal distribution, cf. e.g. [20].

Definition 3.3 We call the nth derivative

$$D_n(s) := C^{(n)}(s), \; s \in \mathbb{R}^+ \tag{8}$$

the *nth Bachelier greek* for $n \in \mathbb{N}$.

$D_n(s)$ represents the nth order sensitivity of the option price with respect to changes of the stock price at time 0. Such sensitivities are often referred to as greeks, which is why we call them Bachelier greeks here.

The summands in our approximation are expressed in terms of Bell polynomials, named after [2].

Definition 3.4 For $k, n \in \mathbb{N}$ with $k \leq n$ and $(x_i)_{i=1,\dots,n-k+1} \in \mathbb{R}^{n-k+1}$, the *incomplete Bell polynomials* $B_{n,k}$ are defined as

$$B_{n,k}((x_i)_{i=1,\dots,n-k+1}) := \sum \frac{n!}{j_1! \cdots j_{n-k+1}!} \prod_{i=1}^{n-k+1} \left(\frac{x_i}{i!}\right)^{j_i},$$

where the sum is taken over all $(j_1, \dots, j_{n-k+1}) \in \mathbb{N}^{n-k+1}$ such that $\sum_{i=1}^{n-k+1} j_i = k$ and $\sum_{i=1}^{n-k+1} i j_i = n$.

Remark 3.5 These polynomials come into play here because they appear in Faà di Bruno's formula on higher order derivatives:

$$\frac{d^n}{dx^n} f(g(x)) = \sum_{k=1}^{n} f^{(k)}(g(x)) B_{n,k}\left(g'(x), g''(x), \dots, g^{(n-k+1)}(x)\right), \tag{9}$$

cf. e.g. [5].

The Bachelier greeks can be expressed explicitly in terms of Bell polynomials.

Lemma 3.6 *For $n \geq 1$ we have*

$$D_n(s) = C^{(n)}(s) = \left(\sqrt{\tilde{\sigma}^2 T}\right)^{1-n} \varphi^{(n-2)}\left(\frac{s - \tilde{K}}{\sqrt{\tilde{\sigma}^2 T}}\right), \tag{10}$$

$$\varphi^{(n)}(x) = \varphi(x) \sum_{k=1}^{n} B_{n,k}(-x, -1, 0, \ldots, 0), \tag{11}$$

where we set $\varphi^{(0)} := \varphi$ and $\varphi^{(-1)} := \Phi$.

3.3 Main Result

We can now express $q_n(1)$ in terms of the above quantities.

Theorem 3.7 *The nth order approximation $q_n(1)$ of the price function $q(1)$ has the representation*

$$q_n(0) = C(0) + \sum_{k=1}^{n} \sum_{j=1}^{k} \frac{T^j}{k!} D_{k+2j}(0) B_{k,j}\left(\left(\frac{\kappa_{2i+4}^X}{(i+1)(i+2)}\right)_{i=1,\ldots,k-j+1}\right).$$

Similarly as in [9] our approximation contains only cumulants of $[X, X]$ resp. X and Bachelier greeks.

4 Numerical Illustration

4.1 Merton Model

We want to assess the accuracy of our approximation numerically in a case study. We consider the Merton model [16], where the Lévy process X is given by a Brownian motion with drift plus an independent compound Poisson process exhibiting Gaussian jumps. Besides being popular in the literature, this model allows for semi-explicit computation of the option price under consideration. Specifically, we have

$$X_t = \gamma t + \sigma W_t + \sum_{k=1}^{N_t} J_k, \; t \in \mathbb{R}^+$$

where $\sigma > 0$, W is a standard Brownian motion, $J_1, J_2, \ldots$ are i.i.d. $N(\nu, \tau^2)$-distributed random variables, and N is a Poisson process with intensity $\alpha > 0$ such

Table 1 Merton model parameters

γ	σ	α	ν	τ
−0.0548	0.280	39.0	−0.00165	0.0457

that $W, N, J_1, J_2, \ldots$ are all independent. The parameters are chosen such that

$$\begin{aligned}
\mathrm{Var}\,(X_1) &= 0.4^2, \\
\mathrm{Skew}\,(X_1) := \frac{E\big((X_1 - E(X_1))^3\big)}{\mathrm{Var}\,(X_1)^{3/2}} &= \frac{0.1}{\sqrt{250}}, \\
\mathrm{ExKurt}\,(X_1) := \frac{E\big((X_1 - E(X_1))^4\big)}{\mathrm{Var}\,(X_1)^2} - 3 &= \frac{5}{250},
\end{aligned}$$

which is in the range of empirical plausible values (cf. e.g. [6, Table 4]), at least if one agrees that risk neutral parameters should not deviate too strongly from statistical ones. The first moment is determined by the martingale requirement $E(\exp(X_1)) = 1$ and amounts to $E(X_1) = -0.08$ in our case. As we have five model parameters, we eliminate the additional degree of freedom by setting the variance arising from the jump component as 49 % of the overall variance of X, following the choice in [9, 10]. The Merton parameters corresponding to this choice are listed in Table 1. The Laplace exponent of X is given by

$$\psi(z) = \gamma z + \frac{\sigma^2}{2} z^2 + \alpha\left(\exp\left(\nu z + \frac{\tau^2}{2} z^2\right) - 1\right).$$

The quadratic variation of X amounts to

$$[X, X]_t = \sigma^2 t + \sum_{k=1}^{N_t} J_k^2, \; t \in \mathbb{R}^+.$$

Since J_1^2/τ^2 is a non-central χ_1^2-distributed random variable, the Laplace transform of J_1^2 equals

$$\varphi_{J_1^2}(z) = \frac{\exp(\frac{z\nu^2}{1-2z\tau^2})}{\sqrt{1-2z\tau^2}}.$$

The Laplace transform of $[X, X]$ is given by

$$\varphi_{[X,X]}(z) = \exp(\sigma^2 z + \alpha(\varphi_{J_1^2}(z) - 1)) \tag{12}$$

because $[X, X]$ is a compound Poisson process whose jumps are distributed as J_1^2. The desired option price equals

$$E\big(([X,X]_T-K)^+\big)=\frac{1}{\pi}\int_0^\infty \mathrm{Re}\left(\varphi_{[X,X]_T}(-R-iy)\frac{e^{-K(R+iy)}}{(R+iy)^2}\right)dy.$$

This follows as in (Lemma 6.3 [14]) with $R>0$ in order to obtain the call rather than the put. The Laplace transform (12) allows for an analytic extension to $U:=(-\infty,1/(2\tau^2))+i\mathbb{R}$, which we denote by $\varphi_{[X,X]}$ as well. Therefore, the moment condition Assumption 2.1(1) holds for $R<1/(4\tau^2)$. However, Assumption 2.1(2) is not satisfied for this model because X has finite jump activity. Nevertheless, the right-hand side of (5) still makes sense for this model. Therefore we examine it in our small numerical study in the subsequent section.

4.2 Discussion of the Numerical Results

We consider maturities $T=1/48, 1/12, 1/4, 1/2$, and 1, measured in years. Our choice of strikes $K=0.14T, 0.16T, 0.18T$ is motivated by the fact that $E([X,X]_T)=0.16T$ for our parameters. Table 2 shows exact and approximate prices of a call option on the quadratic variation in the above Merton model. We observe that the zeroth-order approximation corresponding to a call option in a suitable parametrized Bachelier model has an average relative error of 8.3 % over all strikes and maturities,

Table 2 Exact and approximate prices of a call option on the quadratic variation in a Merton model with normal jumps for $E(X_1)=-0.08$, Var $(X_1)=0.4^2$, $\mathrm{Skew}(X_1)=0.1/\sqrt{250}$ excess kurtosis $\mathrm{Exkurt}(X_1)=5/250$, for varying strike K and maturity T

T	K	$q(1)$	$q_0(1)$	$q_1(1)$	$q_2(1)$	$q_5(1)$
$\frac{1}{48}$	0.00292	0.00120	0.00152	0.00143	0.00121	0.00116
	0.00333	0.00107	0.00130	0.00130	0.00107	0.00105
	0.00375	0.00097	0.00110	0.00119	0.00097	0.00095
$\frac{1}{12}$	0.01167	0.00322	0.00352	0.00335	0.00323	0.00322
	0.01333	0.00249	0.00261	0.00261	0.00249	0.00249
	0.015	0.00191	0.00186	0.00203	0.00191	0.00191
$\frac{1}{4}$	0.035	0.00708	0.00745	0.00717	0.00709	0.00709
	0.04	0.00444	0.00451	0.00451	0.00445	0.00445
	0.045	0.00265	0.00245	0.00272	0.00265	0.00265
$\frac{1}{2}$	0.07	0.01216	0.01259	0.01223	0.01218	0.01217
	0.08	0.00633	0.00638	0.00638	0.00634	0.00634
	0.09	0.00289	0.00259	0.00295	0.00289	0.00289
1	0.14	0.02185	0.02234	0.02192	0.02189	0.02189
	0.16	0.00897	0.00902	0.00902	0.00899	0.00899
	0.18	0.00271	0.00234	0.00275	0.00272	0.00272

Table 3 Exact and approximate prices of a call option on the quadratic variation in a Merton model with normal jumps for $E(X_1) = -0.08$, Var $(X_1) = 0.4^2$, Skew$(X_1) = 0$, excess kurtosis Exkurt$(X_1) = 15/250$, for varying strike K and maturity T

T	K	$q(1)$	$q_0(1)$	$q_1(1)$	$q_2(1)$	$q_5(1)$
$\frac{1}{48}$	0.00292	0.00146	0.00247	0.00232	0.00115	0.00054
	0.00333	0.00139	0.00226	0.00226	0.00110	0.00060
	0.00375	0.00133	0.00205	0.00221	0.00104	0.00064
$\frac{1}{12}$	0.01167	0.00445	0.00540	0.00510	0.00450	0.00440
	0.01333	0.00391	0.00451	0.00451	0.00394	0.00387
	0.015	0.00344	0.00373	0.00403	0.00344	0.00341
$\frac{1}{4}$	0.035	0.00967	0.01057	0.01006	0.00970	0.00967
	0.04	0.00747	0.00782	0.00782	0.00748	0.00747
	0.045	0.00574	0.00557	0.00608	0.00573	0.00574
$\frac{1}{2}$	0.07	0.01578	0.01677	0.01606	0.01580	0.01578
	0.08	0.01082	0.01106	0.01106	0.01082	0.01082
	0.09	0.0722	0.00677	0.00747	0.00721	0.00722
1	0.14	0.02649	0.02763	0.02670	0.02650	0.02649
	0.16	0.01547	0.01564	0.01564	0.01547	0.01547
	0.18	0.00837	0.00763	0.00856	0.00837	0.00837

whereas the second order approximation with an average relative error of 0.2 % turns out to be quite accurate. The error stays approximately the same for the fifth-order and higher approximations. So the approximation $q_2(1)$ seems to be good enough for practical purposes.

One should note, however, that the excellent precision of our approximation is due to the fact that the compensated quadratic variation process $L = [X, X] - E([X, X])$ is reasonably close to a Brownian motion in the sense that its higher-order cumulants are rather small. Table 3 contains the values for higher excess curtosis. We observe that the accuracy decreases in particular for short time to maturity, to the point that it does not provide reasonable values for a time horizon of one week. A thorough comparative study is beyond the scope of this paper. In another direction, one could consider a similar approximation for the effect of time discretization, using a suitably parametrized χ^2-distribution, cf. [15] for leading-order asymptotics. These topics are left for future research.

5 Proofs

The key idea in order to compute $q_n(1)$ is to represent $q(\lambda)$ as

$$q(\lambda) = \int_{R-i\infty}^{R+i\infty} h(\lambda, z)dz$$

with a suitable function $h : [0, 1] \times (R + i\mathbb{R}) \to \mathbb{C}$. Then we interchange differentiation with respect to λ and integration with respect to z and calculate

$$\int_{R-i\infty}^{R+i\infty} \frac{\partial^n}{\partial \lambda^n} h(\lambda, z) dz$$

explicitly.

Lemma 5.1 *We have the integral representation*

$$g(x) = (x - \tilde{K})^+ = \int_{R-i\infty}^{R+i\infty} e^{zx} p(z) dz$$

with

$$p(z) := \frac{e^{-\tilde{K}z}}{2\pi i z^2}.$$

Proof A straightforward calculation shows that

$$\mathcal{L}[g; z] := \int_{-\infty}^{\infty} e^{-zx} g(x) dx = 2\pi i p(z)$$

for $z \in R + i\mathbb{R}$. The Bromwich inversion theorem for the bilateral Laplace transform (cf. [18, Theorem 9.11]) yields

$$g(x) = \frac{1}{2\pi i} \int_{R-i\infty}^{R+i\infty} \mathcal{L}[g; z] e^{zx} dz$$

and hence the claim. □

Lemma 5.2 *For the Lévy measure ν^L of L we have*

$$\int_{[1,\infty)} e^{2Rx} \nu^L(dx) < \infty$$

and

$$\int_{(0,1)} x \nu^L(dx) < \infty.$$

Proof The first statement is equivalent to Assumption 2.1(1), cf. [19, Theorem 25.3 and Proposition 25.4]. The second holds because L is of finite variation, cf. [19, Theorem 21.9]. □

Lemma 5.3 *There are constants $c_1, c_2 > 0$ such that $Re(\psi^\lambda)(z) \le c_1 - c_2|z|^\gamma$ for any $\lambda \in [0, 1]$, $z \in R + i\mathbb{R}$.*

Proof This follows from [10, Lemma 5.7.5] and its proof. □

Lemma 5.4 *For any $c > 0$, $n \in \mathbb{N}$, the function $z \mapsto |z^n e^{cz^2}|$ is bounded on $R + i\mathbb{R}$.*

Proof This follows directly from $\operatorname{Re}(z^2) = R^2 - \operatorname{Im}(z)^2$. □

Lemma 5.5 *For any $c > 0$, $n \in \mathbb{N}$, we have that $z \mapsto |z^n e^{\psi^\lambda(z)T}|$ is uniformly bounded for $\lambda \in [0, 1]$ and $z \in R + i\mathbb{R}$.*

Proof This follows from Lemma 5.3. □

Lemma 5.6 *The Bachelier greeks $D_n(s)$ can be written as*

$$D_n(s) = \int_{R-i\infty}^{R+i\infty} z^n e^{zs+\frac{1}{2}\tilde{\sigma}^2 T z^2} p(z)dz.$$

Proof Note that

$$\begin{aligned}
C(s) &= \int g(s + \sqrt{\tilde{\sigma}^2 T} x)\varphi(x)dx \\
&= \int \int_{R-i\infty}^{R+i\infty} e^{z(s+\sqrt{\tilde{\sigma}^2 T} x)} p(z)dz\varphi(x)dx \\
&= \int_{R-i\infty}^{R+i\infty} \int e^{z(s+\sqrt{\tilde{\sigma}^2 T} x)} \varphi(x)dxp(z)dz \\
&= \int_{R-i\infty}^{R+i\infty} e^{zs+\frac{1}{2}\tilde{\sigma}^2 T z^2} p(z)dz.
\end{aligned}$$

Taking the nth derivative yields

$$\begin{aligned}
D_n(s) &= \frac{\partial^n}{\partial s^n} \int_{R-i\infty}^{R+i\infty} e^{zs+\frac{1}{2}\tilde{\sigma}^2 T z^2} p(z)dz \\
&= \int_{R-i\infty}^{R+i\infty} \frac{\partial^n}{\partial s^n} e^{zs+\frac{1}{2}\tilde{\sigma}^2 T z^2} p(z)dz \\
&= \int_{R-i\infty}^{R+i\infty} z^n e^{zs+\frac{1}{2}\tilde{\sigma}^2 T z^2} p(z)dz.
\end{aligned}$$

The application of Fubini's theorem and interchanging differentiation and integration is possible due to Lemma 5.4. □

Lemma 5.7 *For any $n \geq 1$ we have*

$$\int x^n e^{Rx} \nu^L(dx) < \infty.$$

Proof This follows from Lemma 5.2 because $\int_{[1,\infty)} e^{2Rx}\nu^L(dx) < \infty$ implies that also $\int_{[1,\infty)} x^n e^{Rx}\nu^L(dx) < \infty$ and $\int_{(0,1)} x\nu^L(dx) < \infty$ yields $\int_{(0,1)} x^n e^{Rx}\nu^L(dx) < \infty$. □

Lemma 5.8 *$\lambda \to \psi^\lambda(z)$ is infinitely often differentiable with*

$$\frac{\partial^n}{\partial\lambda^n}\psi^\lambda(z) = \frac{1}{2}z^{n+2}\int\int_0^1 x^{n+2}e^{s\lambda xz}s^{n-1}(1-s)^2(n+\lambda zxs)ds\nu^L(dx).$$

In particular,

$$\left.\frac{\partial^n}{\partial\lambda^n}\psi^\lambda(z)\right|_{\lambda=0} = \frac{\kappa_{n+2}^L}{(n+1)(n+2)}z^{n+2}.$$

Proof From the definition of L^λ in (3) it follows immediately that its Laplace exponent ψ^λ is given in terms of ψ^1 by

$$\psi^\lambda(z) = \frac{1}{\lambda^2}\psi^1(\lambda z).$$

Since L is a martingale without Brownian motion part, the Laplace exponent of L equals

$$\psi^1(z) = \int (e^{zx} - 1 - zx)\nu^L(dx),$$

which implies

$$\psi^\lambda(z) = \int \frac{1}{\lambda^2}\left(e^{\lambda zx} - 1 - \lambda zx\right)\nu^L(dx).$$

From the Taylor expansion with integral remainder term

$$e^{\lambda zx} = 1 + \lambda zx + \frac{1}{2}(\lambda zx)^2 + \frac{1}{2}(\lambda zx)^3\int_0^1 e^{s\lambda zx}(1-s)^2 ds$$

it follows that

$$\psi^\lambda(z) = \frac{1}{2}z^2\int x^2\nu^L(dx) + \frac{1}{2}z^3\int\int_0^1 \lambda x^3 e^{s\lambda zx}(1-s)^2 ds\nu^L(dx) \qquad (13)$$

for $\lambda \in (0, 1]$. Since $\tilde{\sigma}^2 = \mathrm{Var}\,(L_1) = \int x^2\nu^L(dx)$, (13) holds for $\lambda = 0$ as well. The integrand in (13) is obviously infinitely often partially differentiable with respect to λ. The previous lemma and iterated differentiation under the integral sign yield after straightforward calculation

$$\frac{\partial^n}{\partial\lambda^n}\psi^\lambda(z) = \frac{1}{2}z^{n+2}\int\int_0^1 x^{n+2}e^{s\lambda xz}s^{n-1}(1-s)^2(n+\lambda zxs)ds\nu^L(dx)$$

for $\lambda \in [0, 1]$, $z \in R + i\mathbb{R}$, $n \in \mathbb{N}$ and in particular

$$\frac{\partial^n}{\partial\lambda^n}\psi^\lambda(z)\Big|_{\lambda=0} = \frac{z^{n+2}}{(n+1)(n+2)}\int x^{n+2}\nu^L(dx) = \frac{z^{n+2}}{(n+1)(n+2)}\kappa^L_{n+2}.$$

□

Lemma 5.9

$$\left|\frac{\partial^n}{\partial\lambda^n}\psi^\lambda(z)\right| \leq c(1+|z|^{n+3}), \quad z \in R + i\mathbb{R}$$

with some constant $c < \infty$ that may depend on n but not on $\lambda \in [0, 1]$ and z.

Proof This follows from the explicit representation in the previous lemma. □

Lemma 5.10 $\lambda \to e^{\psi^\lambda(z)T}$ *is infinitely often differentiable with*

$$\left|\frac{\partial^n}{\partial\lambda^n}e^{\psi^\lambda(z)T}\right| \leq ce^{\mathrm{Re}(\psi^\lambda(z)T)}(1+|z|^{n+3})^n, \quad z \in R + i\mathbb{R}$$

for some constant $c < \infty$ that may depend on n but not on λ and z. Moreover, we have

$$\frac{\partial^n}{\partial\lambda^n}e^{\psi^\lambda(z)T}\Big|_{\lambda=0} = e^{\psi^0(z)T}\sum_{k=1}^n z^{n+2k}T^k B_{n,k}\left(\left(\frac{\kappa^X_{2i+4}}{(i+1)(i+2)}\right)_{i=1,\dots,n-k+1}\right).$$

Proof The statement follows from Lemmas 5.8, 5.9 and Faà di Bruno's formula (9). □

Lemma 5.11

$$\int_{R-i\infty}^{R+i\infty}\sup_{\lambda\in[0,1]}\left|\frac{\partial^n}{\partial\lambda^n}e^{\psi^\lambda(z)T}p(z)\right|dz < \infty$$

Proof Lemma 5.10 and $E\big(\exp(zL^\lambda_T)\big) < \infty$ yield that $z \mapsto \sup_{\lambda\in[0,1]}\left|\frac{\partial^n}{\partial\lambda^n}\exp(\psi^\lambda(z)T)\right|$ is bounded, which yields the claim because $\int_{R-i\infty}^{R+i\infty}|p(z)|dz < \infty$. □

Lemma 5.12 *q is infinitely often differentiable on* $[0, 1]$ *with*

$$q^{(n)}(0) = \sum_{k=1}^n T^k D_{n+2k}(0)B_{n,k}\left(\left(\frac{\kappa^X_{2i+4}}{(i+1)(i+2)}\right)_{i=1,\dots,n-k+1}\right).$$

Proof Lemma 5.1 together with Fubini's theorem and Lemma 5.5 yield

$$\begin{aligned}q(\lambda) &= E\left(\int_{R-i\infty}^{R+i\infty} e^{zL_T^\lambda} p(z)dz\right)\\ &= \int_{R-i\infty}^{R+i\infty} E\left(e^{zL_T^\lambda}\right) p(z)dz\\ &= \int_{R-i\infty}^{R+i\infty} e^{\psi^\lambda(z)T} p(z)dz.\end{aligned}$$

Lemma 5.11 ensures that we can interchange differentiation and integration. Hence

$$q^{(n)}(\lambda) = \int_{R-i\infty}^{R+i\infty} \frac{\partial^n}{\partial\lambda^n} e^{\psi^\lambda(z)T} p(z)dz$$

which, using Lemma 5.10, yields

$$\begin{aligned}q^{(n)}(0) &= \int_{R-i\infty}^{R+i\infty} \frac{\partial^n}{\partial\lambda^n} e^{\psi^\lambda(z)T}\Big|_{\lambda=0} p(z)dz\\ &= \int_{R-i\infty}^{R+i\infty} e^{\psi^0(z)T}\left(\sum_{k=1}^{n} z^{n+2k}T^k B_{n,k}\left(\left(\frac{\kappa_{2i+4}^X}{(i+1)(i+2)}\right)_{i=1,\dots,n-k+1}\right)\right)p(z)dz.\end{aligned}$$

Since

$$\int_{R-i\infty}^{R+i\infty} z^{n+2k} e^{\psi^0(z)T} p(z)dz = D_{n+2k}(0)$$

by Lemma 5.6, we are done. □

Proof (Proof of Lemma 2.2) From (13) it follows that $\lim_{\lambda\to 0} e^{\psi^\lambda(iu)} = e^{-\frac{1}{2}\sigma^2u^2}$ for any $u \in \mathbb{R}$. Using Lévy's continuity theorem (e.g. [19, Proposition 2.5(vii)]) we conclude that the univariate marginals of L^λ converge to the univariate marginals of $\tilde{\sigma}W$ as $\lambda \to 0$. By [13, Corollary VII.3.6] this implies convergence of the whole process. □

Proof (Proof of Lemma 2.3) This is stated in Lemma 5.12. □

Proof (Proof of Lemma 3.6) (10) follows from (6) and $\alpha'(x) = \Phi(x)$. Faá di Bruno's formula (9) yields (11). □

Proof (Proof of Theorem 3.7) This follows from Lemma 5.12. □

Acknowledgements The authors thank two anonymous referees for their careful reading and helpful comments.

References

1. Barndorff-Nielsen, O., Shephard, N.: Non-Gaussian Ornstein-Uhlenbeck-based models and some of their uses in financial economics. J. Roy. Stat. Soc. B **63**, 167–241 (2001)
2. Bell, E.: Partition polynomials. Ann. Math. **29**, 38–46 (1927)
3. Benth, F., Groth, M., Kufakunescu, R.: Valuing volatility and variance swaps for a non-Gaussian Ornstein-Uhlenbeck stochastic volatility model. Appl. Math. Financ. **14**, 347–363 (2007)
4. Broadie, D., Jain, A.: The effects of jumps and discrete sampling on volatility and variance swaps. Int. J. Theor. Appl. Financ. **11**, 761–797 (2008)
5. Craik, A.: Prehistory of Faà di Bruno's formula. Am. Math. Mon. **112**, 119–130 (2005)
6. Carr, P., Geman, H., Madan, D.B., Yor, M.: The fine structure of asset returns: an empirical investigation. J. Bus. **75**, 305–332 (2002)
7. Carr, P., Geman, H., Madan, D.B., Yor, M.: Pricing options on realized variance. Financ. Stochast. **9**, 453–475 (2005)
8. Carr, P., Lee, R.: Realized volatility and variance: Options via swaps. RISK **20**, 76–83 (2007)
9. Černý, A., Denkl, S., Kallsen, J.: Hedging in Lévy models and the time step equivalent of jumps. Preprint (2013)
10. Denkl, S.: Second-order approximations to pricing and hedging in presence of jumps and stochastic volatility. Ph.D. dissertation, Kiel (2013)
11. Friz, P., Gatheral, J.: Valuation of volatility derivatives as an inverse problem. Quant. Financ. **5**, 531–542 (2005)
12. Heston, S.: A closed-form solution for options with stochastic volatility with applications to bond and currency options. Rev. Finan. Stud. **6**, 327–343 (1993)
13. Jacod, J., Shiryaev, A.: Limit Theorems for Stochastic Processes. second edn. Springer, Berlin (2013)
14. Kallsen, J., Muhle-Karbe, J., Voss, M.: Pricing options on variance in affine stochastic volatility models. Math. Financ. **21**, 73–98 (2011)
15. Keller-Ressel, M., Muhle-Karbe, J.: Asymptotic and exact pricing of options on variance. Financ. Stochast. **17**, 107–133 (2013)
16. Merton, R.: Option pricing when underlying stock returns are discontinuous. J. Financ. Econ. **3**, 125–144 (1976)
17. Ménassé, C., Tankov, P.: Asymptotic indifference pricing in exponential Lévy models. Preprint (2015)
18. Rudin, W.: Real and Complex Analysis, third edn. McGraw-Hill (1987)
19. Sato, K.: Lévy Processes and Infinitely Divisible Distribution. Cambridge University Press, Cambridge (1999)
20. Schachermayer, W., Teichmann, J.: How close are the option pricing formulas of Bachelier and Black-Merton-Scholes? Math. Financ. **18**, 155–170 (2008)
21. Sepp, A.: Pricing options on realized variance in the Heston model with jumps in returns and volatility. J. Comput. Financ. **11**, 33–70 (2008)
22. Sepp, A.: Pricing options on realized variance in the Heston model with jumps in returns and volatility II: an approximate distribution of the discrete variance. J. Comput. Financ. **16**, 3–32 (2012)
23. Zheng, W., Kwok, Y.: Saddlepoint approximation methods for pricing derivatives on discrete realized variance. Appl. Math. Financ. **21**, 1–31 (2014)

Discrete-Time Quadratic Hedging of Barrier Options in Exponential Lévy Model

Aleš Černý

Abstract We examine optimal quadratic hedging of barrier options in a discretely sampled exponential Lévy model that has been realistically calibrated to reflect the leptokurtic nature of equity returns. Our main finding is that the impact of hedging errors on prices is several times higher than the impact of other pricing biases studied in the literature.

Keywords Barrier option · Quadratic hedging · Lévy model

1 Introduction

We study quadratic hedging and pricing of European barrier options with a particular focus on the magnitude of risk of optimal hedging strategies. In a discretely sampled exponential Lévy model, calibrated to reflect the leptokurtic nature of equity returns, we compute the hedging error of the optimal strategy and evaluate prices that yield reasonable risk-adjusted performance for the hedger. We also confirm what traders already know empirically, namely that the hedging risk of barrier options substantially outstrips that of plain vanilla options.

European barrier options are derivative contracts based on standard European calls or puts with the exception that the option becomes active (or inactive) when the stock price hits a prespecified barrier before the maturity of the option. Options activated in this way are called knock-ins; those deactivated are called knock-outs.

Under the assumptions of the Black–Scholes model barrier options have been valued first by [32] and in more detail by [33]. Early literature on numerical evaluation of barrier option prices concentrates on slow convergence of binomial method, which is due to the difference between the *nominal barrier* specified in the option contract and the *effective barrier* implied by the position of nodes in the stock price lattice. This discrepancy, if not properly controlled, may lead to sizeable mispricing, especially

A. Černý (✉)
Cass Business School, City University London, 106 Bunhill Row, London EC1Y 8TZ, UK
e-mail: Ales.Cerny.1@city.ac.uk

© Springer International Publishing Switzerland 2016

J. Kallsen and A. Papapantoleon (eds.), *Advanced Modelling in Mathematical Finance*,
Springer Proceedings in Mathematics & Statistics 189,
DOI 10.1007/978-3-319-45875-5_12

for options whose barrier is close to the initial stock price. [4, 34] suggest better positioning of nodes in binomial and trinomial lattices to minimize the discrepancy between nominal and effective barrier, whereas [14] propose interpolation between two adjacent values of the effective barrier. [21] devise an adaptive mesh allowing for more nodes (and shorter time steps) around the barrier.

The papers above are concerned with continuously-monitored barriers in the Black-Scholes model. Discrete monitoring, too, can have significant impact on option valuation, and, unlike the continuous monitoring case, does not allow for a simple closed form pricing formula, cf. [22]. A simple asymptotic correction, which works well for barriers not too close to initial stock price, was developed by [5]. For barriers in close proximity of the stock price the Markov chain representation of stock prices developed by [15] is more appropriate. Other papers dealing with discrete monitoring in the log-normal framework include [23, 25, 28, 29, 38]. [1] describe a systematic way of handling discretization errors by means of quadrature. There is also an extensive literature on barrier option pricing by Monte Carlo simulation which we will not touch upon in this paper.

The models discussed above are complete in the sense that one can devise a self-financing trading strategy that perfectly replicates the barrier option. In practice, however, one encounters considerable difficulties in maintaining a delta-neutral position when close to the barrier. This has motivated study of static replication of barrier options with plain vanilla options. [8] use the reflection principle known from barrier option pricing combined with so-called put–call symmetry to write a down-and-out call as a sum of a long call and a short put. Their methodology is to some extent model-free but it only works if the market is complete and if the aforementioned symmetry holds, requiring that risk-free rate be equal to the dividend rate in addition to a certain symmetry of local volatilities. [6] analyze static super- and sub-replication. The latter results are completely model-free at the cost of generating price bounds that are potentially very wide. Other papers on static replication include [2, 7, 12, 13].

Several studies allow for parametric departures from the Black–Scholes model. [16, 26] use Bates' stochastic volatility jump-diffusion model while [30] allows for IID jumps. Several numerical approaches now exist for dealing with a wide class of (possibly infinite variation) Lévy models, see [19, 20].

The paper is organized as follows. In Sect. 2 we specify the theoretical model, describe its calibration and computation of optimal strategies. Section 3 provides economic analysis of the numerical results and Sect. 4 explains the relationship between barrier option prices and hedger's risk-adjusted performance.

2 The Model

We have at our disposal nominal log returns on FT100 equity index in the period January 1st, 1993 to December 31st, 2002, sampled at a 1 min interval. Eventually we wish to say something about optimal hedging of barrier options in a model with rebalancing frequency $\Delta \in [5 \text{ min}, 1 \text{ day}]$ and a daily monitoring of the barrier. We will assume independence and time homogoneity of underlying asset returns at any given rebalancing frequency. This is not to say that stochastic volatility is unimportant in practice, instead we may think of the IID assumption as a useful limiting case when the (unobserved) volatility state changes either very slowly or very quickly. In this view of the world the leptokurtic nature of returns is a source of risk that does not vanish even after stochastic volatility has been factored in appropriately.

The analysis is performed under two self-imposed constraints. The first is to use the available data in a non-parametric way and the second is to perform all numerical analysis in a multinomial lattice.

In these circumstances there are essentially two strategies for calibrating the stock price process. One option is to simply take the data series sampled at time interval Δ, generate a discretized distribution of returns and construct a multinomial lattice using this distribution. An alternative is to consider an underlying continuous-time model from which the daily or hourly returns are extracted. [17] argue that equity return data display sufficient amount of time consistency for such an approach to make sense. The underlying model is then necessarily a geometric Lévy model, cf. Lemma 4.1 in [9]. Such approach also offers an alternative avenue to obtaining asymptotics as Δ tends to zero by studying quadratic hedging for barrier options directly in the underlying Lévy model—a task which at present is still outstanding and well beyond the scope of this paper.

2.1 Calibration

We take the original log return data sampled at $\Delta_0 = 1$ min intervals and construct an equidistantly spaced sequence $m_0 < m_1 < \ldots < m_{N+1}$ with spacing δ, such that m_N is the highest and m_1 the lowest log return in the sample. We set $N = 1000$. We then identify the frequency of log returns in each interval of length δ centred on m_j, $j = 0, \ldots, N + 1$ and store this information in the vector $\{f_j\}_{j=0}^{N+1}$. We construct an empirical Lévy measure F_{raw} as an absolutely continuous measure with respect to the Lebesgue measure on $\mathbb{R}$

$$F_{\text{raw}}(dx) = \frac{\hat{f}(x)}{\Delta_0} dx,$$

where $\hat{f} = f$ at the points m_j, $\hat{f} = 0$ outside (m_0, m_{N+1}) and elsewhere $\hat{f}$ is obtained as a linear interpolation of f. This construction is motivated by an asymptotic result that links transition probability measure of a Lévy process to its Lévy measure over short time horizons, see [35], Corollary 8.9.

In the next step we normalize the empirical characteristic function of log returns to achieve a pre-specified annualized mean μ and volatility σ. Since the raw empirical Lévy process is square-integrable and therefore a special semimartingale we will use the (otherwise forbidden) truncation function $h(x) = x$. We will construct the log return process by setting

$$\ln S = \ln S_0 + \mu t + \frac{\sigma x}{\sigma_{\text{raw}}} * (J^{L_{\text{raw}}} - \nu^{L_{\text{raw}}}),$$

$$\sigma_{\text{raw}}^2 = \int_{\mathbb{R}} x^2 F_{\text{raw}}(dx),$$

where $J^{L_{\text{raw}}}$ is the jump measure of a Lévy process with Lévy measure F_{raw}, $\nu^{L_{\text{raw}}}$ is its predictable compensator and $*$ denotes a certain stochastic integral as defined in [27]. II.1.27. This yields

$$\kappa(u) := \mu u + \int_{\mathbb{R}} (\mathrm{e}^{ux} - 1 - ux) F(\mathrm{d}x), \tag{1}$$

$$F(G) := \int_{\mathbb{R}} 1_G \left(\frac{\sigma x}{\sigma_{\text{raw}}} \right) F_{\text{raw}}(dx). \tag{2}$$

We fix the annualized volatility of log returns at $\sigma = 0.2$, but to check the robustness of our results we allow the mean log return to take 2 different values $\mu \in \{-0.1, 0.1\}$, the first representing a bear market and the second representing a bull market.

Instead of the non-parametric calibration procedure above one could instead estimate a model from a convenient parametric family, such as the generalized hyperbolic family, as outlined in [18]. The parametric route offers in some special cases an explicit expression for the log return density at all time horizons which avoids the need for numerical inversion of the characteristic function employed below.

2.2 *Multinomial Lattice*

If Z denotes the log return on time horizon Δ its characteristic function is of the form

$$\mathrm{E}\left[\exp(\mathrm{i}\nu Z)\right] = \mathrm{e}^{\kappa(\mathrm{i}\nu)\Delta},$$

where the cumulant generating function κ is given by Eqs. (1) and (2). Provided that Z has no atom at z the cumulative distribution is given by the inverse Fourier formula, see [35], 2.5xi,

$$P(Z \le z) = \mathscr{H}(c) - \frac{1}{2\pi} \lim_{l \to \infty} \int_{-l}^{l} \frac{\mathrm{e}^{\kappa(\mathrm{i}\lambda - c)\Delta - z(\mathrm{i}\lambda - c)}}{\mathrm{i}\lambda - c} d\lambda,$$

where c is an arbitrarily chosen real number[1] and $\mathscr{H}$ is a step function,

$$\mathscr{H}(x) = \begin{cases} 0 \text{ for } x > 0 \\ \frac{1}{2} \text{ for } x = 0 \\ 1 \text{ for } x < 0 \end{cases}.$$

We now define a discretized distribution of log returns to populate our lattice. The discretized random variable $\hat{Z}$ will take values

$$\hat{z}_j = j\eta \text{ with } j \in [-n_{\text{down}}, n_{\text{up}}] \cap \mathbb{Z},$$

where $n_{\text{down}}, n_{\text{up}}$ are the smallest numbers in $\mathbb{N}$ such that $P(Z \leq -n_{\text{down}}\eta) \leq \alpha$ and $F_Z(Z \leq n_{\text{up}}\eta) \geq 1 - \alpha$, respectively. We use the values $\eta = 0.0005$ and $\alpha = 10^{-5}$. For comparison, the corresponding value of η in [15] with 1001 price nodes is 0.0089. Table 5 shows the number of standard deviations.

The transition probabilities corresponding to different values of log return are defined by

$$\begin{aligned} \hat{p}_j &:= P(Z \leq (j + 1/2)\,\eta) - P(Z \leq (j - 1/2)\,\eta) \text{ for } j \in \left(-n_{\text{down}}, n_{\text{up}}\right) \cap \mathbb{Z}, \\ \hat{p}_j &:= P(Z \leq (j + 1/2)\,\eta) \text{ for } j = -n_{\text{down}}, \\ \hat{p}_j &:= 1 - P(Z \leq (j - 1/2)\,\eta) \text{ for } j = n_{\text{up}}. \end{aligned}$$

To limit the effect of the discretization errors arising from an arbitrary position of the barrier we limit computations to barrier levels that satisfy $\ln B - \ln S \in (\mathbb{Z} + 1/2)\eta$ and use interpolation otherwise.

2.3 Optimal Hedging

In the multinomial lattice constructed above we compute the optimal hedging strategy and the minimal hedging error according to the following theorem.

Theorem 1 *Suppose that there is an $\mathscr{F}_n$-measurable contingent claim H such that $\Xi[H^2] < \infty$. In the absence of transaction costs the dynamically optimal hedging strategy φ solving*

$$\inf_{\vartheta} \mathrm{E}[\left(G_n^{x,\vartheta} - H\right)^2],$$

[1] In numerical calculations with a fixed value of z we choose c so as to minimize the value of the integrand at $\lambda = 0$, see [36], Eq. (3).

subject to ϑ_i being $\mathscr{F}_i$-measurable with G being the value of a self-financing portfolio,

$$\begin{aligned} G_i^{x,\vartheta} &= RG_{i-1}^{x,\vartheta} + \vartheta_{i-1}(S_i - RS_{i-1}), \\ G_0^{x,\vartheta} &= x, \end{aligned}$$

is given by

$$\begin{aligned} \varphi_i &= \xi_i + aR\frac{V_i - G_i^{x,\varphi}}{S_i}, \\ V_i &:= \mathrm{E}_i[(1 - aX_{i+1})V_{i+1}]/(bR), \qquad (3) \\ V_n &:= H, \\ \xi_i &:= \mathrm{Cov}_i\left(V_{i+1}, S_{i+1}\right)/\mathrm{Var}_i\left(S_{i+1}\right) \\ &= \mathrm{E}_i\left[\left(V_{i+1} - RV_i\right)X_{i+1}\right]/\left(S_i\mathrm{E}_i\left[X_{i+1}^2\right]\right), \\ X_i &:= \exp(Z_i) - R, \\ a &:= \mathrm{E}_i\left[X_{i+1}\right]/\mathrm{E}_i\left[X_{i+1}^2\right], \\ b &:= 1 - \left(\mathrm{E}_i\left[X_{i+1}\right]\right)^2/\mathrm{E}_i\left[X_{i+1}^2\right]. \qquad (4) \end{aligned}$$

The hedging performance of the dynamically optimal strategy φ and of the locally optimal strategy ξ is given by

$$\begin{aligned} \mathrm{E}\left[\left(G_n^{x,\varphi} - H\right)^2\right] &= \left(R^2 b\right)^n (x - V_0)^2 + \varepsilon_0^2(\varphi), \\ \mathrm{E}\left[\left(G_n^{x,\xi} - H\right)^2\right] &= \left(R^2\right)^{n-j} (x - V_0)^2 + \varepsilon_0^2(\xi), \\ \varepsilon_0^2(\varphi) &= \sum_{j=0}^{n-1}\left(R^2 b\right)^{n-j-1}\mathrm{E}\left[\psi_j\right], \\ \varepsilon_0^2(\xi) &= \sum_{j=0}^{n-1} R^{2(n-j-1)}\mathrm{E}\left[\psi_j\right], \\ \psi_j &:= \mathrm{E}_j\left[\left(RV_j + \xi_j S_j X_{j+1} - V_{j+1}\right)^2\right] \\ &= \mathrm{Var}_j\left(V_{j+1}^2\right) - \frac{\left(\mathrm{Cov}_j\left(S_{j+1}, V_{j+1}\right)\right)^2}{\mathrm{Var}_j\left(S_{j+1}\right)}. \qquad (5) \end{aligned}$$

Proof See [9], Theorem 3.3. □

3 Numerical Results

We first fix the rebalancing period to $\varDelta = 1$ day and examine the behaviour of hedging errors across maturities, strikes and barrier levels. We then analyze the asymptotics of the hedging error as the rebalancing interval $\varDelta$ approaches 0, keeping the monitoring frequency of the barrier constant. We do so initially for a range of strikes and barrier levels with rebalancing interval $\varDelta = 1$ h and then with fixed strike and barrier level we examine asymptotics going down to $\varDelta = 5$ min.

3.1 Effect of Barrier Position, Maturity and Drift

We consider an up-and-out European call and two maturity dates: 1 and 6 months. During a detailed preliminary analysis we have found that changes in risk-free rate have a very small impact upon hedging errors and therefore we fix the risk-free rate in all computations to $r = 0$. Volatility is normalized to $\sigma = 0.2$ as explained in Sect. 2.1, while the drift takes two values $\mu \in \{-0.1, 0.1\}$. The time units reflect trading time; specifically we assume there are 8 h in a day and 250 days in a year. To be able to compare the size of hedging error across maturities we measure the position of the barrier and of the striking price relative to the initial stock price in terms of their Black–Scholes delta.

For each set of parameters we report five quantities: (i) the Black–Scholes price of a continuously monitored option C, (ii) the Black–Scholes price of a daily monitored option[2] $\hat{V}$, (iii) the standard deviation of the hedging error in a discretely rebalanced Black-Scholes model $\hat{\varepsilon}_0$ obtained from (5) using multinomial approximation of Black-Scholes normal transition probabilities[3] with daily monitoring and daily or hourly rebalancing; (iv) the mean value process V obtained from (3) and (4) using multinomial approximation of Lévy transition probabilities; and (v) the standard deviation of the unconditional expected squared hedging error ε_0 obtained from (5) using multinomial Lévy transition probabilities. The barrier of an up-and-out call has to be above the stock price for the option to be still alive, we therefore parametrize the delta of the barrier by values starting at[4] 10^{-100} and going up to 0.49. The deltas of the striking price range between 0.01 and 0.99. Numerical results for different values of $\varDelta$, T and μ are shown in Tables 1, 2, 3 and 4.

[2]Computation of the discretely monitored option price in Black–Scholes model follows the methodology of [15]. Effectively, the calculation is the same as for V in the empirical model, but the multinomial transition probabilities approximate the Black-Scholes risk-neutral distribution $\mathcal{N}\left(\left(r - \sigma^2/2\right)\varDelta, \sigma^2\varDelta\right)$.

[3]Objective probability distribution of log returns in the Black-Scholes model is $\mathcal{N}\left(\left(\mu - \sigma^2/2\right)\varDelta, \sigma^2\varDelta\right)$.

[4]The barrier with delta of 10^{-100} is so high that the corresponding results are, for all intents and purposes, indistinguishable from a plain vanilla option.

Table 1 Mean value and hedging error for a daily monitored up-and-out call option. $T = 1$ month, $\Delta = 1$ day, $\mu = 0.1$, $r = 0$. For each strike and barrier level we report 5 values: (i) Black–Scholes value of continuously monitored option, (ii) mean value for normally distributed log returns and discretely (daily) monitored option, (iii) hedging error corresponding to (ii); (iv) the mean value process V_0 for the empirical distribution of log returns(discrete monitoring); (v) standard deviation of the unconditional hedging error corresponding to (iv). Strike and barrier levels are parametrized by the Black–Scholes delta of their position

		Barrier (delta/level)					
Strike		1E-100	0.01	0.10	0.30	0.45	0.49
Delta	Level	343.8	114.6	107.9	103.3	100.9	100.3
0.01	114.6	0.019					
		0.019					
		0.104					
		0.020					
		0.122					
0.1	107.9	0.268	0.151				
		0.268	0.172				
		0.286	0.417				
		0.267	**0.170**				
		0.326	**0.443**				
0.3	103.3	1.071	0.874	0.182			
		1.071	0.916	0.255			
		0.408	0.734	0.521			
		1.066	**0.910**	**0.257**			
		0.469	**0.809**	**0.545**			
0.45	100.9	1.900	1.663	0.608	0.023		
		1.900	1.716	0.752	0.052		
		0.430	0.889	0.839	0.212		
		1.894	**1.709**	**0.759**	**0.053**		
		0.491	**0.960**	**0.912**	**0.223**		
0.49	100.3	2.162	1.915	0.767	0.044	0.000	
		2.162	1.971	0.930	0.089	0.001	
		0.427	0.913	0.931	0.290	0.020	
		2.156	**1.964**	**0.938**	**0.092**	**0.001**	
		0.491	**1.007**	**0.991**	**0.300**	**0.020**	
0.75	96.3	4.563	4.247	2.447	0.515	0.054	0.013
		4.563	4.321	2.750	0.754	0.141	0.071
		0.348	1.135	1.471	0.836	0.357	0.252
		4.560	**4.317**	**2.770**	**0.773**	**0.146**	**0.073**
		0.397	**1.222**	**1.610**	**0.915**	**0.379**	**0.265**
0.99	87.5	12.488	12.020	8.764	3.512	0.808	0.262
		12.488	12.134	9.385	4.427	1.598	1.037
		0.065	1.607	2.671	2.138	1.318	1.092
		12.489	**12.134**	**9.429**	**4.491**	**1.632**	**1.050**
		0.077	**1.761**	**2.874**	**2.316**	**1.452**	**1.193**

Table 2 Mean value and hedging error for a daily monitored up-and-out call option. $T = 6$ month, $\Delta = 1$ day, $\mu = 0.1$, $r = 0$. For each strike and barrier level we report 5 values: (i) Black–Scholes value of continuously monitored option, (ii) mean value for normally distributed log returns and discretely (daily) monitored option, (iii) hedging error corresponding to (ii); (iv) the mean value process V_0 for the empirical distribution of log returns(discrete monitoring); (v) standard deviation of the unconditional hedging error corresponding to (iv). Strike and barrier levels are parametrized by the Black–Scholes delta of their position

		Barrier (delta/level)					
Strike		1E-100	0.01	0.10	0.30	0.45	0.49
Delta	Level	2071.0	140.5	121.2	108.8	102.8	101.4
0.01	140.5	0.046					
		0.046					
		0.152					
		0.046					
		0.176					
0.1	121.2	0.635	0.364				
		0.635	0.387				
		0.329	0.674				
		0.631	**0.386**				
		0.381	**0.740**				
0.3	108.8	2.514	2.073	0.447			
		2.514	2.118	0.522			
		0.437	1.238	0.711			
		2.506	**2.115**	**0.526**			
		0.506	**1.364**	**0.778**			
0.45	102.8	4.434	3.910	1.475	0.059		
		4.435	3.966	1.622	0.087		
		0.425	1.366	1.214	0.293		
		4.426	**3.963**	**1.632**	**0.088**		
		0.493	**1.505**	**1.335**	**0.316**		
0.49	101.4	5.038	4.493	1.854	0.114	0.000	
		5.039	4.552	2.020	0.157	0.001	
		0.428	1.526	1.226	0.350	0.029	
		5.030	**4.549**	**2.032**	**0.160**	**0.001**	
		0.496	**1.679**	**1.351**	**0.380**	**0.030**	
0.75	91.8	10.490	9.812	5.793	1.296	0.163	0.055
		10.491	9.888	6.094	1.525	0.240	0.102
		0.314	1.746	2.040	1.013	0.361	0.234
		10.485	**9.890**	**6.120**	**1.543**	**0.246**	**0.105**
		0.364	**1.921**	**2.251**	**1.118**	**0.398**	**0.258**
0.99	72.6	27.452	26.507	19.689	8.316	2.280	1.036
		27.452	26.618	20.272	9.155	2.972	1.659
		0.048	2.605	3.244	2.244	1.307	0.989
		27.452	**26.628**	**20.326**	**9.217**	**3.012**	**1.689**
		0.056	**2.858**	**3.589**	**2.494**	**1.455**	**1.104**

Table 3 Mean value and hedging error for a daily monitored up-and-out call option. $T = 6$ months, $\Delta = 1$ day, $\mu = -0.1$, $r = 0$. For each strike and barrier level we report 5 values: (i) Black–Scholes value of continuously monitored option, (ii) mean value for normally distributed log returns and discretely (daily) monitored option, (iii) hedging error corresponding to (ii); (iv) the mean value process V_0 for the empirical distribution of log returns(discrete monitoring); (v) standard deviation of the unconditional hedging error corresponding to (iv). Strike and barrier levels are parametrized by the Black–Scholes delta of their position

		Barrier (delta/level)					
Strike		1E-100	0.01	0.10	0.30	0.45	0.49
Delta	Level	2070.99	140.55	121.17	108.82	102.83	101.37
0.01	140.55	0.046					
		0.046					
		0.071					
		0.047					
		0.084					
0.1	121.17	0.635	0.364				
		0.635	0.387				
		0.236	0.351				
		0.637	**0.387**				
		0.270	**0.377**				
0.3	108.82	2.514	2.073	0.447			
		2.514	2.118	0.522			
		0.374	0.612	0.515			
		2.514	**2.116**	**0.524**			
		0.438	**0.705**	**0.546**			
0.45	102.83	4.434	3.910	1.475	0.059		
		4.435	3.966	1.622	0.087		
		0.419	0.755	0.798	0.239		
		4.434	**3.962**	**1.626**	**0.088**		
		0.481	**0.822**	**0.913**	**0.268**		
0.49	101.37	5.038	4.493	1.854	0.114	0.000	
		5.039	4.552	2.020	0.157	0.001	
		0.419	0.761	0.901	0.326	0.026	
		5.037	**4.548**	**2.024**	**0.159**	**0.001**	
		0.490	**0.876**	**0.968**	**0.342**	**0.028**	
0.75	91.79	10.490	9.812	5.793	1.296	0.163	0.055
		10.491	9.888	6.094	1.525	0.240	0.102
		0.380	0.939	1.367	0.866	0.391	0.262
		10.489	**9.883**	**6.104**	**1.538**	**0.245**	**0.105**
		0.435	**1.019**	**1.564**	**0.992**	**0.419**	**0.281**
0.99	72.60	27.452	26.507	19.689	8.316	2.280	1.036
		27.452	26.618	20.272	9.155	2.972	1.659
		0.094	1.199	2.382	2.109	1.300	1.019
		27.452	**26.613**	**20.294**	**9.205**	**3.014**	**1.694**
		0.111	**1.371**	**2.591**	**2.306**	**1.487**	**1.165**

Table 4 Mean value and hedging error for a daily monitored up-and-out call option. $T = 1$ month, $\Delta = 1$ h, $\mu = 0.1$, $r = 0$. For each strike and barrier level we report 5 values: (i) Black–Scholes value of continuously monitored option, (ii) mean value for normally distributed log returns and discretely (daily) monitored option, (iii) hedging error corresponding to (ii); (v) the mean value process V_0 for the empirical distribution of log returns(discrete monitoring); (v) standard deviation of the unconditional hedging error corresponding to (iv). Strike and barrier levels are parametrized by the Black–Scholes delta of their position

		Barrier (delta/level)					
Strike		1E-100	0.01	0.10	0.30	0.45	0.49
Delta	Level	342.09	114.57	107.86	103.25	100.90	100.31
0.01	114.63	0.019					
		0.019					
		0.039					
		0.020					
		0.074					
0.1	107.89	0.268	0.151				
		0.268	0.172				
		0.104	0.214				
		0.268	**0.171**				
		0.197	**0.303**				
0.3	103.26	1.071	0.874	0.182			
		1.071	0.916	0.255			
		0.149	0.380	0.279			
		1.068	**0.912**	**0.257**			
		0.282	**0.547**	**0.381**			
0.45	100.90	1.900	1.663	0.608	0.023		
		1.900	1.716	0.752	0.052		
		0.155	0.453	0.451	0.129		
		1.897	**1.710**	**0.758**	**0.053**		
		0.295	**0.651**	**0.628**	**0.165**		
0.49	100.31	2.162	1.915	0.767	0.044	0.000	
		2.162	1.971	0.930	0.089	0.001	
		0.155	0.478	0.487	0.164	0.017	
		2.159	**1.965**	**0.937**	**0.092**	**0.001**	
		0.295	**0.684**	**0.681**	**0.216**	**0.018**	
0.75	96.33	4.563	4.247	2.447	0.515	0.054	0.013
		4.563	4.321	2.750	0.754	0.141	0.071
		0.126	0.595	0.784	0.451	0.193	0.137
		4.562	**4.317**	**2.767**	**0.771**	**0.145**	**0.073**
		0.240	**0.839**	**1.103**	**0.631**	**0.265**	**0.186**
0.99	87.53	12.488	12.020	8.764	3.512	0.808	0.262
		12.488	12.134	9.385	4.427	1.598	1.037
		0.024	0.883	1.409	1.144	0.734	0.606
		12.489	**12.132**	**9.420**	**4.484**	**1.630**	**1.049**
		0.047	**1.224**	**1.975**	**1.604**	**1.020**	**0.837**

We commence with the base case parameters $\Delta = 1$ day, $\mu = 0.1$ in Tables 1 and 2. The mean value process V coincides to a large extent with the Black-Scholes value of a discretely monitored option. This is a striking result, since the model in which V is computed is substantially incomplete, whereas the reasoning behind $\hat{V}$ relies on continuous rebalancing and perfect replication. For $T = 1$ month (Table 1) the difference between V and $\hat{V}$ is always less than 6.4 cents in absolute value, and in relative terms it is less than 3.6 % across all strikes and barrier levels.

The difference between V and $\hat{V}$ tends to diminish with increasing maturity. For $T = 6$ months (Table 2) the difference between V and $\hat{V}$ is less than 6.1 cents in absolute value, and less than 2.7 % in relative terms. The signs of $V - \hat{V}$ follow a pattern across strikes and barrier levels whereby the difference tends to be negative for very high barrier levels in combination with high strike prices, and to be positive elsewhere.

Let us now turn to the hedging errors. Hedging errors of barrier options (columns 4–8) behave differently to those of plain vanilla options (column 3). The hedging error of plain vanilla options are the largest at the money and become smaller for deep-in and deep-out-of-the-money options. In contrast, the hedging error of an up-and-out barrier option increases with decreasing strike price. This happens because for vanilla options the only source of the hedging error is the non-linearity of option pay-off around the strike price, whereas for barriers the main source of the hedging errors is the barrier itself. The lower the strike the higher the pay-off near the barrier and the higher the hedging errors.

Consider an (at-the-money) plain vanilla option with $T = 1$ month to maturity and strike at 100.3 (see Table 1, column 3). The Black-Scholes value of this option is 2.162, and the standard deviation of the unconditional hedging error is 0.427, due to daily rebalancing. If we consider the empirical distribution of log returns, which exhibits excess kurtosis, the hedging error increases to 0.491. Take now a barrier option with the same strike, and barrier at 107.9. The Black–Scholes price of the barrier option is less than a half at 0.930 but the standard deviation of the hedging error is more than double at 0.931. Thus if selling a plain vanilla option at the Black–Scholes price based on historical volatility is not a profitable enterprise, doing the same for barrier options is positively counterproductive. This conclusion is more pronounced for longer maturities and lower strikes, see Table 2 ($T = 6$ months).

Next we examine the effect of the change in the market direction, by contrasting Table 2 ($\mu = 0.1$) with Table 3 ($\mu = -0.1$). The difference between the Black–Scholes no-arbitrage price of a daily monitored barrier option $\hat{V}$ and the mean value process V remains small. The mean value is higher in the bear market for plain vanilla options (column 3) but it is generally marginally lower for barrier options, with the exception of very low strikes in combination with very low barrier levels. The difference in absolute value is less than two cents and less than 1 % in relative terms (with the exception of the two vanilla option with highest strikes). We conclude that V is largely insensitive to the changes in μ and that the Black–Scholes price $\hat{V}$ is a very good proxy for V.

The change in the market direction has a more dramatic effect on the size of unconditional hedging errors. Recall that the standard deviation of the unconditional error is given as a weighted average of one-period hedging errors,

$$\varepsilon_0^2(\varphi) = \sum_{j=0}^{n-1} \left(R^2 b\right)^{n-j-1} \mathrm{E}\left[\psi_j\right],$$

$$\psi_j = \mathrm{Var}_j\left(V_{j+1}^2\right) - \frac{\left(\mathrm{Cov}_j\left(S_{j+1}, V_{j+1}\right)\right)^2}{\mathrm{Var}_j\left(S_{j+1}\right)},$$

where R and b are close to 1. Since V is largely insensitive to the value of μ the values of ψ (as a function of time, stock price and option status) will very much coincide between the bull and the bear market. What will be different is the *expectation* of ψ.

The instantaneous hedging error ψ arises from two non-linearities in the option pay-off—one around the strike price and one along the barrier. The hedging error along the barrier tends to be more significant unless the barrier is either very far away from the stock price or the option is just about to be knocked out. In a bull market prices rise on average and the barrier, being above the initial stock price, contributes more significantly to $\mathrm{E}\left[\psi_j\right]$. $\mathrm{E}\left[\psi_j\right]$ will also contain more significant contribution from the strike region if the option is initially out of the money. In contrast, in a bear market price falls on average and $\mathrm{E}\left[\psi_j\right]$ will put less weight on the barrier region. It will contain a more significant contribution from the strike price, if the option is in the money to begin with. For barrier deltas equal to 10^{-100} and 0.49 we expect the strike region to dominate and therefore the hedging errors in the bear market to be larger for in-the-money options. This intuition is borne out by the numerical results shown in Tables 2 and 3.

3.2 Asymptotics

Let us now examine the effect of more frequent rebalancing by considering $\Delta = 1$ h (Table 4). Although hedging now occurs *hourly* we maintain the *daily monitoring* frequency of the barrier to make the results comparable with those in Table 1.

In the Black–Scholes model the *standard deviation* of the hedging error for *plain vanilla* options decreases with the square root of rebalancing interval, see [3] and [37]. With hourly rebalancing this implies standard deviation equal to $\sqrt{1/8} \approx 35\,\%$ of the daily figure (with 8-h trading day). The theoretical prediction turns out to be accurate, as can be seen by comparing entries marked ii) in each row of column 3 of Tables 1 and 4 which yields the range 36–38 % across all strikes.

In the empirical Lévy model the standard deviation of the unconditional hedging error of plain vanilla options is seen to decay more slowly, see entries marked (v) in each row of column 3 of Tables 1 and 4. With hourly rebalancing it is in the range 60–62 % of the daily rebalancing figures across all strikes. In this instance the higher

Table 5 Kurtosis as a function of rebalancing interval

	Kurtosis lattice		Kurtosis Lévy			
Δ	log	Level	log	Level	$\frac{n_{\text{down}}\eta}{\sigma}$	$\frac{n_{\text{up}}\eta}{\sigma}$
5 min	**69.11**	**69.05**	**72.54**	**72.51**	**27**	**25**
15 min	25.77	25.76	26.18	26.17	17	16
30 min	14.42	14.42	14.59	14.59	12.75	12
1 h	**8.73**	**8.73**	**8.79**	**8.80**	**10**	**9.5**
2 h	5.86	5.87	5.90	5.90	8	7.75
4 h	4.44	4.44	4.45	4.45	6.75	6.75
1 day (8 h)	**3.72**	**3.72**	**3.72**	**3.73**	**5.75**	**5.75**

frequency of hedging is (partially) offset by higher kurtosis of hourly returns. [10], Sect. 13.7, derives an approximation of the hedging error for leptokurtic returns and shows that rebalancing interval must be multiplied by kurtosis minus one to obtain the correct scaling of hedging errors. In our case Table 5 shows the kurtosis of daily returns is 3.72 and the kurtosis of hourly returns is 8.73, thus we should expect hourly errors to equal $\sqrt{1/8 \times 7.73/2.72} \approx 60\,\%$ of the daily errors which matches the actual range of 60–62 % mentioned earlier.

Table 5 compares the kurtosis of returns and log returns in the calibrated Lévy model with the kurtosis achieved in its multinomial lattice approximation. The last two columns show the number of standard deviations of one-period log return (rounded up to the nearest quarter) corresponding to the 10^{-5} and $1 - 10^{-5}$ quantiles of the one-period log return distribution. This is the range represented by the lattice approximation of the Lévy process. As an aside, we observe that the lattice begins to struggle to approximate the kurtosis of the Lévy process well at the 5-min rebalancing interval.

For *barrier options* (columns 4–8 of Tables 1 and 4) the Black-Scholes situation is more complicated because part of the error is caused by the barrier itself and this part has different Δ-asymptotics. Conjecturing that the barrier contributes an error whose *variance* is proportional to the square root of rebalancing interval, see [24], and assuming that fraction α of the error is generated by the strike region and the rest by the barrier, the approximate expression for the hourly total error as a fraction of daily error would read

$$\sqrt{0.35\alpha + \sqrt{0.35}(1-\alpha)}. \tag{6}$$

For barrier options in columns 4 and 5 of Tables 1 and 4 the percentage reduction in hedging error in the Black-Scholes model stands between 51 and 54 % which implies α values in formula (6) between 0.25 and 0.4. Variability of α is to be expected since the relative importance of the two types of errors will depend on barrier and strike levels.

Table 6 Mean value V_0 and unconditional standard deviation of the hedging error ε_0 for parameter values $T = 1, S_0 = 100, B = 107.9, K = 103.3$

	Black-Scholes		Empirical Lévy	
Δ	$\hat{V}_0$	$\hat{\varepsilon}_0$	V_0	ε_0
5 min	0.2525	0.142	0.2548	0.330
15 min	0.2535	0.191	0.2556	0.345
30 min	0.2535	0.231	0.2558	0.360
1 h	0.2536	0.282	0.2559	0.380
2 h	0.2537	0.345	0.2560	0.421
4 h	0.2537	0.424	0.2560	0.470
8 h	0.2538	0.519	0.2561	0.548

One can conjecture that for barrier options in the presence of excess kurtosis the formula (6) will remain the same, only the time scaling factor will be adjusted for excess kurtosis from 0.35 to 0.6 as in the case of plain vanilla options. We thus expect the ratio of hourly to daily errors in the Lévy model to be

$$\sqrt{0.6\alpha + \sqrt{0.6}(1 - \alpha)}. \tag{7}$$

With α in the range 0.25–0.4 heuristic (7) predicts error reduction in the range 71–74 % while the actual figures from columns 4 and 5 of Tables 1 and 4 yield the range 68–70 %, which for practical purposes is a perfectly adequate approximation.

Table 6 provides 5-min error data for one specific strike/barrier combination corresponding to $\alpha = 0.25$. It reports the hedging error ε_0 obtained from (5) and the mean value V_0 obtained from (3) and (4) using the multinomial approximation of the empirical Lévy process and analogous quantitities $\hat{\varepsilon}_0$ and $\hat{V}_0$ obtained from a multinomial approximation of the Black-Scholes model.

The Black-Scholes 5-min time scaling factor is $1/(8 \times 12) = 1/96 = 0.0104$ and the heuristic (6) yields error reduction ratio of

$$\sqrt{0.0104 \times 0.25 + \sqrt{0.0104}(1 - 0.25)} \approx 28\,\%,$$

while in Table 6 we find this ratio to be $0.142/0.519 \approx 27\,\%$. The leptokurtic empirical 5-min distribution leads to the time scaling factor of $68.05/2.72/96 \approx 0.26$ hence the 5-min empirical error is predicted to be

$$\sqrt{0.26 \times 0.25 + \sqrt{0.26}(1 - 0.25)} \approx 67\,\%$$

of the daily error. The actual figure in Table 6 is $0.33/0.548 \approx 60\,\%$. For practical purposes this is again an acceptable approximation.

Our exploratory analysis above points to two open questions in this area of research: (1) calculation of explicit asymptotic expression for hedging error of barrier options in discretely rebalanced Black-Scholes model analogous to the formula of [24] for path-independent options; (2) asymptotic formula for hedging error of barrier options in a continuously rebalanced Lévy model with small jumps. There is a good reason to believe that (1) and (2) are closely linked because similar link has already been established for plain vanilla options, see [11].

4 Sharpe Ratio Price Bounds

In this model, as in reality, the sale of an option and subsequent hedging is a risky activity. If one sells an option at its Black-Scholes value corresponding to historical volatility one effectively enters into an investment with zero mean and non-zero variance. In addition this investment is by construction uncorrelated with the stock returns. To make option trading profitable the trader must aim for a certain level of risk-adjusted returns, which implies selling derivatives above their Black–Scholes value. The question then arises as to what is a sensible measure of risk-adjusted returns and what is a sensible level of compensation for the residual risk.

[10] proposes to measure profitability of investment by its certainty equivalent growth rate adjusted for investor's risk aversion. When this measure is applied to mean-variance preferences, it yields a one-to-one relationship with the ex-ante Sharpe ratio of the investment strategy. Thus, in the present context, the unconditional Sharpe ratio appears as a natural measure of risk-adjusted returns.

It is well known that the square of maximal Sharpe ratio available by trading in two uncorrelated assets equals the sum of squared Sharpe ratios of the individual assets. Since the hedged option position is uncorrelated with the stock we can regard the Sharpe ratio of the hedged position as a meaningful measure of *incremental* performance (i.e. performance over and above optimal investment in the stock).

Suppose that the trader targets a certain level of annualized incremental Sharpe ratio h (say $h = 0.5$). Assuming that he or she can sell the option at price $\tilde{C}$ above the mean value V_0 the resulting Sharpe ratio of the hedged option position equals

$$\frac{e^{rT}(\tilde{C} - V_0)}{\varepsilon_0}.$$

If T is maturity in years the trader should look for a price $\tilde{C}$ such that

$$\frac{e^{rT}(\tilde{C} - V_0)}{\varepsilon_0} = h\sqrt{T},$$

which yields

$$\tilde{C} = V_0 + e^{-rT} h\sqrt{T}\varepsilon_0. \tag{8}$$

For plain vanilla options the price adjustment corresponding to annualized incremental Sharpe ratio of 1 gives rise to a gap between implied volatility and historical volatility of about 150 basis points, robustly across maturities and strikes. If the same price adjustment is performed for barrier options its magnitude is as important as, and often several times dominates, the price adjustment due to discrete monitoring. The fraction $\frac{\sqrt{T}\varepsilon_0}{V_0}$ is reported in Tables 7 and 8.

One obvious conclusion to draw from formula (8) is that prices in an incomplete market are likely to contain both a linear (V_0) and a non-linear (ε_0) component. The prevailing market practice is to use just the linear part V_0 for calibration which often requires distorting the historical distribution of returns to match observed market prices across strikes and maturities. For example, in their calibration of plain vanilla option prices [31] report historical annualized excess kurtosis at 0.002 but risk-neutral excess kurtosis at 0.18 which is a level that the variance-optimal martingale measure

Table 7 Risk premium as a percentage of mean value for up-and-out call. $T = 1$ month, $\Delta = 1$ day, $\mu = 0.1$, $r = 0$. Strike and barrier levels are parametrized by the Black–Scholes delta of their position

		Barrier (delta/level)					
Strike		1E-100	0.01	0.10	0.30	0.45	0.49
Delta	Level	343.8 (%)	114.6 (%)	107.9 (%)	103.3 (%)	100.9 (%)	100.3 (%)
0.01	114.6	177					
0.10	107.9	35	75				
0.30	103.3	13	26	61			
0.45	100.9	7	16	35	121		
0.49	100.3	7	15	30	94	491	
0.75	96.3	3	8	17	34	75	104
0.99	87.5	0.2	4	9	15	26	33

Table 8 Risk premium as a percentage of mean value for up-and-out call. $T = 6$ months, $\Delta = 1$ day, $\mu = 0.1$, $r = 0$. Strike and barrier levels are parametrized by the Black–Scholes delta of their position

		Barrier (delta/level)					
Strike		1E-100	0.01	0.10	0.30	0.45	0.49
Delta	Level	2071.0 (%)	140.5 (%)	121.2 (%)	108.8 (%)	102.8 (%)	101.4 (%)
0.01	114.6	111					
0.10	107.9	17	55				
0.30	103.3	6	19	43			
0.45	100.9	3	11	24	103		
0.49	100.3	3	11	19	69	665	
0.75	96.3	1	6	11	21	47	71
0.99	87.5	0	3	5	8	14	19

that generates V_0 simply cannot reach. This phenomenon gets worse in the presence of exotic options. Formula (8) offers a flexible alternative that may offer better fit of model dynamics to historical return distributions and at the same time provide closer calibration to market prices thanks to the non-linear term ε_0 which has very different characteristics for different types of exotic options, as we have seen in the previous section.

5 Conclusions

In place of conclusions a personal confession. At around 2002 I was performing numerical experiments somewhat similar to the ones presented here, but without the underlying Lévy structure, just purely driven by empirical data and with plain vanilla options. In the process of doing so I convinced myself that Lévy models, which emerge as continuous-time limits of multinomial lattices, are the key mathematical tool to describe market incompleteness. I wrote to Ernst Eberlein, out of the blue, to ask whether I might join his group for a few months to learn properly about this exciting and for me completely new and difficult theory. The result were two stimulating months in Freiburg in early 2004 and a lifetime of mathematical inspiration. Thank you Ernst!

References

1. Andricopoulos, A.D., Widdicks, M., Duck, P.W., Newton, D.P.: Universal option valuation using quadrature methods. J. Financ. Econ. **67**(3), 447–471 (2003)
2. Bowie, J., Carr, P.: Static simplicity. Risk **7**(9), 45–49 (1994)
3. Boyle, P.P., Emanuel, D.: Discretely adjusted option hedges. J. Financ. Econ. **8**, 259–282 (1980)
4. Boyle, P.P., Lau, S.: Bumping up against the barrier with the binomial method. J. Deriv. **1**(4), 6–14 (1994)
5. Broadie, M., Glasserman, P., Kou, S.: A continuity correction for discrete barrier options. Math. Financ. **7**(4), 325–348 (1997)
6. Brown, H., Hobson, D., Rogers, L.C.G.: Robust hedging of barrier options. Math. Financ. **11**(3), 285–314 (2001)
7. Carr, P., Chou, A.: Breaking barriers. Risk **10**(9), 139–146 (1997)
8. Carr, P., Ellis, K., Gupta, V.: Static hedging of exotic options. J. Financ. **53**(3), 1165–1191 (1998)
9. Černý, A.: Optimal continuous-time hedging with leptokurtic returns. Math. Financ. **17**(2), 175–203 (2007)
10. Černý, A.: Mathematical Techniques in Finance: Tools for Incomplete Markets, 2nd edn. Princeton University Press (2009)
11. Černý, A., Denkl, S., Kallsen, J.: Hedging in Lévy models and time step equivalent of jumps (2013). ArXiv preprint http://arxiv.org/pdf/1309.7833
12. Derman, E., Ergener, D., Kani, I.: Forever hedged. Risk **7**(9), 139–145 (1994)
13. Derman, E., Ergener, D., Kani, I.: Static options replication. J. Deriv. **2**, 78–95 (1995)
14. Derman, E., Kani, I., Ergener, D., Bardhan, I.: Enhanced numerical methods for options with barriers. Finan. Anal. J. **51**(6), 65–74 (1995)

15. Duan, J.C., Dudley, E., Gauthier, G., Simonato, J.G.: Pricing discretely monitored barrier options by a Markov chain. J. Deriv. **10**(4), 9–31 (2003)
16. Dupont, D.Y.: Hedging barrier options: Current methods and alternatives. Economics Series 103, Institute for Advanced Studies (2001)
17. Eberlein, E., Özkan, F.: Time consistency of Lévy models. Quan. Financ. **3**, 40–50 (2003)
18. Eberlein, E., Prause, K.: The generalized hyperbolic model: Financial derivatives and risk measures. In: Geman, H., Madan, D., Pliska, S.R., Vorst, T. (eds.) Mathematical Finance: Bachelier Congress 2000, pp. 245–267. Springer (2002)
19. Fang, F., Oosterlee, C.W.: Pricing early-exercise and discrete barrier options by Fourier-cosine series expansions. Numer. Math. **114**(1), 27–62 (2009). doi:10.1007/s00211-009-0252-4, http://dx.doi.org/10.1007/s00211-009-0252-4
20. Feng, L., Linetsky, V.: Pricing discretely monitored barrier options and defaultable bonds in Lévy process models: a fast Hilbert transform approach. Math. Financ. **18**(3), 337–384 (2008)
21. Figlewski, S., Gao, B.: The adaptive mesh model: A new approach to efficient option pricing. J. Financ. Econ. **53**(3), 313–351 (1999)
22. Fusai, G., Abrahams, I., Sgarra, C.: An exact analytical solution for discrete barrier options. Financ. Stochast. **10**(1), 1–26 (2006)
23. Fusai, G., Recchioni, M.: Analysis of quadrature methods for pricing discrete barrier options. J. Econ. Dyn. Control **31**(3), 826–860 (2007)
24. Gobet, E., Temam, E.: Discrete time hedging errors for options with irregular payoffs. Finan. Stochast. **5**(3), 357–367 (2001). doi:10.1007/PL00013539, http://dx.doi.org/10.1007/PL00013539
25. Hörfelt, P.: Pricing discrete European barrier options using lattice random walks. Math. Financ. **13**(4), 503–524 (2003)
26. İlhan, A., Sircar, R.: Optimal static-dynamic hedges for barrier options. Math. Financ. **16**(2), 359–385 (2006)
27. Jacod, J., Shiryaev, A.N.: Limit theorems for stochastic processes, Grundlehren der Mathematischen Wissenschaften [Fundamental Principles of Mathematical Sciences], vol. 288, second edn. Springer, Berlin (2003)
28. Kou, S.G.: On pricing of discrete barrier options. Statistica Sinica **13**(4), 955–964 (2003)
29. Kuan, G., Webber, N.: Valuing discrete barrier options on a Dirichlet lattice. FORC Preprint 140/04, University of Warwick (2004)
30. Leisen, D.: Valuation of barrier options in a Black-Scholes setup with jump risk. Eur. Finance Rev. **3**(3), 319–342 (1999)
31. Madan, D., Carr, P., Chang, E.: The variance gamma process and option pricing. Eur. Finance Rev. **2**(1), 79–105 (1998)
32. Merton, R.C.: Theory of rational option pricing. Bell J. Econ. Manag. Sci. **4**(1), 141–183 (1973)
33. Reiner, E., Rubinstein, M.: Breaking down the barriers. Risk **4**(8), 28–35 (1991)
34. Ritchken, P.H.: On pricing barrier options. J. Deriv. **3**(2), 19–28 (1995)
35. Sato, K.: Lévy Processes and Infinitely Divisible Distributions. Cambridge University Press (1999)
36. Strawderman, R.L.: Computing tail probabilities by numerical Fourier inversion: The absolutely continuous case. Statistica Sinica **14**(1), 175–201 (2004)
37. Toft, K.B.: On the mean-variance tradeoff in option replication with transactions costs. J. Finan. Quant. Anal. **31**(2), 233–263 (1996)
38. Wei, J.: Valuation of discrete barrier options by interpolations. J. Deriv. **6**(1), 51–73 (1998)

Forward Exponential Indifference Valuation in an Incomplete Binomial Model

M. Musiela, E. Sokolova and T. Zariphopoulou

Abstract We introduce and construct indifference prices under exponential forward performance criteria in an incomplete binomial model. We propose a pricing algorithm, which is iterative and yields the price in two sub-steps, locally in time. At the beginning of each period, an intermediate payoff is produced which is non-linear and replicable, and, in turn, it is priced by arbitrage in the second sub-step. The indifference price is thus constructed via an iterative non-linear pricing operator, which also involves a martingale measure. The latter turns out to minimize the reverse relative entropy. Properties of the forward prices are discussed as well as differences with their classical counterparts.

Keywords Forward performance processes · Indifference pricing · Reverse relative entropy

1 Introduction

We introduce, construct and study indifference prices in an incomplete binomial model under forward performance criteria. Such criteria, proposed by two of the

---

The authors would like to thank the Oxford-Man Institute for its hospitality and support, B. Angoshtari and two anonymous referees for their valuable comments and suggestions.

M. Musiela · T. Zariphopoulou (✉)
Oxford-Man Institute, University of Oxford, Oxford, UK
e-mail: zariphop@math.utexas.edu

M. Musiela
e-mail: marek.musiela@oxford-man.ox.ac.uk

E. Sokolova
CDF SUEZ, Houston, USA
e-mail: katusha_s@yahoo.com

T Zariphopoulou
Mathematics and IROM, The University of Texas at Austin, Austin, USA

© Springer International Publishing Switzerland 2016
J. Kallsen and A. Papapantoleon (eds.), *Advanced Modelling in Mathematical Finance*,
Springer Proceedings in Mathematics & Statistics 189,
DOI 10.1007/978-3-319-45875-5_13

authors (see, among others, [12, 15]), complement the traditional expected utility ones by allowing for dynamic adaptation of risk preferences as the market evolves. We refer the reader to, among others, [15–17, 19] for an overview on the forward performance approach.

The binomial model we consider is more general than the ones studied in the traditional exponential indifference valuation literature, for it includes a non-traded stochastic factor that affects not only the claim's payoff (as it is the case, among others, in [1, 4, 10, 11, 23, 24]) but, also, the transition probability and/or the values of the traded stock. This extension is crucial in incorporating models with stochastic investment opportunity sets. Binomial models of this kind were analyzed in the classical setting in [9, 18] for power and exponential utilities, respectively.

We first construct a forward performance process for the incomplete model herein and, in turn, analyze the associated indifference prices. We focus on a criterion of exponential type (cf. (13)) since exponential risk preferences have been predominantly used in indifference valuation.

The main contribution is the construction of a valuation algorithm for the forward indifference prices. We show that, for a claim written at time 0 and maturing at t, its price $\nu_s(C_t)$, $s = 0, 1, \ldots, t$, satisfies

$$\nu_s(C_t) = \mathcal{E}_{\mathbb{Q}^*}^{(s,s+1)}(\nu_{s+1}(C_t)) := E_{\mathbb{Q}^*}\left(\frac{1}{\gamma}\ln E_{\mathbb{Q}^*}\left(e^{\gamma\,\nu_{s+1}(C_t)}\middle|\,\mathcal{F}_s \vee \mathcal{F}_{s+1}^S\right)\middle|\,\mathcal{F}_s\right),$$

where $\mathcal{F}_s$ and $\mathcal{F}_s^S$ are the filtrations generated by both the stock and the stochastic factor, and the stock, respectively, and $\mathbb{Q}^*$ an appropriately chosen martingale measure.

Therefore, the price is constructed iteratively,

$$\nu_s(C_t) = \mathcal{E}_{\mathbb{Q}^*}^{(s,t)}(C_t) := \mathcal{E}_{\mathbb{Q}^*}^{(s,s+1)}\left(\mathcal{E}_{\mathbb{Q}^*}^{(s+1,s+2)} \ldots \mathcal{E}_{\mathbb{Q}^*}^{(t-1,t)}(C_t)\right).$$

Each price iteration has two sub-steps. In the first, the intermediate payoff

$$\mathcal{C}^{(s,s+1)}(\nu_{s+1}(C_t)) := \frac{1}{\gamma}\ln E_{\mathbb{Q}^*}\left(e^{\gamma\,\nu_{s+1}(C_t)}\middle|\,\mathcal{F}_s \vee \mathcal{F}_{s+1}^S\right) \tag{1}$$

is produced, which is non-linear and replicable. In turn, its arbitrage-free price yields, in the second sub-step, the indifference price,

$$\nu_s(C_t) = E_{\mathbb{Q}^*}\left(\mathcal{C}^{(s,s+1)}(\nu_{s+1}(C_t))\middle|\,\mathcal{F}_s\right). \tag{2}$$

Central role plays the emerging pricing measure $\mathbb{Q}^*$, which turns out to be a martingale one that minimizes the reverse relative entropy (see Proposition 7). Moreover, it has the property that the conditional distribution of the stochastic factor, given the information on the traded stock, remains the same as the one under the historical measure (see (36)), in that, for $s = 1, 2, \ldots, t$,

$$\mathbb{Q}^*\left(Y_s\,|\mathcal{F}_{s-1} \vee \mathcal{F}_s^S\right) = \mathbb{P}\left(Y_s\,|\mathcal{F}_{s-1} \vee \mathcal{F}_s^S\right).$$

The forward indifference prices have intuitively pleasing properties. Among others, we show that the above intermediate payoff $\mathcal{C}^{(s,s+1)}\left(\nu_{s+1}\left(C_t\right)\right)$ provides a direct analogue of the traditional certainty equivalent (cf. (51)). Namely, consider the non-linear payoff of certainty equivalent type

$$CE^{(s,s+1)}\left(\nu_{s+1}\left(C_t\right)\right) := -U_{s+1}^{(-1)}\left(E_{\mathbb{P}}\left(U_{s+1}\left(-\nu_{s+1}\left(C_t\right)\right)\middle|\mathcal{F}_s \vee \mathcal{F}_{s+1}^S\right)\right), \quad (3)$$

where $\mathbb{P}$ is the historical measure, U_{s+1} the forward performance process and $U_{s+1}^{(-1)}$ its spatial inverse. We establish that it coincides with the above payoff,

$$CE^{(s,s+1)}\left(\nu_{s+1}\left(C_t\right)\right) = \mathcal{C}^{(s,s+1)}\left(\nu_{s+1}\left(C_t\right)\right).$$

As a result, the forward indifference price can be represented as the arbitrage-free price of an appropriately chosen conditional certainty equivalent for each valuation period.

We also show that the single-period conditional distribution of the pricing measure $\mathbb{Q}^*$ depends exclusively on the associated single-period conditional risk neutral and historical probabilities (see (32), (33)). This, together with the form of $\mathcal{E}_{\mathbb{Q}^*}^{(s,s+1)}$ above, highlight the essential features of the indifference valuation under forward exponential criteria: the price is constructed by "single-period" operations—both in terms of the pricing functional and the involved pricing measure—which are repeated from one-period to the next with single-period adjustments of the conditional risk neutral and historical probabilities. Furthermore, all three pricing ingredients, $\mathbb{Q}^*$, $\mathcal{E}_{\mathbb{Q}^*}^{(s,s+1)}$ and $\mathcal{E}_{\mathbb{Q}^*}^{(s,t)}$, are independent of the maturity of the claim. Finally, because forward performance criteria are defined for all times, sequentially from one period to the next as the market moves (cf. (17)), one can price claims that arrive at later times with arbitrary maturities (see discussion below Corollary 13).

Note that most of these properties fail in the classical setting, where prices are defined in terms of expected utility from terminal wealth in a chosen horizon, say $[0, T]$. Indeed, while the forms of the corresponding single- and multi-step pricing functionals $\mathcal{E}_{\mathbb{Q}_T^{me}}^{(s,s+1)}$ and $\mathcal{E}_{\mathbb{Q}_T^{me}}^{(s,t)}$ are similar to the ones herein (see, [10, 23, 24] for complete markets and a claim written only on the nontraded asset, and [18] for a model like the one herein), the choice of the horizon strongly affects the pricing measure $\mathbb{Q}_T^{me}$, which is the minimal relative entropy one [2, 5, 21]. Moreover, its conditional distribution does not have the aforementioned local features that $\mathbb{Q}^*$ has. From the indifference valuation perspective, once the investment horizon is (pre)chosen, no new claim arriving at a future time, that was not known a priori when the original investment horizon was set up, and maturing beyond T can be priced.

The paper is organized as follows. In Sect. 2, we present the model and its forward investment performance process, and propose an example of exponential type. In Sect. 3, we introduce the forward indifference price and in Sect. 4 we construct the associated pricing algorithm. In Sect. 5, we present various properties of the prices and discuss differences with their classical counterparts.

2 The Model and Its Forward Performance Criteria

We start with the probabilistic setup of the incomplete multi-period binomial model. There are two traded assets, a riskless bond and a stock. The bond is assumed to offer zero interest rate.

The values of the stock are denoted by S_t, $t = 1, 2, \ldots$ with $S_0 > 0$. We define the random variables

$$\xi_t = \frac{S_t}{S_{t-1}}, \ \xi_t = \xi_t^d, \ \xi_t^u \ \text{with} \ 0 < \xi_t^d < 1 < \xi_t^u. \tag{4}$$

Incompleteness comes from a non-traded stochastic factor. Its levels, denoted by Y_t, $t = 0, 1, \ldots,$ satisfy $Y_t \neq 0$. We introduce the random variables

$$\eta_t = \frac{Y_t}{Y_{t-1}}, \ \eta_t = \eta_t^d, \ \eta_t^u \ \text{with} \ 0 < \eta_t^d < \eta_t^u. \tag{5}$$

We then view $\{(S_t, Y_t) : t = 0, 1, \ldots\}$ as a two-dimensional stochastic process defined on a probability space $(\Omega, \mathcal{F}, (\mathcal{F}_t), \mathbb{P})$. The filtration $\mathcal{F}_t$ is generated by S_i and Y_i, or, equivalently, by the random variables ξ_i and η_i, for $i = 0, 1, \ldots, t$. We also consider the filtration $\mathcal{F}_t^S$ generated only by S_i, $i = 0, 1, \ldots, t$. The real (historical) probability measure on Ω and $\mathcal{F}$ is denoted by $\mathbb{P}$.

We introduce the sets

$$A_t = \{\omega : \xi_t(\omega) = \xi_t^u\} \quad \text{and} \quad B_t = \{\omega : \eta_t(\omega) = \eta_t^u\}, \tag{6}$$

and assign the single-period conditional probabilities $\mathbb{P}(A_t B_t | \mathcal{F}_{t-1})$, $\mathbb{P}(A_t B_t^c | \mathcal{F}_{t-1}), \mathbb{P}(A_t^c B_t | \mathcal{F}_{t-1}), \mathbb{P}(A_t^c B_t^c | \mathcal{F}_{t-1})$, for $t = 1, 2, \ldots$.

Throughout, we will be using the notation AB to denote the intersection $A \cap B$ of sets A and B. We will be also using the notations "$Z \in \mathcal{F}_t$" or "Z is $\mathcal{F}_t$-mble" interchangeably to state that a generic random variable Z is $\mathcal{F}_t$-measurable.

An investor starts at $t = 0$ with endowment $X_0 = x$, $x \in \mathbb{R}$, and trades between the stock and the bond, following self-financing strategies. The number of shares of stock held in his portfolio over the time interval $[t-1, t)$, $t = 1, 2, \ldots,$ is denoted by α_t. The set of *admissible* policies is denoted by $\mathcal{A}$ and consists of all sequences $\alpha = \{\alpha_1, \alpha_2, \ldots, \alpha_t, \ldots\}$, where each term α_t is a real-valued $\mathcal{F}_{t-1}$-mble random variable.

The investor's wealth is, then, given, for $t = 1, 2, \ldots$, by

$$X_t^\alpha = x + \sum_{i=1}^{t} \alpha_i \, \Delta S_i, \tag{7}$$

where the price increment $\Delta S_i = S_i - S_{i-1}$.

The performance of the various investment strategies is measured via a stochastic criterion, the so-called *forward performance* process, which measures the output of admissible portfolios and gives a selection criterion as follows: a strategy is deemed optimal if it generates a wealth process whose average performance is maintained over time. Specifically, the average performance of this strategy, at any future date, conditionally on today's information preserves the performance of this strategy up until today. Any strategy that fails to maintain the average performance over time is, then, sub-optimal. We formalize this more rigorously below.

Definition 1 An $\mathcal{F}_t$-adapted process $U_t(x)$ is a forward performance process if, for $t = 0, 1, \ldots,$

(i) the mapping $x \to U_t(x)$, $x \in \mathbb{R}$, is strictly increasing and strictly concave,

(ii) for each $\alpha \in \mathcal{A}$,

$$U_t\left(X_t^{\alpha}\right) \geq E_{\mathbb{P}}\left(U_{t+1}\left(X_{t+1}^{\alpha}\right)\middle|\, \mathcal{F}_t\right), \tag{8}$$

(iii) there exists $\alpha^* \in \mathcal{A}$ for which

$$U_t\left(X_t^{\alpha^*}\right) = E_{\mathbb{P}}\left(U_{t+1}\left(X_{t+1}^{\alpha^*}\right)\middle|\, \mathcal{F}_t\right). \tag{9}$$

The concept of forward performance process was introduced by two of the authors in [12] for the binomial model at hand in a single-period setting. It was subsequently extended to Itô-diffusion markets, and we refer the reader, among others, to [14, 15, 17, 19], and references therein.

Characterizing the entire family of forward performance processes remains an open question and is being currently investigated by the authors and others. In the case of Itô-diffusion markets, a stochastic PDE was derived in [16] for the forward performance process. The novel element therein is the forward performance volatility process, which is an investor-specific input. As a result, forward performance processes are not in general unique.

Special classes of volatilities were proposed in [15], which can be interpreted as zero-volatility cases for alternative market settings under a different numeraire and/or market views. More recent works on the forward SPDE include [3, 8, 19, 20, 22]. For a complete study of the zero-volatility case see [17].

Herein, we study discrete-time forward processes and focus on analyzing the associated indifference prices. Because in the classical expected utility framework such prices have been constructed primarily for exponential risk preferences, we are interested in a similar class of criteria as well.

2.1 An Exponential Forward Performance Process

We look for a forward performance process of the form

$$U_t(x) = -e^{-\gamma x + H_{0,t}}, \; x \in \mathbb{R} \text{ and } \gamma > 0,$$

for an appropriately chosen process $H_{0,t}$, satisfying $H_{0,0} = 0$ and $H_{0,t} \in \mathcal{F}_t$, $t = 1, 2, \ldots$.

As mentioned earlier, forward performance processes are not unique, for they depend critically on the choice of their volatility process. Herein, we focus on a forward process of the above form which, as we show below, turns out to also be decreasing in time, for each x. We choose to start with this class of discrete-time forward criteria because they provide the simplest direct extension of the zero-volatility case in Itô-diffusion markets, which also turn out to be time-monotone processes (see [17]).

For general semimartingale markets, exponential forward processes were analyzed in [27], and subsequently used for the construction of maturity-independent entropic risk measures in [26].

We proceed with some auxiliary results. For $t = 1, 2, \ldots$, we denote by $\mathcal{Q}_t$ the set of equivalent martingale measures defined on $\mathcal{F}_t$. We also denote (with a slight abuse of notation) by $\mathbb{Q}$ its generic element and recall the conditional risk neutral probabilities

$$q_t = \mathbb{Q}\left(A_t \middle| \mathcal{F}_{t-1}\right) = \frac{1 - \xi_t^d}{\xi_t^u - \xi_t^d}, \tag{10}$$

with A_t and ξ_t^d, ξ_t^u as in (6) and (4).

Definition 2 The process h_t, $t = 1, 2, \ldots$, is defined by

$$h_t = q_t \ln \frac{q_t}{\mathbb{P}\left(A_t \middle| \mathcal{F}_{t-1}\right)} + (1 - q_t) \ln \frac{1 - q_t}{\mathbb{P}\left(A_t^c \middle| \mathcal{F}_{t-1}\right)}, \tag{11}$$

with q_t and A_t as in (10) and (6), respectively.

Note that actually $h_t \in \mathcal{F}_{t-1}$ and, moreover,

$$e^{-h_t} = \left(\frac{\mathbb{P}\left(A_t \middle| \mathcal{F}_{t-1}\right)}{q_t}\right)^{q_t} \left(\frac{\mathbb{P}\left(A_t^c \middle| \mathcal{F}_{t-1}\right)}{1 - q_t}\right)^{1-q_t}. \tag{12}$$

We present one of the main results next.

Theorem 3 *Let h as in (11) and $\gamma > 0$. Then, for $t = 1, 2, \ldots$ and $x \in \mathbb{R}$, the process*

$$U_t(x) = -e^{-\gamma x + \sum_{i=1}^{t} h_i}, \tag{13}$$

with $U_0(x) = -e^{-\gamma x}$, is a forward performance.

The policy given, for $i = 1, 2, \ldots, t$, by

$$\alpha_i^* = \frac{1}{\gamma S_{i-1}\left(\xi_i^u - \xi_i^d\right)} \ln \frac{\left(\xi_i^u - 1\right) \mathbb{P}\left(A_i \middle| \mathcal{F}_{i-1}\right)}{\left(1 - \xi_i^d\right) \mathbb{P}\left(A_i^c \middle| \mathcal{F}_{i-1}\right)} \tag{14}$$

is optimal and generates the optimal wealth process

$$X_t^* = x + \frac{1}{\gamma} \sum_{i=1}^{t} \frac{\xi_i - 1}{\left(\xi_i^u - \xi_i^d\right)} \ln \frac{\left(\xi_i^u - 1\right) \mathbb{P}\left(A_i \middle| \mathcal{F}_{i-1}\right)}{\left(1 - \xi_i^d\right) \mathbb{P}\left(A_i^c \middle| \mathcal{F}_{i-1}\right)}. \tag{15}$$

We first present the following auxiliary result.

Lemma 4 *For* $i = 1, 2, \ldots,$ *and* h_i *as in (11), we have*

$$\sup_{\alpha_i \in \mathcal{F}_{i-1}} E_{\mathbb{P}}\left(-e^{-\gamma \alpha_i \Delta S_i} | \mathcal{F}_{i-1}\right) = -e^{-h_i}, \tag{16}$$

with the maximum occurring at α_i^* *given in (14).*

Proof We have, with A_i as in (6),

$$E_{\mathbb{P}}\left(-e^{-\gamma \alpha_i \Delta S_i} \middle| \mathcal{F}_{i-1}\right)$$
$$= E_{\mathbb{P}}\left(-e^{-\gamma \alpha_i S_{i-1}(\xi_i^u - 1)} \mathbf{1}_{A_i} \middle| \mathcal{F}_{i-1}\right) + E_{\mathbb{P}}\left(-e^{-\gamma \alpha_i S_{i-1}(\xi_i^d - 1)} \mathbf{1}_{A_i^c} \middle| \mathcal{F}_{i-1}\right)$$
$$= -\left(e^{-\gamma \alpha_i S_{i-1}(\xi_i^u - 1)} \mathbb{P}\left(A_i \middle| \mathcal{F}_{i-1}\right) + e^{-\gamma \alpha_i S_{i-1}(\xi_i^d - 1)} \mathbb{P}\left(A_i^c \middle| \mathcal{F}_{i-1}\right)\right),$$

where we used the measurability properties of the involved quantities. Direct differentiation yields that the optimum occurs at (14). Then, the first term above becomes

$$-e^{-\gamma \alpha_i^* S_{i-1}(\xi_i^u - 1)} \mathbb{P}\left(A_i \middle| \mathcal{F}_{i-1}\right) = -\left(e^{\gamma \alpha_i^* S_{i-1}(\xi_i^u - \xi_i^d)}\right)^{-\frac{\xi_i^u - 1}{\xi_i^u - \xi_i^d}} \mathbb{P}\left(A_i \middle| \mathcal{F}_{i-1}\right)$$

$$= -\left(\frac{\left(\xi_i^u - 1\right) \mathbb{P}\left(A_i \middle| \mathcal{F}_{i-1}\right)}{\left(1 - \xi_i^d\right) \mathbb{P}\left(A_i^c \middle| \mathcal{F}_{i-1}\right)}\right)^{-\frac{\xi_i^u - 1}{\xi_i^u - \xi_i^d}} \mathbb{P}\left(A_i \middle| \mathcal{F}_{i-1}\right)$$

$$= -\left(\frac{1 - q_i}{q_i}\right)^{-(1-q_i)} \left(\frac{\mathbb{P}\left(A_i \middle| \mathcal{F}_{i-1}\right)}{\mathbb{P}\left(A_i^c \middle| \mathcal{F}_{i-1}\right)}\right)^{-(1-q_i)} \mathbb{P}\left(A_i \middle| \mathcal{F}_{i-1}\right)$$

$$= -\left(\frac{q_i}{1 - q_i}\right)^{1-q_i} \left(\mathbb{P}\left(A_i \middle| \mathcal{F}_{i-1}\right)\right)^{q_i} \left(\mathbb{P}\left(A_i^c \middle| \mathcal{F}_{i-1}\right)\right)^{1-q_i},$$

where we used (10). Similarly,

$$-e^{-\gamma \alpha_i^* S_{i-1}(\xi_i^d - 1)} \mathbb{P}\left(A_i^c \middle| \mathcal{F}_{i-1}\right)$$

$$= -\left(\frac{1-q_i}{q_i}\right)^{q_i} \left(\mathbb{P}\left(A_i \middle| \mathcal{F}_{i-1}\right)\right)^{q_i} \left(\mathbb{P}\left(A_i^c \middle| \mathcal{F}_{i-1}\right)\right)^{1-q_i}.$$

Therefore,

$$E_{\mathbb{P}}\left(-e^{-\gamma\alpha_i^*\Delta S_i} \middle| \mathcal{F}_{i-1}\right)$$

$$= -\left(\mathbb{P}\left(A_i \middle| \mathcal{F}_{i-1}\right)\right)^{q_i} \left(\mathbb{P}\left(A_i^c \middle| \mathcal{F}_{i-1}\right)\right)^{1-q_i} \left(\left(\frac{q_i}{1-q_i}\right)^{1-q_i} + \left(\frac{1-q_i}{q_i}\right)^{q_i}\right)$$

$$= -\left(\frac{\mathbb{P}\left(A_i \middle| \mathcal{F}_{i-1}\right)}{q_i}\right)^{q_i} \left(\frac{\mathbb{P}\left(A_i^c \middle| \mathcal{F}_{i-1}\right)}{q_i}\right)^{1-q_i} (q_i + (1-q_i)) = -e^{-h_i},$$

where we used (12). ■

We continue with the proof of Theorem 3.

Proof Requirement (i) in Definition 1 follows directly. Next, we establish (8). Using (7) and (13), we need to show that, for $t \geq 0$ and $\alpha_i \in \mathcal{F}_{i-1}$, $i = 1, \ldots, t+1$, and $x \in \mathbb{R}$,

$$-e^{-\gamma\left(x+\Sigma_{i=1}^t \alpha_i \Delta S_i\right)+\Sigma_{i=1}^t h_i} \geq E_{\mathbb{P}}\left(-e^{-\gamma\left(x+\Sigma_{i=1}^{t+1}\alpha_i\Delta S_i\right)+\Sigma_{i=1}^{t+1}h_i} \,|\mathcal{F}_t\right).$$

The above inequality reduces to

$$E_{\mathbb{P}}\left(-e^{-\gamma\alpha_{t+1}\Delta S_{t+1}} \,|\mathcal{F}_t\right) \leq -e^{-h_{t+1}},$$

and we easily conclude using Lemma 4.

To show (9) we work as follows. Let X_t^*, $t = 0, 1, \ldots,$ given by (15). We need to establish

$$-e^{-\gamma X_t^* + \Sigma_{i=1}^t h_i} = E_{\mathbb{P}}\left(-e^{-\gamma X_{t+1}^* + \Sigma_{i=1}^{t+1} h_i} \,|\mathcal{F}_t\right).$$

Using that $X_{t+1}^* = X_t^* + \alpha_{t+1}^* \Delta S_{t+1}$ and the measurability of the involved quantities, the above equality simplifies to $E_{\mathbb{P}}\left(e^{-\gamma\alpha_{t+1}^*\Delta S_{t+1}+h_{t+1}} \,|\mathcal{F}_t\right) = 1$, and we conclude using Lemma 4 once more.

We note how $U_t(x)$ is constructed from one period to the next: at each time t,

$$U_t(x) = U_{t-1}(x)\, e^{h_t}$$

$$= U_{t-1}(x) \left(\frac{q_t}{\mathbb{P}\left(A_t \middle| \mathcal{F}_{t-1}\right)}\right)^{q_t} \left(\frac{1-q_t}{\mathbb{P}\left(A_t^c \middle| \mathcal{F}_{t-1}\right)}\right)^{1-q_t}, \tag{17}$$

(cf. (13) and (11)). In other words, to construct $U_t(x)$, we need $U_{t-1}(x)$ and the single-period conditional risk neutral and historical probabilities q_t and $\mathbb{P}(A_t | \mathcal{F}_{t-1})$, measuring the movement of the traded asset for the next upcoming period only, conditionally on today's information. In other words, the forward process is constructed by progressive, forward in time, "single-period" model updates. Moreover, the forward performance process incorporates the market information from initial time 0 up to current time t, "path-by-path", as the term $H_{0,t} = e^{\Sigma_{i=1}^{t} h_i}$ indicates. Thus, $U_t(x)$ evolves in perfect alignment with the market, forward in time.

This is not the case in the classical expected utility framework. For a trading horizon, say $[0, T]$, the classical value process is of the form $V_{t,T}(x) = -e^{-\gamma x + \mathcal{H}_{t,T}}$, with $\mathcal{H}_{t,T}$ being the aggregate minimal entropy conditionally on $\mathcal{F}_t$, till the end of the investment horizon $[t, T]$ (see, for example, [21]). Therefore, for any time $t \in [0, T]$, its construction requires the model specification for the entire remaining investment time $[t, T]$, and incorporates the market information in a much coarser manner, through the term $\mathcal{H}_{t,T}$ associated with the average aggregate relative entropy from t to T, conditionally on $\mathcal{F}_t$.

3 Forward Exponential Indifference Valuation

In this section, we recall the notion of the writer's forward exponential indifference price and provide an iterative algorithm for its construction. Such prices were first introduced in [12] (see, also, [11]) for European claims in a single period model. They were subsequently studied in diffusion models with stochastic volatility in [13], and for American-type claims in [7].

Herein, we consider a generic claim, written on *both* the traded stock and the non-traded factor, say at time t_0, taken for simplicity to be $t_0 = 0$. The claim matures at $t > 0$ yielding payoff C_t, represented as an $\mathcal{F}_t$-mble random variable.

For convenience, we eliminate the "exponential" terminology and also occasionally rewrite some quantities for the reader's convenience.

Definition 5 Consider a claim, written at time $t_0 = 0$ and yielding at $t > 0$ payoff $C_t \in \mathcal{F}_t$. Let U_s, $s = 0, 1, \ldots, t$, be the forward performance process given by

$$U_s(x) = -e^{-\gamma x + \Sigma_{i=1}^{s} h_i}$$

and h as in Definition 1 (cf. (13) and (11)).

For $s = 0, 1, \ldots, t-1$, the *writer's forward indifference price* is defined as the amount $\nu_s(C_t) \in \mathcal{F}_s$ such that, for all wealth levels $x \in \mathbb{R}$,

$$U_s(x) = \sup_{\alpha_{s+1},\ldots,\alpha_t} E_{\mathbb{P}}\left(U_t\left(x + \nu_s(C_t) + \Sigma_{i=s+1}^{t} \alpha_i \Delta S_i - C_t\right) \middle| \mathcal{F}_s\right), \quad (18)$$

with $\alpha_i \in \mathcal{F}_{i-1}$, $i = s+1, \ldots, t$, and $\nu_t(C_t) = C_t$.

Similarly to the classical setting, the above indifference pricing condition reflects the indifference of the writer between two scenaria: start at s with wealth x and trade optimally till t without taking the claim in consideration, or start at s with wealth x and also accept the liability $\nu_s(C_t)$, then trade optimally (with initial wealth $x+\nu_s(C_t)$) till t and also fulfill the liability C_t, at time t.

For the reader's convenience, we start with the construction of the indifference price $\nu_{t-1}(C_t)$, just one period before maturity. Its form will motivate the upcoming choices of the pricing functionals as well as the specification of the emerging pricing measure, for all previous times.

Lemma 6 *At time $t-1$, the indifference price $\nu_{t-1}(C_t)$ is given by*

$$\nu_{t-1}(C_t) = q_t \frac{1}{\gamma} \ln\left(\frac{E_{\mathbb{P}}\left(e^{\gamma C_t}\mathbf{1}_{A_t}\middle| \mathcal{F}_{t-1}\right)}{\mathbb{P}\left(A_t\middle| \mathcal{F}_{t-1}\right)}\right) + (1-q_t)\frac{1}{\gamma}\ln\frac{E_{\mathbb{P}}\left(e^{\gamma C_t}\mathbf{1}_{A_t^c}\middle| \mathcal{F}_{t-1}\right)}{\mathbb{P}\left(A_t^c\middle| \mathcal{F}_{t-1}\right)}, \tag{19}$$

with q_t and A_t as in (10) and (6).

Proof We need to show that for $x\in\mathbb{R}$,

$$U_{t-1}(x) = \sup_{\alpha_t\in\mathcal{F}_{t-1}} E_{\mathbb{P}}\left(-e^{-\gamma(x+\nu_{t-1}(C_t)+\alpha_t\Delta S_t - C_t)+\Sigma_{i=1}^{t} h_i}\middle| \mathcal{F}_{t-1}\right),$$

with $\nu_{t-1}(C_t)$ as in (19). Using (13) and the measurability of the involved quantities, the above reduces to showing

$$\sup_{\alpha_t\in\mathcal{F}_{t-1}} E_{\mathbb{P}}\left(-e^{-\gamma(\nu_{t-1}(C_t)+\alpha_t\Delta S_t - C_t)+h_t}\middle| \mathcal{F}_{t-1}\right) = 1. \tag{20}$$

We have,

$$E_{\mathbb{P}}\left(-e^{-\gamma(\alpha_t\Delta S_t - C_t)}\middle| \mathcal{F}_{t-1}\right)$$

$$= E_{\mathbb{P}}\left(-e^{-\gamma\alpha_t S_{t-1}(\xi_t^u-1)}e^{\gamma C_t}\mathbf{1}_{A_t}\middle| \mathcal{F}_{t-1}\right) + E_{\mathbb{P}}\left(-e^{-\gamma\alpha_t S_{t-1}(\xi_t^d-1)}e^{\gamma Z}\mathbf{1}_{A_t^c}\middle| \mathcal{F}_{t-1}\right)$$

$$= -\left(e^{-\gamma\alpha_t S_{t-1}(\xi_t^u-1)}E_{\mathbb{P}}\left(e^{\gamma C_t}\mathbf{1}_{A_t}\middle| \mathcal{F}_{t-1}\right) + e^{-\gamma\alpha_t S_{t-1}(\xi_t^d-1)}E_{\mathbb{P}}\left(e^{\gamma C_t}\mathbf{1}_{A_t^c}\middle| \mathcal{F}_{t-1}\right)\right)$$

$$= -\left(e^{-\gamma\alpha_t S_{t-1}(\xi_t^u-1)}Z_{t-1}^1 + e^{-\gamma\alpha_t S_{t-1}(\xi_t^d-1)}Z_{t-1}^2\right),$$

with the random variables Z_{t-1}^1, Z_{t-1}^2 defined as

$$Z_{t-1}^1 := E_{\mathbb{P}}\left(e^{\gamma C_t}\mathbf{1}_{A_t}\middle| \mathcal{F}_{t-1}\right) \quad\text{and}\quad Z_{t-1}^2 := E_{\mathbb{P}}\left(e^{\gamma C_t}\mathbf{1}_{A_t^c}\middle| \mathcal{F}_{t-1}\right). \tag{21}$$

The optimum above occurs at

$$\alpha_t^{*,C_t} = \frac{1}{\gamma S_{t-1}\left(\xi_t^u-\xi_t^d\right)}\ln\frac{\left(\xi_t^u-1\right)Z_{t-1}^1}{\left(1-\xi_t^d\right)Z_{t-1}^2} = \frac{1}{\gamma S_{t-1}\left(\xi_t^u-\xi_t^d\right)}\ln\frac{(1-q_t)\,Z_{t-1}^1}{q_t Z_{t-1}^2}.$$

In turn,

$$E_{\mathbb{P}}\left(-e^{-\gamma\left(\alpha_t^{*,C_t}\Delta S_t - C_t\right)}\middle|\,\mathcal{F}_{t-1}\right)$$

$$= -\left(\left(\frac{1-q_t}{q_t}\frac{Z_{t-1}^1}{Z_{t-1}^2}\right)^{-(1-q_t)} Z_{t-1}^1 + \left(\frac{1-q_t}{q_t}\frac{Z_{t-1}^1}{Z_{t-1}^2}\right)^{q_t} Z_{t-1}^2\right)$$

$$= -\left(\frac{Z_{t-1}^1}{q_t}\right)^{q_t}\left(\frac{Z_{t-1}^2}{1-q_t}\right)^{1-q_t}.$$

Therefore,

$$E_{\mathbb{P}}\left(-e^{-\gamma\left(\alpha_t^{*,C_t}\Delta S_t - C_t\right)}\middle|\,\mathcal{F}_{t-1}\right) = -\exp\left(\ln\left(\left(\frac{Z_{t-1}^1}{q_t}\right)^{q_t}\left(\frac{Z_{t-1}^2}{1-q_t}\right)^{1-q_t}\right)\right)$$

$$= -\exp\left(q_t\ln\frac{Z_{t-1}^1}{q_t} + (1-q_t)\ln\frac{Z_{t-1}^2}{1-q_t}\right).$$

Next, observe that

$$q_t\ln\frac{Z_{t-1}^1}{q_t} + (1-q_t)\ln\frac{Z_{t-1}^2}{1-q_t}$$

$$= q_t\ln\frac{E_{\mathbb{P}}\left(e^{\gamma C_t}\mathbf{1}_{A_t}\middle|\,\mathcal{F}_{t-1}\right)}{q_t} + (1-q_t)\ln\frac{E_{\mathbb{P}}\left(e^{\gamma C_t}\mathbf{1}_{A_t^c}\middle|\,\mathcal{F}_{t-1}\right)}{1-q_t}$$

$$= q_t\ln\left(\frac{E_{\mathbb{P}}\left(e^{\gamma C_t}\mathbf{1}_{A_t}\middle|\,\mathcal{F}_{t-1}\right)}{\mathbb{P}\left(A_t\middle|\,\mathcal{F}_{t-1}\right)}\right) + (1-q_t)\ln\frac{E_{\mathbb{P}}\left(e^{\gamma C_t}\mathbf{1}_{A_t^c}\middle|\,\mathcal{F}_{t-1}\right)}{\mathbb{P}\left(A_t^c\middle|\,\mathcal{F}_{t-1}\right)}$$

$$-\left(q_t\ln\frac{\mathbb{P}\left(A_t\middle|\,\mathcal{F}_{t-1}\right)}{q_t} + (1-q_t)\ln\frac{\mathbb{P}\left(A_t^c\middle|\,\mathcal{F}_{t-1}\right)}{1-q_t}\right)$$

$$= \gamma\,\nu_{t-1}(C_t) - h_t.$$

Therefore,

$$\sup_{\alpha_t\in\mathcal{F}_{t-1}} E_{\mathbb{P}}\left(-e^{-\gamma\left(\alpha_t^{*,C_t}\Delta S_t - C_t\right)}\middle|\,\mathcal{F}_{t-1}\right) = -e^{\gamma\,\nu_{t-1}(C_t) - h_t},$$

and (20) follows. ■

Next, we make the following key observations. First, let us define the random variable

$$\mathcal{C}^{(t-1,t)}(C_t) := \frac{1}{\gamma}\ln\frac{E_{\mathbb{P}}\left(e^{\gamma C_t}\middle|\mathcal{F}_{t-1}\right)}{\mathbb{P}(A_t|\mathcal{F}_{t-1})}\mathbf{1}_{A_t} + \frac{1}{\gamma}\ln\frac{E_{\mathbb{P}}\left(e^{\gamma C_t}\middle|\mathcal{F}_{t-1}\right)}{\mathbb{P}(A_t^c|\mathcal{F}_{t-1})}\mathbf{1}_{A_t^c}, \tag{22}$$

and observe that $\mathcal{C}^{(t-1,t)}(C_t) \in \mathcal{F}_t^S$. In particular, it can be expressed as

$$\mathcal{C}^{(t-1,t)}(C_t) = \frac{1}{\gamma}\ln E_{\mathbb{P}}\left(e^{\gamma C_t}\middle|\mathcal{F}_{t-1}\vee\mathcal{F}_t^S\right). \tag{23}$$

In turn, observe that (19) yields that the indifference price is the *conditional expectation* of $\mathcal{C}^{(t-1,t)}(C_t)$ under *any* martingale measure, namely, for all $\mathbb{Q}\in\mathcal{Q}_t$,

$$\nu_{t-1}(C_t) = E_{\mathbb{Q}}\left(\mathcal{C}^{(t-1,t)}(C_t)\middle|\mathcal{F}_{t-1}\right). \tag{24}$$

What the above tells us is that the indifference price $\nu_{t-1}(C_t)$ is constructed via a two-step pricing procedure. In the first step, the claim's payoff C_t is "*distorted*", conditionally on $\mathcal{F}_{t-1}\vee\mathcal{F}_t^S$, and the intermediate payoff $\mathcal{C}^{(t-1,t)}(C_t)$ is created. This payoff is nonlinear and $\mathcal{F}_t^S$-mble. In the second step, the indifference price is produced as the *arbitrage-free* price of this intermediate payoff $\mathcal{C}^{(t-1,t)}(C_t)$.

Note that if $C_t\in\mathcal{F}_t^S$, then, $\mathcal{C}^{(t-1,t)}(C_t) = C_t$ and, naturally, $\nu_{t-1}(C_t) = E_{\mathbb{Q}}(C_t|\mathcal{F}_{t-1})$. In general, the price at t − 1 can be represented as the non-linear expression

$$\nu_{t-1}(C_t) = E_{\mathbb{Q}}\left(\frac{1}{\gamma}\ln E_{\mathbb{P}}\left(e^{\gamma C_t}\middle|\mathcal{F}_{t-1}\vee\mathcal{F}_t^S\right)\middle|\mathcal{F}_{t-1}\right),$$

which involves an *inner non-linear* expression under the *historical* measure, and an *outer conditional expectation* under any *martingale* measure.

Next, we pose the question whether we can actually express the price $\nu_{t-1}(C_t)$ as

$$\nu_{t-1}(C_t) = \mathcal{E}_{\mathbb{Q}^*}^{(t-1,t)}(C_t),$$

for an appropriate *indifference pricing non-linear functional* and a *specific* martingale measure (not necessarily unique) $\mathbb{Q}^*\in\mathcal{Q}_t$. This will provide an intuitively pleasing non-linear analogue of forward indifference prices to their arbitrage-free counterparts.

To this end, observe that the values of the payoff $\mathcal{C}^{(t-1,t)}(C_t)$ suggest that we should seek a martingale measure $\mathbb{Q}^*\in\mathcal{Q}_t$ such that

$$\frac{E_{\mathbb{P}}\left(e^{\gamma C_t}\mathbf{1}_{A_t}\middle|\mathcal{F}_{t-1}\right)}{\mathbb{P}(A_t|\mathcal{F}_{t-1})} = \frac{E_{\mathbb{Q}^*}\left(e^{\gamma C_t}\mathbf{1}_{A_t}\middle|\mathcal{F}_{t-1}\right)}{\mathbb{Q}^*(A_t|\mathcal{F}_{t-1})}$$

and

$$\frac{E_{\mathbb{P}}\left(e^{\gamma C_t}\mathbf{1}_{A_t^c}\middle|\mathcal{F}_{t-1}\right)}{\mathbb{P}(A_t^c|\mathcal{F}_{t-1})} = \frac{E_{\mathbb{Q}^*}\left(e^{\gamma C_t}\mathbf{1}_{A_t^c}\middle|\mathcal{F}_{t-1}\right)}{\mathbb{Q}^*(A_t^c|\mathcal{F}_{t-1})}.$$

We, then, see that it suffices for the candidate measure $\mathbb{Q}^*$ to satisfy

$$\frac{\mathbb{Q}^*\left(A_t B_t \middle| \mathcal{F}_{t-1}\right)}{q_t} = \frac{\mathbb{P}\left(A_t B_t \middle| \mathcal{F}_{t-1}\right)}{\mathbb{P}\left(A_t \middle| \mathcal{F}_{t-1}\right)}, \quad \frac{\mathbb{Q}^*\left(A_t B_t^c \middle| \mathcal{F}_{t-1}\right)}{q_t} = \frac{\mathbb{P}\left(A_t B_t^c \middle| \mathcal{F}_{t-1}\right)}{\mathbb{P}\left(A_t \middle| \mathcal{F}_{t-1}\right)} \tag{25}$$

$$\frac{\mathbb{Q}^*\left(A_t^c B_t \middle| \mathcal{F}_{t-1}\right)}{1-q_t} = \frac{\mathbb{P}\left(A_t^c B_t \middle| \mathcal{F}_{t-1}\right)}{\mathbb{P}\left(A_t^c \middle| \mathcal{F}_{t-1}\right)}, \quad \frac{\mathbb{Q}^*\left(A_t^c B_t^c \middle| \mathcal{F}_{t-1}\right)}{1-q_t} = \frac{\mathbb{P}\left(A_t^c B_t^c \middle| \mathcal{F}_{t-1}\right)}{\mathbb{P}\left(A_t^c \middle| \mathcal{F}_{t-1}\right)}. \tag{26}$$

In turn, we observe that, under such $\mathbb{Q}^*$, the intermediate payoff $\mathcal{C}^{(t-1,t)}(C_t)$ retains its form, in that under *both* $\mathbb{P}$ and $\mathbb{Q}^*$,

$$\mathcal{C}^{(t-1,t)}(C_t) = \frac{1}{\gamma}\ln E_{\mathbb{Q}^*}\left(e^{\gamma C_t}\middle| \mathcal{F}_{t-1} \vee \mathcal{F}_t^S\right) = \frac{1}{\gamma}\ln E_{\mathbb{P}}\left(e^{\gamma C_t}\middle| \mathcal{F}_{t-1} \vee \mathcal{F}_t^S\right).$$

We then see that if we define the non-linear pricing functional

$$\mathcal{E}_{\mathbb{Q}^*}^{(t-1,t)}(Z) := E_{\mathbb{Q}^*}\left(\mathcal{C}^{(t-1,t)}(Z)\middle| \mathcal{F}_t\right) = E_{\mathbb{Q}^*}\left(\frac{1}{\gamma}\ln E_{\mathbb{Q}^*}\left(e^{\gamma Z}\middle| \mathcal{F}_{t-1} \vee \mathcal{F}_t^S\right)\middle| \mathcal{F}_{t-1}\right), \tag{27}$$

for a generic random variable $Z \in \mathcal{F}_t$, we can actually express the indifference price at $t-1$ in the desired concise form

$$\nu_{t-1}(C_t) = \mathcal{E}_{\mathbb{Q}^*}^{(t-1,t)}(C_t). \tag{28}$$

Observe that the measure $\mathbb{Q}^*$ is used in *both* expectations in (27) since the outer one is applied to an $\mathcal{F}_t^S$−mble random variable.

Notice that despite the fact that both forward random functionals U_t and U_{t-1}, entering in the derivation of $\nu_{t-1}(C_t)$, are path-dependent through the terms $\Sigma_{i=1}^t h_i$ and $\Sigma_{i=1}^{t-1} h_i$ appearing in their exponents (cf. (13)), the indifference price $\nu_{t-1}(C_t)$ takes a substantially simplified "single-period" form.

Furthermore, the involved conditional probabilities of the emerging pricing measure $\mathbb{Q}^*$ have also "single-period" dependence, since they are determined *exclusively* by q_t and $\mathbb{P}\left(A_t \middle| \mathcal{F}_{t-1}\right)$ (cf. (25), (26)).

A natural question arises, given the nonlinearity of $\mathcal{C}^{(t-1,t)}(C_t)$, whether it can be interpreted as a certainty equivalent of some form. In Sect. 4, we show that this is indeed the case. Specifically, we establish that

$$\mathcal{C}^{(t-1,t)}(C_t) = -U_t^{(-1)}\left(E_{\mathbb{P}}\left(U_t(-C_t)\middle|\mathcal{F}_{t-1} \vee \mathcal{F}_t^S\right)\right).$$

This also yields a natural interpretation of $\nu_{t-1}(C_t)$ as the arbitrage-free price of a payoff with certainty equivalent characteristics.

In the next section, we show how to extend the above constructions and interpretations to all previous periods $t-2, t-3, \ldots,$ define analogous to (23), (24) pricing functionals and specify a pricing measure from the martingale ones satisfying similar to (25), (26) properties.

4 The (Writer's) Forward Indifference Pricing Algorithm

Motivated by the form of the indifference price $\nu_{t-1}(C_t)$ in (28), we seek an analogous price representation,

$$\nu_s(C_t) = \mathcal{E}_{\mathbb{Q}^*}^{(s,t)}(C_t),$$

for an appropriate chosen multi-period valuation functional $\mathcal{E}_{\mathbb{Q}^*}^{(s,t)}$ and a pricing measure $\mathbb{Q}^*$, for $s = 0, 1, \ldots, t$. As for the case $s = t-1$, the main challenge is how to incorporate the path dependence of the forward functionals U_s and U_t (appearing in Definition 5) coming from the terms $\Sigma_{i=1}^{s} h_i$ and $\Sigma_{i=1}^{t} h_i$ (cf. (13)).

We propose such a multi-period pricing functional of an iterative form, $\mathcal{E}_{\mathbb{Q}^*}^{(s,t)}(\cdot) = \mathcal{E}_{\mathbb{Q}^*}^{(s,s+1)}\left(\mathcal{E}_{\mathbb{Q}^*}^{(s+1,s+2)} \ldots \mathcal{E}_{\mathbb{Q}^*}^{(t-1,t)}(\cdot)\right)$, with the single-period pricing functionals resembling (27). We also show that the appropriate pricing measure $\mathbb{Q}^*$ is a martingale one that has similar to (25), (26) local properties. Furthermore, we prove that it actually minimizes the reverse relative entropy in $[0, t]$ over all martingale measures defined on $\mathcal{F}_t$.

4.1 *The Forward Indifference Pricing Measure, and the Single- and Multi-step Valuation Functionals*

We start with some introductory results and notation. For $t = 1, 2, \ldots,$ recall that $\mathcal{Q}_t$ is the set of equivalent martingale measures and $\mathbb{Q}$ its generic element.

For $s = 1, 2, \ldots, t$, we have

$$\mathbb{Q}\left(\xi_s \in \left\{\xi_s^d, \xi_s^u\right\}, \eta_s \in \left\{\eta_s^d, \eta_s^u\right\} | \mathcal{F}_{s-1}\right) = \mathbb{Q}\left(A_s B_s | \mathcal{F}_{s-1}\right) \mathbf{1}_{A_s B_s}$$

$$+\mathbb{Q}\left(A_s B_s^c \middle| \mathcal{F}_{s-1}\right) \mathbf{1}_{A_s B_s^c} + \mathbb{Q}\left(A_s^c B_s \middle| \mathcal{F}_{s-1}\right) \mathbf{1}_{A_s^c B_s} + \mathbb{Q}\left(A_s^c B_s^c \middle| \mathcal{F}_{s-1}\right) \mathbf{1}_{A_s^c B_s^c},$$

and, similarly, for the historical measure,

$$\mathbb{P}\left(\xi_s \in \left\{\xi_s^d, \xi_s^u\right\}, \eta_s \in \left\{\eta_s^d, \eta_s^u\right\} | \mathcal{F}_{s-1}\right) = \mathbb{P}\left(A_s B_s | \mathcal{F}_{s-1}\right) \mathbf{1}_{A_s B_s}$$

$$+\mathbb{P}\left(A_s B_s^c \middle| \mathcal{F}_{s-1}\right) \mathbf{1}_{A_s B_s^c} + \mathbb{P}\left(A_s^c B_s \middle| \mathcal{F}_{s-1}\right) \mathbf{1}_{A_s^c B_s} + \mathbb{P}\left(A_s^c B_s^c \middle| \mathcal{F}_{s-1}\right) \mathbf{1}_{A_s^c B_s^c},$$

with ξ_s, η_s, A_s, B_s as in (4), (5) and (6).

We will be using the *condensed* notation for the conditional distributions

$$\mathbb{Q}\left(\xi_s, \eta_s | \mathcal{F}_{s-1}\right) \triangleq \mathbb{Q}\left(\xi_s \in \left\{\xi_s^d, \xi_s^u\right\}, \eta_s \in \left\{\eta_s^d, \eta_s^u\right\} | \mathcal{F}_{s-1}\right) \tag{29}$$

and

$$\mathbb{P}\left(\xi_s, \eta_s | \mathcal{F}_{s-1}\right) \triangleq \mathbb{P}\left(\xi_s \in \left\{\xi_s^d, \xi_s^u\right\}, \eta_s \in \left\{\eta_s^d, \eta_s^u\right\} | \mathcal{F}_{s-1}\right). \tag{30}$$

Next, we seek a martingale measure that *minimizes,* for $s = 1, 2, \ldots, t$, the conditional expectation

$$- E_{\mathbb{P}} \left(\ln \frac{\mathbb{Q}(\xi_s, \eta_s | \mathcal{F}_{s-1})}{\mathbb{P}(\xi_s, \eta_s | \mathcal{F}_{s-1})} \middle| \mathcal{F}_{s-1} \right). \tag{31}$$

For this, we need to find the minimizers of

$$- \left(\mathbb{P}(A_s B_s | \mathcal{F}_{s-1}) \ln \frac{\mathbb{Q}(A_s, B_s | \mathcal{F}_{s-1})}{\mathbb{P}(A_s, B_s | \mathcal{F}_{s-1})} + \mathbb{P}(A_s B_s^c | \mathcal{F}_{s-1}) \ln \frac{\mathbb{Q}(A_s B_s^c | \mathcal{F}_{s-1})}{\mathbb{P}(A_s B_s^c | \mathcal{F}_{s-1})} \right.$$

$$\left. + \mathbb{P}(A_s^c B_s | \mathcal{F}_{s-1}) \ln \frac{\mathbb{Q}(A_s^c B_s | \mathcal{F}_{s-1})}{\mathbb{P}(A_s^c B_s | \mathcal{F}_{s-1})} + \mathbb{P}(A_s^c B_s^c | \mathcal{F}_{s-1}) \ln \frac{\mathbb{Q}(A_s^c B_s^c | \mathcal{F}_{s-1})}{\mathbb{P}(A_s^c B_s^c | \mathcal{F}_{s-1})} \right).$$

Direct calculations yield the above quantity is minimized if one chooses

$$\frac{\mathbb{Q}^*(A_s B_s | \mathcal{F}_{s-1})}{q_s} = \frac{\mathbb{P}(A_s B_s | \mathcal{F}_{s-1})}{\mathbb{P}(A_s | \mathcal{F}_{s-1})}, \quad \frac{\mathbb{Q}^*(A_s B_s^c | \mathcal{F}_{s-1})}{q_s} = \frac{\mathbb{P}(A_s B_s^c | \mathcal{F}_{s-1})}{\mathbb{P}(A_s | \mathcal{F}_{s-1})}, \tag{32}$$

$$\frac{\mathbb{Q}^*(A_s^c B_s | \mathcal{F}_{s-1})}{1 - q_s} = \frac{\mathbb{P}(A_s^c B_s^c | \mathcal{F}_{s-1})}{\mathbb{P}(A_s^c | \mathcal{F}_{s-1})}, \quad \frac{\mathbb{Q}^*(A_s^c B_s^c | \mathcal{F}_{s-1})}{1 - q_s} = \frac{\mathbb{P}(A_s^c B_s^c | \mathcal{F}_{s-1})}{\mathbb{P}(A_s^c | \mathcal{F}_{s-1})}. \tag{33}$$

Indeed, observe that the function $f(z) = -\left(\left(\ln \frac{z}{\alpha}\right)\alpha + \left(\ln \frac{c-z}{\beta}\right)\beta\right)$, with $\alpha, \beta, c \in (0, 1)$ achieves for $z \in [0, c]$ a minimum at the point $z^* = \frac{\alpha}{\alpha+\beta} c$. Applying this for the triplets $(\alpha, \beta, c) = \left(\mathbb{P}(A_s B_s | \mathcal{F}_{s-1}), \mathbb{P}(A_s B_s^c | \mathcal{F}_{s-1}), q_s\right)$ and $(\alpha, \beta, c) = \left(\mathbb{P}(A_s^c B_s | \mathcal{F}_{s-1}), \mathbb{P}(A_s^c B_s^c | \mathcal{F}_{s-1}), 1 - q_s\right)$, respectively, we conclude.

We then consider a martingale measure $\mathbb{Q}^*$, defined on $\mathcal{F}_t$, satisfying, for $t = 1, 2, \ldots$, the conditional properties (32) and (33), and we claim that it is well defined. Indeed, for $t = 1, 2, \ldots$, we have

$$\mathbb{Q}^*\left(\xi_1 \in \{\xi_1^d, \xi_1^u\}, \ldots, \xi_t \in \{\xi_t^d, \xi_t^u\}, \eta_1 \in \{\eta_1^d, \eta_1^u\}, \ldots, \eta_t \in \{\eta_t^d, \eta_t^u\}\right) \tag{34}$$

$$= \prod_{s=1}^{t} \mathbb{Q}^*\left(\xi_s \in \{\xi_s^d, \xi_s^u\}, \eta_s \{\eta_s^d, \eta_s^u\} \middle| \mathcal{F}_{s-1}\right) = \prod_{s=1}^{t} \mathbb{Q}^*(\xi_s, \eta_s | \mathcal{F}_{s-1}),$$

with each term being well defined from (32) and (33).

Next, we derive the following characterization result in terms of the reverse relative entropy measure, in which we use the self-evident condensed expressions $\mathbb{Q}(\xi_1, \ldots, \xi_t, \eta_1, \ldots, \eta_t)$ and $\mathbb{P}(\xi_1, \ldots, \xi_t, \eta_1, \ldots, \eta_t)$ to denote the joint distributions of $(\xi_1, \ldots, \xi_t, \eta_1, \ldots, \eta_t)$ under $\mathbb{Q}$ and $\mathbb{P}$, respectively.

Proposition 7 *For $t = 1, 2, \ldots$, let $\mathcal{Q}_t$ be the set of equivalent martingale measures and $\mathbb{Q}^* \in \mathcal{Q}_t$ defined as in (34). Then, the measure $\mathbb{Q}^*$ minimizes the reverse relative* *entropy* $\mathcal{H}_t$, *defined as*

$$\mathcal{H}_t = -E_{\mathbb{P}}\left(\ln \frac{\mathbb{Q}\left(\xi_1, \ldots, \xi_t, \eta_1, \ldots, \eta_t\right)}{\mathbb{P}\left(\xi_1, \ldots, \xi_t, \eta_1, \ldots, \eta_t\right)}\right), \tag{35}$$

for $\mathbb{Q} \in \mathcal{Q}_t$.

Proof First, observe that (35) can be written as

$$\mathcal{H}_t = -\sum_{s=1}^{t} E_{\mathbb{P}}\left(\ln \frac{\mathbb{Q}\left(\xi_s, \eta_s \left|\mathcal{F}_{s-1}\right.\right)}{\mathbb{P}\left(\xi_s, \eta_s \left|\mathcal{F}_{s-1}\right.\right)}\right),$$

and, in turn,

$$\mathcal{H}_t = -\sum_{s=1}^{t} E_{\mathbb{P}}\left(E_{\mathbb{P}}\left(\ln \frac{\mathbb{Q}\left(\xi_s, \eta_s \left|\mathcal{F}_{s-1}\right.\right)}{\mathbb{P}\left(\xi_s, \eta_s \left|\mathcal{F}_{s-1}\right.\right)}\right) \middle| \mathcal{F}_{s-1}\right),$$

and we easily conclude. ■

The next result shows a key property of the measure $\mathbb{Q}^*$. It also provides equalities (37), (38) which will play a main role in the construction of the forward indifference prices. Its proof follows easily.

Proposition 8 *(i) The martingale measure* $\mathbb{Q}^*$ *defined as in (34) satisfies, for* $t = 1, 2, \ldots,$ *and* $s = 1, 2, \ldots, t$,

$$\mathbb{Q}^*\left(Y_s \left|\mathcal{F}_{s-1} \vee \mathcal{F}_s^S\right.\right) = \mathbb{P}\left(Y_s \left|\mathcal{F}_{s-1} \vee \mathcal{F}_s^S\right.\right), \tag{36}$$

with the stochastic factor Y_s *given in (5).*

(ii) Moreover, if Z *is an* $\mathcal{F}_s$*-mble random variable and* A_s *as in (6), we have*

$$\frac{E_{\mathbb{P}}\left(Z\mathbf{1}_{A_s} \left|\mathcal{F}_{s-1}\right.\right)}{\mathbb{P}\left(A_s \left|\mathcal{F}_{s-1}\right.\right)} = \frac{E_{\mathbb{Q}^*}\left(Z\mathbf{1}_{A_s} \left|\mathcal{F}_{s-1}\right.\right)}{\mathbb{Q}^*\left(A_s \left|\mathcal{F}_{s-1}\right.\right)} \tag{37}$$

and

$$\frac{E_{\mathbb{P}}\left(Z\mathbf{1}_{A_s^c} \left|\mathcal{F}_{s-1}\right.\right)}{\mathbb{P}\left(A_s^c \left|\mathcal{F}_{s-1}\right.\right)} = \frac{E_{\mathbb{Q}^*}\left(Z\mathbf{1}_{A_t^c} \left|\mathcal{F}_{s-1}\right.\right)}{\mathbb{Q}^*\left(A_s^c \left|\mathcal{F}_{s-1}\right.\right)}. \tag{38}$$

We introduce the following single- and multi-period forward pricing functionals.

Definition 9 For $t > 0$, let $\mathbb{Q}^*$ be the martingale measure as in (36) and, for $s = 0, 1, \ldots, t-1$, let Z be an $\mathcal{F}_{s+1}$ -mble random variable in $(\Omega, \mathcal{F}, \mathbb{P})$. We define

(i) the *single-step forward price functional*

$$\mathcal{E}_{\mathbb{Q}^*}^{(s,s+1)}(Z) = E_{\mathbb{Q}^*}\left(\frac{1}{\gamma}\ln E_{\mathbb{Q}^*}\left(e^{\gamma Z}\,|\mathcal{F}_s \vee \mathcal{F}_{s+1}^S\right)|\mathcal{F}_s\right) \tag{39}$$

and,

(ii) the *multi-step forward price functional*, $0 \leq s < s' \leq t$,

$$\mathcal{E}_{\mathbb{Q}^*}^{(s,s')}(Z) = \mathcal{E}_{\mathbb{Q}^*}^{(s,s+1)}(\mathcal{E}_{\mathbb{Q}^*}^{(s+1,s+2)}(\ldots\mathcal{E}_{\mathbb{Q}^*}^{(s'-1,s')}(Z))). \tag{40}$$

Remark 10 We *caution* the reader that, in general, for $s' > s+1$ and $Z \in \mathcal{F}_{s'}$,

$$\mathcal{E}_{\mathbb{Q}^*}^{(s,s')}(Z) \neq E_{\mathbb{Q}^*}\left(\frac{1}{\gamma}\ln E_{\mathbb{Q}^*}\left(e^{\gamma Z}\,|\mathcal{F}_s \vee \mathcal{F}_{s'}^S\right)|\mathcal{F}_s\right). \tag{41}$$

The reader familiar with existing indifference pricing algorithms (see, among others, [1, 10, 11, 23, 24]), might find the form of $\mathcal{E}_{\mathbb{Q}^*}^{(s,s+1)}$ and of $\mathcal{E}_{\mathbb{Q}^*}^{(s,s')}$ identical to the ones appearing in these references. This is *not*, however, the case. The results herein are, not only, derived for entirely different risk preference criteria but, also, for more general incomplete market environments, since the nested market model (bond and stock) is incomplete. Moreover, the involved measure is not the minimal entropy one but, rather, the minimal reverse entropy measure.

Another difference, as we show in Proposition 16, is that $\mathcal{E}_{\mathbb{Q}^*}^{(s,s+1)}$ provides an intuitively pleasing *direct analogue* of the arbitrage-free price of a conditional certainty equivalent, while in the classical exponential utility such analogy fails.

The following auxiliary result will be used repeatedly in the construction of the forward pricing algorithm.

Lemma 11 *Let $t > 0$, $s = 0, 1, \ldots, t-1$, and $\mathcal{E}_{\mathbb{Q}^*}^{(s,s+1)}$ be as in (39). Then, if Z is an $\mathcal{F}_{s+1}$-mble random variable,*

$$\sup_{\alpha_{s+1}\in\mathcal{F}_s} E_{\mathbb{P}}\left(-e^{-\gamma(\alpha_{s+1}\Delta S_{s+1}-Z)+h_{s+1}}\middle|\,\mathcal{F}_s\right) = -e^{\gamma\,\mathcal{E}_{\mathbb{Q}^*}^{(s,s+1)}(Z)},$$

with h_s as in (11).

Proof The proof follows by analogous arguments as the ones used to show Lemma 6. For this, we only highlight the main steps. We have

$$E_{\mathbb{P}}\left(-e^{-\gamma(\alpha_{s+1}\Delta S_{s+1}-Z)}\middle|\,\mathcal{F}_s\right) = -\left(e^{-\gamma\alpha_{s+1}S_s(\xi_{s+1}^u-1)}Z_s^1 + e^{-\gamma\alpha_{s+1}S_s(\xi_{s+1}^d-1)}Z_s^2\right),$$

with $Z_s^1 := E_{\mathbb{P}}\left(e^{\gamma Z}\mathbf{1}_{A_{s+1}}\middle|\,\mathcal{F}_s\right)$ and $Z_s^2 := E_{\mathbb{P}}\left(e^{\gamma Z}\mathbf{1}_{A_{s+1}^c}\middle|\,\mathcal{F}_s\right)$. The optimum occurs at the point

$$\alpha_{s+1}^{*,Z} = \frac{1}{\gamma\,S_s\left(\xi_{s+1}^u - \xi_{s+1}^d\right)}\ln\frac{(1-q_{s+1})\,Z_s^1}{q_{s+1}Z_s^2},$$

at which we have

$$E_{\mathbb{P}}\left(-e^{-\gamma(\alpha^*_{s+1}\Delta S_{s+1}-Z)}\Big|\,\mathcal{F}_s\right)=-\left(\frac{Z^1_s}{q_{s+1}}\right)^{q_{s+1}}\left(\frac{Z^2_s}{1-q_{s+1}}\right)^{1-q_{s+1}}$$

$$=-\exp\left(q_{s+1}\ln\frac{Z^1_s}{q_{s+1}}+(1-q_{s+1})\ln\frac{Z^2_s}{1-q_{s+1}}\right).$$

Working as in the proof of Lemma 6, we express the above quantity with respect to the $\mathbb{Q}^*$ measure,

$$q_{s+1}\ln\frac{Z^1_s}{q_{s+1}}+(1-q_{s+1})\ln\frac{Z^2_s}{1-q_{s+1}}$$

$$=q_{s+1}\ln\frac{E_{\mathbb{Q}^*}\left(e^{\gamma Z}\mathbf{1}_{A_{s+1}}\big|\,\mathcal{F}_s\right)}{\mathbb{Q}^*\left(A_{s+1}\big|\,\mathcal{F}_s\right)}+(1-q_{s+1})\ln\frac{E_{\mathbb{Q}^*}\left(e^{\gamma Z}\mathbf{1}_{A^c_{s+1}}\big|\,\mathcal{F}_s\right)}{\mathbb{Q}^*\left(A^c_{s+1}\big|\,\mathcal{F}_s\right)}-h_{s+1},\quad(42)$$

where, we used the definition of h_{s+1}, the measurability of Z and the second part of Proposition 4, and the definition of $\mathcal{E}^{(s,s+1)}_{\mathbb{Q}^*}$. We easily conclude. ■

We are now ready to present the forward indifference pricing algorithm.

Theorem 12 *Consider a claim, introduced at time $t_0=0$, yielding at time $t>0$, payoff $C_t\in\mathcal{F}_t$, and $\nu_s(C_t)$ be defined as in (18). Let, also, $\mathbb{Q}^*$ be as in (36), and $\mathcal{E}^{(s,s+1)}_{\mathbb{Q}^*}$ and $\mathcal{E}^{(s,s')}_{\mathbb{Q}^*}$ as in (39) and (40), respectively. The following statements hold:*

(i) The forward indifference price, $\nu_s(C_t)$, is given, for $s=0,1,\ldots,t-1$, by the iterative algorithm

$$\nu_t(C_t)=C_t,$$

$$\nu_s(C_t)=\mathcal{E}^{(s,s+1)}_{\mathbb{Q}^*}(\nu_{s+1}(C_t))\quad(43)$$

$$=E_{\mathbb{Q}^*}\left(\frac{1}{\gamma}\ln E_{\mathbb{Q}^*}\left(e^{\gamma\,\nu_{s+1}(C_t)}\,|\mathcal{F}_s\vee\mathcal{F}^S_{s+1}\right)|\mathcal{F}_s\right).$$

(ii) The forward indifference price process $\nu_s(C_t)$ is $\mathcal{F}_s$-mble and satisfies, for $s=0,1,\ldots,t-1$,

$$\nu_s(C_t)=\mathcal{E}^{(s,t)}_{\mathbb{Q}^*}(C_t).\quad(44)$$

(iii) The forward indifference price algorithm is consistent across time in that, for $0\le s\le s'<t$, the semigroup property

$$\nu_s(C_t)=\mathcal{E}^{(s,s')}_{\mathbb{Q}^*}(\mathcal{E}^{(s',t)}_{\mathbb{Q}^*}(C_t))\quad(45)$$

$$=\mathcal{E}^{(s,s')}_{\mathbb{Q}^*}(\nu_{s'}(C_t))=\nu_s(\mathcal{E}^{(s',t)}_{\mathbb{Q}^*}(C_t))$$

holds.

Proof Assertions (43) and (44) were proved in Lemma 6 for $s = t - 1$.

To show (43) for $s = t - 2$, we first observe that representation (13) yields with repeated use of Lemma 11 and (13), and of $\alpha_{t-1} \in \mathcal{F}_{t-2}, \alpha_t \in \mathcal{F}_{t-1}$,

$$\sup_{\alpha_{t-1},\alpha_t} E_{\mathbb{P}}\left(U_t(X_t - C_t)|\mathcal{F}_{t-2}\right) = \sup_{\alpha_{t-1}} E_{\mathbb{P}}\left(\sup_{\alpha_t} E_{\mathbb{P}}\left(U_t(X_t - C_t)|\mathcal{F}_{t-1}\right)|\mathcal{F}_{t-2}\right)$$

$$= \sup_{\alpha_{t-1}} E_{\mathbb{P}}\left(e^{-\gamma(x+\alpha_{t-1}\Delta S_{t-1})+\Sigma_{i=1}^{t-1}h_i} \sup_{\alpha_t} E_{\mathbb{P}}\left(-e^{-\gamma(\alpha_t\Delta S_t - C_t)+h_t}|\mathcal{F}_{t-1}\right)|\mathcal{F}_{t-2}\right)$$

$$= \sup_{\alpha_{t-1}} E_{\mathbb{P}}\left(e^{-\gamma(x+\alpha_{t-1}\Delta S_{t-1})+\Sigma_{i=1}^{t-1}h_i}\left(-e^{\gamma\mathcal{E}_{\mathbb{Q}^*}^{(t-1,t)}(C_t)}\right)|\mathcal{F}_{t-2}\right)$$

$$= e^{\Sigma_{i=1}^{t-2}h_i} \sup_{\alpha_{t-1}} E_{\mathbb{P}}\left(-e^{-\gamma(x+\alpha_{t-1}\Delta S_{t-1}-\mathcal{E}_{\mathbb{Q}^*}^{(t-1,t)}(C_t))+h_{t-1}}|\mathcal{F}_{t-2}\right)$$

$$= -e^{-\gamma\left(x-\mathcal{E}_{\mathbb{Q}^*}^{(t-2,t-1)}\left(\mathcal{E}_{\mathbb{Q}^*}^{(t-1,t)}(C_t)\right)\right)+\Sigma_{i=1}^{t-2}h_i} = U_{t-2}\left(x - \mathcal{E}_{\mathbb{Q}^*}^{(t-2,t-1)}\left(\mathcal{E}_{\mathbb{Q}^*}^{(t-1,t)}(C_t)\right)\right)$$

$$= U_{t-2}\left(x - \mathcal{E}_{\mathbb{Q}^*}^{(t-2,t)}(C_t)\right).$$

The rest of the assertions follows along similar albeit tedious arguments. ∎

We conclude with the case of multiple claims. Before we present the general result, let us consider the simple case of two claims, written at $t = 0$ and maturing at $t - 1$ and t, yielding payoffs $C_{t-1} \in \mathcal{F}_{t-1}$ and $C_t \in \mathcal{F}_t$, respectively.

Then, since $C_{t-1} \in \mathcal{F}_t$, we have

$$\nu_{t-1}(C_{t-1} + C_t) = \mathcal{E}_{\mathbb{Q}^*}^{(t-1,t)}(C_{t-1} + C_t)$$

$$= E_{\mathbb{Q}^*}\left(\frac{1}{\gamma}\ln E_{\mathbb{Q}^*}\left(e^{\gamma(C_{t-1}+C_t)}|\mathcal{F}_{t-1} \vee \mathcal{F}_t^S\right)|\mathcal{F}_{t-1}\right) = C_{t-1} + \nu_{t-1}(C_t).$$

Trivially, one may view $C_{t-1} + \nu_{t-1}(C_t)$ as a new claim maturing at time $t - 1$, and, in turn, price it iteratively for $s = t - 2, t - 3, \ldots, 0$. The assumption that all claims are written at time $t = 0$ can be easily removed. Note, however, what in both cases (i.e. common or varying inscription times), the market model needs to be specified at time 0 till the longest a priori known maturity.

Corollary 13 *Let $t = 1, 2, \ldots$ and $s = 0, 1, \ldots, t - 1$. Consider claims $C_s, \ldots, C_j, \ldots C_t$ with $C_j \in \mathcal{F}_j$, $j = s, \ldots, t$, written at $t = 0$. The forward indifference price, $\nu_s(\Sigma_{j=s}^t C_j)$, is given, for $s = 0, 1, \ldots, t - 1$, by the iterative algorithm*

$$\nu_t(C_t) = C_t,$$

$$\nu_s(\Sigma_{j=s}^t C_j) = C_s + \mathcal{E}_{\mathbb{Q}^*}^{(s,s+1)}(C_{s+1} + \nu_{s+1}(\Sigma_{j=s+2}^t C_j))$$

$$= C_s + E_{\mathbb{Q}^*}\left(\frac{1}{\gamma}\ln E_{\mathbb{Q}^*}\left(e^{\gamma\left(C_{s+1}+\nu_{s+1}(\Sigma_{j=s+2}^t C_j)\right)}|\mathcal{F}_s \vee \mathcal{F}_{s+1}^S\right)|\mathcal{F}_s\right).$$

(ii) The forward indifference price process $\nu_s(\Sigma_{j=s}^t C_j)$ *is* $\mathcal{F}_s$*-mble and satisfies, for* $s = 0, 1, \ldots, t,$

$$\nu_s(\Sigma_{j=s}^t C_j)$$

$$= C_s + \mathcal{E}_{\mathbb{Q}^*}^{(s,s+1)}\left(C_{s+1} + \mathcal{E}_{\mathbb{Q}^*}^{(s+1,s+2)}\left(C_{s+2} + \ldots \mathcal{E}_{\mathbb{Q}^*}^{(t-1,t)}(C_t)\right)\right).$$

An interesting case arises when there is *no* a priori knowledge at initial time 0 about all incoming claims and their maturities.

For example, consider the case that a single claim, C_t, is written at time 0 that matures at time t, but it is not known whether additional claims will arrive. Then, at time $s \in (0, t]$, a new claim, say $\tilde{C}_{t'}$, arrives with expiration t'. If $t' < t$, then its valuation is easily accommodated by the above Corollary.

If, however, $t' > t$, then one first needs to specify at time s the market model for the period $(t, t']$, and, in turn, employ the forward exponential criterion for times $t+1, t+2, \ldots, t'$, and price by indifference. This can be readily done, however, since the forward process can be defined for all times, sequentially forward in time.

Note that in the traditional expected utility framework, such flexibility does *not* exist. Indeed, once the investment horizon $[0, t]$ is prespecified at time 0, only claims maturing at times up to t can be priced. Any claim arriving later and with maturity beyond t cannot be priced, because the expected utility problem cannot be extended beyond t unless time-consistency is violated.

5 Properties of Forward Exponential Indifference Prices

The forward indifference pries is constructed via the *optimal* behavior of the investor with and without the claim in consideration. As such, it incorporates and reflects the individual risk preferences. Due to the exponential choice, it is independent of the investor's wealth.

- **Time consistency**
  The forward pricing operator $\mathcal{E}_{\mathbb{Q}^*}^{(s,t)}$ is time consistent, in that the price at any intermediate time, say s, can be thought as the price of a claim equal to the corresponding indifference price at a future time s', namely,

$$\nu_s(C_t) = \nu_s(\nu_{s'}(C_t)), \quad 0 \le s \le s' \le t.$$

  This property is reflected in (45).

- **Scaling and monotonicity properties**
The following properties of $\mathcal{E}_{\mathbb{Q}^*}^{(s,s+1)}(Z)$, $Z, Z' \in \mathcal{F}_{s+1}$ hold:

(i) The mapping $\gamma \to \mathcal{E}_{\mathbb{Q}^*}^{(s,s+1)}(Z;\gamma)$ is increasing and continuous, and

$$\lim_{\gamma\to 0^+} \mathcal{E}_{\mathbb{Q}^*}^{(s,s+1)}(Z;\gamma) = E_{\mathbb{Q}^*}(Z) \text{ and } \lim_{\gamma\to\infty} \mathcal{E}_{\mathbb{Q}^*}^{(s,s+1)}(Z;\gamma) = E_{\mathbb{Q}^*} \|Z\|_{L^{\infty}_{\mathbb{Q}^*\{\cdot|S_s\}}}.$$

Moreover,

$$\lim_{\gamma\to 0} \frac{\partial}{\partial\gamma} \mathcal{E}_{\mathbb{Q}^*}^{(s,s+1)}(Z;\gamma) = \frac{1}{2} E_{\mathbb{Q}^*}\left(Var_{\mathbb{Q}^*}(Z\,|S_s)\right),$$

and, thus,

$$\mathcal{E}_{\mathbb{Q}^*}^{(s,s+1)}(Z;\gamma) = E_{\mathbb{Q}^*}(Z) + \frac{1}{2}\gamma\, E_{\mathbb{Q}^*}\left(Var_{\mathbb{Q}^*}(Z\,|S_s)\right) + o(\gamma).$$

The assertions follow by routine arguments, and their proof is omitted.
(iii) For $\alpha \in (0,1)$, Hölder's inequality gives

$$\mathcal{E}_{\mathbb{Q}^*}^{(s,s+1)}(\alpha Z + (1-\alpha)Z') \le \alpha\mathcal{E}_{\mathbb{Q}^*}^{(s,s+1)}(Z) + (1-\alpha)\mathcal{E}_{\mathbb{Q}^*}^{(s,s+1)}(Z').$$

(iv) For $\alpha > 1$, Jensen's inequality yields

$$\alpha\mathcal{E}_{\mathbb{Q}^*}^{(s,s+1)}(Z) \le \mathcal{E}_{\mathbb{Q}^*}^{(s,s+1)}(\alpha Z),$$

and the reverse inequality for $\alpha \in (0,1)$.
(v) Let $Z = \tilde{Z} + \bar{Z}$, such that $\tilde{Z} \in \mathcal{F}_{s+1}$ and $\bar{Z} \in \mathcal{F}_{s+1}^S$. Then,

$$\mathcal{E}_{\mathbb{Q}^*}^{(s,s+1)}(Z) = \mathcal{E}_{\mathbb{Q}^*}^{(s,t)}\left(\tilde{Z}\right) + E_{\mathbb{Q}^*}\left(\bar{Z}|\,\mathcal{F}_s\right).$$

- **A two-step iterative construction**

The forward indifference price is constructed via an iterative pricing scheme which starts at the claim's maturity and is applied backwards in time in (43). The scheme has local and dynamic properties.

Dynamically, at each time interval, say $(s, s+1)$, the price $\nu_s(C_t)$ is computed via the single-step forward price functional $\mathcal{E}_{\mathbb{Q}^*}^{(s,s+1)}$, applied to the end of the period payoff. The latter turns out to be the indifference price $\nu_{s+1}(C_t)$, as discussed earlier. The functional $\mathcal{E}_{\mathbb{Q}^*}^{(s,s+1)}$ is independent of the specific payoff.

Locally, the pricing role of $\mathcal{E}_{\mathbb{Q}^*}^{(s,s+1)}$ is similar to its single-period counterpart, developed in [10], in that it is non-linear and produces the price in two sub-steps. In the first sub-step, the end of the period payoff $C^{(s,s+1)}(\nu_{s+1}(C_t))$ is distorted and produces an intermediate payoff, say $\mathcal{C}^{(s,s+1)}(\nu_{s+1}(C_t))$, given by

$$\mathcal{C}^{(s,s+1)}\left(\nu_{s+1}(C_t)\right)=\frac{1}{\gamma}\ln E_{\mathbb{Q}^*}\left(e^{\gamma\,\nu_{s+1}(C_t)}\left|\mathcal{F}_s\vee\mathcal{F}_{s+1}^S\right.\right). \tag{46}$$

This payoff is replicable and is, in turn, priced by expectation, yielding

$$\nu_s(C_t)=E_{\mathbb{Q}^*}(\mathcal{C}^{(s,s+1)}\left(\nu_{s+1}(C_t)\right)|\mathcal{F}_s). \tag{47}$$

In the first step, the conditioning is with regards to $\mathcal{F}_s\vee\mathcal{F}_{s+1}^S$ while, in the second, it is only with respect to $\mathcal{F}_s$.

- **Analogies with the static certainty equivalent**

The classical certainty equivalent is a static pricing rule, yielding the price of a generic claim, say Z, as

$$CE\,(Z)=-u^{(-1)}\left(E_{\mathbb{P}}\left(u\left(-Z\right)\right)\right), \tag{48}$$

for a concave and increasing utility function u (see, for example, [6]). Notice that, in contrast to the indifference prices, the above price is derived in the *absence* of any trading activity. Notice, also, that the measure appearing above is the historical probability measure and not any martingale one.

Given that the forward price is constructed taking into account the investor's risk preferences, shall one expect that they would provide *multi-period analogues* of the static certainty equivalent rule? This is not obvious and, as a matter of fact, such analogy fails in the classical setting.

In seeking a multi-period analogue of (48), it is natural to assume that the role of u and $u^{(-1)}$ will be played by the process $U_t\,(x)$ and its spatial inverse $U_t^{(-1)}\,(x)$, with the latter given, for $t=1,2,\ldots,$ by

$$U_t^{(-1)}\,(x)=-\frac{1}{\gamma}\ln\left(-x\right)+\frac{1}{\gamma}\sum_{i=1}^{t}h_i, \tag{49}$$

and $U_0^{(-1)}\,(x)=-\frac{1}{\gamma}\ln\left(-x\right)$, for $x\in\mathbb{R}^-$ and h as in (11).

We now consider an analogue of the certainty equivalent, defined, for $Z\in\mathcal{F}_{s+1}$, as

$$CE^{(s,s+1)}\,(Z):=-U_{s+1}^{(-1)}\left(E_{\mathbb{P}}\left(U_{s+1}\left(-Z\right)\left|\mathcal{F}_s\vee\mathcal{F}_{s+1}^S\right.\right)\right). \tag{50}$$

We remind the reader of $\mathcal{C}^{(s,s+1)}\left(\nu_{s+1}(C_t)\right)$ introduced in (1).

Lemma 14 *Let $t>0$ and $s=0,1,\ldots,t$, and $\mathbb{Q}^*$ be the forward indifference pricing measure. Then, for any $Z\in\mathcal{F}_{s+1}$, the following assertions hold:*

(i) The dynamic certainty equivalent $CE^{(s,s+1)}\,(Z)$ satisfies

$$CE^{(s,s+1)}\,(Z)=\frac{1}{\gamma}\ln E_{\mathbb{Q}^*}\left(e^{\gamma Z}\left|\mathcal{F}_s\vee\mathcal{F}_{s+1}^S\right.\right). \tag{51}$$

(ii) Moreover, $CE^{(s,s+1)}(Z)$ *is invariant under* $\mathbb{P}$ *and* $\mathbb{Q}^*$*, namely,*

$$-U_{s+1}^{(-1)}\left(E_{\mathbb{P}}\left(U_{s+1}(-Z)\left|\mathcal{F}_s \vee \mathcal{F}_{s+1}^S\right.\right)\right) = -U_{s+1}^{(-1)}\left(E_{\mathbb{Q}^*}\left(U_{s+1}(-Z)\left|\mathcal{F}_s \vee \mathcal{F}_{s+1}^S\right.\right)\right).$$

Proof To establish (51), we first observe that, under the measure $\mathbb{Q}^*$,

$$E_{\mathbb{Q}^*}\left(U_{s+1}(-Z)\left|\mathcal{F}_s \vee \mathcal{F}_{s+1}^S\right.\right) = E_{\mathbb{Q}^*}\left(-e^{\gamma Z + \Sigma_{i=1}^{s+1} h_i}\middle|\mathcal{F}_s \vee \mathcal{F}_{s+1}^S\right)$$

$$= -e^{\Sigma_{i=1}^{s+1} h_i} E_{\mathbb{Q}^*}\left(e^{\gamma Z}\middle|\mathcal{F}_s \vee \mathcal{F}_{s+1}^S\right),$$

where we used that $\Sigma_{i=1}^{s+1} h_i$ is $\mathcal{F}_s$-mble. In turn, the forms of (13) and (49) yield

$$-U_{s+1}^{(-1)}\left(E_{\mathbb{Q}^*}\left(U_{s+1}(-Z)\left|\mathcal{F}_s \vee \mathcal{F}_{s+1}^S\right.\right)\right)$$

$$= \frac{1}{\gamma}\ln\left(e^{\Sigma_{i=1}^{s+1} h_i} E_{\mathbb{Q}^*}\left(e^{\gamma Z}\middle|\mathcal{F}_s \vee \mathcal{F}_{s+1}^S\right)\right) - \frac{1}{\gamma}\sum_{i=1}^{s+1} h_i$$

$$= \frac{1}{\gamma}\ln E_{\mathbb{Q}^*}\left(e^{\gamma Z}\middle|\mathcal{F}_s \vee \mathcal{F}_{s+1}^S\right).$$

Using property (36), however, we have that

$$\frac{1}{\gamma}\ln E_{\mathbb{Q}^*}\left(e^{\gamma Z}\middle|\mathcal{F}_s \vee \mathcal{F}_{s+1}^S\right) = \frac{1}{\gamma}\ln E_{\mathbb{P}}\left(e^{\gamma Z}\middle|\mathcal{F}_s \vee \mathcal{F}_{s+1}^S\right),$$

and the rest of the proof follows easily. ■

The above results yield the following representation of the forward indifference price.

Proposition 15 *Consider a claim* C_t *at time* 0 *and yielding payoff* C_t *at time* $t > 0$*. For* $s = 0, 1, \ldots, t$*, its forward indifference price* $\nu_s(C_t)$ *is given as the arbitrage-free price of the conditional certainty equivalent (cf. (50)) of the indifference price at the end of the period, namely,*

$$\nu_s(C_t) = E_{\mathbb{Q}^*}\left(CE^{(s,s+1)}(\nu_{s+1}(C_t))\middle|\mathcal{F}_s\right).$$

- **The pricing measure**

As we have already established, the pricing measure $\mathbb{Q}^*$ is the one that minimizes the reverse relative entropy (cf. Proposition 7). It has the intuitively pleasing property (36), in that, for each period $[s-1, s)$, the conditional on $\mathcal{F}_{s-1} \vee \mathcal{F}_s^S$ distribution of the stochastic factor Y_s is the *same* under both $\mathbb{P}$ and $\mathbb{Q}^*$.

- **Dependence on the maturity of the claim**

The forward pricing functionals $\mathcal{E}_{\mathbb{Q}^*}^{(s,s+1)}$ and $\mathcal{E}_{\mathbb{Q}^*}^{(s,t)}$ are independent of the claim's maturity. Indeed, neither their form nor the involved measure depend on the time t that the claim matures. This does not mean that the price is independent of the claim's maturity, an obvious wrong conclusion. Rather, it says that the forward pricing operator *per se* does not depend on the specific maturity.

This setting is very much aligned with the one in complete markets where the pricing operator, given by the conditional expectation of the (discounted) payoff, is independent of the claim's maturity.

- **Comparison with the traditional exponential utility valuation**

We conclude commenting on some distinct features of the forward and classical exponential indifference prices. To make the notation more familiar with the traditional setting, we assume that the claim matures at time T and that we consider the classical expected utility problem in $[0, T]$ with utility $U_T(x) = -e^{-\gamma x}$, $x \in \mathbb{R}$, $\gamma > 0$.

We recall that the case of a binomial model with exponential preferences in which a claim is written exclusively on a non-traded asset but in a complete nested (stock and bond) market model was studied in [10, 11, 23, 24]. These results were subsequently generalized by the authors in [18] for a setting like the one herein. Similar results for power utilities were analyzed in [9].

Let us denote by $\mu_{s,T}(C_T)$, $s = 1, 2, \ldots, T-1$, the traditional exponential indifference price of the claim C_T and by $V_{s,T}(x)$ the associated value function process. There are several differences between the prices $\mu_{s,T}(C_T)$ and $\nu_t(C_T)$. As it was shown in [18], the classical price is also computed iteratively,

$$\mu_{s,T}(C_T) = E_{\mathbb{Q}_T^{me}}\left(\frac{1}{\gamma}\ln E_{\mathbb{Q}_T^{me}}\left(e^{\gamma\mu_{s+1,T}(C_T)}\middle|\mathcal{F}_s \vee \mathcal{F}_{s+1}^S\right)\middle|\mathcal{F}_s\right),$$

where $\mu_{s+1,T}(C_T)$ is the indifference price of the claim at the end of the period $(s, s+1]$.

The measure $\mathbb{Q}_T^{me}$ is the minimal relative entropy one and its density depends crucially on the horizon choice T, while this not the case with $\mathbb{Q}^*$. As a result, the form of $\mu_{s,T}(\cdot)$ also depends on the horizon choice, while the form of $\nu_t(\cdot)$ does not.

Another difference, is that the classical price has no natural interpretation as the arbitrage-free price of a dynamic conditional certainty equivalent. Indeed, it can be shown[1] that, if Z is $\mathcal{F}_{s+1}$-mble, then

$$\frac{1}{\gamma}\ln E_{\mathbb{Q}_T^{me}}\left(e^{\gamma Z}\middle|\mathcal{F}_s \vee \mathcal{F}_{s+1}^S\right) \neq -V_{s+1,T}^{(-1)}\left(E_{\mathbb{P}}\left(V_{s+1,T}(-Z)\middle|\mathcal{F}_s \vee \mathcal{F}_{s+1}^S\right)\right).$$

[1] The technical arguments are rather tedious and are available upon request. They will also appear in [25].

Finally, as discussed at the end of Sect. 4, the forward indifference valuation mechanism is applicable for claims arriving at arbitrary future times, known or not a priori. This is because the forward criterion can be defined sequentially as time progresses and the market evolves. This is not the case, however, in the classical setting.

A detailed comparative study between the traditional and forward exponential indifference prices, and their respective measures is being carried out by two of the authors in [25].

References

1. Becherer, D.: Rational hedging and valuation of integrated risks under constant absolute risk aversion. Insur.: Math. Econ. **33**, 1–28 (2003)
2. Delbaen, F., Grandits, P., Rheinlander, T., Samperi, D., Schweizer, M., Stricker, C.: Exponential hedging and entropic penalties. Math. Financ. **12**, 99–123 (2002)
3. Karoui, E.l., M'rad, M.: An exact connection between two solvable SDE and a non-linear utility stochastic PDE, SIAM J. Finan. Math. **4**(1), 697–736 (2014)
4. Elliott, R., Van der Hoek, J.: Indifference paradigm for discrete models: Duality methods. In: R. Carmona (ed.) Indifference Pricing. Princeton University Press, (abbreviated title: Duality methods), pp. 321–381 (2009)
5. Frittelli, M.: The minimal entropy martingale measure and the valuation problem in incomplete markets. Math. Financ. **10**, 39–52 (2000)
6. Gerber, H.: An Introduction to Mathematical Risk Theory. Huebner (1979)
7. Leung, T., Sircar, R., Zariphopoulou, T.: Forward indifference valuation of American options. Stochastics **84**(5–6), 751–770 (2012)
8. Liang, G., Zariphopoulou, T.: Representation of homothetic forward performance processes via ergodic and infinite horizon quadratic BSDE in stochastic factor models, submitted for publication (2015)
9. Malamud, S., Trubowitz, E., Wuethrich, M.: Indifference pricing for power utilities. Math. Financ. Econ. **7**(3), 247–280 (2013)
10. Musiela, M., Zariphopoulou, T.: A valuation algorithm for indifference pricing in incomplete markets. Financ. Stochast. **8**, 229–239 (2004)
11. Musiela, M., Zariphopoulou, T.: Derivatives management, investment pricing and the term structure of exponential utilities: The case of binomial model. In: Carmona, R. (ed.) Indifference Pricing. Princeton University Press, (abbreviated title: The single period model), pp. 3–41 (2009)
12. Musiela, M., Zariphopoulou, T.: The backward and forward dynamic utilities and the associated pricing systems: The case study of the binomial model, preprint (2003)
13. Musiela, M., Zariphopoulou, T.: Investment and valuation under backward and forward dynamic exponential utilities in a stochastic factor model, Advances in Mathematical Finance, Applied and Numerical Harmonic Analysis Series, pp. 303–334 (2006)
14. Musiela, M., Zariphopoulou, T.: Optimal asset allocation under forward exponential criteria. In: Kurtz, G. (ed.) Markov Processes and Related Topics: A Festschrift for Thomas, IMS Collections, Institute of Mathematical. Statistics, vol. 4, pp. 285–300 (2008)
15. Musiela, M., Zariphopoulou, T.: Portfolio choice under dynamic investment performance criteria. Quan. Financ. **9**, 161–170 (2009)
16. Musiela, M., Zariphopoulou, T.: Stochastic partial differential equations in portfolio choice. In: Contemporary Quantitative Finance: Essays in Honour of E. Platen, C. Chiarella and A. Novikov, (eds.), pp. 195–215 (2010)

17. Musiela, M., Zariphopoulou, T.: Portfolio choice under space-time monotone performance criteria. SIAM J. Finan. Math. 326–365 (2010)
18. Musiela, M., Sokolova, E., Zariphopoulou, T.: Indifference valuation in incomplete binomial models. Math. Action **3**(2), 1–36 (2010)
19. Natodchiy, S., Tehranchi, M.: Optimal investment for all time horizons and Martin boundary of space-time diffusions. Math. Financ. Published on line, (2015)
20. Nadtochiy, S., Zariphopoulou, T.: A class of homothetic forward investment performance processes with non-zero volatility. In: Kabanov, Y. et al. (Eds.) Inspired by Finance: The Musiela Festschrift, pp. 475–505 (2013)
21. Rouge, R., El Karoui, N.: Pricing via utility maximization and entropy. Math. Financ. **10**, 259–276 (2000)
22. Shkolnikov, M., Sircar, R., and Zariphopoulou, T.: Asymptotic analysis of forward performance processes in incomplete markets and their ill-posed HJB equations. SIAM J. Finan. Math. (to appear)
23. Smith, J., McCardle, K.F.: Valuing oil properties: integrating option pricing and decision analysis approaches. Oper. Res. **46**(2), 198–217 (1998)
24. Smith, J.E., Nau, R.F.: Valuing risky projects: option pricing and analysis. Manage. Sci. **41**(5), 795–816 (1995)
25. Sokolova, E., Zariphopoulou, T.: On the minimal entropy and reverse entropy measures in incomplete binomial models and the associated indifference prices, in preparation (2016)
26. Zariphopoulou, T., Žitković, G.: Maturity-independent risk measures. SIAM J. Finan. Math. **1**, 266–288 (2010)
27. Žitković, G.: A dual characterization of self-generation and exponential forward performances. Ann. Appl. Probab. **19**(6), 2176–2210 (2009)

Almost Surely Optimal Portfolios Under Proportional Transaction Costs

Mark-Roman Feodoria and Jan Kallsen

Abstract In frictionless markets there typically exists a portfolio whose long-term growth rate of wealth almost surely dominates that of any other portfolio. In this note we show that this continues to hold in a Black-Scholes-type market with proportional transaction costs.We heavily rely on results from Gerhold et al. (Financ Stochast 17:325–354 2013 [7]), who determine a portfolio maximizing the *expected* long-term growth rate of wealth in the same setup.

Keywords Portfolio optimization · Proportional transactions costs · Shadow prices

MSC subject classification (2010): 91G20 · 60G51 · 90C59

1 Introduction

Portfolio optimization is one of the oldest problems in Mathematical Finance and it has been considered in manifold contexts and variations. A striking classical result states that in generic frictionless markets there exists a self-financing dynamic portfolio φ whose long-term growth rate of wealth

$$\limsup_{T\to\infty} \frac{1}{T} \log(V_T(\varphi))$$

almost surely dominates that of any competing investment strategy, cf. e.g. [17, Theorem 3.10.1] or [8, Lemma 5.3] for markets with jumps. Here, $V_t(\varphi) = v_0 + \int_0^t \varphi_s dS_s$ represents the value at time t of a portfolio φ which has initial value

M.-R. Feodoria · J. Kallsen (✉)
Mathematisches Seminar, Christian-Albrechts-Universität zu Kiel,
Westring 383, 24098 Kiel, Germany
e-mail: kallsen@math.uni-kiel.de

M.-R. Feodoria
e-mail: feodoria@math.uni-kiel.de

© Springer International Publishing Switzerland 2016

J. Kallsen and A. Papapantoleon (eds.), *Advanced Modelling in Mathematical Finance*,
Springer Proceedings in Mathematics & Statistics 189,
DOI 10.1007/978-3-319-45875-5_14

v_0 and trades assets $S = (S^0, \ldots, S^d)$. This optimal portfolio can be obtained by solving the Merton problem for logarithmic utility or, more specifically, by computing the portfolio maximizing the expected logarithmic utility of terminal wealth. This *numeraire portfolio* is known not to depend on the time horizon and it shares a number of other interesting properties, cf. e.g. [15, 18] and the references therein.

In the presence of proportional transaction costs the solution to the Merton problem with logarithmic utility does depend on the time horizon T. It may therefore be less obvious whether there exists a portfolio dominating any other in the long run. Nevertheless, a natural candidate is provided by the portfolio maximizing the *expected* long-term growth rate of wealth. The latter has been determined by [7, 21] for a Black-Scholes type market with two assets. Based on the results in [7] we show that this portfolio dominates any other's long-term growth rate almost surely and not just in expectation. It turns out that the optimal growth rate is deterministic and can be computed explicitly, again based on the results of [7].

The proof of almost-sure optimality relies on the concept of shadow prices, introduced by [4, 11] and applied in many papers involving proportional transaction costs. It relates the market with transaction costs to a fictitious frictionless market with the same optimal portfolio. For our purposes, this concept turns out to be particularly powerful. Indeed, it allows to reduce the present problem of almost-sure optimality to the classical statement for frictionless markets.

The paper is organized as follows. In Sect. 2 we summarize main results of [7]. Subsequently, we prove the almost sure optimality of the strategy put forward in [7]. In Sect. 4 we verify that the almost surely optimal growth rate coincides with the optimal *expected* growth rate of [7], in parallel to the frictionless case.

2 Trading with Proportional Transaction Costs

We consider a market consisting of a bond with constant interest rate and a stock S following geometric Brownian motion. By switching to discounted prices we may assume the bond price to be constant and equal to 1. The ask price of the stock is modelled as

$$S_t = S_0 \exp\Big(\Big(\mu - \frac{\sigma}{2}\Big)t + \sigma W_t\Big) \tag{1}$$

with constants $S_0, \mu, \sigma > 0$ and standard Brownian motion W. The bid price, on the other hand, is assumed to equal $(1 - \lambda)S$ for some constant $\lambda \in (0, 1)$ representing transaction costs. We set $\theta := \mu/\sigma^2$.

The investor is assumed to enter the market with an initial endowment of v_0 bonds and no shares of stock. Dynamic trading is represented by $\mathbb{R}^2$-valued predictable processes (φ^0, φ) of finite variation. Here φ^0_t, φ_t denote the number of bonds resp. shares of stock at time t. A trading strategy (φ^0, φ) is naturally called *self-financing* if

$$\varphi^0_t = v_0 + \int_0^t (1 - \lambda)S_s d\varphi^{\downarrow}_s - \int_0^t S_s d\varphi^{\uparrow}_s, \quad t \in \mathbb{R}_+, \tag{2}$$

where we write $\varphi = \varphi^{\uparrow} - \varphi^{\downarrow}$ with increasing predictable processes $\varphi^{\uparrow}, \varphi^{\downarrow}$ which do not grow at the same time. A self-financing strategy (φ^0, φ) is *admissible* if its liquidation wealth process

$$V_t(\varphi^0, \varphi) := \varphi_t^0 + \varphi_t^+(1-\lambda)S_t - \varphi_t^- S_t, \quad t \geq 0$$

is almost surely nonnegative. By setting φ^0 as in (2) we can and will identify any predictable process φ of finite variation with the corresponding self-financing strategy (φ^0, φ). Accordingly, we call φ admissible if $V_t(\varphi) := V_t(\varphi^0, \varphi)$, $t \in \mathbb{R}_+$ is nonnegative.

An admissible strategy φ is called *log-optimal* for time horizon $T \in \mathbb{R}_+$ if it maximizes

$$\psi \mapsto \mathbb{E}\left(\log(V_T(\psi))\right)$$

over all admissible strategies ψ. As a natural counterpart for $T \to \infty$, an admissible strategy φ is *expected growth-optimal* if it maximizes

$$\psi \mapsto \limsup_{T\to\infty} \frac{1}{T}\mathbb{E}\left(\log(V_T(\psi))\right)$$

over all admissible ψ. The factor $1/T$ is motivated by the fact that wealth typically grows exponentially in time.

In the following, the corresponding concepts for frictionless markets with some semimartingale price process $\tilde{S}$ will play a role as well. We call a predictable $\tilde{S}$-integrable process φ $\tilde{S}$-admissible if its wealth process

$$\tilde{V}_t(\varphi) := v_0 + \int_0^t \varphi_s d\tilde{S}_s, \quad t \in \mathbb{R}_+$$

stays nonnegative. An $\tilde{S}$*-admissible* strategy is *log-optimal* for the frictionless market $\tilde{S}$ if, for any time horizon $T \in \mathbb{R}_+$, it maximizes

$$\psi \mapsto \mathbb{E}\left(\log\left(v_0 + \int_0^T \psi_t d\tilde{S}_t\right)\right) \tag{3}$$

over all $\tilde{S}$-admissible strategies ψ. It is well known that such a strategy typically exists for frictionless markets, i.e. the optimizer of (3) does not depend on the time horizon T. If $\tilde{S}$ coincides with the above geometric Brownian motion S, the optimal fraction of wealth to be invested in the stock equals the *Merton ratio* θ.

Let us turn back to the market S with transaction costs. As in related studies [2, 5–7, 9, 10, 12–14, 19], a key role in the analysis will be played by shadow prices. For the present problem the following version from [7] is needed.

Definition 2.1 (*Shadow price*) A *shadow price* for the bid-ask processes $(1-\lambda)S$, S is a continuous semimartingale $\tilde{S}$ with $(1-\lambda)S \leq \tilde{S} \leq S$ such that the log-optimal portfolio φ for the frictionless market with price process $\tilde{S}$ exists, is of finite variation and the number of shares φ increases (resp. decreases) only on the set $\{\tilde{S} = S\}$ (resp. $\{\tilde{S} = (1-\lambda)S\}$). Put differently, the corresponding bond investment $\tilde{\varphi}^0 := \tilde{V}(\varphi) - \varphi\tilde{S}$ satisfies (2).

We summarize a few results from [7]. In that paper, a shadow price process is constructed explicitly. The log-optimal portfolio corresponding to this shadow asset turns out to be expected growth-optimal for the original market with bid-ask processes $(1-\lambda)S$, S.

Proposition 2.2 *There exists a shadow price* $\tilde{S}$.

Proof [7, Corollary 5.2] □

Corollary 2.3 *Let* $\tilde{S}$ *be a shadow price such that both the corresponding log-optimal portfolio* φ *and its bond investment* $\varphi^0 := \tilde{V}(\varphi) - \varphi\tilde{S}$ *are nonnegative. Then*

$$\mathbb{E}\left(\log(V_T(\varphi))\right) \geq \mathbb{E}\left(\log(V_T(\psi))\right) + \log(1-\lambda)$$

for any admissible strategy ψ. *Moreover,* φ *is expected growth-optimal for the bid-ask processes* $(1-\lambda)S$, S.

Proof [7, Corollary 1.9] □

Corollary 2.4 *Let* $\tilde{S}$ *be a shadow price with corresponding log-optimal portfolio* φ.

1. *If* $\theta \in (0, 1]$, *then* φ *is expected growth-optimal for the bid-ask processes* $(1-\lambda)S$, S.
2. *If* $\theta \in (1, \infty)$, *there exists* $\lambda_0 > 0$ *such that for all* $\lambda \in (0, \lambda_0)$ *strategy* φ *is expected growth-optimal for the bid-ask processes* $(1-\lambda)S$, S.

Proof 1. By [7, Theorem 5.1] (resp. the proof of [7, Corollary 5.2] for $\theta = 1$) we have $\varphi, \varphi^0 \geq 0$. Now we can apply Corollary 2.3.
2. [7, Lemma 5.3]

□

As is known from related maximization problems under proportional transaction costs, the optimal portfolio remains untouched most of the time and is adjusted infinitesimally whenever it deviates too strongly from the frictionless target. In the present setup, this can be expressed in terms of the fraction of wealth invested in the stock. More specifically, let

$$\pi_t := \frac{\varphi_t S_t}{\varphi_t^0 + \varphi_t S_t}$$

denote the fraction of *book wealth* held in the risky asset, where φ_t^0 denotes the riskless investment from (2). According to [7, Section 5], the optimal strategy from Corollary 2.4 is to keep this fraction in the interval

$$[\underline{\pi}, \overline{\pi}] := \left[\frac{1}{1+c}, \frac{1}{1+c/\bar{s}}\right], \tag{4}$$

where c denotes the unique root of the function

$$f(c) = \begin{cases} \left(\frac{c}{(2\theta-1+2c\theta)(2-2\theta-c(2\theta-1))}\right)^{\frac{1-\theta}{\theta-1/2}} - \frac{1}{1-\lambda}(2\theta-1+2c\theta)^2 & \text{if } \theta \in (0,\infty)\setminus\{\frac{1}{2},1\}, \\ \exp\left(\frac{c^2-1}{c}\right) - \frac{1}{1-\lambda}c^2 & \text{if } \theta = \frac{1}{2} \end{cases}$$

in the interval $(\frac{1-\theta}{\theta}, \infty)$ if $\theta \in (0, \frac{1}{2}]$, in the interval $(\frac{1-\theta}{\theta}, \frac{1-\theta}{\theta-1/2})$ if $\theta \in (\frac{1}{2}, 1)$, resp. in the interval $(\frac{1-\theta}{\theta}, 0)$ if $\theta > 1$, and $\bar{s}$ is defined as

$$\bar{s} := \begin{cases} \left(\frac{c}{(2\theta-1+2c\theta)(2-2\theta-c(2\theta-1))}\right)^{\frac{1}{2\theta-1}} & \text{if } \theta \in (0,\infty)\setminus\{\frac{1}{2},1\}, \\ \exp\left(\frac{c^2-1}{c}\right) & \text{if } \theta = \frac{1}{2}. \end{cases}$$

One could also consider the fraction of *liquidation wealth* held in the risky asset, i.e.

$$\pi_t^L := \frac{\varphi_t(1-\lambda)S_t}{V_t(\varphi)}$$

for positive $\varphi(t)$. A straightforward computation yields that (4) turns into the corresponding interval

$$\left[\frac{1}{1+c/(1-\lambda)}, \frac{1}{1+c/((1-\lambda)\bar{s})}\right]$$

for π^L.

In [7] we can find a statement on the optimal expected growth rate as well:

Proposition 2.5 (Optimal expected growth rate) *The optimal expected long-term growth rate equals*

$$\begin{aligned}\delta &:= \limsup_{T\to\infty} \frac{1}{T}\mathbb{E}\big(\log(V_T(\varphi))\big) \\ &= \limsup_{T\to\infty} \frac{1}{T}\mathbb{E}\big(\log(\tilde{V}_T(\varphi))\big) \\ &= \begin{cases} \frac{(2\theta-1)\sigma^2\bar{s}}{2(1+c)(\bar{s}+(-2-c+2\theta(1+c))\bar{s}^{2\theta})} & \textit{for } \theta \in (0,\infty)\setminus\{\frac{1}{2},1\}, \\ \frac{\sigma^2}{2(1+c)(1+c-\log(\bar{s}))} & \textit{for } \theta = \frac{1}{2}, \\ \frac{\sigma^2}{2} & \textit{for } \theta = 1. \end{cases}\end{aligned}$$

Proof [7, Proposition 5.4 and Remark 5.5] □

In the limit of small transaction costs, the bounds (4) and the expected growth rate δ simplify considerably:

Proposition 2.6 (Asymptotics) *Let $\theta \in (0, \infty) \setminus \{1\}$. In the limit $\lambda \to 0$ we have*

$$\underline{\pi} = \theta - \left(\frac{3}{4}\theta^2(1-\theta)^2\lambda\right)^{1/3} + O(\lambda),$$
$$\overline{\pi} = \theta + \left(\frac{3}{4}\theta^2(1-\theta)^2\lambda\right)^{1/3} + O(\lambda)$$

for the bounds in (4) and

$$\delta = \frac{\mu^2}{2\sigma^2} - \left(\frac{3\sigma^3}{\sqrt{128}}\theta^2(1-\theta)^2\lambda\right)^{2/3} + O(\lambda^{4/3}) \tag{5}$$

for the optimal expected long-term growth rate.

Proof [7, Corollary 6.2 and Proposition 6.3] □

Remark 2.7 For later use we remark that, unless $\theta = 1$, the shadow price in Proposition 2.2 is of the form

$$d\tilde{S}_t = \tilde{S}_t\left(\tilde{\mu}(Y_t)dt + \tilde{\sigma}(Y_t)dW_t\right), \quad \tilde{S}_0 = S_0, \tag{6}$$

where

1. Y is a positively recurrent one-dimensional diffusion with state space $I := [1 \wedge \bar{s}, 1 \vee \bar{s}]$,
2. $\tilde{\mu}$ and $\tilde{\sigma}$ are positive continuous functions on I,
3. $\delta = \int \frac{\tilde{\mu}(s)^2}{2\tilde{\sigma}(s)^2} d\nu(s)$, where ν denotes the stationary distribution of Y.

For $\theta = 1$ the process $\tilde{S} = S$ is a shadow price.

Proof [7, Section 5] □

3 Almost Sure Growth Optimality

Our first main result concerns almost sure growth optimality in the following sense.

Definition 3.1 An admissible strategy φ is called *almost surely growth-optimal* if

$$\limsup_{T\to\infty} \frac{1}{T}\log(V_T(\psi)) \leq \limsup_{T\to\infty} \frac{1}{T}\log(V_T(\varphi))$$

almost surely for any admissible ψ.

Similarly to [7, Corollary 1.9] we have the following results, which do not use the specific model (1) for the stock S.

Proposition 3.2 *Let $\tilde{S}$ be a shadow price with corresponding log-optimal portfolio φ. If $V(\varphi)$ is nonnegative and*

$$\lim_{T\to\infty} \frac{1}{T} \log\left(\frac{V_T(\varphi)}{\tilde{V}_T(\varphi)}\right) = 0 \quad \textit{almost surely,} \tag{7}$$

then φ is almost surely growth-optimal.

Proof Due to $(1-\lambda)S \le \tilde{S} \le S$ we have $\tilde{V}(\varphi) \ge V(\varphi)$. This yields

$$\frac{1}{T}\log(\tilde{V}_T(\varphi)) \ge \frac{1}{T}\log(V_T(\varphi)) = \frac{1}{T}\log(\tilde{V}_T(\varphi)) + \frac{1}{T}\log\left(\frac{V_T(\varphi)}{\tilde{V}_T(\varphi)}\right)$$

and hence

$$\limsup_{T\to\infty} \frac{1}{T}\log(\tilde{V}_T(\varphi)) = \limsup_{T\to\infty} \frac{1}{T}\log(V_T(\varphi)) \quad \text{a.s.} \tag{8}$$

Let ψ be an admissible strategy. Since $\tilde{V}(\psi) \ge V(\psi)$, we have that ψ is $\tilde{S}$-admissible as well (cf. the proof of [7, Proposition 1.8]). From the log-optimality of φ in the frictionless market $\tilde{S}$ we obtain with [17, Theorem 3.10.1] resp. [8, Lemma 5.3]

$$\limsup_{T\to\infty} \frac{1}{T}\log(\tilde{V}_T(\varphi)) \ge \limsup_{T\to\infty} \frac{1}{T}\log(\tilde{V}_T(\psi)) \ge \limsup_{T\to\infty} \frac{1}{T}\log(V_T(\psi)).$$

Together with (8) the assertion follows. □

Corollary 3.7 *Let $\tilde{S}$ be a shadow price with corresponding log-optimal portfolio φ. If φ and φ^0 from (2) are nonnegative, then φ is almost surely growth-optimal.*

Proof Since

$$\tilde{V}_t(\varphi) \ge V_t(\varphi) \ge (1-\lambda)\tilde{V}_t(\varphi) \tag{9}$$

(cf. the proof of [7, Corollary 1.9]), the statement follows from Proposition 3.2 □

Coming back to the Black-Scholes price processes and using the shadow price from Proposition 2.2 we obtain the following corollaries.

Corollary 3.8 *Assume $\lambda < \lambda_0$ if $\theta > 1$, with λ_0 as in the proof of [7, Lemma 5.3]. Let $\tilde{S}$ be the shadow price from Proposition 2.2 with corresponding log-optimal portfolio φ. Then φ is almost surely growth-optimal.*

Proof Case $\theta \le 1$: By [7, Theorem 5.1] (resp. the proof of [7, Corollary 5.2] for $\theta = 1$) we have $\varphi^0, \varphi \ge 0$. The assertion follows from Corollary 3.7.

Case $\theta > 1$: From the proof of [7, Lemma 5.3] it follows that

$$\tilde{V}_t(\varphi) \ge V_t(\varphi) \ge K\tilde{V}_t(\varphi) \tag{10}$$

for some $K > 0$, which yields the claim by Proposition 3.2. □

4 Optimal Growth Rate

As our second main result we want to show that the long-term growth rate of wealth is actually deterministic and hence coincides with the expected long-term growth rate δ of Proposition 2.5. As may be expected, ergodicity plays a key role in this context. For a related statement in the frictionless case, cf. [17, Corollary 3.10.2].

Theorem 4.1 *Suppose that $\lambda < \lambda_0$ if $\theta > 1$. The optimal growth rate in Corollary 3.8 coincides with the optimal expected growth rate in Proposition 2.5, i.e.*

$$\lim_{T\to\infty} \frac{1}{T}\log(V_T(\varphi)) = \delta.$$

Proof Suppose that $\theta \neq 1$. As a first step, we show the assertion for the shadow wealth process $\tilde{V}(\varphi)$. Since the shadow price is of the form (6), the log-optimal *fraction of wealth* equals

$$\tilde{\pi}_t := \frac{\tilde{\mu}(Y_t)}{\tilde{\sigma}^2(Y_t)}, \quad t \geq 0$$

by [16, Example 6.4], i.e. the corresponding log-optimal portfolio satisfies $\varphi_t = \tilde{\pi}_t \tilde{V}_t(\varphi)/\tilde{S}_t$. This implies

$$\tilde{V}_T(\varphi) = v_0 \mathscr{E}\left(\int_0^{\cdot} \frac{\tilde{\pi}_t}{\tilde{S}_t} d\tilde{S}_t\right)_T = v_0 \exp\left(\int_0^T \frac{1}{2}\left(\frac{\tilde{\mu}(Y_t)}{\tilde{\sigma}(Y_t)}\right)^2 dt + \int_0^T \frac{\tilde{\mu}(Y_t)}{\tilde{\sigma}(Y_t)} dW_t\right)$$

and hence

$$\frac{1}{T}\log(\tilde{V}_T(\varphi)) = \frac{1}{T}\log(v_0) + \frac{1}{T}\int_0^T \frac{1}{2}\left(\frac{\tilde{\mu}(Y_t)}{\tilde{\sigma}(Y_t)}\right)^2 dt + \frac{1}{T}\int_0^T \frac{\tilde{\mu}(Y_t)}{\tilde{\sigma}(Y_t)} dW_t. \quad (11)$$

Since the function $f := \tilde{\mu}^2/(2\tilde{\sigma}^2)$ is bounded on I, the ergodic theorem [1, II.35] and Remark 2.7 yield

$$\lim_{T\to\infty} \frac{1}{T}\int_0^T f(Y_s)ds = \int f(y)d\nu(y) = \delta \quad \text{a.s.} \quad (12)$$

The process

$$M_t := \int_0^t \frac{\tilde{\mu}(Y_s)}{\tilde{\sigma}(Y_s)} dW_s, \quad t \geq 0$$

is a continuous local martingale with quadratic variation $[M]_t = \int_0^t 2f(Y_s)^2 ds, t \geq 0$. Since f is bounded away from zero, we have $aT \leq [M]_T \leq bT, T \geq 0$ for some $a, b \in (0, \infty)$. From the law of large numbers for continuous local martingales [20, Exercise V.1.16] we obtain $M_T/[M]_T \to 0$ and hence $M_T/T \to 0$ almost surely for $T \to \infty$. Together with (11, 12) we conclude that

$$\lim_{T\to\infty} \frac{1}{T} \log(\tilde{V}_T(\varphi)) = \delta.$$

The assertion follows now from the fact that $V_T(\varphi)$ and $\tilde{V}_T(\varphi)$ differ at most by a constant factor, cf. (9) resp. (10).

The case $\theta = 1$ is of course similar to [7, Remark 5.5]. Since the entire wealth is invested in stock, we have

$$\log(V_T(\varphi)) = \log(v_0) + \frac{\sigma^2}{2} T + \sigma W_T,$$

which yields the long-term growth rate $\sigma^2/2$ by the strong law of large numbers for standard Brownian motion. □

From Theorem 4.1 and Proposition 2.6 we immediately obtain an asymptotic expansion of the almost sure long-term growth rate for small transaction costs.

The existence of an explicitly computable portfolio which surely dominates any other in the long run may be viewed as a both beautiful and extremely useful mathematical result. But as is well known, it faces severe obstacles in practice. Firstly, the excess drift rate μ is typically small compared to the standard deviation σ. This means that it may take a long time even to beat the bond-only investment strategy with, say, 95 % probability. Put differently, *long term* should rather be interpreted as centuries rather than years. In addition, the frictionless target θ depends linearly on μ which, again since it is small compared to σ, is very hard to estimate in any reliable way. In the presence of limited past data or instationary parameters, it may even be debatable whether the stock's excess drift rate μ is positive at all.

Leaving these disenchanting facts aside, let us finish with a simple numerical example in order to illustrate the results in this paper. We consider a stock with yearly volatility $\sigma = 20\,\%$ and excess drift rate $\mu = 2\,\%$. The *frictionless optimal excess growth rate* $\mu^2/(2\sigma^2) = 0.5\,\%$ seems surprisingly small but should be contrasted with the fact that the stock's long-term *growth rate* (namely $\mu - \sigma^2/2$) vanishes—in spite of its positive *drift rate* $\mu = 2\,\%$. The optimal fraction of wealth invested in the stock equals $\theta = 1/2$, which, due to the factor $\theta^2(1-\theta)^2$ in (5), seems to be a rather unpleasant parameter value if we introduce transaction costs or taxes.

If we consider transaction costs of $\lambda = 1\,\%$, the asymptotic no-trade region from Proposition 2.6 equals [0.42, 0.58], i.e. the investor tries to keep the fraction of wealth invested in stock between 42 % and 58 %. According to the asymptotic formula (5), the frictionless optimal excess growth rate of 0.5 % is lowered by the presence of transaction costs to approximately 0.49 %. In other words, even in the unfavourable case $\mu = 2\,\%$, $\theta = 1/2$ the effect of transaction costs on the optimal long-term growth rate appears to be rather small. For an early reference to related observations cf. [3].

Acknowledgements The authors thank Johannes Muhle-Karbe for fruitful comments and discussions. Moreover, they acknowledge financial support through *DFG-Sachbeihilfe KA 1682/4-1*.

References

1. Borodin, A., Salminen, P.: Handbook of Brownian motion: Facts and Formulae. Springer (2002)
2. Choi, J., Sirbu, M., Zitkovic, G.: Shadow prices and well-posedness in the problem of optimal investment and consumption with transaction costs. SIAM J. Control Optim. **51**, 4414–4449 (2013)
3. Constantinides, G.: Capital market equilibrium with transaction costs. J. Political Econ. 842–862 (1986)
4. Cvitanić, J., Karatzas, I.: Hedging and portfolio optimization under transaction costs: a martingale approach. Math. Financ. **6**, 133–165 (1996)
5. Gerhold, S., Guasoni, P., Muhle-Karbe, J., Schachermayer, W.: Transaction costs, trading volume, and the liquidity premium. Financ. Stochast. **18**, 1–37 (2014)
6. Gerhold, S., Muhle-Karbe, J., Schachermayer, W.: Asymptotics and duality for the Davis and Norman problem. Stochastics **84**, 625–641 (2012)
7. Gerhold, S., Muhle-Karbe, J., Schachermayer, W.: The dual optimizer for the growth-optimal portfolio under transaction costs. Financ. Stochast. **17**, 325–354 (2013)
8. Goll, T., Kallsen, J.: A complete explicit solution to the log-optimal portfolio problem. Ann. Appl. Probab. **13**, 774–799 (2003)
9. Guasoni, P., Muhle-Karbe, J.: Long horizons, high risk aversion, and endogenous spreads. Math. Financ. **25**, 724–753 (2015)
10. Herczegh, A., Prokaj, V.: Shadow price in the power utility case. Ann. Appl. Probab. **25**, 2671–2707 (2015)
11. Jouini, E., Kallal, H.: Martingales and arbitrage in securities markets with transaction costs. J. Econ. Theory **66**, 178–197 (1995)
12. Kallsen, J., Muhle-Karbe, J.: On using shadow prices in portfolio optimization with transaction costs. Ann. Appl. Probab. **20**, 1341–1358 (2010)
13. Kallsen, J., Muhle-Karbe, J.: The General structure of optimal investment and consumption with small transaction costs. Math. Financ. (2013)
14. Kallsen, J., Muhle-Karbe, J.: Option pricing and hedging with small transaction costs. Math. Financ. **25**, 702–723 (2015)
15. Karatzas, I., Kardaras, C.: The numéraire portfolio in semimartingale financial models. Financ. Stochast. **11**, 447–493 (2007)
16. Karatzas, I., Lehoczky, J., Shreve, S., Xu, G.: Martingale and duality methods for utility maximization in an incomplete market. SIAM J. Control Optim. **29**, 702–730 (1991)
17. Karatzas, I., Shreve, S.: Methods of Mathematical Finance. Springer, Berlin (1998)
18. Korn, R., Schäl, M.: On value preserving and growth optimal portfolios. Math. Methods Oper. Res. **50**, 189–218 (1999)
19. Kühn, C., Stroh, M.: Optimal portfolios of a small investor in a limit order market: a shadow price approach. Math. Financ. Econ. **3**, 45–72 (2010)
20. Revuz, D., Yor, M.: Continuous Martingales and Brownian Motion. Springer, Berlin, third edition (1999)
21. Taksar, M., Klass, M., Assaf, D.: A diffusion model for optimal portfolio selection in the presence of brokerage fees. Math. Oper. Res. **13**, 277–294 (1988)

On the Optimal Investment

José Manuel Corcuera, José Fajardo and Olivier Menouken Pamen

Abstract In 1988 Dybvig introduced the payoff distribution pricing model (PDPM) as an alternative to the capital asset pricing model (CAPM). Under this new paradigm agents preferences depend on the probability distribution of the payoff and for the same distribution agents prefer the payoff that requires less investment. In this context he gave the notion of efficient payoff. Both approaches run parallel to the theory of choice of von Neumann and Morgenstern [17], known as the Expected Utility Theory and posterior axiomatic alternatives. In this paper we consider the notion of optimal payoff as that maximizing the terminal position for a chosen preference functional and we investigate the relationship between both concepts, optimal and efficient payoffs, as well as the behavior of the efficient payoffs under different market dynamics. We also show that path-dependent options can be efficient in some simple models.

Keywords Expected utility · Prospect theory · Risk aversion · Law invariant preferences · Growth optimal portfolio · Portfolio numeraire

JEL-Classification 60H30 · 91B16 · 91B25

The work of J.M. Corcuera is supported by the Spanish grant MTM2013-40782-P.
J. Fajardo thanks financial support from CNPq-Brazil.
The research of Menoukeu-Pamen has received partial funding from the European Union's Seventh Framework Programme for research, technological development and demonstration under grant agreement no 318984-RARE.

J.M. Corcuera (✉)
Universitat de Barcelona, Gran Via de Les Corts Catalanes, 585, 08007 Barcelona, Spain
e-mail: jmcorcuera@ub.edu

J. Fajardo
EBAPE, Getulio Vargas Foundation, Rio de Janeiro, Brazil
e-mail: jose.fajardo@fgv.br

O.M. Pamen
IFAM, Department of Mathematical Sciences, University of Liverpool, Liverpool, UK
e-mail: Menoukeu@liv.ac.uk

© Springer International Publishing Switzerland 2016
J. Kallsen and A. Papapantoleon (eds.), *Advanced Modelling in Mathematical Finance*, Springer Proceedings in Mathematics & Statistics 189,
DOI 10.1007/978-3-319-45875-5_15

1 Introduction

The capital asset pricing model (CAPM) can be seeing as an approach to investment analysis based on the following simple assumptions:

Agents preferences depend only on the mean and variance of the payoff.

Between two payoffs with equal variance an agent will choose the one with higher return.

In 1988 Dybvig introduced the payoff distribution pricing model (PDPM) as an alternative to CAPM. His goal was to find another alternative to evaluate investment performance. He assumed that agents preferences depend on the probability distribution of the payoff and for the same distribution agents prefer the payoff that requires less investment.

Both approaches run parallel to the axiomatic theory of choice of Neumann-Morgenstern [17] and the posterior axiomatic alternatives; see for example Föllmer and Schied [11].

The Neumann-Morgenstern [17] axiomatic theory together with the inclusion of risk aversion lead us to the expected utility theory (EUT).

The optimal payoff consists in choosing a payoff in such a way that we obtain the largest expected utility of the payoff for a fixed investment.

Alternatives to EUT are based on modifications or elimination of the *independence axiom.* The independence axiom of the EUT says the following:

A preference relation $\succ$ on a set of probability distributions $\mathcal{X}$ satisfies the independence axiom if for all $\mu, \nu \in \mathcal{X}$, $\mu \succ \nu$ implies

$$\alpha\mu + (1-\alpha)\tau \succ \alpha\nu + (1-\alpha)\tau$$

for all $\tau \in \mathcal{X}$ and $\alpha \in (0, 1]$.

Many examples or paradoxes show that this axiom or principle is not followed by real agents. The following example is a well known paradox where the independence axiom is violated.

Example 1 (*Allais' paradox*) You have to choose between:

$$\mu_1 = 0.33\delta_{2500} + 0.66\delta_{2400} + 0.01\delta_0,$$
$$\mu_2 = \delta_{2400}$$

and later between

$$\nu_1 = 0.33\delta_{2500} + 0.67\delta_0,$$
$$\nu_2 = 0.34\delta_{2400} + 0.66\delta_0.$$

Allais showed that for 66 % of people $\mu_2 \succ \mu_1$ and $\nu_1 \succ \nu_2$. However $\frac{1}{2}(\mu_2 + \nu_1) = \frac{1}{2}(\mu_1 + \nu_2)$ and this violates the independence axiom. In fact if the independence is true and $\mu_2 \succ \mu_1$ and $\nu_1 \succ \nu_2$ we have

$$\alpha\mu_2 + (1-\alpha)\nu_1 \succ \alpha\mu_1 + (1-\alpha)\nu_1 \succ \alpha\mu_1 + (1-\alpha)\nu_2,$$

and taking $\alpha = 1/2$ we obtain

$$\frac{\mu_2+\nu_1}{2} \succ \frac{\mu_1+\nu_2}{2}.$$

The Dual Theory of Choice (DTC) [24] or the Cumulative Prospect Theory (CPT) (see Kahneman-Tverski [14] and Tverski-Kahneman [22]) are some of the alternatives to EUT. Both propose that the optimality of a payoff is a functional of its law. For instance Yaari proposed a preference functional of the form

$$V(X) = \int_0^1 h(1-t)F_X^{-1}(t)\mathrm{d}t,$$

where h: $[0, 1] \longmapsto \mathbb{R}_+$ (distortion function). In the CPT

$$\begin{aligned} V(X) = & \int_0^1 h_1(1-t)u_1\left(\left(F_X^{-1}(t) - x_0\right)_+\right) \mathrm{d}t \\ & - \int_0^1 h_2(t)u_2\left(\left(F_X^{-1}(t) - x_0\right)_-\right) \mathrm{d}t, \end{aligned}$$

with h_1, h_2 distortion functions and u_1 concave and u_2 convex, $x_0 \in \mathbb{R}$ is a reference level where consumers pass from being risk adverse to being risk takers. These functionals are particular cases of

$$V(X) = \int_0^1 L(t, F_X^{-1}(t))\mathrm{d}t.$$

The EUT is included in the previous framework with

$$V(X) = \mathbb{E}\left(u(X)\right) = \int_0^1 u(F_X^{-1}(t))\mathrm{d}t.$$

In this work we investigate the relationship between the concepts of efficient and optimal payoffs. In addition we study the behavior of the efficient portfolio for various derivatives and different assets' price dynamics.

The paper is organised as follows: Sect. 2 contains preliminary results on expected utility theory and payoff distribution pricing model. Section 3 studies efficient payoffs and law invariant preferences. Section 4 is devoted to efficient payoffs in a dynamic setting while Sect. 5 investigates conditional efficient payoffs.

2 EUT and PDPM

We start this section by recalling the definition of a utility function.

Definition 1 A utility function is map $u : \mathbb{R} \to \mathbb{R} \cup \{-\infty\}$, which is strictly increasing and continuous on $\{u > -\infty\}$, of class $\mathbb{C}^2$ and strictly concave in the interior of $\{u > -\infty\}$, and such that marginal utility tends to zero when wealth tends to infinity, i.e.,

$$u'(\infty) := \lim_{x\to\infty} u'(x) = 0.$$

Let us denote the interior of $\{u > -\infty\}$ by $\text{dom}(u)$. We will only consider the two following cases:

Case 1 $\text{dom}(u) = (0, \infty)$ and u satisfies

$$u'(0) := \lim_{x\to 0^+} u'(x) = \infty.$$

Case 2 $\text{dom}(u) = \mathbb{R}$ and u satisfies

$$u'(-\infty) := \lim_{x\to -\infty} u'(x) = \infty.$$

The HARA utility functions $u(x) = \frac{x^{1-p}}{1-p}$ for $p \in \mathbb{R}_+\backslash\{0, 1\}$ and the logarithmic utility $u(x) = \log(x)$ are important examples of **Case** 1 and the exponential utility function $u(x) = -\frac{1}{\alpha}e^{-\alpha x}$ is a typical example of **Case** 2.

Let us fix a pricing measure $\mathbb{Q}$. Given $w_0 > 0$ and a utility function u, we want to find a payoff X, with initial value w_0, that maximizes $\mathbb{E}(u(X))$ that is we consider the following optimization problem

$$\max\left\{\mathbb{E}(u(X)) : \mathbb{E}_{\mathbb{Q}}(X) = w_0\right\}. \tag{1}$$

Such X ifsimplicity we consider that interest rates are zero.

Proposition 1 *The optimal payoff is a decreasing function of* $\frac{d\mathbb{Q}}{d\mathbb{P}}$.

Proof The corresponding Lagrangian for (1) is

$$\mathbb{E}(u(X)) - \lambda\mathbb{E}_{\mathbb{Q}}(X - w_0) = \mathbb{E}\left(u(X) - \lambda\left(X\frac{d\mathbb{Q}}{d\mathbb{P}} - w_0\right)\right).$$

Then, the obvious candidate to be the optimal terminal wealth is

$$X^* := \left(u'\right)^{-1}\left(\lambda\frac{d\mathbb{Q}}{d\mathbb{P}}\right), \tag{2}$$

where λ is the solution of the equation $\mathbb{E}_{\mathbb{Q}}\left[\left(u'\right)^{-1}\left(\lambda\frac{d\mathbb{Q}}{d\mathbb{P}}\right)\right] = w_0$. The existence of X^* follows from the fact that u is strictly concave, so $\left(u'\right)^{-1}(\cdot)$ is a strictly decreasing, and λ is positive and u' takes values on $\mathbb{R}_+$ (in both cases 1 and 2). To see the optimality of X^* we can consider another payoff X and we obtain that

$$\begin{aligned}&\mathbb{E}(u(X)) - \lambda\mathbb{E}_{\mathbb{Q}}(X - w_0) - \left(\mathbb{E}(u(X^*)) - \lambda\mathbb{E}_{\mathbb{Q}}\left(X^* - w_0\right)\right)\\ &= \mathbb{E}\left(u(X) - u(X^*)) - \lambda\left(X - X^*\right)\frac{d\mathbb{Q}}{d\mathbb{P}}\right)\\ &= \frac{1}{2}\mathbb{E}\left(u''(\tilde{X})\left(X - X^*\right)^2\right) \leq 0,\end{aligned}$$

where $\tilde{X}$ is in between X and X^*. Since u is strictly concave, (a.s.) uniqueness follows. ■

Suppose that $Y = \left(u'\right)^{-1}\left(\lambda\frac{d\mathbb{Q}}{d\mathbb{P}}\right)$ is the payoff of certain contract, then this payoff is better than any other payoff X *with the same law* as Y if the risk neutral measure used to price derivatives is $\mathbb{Q}$ and the utility function that we choose is u. Then *a fortiori*

$$\mathbb{E}_{\mathbb{Q}}(X) \geq \mathbb{E}_{\mathbb{Q}}(Y).$$

In fact we have that

$$\mathbb{E}(u(Y)) = \mathbb{E}(u(X)),$$

so if $\mathbb{E}_{\mathbb{Q}}(Y) - \mathbb{E}_{\mathbb{Q}}(X) = h > 0$, we will have that $\mathbb{E}_{\mathbb{Q}}(X + h) = w_0$ and $\mathbb{E}(u(X + h)) > \mathbb{E}(u(Y))$ contradicting the optimality of Y. So among the payoffs with the same law as Y, Y is the payoff with the lowest price. This is the idea of efficient payoff introduced by Dybvig [8] and further developed in Dybvig [9]. Recently a systematic study of efficient payoffs in different contexts has been done by Bernard et al. [2] and Von Hammerstein et al. [13] under the name of cost-efficient payoffs. Here we shall use the term *efficient payoff* for brevity.

Definition 2 A payoff Y is said to be an *efficient payoff if any other payoff* X with the same law is more expensive.

Therefore, we have proved, in the previous paragraph, the following proposition.

Proposition 2 *The optimal payoff w.r.t. the utility function u is an efficient payoff.*

Suppose $Y = \left(u'\right)^{-1}\left(\lambda\frac{d\mathbb{Q}}{d\mathbb{P}}\right)$, and that u is as in **Case** 1 (a similar discussion can be done for **Case** 2), let $h : \mathbb{R}_+ \to \mathbb{R}$ be a non decreasing $\mathbb{C}^1$ function with $h(0) = 0$ and define $Z := h\left(\left(u'\right)^{-1}\left(\lambda\frac{d\mathbb{Q}}{d\mathbb{P}}\right)\right)$. Then we wonder if Z is an optimal payoff w.r.t. another utility function. Let V be such utility function, that is, it must satisfy

$$\left(V'\right)^{-1}\left(\lambda\frac{d\mathbb{Q}}{d\mathbb{P}}\right) = h\left(\left(u'\right)^{-1}\left(\lambda\frac{d\mathbb{Q}}{d\mathbb{P}}\right)\right).$$

Therefore it is sufficient to have that $V(\cdot)$ is a primitive function of $u'(h^{-1}(\cdot))$. Hence $h(Y)$ is an efficient payoff by the argument in the paragraph before Definition 2. As a consequence, if we want to create efficient payoffs with a fixed distribution function $F\colon \mathbb{R}_+ \to [0,1)$ and we assume that $\frac{\mathrm{d}\mathbb{Q}}{\mathrm{d}\mathbb{P}}$ is a continuous random variable, then this efficient payoff is given by

$$F^{-1}\left(1 - F_{\frac{\mathrm{d}\mathbb{Q}}{\mathrm{d}\mathbb{P}}}\left(\frac{1}{\lambda}u'(Y)\right)\right) = F^{-1}\left(1 - F_{\frac{\mathrm{d}\mathbb{Q}}{\mathrm{d}\mathbb{P}}}\left(\frac{\mathrm{d}\mathbb{Q}}{\mathrm{d}\mathbb{P}}\right)\right),$$

where it is assumed that $F^{-1} \in \mathbb{C}^1$, and $F_{\frac{\mathrm{d}\mathbb{Q}}{\mathrm{d}\mathbb{P}}}(\cdot)$ denotes the distribution function of $\frac{\mathrm{d}\mathbb{Q}}{\mathrm{d}\mathbb{P}}$. This efficient payoff is also an optimal payoff w.r.t. a utility function $V(\cdot)$ (belonging to **Case** 1) which is a primitive function of $\lambda F^{-1}_{\frac{\mathrm{d}\mathbb{Q}}{\mathrm{d}\mathbb{P}}}(1 - F(\cdot))$. The factor λ can obviously be omitted. We have derived the following result:

Proposition 3 *Assume that $\frac{\mathrm{d}\mathbb{Q}}{\mathrm{d}\mathbb{P}}$ has a continuous distribution and that F is a smooth distribution function, such that $F^{-1} \in \mathbb{C}^1$. Then*

$$X := F^{-1}\left(1 - F_{\frac{\mathrm{d}\mathbb{Q}}{\mathrm{d}\mathbb{P}}}\left(\frac{\mathrm{d}\mathbb{Q}}{\mathrm{d}\mathbb{P}}\right)\right)$$

*is an efficient payoff. X is also an optimal payoff w.r.t. a utility function (belonging to **Case** 1 or **Case** 2) $V(\cdot)$ which is a primitive function of $F^{-1}_{\frac{\mathrm{d}\mathbb{Q}}{\mathrm{d}\mathbb{P}}}(1 - F(\cdot))$.*

Example 2 It is easy to see that when F and $F_{\log\frac{\mathrm{d}\mathbb{Q}}{\mathrm{d}\mathbb{P}}}$ are Gaussian the corresponding utility funcion is the exponential utility. In fact, if $F_{\log\frac{\mathrm{d}\mathbb{Q}}{\mathrm{d}\mathbb{P}}}(z) = \Phi\left(\frac{z-\mu}{\sigma}\right)$ and $F(u) = \Phi\left(\frac{u-\alpha}{\gamma}\right)$, where $\Phi(\cdot)$ cumulative distribution function of the standard normal distribution, then

$$F^{-1}_{\frac{\mathrm{d}\mathbb{Q}}{\mathrm{d}\mathbb{P}}}(1 - F(u)) = \exp\left\{\mu - \frac{\sigma}{\gamma}(u - \alpha)\right\},$$

and a primitive function, up to multiplicative constants, is given by

$$V(u) := -\frac{\gamma}{\sigma}\exp\left\{-\frac{\sigma}{\gamma}u\right\}, u \in \mathbb{R}.$$

As we shall see later this smoothness condition on F can be relaxed. The relationship between efficient and optimal payoffs has also been studied in a recent paper by Bernard et al. [5].

2.1 Inefficiency of Path Dependent Options

In 1988 Dybvig wrote a paper entitled: Inefficient Dynamic Portfolio or How to Throw Away a Million Dollars in the Stock Market [9]. The title suggests a general or universal result about investment in stock markets. His claim is that path dependent options are inefficient in the sense that we can have a payoff depending only of the final price of the stock, say S_T, with higher terminal utility and the same initial price. Vanduffel et al. [23] obtained the same inneficiency result in a Lévy market model and Kassberger-Liebmann [15] explained when this phenomenon happens. The following simple lemma and theorem clarify the situation.

Lemma 1 *Let $X \geq 0$ be a payoff. Consider a model in which the risk neutral probability $\mathbb{Q}$ satisfies*

$$\frac{\mathrm{d}\mathbb{Q}}{\mathrm{d}\mathbb{P}} \in \sigma(S_T).$$

Then

$$\mathbb{E}_{\mathbb{Q}}(X|S_T) = \mathbb{E}(X|S_T).$$

Proof First, set $Z := \mathbb{E}_{\mathbb{Q}}(X|S_T)$, by definition of the conditional expectation:

$$\mathbb{E}_{\mathbb{Q}}(YZ) = \mathbb{E}_{\mathbb{Q}}(YX) \text{ for all } Y \geq 0,\ Y \in \sigma(S_T),$$

then

$$\mathbb{E}_{\mathbb{Q}}(YZ) = \int_{\Omega} YZ\mathrm{d}\mathbb{Q} = \int_{\Omega} Y\frac{\mathrm{d}\mathbb{Q}}{\mathrm{d}\mathbb{P}}Z\mathrm{d}\mathbb{P} = \int_{\Omega} \bar{Y}Z\mathrm{d}\mathbb{P} = \int_{\Omega} \bar{Y}X\mathrm{d}\mathbb{P},$$

with $\bar{Y} \geq 0$ and $\bar{Y} \in \sigma(S_T)$ arbitrary, so $Z = \mathbb{E}(X|S_T)$. ■

Theorem 1 *If the risk neutral probability satisfies $\frac{\mathrm{d}\mathbb{Q}}{\mathrm{d}\mathbb{P}} \in \sigma(S_T)$, and the savings account is deterministic, path-dependent payoffs are* dominated, *in the sense that there is another payoff with the same initial price and more terminal utility.*

Proof Given a payoff X, define $\bar{X}$ by $\bar{X} := \mathbb{E}_{\mathbb{Q}}(X|S_T)$. Then, the price is the same, since the savings account $(B_t)_{t\geq 0}$ is deterministic,

$$\mathbb{E}_{\mathbb{Q}}\left(\frac{X}{B_T}\right) = \mathbb{E}_{\mathbb{Q}}\left(\frac{1}{B_T}\mathbb{E}_{\mathbb{Q}}(X|S_T)\right).$$

Now, by Lemma 1

$$\bar{X} = \mathbb{E}_{\mathbb{Q}}(X|S_T) = \mathbb{E}(X|S_T),$$

and given a utility function u

$$\mathbb{E}(u(\bar{X})) = \mathbb{E}(u(\mathbb{E}(X|S_T))) \geq \mathbb{E}(\mathbb{E}(u\,(X)\,|S_T)) = \mathbb{E}(u\,(X)),$$

where the inequality follows from Jensen's inequality since u is concave. ■

However, as shown in Example 4, the condition $\frac{\mathrm{d}\mathbb{Q}}{\mathrm{d}\mathbb{P}} \in \sigma(S_T)$ is not satisfied in some simple models and the claim of Dybvig is not true in such cases. In the next section we consider a more general frame that includes EUT.

3 Efficient Payoffs and Law Invariant Preferences

Definition 3 A preference functional $V(X) : L^\infty \to \mathbb{R}$ is called

1. monotone if $X \geq Y$ a.s. implies $V(X) \geq V(Y)$,
2. law invariant if $V(X) = V(Y)$ whenever $X \overset{d}{\sim} Y$.

EUT, DTC and CPT use monotone and law invariant functionals and this law invariance is in agreement with the Dybvig approach.

Here we follow Carlier-Dana [6]. Choose an agent with preference functional V (strictly monotone and law invariant) and initial wealth w_0. Consider the optimization problem

$$\sup\left\{V(X), \mathbb{E}_{\mathbb{Q}}(X) = w_0, X \in L_+^\infty\right\}, \tag{3}$$

where $\mathbb{Q}$ is the pricing measure and let the interest rate be zero. Further, assume that $\psi := \frac{\mathrm{d}\mathbb{Q}}{\mathrm{d}\mathbb{P}}$ has continuous distribution function F_ψ.

Set

$$\mathcal{A} := \{x : (0, 1) \to \mathbb{R}_+, x \text{ is increasing and right continuous}\},$$

and define $v(x) := V(x(U))$ where U is a uniform distribution on $(0, 1)$. Note that $V(X) = v(F_X^{-1})$. Consider now X of the form

$$X = F_X^{-1}(1 - F_\psi(\psi)) = x(1 - F_\psi(\psi)), x \in \mathcal{A}. \tag{4}$$

Then the optimisation problem (3) is equivalent to

$$\sup\left\{v(x), x \in \mathcal{A}, x \text{ bounded}, \int_0^1 F_\psi^{-1}(1-t)x(t)\mathrm{d}t = w_0\right\}. \tag{5}$$

The condition (4) is not a restriction. In fact the solution to the optimal investment has to be in the set of efficient payoffs.

Theorem 2 *Given two random variables X, Y we have*

$$\mathbb{E}(F_X^{-1}(1-U)F_Y^{-1}(U)) \leq \mathbb{E}(XY) \leq \mathbb{E}(F_X^{-1}(U)F_Y^{-1}(U)),$$

where U is a uniform distribution on $(0, 1)$.

So

$$\mathbb{E}_{\mathbb{Q}}(X) = \mathbb{E}\left(\frac{\mathrm{d}\mathbb{Q}}{\mathrm{d}\mathbb{P}} X\right) = \mathbb{E}\left(\psi X\right) = \mathbb{E}(F_{\psi}^{-1}(1-U)F_X^{-1}(U)).$$

Proof By the formula of Hoeffding (see Lemma 2 in Lehmann [16])

$$\begin{aligned}\mathrm{Cov}\,(X,Y) &= \mathbb{E}\,(XY) - \mathbb{E}\,(X)\mathbb{E}(Y)\\ &= \int_{\mathbb{R}}\int_{\mathbb{R}} \left(F_{X,Y}(x,y) - F_X(x)F_Y(y)\right)\mathrm{d}x\mathrm{d}y.\end{aligned}$$

So, the minimum of $\mathbb{E}\,(XY)$, for fixed F_X and F_Y, is obtained when $F_{X,Y}$ is minimum and this minimum is given by the Fréchet lower bound for $F_{X,Y}$ fixed F_X and F_Y (see Fréchet [12]):

$$\min_{F_X(\cdot)=g(\cdot),\,F_Y(\cdot)=h(\cdot)} F_{X,Y}(x,y) = \max(g(x)+h(y)-1,0),$$

and this bound is reached if we take

$$(X,Y) = (F_X^{-1}(1-U), F_Y^{-1}(U)).$$

This is the approach in Bernard et al. [2] to prove the result. Another way of proving it is by using the Hardy-Littlewood inequalities directly (see for instance Theorem A.24 in Föllmer and Schied [11]). ■

Note that if Y is continuous, we can choose $U = F_Y(Y)$ and we can write the random variable

$$\bar{X} := F_X^{-1}(1-U) = F_X^{-1}(1-F_Y(Y)) = \bar{x}(1-F_Y(Y)),\, \bar{x} \in \mathcal{A}.$$

Note that we have solved the problem

$$\min\left\{\mathbb{E}_{\mathbb{Q}}\,(X) : X \sim F\right\},$$

and its solution is given by $X = F^{-1}(1-F_{\psi}(\psi)) = F^{-1}\left(1 - F_{\frac{\mathrm{d}\mathbb{Q}}{\mathrm{d}\mathbb{P}}}\left(\frac{\mathrm{d}\mathbb{Q}}{\mathrm{d}\mathbb{P}}\right)\right)$. Hence we have the following proposition.

Proposition 4 *The optimal payoff w.r.t. a law invariant and monotone functional $V(X)$ and initial wealth w_0, is the efficient payoff with distribution function F that satisfies*

$$F^{-1} = \arg\max_{x\in\mathcal{I}} V(x(U)),$$

where $\mathcal{I} = \left\{x : (0,1) \to \mathbb{R}_+,\, x \text{ increasing, right continuous and bounded}, \int_0^1 F_{\psi}^{-1}(1-t)x(t)\mathrm{d}t = w_0\right\}$ *and U is a uniform distribution on* $(0,1)$.

It is interesting to notice that we have not assumed any additional condition on the preference functional except the monotonicity and the law invariance. Then we cannot in general guarantee the existence of the solution to the problem (3). In the case that

$$v(x) = \int_0^1 h(1-t)u(x(t))\mathrm{d}t,$$

where u is a utility function. We also have the following theorem:

Theorem 3 *(Carlier-Dana [6]) The optimal payoff is an efficient payoff with an inverse distribution function* F^{-1} *that is strictly decreasing iff* F_{ψ}^{-1}/h *is strictly increasing. If* F_{ψ}^{-1}/h *is not increasing there are ranges of values of the pricing density for which* F^{-1} *is constant. If* F_{ψ}^{-1}/h *is decreasing then* F^{-1} *is constant.*

Let us stress that the problem

$$\min \left\{ \mathbb{E}_{\mathbb{Q}}(X) : X \sim F \right\},$$

is exactly what Dybvig considered. That is, for a given distribution of the payoffs; what is the cheapest one? This payoff is the efficient payoff that we defined in the previous section. We have seen that they have the form

$$X = F^{-1}\left(1 - F_{\frac{\mathrm{d}\mathbb{Q}}{\mathrm{d}\mathbb{P}}}\left(\frac{\mathrm{d}\mathbb{Q}}{\mathrm{d}\mathbb{P}}\right)\right).$$

Theorem 4 *A payoff* X *is efficient iff it is a decreasing function of* $\frac{\mathrm{d}\mathbb{Q}}{\mathrm{d}\mathbb{P}}$.

Proof If X is efficient then $X = h\left(\frac{\mathrm{d}\mathbb{Q}}{\mathrm{d}\mathbb{P}}\right)$ with $h = F^{-1}\left(1 - F_{\frac{\mathrm{d}\mathbb{Q}}{\mathrm{d}\mathbb{P}}}(\cdot)\right)$ that is decreasing, on the other hand if $X = h\left(\frac{\mathrm{d}\mathbb{Q}}{\mathrm{d}\mathbb{P}}\right)$ with h decreasing then

$$\begin{aligned} F_X(x) &= 1 - \mathbb{P}(X > x) = 1 - \mathbb{P}\left(h\left(\frac{\mathrm{d}\mathbb{Q}}{\mathrm{d}\mathbb{P}}\right) > x\right) \\ &= 1 - \mathbb{P}\left(\frac{\mathrm{d}\mathbb{Q}}{\mathrm{d}\mathbb{P}} < h^{-1}(x)\right) = 1 - F_{\frac{\mathrm{d}\mathbb{Q}}{\mathrm{d}\mathbb{P}}}\left(h^{-1}(x)\right), \end{aligned}$$

so

$$F_X(h(y)) = 1 - F_{\frac{\mathrm{d}\mathbb{Q}}{\mathrm{d}\mathbb{P}}}\left(h^{-1}(h(y))\right) = 1 - F_{\frac{\mathrm{d}\mathbb{Q}}{\mathrm{d}\mathbb{P}}}(y)$$

and

$$X = F_X^{-1}\left(1 - F_{\frac{\mathrm{d}\mathbb{Q}}{\mathrm{d}\mathbb{P}}}\left(\frac{\mathrm{d}\mathbb{Q}}{\mathrm{d}\mathbb{P}}\right)\right).$$

∎

In the following examples, that can be found in Bernard et al. [2], we illustrate the efficiency or not of the payoff of certain derivatives and the case they are not, we find their corresponding efficient payoff.

Example 3 Consider the Black-Scholes market model, $\mathrm{d}S_t = S_t\,(\mu \mathrm{d}t + \sigma \mathrm{d}W_t)$ and

$$\mathrm{d}B_t = rB_t\mathrm{d}t.$$

Then

$$\frac{\mathrm{d}\mathbb{Q}}{\mathrm{d}\mathbb{P}} = \exp\left\{\frac{r-\mu}{\sigma}W_T - \frac{1}{2}\left(\frac{r-\mu}{\sigma}\right)^2 T\right\},$$

and

$$S_T = S_0 \exp\left\{\left(\mu - \frac{1}{2}\sigma^2\right)T + \sigma W_T\right\}.$$

Hence

$$\frac{\mathrm{d}\mathbb{Q}}{\mathrm{d}\mathbb{P}} = CS_T^{\frac{r-\mu}{\sigma^2}},$$

where C is a constant that depends on T. Then if we assume a bullish market: $\mu > r$, $\frac{\mathrm{d}\mathbb{Q}}{\mathrm{d}\mathbb{P}}$ is a decreasing function of S_T. So, any efficient payoff has to be an increasing function of S_T. In this context, the payoffs

$$X_1 = (K - S_T)_+,\ X_2 = K - S_T$$

are not efficient since they are decreasing functions of S_T. Now

$$\begin{aligned}\log\frac{1}{S_T} &\overset{d}{=} -\log S_0 - \left(\mu - \frac{1}{2}\sigma^2\right)T + \sigma W_T\\ &= \log S_T - 2\log S_0 - 2\left(\mu - \frac{1}{2}\sigma^2\right)T.\end{aligned}$$

That is

$$S_T \overset{d}{=} \frac{c}{S_T},$$

with $c = S_0^2 e^{(2\mu-\sigma^2)T}$. As a consequence the corresponding efficient payoffs of a put option and a short forward are respectively,

$$\bar{X}_1 = \left(K - \frac{c}{S_T}\right)_+ = \frac{K}{S_T}\left(S_T - \frac{c}{K}\right)_+,\ \bar{X}_2 = K - \frac{c}{S_T}.$$

and the corresponding prices of the original and efficient payoffs are:

Short forward contract: $Ke^{-rT} - S_0$; efficient: $Ke^{-rT} - S_0e^{(\mu-r)T}$

Put option : $Ke^{-rT}\Phi(d_-) - S_0\Phi(d_+)$;
Efficient:

$$Ke^{-rT}\Phi\left(d_- - \frac{2(\mu-r)\sqrt{T}}{\sigma}\right) - S_0 e^{(\mu-r)T}\Phi\left(d_+ - \frac{2(\mu-r)\sqrt{T}}{\sigma}\right),$$

with $d_\pm := \dfrac{\log\frac{K}{S_0} - (r \pm \frac{1}{2}\sigma^2)T}{\sigma\sqrt{T}}$.

Note that efficient prices depend on μ, so their estimation can be difficult.

Example 4 Consider the path-dependent payoff

$$X_3 = \left(e^{\frac{1}{T}\int_0^T \log(S_t)\mathrm{d}t} - K\right)_+ .$$

It can be shown that, under a Black-Scholes model, the efficient payoff is

$$\bar{X}_3 = c\left(S_T^{1/\sqrt{3}} - \frac{K}{c}\right)_+, \quad c = S_0^{1-\frac{1}{\sqrt{3}}} e^{\left(\frac{1}{2}-\frac{1}{\sqrt{3}}\right)(\mu-\frac{1}{2}\sigma^2)T}.$$

This is in agreement with Theorem 1: path dependent options have inefficient payoffs if $\frac{\mathrm{d}\mathbb{Q}}{\mathrm{d}\mathbb{P}} = CS_T^{\frac{r-\mu}{\sigma^2}}$. However if we a assume that the stock S evolves as

$$\mathrm{d}S_t = S_t\left(\mu_t \mathrm{d}t + \sigma_t \mathrm{d}W_t\right),$$

and the savings bank account as

$$\mathrm{d}B_t = r_t B_t \mathrm{d}t,$$

with μ_t, σ_t, r_t deterministic and càdlàg, then

$$\frac{\mathrm{d}\mathbb{Q}}{\mathrm{d}\mathbb{P}} = \exp\left\{\int_0^T \frac{r_t-\mu_t}{\sigma_t}\mathrm{d}W_t - \frac{1}{2}\int_0^T \left(\frac{r_t-\mu_t}{\sigma_t}\right)^2 \mathrm{d}t\right\},$$

so

$$\begin{aligned}
\frac{\mathrm{d}\mathbb{Q}}{\mathrm{d}\mathbb{P}} &= \exp\left\{\int_0^T \frac{r_t-\mu_t}{\sigma_t}\mathrm{d}W_t - \frac{1}{2}\int_0^T \left(\frac{r_t-\mu_t}{\sigma_t}\right)^2 \mathrm{d}t\right\} \\
&= \exp\left\{\int_0^T \frac{r_t-\mu_t}{\sigma_t^2}\frac{\mathrm{d}S_t}{S_t} - \frac{1}{2}\int_0^T \frac{r_t^2-\mu_t^2}{\sigma_t^2}\mathrm{d}t\right\} \\
&= C_T \exp\left\{\int_0^T \frac{r_t-\mu_t}{\sigma_t^2}\frac{\mathrm{d}S_t}{S_t}\right\}.
\end{aligned}$$

Then, any payoff that is a decreasing function of

$$V_T = \exp\left\{\int_0^T \frac{r_t - \mu_t}{\sigma_t^2} \frac{\mathrm{d}S_t}{S_t}\right\}$$

will be efficient. Consider for instance a put option

$$(K - S_T)_+$$

$$\log S_T \sim N\left(\int_0^T \left(\mu_t - \frac{1}{2}\sigma_t^2\right) \mathrm{d}t, \int_0^T \sigma_t^2 \mathrm{d}t\right) := N\left(m_T, v_T^2\right)$$

and

$$\log V_T \sim N\left(\int_0^T \frac{\mu_t (r_t - \mu_t)}{\sigma_t^2} \mathrm{d}t, \int_0^T \left(\frac{r_t - \mu_t}{\sigma_t}\right)^2 \mathrm{d}t\right) := N\left(a_T, b_T^2\right),$$

in such a way that an optimal payoff is

$$\left(K - V_T^{\frac{v_T}{b_T}} e^{\frac{v_T}{b_T}(m_T - a_T)}\right)_+,$$

since

$$V_T^{\frac{v_T}{b_T}} e^{\frac{v_T}{b_T}(m_T - a_T)} \stackrel{d}{=} S_T$$

and $K - V_T^{\frac{v_T}{b_T}} e^{\frac{v_T}{b_T}(m_T - a_T)}$ is a decreasing function of V_T. In this situation a path dependent option is better than a vanilla option! contrarily to what the title of Dybvig [9] suggests, as explained in Sect. 2.1.

4 Efficient Payoffs in a Dynamic Setting

Here we follow Becherer [1]. Consider the set of strictly positive self-financing portfolios with initial value one:

$$\mathcal{N} := \left\{N > 0 : N_t = 1 + \int_0^t \varphi_u \mathrm{d}S_u\right\}.$$

$N \in \mathcal{N}$ is said to be the *numeraire portfolio* (NP) if, for all $V \in \mathcal{N}$, V/N is a supermartingale (w.r.t. the probability measure $\mathbb{P}$). We say that an element of $\mathcal{N}$ is the *growth-optimal portfolio* (GOP) if it solves the maximization problem

$$u := \sup_{V\in\mathcal{N}} \mathbb{E}\left(\log V_T\right).$$

We have the following important results.

Theorem 5 *Assume $u < \infty$. Then the numeraire portfolio and the growth-optimal portfolio are the same.*

Proof See Proposition 4.3 in Becherer [1]. ■

Theorem 6 *If the market is complete the numeraire portfolio is given by*

$$N_t = \mathbb{E}\left(\left.\frac{\mathrm{d}\mathbb{P}}{\mathrm{d}\mathbb{Q}}\right|\mathcal{F}_t\right), 0 \le t \le T,$$

with $\mathcal{F}_t := \sigma(S_u, 0 \le u \le t)$.

Proof See Example 1 in Becherer [1]. ■

In the Black-Scholes model

$$\begin{aligned} N_t &= \exp\left\{-\frac{r-\mu}{\sigma}W_t + \frac{1}{2}\left(\frac{r-\mu}{\sigma}\right)^2 t\right\} \\ &= \exp\left\{-\frac{r-\mu}{\sigma}\tilde{W}_t - \frac{1}{2}\left(\frac{r-\mu}{\sigma}\right)^2 t\right\}, \end{aligned}$$

where $\tilde{W}$ is $\mathbb{Q}$-Brownian motion. We have seen that any efficient payoff can be written as a decreasing function of $\frac{\mathrm{d}\mathbb{Q}}{\mathrm{d}\mathbb{P}}$ and consequently as an increasing function of the final value of the numeraire portfolio N_T, say $\tilde{X} = h(N_T)$.

Then the (discounted) value of the replicating portfolio is given by

$$\tilde{V}_t = \mathbb{E}\left(\tilde{X}|\mathcal{F}_t\right) = \mathbb{E}\left(h(N_T)|\mathcal{F}_t\right) = \mathbb{E}\left(h\left(\frac{N_T}{N_t}x\right)\right)\Big|_{x=N_t} =: g(t, N_t),$$

from which (under smoothness assumptions on g), we get

$$\mathrm{d}\tilde{V}_t = \partial_2 g(t, N_t)\mathrm{d}N_t.$$

Hence V is a *locally optimal portfolio* in the sense that it has the largest discounted drift given a diffusion coefficient (Platen [18]) and

$$\frac{\partial_2 g(t, N_t)N_t}{\tilde{V}_t}$$

can be interpreted as a *risk aversion coefficient*.

If the market is incomplete, one uses the numeraire portfolio to get arbitrage free prices of a payoff X by

$$\mathbb{E}\left(\frac{X}{N_T}\right).$$

The latter is referred as the *benchmark approach* where the numeraire is chosen in such a way that the corresponding risk-neutral measure coincides with the historical one (see Platen and Heath [19]). In this case a payoff X is efficient iff X is an increasing function of N_T as above, but if we use a pricing measure $\mathbb{Q}$ a payoff X will be efficient iff it is a decreasing function of $\frac{\mathrm{d}\mathbb{Q}}{\mathrm{d}\mathbb{P}}$. In the continuous case both approaches coincide if we use the minimal martingale measure (see Schweizer [20]).

If we consider an exponential Lévy model for S:

$$\mathrm{d}S_t = S_{t-}\mathrm{d}Z_t,\, S_0 > 0,$$

where Z is a Lévy process with characteristics (d, c^2, ν) (with jumps strictly greater than -1) and the pricing measure $\mathbb{Q}$ is such that Z is a $\mathbb{Q}$-Lévy process it can be seen (see Corcuera et al. [7]) that

$$\frac{\mathrm{d}\mathbb{Q}}{\mathrm{d}\mathbb{P}} = aS_T^b e^{V_T},\ a > 0,\, b \in \mathbb{R} \text{ (that depends on } \mathbb{Q}),$$

and

$$V_T = \int_{-\infty}^{\infty} (\log H(x) - b\log(1+x))\, \tilde{M}((0,t], \mathrm{d}x).$$

with $H(x) = \frac{\mathrm{d}\tilde{\nu}}{\mathrm{d}\nu}(x)$ and where $\tilde{M}((0,t], \mathrm{d}x)$ is the compensated Poisson random measure associated with Z. Tilde indicates the characteristics w.r.t. $\mathbb{Q}$ (see Corcuera et al. [7] for more details).

In such cases an efficient payoff is an increasing function of $S_T^b e^{V_T}$ and only in the case that $V_T \equiv 0$ efficient payoffs are a monotone function of S_T. It corresponds to the case that $\mathbb{Q}$ is the Esscher measure, see von Hammerstein et al. [13].

The benchmark approach coincides with the pricing measure approach when

$$H(x) = \frac{1}{1-bx}, \quad \text{and}$$

$$c^2 b + d - r + b\int_{-\infty}^{\infty} \frac{x^2}{1-bx}\mathrm{d}\nu(x) = 0,$$

since in this case the optimal terminal wealth corresponding to the log-utility can be replicated by using stocks and bonds (see Corcuera et al. [7], Example 4.1).

It will be also interesting to include optimal consumption problem in this context, as for example it is done in Fajardo [10].

5 Conditional Efficient Payoffs

Reducing the importance of a payoff to its law is quite controversial. For instance when one buys a Call option he/she is buying a right to buy a stock at a certain price and this is lost if he/she takes another payoff with the same law but with different values. There are many other examples that suggest that, if there is no perfect correlation, the investor would like a fixed dependency w.r.t. some special payoff.

This approach was introduced by Takahashi-Yamamoto [21]. See also Bernard et al. [3] and Bernard et al. [4].

Suppose a benchmark payoff Y is given, and that the investor wishes to invest in another payoff X with a joint distribution (X, Y) fixed. In other words two payoffs X and Γ are equivalent if $(X, Y) \sim (\Gamma, Y)$, or, equivalently, if $X|Y = y \sim \Gamma|Y = y$ for all y. So, one wants to solve the problem

$$\min_{(X,Y)\sim F_{X,Y}} \mathbb{E}_{\mathbb{Q}}(X). \tag{6}$$

Firstly, given Z, we can find a function $g(Z, Y)$ such that $(X, Y) \sim (g(Z, Y), Y)$. In fact, if we assume that $F_{Z|Y}(z|y)$ is continuous, then, conditionally on $Y = y$, $F_{Z|Y}(Z|y) \sim U(0, 1)$ (note that the random variable $F_{Z|Y}(Z|Y)$ is, therefore, independent of Y) and $F^{-1}_{X|Y}(F_{Z|Y}(Z|y)|y)$ (where $F^{-1}_{X|Y}(\cdot|y)$ is the pseudo-inverse of $F_{X|Y}(\cdot|y)$) will be a random variable such that conditionally on $Y = y$ has the same law as X, then

$$(F^{-1}_{X|Y}(F_{Z|Y}(Z|Y)|Y), Y) \sim (X, Y)$$

and the function we are looking for is $g(z, y) = F^{-1}_{X|Y}(F_{Z|Y}(z|y)|y)$.

Now we can solve the optimization problem (6). We know that

$$\mathbb{E}_{\mathbb{Q}}(X) = \mathbb{E}\left(\frac{\mathrm{d}\mathbb{Q}}{\mathrm{d}\mathbb{P}}X\right),$$

so, since the law of X and $\frac{\mathrm{d}\mathbb{Q}}{\mathrm{d}\mathbb{P}}$ are fixed, if $X \sim h\left(\frac{\mathrm{d}\mathbb{Q}}{\mathrm{d}\mathbb{P}}\right)$ for some decreasing function h, we reach the lower bound for $\mathbb{E}\left(\frac{\mathrm{d}\mathbb{Q}}{\mathrm{d}\mathbb{P}}X\right)$. But we have to fix the conditional law, that is, we need that $(X, Y) \sim \left(h\left(\frac{\mathrm{d}\mathbb{Q}}{\mathrm{d}\mathbb{P}}\right), Y\right)$. Then, according to the previous step, we can take $h\left(\frac{\mathrm{d}\mathbb{Q}}{\mathrm{d}\mathbb{P}}\right) = g\left(\frac{\mathrm{d}\mathbb{Q}}{\mathrm{d}\mathbb{P}}, Y\right)$.

In fact we are solving the conditional problem: in the set of random variables X such that $X|Y = y$ is fixed, we solve the problem

$$\min_{X|Y=y\sim F_{X|Y}} \mathbb{E}\left(\frac{\mathrm{d}\mathbb{Q}}{\mathrm{d}\mathbb{P}}X \,\middle|\, Y = y\right)$$

and the solution is $F^{-1}_{X|Y}\left(F_{\frac{\mathrm{d}\mathbb{Q}}{\mathrm{d}\mathbb{P}}|Y}\left(\frac{\mathrm{d}\mathbb{Q}}{\mathrm{d}\mathbb{P}}\,\middle|\, y\right)\middle|\, y\right) = g\left(\frac{\mathrm{d}\mathbb{Q}}{\mathrm{d}\mathbb{P}}, y\right)$. Consequently

$$\min_{X|Y=y \sim F_{X|Y}} \mathbb{E}\left(\frac{d\mathbb{Q}}{d\mathbb{P}}X\right) = \mathbb{E}\left(\frac{d\mathbb{Q}}{d\mathbb{P}}\Gamma\right),$$

with $\Gamma = g\left(\frac{d\mathbb{Q}}{d\mathbb{P}}, Y\right)$. Three elements interact in the expression: the conditional law of X given Y, the price state density $\frac{d\mathbb{Q}}{d\mathbb{P}}$ and Y.

An additional reason to consider conditional efficient payoffs could be the existence of privileged information about a certain payoff Y. This might be object for future research.

Acknowledgements This research was carried out at CAS—Centre for Advanced Study at the Norwegian Academy of Science and Letters, Research group SEFE.

References

1. Becherer, D.: The numeraire portfolio for unbounded semimartingales. Financ. Stochast. **5**(3), 327–341 (2001)
2. Bernard, C., Boyle, P.P., Vanduffel, S.: Explicit representation of cost-efficient strategies. Finance **35**(2), 5–55 (2014a)
3. Bernard, C., Rüschendorf, L., Vanduffel, S.: Optimal claims with fixed payoff structure. J. Appl. Probab. **51**, 175–188 (2014b)
4. Bernard, C., Moraux, F., Rüschendorf, L., Vanduffel, S.: Optimal payoffs under state-dependent preferences. Quant. Financ. **15**(7), 1157–1173 (2015a)
5. Bernard, C., Chen, J.S., Vanduffel, S.: Rationalizing investors' choices. J. Math. Econ. **59**, 10–23 (2015)
6. Carlier, G., Dana, R.A.: Optimal demand for contingent claims when agents have law invariant utilities. Math. Financ. **21**(2), 169–201 (2011)
7. Corcuera, J.M., Guerra, J., Nualart, D., Schoutens, W.: Optimal investment in a Lévy market. Appl. Math. Optim. **53**(3), 279–309 (2006)
8. Dybvig, P.H.:Distributional analysis of portfolio choice. J. Bus. **61**(3), 369–393 (1988)
9. Dybvig, P.H.: Inefficient dynamic portfolio strategies or how to throw away a million dollars in the stock market. Rev. Finan. Stud. **1**(1), 67–88 (1988b)
10. Fajardo Barbachan, J.: Optimal consumption and investment with Lévy processes. Revista Brasileira de Economia **57**(4), 825–848 (2003)
11. Föllmer, H., Schied, A.: Stochastic finance: an introduction in discrete time. Walter de Gruyter (2011)
12. Fréchet, M.: Généralisation du théoreme des probabilités totales. Fundam. Math. **1**(25), 379–387 (1935)
13. Von Hammerstein, E.A., Lütkebohmert, E., Rüschendorf, L., Wolf, V.: Optimality of payoffs in Lévy models. Int. J. Theor. Appl. Financ. **17**(06), (2014)
14. Kahneman, D., Tversky, A.:Prospect theory: An analysis of decision under risk. Econometrica: J. Econometric Soc. 263–291 (1979)
15. Kassberger, S., Liebmann, T.: When are path-dependent payoffs suboptimal? J. Banking Financ. **36**(5), 1304–1310 (2012)
16. Lehmann, E.L.: Some concepts of dependence. Ann. Math. Stat. **35**(5), 1137–1153 (1966)
17. von Neumann, J., Morgenstern, O.: Theory of Games and Economic Behavior, 2nd edn. Princenton University Press, Princenton NJ (1947)
18. Platen, E.: Arbitrage in continuous complete markets. Adv. Appl. Probab. **34**(3), 540–558 (2002)

19. Platen, E., Heath, D.: A benchmark approach to quantitative finance. Springer Science & Business Media (2006)
20. Schweizer, M.: A minimality property of the minimal martingale measure. Stat. Probab. Lett. **42**(1), 27–31 (1999)
21. Takahashi, A., Yamamoto, K.: Generating a target payoff distribution with the cheapest dynamic portfolio: an application to hedge fund replication. Quant. Financ. **13**(10), 1559–1573 (2013)
22. Tversky, A., Kahneman, D.: Advances in prospect theory: Cumulative representation of uncertainty. J. Risk Uncertainty **5**(4), 297–323 (1992)
23. Vanduffel, S., Chernih, A., Maj, M., Schoutens, W.: A note on the suboptimality of path-dependent pay-offs in Lévy markets. Appl. Math. Financ. **16**(4), 315–330 (2009)
24. Yaari, M.E.: The dual theory of choice under risk. Econometrica: J. Econometric Soc. 95–115 (1987)

Construction and Hedging of Optimal Payoffs in Lévy Models

Ludger Rüschendorf and Viktor Wolf

Abstract The construction of lowest cost strategies for a given payoff has found considerable interest in recent literature and it has been shown in applications to real market data, that cost savings associated with these cost-efficient strategies can be quite substantial. In this paper we provide for a variety of options in the frame of Lévy models cost-efficient counterparts and determine the efficiency loss (resp. gain) in applications to several sets of market data. We discuss specific effects of the cost-efficient payoffs for a series of standard options like puts, calls, self-quanto puts and straddles and butterfly spread options, and develop their pricing. We obtain several new results on dependence of the magnitude of the efficiency loss on various model and option parameters. We show that the cost-efficient payoffs behave improved compared to the standard payoffs concerning hedging properties. We provide concrete hedging simulation schemes for various cost-efficient options. The results of the paper show that cost-efficient payoffs may lead to considerable reduction of cost in markets with pronounced trend.

Keywords Cost efficient payoffs · Levy model · Delta hedging · Esscher measure

1 Introduction to Cost-Efficient Payoffs

The concept of distributional analysis of portfolio choice has been introduced by Dybvig [8]. In a market model $(\Omega, \mathscr{F}, (\mathscr{F}_t)_{0\leq t\leq T}, P)$ with finite time horizon $[0, T]$ let $S = (S_t)_{0\leq t\leq T} \in \mathbb{R}^d$ be a price model for d stocks and $(Z_t)_{0\leq t\leq T}$ a pricing density rendering the discounted process $(e^{-rt}S_tZ_t)_{0\leq t\leq T}$ a P-martingale. The cost of a strategy with terminal payoff X_T then is given by the discounted expected payoff

$$c(X_T) = E[e^{-rT}Z_TX_T]. \tag{1}$$

L. Rüschendorf (✉) · V. Wolf
Department of Mathematical Stochastics, University of Freiburg,
Eckerstraße 1, 79104 Freiburg, Germany
e-mail: ruschen@stochastik.uni-freiburg.de

© Springer International Publishing Switzerland 2016

J. Kallsen and A. Papapantoleon (eds.), *Advanced Modelling in Mathematical Finance*,
Springer Proceedings in Mathematics & Statistics 189,
DOI 10.1007/978-3-319-45875-5_16

For a given payoff distribution G a strategy with terminal payoff $\underline{X}_T$ with distributed payoff G (i.e. $\underline{X}_T \sim G$) is called *cost-efficient* if it minimizes the cost i.e.

$$c(\underline{X}_T) = \min_{X_T \sim G} c(X_T). \tag{2}$$

A strategy with payoff $\overline{X}_T \sim G$ is called *most-expensive* if

$$c(\overline{X}_T) = \max_{X_T \sim G} c(X_T). \tag{3}$$

The difference of the costs $\ell(X_T) = c(X_T) - c(\underline{X}_T)$ is called the *efficiency loss* of X_T.

The following characterization of cost-efficient payoffs has been stated in various generality in a series of papers including Dybvig [8, 9], Jouini and Kallal [13], Dana [7], Schied [17], Burgert and Rüschendorf [6], Bernard and Boyle [1], Bernard et al. [2], Vanduffel et al. [19, 20], and Rüschendorf [14].

Theorem 1 (cost-efficient payoffs)

(a) For a given payoff distribution G it holds that

$$c(\underline{X}_T) = e^{-rT} \int_0^1 G^{-1}(u) F_{Z_T}^{-1}(1-u)du. \tag{4}$$

(b) A payoff $\underline{X}_T \sim G$ is cost-efficient if and only if $\underline{X}_T$ and Z_T are antimonotonic. $\overline{X}_T \sim G$ is most expensive if and only if $\overline{X}_T$ and Z_T are comonotonic.

(c) If F_{Z_T} is continuous then the cost-efficient resp. most expensive payoffs are given by

$$\underline{X}_T = G^{-1}(1 - F_{Z_T}(Z_T)) \quad \textit{resp.} \quad \overline{X}_T = G^{-1}(F_{Z_T}(Z_T)). \tag{5}$$

Theorem 1 has been applied in several papers to determine cost-efficient payoffs in particular in the context of the Samuelson model as well as in some classes of exponential Lévy processes (see Bernard et al. [2], Vanduffel et al. [19], Hammerstein et al. [11]) and has been applied to real market data. In the context of Lévy models $S_t = S_0 e^{L_t}$ with driving Lévy process $L = (L_t)$ the results have been mainly based on the Esscher measure defined by the pricing density

$$Z_t^{\overline{\theta}} = \frac{e^{\overline{\theta} L_t}}{M_{L_t(\overline{\theta})}} \tag{6}$$

where M_{L_t} denotes the moment generating function of L_t and $\overline{\theta}$, the Esscher parameter, is a solution to the equation

$$e^r = \frac{M_{L_1}(\overline{\theta} + 1)}{M_{L_1}(\overline{\theta})}. \tag{7}$$

Condition 7 implies that the Esscher measure $Q^{\overline{\theta}} = Z_T^{\overline{\theta}} P$ is a risk neutral measure for the discounted stock price process $(e^{-rt} S_t)_{0 \le t \le T}$. It has the pleasant property that w.r.t. $Q^{\overline{\theta}}$ L remains a Lévy process with modified parameters.

For exponential Lévy models one gets a simpler representation of efficient strategies and for the cost bounds.

Proposition 2 (cost-efficient payoffs in Lévy models) *Let $(L_t)_{t \ge 0}$ be a Lévy process with continuous distribution function F_{L_T} at maturity $T > 0$, and assume that a solution $\overline{\theta}$ of* (7) *exists.*
If $\overline{\theta} < 0$, the cost-efficient payoff $\underline{X}_T$ and the most-expensive payoff $\overline{X}_T$ with distribution function G are given by

$$\underline{X}_T = G^{-1}(F_{L_T}(L_T)) \textit{ and } \overline{X}_T = G^{-1}(1 - F_{L_T}(L_T)) \textit{ and}$$

$$E[e^{-rT} Z_T^{\overline{\theta}} X_T] \ge E[e^{-rT} Z_T^{\overline{\theta}} \underline{X}_T] = \frac{1}{M_{L_T}(\overline{\theta})} \int_0^1 e^{\overline{\theta} F_{L_T}^{-1}(1-y) - rT} G^{-1}(1-y)\, dy.$$

If $\overline{\theta} > 0$, the cost-efficient and the most-expensive payoffs are given by

$$\underline{X}_T = G^{-1}(1 - F_{L_T}(L_T)) \textit{ and } \overline{X}_T = G^{-1}(F_{L_T}(L_T)) \textit{ and}$$

$$E[e^{-rT} Z_T^{\overline{\theta}} X_T] \ge E[e^{-rT} Z_T^{\overline{\theta}} \underline{X}_T] = \frac{1}{M_{L_T}(\overline{\theta})} \int_0^1 e^{\overline{\theta} F_{L_T}^{-1}(y) - rT} G^{-1}(y)\, dy.$$

$\underline{X}_T$ and $\overline{X}_T$ are almost surely unique.

In Lévy models the market is bullish i.e. $E\frac{S_t}{S_0} > e^{rt}$ iff $\overline{\theta} < 0$ and the market is bearish iff $\overline{\theta} > 0$ (see Proposition 2.2 in Hammerstein et al. [11]). Furthermore, for $\overline{\theta} < 0$ a payoff X_T is cost-efficient iff X_T is an increasing function in L_T and for $\overline{\theta} > 0$, X_T is cost-efficient iff X_T is a decreasing function of L_T. In particular a put is inefficient in increasing markets where $\overline{\theta} < 0$ and a call is inefficient in decreasing markets ($\overline{\theta} > 0$). It is shown (for some examples) that the magnitude of efficiency loss is increasing in the magnitude of the trend in the market described by $|\overline{\theta}|$. As a consequence one gets that path dependent options are not cost-efficient and thus can be improved by cost-efficient options.

The main aim in this paper is to determine cost-efficient payoffs for several classes of monotone and nonmonotone options in Lévy models and thus to present a set of examples showing that the method of cost-efficiency can be used in a great variety of applications. We also extend the known results to describe the magnitude of the efficiency loss in dependence on the model parameters and on the hedging costs for the efficient payoff in comparison to the underlying payoffs. We apply and test the results for several real market data modeled by Lévy processes; in particular we consider the normal inverse Gaussian (*NIG*), the variance Gaussian (*VG*) and the normal model and consider two increasing and two decreasing markets. As a result we find that cost-efficient payoffs may lead to considerable reduction of costs. The magnitude of the efficiency loss depends essentially on the magnitude of trend in the market described by the absolute value of the Esscher parameter $|\overline{\theta}|$. Cost efficient payoffs like all other payoffs are typically not attainable in the incomplete Lévy models. We show however in several examples that cost-efficient payoffs have an improved behaviour concerning hedging compared to the basic payoffs. In particular cost-efficient payoffs do not need extra hedging costs compared to the basic payoffs. The results in this paper are given in the case of real markets and for the case of pricing by the Esscher martingale measure. Some extensions to the multivariate case and to further pricing principles are given in Rüschendorf and Wolf [15, 16]. Puts are inefficient in increasing markets when $\overline{\vartheta} < 0$. Nevertheless investors buy puts, not for their distributional characteristics but for the fact that they provide value when the markets fall. This observation shows that some investors have state-dependent constraints. In recent literature the cost-efficient approach has been extended to deal with such state-dependent constraints (see Bernard et al. [3, 4]). For several details in this paper we refer to the dissertation Wolf [21].

2 Lévy Models and Some Classes of Markets Data

As in Hammerstein et al. [11] we focus in this paper on the modeling of market data by three types of Lévy processes, the *NIG*, the *VG* and the normal model. We apply this modeling to market data of 4 stocks (Volkswagen, Allianz, ThyssenKrupp and E.ON). Two of the stock prices are increasing, two of them are decreasing in the observed period. In spite of the fact that the first two models lead to a better fit of the market data it turns out in the examples that the form of the efficient payoffs is largely independent of the chosen model and the magnitude of the efficiency loss is of similar size in all models and examples of options considered.

2.1 Lévy Models

In this subsection we give a short description of the Lévy models used in the applications in this paper. For a detailed introduction to these models and their role in financial modeling we refer to Eberlein [10], Schoutens [18].

NIG-Model

The NIG model is obtained as a special case of the generalized hyperbolic model $GH(\lambda, \alpha, \beta, \delta, \mu)$ by choosing $\lambda = -\frac{1}{2}$ and can be obtained as a normal mean-variance mixture with an inverse Gaussian mixing distribution. More specifically, if $X \sim NIG(\alpha, \beta, \delta, \mu)$, then X can be represented as

$$X \stackrel{d}{=} \mu + \beta Z + \sqrt{Z} W, \tag{8}$$

where $\mu \in \mathbb{R}$, $W \sim N(0, 1)$, and $Z \sim IG(\delta, \sqrt{\alpha^2 - \beta^2})$ is an inverse Gaussian distributed random variable with $\delta > 0$ and $0 \le |\beta| < \alpha$ that is independent of W. This representation also entails that the infinite divisibility of the mixing inverse Gaussian distribution transfers to the *NIG* mixture distribution, thus there exists a Lévy process $(L_t)_{t \ge 0}$ with $\mathscr{L}(L_1) = NIG(\alpha, \beta, \delta, \mu)$. The Lebesgue density $d_{NIG(\alpha,\beta,\delta,\mu)}$ is given by

$$\begin{aligned} d_{NIG(\alpha,\beta,\delta,\mu)}(x) &= \int_0^\infty d_{N(\mu+\beta y)}(x) d_{IG(\delta,\sqrt{\alpha^2-\beta^2})}(y)\, \mathrm{d}y \\ &= \mathsf{n}(\alpha, \beta, \delta) \frac{K_1(\alpha\sqrt{\delta^2 + (x-\mu)^2})}{\sqrt{\delta^2 + (x-\mu)^2}} e^{\beta(x-\mu)}, \end{aligned} \tag{9}$$

where K_1 is the modified Bessel function of third kind with index 1, and the norming constant $\mathsf{n}(\alpha, \beta, \delta)$ is given by

$$\mathsf{n}(\alpha, \beta, \delta) = \frac{\alpha\delta}{\pi} e^{\delta\sqrt{\alpha^2-\beta^2}}.$$

The corresponding moment generating function $M_{NIG(\alpha,\beta,\delta,\mu)}$ is of the form

$$M_{NIG(\alpha,\beta,\delta,\mu)}(u) = e^{u\mu + \delta\left(\sqrt{\alpha^2-\beta^2} - \sqrt{\alpha^2-(\beta+u)^2}\right)} \tag{10}$$

which is defined for all $u \in (-\alpha - \beta, \alpha - \beta)$. The Esscher parameter $\overline{\theta}$ of the risk neutral Esscher measure $Q^{\overline{\theta}}$, i.e. the solution of (7) (if it exists) is given by

$$\overline{\theta}_{NIG} = -\frac{1}{2} - \beta + \frac{r-\mu}{\delta}\sqrt{\frac{\alpha^2}{1 + (\frac{r-\mu}{\delta})^2} - \frac{1}{4}}.$$

Note that (L_t) remains a *NIG* Lévy process under $Q^{\overline{\theta}}$ with parameter β replaced by $\beta + \overline{\theta}$, $\overline{\theta} = \overline{\theta}_{NIG}$, i.e. w.r.t. $Q^{\overline{\theta}}$ holds: $L_1 \stackrel{d}{=} NIG(\alpha, \beta + \overline{\theta}, \delta, \mu)$.

Variance-Gamma Model

A Variance-Gamma distributed random variable $X \sim VG(\lambda, \alpha, \beta, \mu)$ can be represented as a normal mean-variance mixture as in Eq. (8), but in this case the mixing variable $Z \sim \Gamma(\lambda, \frac{\alpha^2-\beta^2}{2})$ is Gamma distributed with shape parameter $\lambda > 0$ and scale parameter $\frac{\alpha^2-\beta^2}{2}$ where $0 \le |\beta| < \alpha$. Again, the infinite divisibility of $\Gamma(\lambda, \frac{\alpha^2-\beta^2}{2})$ transfers to $VG(\lambda, \alpha, \beta, \mu)$. Analogously as above the corresponding Lebesgue density $d_{VG(\lambda,\alpha,\beta,\mu)}$ is given by

$$d_{VG(\lambda,\alpha,\beta,\mu)}(x) = \mathsf{m}(\lambda, \alpha, \beta)|x-\mu|^{\lambda-\frac{1}{2}} K_\lambda(\alpha|x-\mu|)e^{\beta(x-\mu)} \tag{11}$$

with the norming constant

$$\mathsf{m}(\lambda, \alpha, \beta) = \frac{(\alpha^2-\beta^2)^\lambda}{\sqrt{\pi}(2\alpha)^{\lambda-\frac{1}{2}}\Gamma(\lambda)},$$

and the moment generating function is of the form

$$M_{VG(\lambda,\alpha,\beta,\mu)}(u) = e^{u\mu}\frac{\mathsf{m}(\lambda, \alpha, \beta)}{\mathsf{m}(\lambda, \alpha, \beta+u)} = e^{u\mu}\Big(\frac{\alpha^2-\beta^2}{\alpha^2-(\beta+u)^2}\Big)^\lambda \tag{12}$$

which is defined for all $u \in (-\alpha-\beta, \alpha-\beta)$. Here we have

$$\lim_{u\to\pm\alpha-\beta} M_{VG(\lambda,\alpha,\beta,\mu)}(u) = \infty,$$

and as consequence the condition $2\alpha > 1$ is sufficient to guarantee a unique solution $\overline{\theta}$ of Eq. (7) in the VG case. Some lengthy calculations (see Wolf [21]) show that the Esscher parameter $\overline{\theta}$, i.e. the solution of (7), is given by

$$\overline{\theta}_{VG} = \begin{cases} -\frac{1}{1-e^{-\frac{r-\mu}{\lambda}}} - \beta + \mathrm{sign}(r-\mu)\sqrt{\frac{e^{-\frac{r-\mu}{\lambda}}}{\left(1-e^{-\frac{r-\mu}{\lambda}}\right)^2} + \alpha^2}, & r \neq \mu, \\ -\frac{1}{2} - \beta, & r = \mu. \end{cases} \tag{13}$$

For $L_t \sim VG(\lambda t, \alpha, \beta, \mu t)$ the law of L_t under the Esscher martingale measure is again Variance-Gamma distributed $L_t \sim VG(\lambda t, \alpha, \beta + \overline{\theta}_{VG}, \mu t)$.

Samuelson Model

The classical benchmark model which also is at the basis of the Black–Scholes theory is to assume that the stock price process $(S_0 e^{L_t})_{t\geq 0}$ follows a geometric Brownian motion. In this case, the driving Lévy process is given by

$$L_t = \big(\mu - \frac{\sigma^2}{2}\big)t + \sigma B_t,\ t > 0$$

where $(B_t)_{t\geq 0}$ is a standard Brownian motion under the physical measure P, μ is the drift and σ the volatility parameter. Here we have $\mathscr{L}(L_t) = N((\mu - \frac{\sigma^2}{2})t, \sigma^2 t)$, its Lebesgue density is given by

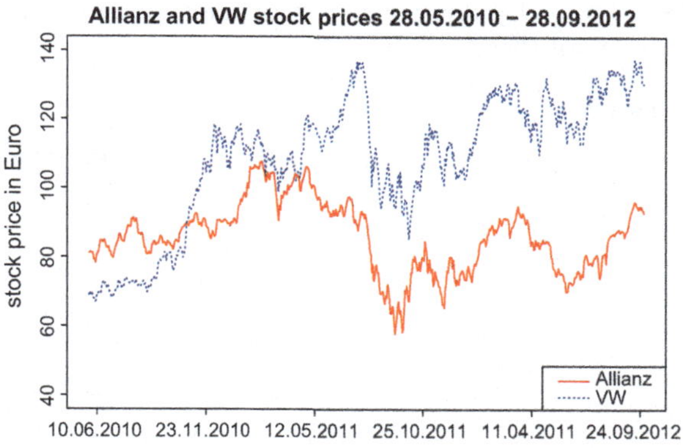

Fig. 1 Daily closing prices of Allianz and Volkswagen used for parameter estimation

$$d_{L_t}(x) = \frac{1}{\sqrt{2\pi\sigma^2 t}} e^{-\frac{1}{2}\frac{(x-(\mu-\frac{\sigma^2}{2})t)^2}{\sigma^2 t}},$$

and the moment generating function of L_1 equals $M_{N(\mu-\frac{\sigma^2}{2},\sigma^2)}(u) = e^{u(\mu-\frac{\sigma^2}{2})+\frac{\sigma^2 u^2}{2}}$. The Esscher parameter $\overline{\theta}$ is a solution of $e^r = \frac{M_{L_1}(\overline{\theta}_N+1)}{M_{L_1}(\overline{\theta}_N)} = e^{\mu+\overline{\theta}_N\sigma^2}$ and is given by $\overline{\theta}_N = \frac{r-\mu}{\sigma^2}$.

Under the Esscher martingale measure $Q^{\overline{\theta}}$ holds $L_t \sim N((r-\frac{\sigma^2}{2})t, \sigma^2 t)$.

2.2 *Modeling of Market Data*

We apply the Lévy models from Sect. 2.1 to model German stock price data for Allianz and Volkswagen and for E.ON and ThyssenKrupp from May 28, 2010, to September 28, 2012, which are shown in Fig. 1 and in Fig. 2 respectively. The estimated parameters and the corresponding Esscher parameter from the daily log-returns of Allianz and Volkswagen are given in Table 1 and of E.ON and ThyssenKrupp in Table 2.

The fitted densities in the three Lévy models are displayed in Fig. 3 for Allianz and Volkswagen and in Fig. 4 for E.ON and ThyssenKrupp. It stands out that the normal density curve fits worse than the *NIG* and *VG* density. The interest rate used to calculate the Esscher parameter $\overline{\theta}$ is $r = 4.2027 \times 10^{-6}$ which corresponds to the continuously compounded daily-Euribor rate of October 1, 2012. From the estimated Esscher parameter we find that the Allianz and Volkswagen data have a positive trend while the ThyssenKrupp and E.ON data have a slight negative trend.

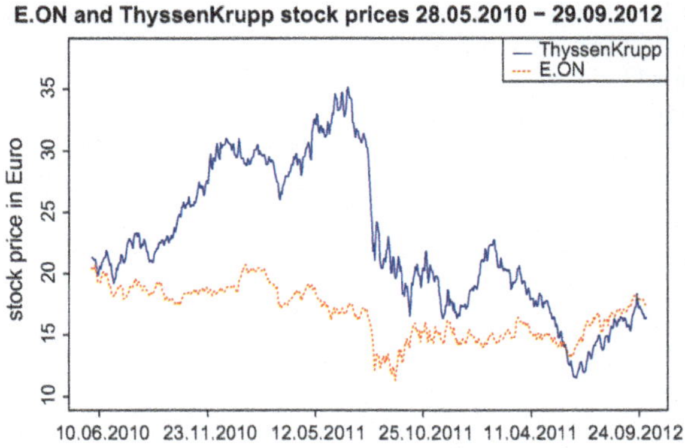

Fig. 2 Daily closing prices of E.ON and ThyssenKrupp used for parameter estimation

Table 1 Estimated parameters from daily log-returns of Allianz and Volkswagen for the NIG, the VG, and the Samuelson model

Allianz	λ	α	β	δ	μ	$\overline{\theta}$
NIG	−0.5	35.01998	−0.36857	0.01478	0.000376	−1.01266
VG	1.03086	72.01061	0.55168	0.0	1.941×10^{-8}	−1.04116
Normal	$\mu = 4.2757 \times 10^{-4}$, $\sigma = 0.02026$					−1.03143
Volkswagen	λ	α	β	δ	μ	$\overline{\theta}$
NIG	−0.5	48.85903	−0.84151	0.02313	0.001451	−2.70867
VG	1.60198	82.94782	−2.16537	0.0	0.00206	−2.73948
Normal	$\mu = 0.00129$, $\sigma = 0.02162$					−2.74475

Table 2 Estimated parameters from daily log-returns of E.ON and ThyssenKrupp for the NIG, the VG, and the Samuelson model

E.ON	λ	α	β	δ	μ	$\overline{\theta}$
NIG	−0.5	44.831	−0.639	0.016	-5.25×10^{-5}	0.297816
VG	1.276	86.399	−0.63	0.0	-6.17×10^{-5}	0.322992
Normal	$\mu = -0.0001$, $\sigma = 0.018878$					0.293082
ThyssenKrupp	λ	α	β	δ	μ	$\overline{\theta}$
NIG	−0.5	42.01665	−2.08815	0.02554	0.000846	0.203533
VG	1.43896	69.05434	−0.92983	0.0	0.000137	0.210135
Normal	$\mu = -0.000128$, $\sigma = 0.02447$					0.220797

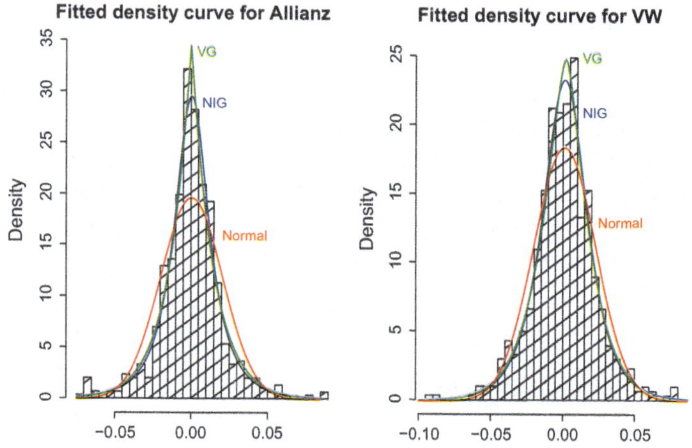

Fig. 3 Fitted density curves for Allianz and Volkswagen

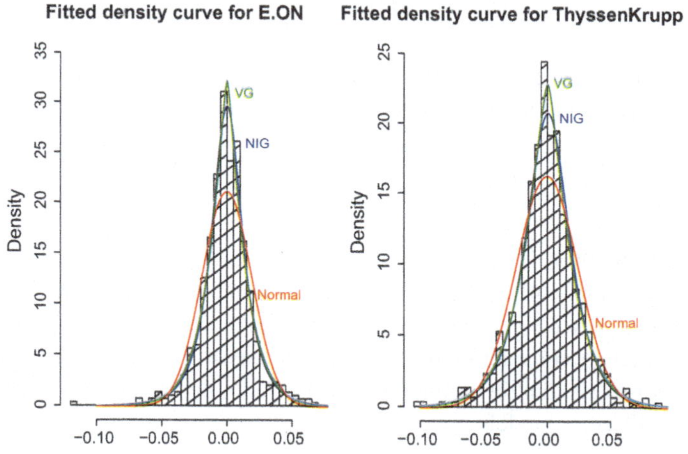

Fig. 4 Fitted density curves for E.ON and ThyssenKrupp

3 Cost-Efficient Payoffs for Monotone and Nonmonotone Options

In this section we apply the results on cost-efficiency to a series of options in the class of Lévy models in Sect. 2 and apply them to the market data introduced and modeled in Sect. 2.2. The examples give an impression on the magnitude of efficiency loss in terms of the parameters and shows that this methodology is also applicable to the improvement of nonmonotone options where calculations typically have to be done numerically.

3.1 Put Options

(Long) put options are inefficient in increasing markets where $\overline{\theta} < 0$ (see Hammerstein et al. [11]). Thus calculation of cost-efficient options is only of interest in the Volkswagen (VW), Allianz (Al) examples. We start with an example which was already analyzed in Hammerstein et al. [11]. We give a short presentation of these results, extend them in various respects and compare with the following options.

For a put option with strike K and maturity $T > 0$, i.e.

$$X_T^{\text{Put}} = (K - S_T)_+ = (K - S_0 e^{L_T})_+ \tag{14}$$

the payoff distribution is given by

$$G_{\text{Put}}(x) = P\big(X_T^{\text{Put}} \le x\big) = \begin{cases} 1, & \text{if } x \ge K, \\ 1 - F_{L_T}\big(\ln\big(\frac{K-x}{S_0}\big)\big), & \text{if } 0 \le x < K, \\ 0, & \text{if } x < 0. \end{cases} \tag{15}$$

Applying Proposition 2 for $\overline{\theta} < 0$ the cost-efficient payoff that generates the same distribution G_{Put} as the long put is given by

$$\underline{X}_T^{\text{Put}} = G_{\text{Put}}^{-1}(F_{L_T}(L_T)) = \big(K - S_0 e^{F_{L_T}^{-1}(1 - F_{L_T}(L_T))}\big)_+ \tag{16}$$

with payoff function $\underline{\omega}^{\text{Put}}(y) := \big(K - S_0 e^{F_{L_T}^{-1}(1 - F_{L_T}(\ln(\frac{y}{S_0})))}\big)_+$.
Fig. 5 displays the payoff X_T^{Put} of a long put option on one Allianz stock with strike $K = 98$ and maturity $T = 23$ days, and its cost-efficient counterparts $\underline{X}_T^{\text{Put}}$ for the three Lévy models under consideration. Although the payoff profiles look quite similar, a closer look reveals that the optimal payoff is model-dependent and slightly varies between the different models.

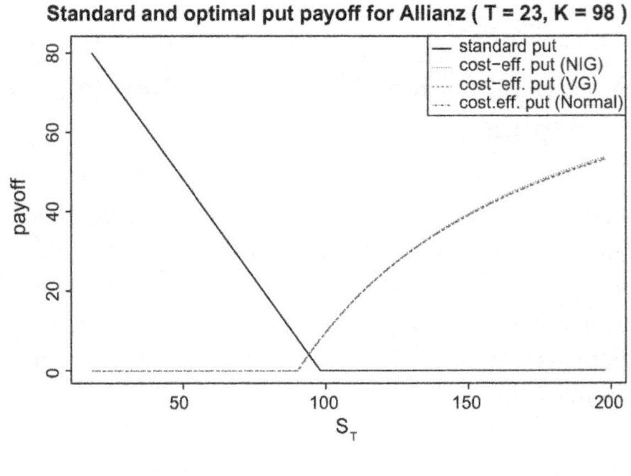

Fig. 5 Classical put and its cost-efficient counterparts for Allianz. $S_0 = 93.42$, closing price October 1, 2012, $\overline{\vartheta} < 0$

Table 3 Comparison of the cost of a long put option on Allianz and Volkswagen, resp., and the corresponding cost-efficient payoffs in different Lévy models. $S_0 = 93.42$, $K = 98$, $T = 23$ for Allianz and $S_0 = 130.55$, $K = 135$, $T = 23$ for Volkswagen

Allianz	$c(X_T^{\text{Put}})$	$c(\underline{X}_T^{\text{Put}})$	Efficiency loss in %
NIG	6.4495	5.2825	18.09
VG	6.3681	5.2270	17.92
Normal	6.4324	5.2683	18.10
Volkswagen	$c(X_T^{\text{Put}})$	$c(\underline{X}_T^{\text{Put}})$	Efficiency loss in %
NIG	8.0064	4.0871	48.95
VG	7.9765	4.0603	49.10
Normal	7.9909	4.0749	49.01

Note that the cost-efficient long put payoff is increasing and is bounded by K as is its vanilla counterpart, as follows from

$$\lim_{S_T\to\infty} \underline{X}_T^{\text{Put}} = \lim_{S_T\to\infty} \left(K - S_0 e^{F_{L_T}^{-1}(1-F_{L_T}(\ln(\frac{S_T}{S_0})))}\right)_+ = K.$$

In Table 3 we compare the cost of a long put option on Allianz and Volkswagen with their cost-efficient counterparts for the Lévy models discussed in Sect. 2. All computations are based on the estimated parameters given in Table 1 above. The initial stock prices S_0 of Allianz resp. Volkswagen are the closing prices at October 1, 2012, and the time to maturity is chosen to be $T = 23$ trading days, meaning that the put options mature on November 1, 2012. According to Proposition 2, the cost of the efficient put can be calculated by

$$c(\underline{X}_T^{\text{Put}}) = E\left[e^{-rT} Z_T^{\overline{\theta}} \underline{X}_T^{\text{Put}}\right] = \frac{1}{M_{dist}(\overline{\theta})} \int_0^1 e^{\overline{\theta} F_{dist}^{-1}(1-y)-rT} \left(K - S_0 e^{F_{dist}^{-1}(y)}\right)_+ \mathrm{d}y \quad (17)$$

where *dist* is $NIG(\alpha, \beta, \delta T, \mu T)$, $VG(\lambda T, \alpha, \beta, \mu T)$, or $N((\mu - \frac{\sigma^2}{2})T, \sigma^2 T)$.

The cost $c(X_T^{\text{Put}})$ of the vanilla put in the *NIG* model is given by

$$\begin{aligned} c(X_T^{\text{Put}}) &= E_{\overline{\theta}}\left[e^{-rT}(K - S_T)_+\right] \\ &= e^{-rT} \int_{-\infty}^{\ln(K/S_0)} (K - S_0 e^x) Z_T^{\overline{\theta},x}\, d_{NIG(\alpha,\beta,\delta T,\mu T)}(x)\, \mathrm{d}x \\ &= e^{-rT} K F_{NIG(\alpha,\beta+\overline{\theta},\delta T,\mu T)}\left(\ln\left(\tfrac{K}{S_0}\right)\right) - S_0 F_{NIG(\alpha,\beta+\overline{\theta}+1,\delta T,\mu T)}\left(\ln\left(\tfrac{K}{S_0}\right)\right), \end{aligned} \quad (18)$$

where $Z_T^{\overline{\theta},x} = \frac{e^{\overline{\theta}x}}{M_{NIG(\alpha,\beta,\delta T,\mu T)}(\overline{\theta})}$. For the *VG* model one analogously obtains

$$c(X_T^{\text{Put}}) = e^{-rT} K F_{VG(\lambda T,\alpha,\beta+\overline{\theta},\mu T)}\left(\ln\left(\tfrac{K}{S_0}\right)\right) - S_0 F_{VG(\lambda T,\alpha,\beta+\overline{\theta}+1,\mu T)}\left(\ln\left(\tfrac{K}{S_0}\right)\right). \quad (19)$$

In the Samuelson model, $c(X_T^{\text{Put}})$ is calculated by the well-known Black–Scholes put price formula.

In an exponential Lévy model with Lévy process $L_T \overset{d}{=} NIG(\alpha, \beta, \delta T, \mu T)$ or $L_T \overset{d}{=} VG(\lambda T, \alpha, \beta, \mu T)$ and with Esscher parameter $\overline{\theta}$ we have

$$c(X_T^{\text{Put}}) = e^{-rT} K F_{L_T^{\overline{\theta}}}\Big(\ln\Big(\frac{K}{S_0}\Big)\Big) - S_0 F_{L_T^{\overline{\theta}+1}}\Big(\ln\Big(\frac{K}{S_0}\Big)\Big), \tag{20}$$

since $L_T^{\overline{\theta}+k} \overset{d}{=} NIG(\alpha, \beta + \overline{\theta} + k, \delta T, \mu T)$ or $VG(\lambda T, \alpha, \beta + \overline{\theta} + k, \mu T)$, $k = 0, 1$. If $L_T \overset{d}{=} N((\mu - \frac{\sigma^2}{2})T, \sigma^2 T)$, then we have

$$c(X_T^{\text{Put}}) = e^{-rT} K \Phi(h) - S_0 \Phi(h - \sigma\sqrt{T}), \tag{21}$$

where $h = \frac{1}{\sigma\sqrt{T}}(\ln(\frac{K}{S_0}) - (r - \frac{\sigma^2}{2})T)$.

For symmetric Lévy processes which fulfill $L_T \overset{d}{=} vT - L_T$, where $v \in \mathbb{R}$, the cost of the cost-efficient put (17) can be evaluated without any integration. This symmetry holds true in particular in the normal case. Numerical computations of prices, then, become a lot easier. The price formula for the cost-efficient put in the Samuelson model is given in Bernard et al. [2, Sect. 5.2]. Some similar calculations yield in the Lévy case the following result (for details see Wolf [21, Proposition 5.25]).

Proposition 3 (Price of efficient puts in symmetric Lévy models) *Let X_T^{Put} be the payoff of a vanilla put option with strike K, maturity $T > 0$. Suppose $(L_t)_{t\geq 0}$ is a Lévy process such that $L_T \overset{d}{=} vT - L_T$. If $\overline{\theta}$ is an Esscher parameter, then the cost of the cost-efficient put $\underline{X}_T^{\text{Put}}$ w.r.t. the Esscher measure is given by $c(X_T^{\text{Put}})$ if $\overline{\theta} > 0$ and by*

$$\begin{aligned} c(\underline{X}_T^{\text{Put}}) &= e^{-rT} K \Big(1 - F_{L_T^{\overline{\theta}}}\Big(\ln\Big(\frac{S_0}{K}\Big) + vT\Big)\Big) \\ &\quad - e^{-(r-v)T} S_0 \frac{M_{L_T}(\overline{\theta} - 1)}{M_{L_T}(\overline{\theta})} \Big(1 - F_{L_T^{\overline{\theta}-1}}\Big(\ln\Big(\frac{S_0}{K}\Big) + vT\Big)\Big) \end{aligned} \tag{22}$$

if $\overline{\theta} < 0$. Here $L_T^{\overline{\theta}}$ denotes the Lévy process at maturity under the Esscher measure $Q^{\overline{\theta}}$. In particular, in the Samuelson model we have that $L_T \overset{d}{=} 2(\mu - \frac{\sigma^2}{2})T - L_T$. Thus for $\overline{\theta} < 0$

$$c(\underline{X}_T^{\text{Put}}) = e^{-rT} K \Phi(\underline{h}) - e^{2(\mu - r)T} S_0 \Phi(\underline{h} - \sigma\sqrt{T}) \tag{23}$$

where $\underline{h} = \frac{1}{\sigma\sqrt{T}}\big(\ln(\frac{K}{S_0}) - (\mu - \frac{\sigma^2}{2})T + (r - \mu)T\big)$.

The results from Table 3 show that the savings from choosing the cost-efficient strategies can be quite large: For Allianz, the cost of the efficient put is less than 83 % of the price of the plain vanilla put, and in case of Volkswagen the vanilla put is almost twice as expensive as the efficient put. The great differences in the efficiency losses of the Allianz and Volkswagen puts may seem somewhat surprising at first glance because the stock price to strike ratio $\frac{S_0}{K}$ is roughly the same in both cases (0.953 for Allianz and 0.967 for Volkswagen). But the difference is induced by the greater magnitude of positive trend in the VW data compared to Allianz as seen from Table 2. The value of $|\overline{\theta}|$ for Volkswagen is more than 2.5 times as large as that of Allianz. For each stock itself the efficiency losses obtained under the different Lévy models are of almost the same size and thus seem to be widely model-independent.

In contrast to the latter static formulas and results we also consider time dynamic behaviour of the cost of the cost-efficient payoff. Therefore, we keep the payoff function $\underline{\omega}^{\text{Put}}(y) = \left(K - S_0 e^{F_{L_T}^{-1}(1-F_{L_T}(\ln(\frac{y}{S_0})))}\right)_+$ of the cost-efficient long put fixed within the trading period $[0, T]$. The payoff function $\underline{\omega}^{\text{Put}}$ depends on S_0 which becomes a location parameter in this context. In consequence, the price at time $t < T$ of a cost-efficient long put with maturity T is given by

$$c_t(\underline{X}_T^{\text{Put}}) = e^{-r(T-t)} E\left[Z^{\overline{\theta}}_{T-t}\left(K - S_0 e^{F_{L_T}^{-1}\left(1-F_{L_T}\left(\ln(\frac{y}{S_0})+L_{T-t}\right)\right)}\right)_+\right]\Big|_{y=S_t}. \tag{24}$$

In Fig. 6 we notice that during some time in the trading period $[0, T]$ the cost of the cost-efficient long put exceeds the cost of the plain vanilla long put. For the case of the Allianz stock this is even beneficial for writers of the cost-efficient long put $\underline{X}_T^{\text{Put}}$ since at maturity, November 1, 2012, the higher price corresponds to a higher payout. However, we could also have the reverse situation. In other words, an initially optimal strategy may become less profitable as its vanilla counterpart if the market scenario significantly changes in between.

Although, the cost-efficient put behaves like a modified call, i.e. it is increasing in L_T, both X_T^{Put} and $\underline{X}_T^{\text{Put}}$ end up in the money, whereas a plain vanilla call would expire worthless. But besides this abnormal behaviour the progression of the cost of the long put and its cost-efficient counterpart exhibit similar price behaviour as one would expect from vanilla long call and put options.

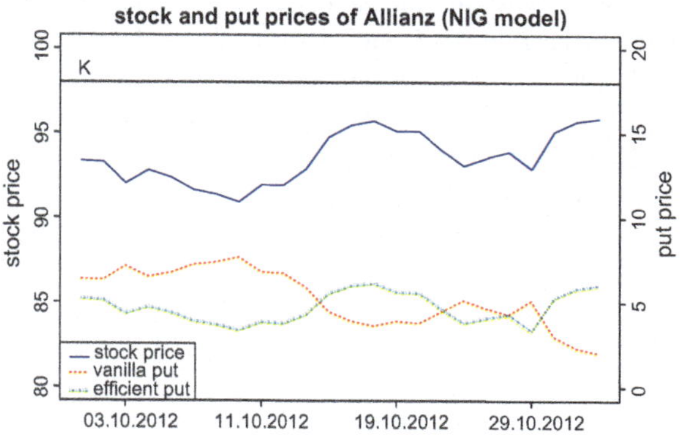

Fig. 6 Stock and put prices along the period $[0, T]$ for Allianz strike $K = 98$, maturity $T = 23$ days

Remark 4 (**Short put option**) Similarly for the short put $X_T^{-\text{Put}} = -(K - S_0 e^{L_T})_+$ which is inefficient for $\underline{\theta} > 0$ the cost-efficient version is given by

$$\underline{X}_T^{-\text{Put}} = G_{-\text{Put}}^{-1}(1 - F_{L_T}(L_T)) = \big(S_0 e^{F_{L_T}^{-1}(1-F_{L_T}(L_T))} - K\big)_-. \tag{25}$$

and we obtain by simple arguments the following duality relation.
For $\overline{\theta} < 0$ holds

$$\underline{X}_T^{\text{Put}} = -\overline{X}_T^{-\text{Put}} \text{ and } c\big(\underline{X}_T^{\text{Put}}\big) = -c\big(\overline{X}_T^{-\text{Put}}\big). \tag{26}$$

Similarly, if $\overline{\theta} > 0$ we have

$$\underline{X}_T^{-\text{Put}} = -\overline{X}_T^{\text{Put}} \quad \text{as well as} \quad c\big(\underline{X}_T^{-\text{Put}}\big) = -c\big(\overline{X}_T^{\text{Put}}\big).$$

3.2 Call Options

Call options are inefficient in decreasing markets i.e. when $\overline{\theta} > 0$. For the call $X_T^{\text{Call}} = (S_T - K)_+ = (S_0 e^{L_T} - K)_+$ with payoff function $\omega^{\text{Call}}(y) := (y - K)_+$ we obtain the payoff distribution function $G_{\text{Call}} = F_{X_T^{\text{Call}}}$ by

$$G_{\text{Call}}(x) = P\big(X_T^{\text{Call}} \le x\big) = \begin{cases} F_{L_T}\big(\ln\big(\frac{K+x}{S_0}\big)\big), & \text{if } x \ge 0, \\ 0, & \text{if } x < 0. \end{cases} \tag{27}$$

Applying Proposition 2 for $\overline{\theta} > 0$ the cost-efficient payoff that generates the same distribution G_{Call} as the long call option is given by

$$\underline{X}_T^{\text{Call}} = G_{\text{Call}}^{-1}(1 - F_{L_T}(L_T)) = \Big(S_0 e^{F_{L_T}^{-1}(1-F_{L_T}(L_T))} - K\Big)_+ \tag{28}$$

with payoff function $\underline{\omega}^{\text{Call}}(y) := (S_0 e^{F_{L_T}^{-1}(1-F_{L_T}(\ln(\frac{y}{S_0})))} - K)_+$.

Figure 7 displays the payoff X_T^{Call} of a long call option on one ThyssenKrupp stock with strike $K = 16.5$ and maturity $T = 23$ days, and its cost-efficient counterparts $\underline{X}_T^{\text{Call}}$ for the three Lévy models under consideration. As seen before the optimal payoff is model-dependent and slightly varies between the different models.

Next, we state formulas for the cost of the standard call in the three Lévy models. Let $(L_t)_{t\ge 0}$ be a Lévy process, with $\mathscr{L}(L_T) = NIG(\alpha, \beta, \delta T, \mu T)$ or $VG(\lambda T, \alpha, \beta, \mu T)$. If $\overline{\theta}$ is a Esscher parameter for L, then we have

$$c(X_T^{\text{Call}}) = S_0\Big(1 - F_{L_T^{\overline{\theta}+1}}\Big(\ln\Big(\frac{K}{S_0}\Big)\Big)\Big) - e^{-rT}K\Big(1 - F_{L_T^{\overline{\theta}}}\Big(\ln\Big(\frac{K}{S_0}\Big)\Big)\Big), \tag{29}$$

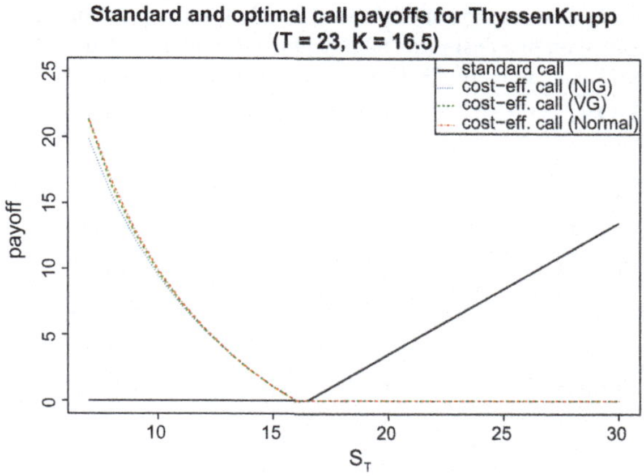

Fig. 7 Classical call and its cost-efficient counterparts for ThyssenKrupp. $S_0 = 16.73$, closing price October 1, 2012, $\overline{\vartheta} > 0$

where $\mathscr{L}(L_T^{\overline{\theta}+k})$ is $NIG(\alpha, \beta + \overline{\theta} + k, \delta T, \mu T)$ or $VG(\lambda T, \alpha, \beta + \overline{\theta} + k, \mu T)$, $k = 0, 1$. If $\mathscr{L}(L_T) = N((\mu - \frac{\sigma^2}{2})T, \sigma^2 T)$, then we have

$$c(X_T^{\text{Call}}) = S_0 \Phi(-h + \sigma\sqrt{T}) - e^{-rT} K \Phi(-h), \tag{30}$$

where $h = \frac{1}{\sigma\sqrt{T}}(\ln(\frac{K}{S_0}) - (r - \frac{\sigma^2}{2})T)$.

Similarly to the case of a put in (14) also for a call a simple formula can be given for cost-efficient calls for symmetric Lévy models which fulfill $L_T \stackrel{d}{=} vT - L_T$, $v \in \mathbb{R}$.

Proposition 5 (Price of efficient calls in symmetric Lévy models)

Suppose $(L_t)_{t\geq 0}$ is a Lévy process such that $\mathscr{L}(L_T) = \mathscr{L}(vT - L_T)$. If $\overline{\theta}$ is a Esscher parameter, then the cost of the cost-efficient call $\underline{X}_T^{\text{Call}}$ equals

$$\begin{aligned} c(\underline{X}_T^{\text{Call}}) =& e^{-(r-v)T} S_0 \frac{M_{L_T}(\overline{\theta} - 1)}{M_{L_T}(\overline{\theta})} F_{L_T^{\overline{\theta}-1}}\left(\ln\left(\frac{S_0}{K}\right) + vT\right) \\ & - e^{-rT} K F_{L_T^{\overline{\theta}}}\left(\ln\left(\frac{S_0}{K}\right) + vT\right) \end{aligned} \tag{31}$$

if $\overline{\theta} > 0$, where $L_T^{\overline{\theta}}$ denotes the Lévy process at maturity under the Esscher measure $Q^{\overline{\theta}}$. In particular, in the Samuelson model we have that $L_T \stackrel{d}{=} 2(\mu - \frac{\sigma^2}{2})T - L_T$, thus

$$c(\underline{X}_T^{\text{Call}}) = e^{2(\mu-r)T} S_0 \Phi(-\underline{h} + \sigma\sqrt{T}) - e^{-rT} K \Phi(-\underline{h}) \tag{32}$$

if $\overline{\theta} > 0$, where $\underline{h} = \frac{1}{\sigma\sqrt{T}}\left(\ln(\frac{K}{S_0}) - (\mu - \frac{\sigma^2}{2})T + (r - \mu)T\right)$.

In Table 4 we compare the cost of a long call option on E.ON and ThyssenKrupp with their cost-efficient counterparts for the Lévy models discussed in Sect. 2. The results from Table 4 show that the savings from choosing the cost-efficient strategies

Table 4 Comparison of the cost of a long call option on E.ON and ThyssenKrupp, resp., and the corresponding cost-efficient payoffs in different Lévy models. $S_0 = 17.48$, $K = 17.24$, $T = 23$ for E.ON and $S_0 = 16.73$, $K = 16.5$, $T = 23$ for ThyssenKrupp

E.ON	$c(X_T^{\text{Call}})$	$c(\underline{X}_T^{\text{Call}})$	Efficiency loss in %
NIG	0.7502	0.7018	6.45
VG	0.7398	0.6893	6.83
Normal	0.7550	0.7073	6.32
Thyssen	$c(X_T^{\text{Call}})$	$c(\underline{X}_T^{\text{Call}})$	Efficiency loss in %
NIG	0.9016	0.8484	5.90
VG	0.8989	0.8443	6.07
Normal	0.8987	0.8418	6.33

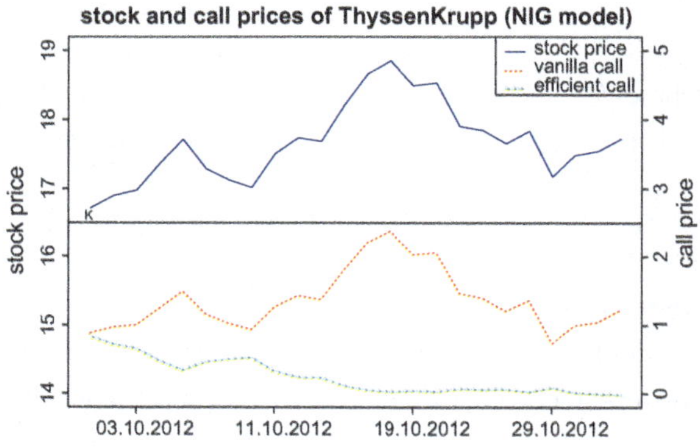

Fig. 8 Stock and call prices along the period $[0, T]$ for ThyssenKrupp, strike $K = 16.5$, maturity $T = 23$ days

are close to each other as is the magnitude of $|\overline{\theta}|$ for ThyssenKrupp and E.ON (compare Table 2). The time dynamic behaviour of the cost-efficient call compared to the standard call is displayed in Fig. 8. There we notice that the formerly bearish market setting of the ThyssenKrupp changed to a bullish market setting, since the drift of the stock price altered its direction from negative to positive (cf. Fig. 2). Moreover, the stock price remains above the strike during the entire trading period $[0, T]$. As the cost-efficient long call behaves like a modified put, it decreases over $[0, T]$ and expires worthless. Indeed, this is an unpropitious example for writers of the cost-efficient long call $\underline{X}_T^{\text{Call}}$.

Again as in the put case we get the following symmetry relation between long and short calls: If $\overline{\theta} > 0$, it holds that

$$\underline{X}_T^{\text{Call}} = -\overline{X}_T^{-\text{Call}} \text{ and } c\big(\underline{X}_T^{\text{Call}}\big) = -c\big(\overline{X}_T^{-\text{Call}}\big).$$

Similarly, if $\overline{\theta} < 0$ we have $\underline{X}_T^{-\text{Call}} = -\overline{X}_T^{\text{Call}}$ as well as $c\big(\underline{X}_T^{-\text{Call}}\big) = -c\big(\overline{X}_T^{\text{Call}}\big)$.

3.3 Self-quanto Calls and Puts

A quanto option is a (typically European) option whose payoff is converted into a different currency or numeraire at maturity at a pre-specified rate, called the quanto-factor. Such products are attractive for speculators and investors who wish to have exposure to a foreign asset, but without the corresponding exchange rate risk. Quanto options are attractive because they shield the purchaser from exchange rate fluctuations. In the special case of a self-quanto option the numeraire is the underlying asset price at maturity itself. The payoff of a long self-quanto call with maturity T and strike price K is

$$X_T^{\text{sqC}} = S_T \cdot (S_T - K)_+ = S_0 e^{L_T} (S_0 e^{L_T} - K)_+$$

which is monotonically increasing in L_T and thus not cost-efficient if $\overline{\theta} > 0$. Its payoff function is then given by $\omega^{\text{sqC}}(y) := y(y-K)_+$. To derive the corresponding distribution function $G_{\text{sqC}} = F_{X_T^{\text{sqC}}}$, observe that the positive solution S_T^* of the quadratic equation $S_T^2 - KS_T = x$, $x > 0$, is given by

$$S_T^* = \frac{K}{2} + \sqrt{\frac{K^2}{4} + x},$$

then $\{S_T^2 - KS_T - x \leq 0\} = \{S_T \leq S_T^*\}$, hence

$$G_{\text{sqC}}(x) = P\big(X_T^{\text{sqC}} \leq x\big) = \begin{cases} F_{L_T}\left(\ln\left(\frac{\frac{K}{2}+\sqrt{\frac{K^2}{4}+x}}{S_0}\right)\right), & \text{if } x \geq 0, \\ 0, & \text{if } x < 0. \end{cases}$$

The generalized inverse is given by

$$G_{\text{sqC}}^{-1}(y) = S_0 e^{F_{L_T}^{-1}(y)} \left(S_0 e^{F_{L_T}^{-1}(y)} - K\right)_+, \quad y \in (0,1). \tag{33}$$

Consequently according to Proposition 2 the cost-efficient strategy for a long self-quanto call in the case $\overline{\theta} > 0$ is,

$$\underline{X}_T^{\text{sqC}} = G_{\text{sqC}}^{-1}(1 - F_{L_T}(L_T)) = S_0 e^{F_{L_T}^{-1}(1-F_{L_T}(L_T))} \left(S_0 e^{F_{L_T}^{-1}(1-F_{L_T}(L_T))} - K\right)_+ \tag{34}$$

with payoff function $\underline{\omega}^{\text{sqC}}(y) := S_0 e^{F_{L_T}^{-1}\left(1-F_{L_T}(\ln(\frac{y}{S_0}))\right)} \left(S_0 e^{F_{L_T}^{-1}\left(1-F_{L_T}(\ln(\frac{y}{S_0}))\right)} - K\right)_+$.

Figure 9 displays the payoff X_T^{sqC} of a long self-quanto call option on one ThyssenKrupp stock with strike $K = 16.5$ and maturity $T = 23$ days, and its cost-efficient counterparts $\underline{X}_T^{\text{sqC}}$ for the three Lévy models under consideration.

The cost of the efficient self-quanto call can be calculated using (33),

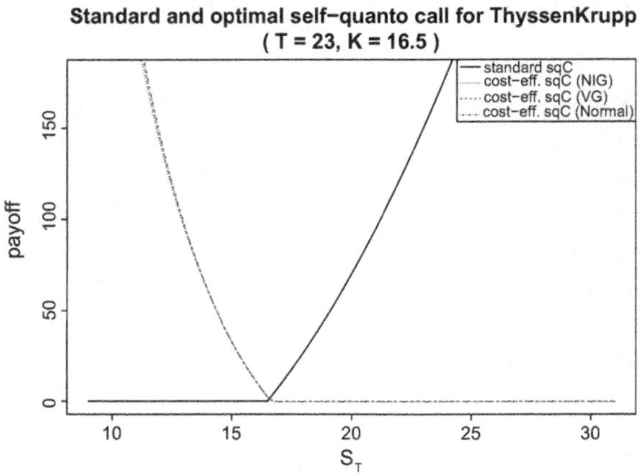

Fig. 9 Classical self-quanto call and its cost-efficient counterparts for ThyssenKrupp. $S_0 = 16.73$

$$c(\underline{X}_T^{\text{sqC}}) = \frac{1}{M_{dist}(\overline{\theta})} \int_0^1 e^{\overline{\theta} F_{dist}^{-1}(y) - rT} S_0 e^{F_{dist}^{-1}(1-y)} \left(S_0 e^{F_{dist}^{-1}(1-y)} - K \right)_+ \mathrm{d}y \tag{35}$$

where *dist* is $NIG(\alpha, \beta, \delta T, \mu T)$, $VG(\lambda T, \alpha, \beta, \mu T)$, or $N((\mu - \frac{\sigma^2}{2})T, \sigma^2 T)$.

If $\overline{\theta}$ is an Esscher parameter and $M_{L_T}(\overline{\theta} + 2) < \infty$, then

$$c(X_T^{\text{sqC}}) \le e^{-rT} E_{\overline{\theta}}[S_T^2] = e^{-rT} S_0^2 \frac{M_{L_T}(\overline{\theta} + 2)}{M_{L_T}(\overline{\theta})} < \infty. \tag{36}$$

In general this holds true for the Samuelson model, since the moment generating function of L_1, $M_{N((\mu - \frac{\sigma^2}{2}), \sigma^2)}(u)$ is defined for all $u \in \mathbb{R}$. For the *NIG* resp. *VG* model

$$\overline{\theta} + 2 \in (-\alpha - \beta, \alpha - \beta) \text{ and } \overline{\theta} \in (-\alpha - \beta, \alpha - \beta) \tag{37}$$

which implies that $\overline{\theta} \in (-\alpha - \beta, \alpha - \beta - 2)$. All estimated parameters from the daily log returns of E.ON and ThyssenKrupp from Table 2 fulfill this condition as well as Eq. (37). We get the following pricing formula.

Proposition 6 (Price of a vanilla self-quanto call)

Let $(L_t)_{t \ge 0}$ be a Lévy process, such that $L_T \overset{d}{=} NIG(\alpha, \beta, \delta T, \mu T)$ or $VG(\lambda T, \alpha, \beta, \mu T)$. If $M_{L_T}(\overline{\theta} + 2) < \infty$, where $\overline{\theta}$ is an Esscher parameter, then we have

$$c(X_T^{\text{sqC}}) = \frac{M_{L_T}(\overline{\theta} + 2)}{M_{L_T}(\overline{\theta} + 1)} S_0^2 \Big(1 - F_{L_T^{\overline{\theta}+2}}\Big(\ln\Big(\frac{K}{S_0}\Big)\Big)\Big) - S_0 K \Big(1 - F_{L_T^{\overline{\theta}+1}}\Big(\ln\Big(\frac{K}{S_0}\Big)\Big)\Big) \tag{38}$$

where $L_T^{\overline{\theta}+k} \overset{d}{=} NIG(\alpha, \beta + \overline{\theta} + k, \delta T, \mu T)$ or $VG(\lambda T, \alpha, \beta + \overline{\theta} + k, \mu T)$, $k = 0, 1, 2$.

If $L_T \overset{d}{=} N((\mu - \frac{\sigma^2}{2})T, \sigma^2 T)$, then we have

$$c(X_T^{\text{sqC}}) = e^{(r+\sigma^2)T} S_0^2 \Phi\big(-h' + \sigma\sqrt{T}\big) - S_0 K \Phi\big(-h'\big), \tag{39}$$

where $h' = \frac{1}{\sigma\sqrt{T}}(\ln(\frac{K}{S_0}) - (r + \frac{\sigma^2}{2})T)$.

For details of the proof we refer to Wolf [21].

Remark 7 (**symmetric Lévy case**) In the case of symmetric Lévy processes where $L_T \stackrel{d}{=} \upsilon T - L_T$ for some $\upsilon \in \mathbb{R}$ the formulas for the cost of efficient self-quanto calls simplify to

$$\begin{aligned} c(\underline{X}_T^{\text{sqC}}) =& e^{\upsilon T} S_0 \Big(e^{-(r-\upsilon)T} S_0 \frac{M_{L_T}(\overline{\theta} - 2)}{M_{L_T}(\overline{\theta})} F_{L_T^{\overline{\theta}-2}} \Big(\ln\Big(\frac{S_0}{K}\Big) + \upsilon T\Big) \\ &- e^{-rT} K \frac{M_{L_T}(\overline{\theta} - 1)}{M_{L_T}(\overline{\theta})} F_{L_T^{\overline{\theta}-1}} \Big(\ln\Big(\frac{S_0}{K}\Big) + \upsilon T\Big)\Big) \end{aligned} \tag{40}$$

where $L_T^{\overline{\theta}}$ denotes the Lévy process at maturity under the Esscher measure $Q^{\overline{\theta}}$.

In particular, in the Samuelson model we have that $L_T \stackrel{d}{=} 2(\mu - \frac{\sigma^2}{2})T - L_T$, thus

$$c(\underline{X}_T^{\text{sqC}}) = e^{2(\mu-r)T} S_0 \Big(e^{-rT} e^{2(\mu+\frac{\sigma^2}{2})T} S_0 \Phi(-h + \sigma\sqrt{T}) - K\Phi(-h)\Big) \tag{41}$$

where $h = \frac{1}{\sigma\sqrt{T}}\big(\ln(\frac{K}{S_0}) - (\mu + \frac{\sigma^2}{2})T + (r - \mu)T\big)$.

We display the cost of a long self-quanto call option on E.ON and ThyssenKrupp with their cost-efficient counterparts for the three Lévy models under consideration in Table 5. Again, we emphasize that the relative efficiency loss of the self-quanto option on E.ON and ThyssenKrupp has almost the same size. The same is true for the corresponding Esscher parameter $\overline{\theta}$.

Utilizing Proposition 2 and the explicit formula of the payoff function $\underline{\omega}^{\text{sqC}}$ the price at time $t < T$ of a cost-efficient long call with maturity T can be computed as

Table 5 Comparison of the cost of a long self-quanto call option on E.ON and ThyssenKrupp, resp., and the corresponding cost-efficient payoffs. $S_0 = 17.48$, $K = 17.24$, $T = 23$ for E.ON and $S_0 = 16.73$, $K = 16.5$, $T = 23$ for ThyssenKrupp

E.ON	$c(X_T^{\text{sqC}})$	$c(\underline{X}_T^{\text{sqC}})$	Efficiency loss in %
NIG	14.6161	13.6394	6.68
VG	14.3741	13.3613	7.05
Normal	14.6988	13.7397	6.53
Thyssen	$c(X_T^{\text{sqC}})$	$c(\underline{X}_T^{\text{sqC}})$	Efficiency loss in %
NIG	17.4182	16.3441	6.17
VG	17.3619	16.2628	6.50
Normal	17.3394	16.1980	6.58

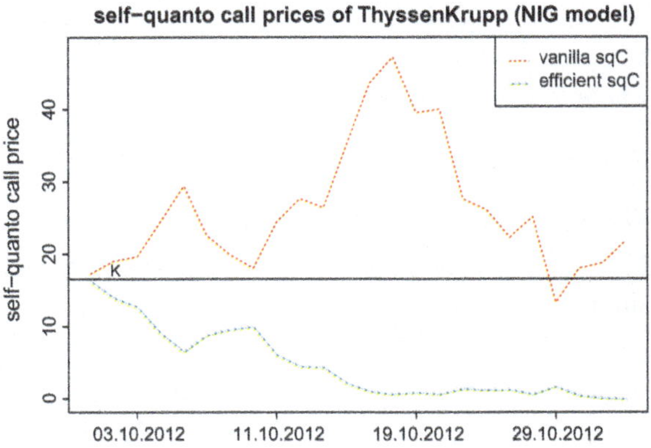

Fig. 10 Cost of a classical self-quanto call and its cost-efficient counterpart for ThyssenKrupp. $S_0 = 16.73$

$$c_t(\underline{X}_T^{\text{sqC}}) = e^{-r(T-t)} E\Big[Z_{T-t}^{\overline{\theta}}\, S_0 e^{F_{L_T}^{-1}\left(1-F_{L_T}\left(\ln(\frac{y}{S_0})+L_{T-t}\right)\right)} \cdot \\ \Big(S_0 e^{F_{L_T}^{-1}\left(1-F_{L_T}\left(\ln(\frac{y}{S_0})+L_{T-t}\right)\right)} - K\Big)_+ \Big] \Big|_{y=S_t}$$

From Fig. 10 the leverage effect of the self-quanto call payoff is clearly recognizable in comparison to the standard long call payoff. The peaks and lows are more pronounced then in the vanilla call case (compare Fig. 8).

Again, we have the symmetry relation $G_{\text{sqC}}^{-1}(y) = -G_{-\text{sqC}}^{-1}(1-y)$ and conclude: If $\overline{\theta} > 0$, then $\underline{X}_T^{\text{sqC}} = -\overline{X}_T^{-\text{sqC}}$ and $c\big(\underline{X}_T^{\text{sqC}}\big) = -c\big(\overline{X}_T^{-\text{sqC}}\big)$.
If $\overline{\theta} < 0$ then $\underline{X}_T^{-\text{sqC}} = -\overline{X}_T^{\text{sqC}}$ as well as $c\big(\underline{X}_T^{-\text{sqC}}\big) = -c\big(\overline{X}_T^{\text{sqC}}\big)$.

3.4 Self-quanto Put Options

A long self-quanto put $X_T^{\text{sqP}} = S_T(K - S_T)_+$ with payoff function $\omega^{\text{sqP}}(y) := y(K - y)_+$ and strike $K > 0$ is designed to profit from moderate decreasing prices of the underlying security. It provides highest outcomes when the price is at $\frac{K}{2}$. Its distribution function can be calculated and is presented in the following formula (42). Since the payoff function ω^{sqP} is not monotone self-quanto puts are not efficient for $\overline{\theta} \neq 0$. The payoff distribution function G_{sqP} can be calculated as

$$G_{\text{sqP}}(x) = \begin{cases} 1, & x \geq \frac{K^2}{4}, \\ 1 - \left(F_{L_T}\left(\ln\left(\frac{\frac{K}{2}+\sqrt{\frac{K^2}{4}-x}}{S_0}\right)\right) - F_{L_T}\left(\ln\left(\frac{\frac{K}{2}-\sqrt{\frac{K^2}{4}-x}}{S_0}\right)\right)\right), & 0 < x < \frac{K^2}{4}, \\ 1 - F_{L_T}\left(\ln(\frac{K}{S_0})\right), & x = 0, \\ 0, & x < 0. \end{cases} \tag{42}$$

Figure 11 displays the payoff X_T^{sqP} of a long self-quanto put option on one Allianz stock with strike $K = 98$ and maturity $T = 23$ days and its cost-efficient counterparts $\underline{X}_T^{\text{sqP}}$ for the three Lévy models under consideration. The bearish market counterpart is illustrated in Fig. 12 which shows the payoff $\underline{X}_T^{\text{sqP}}$ of a cost-efficient long self-quanto put option on one ThyssenKrupp stock with strike $K = 16.5$ and maturity $T = 23$ days. Note, that all three Lévy models generate fairly equal plots.

For the price of a standard and optimal self-quanto put we have similarly as in the call case the following simplified formula.

Proposition 8 (Price of a vanilla self-quanto put) *Let $(L_t)_{t\geq 0}$ be a Lévy process, such that $L_T \stackrel{d}{=} NIG(\alpha, \beta, \delta T, \mu T)$ or $VG(\lambda T, \alpha, \beta, \mu T)$. If $\overline{\theta}$ is a Esscher parameter, then we have*

$$c(X_T^{\text{sqP}}) = S_0 K F_{L_T^{\overline{\theta}+1}}\left(\ln\left(\frac{K}{S_0}\right)\right) - S_0^2 E\left[\frac{e^{(\overline{\theta}+2)L_T}}{M_{L_T}(\overline{\theta}+1)}\mathbb{1}_{\left\{L_T < \ln\left(\frac{K}{S_0}\right)\right\}}\right]$$

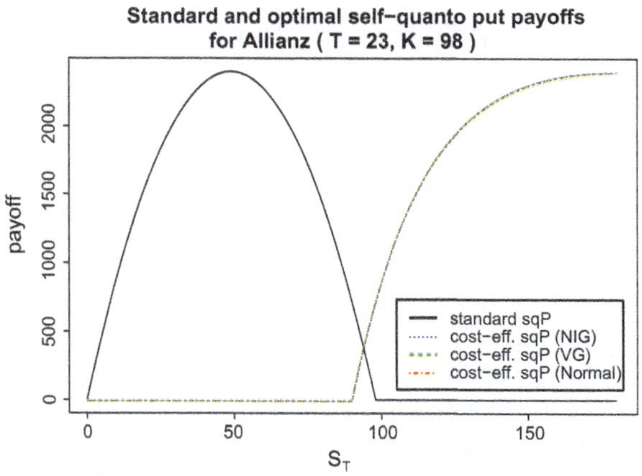

Fig. 11 Classical self-quanto put and its cost-efficient counterparts for Allianz. $S_0 = 93.42$

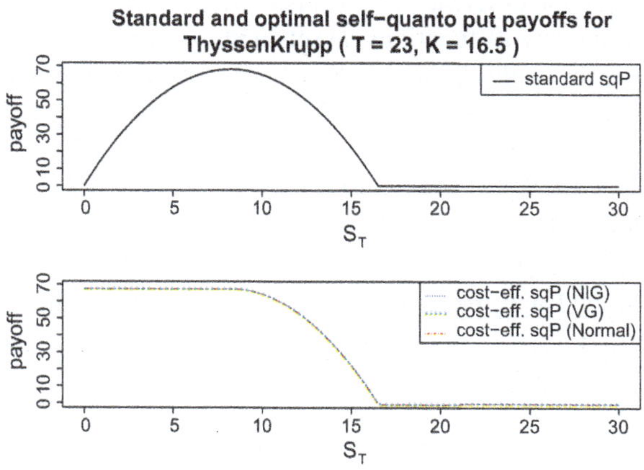

Fig. 12 Classical self-quanto put and its cost-efficient counterparts for ThyssenKrupp. $S_0 = 16.73$

Table 6 Comparison of the cost of a long self-quanto put option on Allianz and ThyssenKrupp resp., and the corresponding cost-efficient payoffs. $S_0 = 93.42$, $K = 98$, $T = 23$ for Allianz and $S_0 = 16.73$, $K = 16.5$ for ThyssenKrupp

Allianz	$c(X_T^{\text{sqP}})$	$c(\underline{X}_T^{\text{sqP}})$	Efficiency loss in %
NIG	547.2179	452.8534	17.24
VG	542.4431	449.5875	17.12
Normal	546.3491	452.2157	17.23
ThyssenKrupp	$c(X_T^{\text{sqP}})$	$c(\underline{X}_T^{\text{sqP}})$	Efficiency loss in %
NIG	9.5988	9.5987	0.001041
VG	9.5737	9.5736	0.001044
Normal	9.5826	9.5825	0.001043

where $L_T^{\overline{\theta}+k} \stackrel{d}{=} NIG(\alpha, \beta + \overline{\theta} + k, \delta T, \mu T)$ *or* $VG(\lambda T, \alpha, \beta + \overline{\theta} + k, \mu T)$, $k = 0, 1, 2$. *If in addition* $M_{L_T}(\overline{\theta} + 2) < \infty$, *then*

$$c(X_T^{\text{sqP}}) = S_0 K F_{L_T^{\overline{\theta}+1}}\left(\ln\left(\frac{K}{S_0}\right)\right) - \frac{M_{L_T}(\overline{\theta}+2)}{M_{L_T}(\overline{\theta}+1)} S_0^2 F_{L_T^{\overline{\theta}+2}}\left(\ln\left(\frac{K}{S_0}\right)\right). \tag{43}$$

For $L_T \stackrel{d}{=} N((\mu - \frac{\sigma^2}{2})T, \sigma^2 T)$ *we have*

$$c(X_T^{\text{sqP}}) = S_0 K \Phi(h) - e^{(r+\sigma^2)T} S_0^2 \Phi(h - \sigma\sqrt{T}) \tag{44}$$

where $h = \frac{1}{\sigma\sqrt{T}}(\ln(\frac{K}{S_0}) - (r + \frac{\sigma^2}{2})T)$.

Table 6 gives the cost of a long self-quanto put option on Allianz with their cost-efficient counterparts for the three Lévy models under consideration. To cover the bearish markets the analogous results for the cost of a long self-quanto put option on ThyssenKrupp are included in Table 6. One observes that for the ThyssenKrupp stock the efficiency loss is insignificantly small. This is due to the fact that equality of the payoff functions on sets with high probability (under the Esscher martingale measure) plus boundedness on the complementary set implies nearly equal cost and, thus small efficiency loss. This is quantified in the next remark in the *NIG* case.

Remark 9 Assume that $L_T \stackrel{d}{=} NIG(\alpha, \beta, \delta T, \mu T)$, then the payoff functions of the long self-quanto put X_T^{sqP} and of the cost-efficient counterpart $\underline{X}_T^{\text{sqP}}$ are both bounded by $C = (\frac{K}{2})^2 = 68.0625$ and are identical on the interval $I = [\ln(\frac{8.25}{S_0}), \ln(\frac{30}{S_0})]$ with probability (w.r.t. Esscher martingale measure) nearly one, $F_{L_T^{\overline{\theta}}}(I) = F_{L_T^{\overline{\theta}}}(\ln(30/S_0)) - F_{L_T^{\overline{\theta}}}(\ln(8.25/S_0)) = 99.9999\,\%$. Hence

$$\begin{aligned}\ell(X_T^{\text{sqP}}) &= e^{-rT}E_{Q^{\bar\theta}}[X_T^{\text{sqP}} - \underline{X}_T^{\text{sqP}}]\\ &= \int_I (X_T^{\text{sqP},x} - \underline{X}_T^{\text{sqP},x})dF_{L_T^{\bar\theta}}(x) + \int_{I^c} (X_T^{\text{sqP},x} - \underline{X}_T^{\text{sqP},x})dF_{L_T^{\bar\theta}}(x)\\ &= \int_{I^c} (X_T^{\text{sqP},x} - \underline{X}_T^{\text{sqP},x})dF_{L_T^{\bar\theta}}(x)\\ &\le \sup_{x\in I^c}(X_T^{\text{sqP},x} - \underline{X}_T^{\text{sqP},x})\cdot F_{L_T^{\bar\theta}}(I^c) \le C\cdot 0.00001 = 0.0001089,\end{aligned}$$

where $X_T^{\text{sqP},x} = \omega^{\text{sqP}}(S_0e^x) = S_0e^x(K - S_0e^x)_+$ and $\underline{X}_T^{\text{sqP},x} = \underline{\omega}^{\text{sqP}}(x) = G_{\text{sqP}}^{-1}(1 - F_{L_T}(\ln(\frac{x}{S_0})))$.

Figure 13 illustrates the nearly equal payoff function of the standard and optimal self-quanto put in the bearish market situation. This leads to almost identical prices during the entire trading period $[0, T]$ in October 2012. Figure 14, shows distinctive similarities to Fig. 6 with more pronounced peaks and lows which is due to the design of the long self-quanto put option. The prices of the efficient options always roughly move in the direction opposite to that of the standard options which reflects the reversed monotonicity properties of the underlying payoff profiles in Fig. 11.

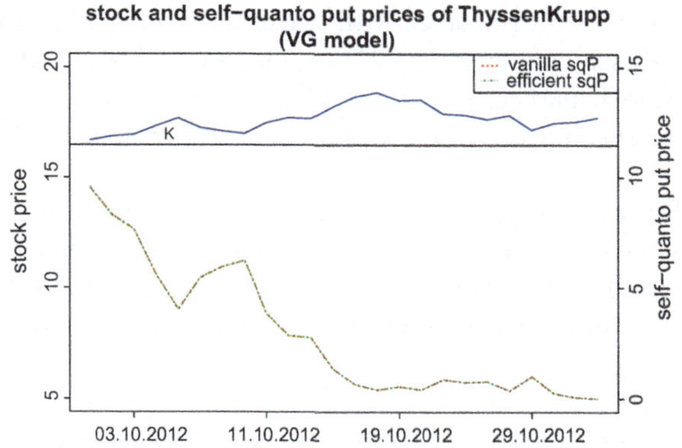

Fig. 13 Evolution of prices of standard and cost-efficient self-quanto put with strike $K = 16.5$ for ThyssenKrupp in the VG model

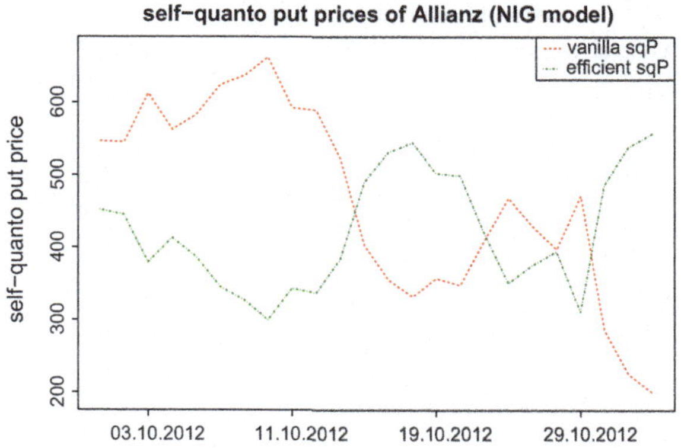

Fig. 14 Evolution of prices of standard and cost-efficient self-quanto put with strike $K = 98$ for Allianz in the NIG model

3.5 Long Straddle Options

A long straddle investment strategy allows the holder to profit based on how much the price of the underlying security moves, regardless of the direction of price movement. A long straddle option X_T^{strdl} is realized by going long in both a call option and a put option on some stock, index or other underlying, i.e. $X_T^{\text{strdl}} = X_T^{\text{Call}} + X_T^{\text{Put}}$. It involves buying the put and call options at the same strike $K > 0$ with the same maturity T. A profit is gained if the underlying price moves a long way from the strike price, either above or below. The payoff distribution function is given in the following lemma.

Lemma 10 *Let $(L_t)_{t\geq 0}$ be a Lévy process with continuous distribution function F_{L_T} at maturity $T > 0$. The distribution function G_{strdl} of the payoff of the long straddle X_T^{strdl} with strike $K > 0$ at maturity T is given by*

$$G_{\text{strdl}}(x) = \begin{cases} F_{L_T}(\ln(\frac{K+x}{S_0})), & x \geq K, \\ F_{L_T}(\ln(\frac{K+x}{S_0})) - F_{L_T}(\ln(\frac{K-x}{S_0})), & 0 \leq x < K, \\ 0, & x < 0. \end{cases} \tag{45}$$

The cost of the cost-efficient long straddle is compared with its vanilla counterpart in Table 7, while in Fig. 15 we contrast the payoffs X_T^{strdl} of a long straddle option on one Volkswagen stock with strike $K = 135$ and maturity $T = 23$ days, and its cost-efficient counterparts $\underline{X}_T^{\text{strdl}}$ for the three Lévy models under consideration. For the bearish markets we present in Fig. 17, the payoff $\underline{X}_T^{\text{strdl}}$ of a cost-efficient long straddle option on one ThyssenKrupp stock with strike $K = 16.5$ and maturity $T = 23$ days.

Regarding Fig. 15 we notice that with increasing stock price at maturity of the Volkswagen the payoff of the cost-efficient long straddle dominates that of the standard long straddle. This difference becomes more and more irrelevant with increasing stock price and takes the greatest value if the stock prices moves around the exercise

Table 7 Comparison of the cost of a long straddle option on Volkswagen and ThyssenKrupp resp., and the corresponding cost-efficient payoffs. $S_0 = 130.55$, $K = 135$, $T = 23$ for Volkswagen and $S_0 = 16.73$, $K = 16.5$ for ThyssenKrupp

Volkswagen	$c(X_T^{\text{strdl}})$	$c(\underline{X}_T^{\text{strdl}})$	Efficiency loss in %
NIG	11.5759	8.9844	22.39
VG	11.5161	8.9239	22.51
Normal	11.5448	8.9722	22.28
ThyssenKrupp	$c(X_T^{\text{strdl}})$	$c(\underline{X}_T^{\text{strdl}})$	Efficiency loss in %
NIG	1.5717	1.5377	2.17
VG	1.5662	1.5312	2.23
Normal	1.5657	1.5293	2.32

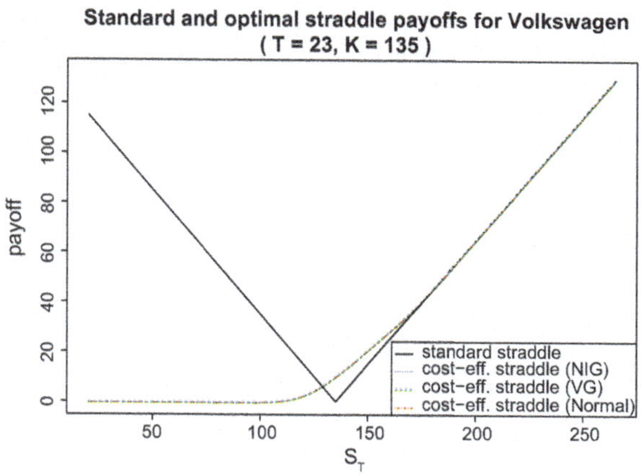

Fig. 15 Payoff functions of a classical straddle option and its cost-efficient counterparts for Volkswagen

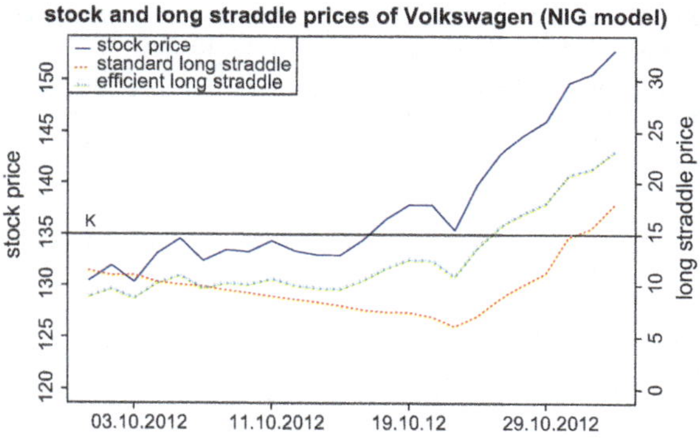

Fig. 16 Evolution of prices of standard and cost-efficient long straddle with strike $K = 135$ for Volkswagen in the NIG model

price. A reverse pattern is depicted in Fig. 17. Thus, the evolution of the prices of standard and cost-efficient long straddle for Volkswagen, Fig. 16, shows that close to maturity, where the stock price rapidly increases, the costs are increasing too, and the cost-efficient long straddle is more expensive than the standard long straddle option. For the ThyssenKrupp stock we have a more complex situation. Here, the trend alters from bearish to bullish within the trading period, thus the stock price increases close to maturity. Hence, the payoff of the cost-efficient long straddle becomes less worthy which is illustrated in Fig. 18.

3.6 *Long Call Butterfly Spread Options*

A long (call) butterfly option strategy is created to earn substantial but limited profits with great probability. It is a limited risk and non-directional financial investment strategy, and due to its design it is a suitable neutral option strategy for low volatility markets. A long butterfly spread is the combination of two long calls C_3 and C_1 with

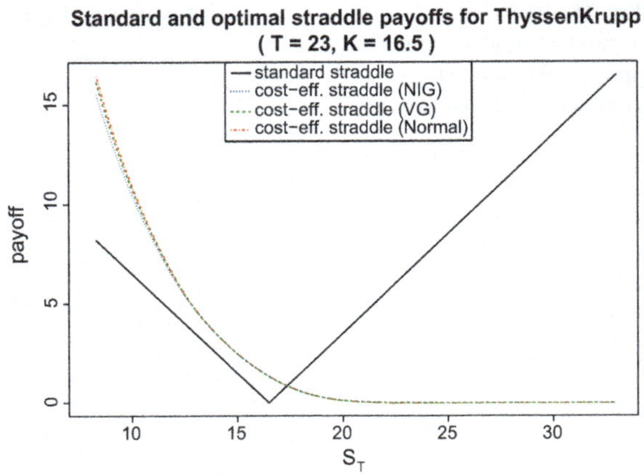

Fig. 17 Payoff functions of a classical straddle option and its cost-efficient counterparts for ThyssenKrupp

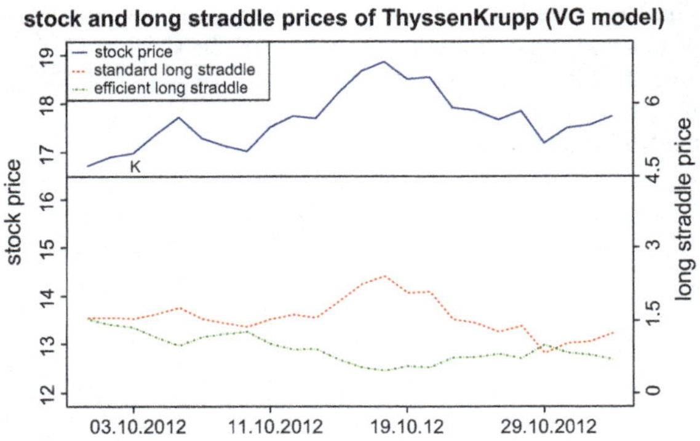

Fig. 18 Evolution of prices of standard and cost-efficient long straddle with strike $K = 16.5$ for ThyssenKrupp in the VG model

strikes $K_3 > K_1 > 0$, and two short calls $-C_2$ with strike $K_2 = \frac{K_1+K_3}{2}$. The payoff X_T^{bfly} of a butterfly spread is given by

$$X_T^{\text{bfly}} = (S_T - K_1)_+ + (S_T - K_3)_+ - 2(S_T - K_2)_+.$$

An investor may take a long butterfly position if he expects that the market is mildly volatile, thus profiting the most if the stock price is at K_2. The payoff distribution function can be calculated and is given by:

$$G_{\text{bfly}}(x) = \begin{cases} 1, & x > K_2 - K_1, \\ 1 - F_{L_T}(\ln(\frac{K_3 - x}{S_0})) + F_{L_T}(\ln(\frac{K_1 + x}{S_0})), & 0 \le x \le K_2 - K_1, \\ 0, & x < 0\,. \end{cases} \tag{46}$$

In Fig. 19 and 21 the payoffs and the prices of of a butterfly spread and its efficient counterpart of one Allianz stock with strikes $K_1 = 94$ and $K_3 = 104$ are presented. For the bearish markets, Fig. 20 shows the payoff $\underline{X}_T^{\text{bfly}}$ of a cost-efficient long butterfly

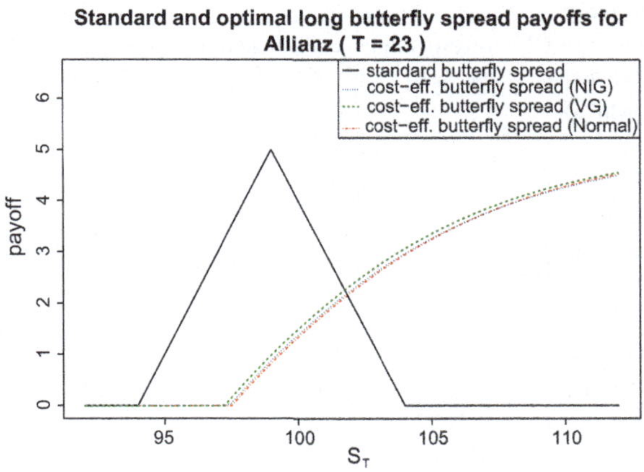

Fig. 19 Payoff functions of a classical long butterfly option and its cost-efficient counterparts for Allianz. $S_0 = 93.42$, $K_1 = 94$ and $K_3 = 104$

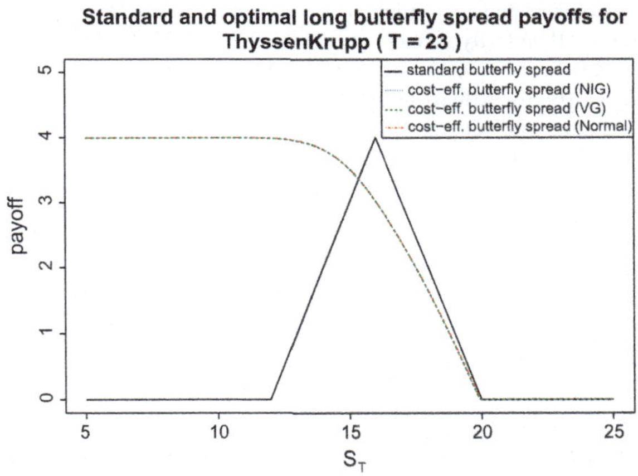

Fig. 20 Payoff functions of a classical long butterfly option and its cost-efficient counterparts for ThyssenKrupp. $S_0 = 16.73$, $K_1 = 12$ and $K_3 = 20$

spread option on one ThyssenKrupp stock with strikes $K_1 = 12$ and $K_3 = 20$. The cost of a long butterfly spread option on Allianz and ThyssenKrupp with their cost-efficient counterparts for the three Lévy models under consideration are considered in Table 8. Again, in case of the ThyssenKrupp stock we notice that the more the payoffs functions resemble on sets with greater mass the smaller becomes the efficiency loss (cf. Figs. 20, 21 and also Remark 9). We see from Fig. 20 that the payoff of the cost-efficient version of the long butterfly spread is dominated if the stock price of the underlying is greater than approximately 15.5. Since the stock price of the ThyssenKrupp is above 15.5 during the entire trading period $0 < t \leq T$ the cost of the standard butterfly spread dominates the cost of the efficient counterpart which can be seen in Fig. 22.

Table 8 Comparison of the cost of a long butterfly spread option on Allianz and ThyssenKrupp resp., and the corresponding cost-efficient payoffs. $S_0 = 93.42$, $K_1 = 94$, $K_3 = 104$, for Volkswagen and $S_0 = 16.73$, $K_1 = 12$, $K_3 = 20$ for ThyssenKrupp

Allianz	$c(X_T^{\text{bfly}})$	$c(\underline{X}_T^{\text{bfly}})$	Efficiency loss in %
NIG	0.8398	0.7739	7.84
VG	0.8475	0.7825	7.67
Normal	0.8349	0.7691	7.87
ThyssenKrupp	$c(X_T^{\text{bfly}})$	$c(\underline{X}_T^{\text{bfly}})$	Efficiency loss in %
NIG	2.4153	2.4008	0.60
VG	2.4210	2.4064	0.60
Normal	2.4196	2.4044	0.63

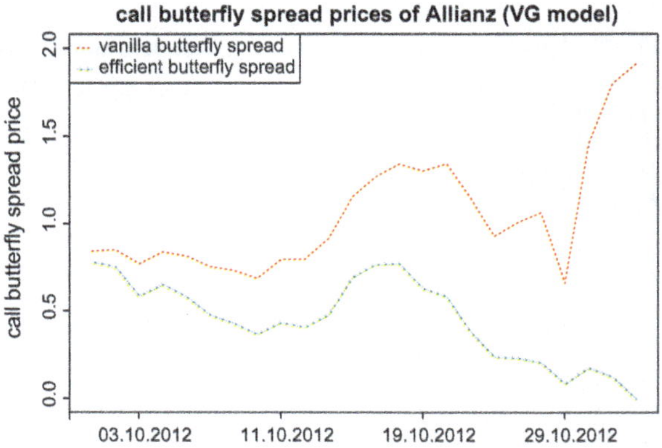

Fig. 21 Evolution of prices of standard and cost-efficient long call butterfly spread with strikes $K_1 = 94$ and $K_3 = 104$ for Allianz in the VG model in October 2012, $T = 23$ days

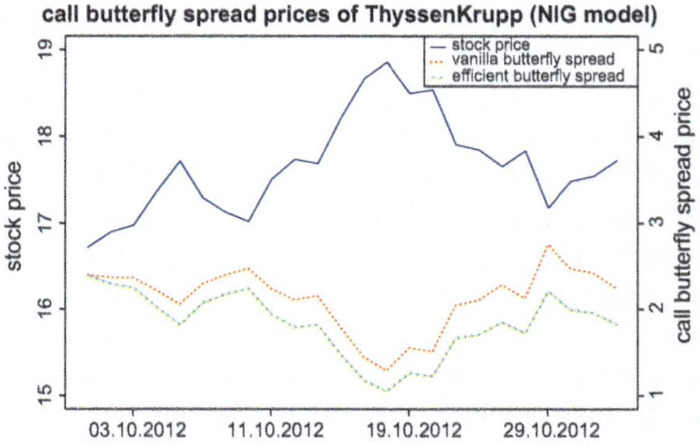

Fig. 22 Evolution of prices of standard and cost-efficient long call butterfly spread with strikes $K_1 = 12$ and $K_3 = 20$ for ThyssenKrupp in the NIG model in October 2012, $T = 23$ days

4 Efficiency Loss for Monotone Payoff Functions

The efficiency loss $\ell(\overline{\theta}) = \ell(\overline{\theta}, \eta) = e^{-rT} E_{\overline{\theta}}(X_T - \underline{X}_T)$ depends on the Esscher parameter $\overline{\theta} = \overline{\theta}(\eta)$ and in particular on the model parameter $\eta = (\eta_1, \ldots, \eta_k)$. In Hammerstein et al. [11] it has been shown that $\ell(\overline{\theta}, \eta)$ is an increasing function in $|\overline{\theta}|$ which leads in particular in the examples of put and call options to the result that the magnitude $|\overline{\theta}|$ of market trend determines the magnitude of the efficiency loss.

In the previous sections we typically reported the relative efficiency loss $\frac{\ell(\overline{\theta})}{c(X_T)} = \ell_r(\overline{\theta})$ which might be more relevant for applications. In this section we study the efficiency loss for plain vanilla puts and calls as well as for self-quanto calls. For vanilla puts as for its cost-efficient counterparts the cost rises with increasing strike price. Hence, it would be interesting to know how the efficiency loss resp. relative efficiency loss behaves when changing the strike. The next theorem confirms that the efficiency loss $\ell(K)$ in case of the put option is increasing in the strike while the relative efficiency loss $\ell_r(K)$ shows an opposite behaviour. This has noticeable consequences for trading put options, when investors are seeking to maximize their (relative) efficiency loss. Related results for the call resp. self-quanto call option are given too.

Theorem 11 (Efficiency loss vs. relative efficiency loss for X_T^{Put}, influence of strike) *Let $(L_t)_{t\geq 0}$ be a Lévy process with continuous and strictly increasing distribution function F_{L_T} at maturity $T > 0$. Suppose X_T^{Put} is the payoff of a long put option with strike $K > 0$ and let $\overline{\theta}$ be an Esscher parameter.*

1. *The efficiency loss, $\ell(K) := c(X_T^{\text{Put}}) - c(\underline{X}_T^{\text{Put}})$ is increasing in K.*
2. *$\frac{\partial}{\partial K} c(\underline{X}_T^{\text{Put}}) \leq \frac{\partial}{\partial K} c(X_T^{\text{Put}})$ and the costs of the standard and efficient put are increasing in K.*
3. *The relative efficiency loss $\ell_r(K) := \frac{\ell(K)}{c(X_T^{\text{Put}})}$ decreases in K.*

Proof 1. If $\overline{\theta} > 0$ then $\underline{X}_T^{\text{Put}} = X_T^{\text{Put}}$ and $\ell(K) \equiv 0$, thus w.l.g., let $\overline{\theta} < 0$. By definition $\ell(K) \geq 0$ for all $K \in \mathbb{R}_+$, thus $\ell : \mathbb{R}_+ \to \mathbb{R}_+$. Define $C := e^{-rT} E_{\overline{\theta}}[S_0 e^{F_{L_T}^{-1}(1-F_{L_T}(L_T))} - S_T]$ and observe that the pair $(X_1^+, X_2^+) := (Z_T^{\overline{\theta}}, S_0 e^{F_{L_T}^{-1}(1-F_{L_T}(L_T))})$ is comonotonic whereas the pair $(X_1^-, X_2^-) := (Z_T^{\overline{\theta}}, S_T)$ is countermonotonic. The marginals $F_{X_1^+} = F_{X_1^-}$ and $F_{X_2^+} = F_{X_2^-}$ are equal, thus, Hoeffdings inequality implies

$$e^{rT} C = \text{Cov}(Z_T^{\overline{\theta}}, S_0 e^{F_{L_T}^{-1}(1-F_{L_T}(L_T))}) - \text{Cov}(Z_T^{\overline{\theta}}, S_T) \geq 0$$

and thus $C \geq 0$. Since the first term is non-negative and the second is non-positive it holds that $C = 0$ if and only if $\text{Cov}(Z_T^{\overline{\theta}}, S_0 e^{F_{L_T}^{-1}(1-F_{L_T}(L_T))}) = \text{Cov}(Z_T^{\overline{\theta}}, S_T) = 0$. Since $Z_T^{\overline{\theta}} = h(S_T)$ for some decreasing function $h : \mathbb{R}_+ \to \mathbb{R}_+$, this can only be true if and only if L_T respectively L_1 is degenerate. By our general assumption this case is excluded and thus $C > 0$. Further, we easily obtain that $\ell(0) = 0$ and we also have

$$\begin{aligned} E_{\overline{\theta}}[(K-a)_+ - (K-b)_+] &= E_{\overline{\theta}}[\min(K,b) - \min(K,a)] \\ &\leq E_{\overline{\theta}}[\min(K,b)] \\ &\leq E_{\overline{\theta}}[b]. \end{aligned}$$

Hence, with the identity $(K-a)_+ = K - \min(K,a)$ for all $K, a \in \mathbb{R}_+$, the dominated convergence theorem yields

$$\begin{aligned} \lim_{K\to\infty} E_{\overline{\theta}}[\min(K,b) - \min(K,a)] &= E_{\overline{\theta}}[\lim_{K\to\infty} \min(K,b)] - E_{\overline{\theta}}[\lim_{K\to\infty} \min(K,a)] \\ &= E_{\overline{\theta}}[b-a] \end{aligned}$$

for all $a, b \in \mathbb{R}_+$, such that $E_{\overline{\theta}}[b], E_{\overline{\theta}}[a] < \infty$. Putting $a = S_T$ and $b = S_0 e^{F_{L_T}^{-1}(1-F_{L_T}(L_T))}$ we have shown that

$$\lim_{K\to\infty} \ell(K) = C > 0.$$

For the proof of 1. it is sufficient to show the existence of a $K^* \in \mathbb{R}_+$ such that ℓ is convex on $[0, K^*)$ and concave on $[K^*, \infty)$. For $\overline{\theta} < 0$ the plain vanilla put is most-expensive. By Proposition 2 the efficiency loss is given by

$$\ell(K) = \frac{1}{M_{L_T}(\overline{\theta})} \int_0^1 e^{\overline{\theta} F_{L_T}^{-1}(1-y)-rT} \big((K - S_0 e^{F_{L_T}^{-1}(1-y)})_+ - (K - S_0 e^{F_{L_T}^{-1}(y)})_+\big)\, dy.$$

Note that $\ell(K)$ is bounded from above by the price $c(X_T^{\text{Put}})$ of the original long put which obviously is finite for all $K \in \mathbb{R}_+$. Moreover, the functions

$$\begin{aligned} f_1(K,y) &= e^{\overline{\theta} F_{L_T}^{-1}(1-y)-rT} (K - S_0 e^{F_{L_T}^{-1}(1-y)})_+ \quad \text{and} \\ f_2(K,y) &= e^{\overline{\theta} F_{L_T}^{-1}(1-y)-rT} (K - S_0 e^{F_{L_T}^{-1}(y)})_+ \end{aligned}$$

are differentiable in K for all $y \in [0,1]$. The points $K = S_0 e^{F_{L_T}^{-1}(1-y)}$, $K = S_0 e^{F_{L_T}^{-1}(y)}$ can be neglected since the left- and right-hand derivatives are bounded. The partial derivatives are

$$\begin{aligned} \frac{\partial}{\partial K} f_1(K,y) &= e^{\overline{\theta} F_{L_T}^{-1}(1-y)-rT} \mathbb{1}_{[0, F_{L_T}(\ln(\frac{K}{S_0})))}(1-y) \quad \text{and} \\ \frac{\partial}{\partial K} f_2(K,y) &= e^{\overline{\theta} F_{L_T}^{-1}(1-y)-rT} \mathbb{1}_{[0, F_{L_T}(\ln(\frac{K}{S_0})))}(y). \end{aligned}$$

It holds that $|\frac{\partial}{\partial K} f_i(K,y)| \leq e^{\overline{\theta} F_{L_T}^{-1}(1-y)-rT}$, $i = 1, 2$. For establishing the integrability of $e^{\overline{\theta} F_{L_T}^{-1}(1-y)-rT}$ observe that

$$\int_0^1 e^{\overline{\theta}F_{L_T}^{-1}(1-y)-rT}\,\mathrm{d}y = \int_0^1 e^{\overline{\theta}F_{L_T}^{-1}(z)-rT}\,\mathrm{d}z$$
$$= \int_{-\infty}^{\infty} e^{\overline{\theta}x-rT} f_{L_T}(x)\,\mathrm{d}x = e^{-rT}M_{L_T}(\overline{\theta}) < \infty,$$

where f_{L_T} denotes the density of L_T which exists and is strictly positive on $\mathbb{R}$ due to our assumptions on F_{L_T}. Hence, we can interchange differentiation and integration and obtain

$$\frac{\partial \ell}{\partial K}(K) = \frac{e^{-rT}}{M_{L_T}(\overline{\theta})}\int_0^{F_{L_T}(\ln(\frac{K}{S_0}))} e^{\overline{\theta}F_{L_T}^{-1}(y)} - e^{\overline{\theta}F_{L_T}^{-1}(1-y)}\,\mathrm{d}y. \tag{47}$$

Differentiating w.r.t. K once again yields

$$\frac{\partial^2}{\partial^2 K}\ell(K) = \frac{e^{-rT}}{M_{L_T}(\overline{\theta})}\left(f_{L_T}\left(\ln\left(\frac{K}{S_0}\right)\right)\right)\frac{1}{K}\left[e^{\overline{\theta}F_{L_T}^{-1}\left(F_{L_T}(\ln(\frac{K}{S_0}))\right)} - e^{\overline{\theta}F_{L_T}^{-1}\left(1-F_{L_T}(\ln(\frac{K}{S_0}))\right)}\right].$$

Thus, $\frac{\partial^2}{\partial^2 K}\ell(K) > 0$ if and only if

$$\overline{\theta}F_{L_T}^{-1}\left(F_{L_T}\left(\ln\left(\frac{K}{S_0}\right)\right)\right) > \overline{\theta}F_{L_T}^{-1}\left(1 - F_{L_T}\left(\ln\left(\frac{K}{S_0}\right)\right)\right),$$

or equivalently, $F_{L_T}(\ln(\frac{K}{S_0})) < 1 - F_{L_T}(\ln(\frac{K}{S_0}))$, since $\overline{\theta} < 0$.
For $K^* := S_0 e^{F_{L_T}^{-1}(0.5)}$ we obtain $K < K^*$ if and only if $F_{L_T}(\ln(\frac{K}{S_0})) < 1 - F_{L_T}(\ln(\frac{K}{S_0}))$. Thus, ℓ is convex on $[0, K^*)$. Analogously, we get $\frac{\partial^2}{\partial^2 K}\ell(K) \leq 0$ if and only if $K \geq K^*$. Thus, ℓ is concave on $[K^*, \infty)$ as consequence we therefore get that ℓ is increasing.

2. This follows directly from the fact that $\frac{\partial}{\partial K}\ell(K) \geq 0$, thus $\frac{\partial}{\partial K}c(\underline{X}_T^{\text{Put}}) \leq \frac{\partial}{\partial K}c(X_T^{\text{Put}})$ and

$$\frac{\partial}{\partial K}c(\underline{X}_T^{\text{Put}}) = \frac{e^{-rT}}{M_{L_T}(\overline{\theta})}\int_0^{F_{L_T}(\ln(\frac{K}{S_0}))} e^{\overline{\theta}F_{L_T}^{-1}(1-y)}\,\mathrm{d}y \geq 0$$

as can be seen from Eq. (47).

3. Note that the function $\ell_{\text{r}}(K)$ is decreasing in K if and only if for all compact intervals $[K_1, K_2]$ with $K_1 < K_2 \in \mathbb{R}_+$ we have $\max_{K\in[K_1,K_2]}\ell_{\text{r}}(K) = \ell_{\text{r}}(K_1)$, or equivalently $K_1 \in \operatorname{argmax}_{K\in[K_1,K_2]}\ell_{\text{r}}(K)$. Since $c(\underline{X}_T^{\text{Put}})$ is an increasing function of K it holds that for all $K_1 < K_2 \in \mathbb{R}_+$ the cost of the efficient put $c(\underline{X}_T^{\text{Put}})$ is minimal at K_1, or equivalently for all $K_1 < K_2 \in \mathbb{R}_+$ we have

$$K_1 \in \underset{K\in[K_1,K_2]}{\operatorname{argmin}}\, c(\underline{X}_T^{\text{Put}}) = \underset{K\in[K_1,K_2]}{\operatorname{argmin}} \big(c(X_T^{\text{Put}}) - \ell(K)\big)$$
$$= \underset{K\in[K_1,K_2]}{\operatorname{argmax}} \Big(\frac{\ell(K)}{c(X_T^{\text{Put}})}\Big) = \underset{K\in[K_1,K_2]}{\operatorname{argmax}}\, \ell_{\text{r}}(K).$$

Also, for $K_1 < K_2 \in \mathbb{R}_+$ it yields that $\max_{K\in[K_1,K_2]} \ell_{\text{r}}(K) = \ell_{\text{r}}(K_1)$, since ℓ_{r} is decreasing in K. Thus, the assertion is proven.

The latter result reveals that potential greater savings in buying an efficient option with higher strike could be annihilated by the higher cost one has to pay for. The reverse holds true for the relative efficiency loss.

Remark 12 (a) **Total savings vs. higher distribution w.r.t. $\leq_{\text{st}}$**
Consider an investor with budget $B \in \mathbb{R}_+$ who aims to buy a put option which generates a payoff X_T^{Put} at maturity $T > 0$. Further, consider put options $X_T^{\text{Put},i}$ with strike $K_i > 0,\ i = 1, 2$ and the efficient counterparts $\underline{X}_T^{\text{Put},i}$ with cost denoted by c_i respectively $\underline{c}_i$. We compute that

$$B = \frac{B}{c_i} \cdot c_i = \frac{B}{c_i}\big(\underline{c}_i + (c_i - \underline{c}_i)\big) = \frac{B}{c_i} \cdot \underline{c}_i + \frac{B}{c_i}(c_i - \underline{c}_i),\ i = 1, 2,$$

that is, $\frac{B}{c_i}(c_i - \underline{c}_i) = B \cdot \ell_{\text{r}}(K_i)$ denotes the total savings when buying $\frac{B}{c_i}$ shares of the efficient put option. By Theorem 11 we see that $B\ell_{\text{r}}(K_2) \leq B\ell_{\text{r}}(K_1)$ if $K_1 < K_2$, i.e. the total savings decrease in the strike, when buying the associated efficient put option. In other words, when choosing the put option with the higher strike K_2 which generates a stochastically larger distribution one has to pass on the amount of $B \cdot (\ell_{\text{r}}(K_1) - \ell_{\text{r}}(K_2)) \geq 0$ of potential savings.

(b) **Bounds for efficiency loss**
The proof of Theorem 11 also establishes the following bound for the efficiency loss of a long put option in bullish markets.

$$0 \leq \ell(X_T^{\text{Put}}) \leq S_0 E_{\overline{\theta}}[e^{F_{L_T}^{-1}(1-F_{L_T}(L_T))-rT} - 1]. \tag{48}$$

Some concrete results on the relative efficiency loss for put options with different strikes on the Allianz and Volkswagen stock can be found in Table 9. These show the decrease of the relative efficiency loss in the strike K for Allianz and Volkswagen. We give analogous results for the long call and the self-quanto call option. The results confirm that the efficiency loss in case of the plain call and self-quanto call option is decreasing and the relative efficiency loss is increasing in the strike, thus, has a reverse behaviour as for the put option. For these examples comparisons of the relative efficiency loss are given in Tables 10 and 11. Although the payoff profile of a plain vanilla call considerably differs from the profile of the self-quanto option, the monotonicity of the relative efficiency loss does not exhibit substantial differences.

Table 9 Relative efficiency loss for a long put option on Allianz and Volkswagen $S_0 = 93.42$, $T = 23$ for Allianz and $S_0 = 130.55$, $T = 23$ for Volkswagen; from Table 1

Efficiency loss in %			
Allianz	$K_1 = 92$	$K_2 = 95$	$K_3 = 98$
NIG	24.01	20.89	18.09
VG	23.86	20.73	17.92
Normal	23.84	20.83	18.10
Volkswagen	$K_1 = 130$	$K_2 = 133$	$K_3 = 135$
NIG	54.94	51.33	48.95
VG	55.09	51.48	49.10
Normal	54.92	51.35	49.01

Table 10 Relative efficiency loss for a long call option on E.ON and ThyssenKrupp $S_0 = 17.48$, $T = 23$ for E.ON and $S_0 = 16.73$, $T = 23$ for ThyssenKrupp

Efficiency loss in %			
E.ON	$K_1 = 17.24$	$K_2 = 19.48$	$K_3 = 20.72$
NIG	6.45	11.04	13.75
VG	6.83	11.66	14.53
Normal	6.32	10.61	13.15
ThyssenKrupp	$K_1 = 16.5$	$K_2 = 18.5$	$K_3 = 20.5$
NIG	5.90	8.74	11.74
VG	6.07	8.96	11.99
Normal	6.33	9.27	12.35

Table 11 Relative efficiency loss for a long self-quanto call option on E.ON and ThyssenKrupp. $S_0 = 17.48$, $T = 23$ for E.ON and $S_0 = 16.73$, $T = 23$ for ThyssenKrupp

Efficiency loss in %			
E.ON	$K_1 = 17.24$	$K_2 = 19.48$	$K_3 = 20.72$
NIG	6.68	11.17	13.85
VG	7.05	11.80	14.67
Normal	6.53	10.72	13.20
ThyssenKrupp	$K_1 = 16.5$	$K_2 = 18.5$	$K_3 = 20.5$
NIG	6.17	8.91	11.86
VG	6.33	9.13	12.11
Normal	6.58	9.43	12.46

Proposition 13 (Efficiency loss vs. relative efficiency loss for X_T^{Call}) *Let $(L_t)_{t\geq 0}$ be a Lévy process with continuous and strictly increasing distribution function F_{L_T} at maturity $T > 0$. Let X_T^{Call} be the payoff of a long call option with strike $K > 0$ and let $\overline{\theta}$ be an Esscher parameter.*

1. *The efficiency loss, as a function of the strike, $\ell(K) := c(X_T^{\text{Call}}) - c(\underline{X}_T^{\text{Call}})$ is decreasing in K.*
2. $\frac{\partial}{\partial K} c(X_T^{\text{Call}}) \leq \frac{\partial}{\partial K} c(\underline{X}_T^{\text{Call}})$ *and the cost of the standard and efficient call are decreasing in K.*
3. *The relative efficiency loss $\ell_r(K) := \frac{\ell(K)}{c(X_T^{\text{Call}})}$ increases in K.*

As for the put option, the latter findings immediately implies the following bound for the efficiency loss of a long call option in bearish markets.

$$0 \leq \ell(X_T^{\text{Call}}) \leq S_0 E_{\overline{\theta}}[1 - e^{F_{L_T}^{-1}(1-F_{L_T}(L_T))-rT}]. \tag{49}$$

Remark 14 (**monotonicity for self-quanto calls**) The monotonicity results in Theorem 11 and Proposition 13 also hold true in the same way for the efficiency loss of self-quanto calls. For details see Wolf [21].

5 Delta Hedging of Cost-Efficient Strategies in Lévy Models

In the following we discuss the *delta hedge*, i.e. the derivative of the cost of a strategy with respect to the underlying for the cost-efficient payoff. Furthermore, concrete hedging simulation schemes are provided for the standard and the cost-efficient put, long call butterfly spread and self-quanto put in the NIG model. Our delta hedging simulation schemes are inspired by the approach of Hull [12, Sect. 14.5]. Moreover, we demonstrate that delta hedging of cost-efficient puts can be efficiently applied in practice and that the obtained hedge errors are usually not greater, but often even smaller that those of the corresponding vanilla puts. Also an alternative delta hedging approach based on a rollover strategy is introduced. The delta hedging strategies obtained by this alternative hedging technique have the potential to outperform the classical ones in this context.

5.1 *Introduction to Delta Hedging*

The Greek delta measures the exposure of a derivative to changes in the value of the underlying. By delta hedging we mean the process of keeping the delta of a portfolio which consists of related financial securities as close to zero as possible. Thus, by delta hedging investors attempt to make their portfolio immune to small changes in the price of the underlying asset in the next small interval of time. If the underlying asset is traded sufficiently liquid in the market, delta hedging is a simple, but nevertheless fairly effective way to cover a risky position and is therefore widely used in practice.

For puts $X_T^{\text{Put}} = (K - S_T)_+$ and calls $X_T^{\text{Call}} = (S_T - K)_+$ with strike K hedging of cost-efficient options with maturity T has already been investigated in Hammerstein et al. [11]. We first restate the main findings in this paper. Recall that the payoff function $\underline{\omega}^{\text{Put}}(y) = (K - S_0 e^{F_{L_T}^{-1}(1-F_{L_T}(\ln(\frac{y}{S_0})))})_+$ of the cost-efficient long put (and call as well) is kept fixed within the trading period $[0, T]$. For $\overline{\theta} < 0$ the price at time $t < T$ of a cost-efficient long put with maturity T is given by

$$c_t(\underline{X}_T^{\text{Put}}) = e^{-r(T-t)} E\Big[Z_{T-t}^{\overline{\theta}} \left(K - S_0 e^{F_{L_T}^{-1}\left(1-F_{L_T}\left(\ln(\frac{y}{S_0})+L_{T-t}\right)\right)}\right)_+\Big]\Big|_{y=S_t}. \tag{50}$$

For $\overline{\theta} > 0$ the price of the cost-efficient call option is given by

$$c_t(\underline{X}_T^{\text{Call}}) = e^{-r(T-t)} E\Big[Z_{T-t}^{\overline{\theta}} \left(S_0 e^{F_{L_T}^{-1}\left(1-F_{L_T}\left(\ln(\frac{y}{S_0})+L_{T-t}\right)\right)} - K\right)_+\Big]\Big|_{y=S_t}. \tag{51}$$

Assuming strictly increasing distribution functions F_{L_t} the Greek delta of a cost-efficient payoff $\underline{X}_T$ with differentiable payoff-function $\underline{w}^X$ such that $\frac{\partial \underline{w}^X}{\partial S_t}\left(S_t e^{L_{T-t}}\right) \in \mathscr{L}^1(Z_{T-t}^{\overline{\theta}+1} P)$, where $\overline{\theta}$ is an Esscher parameter, then is given for $t < T$ by

$$\underline{\Delta}_t^X = \frac{\partial}{\partial S_t} c_t(\underline{\omega}^X(S_T)) = E_{\overline{\theta}+1}\left[\frac{\partial \underline{\omega}^X}{\partial S_t}(S_t e^{L_{T-t}})\right]. \tag{52}$$

For the basic vanilla payoffs the assumptions on $\underline{w}^{X_T}$ are fulfilled and lead to more concrete formulas in the Lévy models considered in this paper; for example for cost-efficient puts one gets

$$\underline{\Delta}_t^{\text{Put}} = S^* \int_{K^*}^{\infty} e^{\overline{\theta}x + F_{L_T}^{-1}(1-F_{L_T}(\ln(\frac{S_t}{S_0})+x))} \frac{d_{L_T}(\ln(\frac{S_t}{S_0})+x)}{d_{L_T}(1-F_{L_T}(\ln(\frac{S_t}{S_0})+x))} d_{L_{T-t}}(x)\,dx, \tag{53}$$

where $K^* := \ln(\frac{S_0}{S_t}) + F_{L_T}^{-1}(1 - F_{L_T}(\ln(\frac{K}{S_0})))$ and $S^* := \frac{S_0}{S_t \cdot M_{L_{T-t}}(\overline{\theta}+1)}$. In cases where the payoff function $\underline{\omega}^X$ of the cost-efficient payoff $\underline{X}_T$ is not explicitly given, as for all options X_T with non-monotone payoff function, the latter result becomes impractical. In such circumstances we utilize the standard approximation $\Delta^X = \frac{\Delta c}{\Delta S}$ for the Greek delta associated to a payoff X_T, where ΔS indicates a small change in the stock price and Δc expresses the corresponding change in the option price. In Fig. 23 the relationship between the cost-efficient self-quanto put price and the underlying stock price is illustrated for Volkswagen in the NIG model at time $t = 0$. The Greek delta $\underline{\Delta}_0^{\text{sqP}}$ of the cost-efficient self-quanto put is the slope of the dotted line.

Using the representation of the NIG density we obtain from the formula for the cost of the vanilla put option in (18):

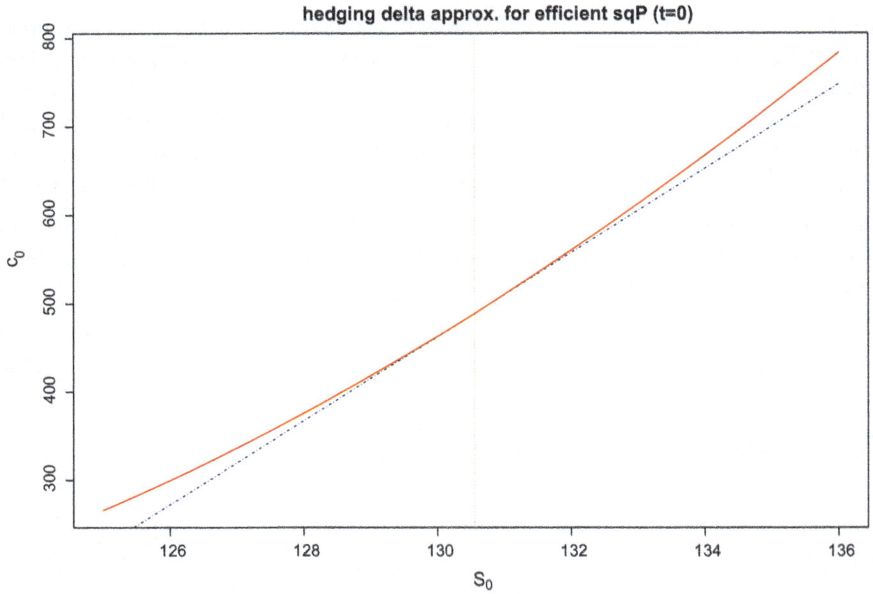

Fig. 23 The relationship between the cost-efficient long self-quanto put price with strike $K = 135$, and the underlying stock price at initial time $t = 0$ for the Volkswagen stock in the NIG model, $T = 23$ days. The vertical dotted line marks the actual initial value $S_0 = 130.55$

$$\begin{aligned}\Delta_t^{\text{Put}} = \frac{\partial c(X_{T-t}^{\text{Put}})}{\partial S_t} &= -\frac{Ke^{-r(T-t)}}{S_t} d_{NIG(\alpha,\beta+\overline{\theta},\delta(T-t),\mu(T-t))}\left(\ln\left(\tfrac{K}{S_t}\right)\right) \\ &\quad - F_{NIG(\alpha,\beta+\overline{\theta}+1,\delta(T-t),\mu(T-t))}\left(\ln\left(\tfrac{K}{S_t}\right)\right) \\ &\quad + d_{NIG(\alpha,\beta+\overline{\theta}+1,\delta(T-t),\mu(T-t))}\left(\ln\left(\tfrac{K}{S_t}\right)\right) \\ &= -\frac{Ke^{-r(T-t)}}{S_t} d_{NIG(\alpha,\beta+\overline{\theta},\delta(T-t),\mu(T-t))}\left(\ln\left(\tfrac{K}{S_t}\right)\right) \\ &\quad - F_{NIG(\alpha,\beta+\overline{\theta}+1,\delta(T-t),\mu(T-t))}\left(\ln\left(\tfrac{K}{S_t}\right)\right) \\ &\quad + \frac{Ke^{-r(T-t)}}{S_t} d_{NIG(\alpha,\beta+\overline{\theta},\delta(T-t),\mu(T-t))}\left(\ln\left(\tfrac{K}{S_t}\right)\right) \\ &= -F_{NIG(\alpha,\beta+\overline{\theta}+1,\delta(T-t),\mu(T-t))}\left(\ln\left(\tfrac{K}{S_t}\right)\right).\end{aligned} \tag{54}$$

Example 15 (**Simulation of Delta hedging**) We investigate the following example: A financial institution has sold for 40,871 € a cost-efficient long put option on 10,000 shares of Volkswagen (cf. Table 3 for prices of a cost-efficient put). We assume that this is a non-dividend paying stock. The trading period is October 2012, thus $S_0 = 130.55$ € and $T = 23$ days. Further, the exercising price is $K = 135$ € and the riskless interest rate equals the continuously compounded daily Euribor rate at October 1, 2012, $r = 4.2027 \times 10^{-6}$. The hedge is supposed to be adjusted every three trading days, i.e. at October 1st, 4th, 8th, 11th and so on. Table 12 provides a

Table 12 Simulation of delta hedging for a cost-efficient long put on 10,000 Volkswagen shares in the NIG model

Days	Stock price	$\underline{\Delta}_t^{\mathrm{Put}}$	Shares purchased/sold	Cost of shares purchased/sold	Cumulative cost (interest cost)
$t = 1$	130.55	0.4082	(+)4082	+532,905.10	+532,905.10 (0.00)
$t = 4$	133.55	0.4765	(+) 683	+ 91,009.75	+ 623,921.57 (6.72)
$t = 7$	133.58	0.4913	(+) 148	+ 19,769.25	+ 643,698.70 (7.87)
$t = 10$	133.50	0.4971	(+) 58	+ 7,743.00	+ 651,449.82 (8.21)
$t = 13$	134.60	0.545	(+) 479	+ 64,473.40	+ 715,931.41 (9.03)
$t = 16$	138.00	0.6829	(+)1379	+190,302.00	+ 906,242.44 (11.43)
$t = 19$	143.00	0.8197	(+)1368	+195,624.00	+1,101,877.87 (13.89)
$t = 22$	149.84	0.8020	(−) 177	− 26,521.68	+1,075,370.08 (9.04)

delta hedging scheme for a cost-efficient long put. Initially, the delta equals $\underline{\Delta}_0^{\mathrm{Put}} = 0.4082$. This means that as soon as the option is written, 532,905.10 € must be borrowed to buy 4,082 shares at a price of S_0. The financial institution encounters interest cost of 6.72 € for the first three trading days. If the delta declines, shares are sold to maintain the hedge implying a reduced cumulative and interest cost. Note, towards the end of the life of the option it is not necessary that the delta of a cost-efficient long put approaches 1.0 when it is apparent that the option will be exercised, since this is typically the case for a standard call only. The optimal long put behaves like a modified call and its corresponding payoff function strongly distinguishes from its vanilla counterpart (compare Fig. 5).

Hedging cost:	
	Total cost at maturity $= 1{,}075{,}379.00$ €
	Long position $= -1{,}227{,}060.00$ €
	Payoff at maturity $\underline{\omega}^{\mathrm{Put}}(S_T) = 181{,}069.50$ €
	Cost of hedging $= 29{,}388.50$ €

On one hand, at maturity the total cost for the hedger adds up to 1,075,379 € plus the payoff $\underline{\omega}^{\mathrm{Put}}(S_T) = 181{,}069.50$ € for the buyer of the optimal put. On the other hand, by selling the long position on the Volkswagen stock the hedger earns $8{,}020 \times 153$ € $= 1{,}227{,}060$ €, thus the cost of the option to the writer equals 29,388.50 € which is 11,482.50 € below the actual price of the option. The performance of the delta hedging gets steadily better as the hedge is monitored more frequently.

The analogous simulation of delta hedging of a standard long put on 10,000 Volkswagen stocks is presented in Table 13. Note that the option closes out of the money. The cost of hedging of the standard long put sums up to 65,108.74 € which is 14955.26 € below the actual price (80,064 €) of the option.

Table 13 Simulation of delta hedging for a long put on 10,000 Volkswagen shares in the NIG model

Days	Stock price	Δ_t^{Put}	Shares purchased/ sold	Cost of shares purchased/ sold	Cumulative cost (interest cost)
$t = 1$	130.55	−0.6050	(−) 6050	−789,827.50	−789,827.50 (0.00)
$t = 4$	133.55	−0.5320	(+) 730	+ 97,272.50	−692,545.04 (9.96)
$t = 7$	133.58	−0.5270	(+) 50	+ 6,678.80	−685,857.51 (8.73)
$t = 10$	133.50	−0.5363	(−) 93	− 12,415.50	−698,264.36 (8.65)
$t = 13$	134.60	−0.4988	(+) 375	+ 50,475.00	−647,780.56 (8.80)
$t = 16$	138.00	−0.3411	(+)1577	+217,626.00	−430,146.39 (8.17)
$t = 19$	143.00	−0.1081	(+)2330	+333,190.00	− 96,950.97 (5.42)
$t = 22$	149.84	−0.0026	(+)1055	+158,081.20	+ 61,130.23 (1.22)

Hedging cost:	Total cost at maturity	= 61,130.74 €
	Short position	= 3,978.00 €
	Payoff at maturity $\omega^{\text{Put}}(S_T)$	= 0.00 €
	Cost of hedging	= 65,108.74 €

We see that delta hedging of cost-efficient options is as complex as for standard options if the numerical techniques are present. From the hedgers point of view it is surely beneficial to provide several (differently priced) delta-hedgeable options with identical payoff distributions to its customers. Further hedging simulations of a long call butterfly spread and its cost-efficient counterpart for ThyssenKrupp and a delta hedging simulation of a self-quanto put and its cost-efficient counterpart for Volkswagen in the NIG model are given in Wolf [21].

5.2 *Alternative Delta Hedging Using Cost-Efficient Strategies*

While in Sect. 5.1 we used cost-efficient strategies at time $t = 0$ and kept the payoff profile fixed up to time T an alternative rollover strategy has been introduced in Hammerstein et al. [11]. In this strategy the cost-efficient payoffs $\underline{X}_{T-t}$ at time t are hedged by a rollover strategy, i.e. an Δ-hedge which reproduces the evolution of the efficient option prices $c(\underline{X}_{T-t})$. This can be regarded as an alternative way to hedge the final payoff X_T. We denote the corresponding hedging deltas by Δ_t^{ro}.

For a cost-efficient put at time t with time to maturity T-t we have

$$\underline{X}_{T-t}^{\text{Put}} = \left(K - S_t e^{F_{L_{T-t}}^{-1}(1-F_{L_{T-t}}(L_{T-t}))}\right)_+. \tag{55}$$

We find

$$\underline{X}_{T-t}^{\text{Put}} \to X_T^{\text{Put}} \quad \text{and} \quad c\left(X_{T-t}^{\text{Put}}\right) - c\left(\underline{X}_{T-t}^{\text{Put}}\right) \to 0 \quad \text{as } t \to T.$$

For $\overline{\theta} < 0$ we have for the alternative delta Δ_t^{roP} of the long vanilla put X_T^{Put} at time t

$$\Delta_t^{\text{roP}} = -\frac{1}{M_{L_{T-t}}(\overline{\theta})} \int_0^{F_{L_{T-t}}\left(\ln\left(\frac{K}{S_t}\right)\right)} e^{\overline{\theta} F_{L_{T-t}}^{-1}(1-y) + F_{L_{T-t}}^{-1}(y) - r(T-t)}\, \mathrm{d}y. \tag{56}$$

For $\overline{\theta} > 0$, the alternative delta Δ_t^{roC} of the long vanilla call X_T^{Call} at time t is

$$\Delta_t^{\text{roC}} = \frac{1}{M_{L_{T-t}}(\overline{\theta})} \int_0^{1-F_{L_{T-t}}\left(\ln\left(\frac{K}{S_t}\right)\right)} e^{\overline{\theta} F_{L_{T-t}}^{-1}(y) + F_{L_{T-t}}^{-1}(1-y) - r(T-t)}\, \mathrm{d}y. \tag{57}$$

Equations (56) and (57) imply that the alternative deltas Δ_t^{roP}, Δ_t^{roC} for the vanilla puts and calls have the same sign as their classical counterparts Δ_t^{Put}, Δ_t^{Call}, which is in line with the intuition. The absolute values of the rollover deltas for calls are smaller than the classical deltas of calls while this is also the case typically for puts.

Comparison of deltas:

(1) For a vanilla call and if $\overline{\theta} > 0$, then for each

$$t \in [0, T) \colon 0 \le \Delta_t^{\text{roC}} \le \Delta_t^{\text{Call}}. \tag{58}$$

(2) In the put case if $\overline{\theta} < 0$ and $F_{L_{T-t}}\left(\ln\left(\frac{K}{S_t}\right)\right) \le q^*$ where $q^* \in \left(\frac{1}{2}, 1\right]$ is the unique positive root of

$$\mathsf{D}_P(q) = \frac{1}{M_{L_{T-t}}(\overline{\theta})} \int_0^q e^{\overline{\theta} F_{L_{T-t}}^{-1}(y) + F_{L_{T-t}}^{-1}(y)} - e^{\overline{\theta} F_{L_{T-t}}^{-1}(1-y) + F_{L_{T-t}}^{-1}(y)}\, \mathrm{d}y \tag{59}$$

in $[0, 1]$, then $\Delta_t^{\text{Put}} \le \Delta_t^{\text{roP}} \le 0$.

For details see Hammerstein et al. [11]. The proof makes use of monotonicity properties of

$$\mathsf{D}_C(q) = \frac{1}{M_{L_{T-t}}(\overline{\theta})} \int_0^q e^{\overline{\theta} F_{L_{T-t}}^{-1}(1-y) + F_{L_{T-t}}^{-1}(1-y)} - e^{\overline{\theta} F_{L_{T-t}}^{-1}(y) + F_{L_{T-t}}^{-1}(1-y)}\, \mathrm{d}y$$

and

$$\mathsf{D}_P(q) = \frac{1}{M_{L_{T-t}}(\overline{\theta})} \int_0^q e^{\overline{\theta} F_{L_{T-t}}^{-1}(y) + F_{L_{T-t}}^{-1}(y)} - e^{\overline{\theta} F_{L_{T-t}}^{-1}(1-y) + F_{L_{T-t}}^{-1}(y)}\, \mathrm{d}y.$$

More precisely it holds under the conditions specified above

(1) For $\overline{\theta} > 0$, $\mathsf{D}_C \ge 0$ in $[0, 1]$, D_C is increasing on $\left[0, \frac{1}{2}\right]$ and decreasing on $\left[\frac{1}{2}, 1\right]$.
(2) For $\overline{\theta} < 0$, $\mathsf{D}_P \ge 0$ in $[0, q^*]$, D_P is increasing on $\left[0, \frac{1}{2}\right]$ and decreasing on $\left[\frac{1}{2}, 1\right]$.

As consequence we obtain for cost-efficient bull resp. bear spread options.

Corollary 16 *Under the assumptions above we have:*

(a) *For cost-efficient and vanilla bull spreads with strikes* $0 < K_1 < K_2$*, holds:* *If* $\overline{\theta} > 0$*, then* $0 \leq \Delta_t^{ro\text{-bull}} \leq \Delta_t^{\text{bull}}$ *for* $F_{L_{T-t}}(\ln(\frac{K_1}{S_t})) > \frac{1}{2}$ *and* $0 \leq \Delta_t^{\text{bull}} \leq \Delta_t^{ro\text{-bull}}$ *for* $F_{L_{T-t}}(\ln(\frac{K_2}{S_t})) < \frac{1}{2}$.
For $\overline{\theta} < 0$ *we have* $\Delta_t^{ro\text{-bull}} = \Delta_t^{\text{bull}}$.

(b) *In the bear spread case, we have* $\Delta_t^{ro\text{-bear}} = \Delta_t^{\text{bear}}$ *for* $\overline{\theta} > 0$.
If $\overline{\theta} < 0$*, then* $\Delta_t^{ro\text{-bear}} \leq \Delta_t^{\text{bear}} \leq 0$ *for* $F_{L_{T-t}}(\ln(\frac{K_1}{S_t})) > \frac{1}{2}$ *and* $\Delta_t^{\text{bear}} \leq \Delta_t^{ro\text{-bear}} \leq 0$ *for* $F_{L_{T-t}}(\ln(\frac{K_2}{S_t})) < \frac{1}{2}$.

Proof (a) Since the vanilla and cost-efficient bull spread coincide for $\overline{\theta} < 0$, the equation $\Delta_t^{\text{ro-bull}} = \Delta_t^{\text{bull}}$ is obvious. Let $\overline{\theta} > 0$ and denote by C_i a call option with strike K_i, $i = 1, 2$, then from the definition of a bull spread we easily arrive at

$$c(X_{T-t}^{\text{bull}}) = c(X_{T-t}^{C_1}) - c(X_{T-t}^{C_2}) \quad \text{and} \quad c(\underline{X}_{T-t}^{\text{bull}}) = c(\underline{X}_{T-t}^{C_1}) - c(\underline{X}_{T-t}^{C_2}).$$

The corresponding deltas are known and equal

$$\Delta_t^{\text{bull}} = \Delta_t^{C_1} - \Delta_t^{C_2} \quad \text{and} \quad \Delta_t^{\text{ro-bull}} = \Delta_t^{\text{roC}_1} - \Delta_t^{\text{roC}_2}.$$

From Eqs. (50) and (51) for $T - t$ it is easily seen that both $\Delta_t^{C_i}$ and $\Delta_t^{\text{roC}_i}$ are decreasing functions in the strike K_i, i.e. $\Delta_t^{C_1} \geq \Delta_t^{C_2}$ and $\Delta_t^{\text{roC}_1} \geq \Delta_t^{\text{roC}_2}$. Thus, we have $\Delta_t^{\text{bull}} \geq 0$ and $\Delta_t^{\text{ro-bull}} \geq 0$. Now, consider the difference of the deltas for cost-efficient and vanilla bull spread which equals

$$\begin{aligned} \Delta_t^{\text{bull}} - \Delta_t^{\text{ro-bull}} &= (\Delta_t^{C_1} - \Delta_t^{C_2}) - (\Delta_t^{\text{roC}_1} - \Delta_t^{\text{roC}_2}) \\ &= (\Delta_t^{C_1} - \Delta_t^{\text{roC}_1}) - (\Delta_t^{C_2} - \Delta_t^{\text{roC}_2}) \\ &= \mathsf{D}_{C_1}(q_1) - \mathsf{D}_{C_2}(q_2) \end{aligned}$$

where $q_i = 1 - F_{L_{T-t}}(\ln(\frac{K_i}{S_t}))$. Since $q_1 > q_2$ we obtain, using the above stated monotonicity properties of D_C, D_P,

$$\Delta_t^{\text{bull}} - \Delta_t^{\text{ro-bull}} \geq 0 \text{ for } q_1 < \frac{1}{2} \text{ and } \Delta_t^{\text{bull}} - \Delta_t^{\text{ro-bull}} \leq 0 \text{ for } q_2 > \frac{1}{2}.$$

This proves the assertion.

(b) Again, the vanilla and cost-efficient bear spread coincide for $\overline{\theta} > 0$, the equation $\Delta_t^{\text{ro-bear}} = \Delta_t^{\text{bear}}$ is clear. Let $\overline{\theta} < 0$ and denote P_i a put option with strike K_i, $i = 1, 2$, then we obtain from the Eqs. (50) and (51) that both $\Delta_t^{P_i}$ and $\Delta_t^{\text{roP}_i}$ are decreasing functions in the strike K_i. This implies completely analogous to the bull spread case that $\Delta_t^{\text{bear}} = \Delta_t^{P_2} - \Delta_t^{P_1} \leq 0$ and $\Delta_t^{\text{ro-bear}} = \Delta_t^{\text{roP}_2} - \Delta_t^{\text{roP}_1} \leq 0$. Moreover, rearranging the difference yields

$$\begin{aligned}\Delta_t^{\text{ro-bear}} - \Delta_t^{\text{bear}} &= (\Delta_t^{\text{roP}_2} - \Delta_t^{\text{roP}_1}) - (\Delta_t^{P_2} - \Delta_t^{P_1}) \\ &= (\Delta_t^{\text{roP}_2} - \Delta_t^{P_2}) - (\Delta_t^{\text{roP}_1} - \Delta_t^{P_1}) \\ &= \mathsf{D}_{P_1}(q_2) - \mathsf{D}_{P_2}(q_1)\end{aligned}$$

where $q_i = F_{L_{T-t}}(\ln(\frac{K}{S_t}))$. Since $q_2 > q_1$ we obtain, as above that,

$$\Delta_t^{\text{ro-bear}} - \Delta_t^{\text{bear}} \geq 0 \text{ for } q_2 < \tfrac{1}{2} \text{ and } \Delta_t^{\text{ro-bear}} - \Delta_t^{\text{bear}} \leq 0 \text{ for } q_1 > \tfrac{1}{2}.$$

Thus, the statement is proven.

5.3 Application to Real Market Data

In the following we illustrate the hedging results by some examples for the put case. We first consider the price evolution $(c(X_{T-t}^{\text{Put}}))_{0\leq t\leq T}$ of a vanilla put and a cost-efficient put $c_t(\underline{X}_T^{\text{Put}})_{0\leq t\leq T}$ on the Allianz and the Volkswagen stock which are assumed to be issued on October 1, 2012, and to mature on November 1, 2012.

Figure 24 shows the prices of the Allianz stock and the corresponding puts with strike $K = 98$ within the aforementioned time period, as well as the values of the deltas $(\Delta_t^{\text{Put}})_{0\leq t\leq T}$ resp. $(\underline{\Delta}_t^{\text{Put}})_{0\leq t\leq T}$ associated to both puts. Here, all calculations are based on the *NIG* model; the *NIG* parameters for Allianz can be found in Table 1. As is obvious from Fig. 24, the price of the cost-efficient put evolves almost exactly in the opposite way as that of the vanilla put. This reflects the fact that the payoff profiles of both puts are, in some sense, reversed to each other (see Fig. 5); the efficient put roughly behaves like a vanilla call. However, the efficient put ends in the money although the price of the Allianz stock remains below the strike price at maturity because its payoff function already takes positive values for some $S_T < K$. The opposite behaviour of the efficient and the vanilla put is also mirrored in the values of the associated deltas. Because the values of the deltas at maturity are not

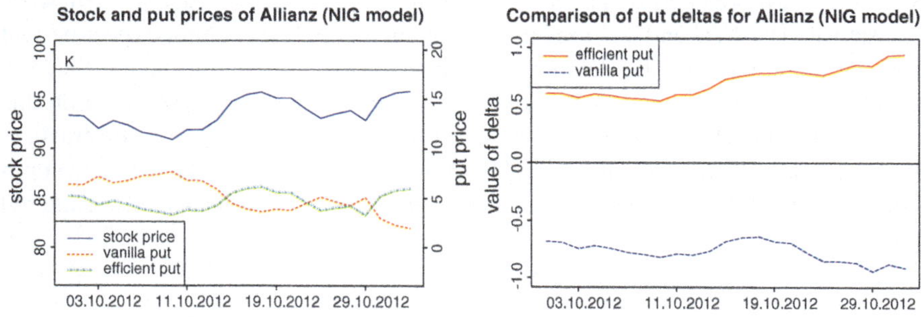

Fig. 24 *Left* Stock price of Allianz and the prices of the associated vanilla resp. efficient put. *Right* Comparison of the deltas of the vanilla and the efficient put on Allianz

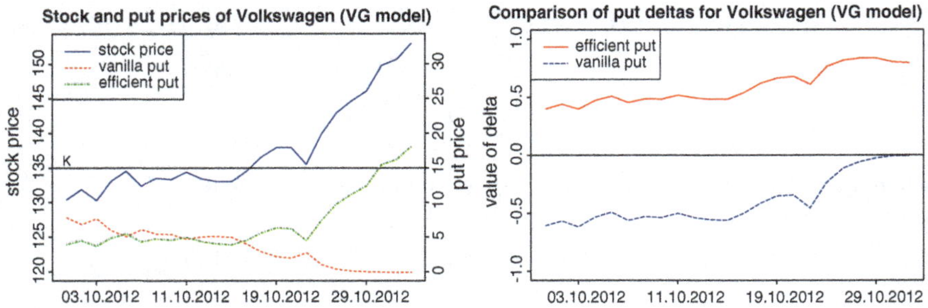

Fig. 25 *Left* Stock price of Volkswagen and the prices of the associated vanilla resp. efficient put. *Right* Comparison of the deltas of the vanilla and the efficient put on Volkswagen

relevant for hedging purposes any more, Fig. 24 only shows the deltas up to one day to maturity, that is, from October 1, 2012, to October 31, 2012. The results obtained for the other two Lévy models (normal and VG) look quite similar and therefore are not plotted here separately. Since the risk-neutral Esscher parameter roughly are of the same size for all three models (see Table 1) and also the put prices and efficiency losses in Table 3 are almost identical, one should not expect greater differences here.

Figure 25 shows the evolution of the prices of the Volkswagen stock and the cost-efficient and vanilla puts on it with strike $K = 135$ as well as the corresponding deltas. Again, the results do not differ much between all three Lévy models under consideration, thus we only show the plots for the VG case. The delta of the vanilla put in this model can be derived analogously as above to be

$$\Delta_t^{\text{Put}} = \frac{\partial c(X_{T-t}^{\text{Put}})}{\partial S_t} = -F_{VG(\lambda(T-t),\alpha,\beta+\overline{\theta}+1,\mu(T-t))}\left(\ln\left(\tfrac{K}{S_t}\right)\right).$$

Note that in this example we have $S_T > K$, therefore the vanilla put expires worthless, and the corresponding delta converges to zero, whereas the efficient put ends deep in the money.

However, computing the put deltas is only one side of the coin, market participants will surely be more interested in how well the hedging strategies based on them work in practice. The *NIG* and *VG* models are incomplete, so one cannot expect perfect hedging there, but also the Samuelson model is only complete in theory. Since in reality just discrete hedging is feasible, one will encounter hedge errors within this framework, too. The magnitude of these errors is, of course, relevant for practical applications. Therefore, we also calculate and compare the hedge errors that occur in delta hedging of the vanilla and efficient puts on Allianz and Volkswagen considered before.

Delta Hedging Strategy

The hedge portfolios are rebalanced daily, hence the portfolio weights δ_t (amount of stock at time t) and b_t (amount of money on the savings account at t) just have to be calculated at the discrete times $t = 0, 1, \dots, T-1$. For the vanilla puts $\delta_t = \Delta_t^{\text{Put}}$, and in case of the efficient puts we have $\delta_t = \underline{\Delta}_t^{\text{Put}}$. Depending on the put type under consideration, we analogously set $c_t = c(X_{T-t}^{\text{Put}})$ or $c_t = c_t(\underline{X}_T^{\text{Put}})$, respectively. At the initial time $t = 0$, the hedge portfolio is set up with the weights δ_0 and $b_0 = -\delta_0 S_0 + c_0$ since the writer of the put obtains c_0 from the buyer, shorts $|\delta_0|$ stocks and deposits all income on his savings account. At time $t > 0$, the value of the portfolio *before* rebalancing is $\delta_{t-1} S_t + e^r b_{t-1}$, and we define the corresponding hedge error by

$$\mathrm{e}_t := c_t - \delta_{t-1} S_t - e^r b_{t-1},$$

so positive hedge errors mean losses. At the end of the trading day, the new weights δ_t and $b_t = c_t - \delta_t S_t$ are chosen to ensure that the value of the portfolio again exactly coincides with the present put price. Using the above definition of e_t, we can alternatively represent b_t in the form

$$b_t = \mathrm{e}_t + e^r b_{t-1} + S_t(\delta_{t-1} - \delta_t).$$

This means that the hedge error is nothing but the amount of money one has to additionally inject in or withdraw from the savings account after adapting the stock position to make the value of the hedge portfolio congruent with the current put price.

Remark 17 In general, the size of the hedge error also depends on the rebalancing frequency and the continuity properties of the payoff function. Our empirical results below show that for standard and efficient puts a daily rebalancing of the portfolio already is sufficient to get a fairly precise approximation to the current option prices. A thorough theoretical analysis of the behaviour of hedge errors resulting from delta and quadratic hedging strategies in exponential Lévy models can be found in Brodén and Tankov [5].

The upper graphs of Fig. 26 display the hedge errors obtained from delta hedging of the different puts on Allianz and Volkswagen. At the beginning, the hedge errors of the efficient and the vanilla puts behave fairly similarly, but with time passing the distinctions increase. This might again be explained by the different shapes of the payoff profiles and the different signs of the corresponding deltas which lead to more pronounced differences in the hedge errors as the time to maturity becomes smaller. The sums $\sum_{t=0}^{22} |\mathrm{e}_t|$ of the absolute hedge errors for Allianz are 1.296 (efficient put) and 1.798 (vanilla put), for Volkswagen we obtain 1.794 (efficient put) resp. 2.252 (vanilla put). This indicates that cost-efficient options can be hedged at least as efficiently as standard options. However, since the prices of vanilla and efficient puts can differ significantly over time, one should not only look at the absolute hedge errors to confirm this assertion, but also take the relative or percentage hedge errors $\tilde{\mathrm{e}}_t := \frac{\mathrm{e}_t}{c_t}$ into account. The values of $\tilde{\mathrm{e}}_t$ for the Allianz puts are shown in the

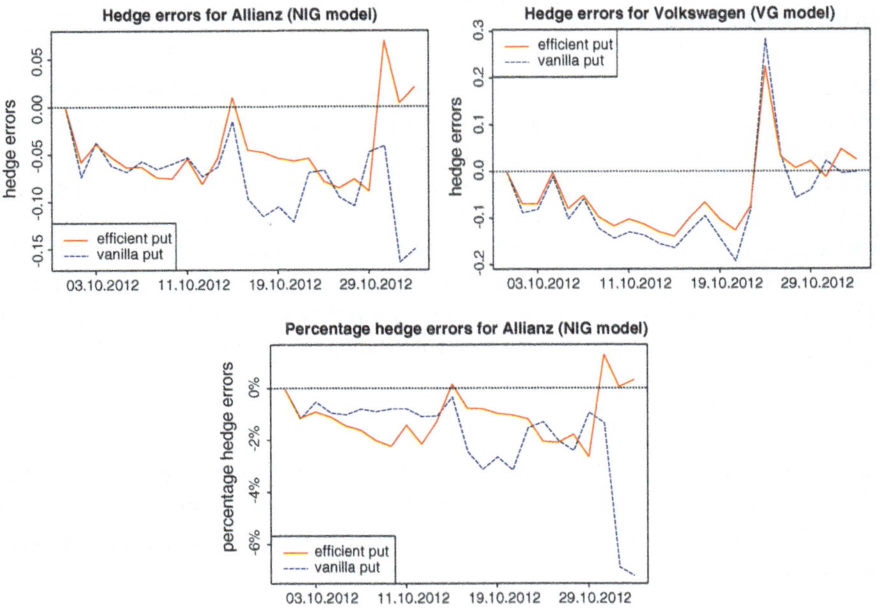

Fig. 26 Delta hedge errors of the efficient and vanilla puts on Allianz with strike $K = 98$ and Volkswagen with strike $K = 135$

lower graph of Fig. 26. For the efficient put, we obtain $\sum_{t=0}^{22} |\tilde{e}_t| = 0.299$, and the corresponding value for the vanilla put is 0.438. Analogous computations for the Volkswagen puts would not make much sense here because there the vanilla put ends up deep out of the money, therefore the $\tilde{e}_t$ would tend to infinity as t approaches T.

In the last part of this section, we compare the alternative hedging strategy for vanilla puts based on the rollover-deltas Δ_t^{roP} with its classical counterpart and investigate if it can provide an efficient and more robust way to hedge the final put payoff $(K - S_T)_+$ as expected from our comparison result. For this purpose, we again consider the vanilla puts on Allianz and Volkswagen with the same strikes and maturity as before, but now contrast the corresponding price processes $(c(X_{T-t}^{\text{Put}}))_{0 \le t \le T}$ with the series $(c(\underline{X}_{T-t}^{\text{Put}}))_{0 \le t \le T}$ of prices of efficient puts which are newly initiated at each day t. Figures 27 and 28 show the stock and put price processes for Allianz in the NIG model and for Volkswagen in the VG model, respectively, as well as a graphical comparison of the associated classical put deltas Δ_t^{Put} and rollover-deltas Δ_t^{roP}. The condition of the comparison results in (58), (59) is fulfilled for all $0 \le t \le T$, the absolute values of the rollover-deltas are always smaller than those of the classical deltas for both stocks.

This indicates that the hedging strategies based on the rollover-deltas may indeed allow for a less expensive way to replicate the final put payoff. The advantage of lower hedging costs might be annihilated by larger hedging errors though. Therefore one has to take these into account before coming to a conclusion. Using some of the

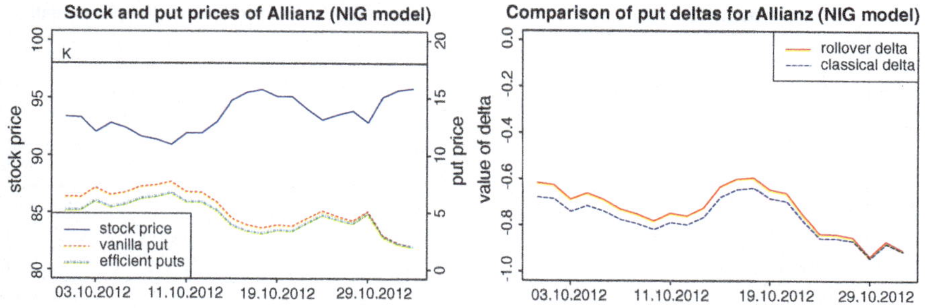

Fig. 27 *Left* Stock price of Allianz and the prices of the associated vanilla resp. efficient puts. *Right* Comparison of the deltas Δ_t^{Put}, Δ_t^{roP} of the vanilla put on Allianz on the *left*

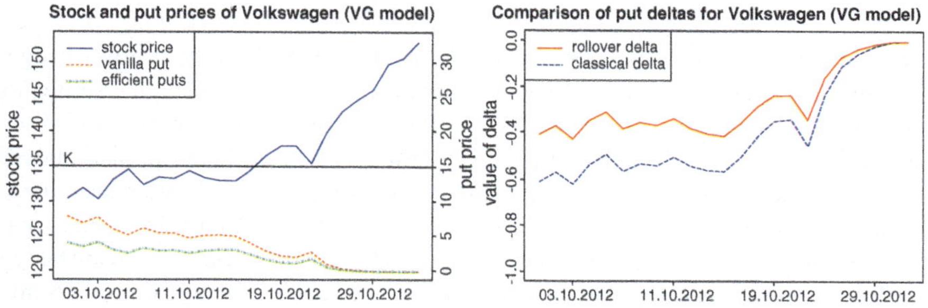

Fig. 28 *Left* Stock price of Volkswagen and the prices of the associated vanilla resp. efficient puts. *Right* Comparison of the deltas Δ_t^{Put}, Δ_t^{roP} of the vanilla put on Volkswagen on the *left*

notations from above, we define the hedge error for the alternative hedging strategy by

$$\mathrm{e}_t := c(\underline{X}_{T-t}^{\text{Put}}) - \Delta_t^{\text{roP}} S_t - e^r b_{t-1}.$$

Observe that we do not use the price $c(X_{T-t}^{\text{Put}})$ of the vanilla put at time t in the above definition although we want to hedge its final payoff. Since the rollover-deltas Δ_t^{roP} are intended to replicate the prices $c(\underline{X}_{T-t}^{\text{Put}})$, and $c(\underline{X}_{T-t}^{\text{Put}}) < c(X_{T-t}^{\text{Put}})$ for all $0 \leq t < T$ because $\overline{\theta} < 0$ here, a comparison of the value of the hedge portfolio at time t with $c(X_{T-t}^{\text{Put}})$ would lead to a systematic overestimation of the hedge error. Moreover, we only consider options of European type here. Therefore it is more important to look at the hedge error at maturity which tells us how precise the hedging strategies can reproduce the final obligation of the writer of the option. At time T, however, we have $c(\underline{X}_{T-T}^{\text{Put}}) = c(X_{T-T}^{\text{Put}}) = (K - S_T)_+$ as pointed out before, so there the hedge error is defined without ambiguity.

We finally take a look at the hedge errors obtained from the two delta hedging strategies for the vanilla puts on Allianz and Volkswagen which are visualized in Fig. 29. For Allianz, the hedge errors e_T at maturity are -0.149 for the classical delta hedge and -0.085 for the alternative rollover-delta hedge, and the sum $\sum_{t=0}^{22} |\mathrm{e}_t|$ of

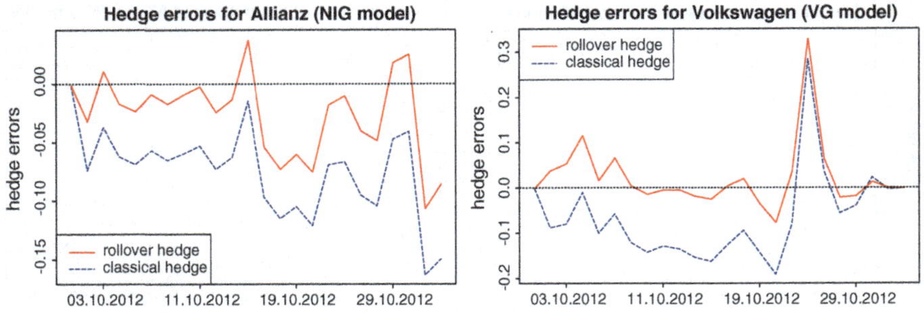

Fig. 29 *Left* Delta hedge errors of the vanilla put on Allianz with strike $K = 98$ in the NIG model. *Right* Delta hedge errors of the vanilla put on Volkswagen with strike $K = 135$ in the VG model

the absolute hedge errors is 1.789 for the classical and 0.802 for the rollover hedge. The final hedge errors e_T for the Volkswagen put are zero for both hedging strategies (which is not so surprising because the vanilla put expires worthless here), and the sums of the absolute hedge errors are 2.252 for the classical and 0.983 for the rollover hedge. This shows that the latter can yield at least comparable and often even more accurate results than the classical delta hedging strategy. In case of the Allianz put, the classical delta hedge tends to superhedge the option, that is, the value of the hedge portfolio is always greater than the option price. The rollover hedge does the same on most days, but produces smaller absolute hedge errors. In view of the comparison results in (58) and (59), we suppose that analogous assertions will also hold for calls and probably also for more complex options.

References

1. Bernard, C., Boyle, P.P.: Explicit representation of cost-efficient strategies. Available at SSRN. http://ssrn.com/abstract=1695543. (2010)
2. Bernard, C., Boyle, P.P., Vanduffel, S.: Explicit representation of cost-efficient strategies. Finance **35**(2), 5–55 (2014a)
3. Bernard, C., Moraux, F., Rüschendorf, L., Vanduffel, S.: Optimal payoffs under state-dependent constraints. Quant. Finance **15**(7), 1157–1173 (2015)
4. Bernard, C., Rüschendorf, L., Vanduffel, S.: Optimal claims with fixed payoff structure. J. Appl. Probab. **51A**(1), 175–188 (2014b)
5. Brodén, M., Tankov, P.: Tracking errors from discrete hedging in exponential Lévy models. Int. J. Theor. Appl. Finance **14**(6), 803–837 (2011)
6. Burgert, C., Rüschendorf, L.: On the optimal risk allocation problem. Stat. Decis. **24**(1), 153–171 (2006)
7. Dana, R.A.: A representation result for concave Schur concave functions. Math. Finance **14**, 613–634 (2005)
8. Dybvig, P.: Distributional analysis of portfolio choice. J. Bus. **61**(3), 369–393 (1988a)
9. Dybvig, P.: Inefficient dynamic portfolio strategies or how to throw away a million dollars in the stock market. Rev. Fin. Stud. **1**(1), 67–88 (1988b)

10. Eberlein, E.: Application of generalized hyperbolic Lévy motions to finance. In: T. Mikosch, S. Resnick (eds.) Lévy Processes: Theory and Applications, pp. 319–337. Birkhäuser Verlag (2001)
11. Hammerstein, E.A.v., Lütkebohmert, E., Rüschendorf, L., Wolf, V.: Optimality of payoffs in Lévy models. Int. J. Theor. Appl. Finance **17**(6), 1450041 [46 pp.] (2014)
12. Hull, J.C.: Options, Futures, and Other Derivatives. Prentice-Hall, 3rd ed. (1997)
13. Jouini, E., Kallal, H.: Efficient trading strategies in the presence of market frictions. Rev. Fin. Stud. **14**(2), 343–369 (2001)
14. Rüschendorf, L.: Risk bounds, worst case dependence and optimal claims and contracts. In: Proceedings of the AFMath Conference, p. 23–36, Brussels (2012)
15. Rüschendorf, L., Wolf, V.: On the method of optimal portfolio choice by cost-efficiency. Appl. Math. Finance. **23**(2), 158–173 (2016)
16. Rüschendorf, L., Wolf, V.: Cost-efficiency in multivariate Lévy models. Depend. Model. **3**(1), 2298–2300 (2015)
17. Schied, A.: On the Neyman-Pearson problem for law-invariant risk measures and robust utility functionals. Ann. Appl. Probab. **14**, 1398–1423 (2004)
18. Schoutens, W.: Lévy Processes in Finance: Pricing Financial Derivatives. Wiley (2003)
19. Vanduffel, S., Ahcan, A., Henrard, L., Maj, M.: An explicit option-based strategy that outperforms Dollar cost averaging. Int. J. Theor. Appl. Fin. **15**(2), 1250013 [19 pp] (2012)
20. Vanduffel, S., Cheng, X., Dhaene, J., Goovaerts, M., Henrard, L.: Optimal approximations for risk measures of sums of log-normals based on conditional expectations. J. Comput. Appl. Math. **221**(1), 202–218 (2008)
21. Wolf, V.: Comparison of Markovian Price Processes and Optimality of Payoffs. Dissertation, Universität Freiburg (2014)

Part IV
Term-Structure Modelling

No Arbitrage Theory for Bond Markets

Irene Klein, Thorsten Schmidt and Josef Teichmann

Abstract We investigate default-free bond markets and relax assumptions on the numéraire, which is typically chosen to be the bank account. Considering numéraires different from the bank account allows us to study bond markets where the bank account process is not a valid numéraire or does not exist at all. We argue that this feature is not the exception, but rather the rule in bond markets when starting with, e.g., terminal bonds as numéraires. Our setting are general càdlàg processes as bond prices, where we employ directly methods from large financial markets. Moreover, we do not restrict price processes to be semimartingales, which allows for example to consider markets driven by fractional Brownian motion. In the core of the article we relate the appropriate no arbitrage assumptions (NAFL), i.e. no asymptotic free lunch, to the existence of an equivalent local martingale measure with respect to the terminal bond as numéraire, and no arbitrage opportunities of the first kind (NAA1) to the existence of a supermartingale deflator, respectively. In all settings we obtain existence of a generalized bank account as a limit of convex combinations of roll-over bonds. The theory is illustrated by several examples.

Keywords Large financial markets · Bond markets · Interest rate theory · Forward measure · Short-rate · Numéraire

---

I. Klein
Department of Statistics and OR, University of Vienna,
Oskar-Morgensternplatz 1, 1090 Vienna, Austria
e-mail: irene.klein@univie.ac.at

T. Schmidt
Mathematical Insitute, University Freiburg, Eckerstr. 1,
79104 Freiburg im Breisgau, Germany
e-mail: thorsten.schmidt@stochastik.uni-freiburg.de

J. Teichmann (✉)
Department of Mathematics, ETH Zurich, Rämistr. 101, 8092 Zurich, Switzerland
e-mail: josef.teichmann@math.ethz.ch

© Springer International Publishing Switzerland 2016

J. Kallsen and A. Papapantoleon (eds.), *Advanced Modelling in Mathematical Finance*,
Springer Proceedings in Mathematics & Statistics 189,
DOI 10.1007/978-3-319-45875-5_17

1 Introduction

Most of the term structure models in the literature are based on the fundamental assumption that bond prices $P(t, T)$ together with a numéraire bank account process $B(t)$ of finite total variation form an arbitrage-free market. Formally speaking this means that we can find an equivalent local martingale measure for the collection of stochastic processes $(B(t)^{-1}P(t, T))_{0\leq t\leq T}$ representing bond prices discounted by the bank account's current value. If we assume additionally that those local martingales are indeed martingales, then we arrive at the famous relationship

$$P(t, T) = E_Q\left[\frac{B_t}{B_T}\middle|\mathscr{F}_t\right] \tag{1}$$

for $0 \leq t \leq T$ with respect to a pricing measure Q. For a comprehensive treatment of this special case see [13]. If we assume alternatively the existence of forward rates, we arrive at the Heath-Jarrow-Morton (HJM) drift condition for the stochastic forward rate process encoding the previous local martingale property (see [3, 20]).

On the other hand, bank account processes are limits of roll-over constructions and therefore only approximately given in real markets. Even if there is a bank account process it is not innocent to take it as numéraire. Therefore we avoid assumptions on even the existence of a bank account process and—in case of existence—we do not assume that it can be chosen as numéraire. This can be compared to the famous BGM market model approach, see [4].

It might appear that this relaxation of assumptions leads to more general but possibly less interesting term structure models. However, we show that term structure models, where the bank account process is not a numéraire, are of some interest and importance.

Let us demonstrate our setting with an example: we generate an arbitrage-free market of zero coupon bonds by the formula

$$P(t, T) = E\left[\frac{S^*_t}{S^*_T}\middle|\mathscr{F}_t\right] \tag{2}$$

for $0 \leq t \leq T \leq T^*$, where S^* denotes a strictly positive process under some appropriate integrability assumptions. The market is indeed free of arbitrage since the (appropriately normalized) density process $\left(\frac{P(t,T^*)}{S^*_t}\right)_{0\leq t\leq T^*}$ yields an equivalent measure Q^*, the T^*-forward measure, such that the discounted price processes $\left(P(t, T^*)^{-1}P(t, T)\right)_{0\leq t\leq T}$ with respect to the terminal bond $P(., T^*)$ are Q^*-martingales.

Consider now the case where $\frac{1}{S^*}$ is a strict local martingale and where the bank account process obtained as a limit of roll-over portfolios is identically equal to 1, see Sect. 7.1 for a natural example. Apparently there is no equivalent pricing measure Q such that the pricing formula (1) holds true with the bank account process $B = 1$, since the term structure is non-trivial by strict local martingality. In the appendix section this example is further analyzed from a change of numéraire perspective, see Appendix 8, in particular a dichotomy between numéraires and bubbles is outlined there.

It is now the aim of this article to understand bond market dynamics under the weaker assumption that there are no arbitrages with respect to a terminal bond numéraire. This is a minimal assumption, which appears to us—in light of the previous example—more appropriate.

From a mathematical point of view we believe that the technology of large financial markets is the right tool to understand the nature of arbitrage in the considered (infinite-dimensional) bond market, since we want to avoid artificially introduced trading strategies. More precisely, we fix a terminal maturity T^* and consider the bond market (for maturities $T \leq T^*$) with respect to the terminal bond as a numéraire. We can only trade a finite number of assets, but we can take more and more of them and so approximate a portfolio with an infinite number of assets. In contrast to, e.g., [3, 8, 14] we do not introduce infinite–dimensional trading strategies but only approximate by finite portfolios, an idea which stems from the theory of large financial markets. As a direct consequence, we avoid pitfalls for measure-valued strategies pointed out in [40]. A second advantage is that we are able to consider markets driven by general càdlàg processes with only a weak regularity in maturity. This extends beyond semimartingale models as considered in the above mentioned articles and in [13].

The structure of the article is as follows: in Sect. 2 and 3 we introduce our model for a bond market with an appropriate interpretation as a large financial market and characterize notions of no arbitrage.

In Sect. 4 we relate the appropriate no arbitrage assumption, which is no free lunch (NFL) on the bond market, to the global existence of an equivalent local martingale measure. Indeed, we can prove the existence of an equivalent local martingale measure for all bonds with maturity $T \leq T^*$ in terms of the bond $P(t, T^*)$ as numéraire. This is in contrast to common bond market models in the literature which often start with the assumption of existence of an equivalent (local) martingale measure, whereas we directly define a notion of no arbitrage and then the existence of a local martingale measure follows.

In Sect. 5 we prove by a Komlós-type argument that under the assumption of NAFL there exists a candidate process for the bank account as a limit of convex combinations of roll-over bonds. This bank account is a supermartingale in terms of the terminal bond.

In Sect. 6 we will see that it is possible to further relax the assumptions on the bond market. If we only assume that the bond market does not allow asymptotic arbitrage opportunities of first kind (AA1) in the sense of large financial markets as in [23], we cannot guarantee the global existence of an equivalent local martingale measure. However, we can prove that there exists a strictly positive supermartingale deflator for each sequence of bonds with maturities that do not induce an AA1. If there exists a dense sequence of maturities in $[0, T^*]$, such that the induced large financial market is free of AA1, then there exists a supermartingale deflator for the bond market with all maturities in $[0, T^*]$. In this relaxed setting we can still show the existence of a generalized bank account as a limit of convex combinations of roll-over bonds, which is a supermartingale in terms of the terminal bond. This section is related to results of Kardaras, see, e.g., [26]

In Sect. 7 we illustrate the setup with four examples: first, we consider the example with optimal growth portfolio being a strict local martingale. Second, a bond market model driven by fractional Brownian motion is studied, where bond prices are not semimartingales, but bond prices in terms of the numéraire are. Third, we consider an extension of the Heath-Jarrow-Morton approach where the bond prices as functions of the maturity are continuous but of unbounded variation such that a short rate does not exist. Fourth, we give an example illustrating possible pitfalls when considering limits of roll-overs as numéraire: in a setting of not uniformly integrable bond prices, the limit of roll-overs does not qualify as numéraire because it reaches zero with probability one.

2 Market Models for Bond Markets

We consider the following model for a bond market. Let $(\Omega, \mathscr{F}, (\mathscr{F}_t)_{t\geq 0}, P)$ be a filtered probability space where the filtration satisfies the usual conditions. For each $T \in [0, \infty)$ we denote by $(P(t, T))_{0\leq t\leq T}$ the price process of a bond with maturity T. For all T, $(P(t, T))_{0\leq t\leq T}$ is a strictly positive càdlàg stochastic process adapted to $(\mathscr{F}_t)_{0\leq t\leq T}$ with $P(T, T) = 1$. We assume that the price process is almost surely right continuous in the second variable, where the nullset does not depend on t, indeed we make

Assumption 1 There is $N \in \mathscr{F}$ with $P(N) = 0$ such that

$$N \supseteq \bigcup_{t\in[0,\infty)} \{\omega : T \to P(t, T)(\omega) \text{ is not right continuous}\}.$$

For a generic process X and a stopping time τ we denote by $(X_t^\tau) = (X_{t\wedge\tau})$ the process stopped at τ.

Assumption 2 We make the following assumption on uniform local boundedness for $P(., T)$ and local boundedness for $P(., T)^{-1}$:

(1) For any T there is $\varepsilon > 0$, an increasing sequence of stopping times $\tau_n \to \infty$ and $\kappa_n \in [0, \infty)$ such that
$$P(t, U)^{\tau_n} \leq \kappa_n,$$
for all $U \in [T, T + \varepsilon)$ and all $t \leq T$.

(2) There exists a nonempty set $\mathscr{T} \subset [0, \infty)$ such that $\left(\frac{1}{P(t,T^*)}\right)_{0\leq t\leq T^*}$ is locally bounded for all $T^* \in \mathscr{T}$.

The set $\mathscr{T}$ denotes the maturities of those bonds which we shall consider as candidate numéraires.

Remark 1 Note that Assumption 2 is fulfilled in the reasonable special case, where $P(., T)$ and $P(., T)^{-1}$ are locally bounded for any T and, for any fixed t, the function $T \mapsto P(t, T)$ is non–increasing. This, for example, holds, if there exists a non-negative short-rate.

In the following assumption we consider a numéraire related to a *terminal maturity* $T^* \in \mathcal{T}$.

Assumption 3 For all finite collections of maturities $T_1 < T_2 < \cdots < T_n \leq T^*$ with $T^* \in \mathcal{T}$ there exists a measure $Q \sim P|_{\mathcal{F}_{T^*}}$ such that $\left(\frac{P(t,T_i)}{P(t,T^*)}\right)_{0\leq t\leq T_i}$ is a local Q-martingale, $i = 1, \ldots, n$.

The measure Q from Assumption 3 is called the T^*-forward-measure for the finite market consisting of bonds $P(., T_i), i = 1, \ldots, n$ and the numéraire $P(., T^*)$.

Remark 2 Note that we do not assume the existence of a short-rate or even a bank account. Moreover, we do not assume that $P(., T)$ is a semimartingale. However, Assumption 3 implies that, for a finite collection of maturities, only bonds in terms of the numéraire $P(., T^*)$ are semimartingales under the objective measure P, because they are local martingales under the equivalent measure Q. Moreover, they are locally bounded because we assumed that $P(., T)$ is locally bounded, for any T, and $P(., T^*)^{-1}$ is locally bounded for $T^* \in \mathcal{T}$.

If there exists a short-rate and an equivalent martingale measure for all discounted bond processes, then Assumption 3 follows immediately.

Lemma 1 *Assume that there exists the locally integrable short-rate process* $(r_t)_{t\geq 0}$ *and let* $B_t := e^{\int_0^t r_s ds}$ *for* $t \geq 0$. *Assume that there exists a measure* Q *such that* $Q|_{\mathcal{F}_t} \sim P|_{\mathcal{F}_t}$ *for* $t \geq 0$ *and such that* $\left(B_t^{-1} P(t, T)\right)_{0\leq t\leq T}$ *is a* Q*–martingale, for all* $T \in [0, \infty)$. *Then, for any finite collection of maturities* $T_1 < \cdots < T_n$, *the measure* Q^{T_n} *with*

$$Z^n := \frac{dQ^{T_n}}{dQ|_{\mathcal{F}_{T_n}}} = \frac{(B_{T_n})^{-1}}{E_Q[B_{T_n}^{-1}]}$$

fulfills Assumption 3.

Proof Let Q^{T_n} be defined as above. We have to show that $\left(\frac{P(t,T_i)}{P(t,T_n)}\right)_{0\leq t\leq T_i}$ is a (local) Q^{T_n}-martingale, which is the case iff $\left(\frac{P(t,T_i)}{P(t,T_n)} \cdot E_Q[Z^n|\mathcal{F}_t]\right)_{0\leq t\leq T_i}$ is a (local) Q-martingale. As $P(T_n, T_n) = 1$ we have that $Z^n = \frac{1}{E_Q[B_{T_n}^{-1}]} \frac{P(T_n,T_n)}{B_{T_n}}$. Hence we get by the martingale property of $B^{-1}P(., T_n)$ that

$$\frac{P(t, T_i)}{P(t, T_n)} E_Q[Z^n|\mathcal{F}_t] = \frac{P(t, T_i)}{P(t, T_n)} \frac{P(t, T_n)}{E_Q[B_{T_n}^{-1}]B_t} = \frac{1}{E_Q[B_{T_n}^{-1}]} \frac{P(t, T_i)}{B_t},$$

which is a Q-martingale. □

3 Bond Markets as Large Financial Markets

Assumption 3 means that for a finite selection of bonds considered with respect to a certain numéraire (the bond with the largest maturity) there exists an equivalent local martingale measure. Our aim will be the following: for a fixed maturity $T^* \in \mathcal{T}$, we aim at finding a measure Q^* such that all bonds with maturity $T \leq T^*$ are local martingales under Q^* in terms of the numéraire $P(t, T^*)$. In Sect. 4 we will present a general theorem.

We introduce a large financial market connected to the bond market. We choose a finite time horizon $T > 0$ as this will be sufficient for our purpose. We start with a short overview of the facts on large financial markets that we will need. Note that we do not present large financial in the full generality of the literature but only in the following *nested* setting. Let $(S_t^n)_{t\in[0,T]}$, $n = 1, 2, \ldots,$ be a sequence of $\mathbb{R}$-valued semimartingales based on a filtered probability space $(\Omega, \mathcal{F}, (\mathcal{F}_t)_{t\in[0,T]}, P)$ where the filtration satisfies the usual assumptions. Moreover let an additional asset $S_t^0 \equiv 1$ be given (this means we assume that all assets are already discounted with respect to some numéraire, so we have that one of the assets, i.e. S^0, equals 1). For each $n \geq 1$ we define a classical market model (referred to as *finite market* n) given by the $\mathbb{R}^{n+1}$–valued semimartingale $\mathbf{S}_t^n = (S_t^0, S_t^1, \ldots, S_t^n)$, $t \in [0, T]$, which describes the price processes of the first $n + 1$ tradable assets.

Let $\mathbf{H}$ be a predictable $\mathbf{S}^n$-integrable process and $(\mathbf{H} \cdot \mathbf{S}^n)_t$ the stochastic integral of $\mathbf{H}$ with respect to $\mathbf{S}^n$. The process $\mathbf{H}$ is an admissible trading strategy if $\mathbf{H}_0 = 0$ and there is $a > 0$ such that $(\mathbf{H} \cdot \mathbf{S}^n)_t \geq -a$, $0 \leq t \leq T$. If the bound from below is $-a$ then $\mathbf{H}$ is called a-admissible. Define

$$\mathbf{K}^n = \{(\mathbf{H} \cdot \mathbf{S}^n)_T : \mathbf{H}\,\text{admissible}\} \text{ and } \mathbf{C}^n = (\mathbf{K}^n - L_+^0) \cap L^\infty. \tag{3}$$

Here $L^0 = L^0(\Omega, \mathcal{F}, P)$ denotes all random variables with values in $\mathbb{R}$ and $L^\infty = L^\infty(\Omega, \mathcal{F}, P)$ is the Banach space of essentially bounded random variables. $\mathbf{K}^n$ can be interpreted as the cone of all replicable claims in the finite market n, and $\mathbf{C}^n$ is the cone of all claims in L^∞ that can be super replicated in this market.

In all the following we will use the notation $E[\cdot]$ for expectation with respect to the original measure P and $E_R[\cdot]$ for expectation with respect to a probability measure R different from P. Define the set $\mathbf{M}_e^n$ of equivalent separating measures for the finite market n as

$$\begin{aligned}\mathbf{M}_e^n &= \{Q \sim P : E_Q[f] \leq 0 \text{ for all } f \in \mathbf{C}^n\} \\ &= \{Q \sim P : E_Q[f] \leq 0 \text{ for all } f \in \mathbf{K}^n\}.\end{aligned} \tag{4}$$

If $\mathbf{S}^n$ is (locally) bounded then $\mathbf{M}_e^n$ consists of all equivalent probability measures such that $\mathbf{S}^n$ is a (local) martingale.

A *large financial market* is the sequence of the finite market models, i.e. the sequence of the market models induced by the $d(n)$-dimensional semimartingales $\mathbf{S}^n$. As a consequence, we cannot trade with an actually infinite number of securities (so that we avoid artificially introduced infinite-dimensional trading strategies), but we can trade in more and more assets and in this way approximate something infinite-dimensional.

We impose the following assumption, which is standard in the theory of large financial markets:

$$\mathbf{M}_e^n \neq \emptyset, \qquad \text{for all } n \in \mathbb{N}. \tag{5}$$

This implies that any no arbitrage condition (such as *no arbitrage*, *no free lunch with vanishing risk*, *no free lunch*) holds for each finite market n.

However, there is still the possibility of various approximations of an arbitrage profit by trading on the sequence of market models. We will need the notions no asymptotic free lunch and no asymptotic free lunch with bounded risk and later on no asymptotic arbitrage of first kind, see Sect. 6.

No asymptotic free lunch (NAFL) is the large financial markets analogue of the classical no free lunch condition (NFL) of Kreps [32]. We will first recall the classical NFL condition here for a finite market n. Let $\mathbf{C}^n$ be defined as in (3).

Definition 1 The condition NFL holds on the finite market n if

$$\overline{\mathbf{C}^n}^* \cap L_+^\infty = \{0\}, \tag{6}$$

where $\overline{\mathbf{C}^n}^*$ denotes the weak-star-closure of $\mathbf{C}^n$.

This means by super replicating claims in an admissible way with a finite number of assets we cannot approximate in a weak-star sense a strictly positive gain.

Now NAFL can be defined in analogous way as the condition NFL but for the whole sequence of sets $\mathbf{C}^n$:

Definition 2 A given large financial market satisfies NAFL if

$$\overline{\bigcup_{n=1}^{\infty} \mathbf{C}^n}^* \cap L_+^\infty = \{0\}.$$

If NAFL holds then it is not possible to approximate a strictly positive profit in a weak-star sense by trading in any finite number of the given assets (although we can use more and more of them).

Remark 3 Note that in the literature the term large financial market is used for a more general concept where each market n is based on a different filtered probability space. When based on one fixed filtered probability space, it is moreover crucial in Definition 2, that in our nested setting we have that, for all $n \geq 1$, $\mathbf{C}^n \subseteq \mathbf{C}^{n+1}$ and hence $\bigcup_{n=1}^{\infty} \mathbf{C}^n$ is a convex cone. In a more general situation the condition NAFL cannot be reduced to the above rather easy form but the general definition that was introduced in [27] has to be used, see also [29]. In our nested setting, we will not have to deal with these technicalities which are common in large financial markets.

Let us now introduce a large financial markets' structure for the bond market introduced in Sect. 2.

Definition 3 Let $T^* \in \mathscr{T}$ where $\mathscr{T}$ is the set from Assumption 2. Fix a sequence $(T_i)_{i \in \mathbb{N}}$ in $[0, T^*]$. Define the $n+1$-dimensional stochastic process $(\mathbf{S}^n) = (S^0, \dots, S^n)$ on $[0, T^*]$ as follows:

$$S_t^i = \begin{cases} \frac{P(t,T_i)}{P(t,T^*)} & \text{for } 0 \le t \le T_i \\ \frac{1}{P(T_i,T^*)} & \text{for } T_i < t \le T^* \end{cases}, \tag{7}$$

for $i = 1, \dots, n$ and $S_t^0 = \frac{P(t,T^*)}{P(t,T^*)} \equiv 1$.

The large financial market consists of the sequence of classical market models given by the $(n+1)$-dimensional stochastic processes $(\mathbf{S}^n)_{t \in [0,T^*]}$ based on the filtered probability space $\left(\Omega, \mathscr{F}, (\mathscr{F}_t)_{t \in [0,T^*]}, P|_{\mathscr{F}_{T^*}}\right)$.

Definition 4 The bond market $(P(t,T))_{0 \le t \le T}$ for $0 \le T \le T^*$ satisfies NAFL if there exists a dense sequence $(T_i)_{i \in \mathbb{N}}$ in $[0, T^*]$, such that the large financial market of Definition 3 satisfies the condition NAFL.

Since all involved semimartingales $\mathbf{S}^n$ are locally bounded due to Assumption 2, it is sufficient to deal with equivalent local martingale measures. Hence, the set $\mathbf{M}_e^n$ from (4) is given as follows:

$$\mathbf{M}_e^n = \{Q^n \sim P|_{\mathscr{F}_{T^*}} : \mathbf{S}^n \text{ local} Q^n - \text{martingale}\}.$$

By Assumption 3 we have that $\mathbf{M}_e^n \neq \emptyset$ for all $n \in \mathbb{N}$, so the standard assumption (5) for large financial markets holds. Note that this also implies that each $\mathbf{S}^n$ is a semimartingale, so this is not a problem in Definition 3.

Remark 4 Obviously Assumption 3 can be weakened if the bond market satisfies condition NAFL. Indeed it is sufficient to assume that all the processes from Definition 4 $(S_t^n)_{0 \le t \le T^*}$, $n \in \mathbb{N}$, are semimartingales. Then the stochastic integrals and therefore the sets $\mathbf{C}^n$ make sense. In this case, the existence of a local martingale measure for a finite number of assets follows by NAFL as $\overline{\bigcup_{n=1}^{\infty} \mathbf{C}^n}^* \cap L_+^\infty = \{0\}$ implies that $\overline{\mathbf{C}^n}^* \cap L_+^\infty = \{0\}$, for all n. Hence NFL holds for all $(S^0, \dots, S^n)$, all n and so Assumption 3 follows.

Remark 5 Note that, in the general case of processes with non-continuous paths, the condition NAFL cannot be replaced by a less technical and more intuitive condition, such as, for example a no asymptotic free lunch condition involving a bounded or vanishing risk, see the examples in [28, 30]. However, after the present paper was finished, it turned out in recent results, that in the nested setting (as also used in the present paper) the condition NAFL is in fact equivalent to a newly defined condition of *no asymptotic free lunch with vanishing risk*, see [7]. It only became clear later on and very recently that the crucial fact here is the nestedness of the large financial market (i.e., $\mathbf{C}^n \subseteq \mathbf{C}^{n+1}$, for all n). The examples of [28, 30] cannot be extended to the nested setting.

4 Global Existence of an Equivalent Local Martingale Measure

The large financial market induced by the bond market provides an adequate framework to analyze existence of an equivalent local martingale measure. For each $T^* \in \mathcal{T}$ with the set $\mathcal{T}$ from Assumption 2, we will find a measure Q^* such that all bond prices with maturity $T \leq T^*$ discounted by the numéraire $P(., T^*)$, are local martingales under Q^*.

In fact, we immediately obtain a measure $Q^* \sim P|_{\mathcal{F}_{T^*}}$ such that $\left(\frac{P(t,T_i)}{P(t,T^*)}\right)_{0\leq t\leq T_i}$ is a local Q^*-martingale for all T_i in the dense subset of maturities of Definition 4. This is just the classical Kreps-Yan result which we state in an abstract version below, for a proof see [38]. It remains to show that the local martingale-property holds for all maturities $T \in [0, T^*]$.

Theorem 4 (Kreps, Yan) *Let C be a convex cone in L^∞ such that $-L^\infty_+ \subseteq C$, C is weak-star-closed and $C \cap L^\infty_+ = \{0\}$. Then there exists g in L^1 such that $g > 0$ a.s. and $E[fg] \leq 0$ for all $f \in C$.*

Theorem 5 *Fix any $T^* \in \mathcal{T}$ and let Assumptions 1, 2 and 3 hold. Then, the bond market satisfies NAFL (see Definition 4), if and only if there exists a measure $Q^* \sim P|_{\mathcal{F}_{T^*}}$ such that $\left(\frac{P(t,T)}{P(t,T^*)}\right)_{0\leq t\leq T}$ is a local Q^*-martingale for all $T \in [0, T^*]$.*

Remark 6 In Theorem 5 we consider the NAFL condition for the large financial market as in Definition 3 with respect to one fixed, dense sequence of T_i in $[0, T^*]$. However, Theorem 5 shows that there is an equivalent measure Q^* such that bond prices for *all* maturities in $[0, T^*]$ discounted by the numéraire are local martingales with respect to Q^*. Hence the choice of the dense sequence in Definition 3 does not play a role, in fact, the general theorem about NAFL by Klein [27] implies that for any such sequence of maturities the corresponding large financial market (induced by the bond market with these maturities) satisfies NAFL.

Proof (Proof of Theorem 5) We denote by $(T_i)_{i\in\mathbb{N}}$ the dense sequence from Definition 4. Consider the large financial market of Definition 3. By Theorem 4 we get for $C = \overline{\bigcup_{n=1}^\infty \mathbf{C}^n}^*$ a $g \in L^1(\Omega, \mathcal{F}_{T^*}, P)$, $g > 0$ such that $E[fg] \leq 0$ for all $f \in C$. Take $\frac{g}{E[g]}$ as the density of a probability measure $Q^* \sim P|_{\mathcal{F}_{T^*}}$. As all $S^i = \frac{P(t,T_i)}{P(t,T^*)}$ are locally bounded this gives that S^i is a local Q^*-martingale. Indeed, choose τ such that $(S^i_{t\wedge\tau})_{0\leq t\leq T_i}$ is bounded, and let $s < t \leq T_i$, $A \in \mathcal{F}_s$. Then we have that $\pm(\mathbb{1}_{]\!]0,\tau]\!]}\mathbb{1}_A\mathbb{1}_{]s,t]} \cdot S^i)_T = \pm\mathbb{1}_A(S^i_{t\wedge\tau} - S^i_{s\wedge\tau}) \in \mathbf{C}^i$. This gives the local martingale property under Q^*.

It remains to show that for any $T < T^*$ which is not an element of the sequence (T_i) we get the local martingale property of $\left(\frac{P(t,T)}{P(t,T^*)}\right)_{0\leq t\leq T}$ with respect to Q^* as well. As the sequence (T_i) is dense in $[0, T^*]$ there exists a subsequence denoted by $(\tilde{T}_i)$ with $\tilde{T}_i \to T$ for $i \to \infty$ (w.l.o.g. assume that $\tilde{T}_i \geq T$ for all i).

Let $\frac{P(t,T)}{P(t,T^*)} := X_t$ and $\frac{P(t,\tilde{T}_i)}{P(t,T^*)} := X_t^i$ for each i. As a consequence of Assumption 2 there exists $\varepsilon > 0$, an increasing sequence of stopping times $\sigma_n \to \infty$ and constants $\kappa_n > 0$ such that for all $U \in [T, T+\varepsilon)$ and all $t \in [0, T]$ we have that

$$\left(\frac{P(t,U)}{P(t,T^*)}\right)^{\sigma_n} \leq \kappa_n.$$

Hence for i large enough, such that $\tilde{T}_i \in [T, T+\varepsilon)$, say $i \geq i_\varepsilon$, we have that

$$X^i_{t\wedge\sigma_n} \leq \kappa_n \quad \text{for all } t \in [0, T]. \tag{8}$$

By the first part of the proof, for any i, X^i is a local Q^*-martingale. So, (8) gives that, for $i \geq i_\varepsilon$, $(X^i)^{\sigma_n}$ is a Q^*-martingale (as it is a bounded local martingale).

Fix $\sigma = \sigma_n$. W.l.o.g. we can replace σ by $\sigma \wedge T$ as we will use it only on $[0, T]$. We will now show that, for all $t \in [0, T]$, we have that, for $i \to \infty$

$$X^i_{t\wedge\sigma} \to X_{t\wedge\sigma} \quad \text{a.s.} \tag{9}$$

This holds iff $P(t\wedge\sigma, \tilde{T}_i) \to P(t\wedge\sigma, T)$ a.s. By right-continuity of $U \to P(t, U)$ it is clear that $P(t, \tilde{T}_i)\mathbb{1}_{\{t<\sigma\}} \to P(t, T)\mathbb{1}_{\{t<\sigma\}}$ a.s.

So it remains to show that $P(\sigma, \tilde{T}_i)\mathbb{1}_{\{t\geq\sigma\}} \to P(\sigma, T)\mathbb{1}_{\{t\geq\sigma\}}$ a.s. Take any $\omega \in \Omega \setminus N$, where N is the nullset of Assumption 1, then we have that $\sigma(\omega) = s$ for some $s \in [0, T]$ and as $\tilde{T}_i \downarrow T$ we get $P(s, \tilde{T}_i)(\omega) \to P(s, T)(\omega)$, and hence $P(\sigma, \tilde{T}_i)(\omega) \to P(\sigma, T)(\omega)$, so (9) holds.

Let $s < t \leq T$. By (9) we have that $X^i_{t\wedge\sigma} \to X_{t\wedge\sigma}$ a.s. for all $t \in [0, T]$. Hence

$$E_{Q^*}[X_{t\wedge\sigma}|\mathscr{F}_s] = E_{Q^*}[\lim_{i\to\infty} X^i_{t\wedge\sigma}|\mathscr{F}_s] = \lim_{i\to\infty} E_{Q^*}[X^i_{t\wedge\sigma}|\mathscr{F}_s] = \lim_{i\to\infty} X^i_{s\wedge\sigma} = X_{s\wedge\sigma},$$

where the second equality follows by dominated convergence as by (8) we have that $0 < X^i_{t\wedge\sigma} \leq \kappa$ for all $i \geq i_\epsilon$. The third equality is the martingale property of $(X^i)^\sigma$ for $i \geq i_\epsilon$. This gives that $(X^\sigma_t)_{0\leq t\leq T}$ is a Q^*-martingale. As this holds for each σ in the localizing sequence, $(X_t)_{0\leq t\leq T}$ is a local Q^*-martingale. □

5 Existence of a Bank Account

It is possible to obtain a candidate process for the bank account by a limit of rolled over bonds as we show now. Throughout this section we assume that all the assumptions of Theorem 5 hold.

Definition 5 Let $0 = t_0^n < t_1^n < \cdots < t_{k_n}^n = T^*$ be a sequence of refining partitions of $[0, T^*]$. Define, for each n, the *roll-over* B^n as follows: $B_0^n = 1$ and

$$B_t^n = \begin{cases} \prod_{i=1}^{j} \frac{1}{P(t_{i-1}^n, t_i^n)} & \text{for } t = t_j^n, j = 1, \dots, k_n \\ B_{t_j^n}^n P(t, t_j^n) & \text{for } t_{j-1}^n < t \le t_j^n, j = 1, \dots, k_n \end{cases}$$

The sequence of these roll-overs can be viewed as a replacement of a bank account even without passing to a limit. This is in the spirit of large financial markets, where one often approximates in a finite way for larger and larger n but one does not actually pass to the limit.

We shall see that one can still pass to the limit by taking convex combinations, which will provide us with the notion of a generalized bank account. First we shall observe some properties of the sequence of roll-overs.

Lemma 2 *There exists a self-financing strategy* $\hat{\mathbf{H}}_t^n = (\hat{H}_t^1, \dots, \hat{H}_t^{k_n})$ *on the market containing the* k_n*-dimensional asset* $\hat{\mathbf{S}}^n(\cdot) = (P(., t_1^n), \dots, P(., t_{k_n}^n))$ *such that* $B_t^n = \langle \hat{\mathbf{H}}_t, \hat{\mathbf{S}}_t \rangle$. *Discounted by the numéraire* $P(t, t_{k_n}^n) = P(t, T^*)$ *this gives an admissible strategy* $\mathbf{H}^n$ *such that* $\frac{B_t^n}{P(t,T^*)} = \frac{1}{P(0,T^*)} + (\mathbf{H}^n \cdot \mathbf{S}^n)_t > 0$, *where* $\mathbf{S}^n$ *is the process* $\hat{\mathbf{S}}^n$ *discounted by the numéraire* $P(t, T^*)$. *In particular,* $\left(\frac{B_t^n}{P(t,T^*)}\right)_{0 \le t \le T^*}$ *is a positive local martingale and hence a supermartingale with respect to the measure* Q^* *of Theorem 5.*

Proof The strategy $\hat{\mathbf{H}}_t^n$ is given as follows. Fix j and let $t_{j-1}^n < t \le t_j^n$, then

$$\hat{H}_t^i = \begin{cases} \prod_{l=1}^{j} \frac{1}{P(t_{l-1}^n, t_l^n)} & \text{for } i = j \\ 0 & \text{for } i \ne j, \end{cases}$$

which is equivalent to

$$\hat{H}_t^i = \sum_{j=1}^{k_n} B_{t_j^n}^n \mathbb{1}_{(t_{j-1}^n, t_j^n]}(t) \delta_{ij},$$

which is predictable since $B_{t_j^n}^n \in \mathscr{F}_{t_{j-1}^n}$.
So we get for $t_{j-1}^n < t \le t_j^n$

$$\begin{aligned} B_t^n &= \langle \hat{\mathbf{H}}_t^n, \hat{\mathbf{S}}_t^n \rangle \\ &= \sum_{i=1}^{k_n} \hat{H}_t^i \hat{S}_t^i = \prod_{l=1}^{j} \frac{1}{P(t_{l-1}^n, t_l^n)} \cdot P(t, t_j^n) \\ &= B_{t_j^n}^n P(t, t_j^n). \end{aligned}$$

This is self-financing as

$$\langle \hat{\mathbf{H}}_{t_{j-1}^n}^n, \hat{\mathbf{S}}_{t_{j-1}^n}^n \rangle = \langle \hat{\mathbf{H}}_t^n, \hat{\mathbf{S}}_{t_{j-1}^n}^n \rangle$$

for $t_{j-1}^n < t \le t_j^n$. Indeed the left hand side equals $B_{t_{j-1}^n}^n P(t_{j-1}^n, t_{j-1}^n) = B_{t_{j-1}^n}^n$ and the right hand side equals $B_{t_j^n}^n P(t_{j-1}^n, t_j^n) = B_{t_{j-1}^n}^n$.

After discounting by the numéraire $P(t, T^*)$ we have the initial investment $\langle \hat{\mathbf{H}}_1, \mathbf{S}_0 \rangle = \frac{1}{P(0,t_1^n)} \frac{P(0,t_1^n)}{P(0,T^*)} = \frac{1}{P(0,T^*)}$. For $t_{j-1}^n < t \le t_j^n$ as $B_{t_{j-1}^n}^n = \langle \hat{\mathbf{H}}_{t_{j-1}^n}, \hat{\mathbf{S}}_{t_{j-1}^n} \rangle = \langle \hat{\mathbf{H}}_t, \hat{\mathbf{S}}_{t_{j-1}^n} \rangle$ the increment equals

$$\begin{aligned} \frac{B_t^n}{P(t,T^*)} - \frac{B_{t_{j-1}^n}^n}{P(t_{j-1}^n, T^*)} &= \sum_{i=1}^{k_n} \hat{H}_t^i \frac{P(t,t_i^n)}{P(t,T^*)} - \sum_{i=1}^{k_n} \hat{H}_t^i \frac{P(t_{j-1}^n, t_i^n)}{P(t_{j-1}^n, T^*)} \\ &= \hat{H}_t^{k_n}(1-1) + \sum_{i=1}^{k_n - 1} \hat{H}_t^i \left(\frac{P(t,t_i^n)}{P(t,T^*)} - \frac{P(t_{j-1}^n, t_i^n)}{P(t_{j-1}^n, T^*)} \right) \\ &= B_{t_j^n}^n \left(\frac{P(t,t_j^n)}{P(t,T^*)} - \frac{P(t_{j-1}^n, t_j^n)}{P(t_{j-1}^n, T^*)} \right). \end{aligned}$$

Summing the increments up we arrive at the stochastic integral

$$\frac{B_t^n}{P(t,T^*)} = \frac{1}{P(0,T^*)} + \sum_{j=1}^{k_n} B_{t_j^n}^n \left(\frac{P(t \wedge t_j^n, t_j^n)}{P(t \wedge t_j^n, T^*)} - \frac{P(t \wedge t_{j-1}^n, t_j^n)}{P(t \wedge t_{j-1}^n, T^*)} \right).$$

As $\frac{B_t^n}{P(t,T^*)} = \frac{1}{P(0,T^*)} + (\mathbf{H}^n \cdot \mathbf{S}^n)_t$ is bounded from below and $\mathbf{S}^n$ is a local Q^*-martingale the discounted roll-over is a Q^*-supermartingale.

The existence of limits for refined roll-overs is apparently delicate. The following theorem is proved by a Komlós-type argument as in [18, Lemma 5.2] and provides us with a generalized bank account, that always exists (under the assumptions of Theorem 5) and which is always a supermartingale with respect to the measure Q^* of Theorem 5.

Theorem 6 *Let $((B_t^n)_{0 \le t \le T^*})$ be the sequence of roll-overs given as in Definition 5. There exists a sequence of convex combinations $\tilde{B}^n \in \mathrm{conv}(B^n, B^{n+1}, \dots)$ and a càdlàg stochastic process $(B_t)_{0 \le t \le T^*}$, in the following called* generalized bank account, *such that*

$$B_t = \lim_{q \downarrow t} \lim_{n \to \infty} \tilde{B}_q^n,$$

with $B_0 \le 1$ and $0 \le B_t < \infty$, for all $t \le T^*$. The generalized bank account has the following properties.

1. *The process $(V_t)_{0 \le t \le T^*}$, where $V_t = \frac{B_t}{P(t,T^*)}$, is a supermartingale with respect to the measure Q^* of Theorem 5.*
2. *If $0 < P(t,T) \le 1$, for all $T \le T^*$, then $P(B_t \ge 1) = 1$, for all $t \le T^*$.*

Remark 7 In general, we can only say that the process $(V_t)_{0 \le t \le T}$ is a supermartingale with respect to Q^* (and not a local martingale), see Sect. 7.4 for an example.

Proof Consider the sequence of roll-overs $M_t^n := \frac{B_t^n}{P(t,T^*)}$ discounted by the numéraire $P(.,T^*)$. By Lemma 2 these processes are supermartingales (and bounded from below by 0) with respect to the measure Q^* of Theorem 5. The existence of a limit of convex combinations of the M^n follows by Lemma 5.2 of [18], we recall the proof here. Let $\mathscr{D} = ([0,T^*] \cap \mathbb{Q}) \cup \{T^*\}$. This is a dense countable subset of $[0,T^*]$. By Lemma A.1.1 of [9] and a diagonalization procedure we find a sequence of processes $\tilde{M}^n \in \operatorname{conv}\left(\frac{B^n}{P(.,T^*)}, \frac{B^{n+1}}{P(.,T^*)}, \dots\right)$ such that, for all $q \in \mathscr{D}$, $\tilde{M}_q^n$ a.s. converges to a random variable V_q' with values in $[0,\infty]$. For each q, we have that $\operatorname{conv}(M_q^n, M_q^{n+1}, \dots)$ is bounded in L^0, as all M^n are positive supermartingales with starting value $\frac{1}{P(0,T^*)}$. Hence for each $\tilde{M} \in \operatorname{conv}(M_q^n, M_q^{n+1}, \dots)$ we have that $E_{Q^*}[|\tilde{M}|] = E_{Q^*}[\tilde{M}] \le \frac{1}{P(0,T^*)}$, so the set of convex combinations is bounded in $L^1(Q^*)$ hence in L^0. Lemma A.1.1 of [9] gives then that $V_q' < \infty$ a.s.

Moreover, for $r < q, r, q \in \mathscr{D}$, by Fatou and the supermartingale property of $\tilde{M}^n$ we have that

$$E_{Q^*}[V_q'|\mathscr{F}_r] = E_{Q^*}[\lim_{n\to\infty} \tilde{M}_q^n|\mathscr{F}_r] \le \liminf_{n\to\infty} E_{Q^*}[\tilde{M}_q^n|\mathscr{F}_r] \le \lim_{n\to\infty} \tilde{M}_r^n = V_r'.$$

Therefore $(V_q')_{q\in\mathscr{D}}$ is a discrete Q^*-supermartingale. By standard arguments (using Doob's Upcrossing Lemma) we get that $(V_t)_{0\le t\le T^*}$ is a càdlàg supermartingale, where, for all $t \in [0,T^*[$,

$$V_t := \lim_{q\downarrow t} V_q',$$

and $V_{T^*} := V_{T^*}'$ (recall that $T^* \in \mathscr{D}$). Note that $V_0 \le \frac{1}{P(0,T^*)}$ as

$$V_0 = \lim_{q\downarrow 0} V_q' = E_{Q^*}[\lim_{q\downarrow 0} V_q'|\mathscr{F}_0] \le V_0' = \frac{1}{P(0,T^*)}. \qquad \square$$

Define now $B_t := V_t P(t,T^*)$, this is a càdlàg process as V_t and $P(t,T^*)$ are càdlàg. As the process $P(t,T^*)$ is right-continuous in t easy computations show that $B_t = \lim_{q\downarrow t}\lim_{n\to\infty}\tilde{B}_q^n$, where $\tilde{B}_q^n = \sum_{i=1}^{k_n}\lambda_i^n B_q^i = P(q,T^*)\tilde{M}_q^n$. By definition $B_0 = P(0,T^*)V_0 \le 1$.

Let now $P(t,T) \le 1$, for all $T \le T^*, t \le T$. Then we see from the definition of the roll-over as product of terms of the form $\frac{1}{P(t_i,t_{i+1})} \ge 1$ that $B_t^n \ge 1$ for all n,t. The same holds for all convex combinations and therefore for the limits as above.

Remark 8 With a view to what it means to be a numéraire (see [10]) we can ask why just terminal bonds qualify as numéraires by default in our setting: the answer is that we could take any other reasonably behaved stochastic process (the inverse has to be locally bounded) and plug it into Assumption 3 instead of $P(.,T^*)$. Conclusions would remain the same, of course with respect to the chosen numéraire. For instance we could think of taking discrete roll-over bonds as numéraires.

6 On the Existence of a Supermartingale Deflator and a Generalized Bank Account

In this section, we relax the assumptions on the bond market and investigate under which conditions there is a supermartingale deflator. This is motivated by the fact that we are lead to supermartingale deflators by the very structure of bond market models. Indeed if we have a non-vanishing generalized bank account and decide to choose it as market numéraire, see Theorem 6, then our theory only provides us with a supermartingale deflator structure.

The results about supermartingale deflators in this section are related to results of Kostas Kardaras, see, e.g., [26]. Consider a large financial market induced by a sequence of semimartingales S^i, $i = 0, 1, \ldots$ on a fixed filtered probability space, where the filtration satisfies the usual conditions, such that $(\mathbf{S}_t^n)_{t\in[0,T^*]} = (S_t^0, S_t^1, \ldots, S_t^n)_{t\in[0,T^*]}$. Recall that $S_t^0 \equiv 1$, i.e. the numéraire has been fixed and prices are discounted by the chosen numéraire. The sets $\mathbf{K}^n$, $\mathbf{C}^n$, $\mathbf{M}_e^n$ are defined as previously. We assume that each finite market satisfies (NFLVR), i.e. (5) holds for all n. In contrast to the previous sections, we do not assume here that the semimartingales are locally bounded. In this case, the set $\mathbf{M}_e^n$ as in (4) consists of all equivalent probability measures Q such that stochastic integrals $(\mathbf{H}^n \cdot \mathbf{S}^n)_t$, $0 \leq t \leq T^*$, with admissible integrands $\mathbf{H}^n$ (i.e. $(\mathbf{H}^n \cdot \mathbf{S}^n)_{T^*} \in \mathbf{K}^n$) are Q-supermartingales. It was shown in [12] that under the condition no free lunch with vanishing risk the set of equivalent sigma-martingale measures for $\mathbf{S}^n$ is dense in the set $\mathbf{M}_e^n$.

The notion *no asymptotic arbitrage of first kind* (NAA1) was introduced in [23].

Definition 6 A large financial market admits an asymptotic arbitrage opportunity of first kind if there exists a subsequence, again denoted by n, and trading strategies $\mathbf{H}^n$ with

1. $(\mathbf{H}^n \cdot \mathbf{S}^n)_t \geq -\varepsilon_n$ for all $t \in [0, T^*]$,
2. $P((\mathbf{H}^n \cdot \mathbf{S}^n)_{T^*} \geq C_n) \geq \alpha$,

for all n, where $\alpha > 0$, $\varepsilon_n \to 0$ and $C_n \to \infty$.

We say that the large financial market satisfies the condition NAA1 if there are no asymptotic arbitrage opportunities of first kind.

The following result for large financial markets provides us with supermartingale deflators for bond markets.

Theorem 7 *Consider the large financial market induced by the sequence of semimartingales* $(\mathbf{S}_t^n)_{0\leq t\leq T^*} = (S_t^0, S_t^1, \ldots, S_t^n)_{t\in[0,T^*]}$, $n = 1, 2, \ldots$ *and assume that (5) holds for all n. Then NAA1 holds, if and only if there exists a strictly positive supermartingale* $(Z_t)_{0\leq t\leq T^*}$ *with* $Z_0 \leq 1$, *such that* $(Z_t(X_t + a))_{0\leq t\leq T^*}$ *is a supermartingale for all processes X with* $X_{T^*} \in \bigcup_{n=1}^{\infty} \mathbf{K}^n$ *where X is a-admissible.*
Moreover, if NAA1 holds, then: if $S_t^i \geq -a$, $0 < t \leq T^*$, *for some* $i \in \mathbb{N}$ *and some* $a \geq 0$, *then* $\left(Z_t(S_t^i + a)\right)_{0\leq t\leq T^*}$ *is a supermartingale.*

The supermartingale Z is called *supermartingale deflator for the large financial market.*

In order to prove Theorem 7 we will use a result from the theory of large financial markets. In this aspect our proof differs from Kardaras' proofs of similar results.

Under the assumption of NAA1 Kabanov and Kramkov proved Theorem 8 in [23] in the complete setting, the most general result can be found in [24]. Note that the general theorem in [30] was only proved under local boundedness assumptions on all processes, but it holds in the general case as well and is equivalent to the result of [24]. We choose to take the setting of [30] as the formulation of NAA1 given there is more convenient for our presentation (and anyway completely equivalent to the formulation in [24]).

We will formulate Theorem 8 for the special case of our nested setting on one fixes probability space $(\Omega, \mathscr{F}, P)$.

Theorem 8 *A large financial market as defined in Sect. 3 satisfies NAA1 if and only if there exists a sequence of probability measures $Q^n \in \mathbf{M}_e^n$ such that $P \lhd (Q^n)$.*

$P \lhd (Q^n)$ means that the measure P is contiguous with respect to the sequence of measures (Q^n), i.e., whenever for a sequence of measurable sets A^n we have that $Q^n(A^n) \to 0$ then $P(A^n) \to 0$. In our case where, for each n, $P \ll Q^n$, the notion of contiguity can be interpreted as a uniform absolute continuity in the following sense: for each $\varepsilon > 0$ there is $\delta > 0$ such that, for all n and $A^n \in \mathscr{F}$ with $Q^n(A^n) < \delta$ we have that $P(A^n) < \varepsilon$.

Let us now proceed with the proof of Theorem 7. In order to apply Theorem 8 we need the following useful lemma.

Lemma 3 *Let Q^n, $n \geq 1$, be a sequence of probability measures and P a probability measure on $(\Omega, \mathscr{F}_{T^*})$ such that $Q^n \sim P$, for all n, and $P \lhd (Q^n)$. Let $Z_t^n = E\left[\frac{dQ^n}{dP}|\mathscr{F}_t\right]$. Then there exists a càdlàg supermartingale $(Z_t)_{0 \leq t \leq T^*}$ with $Z_0 \leq 1$ and a sequence of $\tilde{Z}^n \in \operatorname{conv}(Z^n, Z^{n+1}, \dots)$ such that, for all $t \in [0, T^*]$,*

$$Z_t = \lim_{q \downarrow t} \lim_{n \to \infty} \tilde{Z}_q^n. \tag{10}$$

Moreover $P(Z_t > 0) = 1$, for all t.

Proof Let $\mathscr{D} = ([0, T^*] \cap \mathbb{Q}) \cup \{T^*\}$. The processes $(Z_t^n)_{0 \leq t \leq T^*}$ are positive martingales with $Z_0^n = 1$. As in the proof of Theorem 6 there exists a sequence $\tilde{Z}^n \in \operatorname{conv}(Z^n, Z^{n+1}, \dots)$ such that Z as in (10) is a càdlàg supermartingale, with $0 \leq Z_t < \infty$ for $t \in [0, T^*]$. As $\tilde{Z}_0^n = 1$, for all n, we have that $Z_0 \leq 1$.

It remains to show that, for all t, $P(Z_t > 0) = 1$. We will show that this holds for T^*, which implies the statement for all $t \leq T^*$. Indeed $Z_{T^*} > 0$ a.s. implies $E[Z_{T^*}|\mathscr{F}_t] > 0$ a.s. By the supermartingale property,

$$Z_t \geq E[Z_{T^*}|\mathscr{F}_t] > 0 \quad \text{a.s.}$$

Assume now that for $A = \{Z_{T^*} = 0\}$ we have that $P(A) = \alpha > 0$. As $T^* \in \mathscr{D}$, we have that $\mathbb{1}_A \tilde{Z}^n_{T^*} \to \mathbb{1}_A Z_{T^*} = 0$ a.s. This implies that, for all $\varepsilon > 0$,

$$P(\mathbb{1}_A \tilde{Z}^n_{T^*} > \varepsilon) \to 0. \tag{11}$$

Hence, for $\varepsilon = 2^{-N}$, there is $m_N \uparrow \infty$ such that, for all $n \geq m_N$, $P(\mathbb{1}_A \tilde{Z}^n_{T^*} > 2^{-N}) < 2^{-N}$. Define

$$A_n := A \cap \{\mathbb{1}_A \tilde{Z}^n_{T^*} \leq 2^{-N}\} \qquad \text{for } m_N \leq n < m_{N+1}.$$

For $n \geq m_{N_0}$, such that $2^{-N_0} \leq \frac{\alpha}{2}$ we have that

$$P(A_n) \geq P(A) - P(\mathbb{1}_A \tilde{Z}^n_{T^*} > 2^{-N_0}) \geq \frac{\alpha}{2}. \tag{12}$$

Define the probability measure $\tilde{Q}^n$ by $\frac{d\tilde{Q}^n}{dP} := \tilde{Z}^n_{T^*}$. Then, $\tilde{Q}^n \in \mathbf{M}^n_e$ as the density $Z^n_{T^*}$ is a convex combination of densities of equivalent probability measures $Q^k \in \mathbf{M}^k_e$, $k \geq n$. For $m_N \leq n \leq m_{N+1}$, we have that

$$\tilde{Q}^n(A_n) = E[\tilde{Z}^n_{T^*} \mathbb{1}_A \mathbb{1}_{\{\mathbb{1}_A \tilde{Z}^n_{T^*} \leq 2^{-N}\}}] \leq 2^{-N}. \qquad \square$$

This shows that $\tilde{Q}^n(A_n) \to 0$. As $\tilde{Q}^n \in \text{conv}(Q^n, Q^{n+1}, \ldots)$, there is $k_n \geq n$ with $k_n \to \infty$ such that $Q^{k_n}(A_n) \to 0$, for $n \to \infty$. As $P \lhd (Q^n)$ it is contiguous with respect to any subsequence of (Q^{k_n}) as well, so we should have $P(A_n) \to 0$ which is a contradiction to (12). Hence $Z_{T^*} > 0$ a.s.

Proof (Proof of Theorem 7) Assume that NAA1 holds. Then by Theorem 8 there exists a sequence of probability measures $Q^n \in \mathbf{M}^n_e$ such that $P \lhd (Q^n)$. Take a strictly positive supermartingale Z which is induced by (Q^n) and $\mathscr{D} = ([0, T^*] \cap \mathbb{Q}) \cup \{T^*\}$ as in Lemma 3. For all $q \in \mathscr{D}$, denote $Z'_q := \lim_{n\to\infty} \tilde{Z}^n_q$ where $\tilde{Z}^n$ are the convex combinations as in the proof of Lemma 3. Let X be such that $X_{T^*} \in \mathbf{K}^n$ for some n and $X_t \geq -a$ for all $0 \leq t \leq T^*$, i.e., X is a-admissible. Let $X^a_t = X_t + a$. We will show that $(Z_t X^a_t)_{0 \leq t \leq T^*}$ is a supermartingale. Indeed, let $r < q$, $r, q \in \mathscr{D}$. Define, $\tilde{Q}^n$ by $\frac{d\tilde{Q}^n}{dP} := \tilde{Z}^n_{T^*}$. As shown in the proof of Lemma 3, $\tilde{Q}^n \in \mathbf{M}^n_e$. This implies that $(\tilde{Z}^n_t X^a_t)_{0 \leq t \leq T^*}$ is a supermartingale. Then, by Fatou, we get

$$\begin{aligned} E[Z'_q X^a_q | \mathscr{F}_r] &= E[\lim_{n\to\infty} \tilde{Z}^n_q X^a_q | \mathscr{F}_r] \\ &\leq \liminf_{n\to\infty} E[\tilde{Z}^n_q X^a_q | \mathscr{F}_r] \\ &\leq \lim_{n\to\infty} \tilde{Z}^n_r X^a_r = Z'_r X^a_r. \end{aligned}$$

So $(Z'_q X^a_q)_{q \in \mathscr{D}}$ is a discrete supermartingale. Let now $s < t < T^*$ and $s_k \downarrow s$, $t_j \downarrow t$ for rational s_k, t_j (for $t = T^*$ take $t_j \equiv T^*$). Then we have that

$$\begin{aligned}E[Z_t X_t^a|\mathscr{F}_{s_k}] &= E[\lim_{j\to\infty} Z'_{t_j} X_{t_j}^a|\mathscr{F}_{s_k}]\\ &\le \liminf_{j\to\infty} E[Z'_{t_j} X_{t_j}^a|\mathscr{F}_{s_k}]\\ &\le Z'_{s_k} X_{s_k}^a,\end{aligned}$$

where the equality holds by the definition of Z and by the right continuity of X^a, the first inequality is Fatou, the second inequality is the discrete supermartingale property. The right-continuity of the filtration together with the definition of Z and the right continuity of X^a gives

$$E[Z_t X_t^a|\mathscr{F}_s] = \lim_{k\to\infty} E[Z_t X_t^a|\mathscr{F}_{s_k}] \le \lim_{k\to\infty} Z'_{s_k} X_{s_k}^a = Z_s X_s^a.$$

Hence, ZX^a is a supermartingale.

For the converse, assume that there is a supermartingale deflator for the large financial market. Suppose there exists an asymptotic arbitrage of first kind, that is, there exists a sequence $X_{T^*}^k \in \mathbf{K}^{n_k}$ such that $X_t^k \ge -\varepsilon_k$, $0 \le t \le T^*$, and $P(X_{T^*}^k \ge C_k) \ge \alpha$ with $\varepsilon_k \to 0$ and $C_k \to \infty$. We have that $X^k Z$ are supermartingales, for all k. Hence, as $X_0^k = 0$ and X^k is ε_k-admissible,

$$E[X_T^k Z_T] \le E[(X_T^k + \varepsilon_k) Z_T] \le (X_0^k + \varepsilon_k) Z_0 \le \varepsilon_k. \tag{13}$$

On the other hand, let $A_k := \{X_{T^*}^k \ge C_k\}$. As $Z_0 \le 1$ and by the properties of X^k,

$$E[X_{T^*}^k Z_{T^*}] \ge C_k E[Z_{T^*} \mathbb{1}_{A_k}] - \varepsilon_k E[Z_{T^*}] \ge C_k E[Z_{T^*} \mathbb{1}_{A_k}] - \varepsilon_k. \tag{14}$$

By assumption $P(A_k) \ge \alpha$, for all k. We claim that there exists $\beta > 0$ such that, $P(\{Z_{T^*} > \beta\} \cap A_k) \ge \frac{\alpha}{2}$ for all k. Suppose not, then for each $j \ge 1$ and $\beta = \frac{1}{j}$, there is k_j such that $P(\{Z_{T^*} > \frac{1}{j}\} \cap A_{k_j}) < \frac{\alpha}{2}$ and hence $P(\{Z_{T^*} \le \frac{1}{j}\} \cap A_{k_j}) \ge P(A_{k_j}) - \frac{\alpha}{2} \ge \frac{\alpha}{2}$. Therefore

$$P(Z_{T^*} = 0) = \lim_{j\to\infty} P(Z_{T^*} \le \frac{1}{j}) \ge \liminf_{j\to\infty} P(\{Z_{T^*} \le \frac{1}{j}\} \cap A_{k_j}) \ge \frac{\alpha}{2},$$

a contradiction to the integrability of Z_{T^*} and $Z_{T^*} > 0$ a.s. Equation (14) then implies that

$$E[X_{T^*}^k Z_{T^*}] \ge C_k E[Z_{T^*} \mathbb{1}_{A_k \cap \{Z_{T^*} > \beta\}}] - \varepsilon_k \ge C_k \beta \frac{\alpha}{2} - \varepsilon_k > \varepsilon_k,$$

for k large enough. This gives a contradiction to (13).

We still have to prove the second statement of the theorem. Assume that $S_t^i \ge -a$, $0 \le t \le T^*$ for some i. Define the trivial predictable $H_t = \mathbb{1}_{]0,T^*]}(t)$, then, for $t \le T^*$,

$$X_t = (H \cdot S^i)_t = S_t^i - S_0^i \ge -a - S_0^i.$$

Hence $X_{T^*} \in \mathbf{K}^n$, for $n \geq i$. Therefore $((X_t + a + S_0^i)Z_t)_{0 \leq t \leq T^*}$ is a supermartingale. This immediately gives that $(S_t^i + a)Z_t$ is a supermartingale as $S_t^i + a = X_t + a + S_0^i$. □

In the sequel we apply the results to bond markets: again, for each $T \in [0, T^*]$, $(P(t, T))_{0 \leq t \leq T}$ is a strictly positive càdlàg stochastic process adapted to $(\mathscr{F}_t)_{0 \leq t \leq T}$ with $P(T, T) = 1$. We assume that, for fixed t, the function $T \mapsto P(t, T)$ is almost surely right-continuous. Note, that in this section, we do not have any local boundedness assumptions on $P(t, T)$ or $\frac{1}{P(t,T)}$. We will again have to assume that in the case of a finite number of assets discounted with the numéraire $P(t, T^*)$ we will not have any arbitrage opportunities. This is again a consequence of the following assumption on existence of the T^*-forward measure. Consider any $0 < T_1 < \cdots < T_n \leq T^*$ and define the cone $\mathbf{C}(T_1, \ldots, T_n, T^*)$ as in (3) where S^i is defined as in (7). Note that we do not assume here that $T^* \in \mathscr{T}$.

Assumption 9 For all finite collections of maturities $T_1 < T_2 < \cdots < T_n \leq T^*$ there exists a measure $Q \sim P|_{\mathscr{F}_{T^*}}$ such that for all $f \in \mathbf{C}(T_1, \ldots, T_n, T^*)$ we have that $E_Q[f] \leq 0$.

The condition $E_Q[f] \leq 0$ means that the measure $Q \in \mathbf{M}_e(T_1, \ldots, T_n, T^*)$, where the definition of the set of separating measures is analogous as in (4). Note that Assumption 9 implies the existence of an equivalent sigma-martingale measure for $S^1, \ldots, S^n$ given as in (7), and therefore these processes are semimartingales, see [12]. As in Definition 3, any sequence $(T_i)_{i \in \mathbb{N}}$ in $[0, T^*]$ induces a large financial market.

Definition 7 The bond market satisfies NAA1 w.r.t. a sequence $(T_i)_{i \in \mathbb{N}}$ in $[0, T^*]$ if for the large financial market induced by $(T_i)_{i \in \mathbb{N}}$ there does not exist an asymptotic arbitrage of first kind.

Theorem 10 *Fix a sequence $(T_i)_{i \in \mathbb{N}}$ in $[0, T^*]$. The bond market satisfies NAA1 w.r.t. $(T_i)_{i \in \mathbb{N}}$ if and only if there exists a strictly positive supermartingale deflator $(Z_t)_{0 \leq t \leq T^*}$ for the large financial market induced by $(T_i)_{i \in \mathbb{N}}$. If $(T_i)_{i \in \mathbb{N}}$ is dense in $[0, T^*]$, then $\left(Z_t \frac{P(t,T)}{P(t,T^*)}\right)_{0 \leq t \leq T}$ is a supermartingale for all $T \leq T^*$.*

Proof (Proof of Theorem 10) Everything follows by Theorem 7. In particular, as $\frac{P(t,T_i)}{P(t,T^*)} > 0$, we have that $(\frac{P(t,T_i)}{P(t,T^*)} Z_t)_{0 \leq t \leq T_i}$ is a supermartingale for each T_i. It only remains to show, that $\frac{P(t,T)}{P(t,T^*)} Z_t$ is a supermartingale for each $T \leq T^*$ which is not an element of the dense sequence. Note that for T^* the statement holds, as $\frac{P(t,T^*)}{P(t,T^*)} Z_t = Z_t$ is a supermartingale. Let $T < T^*$. Choose $\tilde{T}_i \downarrow T$ with $\tilde{T}_i$ elements of the dense sequence in $[0, T^*]$. Let $X_t := \frac{P(t,T)}{P(t,T^*)}$ and $X_t^i := \frac{P(t,\tilde{T}_i)}{P(t,T^*)}$. As, for each i, $X_t^i > 0$ a.s., for all t, we get by Theorem 7 that $Z_t X_t^i$ is a supermartingale. Hence

$$E[X_t Z_t | \mathscr{F}_s] = E[\lim_{i \to \infty} X_t^i Z_t | \mathscr{F}_s] \quad \leq \liminf_{i \to \infty} E[X_t^i Z_t | \mathscr{F}_s] \leq \lim_{i \to \infty} X_s^i Z_s = X_s Z_s,$$

as $U \mapsto P(v, U)$ is right-continuous, for each v (and therefore $X_v^i \to X_v$ for $v = s, t$) and by Fatou.

Finally we will show that under the weaker assumptions of this section we will still be able to define a generalized bank account.

Theorem 11 *Let (T_i) be a dense sequence in $[0, T^*]$ such that NAA1 holds. Let $(B_t^n)_{t\in[0,T^*]}$ be the sequence of roll-overs as in Definition 5, where the refining partition $\{t_1^n, \dots, t_{k_n}^n\}$ is chosen such $\bigcup_{n\in\mathbb{N}}\{t_1^n, \dots, t_{k_n}^n\} \subseteq (T_i)$. Then there exists a sequence of convex combinations $\tilde{B}^n \in \operatorname{conv}(B^n, B^{n+1}, \dots)$ and a càdlàg stochastic process $(B_t)_{0\le t\le T^*}$ (the generalized bank account) such that*

$$B_t = \lim_{q\downarrow t}\lim_{n\to\infty} \tilde{B}_q^n,$$

with $B_0 \le 1$ and $0 \le B_t < \infty$, for all $t \le T^*$. The generalized bank account has the following properties.

1. *The process $\left(\frac{Z_t B_t}{P(t,T^*)}\right)_{0\le t\le T^*}$ is a supermartingale.*
2. *If $0 < P(t, T) \le 1$, for all $T \le T^*$, then $P(B_t \ge 1) = 1$, for all $t \le T^*$.*

The interpretation of this Theorem is, that for any refining sequence of partitions, which does not produce an asymptotic arbitrage opportunity of first kind in the induced large financial market, there does exist a generalized bank account. In particular, if the bond market does not allow an asymptotic arbitrage opportunity of first kind for any sequence of maturities in $[0, T^*]$ (for the respective induced large financial market as in Definition 3), then any refining sequence of partitions gives a generalized bank account in the sense of Theorem 11. If, moreover, the bond $P(t, T) \le 1, 0 \le t \le T \le T^*$, then we can say that B_t is bounded from below by 1 a.s. This corresponds to the case, where a non-negative short-rate exists.

Proof By Lemma 2 we have that $X_{T^*}^n := \frac{B_{T^*}^n}{P(t,T^*)} - \frac{1}{P(0,T^*)} \in \mathbf{K}^{m_n}$ for some m_n large enough and $(X_t)_{0\le t\le T^*}$ is $\frac{1}{P(0,T^*)}$-admissible. By NAA1 and Theorem 10 there exists a strictly positive càdlàg supermartingale Z such that, for all n, $(V_t^n)_{t\in[0,T^*]}$ is a supermartingale, where $V_t^n := Z_t \frac{B_t^n}{P(t,T^*)} = Z_t(X_t^n + \frac{1}{P(0,T^*)})$, since all points of the partition defining the roll-over bond are contained in the dense sequence instrumental for the definition of Z. As in the proof of Theorem 6 we get a sequence of convex combination $\tilde{V}_t^n \in \operatorname{conv}(V^n, V^{n+1}, \dots)$ and a càdlàg supermartingale $0 \le V_t < \infty$ such that $V_t = \lim_{q\downarrow t}\lim_{n\to\infty} \tilde{V}_q^n$. Moreover, $V_0 \le 0$ as $V_0 \le \lim_{n\to\infty} \tilde{V}_0^n = 0$. Define

$$B_t = \frac{V_t}{Z_t} P(t, T^*). \qquad \square$$

Similarly as in the proof of Theorem 6, we see that $B_t = \lim_{q\downarrow t}\lim_{n\to\infty} \tilde{B}_q^n$, where $\tilde{B}_q^n$ are the corresponding convex combinations of B_q^n, i.e., $\tilde{B}_q^n = \frac{\tilde{V}_q^n}{Z_q} P(q, T^*)$. (Use the right continuity of $t \mapsto P(t, T)$ and $t \mapsto Z_t$.) Clearly $B_0 = \frac{V_0}{Z_0} P(0, T^*) \le 1$.

7 Examples

7.1 A Strict Local Martingale Deflator

In this section we consider the example touched upon in the introduction in more detail. We follow [21] and place ourselves in the so-called benchmark approach. The fundamental principle in this approach is that pricing is performed under the real-world measure with the growth optimal portfolio, denoted by S^*, chosen as numéraire. A well diversified index such as the S&P 500 can be chosen as proxy for the growth optimal portfolio. As in [5] it will not be necessary to describe the market in detail, as the essential ingredient for term structure modelling is the numéraire S^*.

Let Q denote the objective probability measure. In the benchmark approach fair prices of a payoff X at time $T > 0$ are given by

$$\pi_t(X) = S_t^* E_Q[\frac{X}{S_T^*}|\mathscr{F}_t],$$

and, as a consequence, one obtains bond prices of the form

$$P(t,T) = E_Q\big[\frac{S_t^*}{S_T^*}|\mathscr{F}_t\big]. \tag{15}$$

In reality, the inverse of the S&P 500 accumulation index appears to follow a strict supermartingale rather than a martingale, see [21]. Motivated by this, we give an example where the inverse of the growth optimal portfolio is related to a positive, strict local martingale. Consider the case where

$$\frac{1}{S_t^*} = \frac{A(t)}{\| x + W_t \|^2} =: \xi_t$$

with a positive, deterministic, càdlàg function $A\colon [0,\infty) \mapsto (0,\infty)$, a four-dimensional standard Brownian motion W and $0 \neq x \in \mathbb{R}^4$. Then $(\| x + W_t \|^2)_{t\geq 0}$ is a squared Bessel process of dimension four and its inverse is a strict local martingale. Surprisingly, it turns out that the benchmark approach for term structure modelling can be linked to risk-neutral pricing when the bond with longest available maturity in the market is chosen as numéraire: let $T^* > 0$ denote the longest maturity of bonds available in the market. As we will show, there exists an equivalent probability measure Q^* such that all bond prices with maturity $T \leq T^*$ discounted by the numéraire $P(.,T^*)$ are even martingales under Q^* and we arrive in the situation of Theorem 5. Indeed, let $\alpha = \frac{1}{E_Q[\xi_{T^*}]}$ and define

$$\frac{dQ^*}{dQ} = \alpha \xi_{T^*}.$$

The density process $Z_t^*, 0 \le t \le T^*$, satisfies

$$Z_t^* = \alpha E_Q[\xi_{T^*}|\mathcal{F}_t] = \alpha \xi_t P(t, T^*).$$

Therefore

$$Z_t^* \frac{P(t,T)}{P(t,T^*)} = \alpha E_Q[\xi_T|\mathcal{F}_t],$$

hence $\frac{P(t,T)}{P(t,T^*)}, 0 \le t \le T$, is a martingale with respect to Q^*.

Using Markovianity of W and integrating over the transition density of squared Bessel processes one obtains the following explicit expression for $P(t, T)$.

$$P(t,T) = \frac{A(T)}{A(t)} E_Q\left[\frac{\| x + W_t \|^2}{\| x + W_T \|^2} \,\Big|\, \mathcal{F}_t\right] = \frac{A(T)}{A(t)}\left(1 - e^{-\frac{\|x+W_t\|^{-2}}{2(T-t)}}\right), \tag{16}$$

see Eq. 8.7.17 in [21].

Up to now, there does not exist a bank account in the market. The limit of rolled over bonds is a natural candidate for a bank account which we study in detail now. Consider B_t^n as defined in Definition 5, where we additionally assume that there exists a constant $K \ge 1$ such that, for all n, $\max_{1 \le i \le k_n} |t_i^n - t_{i-1}^n| \le \frac{KT^*}{k_n}$. The limit of the rolled over bonds is computed in the following result.

Lemma 4 *The almost sure limit of the roll-over portfolios can be calculated, i.e.*

$$\lim_{n\to\infty} B_t^n = B_t := \frac{A(0)}{A(t)}$$

for each $t \ge 0$.

Proof Let $0 = t_0^n < t_1^n \cdots < t_{k_n}^n = T^*$ and $\delta_n := \max_{1 \le i \le k_n}(t_i^n - t_{i-1}^n)$. By assumption $\delta_n \le \frac{KT^*}{k_n} \to 0$ for $n \to \infty$. Fix t, then for each n there is $j_n \le k_n$ such that $t_{j_n-1}^n < t \le t_{j_n}^n$. We have that

$$B_t^n = \prod_{i=1}^{j_n} \frac{1}{P(t_{i-1}^n, t_i^n)} P(t, t_{j_n}^n) = \frac{A(0)}{A(t_{j_n}^n)} \prod_{i=1}^{j_n} \left(1 - e^{-\frac{\|x+W_{t_{i-1}^n}\|^{-2}}{2(t_i^n - t_{i-1}^n)}}\right)^{-1} P(t, t_{j_n}^n).$$

By the right continuity of A we have that $A(t_{j_n}^n) \to A(t)$ for $n \to \infty$ and it is clear that $P(t, t_{j_n}^n) \to 1$ for $n \to \infty$. Moreover, as $0 < 1 - e^{-\frac{\|x+W_{t_{i-1}^n}\|^{-2}}{2(t_i^n - t_{i-1}^n)}} \le 1$, for all i, we get

$$\prod_{i=1}^{j_n} \left(1 - e^{-\frac{\|x+W_{t_{i-1}^n}\|^{-2}}{2(t_i^n - t_{i-1}^n)}}\right)^{-1} \ge 1.$$

We will now show that

$$\lim_{n\to\infty}\prod_{i=1}^{j_n}\left(1-e^{-\frac{\|x+W_{t_{i-1}^n}\|^{-2}}{2(t_i^n-t_{i-1}^n)}}\right)^{-1}\leq 1 \quad \text{a.s.},$$

which then implies that $B_t^n \to B_t = \frac{A(0)}{A(t)}$. Let $m_t := \min_{s\leq t}\|x+W_s\|^{-2}$ be the running minimum of the inverse of the squared Bessel process of dimension 4. We have that $m_t > 0$ a.s. By assumption $\delta_n \leq \frac{KT^*}{k_n}$. Hence

$$\prod_{i=1}^{j_n}\left(1-e^{-\frac{\|x+W_{t_{i-1}^n}\|^{-2}}{2(t_i^n-t_{i-1}^n)}}\right)^{-1}\leq\prod_{i=1}^{j_n}\left(1-e^{-\frac{k_n m_t}{2KT^*}}\right)^{-1}$$
$$=\left(1-e^{-\frac{k_n m_t}{2KT^*}}\right)^{-j_n}\leq\left(1-e^{-\frac{k_n m_t}{2KT^*}}\right)^{-k_n},$$

as, clearly, $0 < 1-e^{-\frac{k_n m_t}{2KT^*}} \leq 1$. But as for almost all ω we have that $m_t(\omega) > 0$ and as $(1-e^{-ak_n})^{-k_n} \to 1$ for $n\to\infty$ and $a > 0$ we get that

$$\lim_{n\to\infty}\left(1-e^{-\frac{k_n m_t}{2KT^*}}\right)^{-k_n} = 1 \text{ a.s.}$$

By Theorem 6, $(B_t)_{0\leq t\leq T^*}$ discounted with respect to the numéraire $P(., T^*)$ is a Q^*–supermartingale. Lemma 4 and the definition of the measure Q^* moreover gives that $\left(\frac{B_t}{P(t,T^*)}\right)_{0\leq t\leq T^*}$ is a strict local martingale under Q^* as

$$Z_t^*\frac{B_t}{P(t,T^*)} = \alpha\xi_t B_t = \alpha\frac{A(0)}{\|x+W_t\|^2};$$

hence we cannot choose B as numéraire in this market.

In Fig. 1 we consider the case where $A_t \equiv 1$ and show the term structure given by (16) in comparison to the constant term structure. In Sect. 7.4 we will meet an example where investing in the roll-over account may even lead to a total loss of invested money.

7.2 Bond Markets Driven by Fractional Brownian Motion

The purpose of this section is to illustrate the applicability of our approach beyond semimartingale models. The semimartingale assumption is standard in the literature on bond markets (see for example [13] and the referenced literature therein), while we were able to show that this is in general not necessary. In this regard, we study some models for bond markets driven by fractional Brownian motion. In these models, the bank account, the forward rates or even the bond prices themselves may no longer be semimartingales. However, *discounted* bond prices are martingales, such that an appropriate no-arbitrage condition still holds.

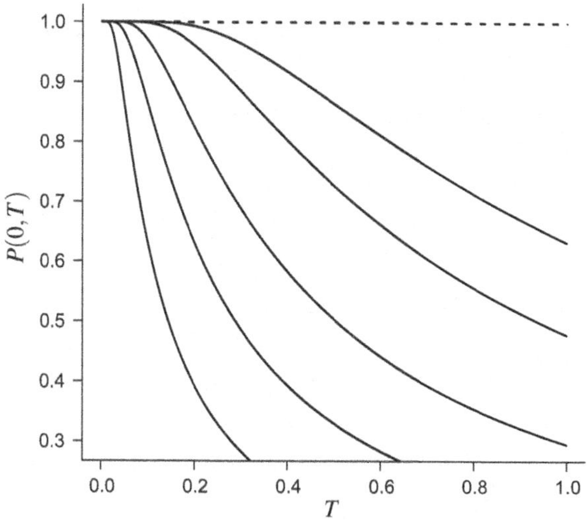

Fig. 1 This figure illustrates the two term structures: the term structure from Eq. (16) with $A_t \equiv 1$, is shown by the *lines* $T \mapsto P(0,T)$ for different $\| x \|^{-2} \in \{0.2, 0.4, 0.7, 1.3, 2\}$. The constant term structure $T \mapsto \tilde{P}(0,T) \equiv 1$ is represented by the *dashed line*

More precisely, we first consider a locally integrable short-rate process which is given by a time-inhomogeneous variant of fractional Brownian motion and show that discounted bond prices turn out to be martingales. Forward rates, however, are not semimartingales in this case which is in contrast to typical forward rate approaches as in [20]. In Remark 9 we consider a bank account directly driven by a fractional Brownian motion. In this case a short-rate does not exist and neither the bank account nor bond prices are semimartingales.

Regarding (1) we need to obtain the conditional distribution of a fractional Brownian motion, which we establish following [36]; see also [17] for related results. A fractional Brownian motion (FBM) with Hurst parameter $H \in (0,1)$ is a zero-mean stationary Gaussian process $Z = Z^H$ with covariance function

$$E[Z_s Z_t] = \frac{1}{2}\big(|s|^{2H} + |t|^{2H} - |s-t|^{2H}\big).$$

For $H = \frac{1}{2}$, Z is a standard Brownian motion. If $H > \frac{1}{2}$ the fractional Brownian motion has long-range dependence. Moreover, for $H \neq \frac{1}{2}$, Z is no longer a semimartingale.

To ease the exposition we consider $H > \frac{1}{2}$ only. Define the right-sided fractional Riemann-Liouville integral of order $\alpha > 0$ by

$$\big(I^{\alpha}_{t-} f\big)(s) := \frac{1}{\Gamma(\alpha)} \int_s^t f(u)(u-s)^{\alpha-1} du, \quad s \in (0,t).$$

I^0 is the identity. The fractional derivative of order $0 < \alpha < 1$ is denoted by $I^{-\alpha}$, i.e.

$$\left(I_{t-}^{-\alpha} f\right)(s) := -\frac{1}{\Gamma(1-\alpha)} \frac{d}{ds} \int_s^t f(u)(u-s)^{-\alpha} du, \quad s \in (0,t).$$

For the further analysis it will be useful to consider $\kappa := H - \frac{1}{2}$ instead of H itself. Fix a finite time horizon $T^* > 0$. Let[1]

$$(K_\kappa f)(s) := c_\kappa s^{-\kappa} \left(I_{T^*-}^{\kappa} (\cdot^\kappa f(\cdot)) \right)(s),$$

with constant $c_\kappa = \sqrt{\frac{\pi\kappa(2\kappa+1)}{\Gamma(1-2\kappa)\sin(\pi\kappa)}}$. The adjoint operator of K_κ is

$$(K_\kappa^* f)(s) := c_\kappa s^{-\kappa} \left(I_{T^*-}^{-\kappa} (\cdot^\kappa f(\cdot)) \right)(s).$$

It turns out that for $H > \frac{1}{2}$ the proper space of deterministic integrands to consider is, see [36],

$$\Lambda_{T^*}^{\kappa} := \left\{ f : \exists \varphi_f \in L^2[0,T^*], \text{ s.t. } f(s) = (K_\kappa^* \varphi_f)(s) \right\}.$$

Then $\Lambda_{T^*}^{\kappa}$ is a Hilbert space with corresponding norm

$$\| f \|_{\Lambda_{T^*}^{\kappa}} := \| K_\kappa f \|_{L^2([0,T^*])} .$$

The integral of $f \in \Lambda_{T^*}^{\kappa}$ w.r.t. the fractional Brownian motion Z is obtained as the limit of $\int f_n dZ$ with elementary f_n s.t. $\| f_n - f \|_{\Lambda_{T^*}^{\kappa}} \to 0$. Of course, for elementary f, say $f = \sum a_i \mathbb{1}_{(s_i,t_i)}$ the integral equals $\int f dZ = \sum a_i (Z_{t_i} - Z_{s_i})$. Letting $k^\kappa(t,s) := (K^\kappa \mathbb{1}_{[0,t]})(s)$ the covariance function of Z has the following representation

$$R_\kappa(t,s) := E[Z_s Z_t] = \int_0^t k^\kappa(t,w) k^\kappa(s,w) dw. \tag{17}$$

For $\kappa = 0$ we obtain $K_\kappa = \text{id}$, i.e. $R_\kappa(t,s) = s \wedge t$ which is the covariance function of a Brownian motion. From the representation in (17) it is immediate that

$$Z_t \stackrel{\mathcal{L}}{=} \int_0^t k^\kappa(t,w) dB_w \tag{18}$$

where B is a standard Brownian motion. This result was already discovered in the seminal work of [33] and leads to the following representation of conditional expectations (Theorem 7.1 in [36]): let $\mathscr{F}_t^Z := \sigma(Z_s : 0 \le s \le t)$. For $0 < s < t$

[1] We write $\cdot^{-\kappa} f(\cdot)$ short for the function $u \mapsto u^{-\kappa} f(s)$.

$$E\big[Z_u|\mathscr{F}_t^Z\big] = Z_t + \int_0^t \psi_u(t,w)dZ_w \tag{19}$$

with

$$\psi_u(t,w) = \psi_u^\kappa(t,w) := \frac{\sin(\pi\kappa)}{\pi} w^{-\kappa}(t-w)^{-\kappa} \int_t^u \frac{z^\kappa(z-t)^\kappa}{z-w} dz.$$

Note that

$$\psi_u(t,w) = w^{-\kappa}(I_{t-}^{-\kappa}(I_{u-}^{\kappa}(\cdot^\kappa \mathbb{1}_{[t,u)})))(w).$$

Proceeding similarly, we are able to compute the conditional covariance of Z.

Lemma 5 *For* $0 < t < u, v$,

$$E[Z_u Z_v|\mathscr{F}_t^Z] = E[Z_u|\mathscr{F}_t^Z] \cdot E[Z_v|\mathscr{F}_t^Z] + \int_t^{u\wedge v} k^\kappa(u,w)k^\kappa(v,w)dw.$$

Proof The proof mainly relies on (18). We have that

$$\begin{aligned} E\Big[\int_0^u k^\kappa(u,w)dB_w \int_0^v k^\kappa(v,w)dB_w|\mathscr{F}_t^B\Big] \\ = \int_0^t k^\kappa(u,w)dB_w \int_0^t k^\kappa(v,w)dB_w \\ + E\Big[\int_t^u k^\kappa(u,w)dB_w \int_t^v k^\kappa(v,w)dB_w|\mathscr{F}_t^B\Big]. \end{aligned} \tag{20}$$

As standard Brownian motion has independent increments, the last expectation is easily computed, leading to the last term in our result. It remains to represent the first addend in terms of Z. Using (18) we obtain

$$E[Z_u|\mathscr{F}_t^Z] = \int_0^t k^\kappa(u,w)dB_w$$

and we conclude.

With these results at hand we are ready to consider bond markets where the short-rate is driven by fractional Brownian motion. Fix a measure $Q \sim P$ and assume that Z is a FBM with parameter κ under Q. As numéraire we consider the bank account $B(t) = \exp(\int_0^t r_u du)$ where the short-rate is given by

$$r_t = \mu(t) + \sigma(t)Z_t, \quad 0 \le t \le T^*, \tag{21}$$

with $\mu : [0, T^*] \mapsto \mathbb{R}^+$ in $L^1[0, T^*]$ and $\sigma : [0, T^*] \mapsto \mathbb{R}^+$ being an element of $\Lambda_{T^*}^\kappa$. Bond prices are given by

$$P(t,T) = E_Q[\frac{B_t}{B_T}|\mathscr{F}_t] = E_Q\Big[\exp\Big(-\int_t^T r_u du\Big)|\mathscr{F}_t\Big], \qquad 0 \le t \le T \le T^*. \tag{22}$$

For $0 \le t \le T \le T^*$, we denote by $\mu^*(t,T) := \int_t^T \mu(u)du$, $\sigma^*(t,T) := \int_t^T \sigma(u)du$, and

$$\gamma^*(t,T) := \frac{1}{2}\int_t^T \int_t^T \int_0^{u\wedge v} \sigma(u)\sigma(v)k^\kappa(u,w)k^\kappa(v,w)\,dw\,du\,dv.$$

The following result gives bond prices and forward rates when the short-rate satisfies (21).

Proposition 1 *Under* (21) *and for* $0 \le t \le T \le T^*$, *the bond prices equal*

$$P(t,T) = \exp\Big[-\mu^*(t,T) - \sigma^*(t,T)Z_t - \int_t^T \int_0^t \sigma(u)\psi_u(t,w)dZ_w\,du + \gamma^*(t,T)\Big],$$

and associated forward rates are given by

$$f(t,T) = \mu(T) + \sigma(T)\Big(Z_t + \int_0^t \psi_T(t,w)dZ_w\Big) - \partial_T\gamma^*(t,T). \tag{23}$$

Proof First, note that $J(t,T) := \int_t^T r_u du$ is a Gaussian process with

$$\begin{aligned} E_Q[J(t,T)|\mathscr{F}_t] &= \int_t^T \Big(\mu(u) + \sigma(u)E_Q[Z_u|\mathscr{F}_t]\Big)du \\ &= \mu^*(t,T) + \int_t^T \sigma(u)(Z_t + \int_0^t \psi_u(t,w)dZ_w)du \\ &= \mu^*(t,T) + \sigma^*(t,T)Z_t + \int_t^T \sigma(u)\int_0^t \psi_u(t,w)dZ_w\,du, \end{aligned}$$

using (19). For the conditional variance of $J(t,T)$ note that

$$\mathrm{Var}[J(t,T)|\mathscr{F}_t] = E_Q\Big[\Big(\int_t^T \sigma(u)\Big(Z_u - E_Q[Z_u|\mathscr{F}_t]\Big)du\Big)^2|\mathscr{F}_t\Big]. \tag{24}$$

By Lemma 5 we obtain

$$E_Q[Z_uZ_v|\mathscr{F}_t] - E_Q[Z_u|\mathscr{F}_t]E_Q[Z_v|\mathscr{F}_t] = \int_t^{u\wedge v} k^\kappa(u,w)k^\kappa(v,w)dw.$$

Inserting this into (24) gives that

$$\mathrm{Var}[J(t,T)|\mathscr{F}_t] = \int_t^T \int_t^T \sigma(u)\sigma(v) \int_t^{u\wedge v} k^{\kappa}(u,w)k^{\kappa}(v,w)dw\,dv\,du.$$

Finally, note that $J(t,T)$ is, conditional on $\mathscr{F}_t$, a Gaussian random variable. Using its Laplace-Stieltjes transform we obtain the claim on bond prices. Observe that bond prices are absolutely continuous in maturity T. Then the expression on forward rates, (23), follows directly from

$$f(t,T) = -\partial_T \log P(t,T)$$

and we conclude.

By (22), discounted bond prices are Q-martingales. Moreover, from Lemma 1 it follows that Assumption (3) holds. Hence Q is an ELMM such that by Theorem 5 NAFL holds.

Remark 9 In [13] semimartingale models are covered while we drop this assumption in our setup. Consider for example the case where $B_t = \exp(\mu(t) + \sigma(t)Z_t)$ with càdlàg functions $\mu : [0,T^*] \to \mathbb{R}$ and $\sigma : [0,T^*] \to \mathbb{R}_{>0}$. Then, for $H \neq \frac{1}{2}$ the bank account is not a semimartingale. Analogously to Proposition 1, bond prices can be computed. A short calculation yields that

$$P(t,T) = E_Q[\frac{B_t}{B_T}|\mathscr{F}_t] = \exp\Big(\mu(t) + \sigma(t)Z_t - \mu(T) - \sigma(T)(Z_t + \int_0^t \psi_T(t,w)dZ_w)$$
$$+ \frac{1}{2}\int_t^T (k^{\kappa}(T,w))^2 dw\Big).$$

Furthermore, $((B_t)^{-1}P(t,T))_{0\leq t\leq T}$ is a Q-martingale for all $T \in [0,T^*]$ while bond prices themselves are no semimartingales if $H \neq \frac{1}{2}$. Again, by Lemma 1 this case is included in our setup.

7.3 An Extension of the HJM Setup

In this section we extend the HJM setup by an additional component which is not absolutely continuous in terms of maturity, such that, in general, a short-rate does not exist in this framework. We point out in Remark 11 that in credit risk a number of examples exist in the literature where the term structure is not even continuous. Note that all these models, however, satisfy our assumption on right-continuity formulated in Assumption 1.

Using Theorem 5 we classify those models which satisfy NAFL by means of a generalised drift condition. The HJM-model is contained as special case.

Fix a finite time horizon T^* and a measure $Q \sim P|_{T^*}$. There are two independent Q-Brownian motions W and V where W is d-dimensional and V is one-dimensional. We consider the filtration $(\mathscr{F}_t)_{t\geq 0}$ given by

$$\mathscr{F}_t = \sigma(W_s : 0 \leq s \leq t, V_u : u \geq 0) \vee N$$

which is the initial enlargement of the natural filtration of W with the full path of V and all P-nullsets N. We assume that bond prices are given by

$$P(t,T) = \exp\Big(- \int_t^T f(t,u) dV(u) - \int_t^T g(t,u) du \Big), \quad 0 \leq t \leq T \leq T^* \quad (25)$$

with families of Itô-processes f and g to be specified below. This includes the HJM-framework if $f \equiv 0$. In the following we characterize when the considered measure Q is an (equivalent) local martingale measure in the sense used in Theorem 5. All models which satisfy NAFL are given by an equivalent change to such a local martingale measure.

For given initial curves $T \mapsto f(0,T)$ and $T \mapsto g(0,T)$ we assume that f and g satisfy

$$f(t,T) = f(0,T) + \int_0^t a(s,T) ds + \int_0^t b(s,T) dW_s, \quad (26)$$

$$g(t,T) = g(0,T) + \int_0^t c(s,T) ds + \int_0^t d(s,T) dW_s, \quad (27)$$

for $0 \leq t \leq T \leq T^*$. Denote by $\mathscr{O}$ the optional sigma-algebra on $\Omega \times \mathbb{R}_+$. We assume the following regularity conditions:

$$a, b, c \text{ and } d \text{ are } \mathscr{O} \otimes \mathscr{B}(\mathbb{R}_+) - \text{measurable}, \quad \text{(HJM1)}$$

$$\int_0^{T^*} \int_0^{T^*} (|a(s,t)| + |c(s,t)|) ds\, dt < \infty, \quad \text{(HJM2)}$$

$$\sup_{0 \leq s \leq t \leq T^*} (\| b(s,t) \| + \| d(s,t) \|) < \infty. \quad \text{(HJM3)}$$

Recall that Q is an equivalent local martingale measure (ELMM) if

$$\left(\frac{P(t,T)}{P(t,T^*)} \right)_{0 \leq t \leq T} \quad \text{is a local martingale for all } T \in [0, T^*].$$

For any $T \leq T^*$ we set

$$A(t,T) := \int_T^{T^*} a(t,u) dV_u, \quad C(t,T) := \int_T^{T^*} c(t,u) du,$$

and similar for b (as for a) and d (as for c).

Proposition 2 *Under (HJM1)–(HJM3), Q is an ELMM iff*

$$0 = A(t,T) + C(t,T) + \frac{1}{2}(\| B(t,T) \|^2 + \| D(t,T) \|^2), \quad for\, t \le T \le T^*, \\ dQ \otimes dt - a.s. \tag{28}$$

Proof The formulation in terms of forward rates in (25) directly gives that

$$Z(t,T) := \frac{P(t,T)}{P(t,T^*)} = \exp\Big(\int_T^{T^*} f(t,u)dV(u) + \int_T^{T^*} g(t,u)du\Big).$$

The dynamics of f, given in (26), implies

$$\begin{aligned}\int_T^{T^*} f(t,u)dV(u) &= \int_T^{T^*} f(0,u)dV(u) \\ &\quad + \int_T^{T^*}\int_0^t a(s,u)ds\, dV(u) + \int_T^{T^*}\int_0^t b(s,u)dW_s\, dV(u) \\ &= \int_T^{T^*} f(0,u)dV(u) + \int_0^t A(s,T)ds + \int_0^t B(s,T)dW_s\end{aligned}$$

by the stochastic Fubini theorem (see, e.g., Theorem 6.2 in [16]). We obtain a similar expression for the second integral. Hence, by the Itô formula,

$$\begin{aligned}dZ(t,T) = Z(t,T)\Big(&A(t,T)dt + B(t,T)dW_t + \frac{1}{2} \| B(t,T) \|^2 dt \\ &+ C(t,T)dt + D(t,T)dW_t + \frac{1}{2} \| D(t,T) \|^2 dt\Big),\end{aligned}$$

for $0 \le t \le T \le T^*$. These processes are local martingales if and only if their drifts vanish. This is equivalent to (28) and we conclude.

Remark 10 The classical HJM-drift condition, i.e. the drift condition for the case $f \equiv 0$, can be obtained as follows: if the limit of the roll-overs $B(t) = \exp(\int_0^t g(s,s)ds)$ qualifies as numéraire, which is equivalent to the assumption that

$$\frac{B(t)P(0,T^*)}{P(t,T^*)}, \quad 0 \le t \le T^*$$

is a true Q-martingale, one can change to the equivalent measure $\tilde{Q}$ where B is taken as numéraire. Considering the dynamics of g, as in (27), under $\tilde{Q}$ then gives the well-known drift condition as in [20].

Example 1 Consider the simple case where $f(t,u) = \int_0^t a(s,u)du + W(t)$. We assume that $a(t,u)$ and $g(t,u)$ are $\mathscr{F}_0$-measurable functions which are bounded and continuous. Moreover, g is differentiable in the first coordinate. The initial term structure is flat, i.e. $a(0,T) = g(0,T) = 0$ for all $T \geq 0$. We have that $b(t,T) = 1$, $c(t,T) = g'(t,T)$ and $d(t,T) = 0$. The drift condition (28) in this setup reads

$$0 = \int_T^{T^*} a(t,u)dV(u) + \int_T^{T^*} g'(t,u)du + \frac{1}{2}(V_{T^*} - V_T)^2. \tag{29}$$

We consider $(V_{T^*} - V_T)^2$ as pathwise stochastic integral and an application of Itô's formula reversely in time gives

$$(V_T - V_{T^*})^2 = \int_{T^*}^{T} 2(V(u) - V(T^*))dV(u) + \int_{T^*}^{T} du.$$

We conclude that (28) holds if and only if $a(t,u) = (V(u) - V(T^*))$ and $g'(t,u) = 1/2$.

Remark 11 (Discontinuous term structures) In some models in credit risk, discontinuous term structures appear in a natural way: consider a company subject to default risk. A model which encompasses both the reduced-form approach and the structural approach was proposed in [2]. The authors model the default time as

$$\tau = \inf\{t \geq 0 : \Gamma_t \geq \xi\}$$

with a non-decreasing process Γ and an independent, positive random variable ξ. If, for example, $\Gamma_t = t$ and ξ is equal to 1 with a probability of $1/2$ and equal to a standard exponential random variable with probability $1/2$, then

$$Q(\tau > T) = \frac{1}{2}(1_{\{T<1\}} + e^{-T}),$$

and as a consequence, for $0 \leq t \leq T$,

$$Q(\tau > T | \tau > t) = \frac{1_{\{T<1\}} + e^{-T}}{1_{\{t<1\}} + e^{-t}}, \tag{30}$$

where Q can either be an equivalent martingale measure or, in the benchmark approach considered in Sect. 7.1, the objective probability measure. Consider a numéraire S^* which is independent of τ. Then, defaultable bond prices are given by

$$P^d(t,T) = E_Q[\frac{S_t^*}{S_T^*} 1_{\{\tau>T\}}|\mathscr{F}_t]$$
$$= 1_{\{\tau>t\}} P(t,T) Q(\tau > T|\tau > t).$$

From (30) it follows that the defaultable term structure $T \mapsto P(t,T)$ has a discontinuity at 1, for $0 \le t < 1$. A detailed study of a general approach to credit risk in extended HJM models can be found in [19].

7.4 On the Supermartingale Property of the Generalized Bank Account

In this section we elaborate on properties of the generalized bank account introduced as a limit of rolled over bonds in Theorem 6. In general, we can only state that the generalized bank account in terms of the terminal numéraire is a supermartingale (not necessarily a local martingale), see Remark 7. We shall illustrate this in the following by a concrete example where, by means of not uniformly integrable martingales, the generalized bank account reaches zero almost surely in finite time, see Lemma 7. As it starts from 1 it is a supermartingale which is not a local martingale.

Besides the technical interest in this example it also has an interesting economic interpretation: it answers the question raised in Sect. 7.1, namely how much money can be lost by investing in the roll-over strategy.

We consider a market with NAFL and denote by Q^* the measure in Theorem 5. Our starting point are bond prices of the form

$$P(t,T) = E_{Q^*}[\frac{N_t}{N_T}|\mathscr{F}_t] \tag{31}$$

with a finite time horizon $T^* = 2$ and the numéraire N, chosen as follows: let $\tau\colon [0,1) \to \mathbb{R}_{\ge 0}$ be an increasing, differentiable time transformation with $\tau(0) = 0$ and $\tau(t) \to \infty$ as $t \to 1$. The numéraire N is given by

$$N(t) := P(t,T^*) = \begin{cases} \exp(W^2_{\tau(t)} - \tau(t)^2) & 0 \le t < 1 \\ 1 & t \in [1,2]. \end{cases}$$

with Q^*-Brownian motion W. Note that N is càdlàg: for any $\epsilon > 0$

$$Q^*(N(t) \le \epsilon) = Q^*(\tau(t)\xi^2 \le \log\epsilon + \tau(t)^2)$$
$$= 2\Phi(\sqrt{\tau(t) + \tau(t)^{-1}\log\epsilon}) - 1$$

for a standard normal random variable ξ. The last expression converges to 1 as $\tau(t) \to \infty$ and existence of left limits of N follows. However, N is not uniformly integrable. The filtration is given by $\mathscr{F}_t := \sigma(W_{\tau(s)} : 0 \le s \le t),\ t \in [0, 2]$, with the usual augmentation by null sets.

We compute the bond prices with the following lemma.

Lemma 6 *For a standard normal random variable ξ, and $a < \frac{1}{2}$ we have that*

$$E[\exp(a\xi^2 + b\xi)] = e^{\frac{b^2}{2(1-2a)}}(1-2a)^{-\frac{1}{2}}.$$

Proof We start by observing that

$$E[\exp(a\xi^2 + b\xi)] = \int \frac{1}{\sqrt{2\pi}} e^{-\frac{x^2(1-2a)}{2}+bx}\,dx. \tag{32}$$

Let $s := (1-2a)^{-1}$. Then

$$\begin{aligned}(1.32) &= \int \frac{1}{\sqrt{2\pi}} e^{-\frac{x^2-2sbx+s^2b^2}{2s}+\frac{sb^2}{2}}\,dx \\ &= e^{\frac{sb^2}{2}} s^{\frac{1}{2}} \int \frac{1}{\sqrt{2\pi s}} e^{-\frac{(x-sb)^2}{2s}}\,dx \\ &= e^{\frac{b^2}{2(1-2a)}}(1-2a)^{-\frac{1}{2}}.\end{aligned}$$

□

Bond prices now can be computed from (31): Note that

$$\begin{aligned}E_{Q^*}[e^{-W_T^2+W_t^2}|\mathscr{F}_t] &= E_{Q^*}[e^{-(W_T-W_t)^2-2W_t(W_T-W_t)}|W_t] \\ &= E_{Q^*}[e^{-(T-t)\xi^2-2W_t\sqrt{T-t}\,\xi}|W_t] \\ &=: \exp(W_t^2 f(t,T) - g(t,T))\end{aligned}$$

where ξ is standard normal, independent of W_t; we obtain $f(t,T) = 2(T-t)(1+2(T-t))^{-1}$ and $g(t,T) = \frac{1}{2}\log(1+2(T-t))$ using Lemma 6. Hence,

$$P(t,T) = \exp(W_{\tau(t)}^2 f_\tau(t,T) - g_\tau(t,T) + \tau^2(T) - \tau^2(t)),$$

$0 \le t \le T < 1$, where we set $f_\tau(t,T) := f(\tau(t), \tau(T)))$ and similarly for g_τ. For $t \ge 1$ the term structure is flat, i.e. $P(t,T) = 1$.

Now we turn to the limit of the roll-over account. Fix $T < 1$ and consider $t_i^n := t_i = \tau^{-1}(iT/n)$. Then

$$B_{t_n}^n = \exp(-\sum_{i=1}^n f_\tau(t_{i-1}, t_i) W_{\tau(t_{i-1})}^2 + \sum_{i=1}^n g_\tau(t_{i-1}, t_i) - \tau^2(t_n)).$$

We have that

$$\exp(\sum_{i=1}^{n} g_\tau(t_{i-1}, t_i)) = \exp(\frac{1}{2}\sum_{i=1}^{n} \log(1 + 2(\tau(t_i) - \tau(t_{i-1}))))$$
$$\to e^{\tau(T)}$$

by Taylor expansion and continuity of τ. Moreover,

$$\sum_{i=1}^{n} f_\tau(t_{i-1}, t_i) W^2_{\tau(t_{i-1})} = 2\sum_{i=1}^{n} \frac{W^2_{\tau(t_{i-1})}}{1 + 2(\tau(t_i) - \tau(t_{i-1}))} (\tau(t_i) - \tau(t_{i-1}))$$
$$\to 2\int_0^T W^2_{\tau(s)} d\tau(s) = 2\int_0^{\tau(T)} W_s^2 ds.$$

Hence,

$$B^n(T) = \exp(-\sum_{i=1}^{n} W^2_{\tau(t_{i-1})} f_\tau(t_{i-1}, t_i) + \sum_{i=1}^{n} g_\tau(t_{i-1}, t_i) - \tau^2(t_n))$$
$$\to \exp(-2\int_0^{\tau(T)} W_s^2 ds + \tau(T) - \tau^2(T)).$$

The discounted limit of the roll-over account turns out to be

$$V(T) = P(T, T^*)^{-1} B(T) = \exp(-W^2_{\tau(T)} - 2\int_0^{\tau(T)} W_s^2 ds + \tau(T))$$
$$= Z(\tau(T)),$$

letting

$$Z(T) := \exp(T - 2\int_0^T W_s^2 ds - W_T^2). \tag{33}$$

We are interested in

$$V(1) = \lim_{T\to 1} Z(\tau(T)) = \lim_{T\to\infty} Z(T).$$

The following lemma shows that $\lim_{T\to\infty} Z(T) = 0$, hence $B(1) = 0$. It turns out that investing in the roll-over strategy leads to the total loss of invested money such that the classical risk-free investment strategy becomes highly risky in this example.

Lemma 7 *Consider Z as in Eq. (7.19). Then Z converges to* 0 *Q^*-almost surely as $T \to \infty$.*

Proof Note that Z is a non-negative local martingale and hence by the supermartingale convergence theorem[2] the limit Z_∞ exists and is in L^1. Moreover, we have

[2] See Theorem 1.3.15 in [25].

that

$$Z_t \le X_t := \exp(t - 2\int_0^t W_s^2 ds), \qquad t \ge 0.$$

We compute the distribution of X_t by P. Lévy's diagonalization procedure. Fix $T > 0$. Using N. Wiener's construction of Brownian motion we obtain

$$W_t = \sum_{k\ge 1} \frac{\sin(k\pi t/T)}{k} \xi_k \sqrt{T}, \quad t \in [0, T]$$

with i.i.d. standard normal $\xi_1, \xi_2, \ldots$ Then

$$\begin{aligned}\int_0^T W_t^2 dt &= \int_0^T \Big[\sum_{k\ge 1} \frac{\sin(k\pi t/T)}{k} \xi_k \sqrt{T}\Big]^2 dt \\ &= \sum_{k,j\ge 1} \frac{T\xi_k\xi_j}{kj} \int_0^T \sin(k\pi t/T)\sin(j\pi t/T)dt \\ &= \sum_{k\ge 1} \frac{T^2\xi_k^2}{2k^2},\end{aligned}$$

by orthogonality of the trigonometric functions, i.e.

$$\int_0^T \sin(k\pi t/T)\sin(j\pi t/T)dt = \mathbb{1}_{\{k=j\}}\frac{T}{2}.$$

Hence, for $u \ge 0$, we obtain with Lemma 6 that

$$\begin{aligned}E\Big[e^{-u\int_0^T W_t^2 dt}\Big] &= \prod_{k\ge 1} E[e^{-u\frac{T^2}{2k^2}\xi_k^2}] \\ &= \Big[\prod_{k\ge 1}(1 + \frac{uT^2}{k^2})\Big]^{-1/2} \\ &= \Big[\frac{\sinh(\pi\sqrt{uT^2})}{\pi\sqrt{uT^2}}\Big]^{-1/2}.\end{aligned}$$

Next, by Fatou's lemma,

$$E[(X_\infty)^u] \le \lim_{T\to\infty} e^{uT}\Big[\frac{\sinh(\pi\sqrt{2uT^2})}{\pi\sqrt{2uT^2}}\Big]^{-1/2}.$$

Note that, for $0 < u < \pi^2/2$,

$$\left[T^{-1}e^{2uT}(e^{\pi\sqrt{2u}T}-e^{-\pi\sqrt{2u}T})\right]^{-1}=\frac{T}{e^{T(\pi\sqrt{2u}-2u)}-e^{-T(\pi\sqrt{2u}+2u)}}\to 0$$

as $T \to \infty$. This shows that for the non-negative random variable X_∞ and $0 < u < \pi^2/2$, $E[(X_\infty)^u] = 0$. By the generalized Markov inequality, for $\epsilon > 0$ and $0 < u < \pi^2/2$,

$$P(X_\infty \geq \epsilon) = P((X_\infty)^u \geq \epsilon^u) \leq \epsilon^{-u} E[(X_\infty)^u] = 0$$

such that $X_\infty = 0$ almost surely and we conclude.

8 Appendix: Change of Numéraire and Bubbles

In this appendix section we outline some basic definitions and conclusions on numéraires and bubbles, since changes of numéraire are used frequently. The goal of this section is to add some possibly new definition to the large literature on these issues, however, no deep results are proved.

In seminal works on the absence of arbitrage in financial markets the numéraire (portfolio) plays a distinguished rôle, see [9]. Additionally in markets with stochastic interest rates, or foreign exchange markets, change of numéraire is an important technique. It turns out that the question which portfolios do qualify as numéraire is surprisingly subtle and often only indirectly solved: usually one characterizes possible changes of numéraire mathematically but *no economically reasonable* properties of numéraire portfolios are laid down (see for instance the seminal work [10], where numéraire portfolios are characterized as maximal, admissible, strictly positive portfolios). We would like to close this small gap in the following paragraphs by providing a simple definition of numéraire portfolios, which can be also mirrored in the world of bubbles and liquidity, and which still makes sense in discrete time and under trading constraints.

Intuitively, a portfolio can be used as numéraire if it is strictly positive and allows for short-selling, i.e. the investor is able to find a reasonable counterparty from whom the portfolio can be borrowed and she sells it then on the market. Short-selling might require arbitrarily high credit lines when the portfolio is to be returned, so the counterparty faces the risk of the investor's bankruptcy. Mathematically speaking this might lead to arbitrages in the virtual world after a change of numéraire (see [10]). Hence some conditions on the behaviour of the short-sold portfolio from below must be imposed. On the other hand we do not want to bound the short-sold portfolio from below by some number, hence the usual admissibility condition is too strong. Instead of admissibility of the short-sold portfolio we require a uniform integrability condition with respect to some equivalent local martingale measure. In other words: we extend the notion of traded portfolios a bit beyond admissibility and call a strictly positive portfolio N a numéraire if N and $-N$ are traded in this extended sense. Such approaches have been successfully investigated in [11] in the context of workable claims, or in [39] via a re-formulation of the Ansel-Stricker framework [1]. We consider here the second approach which seems to us slightly more descriptive, but we could also simply formulate everything in the context of workable claims. Notice that the second definition also makes sense under trading constraints.

We give a precise definition which reflects this insight and which *leads* to the well known change of numéraire formulas, see [10]. Furthermore we relate this intuitive and economically meaningful definition with the notion of bubbles: a positive portfolio is modelled in a bubble state if it does *not* qualify as numéraire. Both concepts will play an important role when it comes to the notion of liquidity in bond markets.

Consider a filtered probability space $(\Omega, \mathscr{F}, (\mathscr{F}_t)_{t\in[0,T]}, P)$, where the filtration satisfies the usual conditions. The price process of traded assets $(\mathbf{X}_t)_{t\in[0,T]} = (X_t^0, \dots, X_t^d)_{t\in[0,T]}$ is a $d+1$-dimensional adapted process with càdlàg trajectories, where at least one process, say X^0, is positive, i.e. $X^0 > 0$. We introduce the process of discounted assets,

$$\mathbf{S} := \left(1, \frac{X^1}{X^0}, \dots, \frac{X^d}{X^0}\right)$$

and assume without loss of generality that we are dealing from now on with a semimartingale S. Let $\mathbf{H}$ be a predictable $\mathbf{S}$-integrable process and denote by $(\mathbf{H}\cdot\mathbf{S})$ the stochastic integral process of $\mathbf{H}$ with respect to $\mathbf{S}$, the *(portfolio) wealth process*. The process $\mathbf{H}$ is called an *a-admissible trading strategy* if there is $a \geq 0$ such that $(\mathbf{H}\cdot\mathbf{S})_t \geq -a$ for all $t \in [0, T]$. A strategy is called admissible if it is a-admissible for some $a \geq 0$. Define

$$\mathbf{K} = \{(\mathbf{H}\cdot\mathbf{S})_T : H \text{ admissible}\} \text{ and } \mathbf{C} = \{g \in L^\infty(P) : g \leq f \text{ for some } f \in K\}.$$

Then $\mathbf{K}$ and $\mathbf{C}$ form convex cones in $L^0(\Omega, \mathscr{F}, P)$.

The condition *no free lunch with vanishing risk* (NFLVR) is the right concept of no arbitrage, since it combines mathematically minimal assumptions with an economically reasonable interpretation, see [9, 12].

Definition 8 The market $\mathbf{S}$ satisfies (NFLVR) if

$$\bar{\mathbf{C}} \cap L_+^\infty(P) = \{0\},$$

where $\bar{\mathbf{C}}$ denotes the closure of C with respect to the norm topology of $L^\infty(P)$.

This means that a free lunch with vanishing risk exists, if there exists a free lunch $f \in L_+^\infty(P)$, which can be approximated by a sequence of portfolio wealth processes $(f_n) = ((\mathbf{H_n}\cdot\mathbf{S})) \in \mathbf{K}$ with $\frac{1}{n}$-admissible integrands $\mathbf{H}_n$, such that

$$\lim_{n\to\infty} \| f - f_n \|_\infty = 0$$

with respect to the norm topology of $L^\infty(P)$. Define the set $\mathbf{M}_e$ of equivalent separating measures as

$$\mathbf{M}_e = \{Q \sim P|_{\mathscr{F}_T} : E_Q[f] \le 0 \text{ for all } f \in \mathbf{K}\}.$$

If $\mathbf{S}$ is (locally) bounded then $\mathbf{M}_e$ consists of all equivalent probability measures such that $\mathbf{S}$ is a (local) martingale.

Having a general change of numéraire theorem in mind it turns out that the concept of admissibility is too strong, since we want to talk about unbounded portfolio wealth processes and their negative to be admissible. Also we want to consider market extensions of the market $\mathbf{S}$ by assets $\mathbf{Y}$. We assume from now on (NFLVR) for the market $\mathbf{S}$. We call assets $\mathbf{Y}$ a market extension of $\mathbf{S}$ if $\mathbf{S}' := (\mathbf{S}, \mathbf{Y})$ satisfies (NFLVR). We define in the sequel a larger class of trading strategies which we call Q-admissible. This is a generalization of admissibility as introduced above, i.e. every admissible strategy is Q-admissible. The definition is in spirit of the results of Strasser in [39].

Definition 9 Fix $Q \in \mathbf{M}_e$. Consider an extension of the original market $\mathbf{S}' := (\mathbf{S}, \mathbf{Y})$ by finitely many assets $\mathbf{Y}$ such that the process $\mathbf{S}'$ is a Q-local martingale. Consider furthermore a predictable, $\mathbf{S}'$-integrable process φ and the sequence of hitting times

$$\sigma_n := \inf\{t \ge 0 : (\varphi \cdot \mathbf{S}')_t \le -n\}, \quad n \ge 1.$$

The trading strategy φ is called *Q-admissible* (such as the corresponding stochastic integral, the wealth process), if

$$\liminf_{n\to\infty} E_Q[(\varphi \cdot \mathbf{S}')^-_{\sigma_n} \mathbb{1}_{\{\sigma_n < \infty\}}] = 0.$$

Define

$$\mathbf{L}^Q = \{x + (\varphi \cdot \mathbf{S}') : x \in \mathbb{R}, \quad \varphi \text{ is } Q\text{-admissible}\}.$$

and

$$\mathbf{L} = \cup_{Q \in \mathbf{M}_e} \mathbf{L}^Q.$$

Remark 12 We extend the set of admissible portfolios but due to Theorem 3 in [39] we do not introduce arbitrages, since every wealth process $(\varphi \cdot \mathbf{S}')$ for a Q-admissible strategy is a supermartingale. We also do not introduce free lunches, since this notion *only* depends on a-admissible strategies.

Remark 13 We could use a less general but more appealing definition of L^Q when we do not allow for a market extension $\mathbf{S}'$. Then numéraires are traded portfolios in the original market $\mathbf{S}$. In our definition all possible price processes for payoffs at time T are added. Notice that we should consider L^Q as set of trading strategies of our market, but not their union, since the union might contain contradictory pricing structures for one payoff.

Now we are in the position to make our intuitive definition of numéraire portfolios precise: a numéraire portfolio is a strictly positive portfolio which allows for short-selling, i.e. the negative of its wealth process is still given by a Q-admissible trading strategy for some $Q \in \mathbf{M}_e$, and hence is an element of $\mathbf{L}$.

Definition 10 A strictly positive process $N \in \mathbf{L}$ with $N_0 = 1$ is called a *strong numéraire* (in discounted terms with respect to S^0), if

$$N \in \mathbf{L}^Q \text{ and } -N \in \mathbf{L}^Q \tag{34}$$

for all $Q \in \mathbf{M}_e$. It is called *weak numéraire* (in discounted terms with respect to S^0), if (34) holds for at least one $Q \in \mathbf{M}_e$, i.e. N and $-N$ are elements of $\mathbf{L}$.

This definition has a clear economic meaning and easy consequences: as it should be, a weak numéraire qualifies as an accounting unit, where the classical change of numéraire technique is possible: there exist an equivalent measure $Q \in \mathbf{M}_e$ under which $N = (1 + (\varphi \cdot \mathbf{S}'))$ is a true Q-martingale.

Theorem 12 *The following statements are equivalent:*

(i) A strictly positive process N with $N_0 = 1$ is a weak numéraire.
(ii) There exists $Q \in \mathbf{M}_e$ such that N is a Q-martingale.

Proof Both directions are easy: if there exists $Q \in \mathbf{M}_e$ such that N is a true Q-martingale, then by adding N to the market $\mathbf{S}$ we obtain an element of $\mathbf{L}^Q$, but due to its uniform integrability $-N \in \mathbf{L}^Q$: hence N is a weak numéraire. If, on the other hand, $N \in \mathbf{L}^Q$ for some $Q \in \mathbf{M}_e$, then N is a Q-supermartingale *together* with $-N$, which in turn means that N is a Q-martingale.

Our definition of a numéraire has a clear relation to bubbles: a portfolio or an asset which does not qualify as numéraire is in a bubble state. Again this very intuitive definition leads to the meanwhile classical definition of a bubble, see [6]. In other words: if an asset S^i is a strict local martingale under any $Q \in \mathbf{M}_e$, $-S^i$ is not Q-admissible and hence it does not qualify as weak numéraire.

Definition 11 A strictly positive process $B \in \mathbf{L}$ is (modelled) in a *strong bubble state* if $-B \notin \mathbf{L}$, i.e. for all $Q \in \mathbf{M}_e$ the wealth process B is a strict local martingale. It is (modelled) in a *weak bubble state* if $-B \notin \mathbf{L}^Q$ for some $Q \in \mathbf{M}_e$, i.e. for this $Q \in \mathbf{M}_e$ the wealth process B is a strict local martingale.

Theorem 13 *A strictly positive portfolio $B \in \mathbf{L}$ with $B_0 = 1$ is in a strong bubble state if and only if B does not qualify as weak numéraire portfolio. A strictly positive portfolio $B \in \mathbf{L}$ with $B_0 = 1$ is in a weak bubble state if and only if B does not qualify as strong numéraire.*

Remark 14 Notice that this notion of bubble, such as the notion of numéraire, depends crucially on the set of trading strategies, which in turn under constraints also leads to notions of bubbles in discrete time. Conditions classifying certain strict local martingales and the relation to bubbles may be found in [34].

Example 2 Consider now again as in the introduction the case where the growth optimal portfolio is exogenously given such that $\frac{1}{S^*}$ is a strict local martingale. Furthermore we assume that the roll-over portfolio leads to the bank account process B equal to 1 even though the term structure will be non-trivial due to the strict local martingale property, see Sect. 7.1 for an example of these properties. In contrast the bank account process 1 does not qualify as numéraire, since $\frac{1}{S^*}$, or equivalently $\frac{1}{P(.,T^*)}$ is a strict local martingale. The bond market with respect to the numéraire $P(., T^*)$ is free of (asymptotic) arbitrage in the classical sense, even if one adds the bank account process as additional traded asset, but we cannot perform a change of numéraire towards the numéraire $B = 1$.

An interesting aspect of the previous example stems from the introduction of virtual term structures related to bank account processes B^n. We think here of the (finite) roll-over processes, i.e., for a sequence of refining partitions $0 = t_0^n < t_1^n < \cdots < t_{k_n}^n = T^*$ of $[0, T^*]$ define, for each n,

$$B_t^n = \begin{cases} \prod_{i=1}^{j} \frac{1}{P(t_{i-1}^n, t_i^n)} & \text{for } t = t_j^n,\ j = 1, \ldots, k_n, \\ B_{t_j^n}^n & \text{for } t_{j-1}^n < t \leq t_j^n,\ j = 1, \ldots, k_n. \end{cases}$$

Notice the difference to Definition 5. We have $\lim_{n\to\infty} B^n = B^\infty = B = 1$ as announced before, see Sect. 7.1 for a concrete example. These virtual term structures can be interpreted as high-liquidity term structures, which one would actually expect in the market if there was enough liquidity in the respective numéraire: this amounts to pricing with the corresponding supermartingale deflator

$$\frac{1}{E_{Q^*}[\frac{B_T^n}{P(T,T^*)}]} \frac{B_T^n}{P(T, T^*)} \frac{1}{B_T^n},$$

which is derived from changing measure by the local martingale density $\frac{B_T^n}{P(T,T^*)}$. When pricing 1 at time T with respect to this deflator we obtain an alternative term structure $\tilde{P}^n(t, T)$

$$\begin{aligned} E_{Q^*}\Big[\frac{B_T^n}{P(T, T^*)} \frac{1}{E_{Q^*}[\frac{B_T^n}{P(T,T^*)}]} \frac{1}{B_T^n}\Big] =& E_{Q^*}\Big[\frac{1}{P(T, T^*)} \frac{1}{E_{Q^*}[\frac{B_T^n}{P(T,T^*)}]}\Big] \\ =& \frac{1}{B_0^n} \tilde{P}^n(0, T), \end{aligned}$$

which yields

$$\tilde{P}^n(0, T) = \frac{B_0^n P(0, T)}{P(0, T^*) E_{Q^*}[\frac{B_T^n}{P(T,T^*)}]} > P(0, T), \qquad (35)$$

for each n, i.e., the virtual term structures show lower interest rates (due to higher liquidity) than $P(t, T)$. In case of B^{∞} we apparently obtain the virtual term structure $\tilde{P}^{\infty}(0, T) = 1$, which corresponds to the highest liquidity virtual term structure, with overnight borrowing at no cost available.

Acknowledgements The authors thank ETH Foundation for its support of this research project. The first and second author thank the Forschungsinstitut Mathematik at ETH Zürich for its generous hospitality.

References

1. Ansel, J.P., Stricker, C.: Couverture des actifs contingent et prix maximum. Ann. Inst. Henri Poincaré **30**, 303–315 (1994)
2. Bélanger, A., Shreve, S.E., Wong, D.: A general framework for pricing credit risk. Math. Finance **14**, 317–350 (2004)
3. Björk, T., Di Masi, G., Kabanov, Y., Runggaldier, W.: Towards a general theory of bond markets. Finance Stoch. **1**, 141–174 (1997)
4. Brace, A., Gatarek, D., Musiela, M.: The market model of interest rate dynamics. Math. Finance **7**(2), 127–155 (1997)
5. Bruti-Liberati, N., Nikitopoulos-Sklibosios, C., Platen, E., Schlögl, E.: Alternative defaultable term structure models. Asia-Pacific Fin. Markets. **16**, 1–31 (2009)
6. Cox, A.M.G., Hobson, D.G.: Local martingales, bubbles and option prices. Finance Stoch. **9**, 477–492 (2005)
7. Cuchiero, C., Klein, I., Teichmann, J.: A new perspective on the fundamental theorem of asset pricing for large financial markets. To appear in Theory of Probability and its Applications. Teor. Veroyatnost. i Primenen. **60**(4), 660–685 (2015)
8. De Donno, M. Pratelli, M.: A theory of stochastic integration for bond markets. Ann. Appl. Prob. **15**:2773–2791 (2005)
9. Delbaen, F., Schachermayer, W.: A general version of the fundamental theorem of asset pricing. Math. Ann. **300**, 463–520 (1994)
10. Delbaen, F., Schachermayer, W.: The no-arbitrage property under a change of numéraire. Stochast. Stochast. Rep. **53**, 213–226 (1995)
11. Delbaen, F. Schachermayer, W.: The Banach space of workable contingent claims in arbitrage theory. Annales de l' I.H.P. **33**, 113–144 (1997)
12. Delbaen, F., Schachermayer, W.: The fundamental theorem of asset pricing for unbounded stochastic processes. Math. Ann. **312**, 215–250 (1998)
13. Döberlein, F., Schweizer, M.: On savings accounts in semimartingale term structure models. Stoch. An. Appl. **19**, 605–626 (2001)
14. Ekeland, I., Taflin, E.: A theory of bond portfolios. Ann. Appl. Probab. **15**, 1260–1305 (2005)
15. Fernholz, R., Karatzas, I.: Relative arbitrage in volatility-stabilized markets. Ann. Fin. **1**, 149–177 (2005)
16. Filipović, D.: Term Structure Models: A Graduate Course. Springer, Berlin, Heidelberg, New York
17. Fink, H., Zähle, M., Klüppelberg, C.: Conditional distributions of processes related to fractional Brownian motion. J. App. Probab. **50**, 166–183 (2013)
18. Föllmer, H., Kramkov, D.: Optional decomposition under constraints. Probab. Theory Rel. Fields **109**, 1–25 (1997)
19. Gehmlich, F., Schmidt, T.: Dynamic defaultable term structure modelling beyond the intensity paradigm. Forthcoming Math. Finance. arXiv:1411.4851

20. Heath, D., Jarrow, R.A., Morton, A.J.: Bond pricing and the term structure of interest rates. Econometrica **60**, 77–105 (1992)
21. Heath, D., Platen, E.: A Benchmark Approach to Quantitative Finance. Springer (2006)
22. Jacod, J., Shiryaev, A.N.: Limit Theorems for Stochastic Processes. Springer (1987)
23. Kabanov, Y., Kramkov, D.: Large financial markets: asymptotic arbitrage and contiguity. Theory Probab. Appl. **39**, 222–228 (1994)
24. Kabanov, Y., Kramkov, D.: Asymptotic arbitrage in large financial markets. Finance Stoch. **2**, 143–172 (1998)
25. Karatzas, I., Shreve, S.: Brownian Motion and Stochastic Calculus. Springer (1988)
26. Kardaras, K.: Generalized supermartingale deflators under limited information. Math. Finance **23**, 186–197 (2013)
27. Klein, I.: A fundamental theorem of asset pricing for large financial markets. Math. Finance **10**, 443–458 (2000)
28. Klein, I.: Free lunch for large financial markets with continuous price processes. Ann. Appl. Probab. **13**, 1494–1503 (2003)
29. Klein, I.: No asymptotic free lunch reviewed in the light of Orlicz spaces. Séminaire de Probabilités XLI **1934**, 443–454 (2008)
30. Klein, I., Schachermayer, W.: Asymptotic Arbitrage arbitrage in non-complete large financial markets. Theory Probab. Appl. **41**, 927–934 (1996)
31. Klein, I., Schachermayer, W.: A quantitative and a dual version of the Halmos-Savage theorem with applications to Mathematical Finance. Ann. Probab. **24**, 867–881 (1996)
32. Kreps, D.M.: Arbitrage and equilibrium in economies with infinitely many commodities. J. Math. Econom. **8**, 15–35 (1981)
33. Mandelbrot, B., Van Ness, J.: Fractional Brownian motions, fractional noises and applications. SIAM Rev. **10**, 422–437 (1968)
34. Mijatović, A., Urusov, M.: On the martingale property of certain local martingales. Probab. Theory Relat. Fields **152**, 1–30 (2012)
35. Pipiras, V., Taqqu, M.: Integration questions related to fractional Brownian motion. Probab. Theory Relat. Fields **118**, 251291 (2000)
36. Pipiras, V., Taqqu, M.: Are classes of deterministic integrands for fractional Brownian motion on an interval complete? Bernoulli **7**, 95–105 (2001)
37. Revuz, D., Yor, M.: Continuous Martingales and Brownian Motion. Springer (1999)
38. Schachermayer, W.: Martingale measures for discrete time processes with infinite horizon. Math. Finance **4**, 25–56 (1994)
39. Strasser, E.: Necessary and sufficient conditions for the supermartingale property of a stochastic integral with respect to a local martingale. Séminaire de Probabilités XXXVII, 385–393, Lecture Notes in Math., 1832, Springer, Berlin (2003)
40. Taflin, E.: Generalized integrands and bond portfolios: pitfalls and counter examples. Ann. Appl. Probab. **21**, 266–282 (2011)

A Unified View of LIBOR Models

Kathrin Glau, Zorana Grbac and Antonis Papapantoleon

Abstract We provide a unified framework for modeling LIBOR rates using general semimartingales as driving processes and generic functional forms to describe the evolution of the dynamics. We derive sufficient conditions for the model to be arbitrage-free which are easily verifiable, and for the LIBOR rates to be true martingales under the respective forward measures. We discuss when the conditions are also necessary and comment on further desirable properties such as those leading to analytical tractability and positivity of rates. This framework allows to consider several popular models in the literature, such as LIBOR market models driven by Brownian motion or jump processes, the Lévy forward price model as well as the affine LIBOR model, under one umbrella. Moreover, we derive structural results about LIBOR models and show, in particular, that only models where the forward price is an exponentially affine function of the driving process preserve their structure under different forward measures.

Keywords LIBOR · Forward price · Semimartingales · LIBOR market models · Lévy forward price models · Affine LIBOR models

2010 Mathematics Subject Classification 91G30 · 60G44

Financial support from the PROCOPE project "Financial markets in transition: mathematical models and challenges" and the Europlace Institute of Finance project "Post-crisis models for interest rate markets" is gratefully acknowledged.

K. Glau
Center for Mathematics, Technical University of Munich,
Parkring 11, 85748 Garching bei München, Germany
e-mail: kathrin.glau@tum.de

Z. Grbac
Laboratoire de Probabilités et Modèles Aléatoires, Université Paris Diderot,
75205 Paris Cedex 13, France
e-mail: grbac@math.univ-paris-diderot.fr

A. Papapantoleon (✉)
Institute of Mathematics, TU Berlin, Strasse des 17. Juni 136, 10623 Berlin, Germany
e-mail: papapan@math.tu-berlin.de

© Springer International Publishing Switzerland 2016

J. Kallsen and A. Papapantoleon (eds.), *Advanced Modelling in Mathematical Finance*,
Springer Proceedings in Mathematics & Statistics 189,
DOI 10.1007/978-3-319-45875-5_18

1 Introduction

The LIBOR and EURIBOR interest rates rank among the most important interest rates worldwide. They are determined on a daily basis by a panel of banks for a number of maturities, while LIBOR is also determined for several currencies. LIBOR and EURIBOR serve as underlying rates for an enormous amount of financial transactions. In 2012, the outstanding values of contracts with LIBOR as reference were estimated at roughly summing up to 300 trillion USD; see Wheatley [32]. Therefore, the development of suitable policies and regulations for the fair calculation of LIBOR and EURIBOR, as well as of mathematical models for the fair evaluation of interest rate products, is essential for the financial industry and also serves the general public interest.

The modeling of the dynamics of LIBOR and EURIBOR rates is a challenging task due to the high dimensionality of the modeled objects. The major difference, from a modeling point of view, between interest rates and stock prices lies in the fact that stock prices are observed at each time point as a single value, once the difference between bid and ask prices is ignored, while interest rates are observed at each time point for several maturities. Moreover, these different rates (for the different maturities) are interdependent. Their joint modeling is indispensable because they jointly enter already the basic interest rate derivatives as underlying rates. In addition, the rates for different period lengths can no longer be derived from simple no-arbitrage relations. Indeed, the financial crisis of 2007–2009 has fundamentally changed the attitude of market participants towards risks in the interbank sector, regarding in particular counterparty and liquidity risk, which have a direct effect on the LIBOR rates for different lending periods; see, for example, Filipović and Trolle [11]. Summarizing, LIBOR modeling presents a challenge to jointly model the rates for different maturities and periods in an arbitrage-free way and such that the resulting pricing formulas are fast and accurately computable for all liquid derivatives, such as caps and swaptions. In this work, we provide a unified mathematical foundation for some of the most important of the existing LIBOR models in the literature. On this basis we gain valuable structural insight in modeling LIBOR rates. In particular we derive sufficient conditions for the validity of mandatory model features such as arbitrage-freeness and investigate those that typically support computational ease.

The seminal articles by Brace et al. [4] and Miltersen et al. [26] introduced the LIBOR Market Model (LMM), that became known also as the BGM model. The celebrity of the model certainly is, at least partly, owed to the fact that the BGM model reproduces the market standard Black's formula for caps. Moreover, the backward construction of LIBOR rates in Musiela and Rutkowski [27] has proven to be extendible beyond models driven by Brownian motion. The article by Eberlein and Özkan [10] introduced one of the first LIBOR models driven by jump processes and also proposed the Lévy-driven forward price model. In Jamshidian [16, 17] LIBOR models driven by general semimartingales were presented. In interest rate modeling, jump processes have several advantages. Firstly, just as in stock price modeling, their distributional flexibility allows to better capture the empirical distributions of logarithmic returns, see for instance Eberlein and Kluge [8, 9]. Secondly, the traditional way to jointly model the rates for different maturities would suggest to introduce one component of a multi-dimensional Brownian motion for each maturity, so as to introduce one stochastic factor for each source of risk. A Lévy process with an infinite jump activity, in contrast, introduces infinitely many sources of risk, already as a one-dimensional process. In this regard, jump processes show their potential to reduce the dimension of the related computational problems for pricing and hedging. However, they also bring along a new level of technical challenges, in particular the measure changes between forward measures become more involved and the backward construction typically requires a more sophisticated justification. Additionally, various extensions of the LIBOR market model to stochastic volatility have also appeared in the literature, cf. Wu and Zhang [33], Belomestny et al. [2] and Ladkau et al. [25]. Recently, a modeling approach under one single forward measure, the terminal measure, has been proposed in Keller-Ressel et al. [21], which is based on affine processes. We refer to Schoenmakers [31] and Papapantoleon [29] for an overview of the modeling approaches and the existing literature. Regarding the post-crisis LIBOR models we refer to Bianchetti and Morini [3] and Grbac and Runggaldier [14].

In view of the high level of technical sophistication that LIBOR models have reached in today's literature and also of the new demands they are faced with, we propose an abstract perspective on LIBOR modeling in order to obtain:

- a unified view on different modeling approaches, such as the LIBOR market models, the Lévy forward price models and the affine LIBOR models;
- transparent conditions that guarantee:
  - positivity of bond prices and arbitrage-freeness—the fundamental model requirements;
  - martingality of the forward prices under their corresponding forward measures, which paves the way for change of numeraire techniques and tractable pricing formulas;
  - structure preservation under different forward measures, a feature that is beneficial in connection with change of numeraire techniques;
- the validity of further desirable model properties that lead to analytically tractable models.

This article is structured as follows: in Sect. 2 we introduce the main modeling objects and formalize model axioms as well as desirable model properties that entail computational tractability. In Sect. 3, we provide two general modeling approaches based on general semimartingales and generic functional forms for the evolution of rates, and derive sufficient conditions for the arbitrage-freeness of the models and for the forward price processes to be uniformly integrable martingales under their corresponding forward measures; positivity of bond prices holds by construction. On this basis we derive conditions that imply positivity of LIBOR rates and ensure computational tractability. As an interesting additional insight we show that essentially only models in which the forward price processes are exponentials of an affine function of a semimartingale are structure preserving under different forward measures. In Sect. 4, we present several LIBOR models in the guise of the general modeling framework and investigate sufficient conditions that lead to further essential model features. Finally, required results from semimartingale theory are derived in the appendix.

2 Axioms and Desirable Properties

Let $(\Omega, \mathcal{F} = \mathcal{F}_{T_*}, \mathbb{F} = (\mathcal{F}_t)_{t\in[0,T_*]}, \mathbb{P}_N)$ denote a complete stochastic basis in the sense of Jacod and Shiryaev [15, Def. I.1.3], where T_* denotes a finite time horizon. Consider a discrete tenor structure $\mathcal{T} := \{0 = T_0 < \ldots < T_N \leq T_*\}$ with $\delta_k = T_k - T_{k-1}$ for $k \in \mathcal{K} := \{1, \ldots, N\}$, and define $\bar{\mathcal{K}} := \mathcal{K} \setminus \{N\}$. We assume that zero-coupon bonds with maturities $T_1, \ldots, T_N$ are traded in the market and denote by $B(t, T_k)$ the time-t price of the zero-coupon bond with maturity T_k, for all $k \in \mathcal{K}$. We associate to each date T_k the *numeraire pair* $(B(\cdot, T_k), \mathbb{P}_k)$, meaning that bond prices discounted by the numeraire $B(\cdot, T_k)$ are $\mathbb{P}_k$-local martingales, for all $k \in \mathcal{K}$. The measures $\mathbb{P}_k$ are then called *forward (martingale) measures*. Moreover, let $\mathcal{M}_{\text{loc}}(\mathbb{P})$ denote the set of local martingales with respect to the measure $\mathbb{P}$.

The *forward LIBOR rate*, denoted by $L(t, T_k)$, is a discretely compounded interest rate determined at time t for the future accrual interval $[T_k, T_{k+1}]$. It is related to bond prices via

$$L(t, T_k) = \frac{1}{\delta_k}\left(\frac{B(t, T_k)}{B(t, T_{k+1})} - 1\right), \qquad t \in [0, T_k], \tag{1}$$

for $k \in \bar{\mathcal{K}}$. The *forward price process* $F(\cdot, T_k, T_n)$ is defined as follows

$$F(t, T_k, T_n) = \frac{B(t, T_k)}{B(t, T_n)}, \qquad t \in [0, T_k \wedge T_n], \tag{2}$$

for all $k, n \in \mathcal{K}$. The forward LIBOR rate $L(t, T_k)$ and the forward price $F(t, T_k, T_{k+1})$ are connected via

$$F(t, T_k, T_{k+1}) = 1 + \delta_k L(t, T_k). \tag{3}$$

We will describe in the sequel several axioms and properties that LIBOR models should possess in order to be economically meaningful on the one hand, and applicable in practice on the other. In particular, we will distinguish between three different groups of attributes. The first group consists of *necessary axioms*, which are needed to build a sound financial model. These are:

(𝔸1) Bond prices are *positive*, i.e. $B(\cdot, T_k) > 0$ for all $k \in \mathcal{K}$;
(𝔸2) The model is *arbitrage-free*, i.e. $\frac{B(\cdot,T_k)}{B(\cdot,T_N)} \in \mathcal{M}_{\text{loc}}(\mathbb{P}_N)$ for all $k \in \mathcal{K}$.

The first axiom is justified since bond prices are traded assets with a positive payoff, thus should have a positive price. The second axiom precludes the existence of arbitrage opportunities and could be equivalently formulated under any forward measure, i.e. the model is arbitrage-free if $\frac{B(\cdot,T_k)}{B(\cdot,T_n)} \in \mathcal{M}_{\text{loc}}(\mathbb{P}_n)$ for all $k \in \mathcal{K}$ and some $n \in \mathcal{K}$; see also Musiela and Rutkowski [28, §14.1.3] and Klein et al. [22].

The second group consists of *tractability properties*, which simplify computations in the model. Out of several possible choices, we will concentrate on the following:

($\mathbb{B}$1) Forward prices are *true martingales*, i.e. $\frac{B(\cdot,T_k)}{B(\cdot,T_N)} \in \mathcal{M}(\mathbb{P}_N)$ for all $k \in \mathcal{K}$.
($\mathbb{B}$2) The model is *structure preserving*, i.e. the semimartingale characteristics of the driving process are transformed in a deterministic way under forward measures.
($\mathbb{B}$3) Each LIBOR rate is a *Markov process* under its corresponding forward measure.
($\mathbb{B}$4) The initial LIBOR rates are direct model inputs.

These properties are not necessary to build an arbitrage-free model, but are very convenient in several aspects. The first property allows to compute option prices as conditional expectations and to relate the forward measures via a density process. Hence, several option pricing formulas can be simplified considerably by changing to a more convenient forward measure. The second property yields that the processes driving each LIBOR rate remain in the same class of processes under each forward measure. Moreover, ($\mathbb{B}$1) combined with ($\mathbb{B}$2) typically allows to derive closed-form or semi-analytical pricing formulas for liquid products such as caps and swaptions. ($\mathbb{B}$3) also allows to simplify certain option pricing problems and to use PDE methods. Additionally, if the initial term structure is a direct input in the model, i.e. ($\mathbb{B}$4) holds, then we avoid using a numerical procedure to fit the currently observed bond prices.

Finally, we shall also discuss the following model property:

($\mathbb{C}$) LIBOR rates are always non-negative.

Until the recent financial crisis, LIBOR rates were always non-negative, hence the possibility of rates becoming negative has been considered as a drawback of a model. As a consequence, several LIBOR models have been designed to produce non-negative LIBOR rates. Nowadays, the quoted LIBOR rates are at extremely low levels and even negative LIBOR rates for several tenors have been reported over longer time periods, which prompts us to take this into account in the modeling. Therefore, it is important to know which models allow for negative rates as well as which of the existing models for positive rates can easily be adapted to allow the rates to go below zero. Moreover, the techniques used to construct non-negative LIBOR rates can often be adapted to model other related non-negative quantities such as spreads in multiple curve models.

3 A Unified Construction of LIBOR Models

Models for the evolution of LIBOR rates are constructed in the literature either using a backward induction approach, where rates are specified successively under different forward measures, or by modeling all rates simultaneously under one measure, typically the terminal forward measure. The former approach has been used for the construction of LIBOR market models and forward price models, while the latter is used for affine LIBOR models and Markov functional models. The aim of this section is to offer a unified construction of LIBOR models by emphasizing the common features in both approaches.

3.1 *Modeling Rates via Backward Induction*

The aim of this section is to formulate sufficient conditions and to present a generic construction of LIBOR (market) models using the backward induction approach. The driving process is a general semimartingale and the functional form of the dynamics is also generic.

The following key observations of Musiela and Rutkowski [27] lie at the heart of the constructions via backward induction:

- A model for the LIBOR rates $(L(\cdot, T_k))_{k\in\bar{\mathcal{K}}}$ is *arbitrage-free* if $L(\cdot, T_k)$ is a $\mathbb{P}_{k+1}$-local martingale for all $k \in \bar{\mathcal{K}}$.
- The *forward measures* $(\mathbb{P}_k)_{k\in\bar{\mathcal{K}}}$ are related via the Radon-Nikodym derivatives

$$\frac{\mathrm{d}\mathbb{P}_k}{\mathrm{d}\mathbb{P}_{k+1}} = \frac{1 + \delta_k L(T_k, T_k)}{1 + \delta_k L(0, T_k)}, \qquad \text{for all } k \in \bar{\mathcal{K}}. \tag{4}$$

Therefore, in order to construct a LIBOR model it suffices to specify the dynamics either of the LIBOR rate $L(\cdot, T_k)$ itself or of the forward price process $F(\cdot, T_k, T_{k+1}) = 1 + \delta L(\cdot, T_k)$ for all $k \in \bar{\mathcal{K}}$, and both choices determine the densities in (4) as well.

Our construction is based on specifying an exponential semimartingale for the dynamics of the forward price process with the following functional form:

$$F(\cdot, T_k, T_{k+1}) = \mathrm{e}^{f^k(\cdot, X)}, \tag{5}$$

where f^k are functions for each $k \in \bar{\mathcal{K}}$ and X is a semimartingale. This approach unifies the construction of the LIBOR market models and the forward price models by appropriate choices of f^k and X that will be discussed in Sect. 4.

Consider an $\mathbb{R}^d$-valued semimartingale $X = (X_t)_{0\leq t\leq T_N}$ on $(\Omega, \mathcal{F}, \mathbb{F}, \mathbb{P}_N)$ and a collection of functions $f^k : [0, T_N] \times \mathbb{R}^d \to \mathbb{R}$ for all $k \in \bar{\mathcal{K}}$, which satisfy the following assumptions:

($\mathbb{LIP}$) The function f^k belongs to $C^{1,2}([0, T_N] \times \mathbb{R}^d)$ and is globally Lipschitz, i.e.

$$|f^k(t, x) - f^k(t, y)| \leq K^k |x - y|,$$

for every $t \in [0, T_N]$ and any $x, y \in \mathbb{R}^d$, where $K^k > 0$ is a constant.

($\mathbb{INT}$) The process X is an $\mathbb{R}^d$-valued semimartingale with absolutely continuous characteristics (b^N, c^N, F^N) under $\mathbb{P}_N$, such that the following conditions hold

$$\int_0^{T_N}\int_{\mathbb{R}^d} \left\{|x|^2 1_{\{|x|\le 1\}} + |x| e^{K|x|} 1_{\{|x|>1\}}\right\} F_t^N(\mathrm{d}x)\mathrm{d}t < C_1 \tag{6}$$

and

$$\int_0^{T_N} \|c_t^N\| \mathrm{d}t < C_2, \tag{7}$$

for some constants $C_1, C_2 > 0$ and $K = \sum_{k=1}^{N-1} K^k$.

We denote by $\|\cdot\|$ the Euclidean norm on $\mathbb{R}^d$ and by $\langle\cdot,\cdot\rangle$ the associated scalar product.

Remark 3.1 The characteristic triplet of the semimartingale X under the forward measure $\mathbb{P}_k$ is denoted by (b^k, c^k, F^k), while the truncation function can always be chosen the identity (i.e. $h(x) = x$) due to (6). Moreover, we use the standard conventions $\sum_\emptyset = 0$ and $\prod_\emptyset = 1$.

Theorem 3.2 *Consider an $\mathbb{R}^d$-valued semimartingale X and functions f^k such that Assumptions ($\mathbb{LIP}$) and ($\mathbb{INT}$) are satisfied for each $k \in \bar{\mathcal{K}}$. Assume that the forward price processes are modeled via*

$$F(t, T_k, T_{k+1}) = e^{f^k(t, X_t)}, \qquad t \in [0, T_k], \tag{8}$$

and the following drift condition *is satisfied*

$$\begin{aligned}
\langle \mathrm{D}f^k(t, X_{t-}), b_t^N\rangle = &-\frac{\mathrm{d}}{\mathrm{d}t} f^k(t, X_{t-}) - \frac{1}{2}\sum_{i,j=1}^{d} \mathrm{D}_{ij}^2 f^k(t, X_{t-})(c_t^N)^{ij} \\
&- \frac{1}{2}\langle \mathrm{D}f^k(t, X_{t-}), c_t^N \mathrm{D}f^k(t, X_{t-})\rangle \\
&- \sum_{j=k+1}^{N-1} \langle \mathrm{D}f^k(t, X_{t-}), c_t^N \mathrm{D}f^j(t, X_{t-})\rangle \qquad (\mathbb{DRIFT}) \\
- \int_{\mathbb{R}^d} &\left\{\left(e^{f^k(t, X_{t-}+x) - f^k(t, X_{t-})} - 1\right)\right. \\
&\left.\times \prod_{j=k+1}^{N-1} e^{f^j(t, X_{t-}+x) - f^j(t, X_{t-})} - \langle \mathrm{D}f^k(t, X_{t-}), x\rangle\right\} F_t^N(\mathrm{d}x)
\end{aligned}$$

for each $k \in \bar{\mathcal{K}}$. Then, the measures $(\mathbb{P}_k)_{k\in\bar{\mathcal{K}}}$ defined via

$$\frac{\mathrm{d}\mathbb{P}_k}{\mathrm{d}\mathbb{P}_{k+1}} = \frac{\mathrm{e}^{f^k(T_k, X_{T_k})}}{\mathrm{e}^{f^k(0, X_0)}} \tag{9}$$

are equivalent forward measures and the forward prices processes $F(\cdot, T_k, T_{k+1})$ are uniformly integrable martingales with respect to $\mathbb{P}_{k+1}$, for each $k \in \bar{\mathcal{K}}$. In particular, the model $(F(\cdot, T_k, T_{k+1}))_{k\in\bar{\mathcal{K}}}$ is arbitrage-free and satisfies Axioms $(\mathbb{A}1)$ and $(\mathbb{A}2)$, as well as Property $(\mathbb{B}1)$.

Proof The statement is proved via backward induction, motivated by the backward construction of LIBOR and forward price models.

First step: We start from the forward price process $F(\cdot, T_{N-1}, T_N)$ whose dynamics are

$$F(t, T_{N-1}, T_N) = \mathrm{e}^{f^{N-1}(t, X_t)}, \qquad t \in [0, T_{N-1}],$$

and examine its properties under the measure $\mathbb{P}_N$. The function f^{N-1} satisfies $(\mathbb{LIP})$ and the process X satisfies $(\mathbb{INT})$, hence the process $f^{N-1}(\cdot, X)$ is an exponentially special semimartingale by Proposition A.2. Using Proposition A.1, we have that $F(\cdot, T_{N-1}, T_N)$ is a $\mathbb{P}_N$-local martingale if the following condition holds:

$$\begin{aligned} &\langle \mathrm{D}f^{N-1}(t, X_{t-}), b_t^N \rangle \\ &= -\frac{\mathrm{d}}{\mathrm{d}t} f^{N-1}(t, X_{t-}) - \frac{1}{2} \sum_{i,j=1}^{d} \mathrm{D}_{ij}^2 f^{N-1}(t, X_{t-}) (c_t^N)^{ij} \\ &\quad - \frac{1}{2} \langle \mathrm{D}f^{N-1}(t, X_{t-}), c^N \mathrm{D}f^{N-1}(t, X_{t-}) \rangle \\ &\quad - \int_{\mathbb{R}^d} \left(\mathrm{e}^{f^{N-1}(t, X_{t-}+x) - f^{N-1}(t, X_{t-})} - 1 - \langle \mathrm{D}f^{N-1}(t, X_{t-}), x \rangle \right) F_t^N(\mathrm{d}x), \end{aligned} \tag{10}$$

which is actually $(\mathbb{DRIFT})$ for $k = N - 1$. Moreover, Proposition A.2 yields that $F(\cdot, T_{N-1}, T_N)$ is even a $\mathbb{P}_N$-uniformly integrable martingale. Therefore, we can use $F(\cdot, T_{N-1}, T_N)$ as a density process to define the measure $\mathbb{P}_{N-1}$ via

$$\left. \frac{\mathrm{d}\mathbb{P}_{N-1}}{\mathrm{d}\mathbb{P}_N} \right|_{\mathcal{F}_\cdot} = \frac{F(\cdot, T_{N-1}, T_N)}{F(0, T_{N-1}, T_N)} = \frac{\mathrm{e}^{f^{N-1}(\cdot, X)}}{\mathrm{e}^{f^{N-1}(0, X_0)}},$$

and the characteristics of the process X under the measure $\mathbb{P}_{N-1}$ are provided by

$$\begin{aligned}
b_t^{N-1} &= b_t^N + c_t^N \mathrm{D} f^{N-1}(t, X_{t-}) \\
&\quad + \int_{\mathbb{R}^d} \left(\mathrm{e}^{f^{N-1}(t,X_{t-}+x)-f^{N-1}(t,X_{t-})} - 1\right) x F_t^N(\mathrm{d}x) \\
c_t^{N-1} &= c_t^N \\
F_t^{N-1}(\mathrm{d}x) &= \mathrm{e}^{f^{N-1}(t,X_{t-}+x)-f^{N-1}(t,X_{t-})} F_t^N(\mathrm{d}x);
\end{aligned}$$

cf. Lemma A.4.

Then, we proceed backwards by considering the 'next' forward price process $F(\cdot, T_{N-2}, T_{N-1})$ with dynamics

$$F(t, T_{N-2}, T_{N-1}) = \mathrm{e}^{f^{N-2}(t,X_t)}, \qquad t \in [0, T_{N-2}], \tag{11}$$

and verifying that subject to ($\mathbb{LIP}$), ($\mathbb{INT}$) and ($\mathbb{DRIFT}$) it is a $\mathbb{P}_{N-1}$-uniformly integrable martingale. Thus, it can be used as a density process to define the measure $\mathbb{P}_{N-2}$.

Next, we provide the general step of the backward induction.

General step: Let $k \in \{1, \dots, N-1\}$ be fixed and consider the process X, the functions $f^{k+1}, \dots, f^{N-1}$ and the measures $\mathbb{P}_{k+1}, \dots, \mathbb{P}_N$ which are defined recursively via

$$\frac{\mathrm{d}\mathbb{P}_{k+1}}{\mathrm{d}\mathbb{P}_{k+2}}\bigg|_{\mathcal{F}_\cdot} = \frac{F(\cdot, T_{k+1}, T_{k+2})}{F(0, T_{k+1}, T_{k+2})} = \frac{\mathrm{e}^{f^{k+1}(\cdot,X)}}{\mathrm{e}^{f^{k+1}(0,X_0)}}.$$

Assume that the forward price processes $F(\cdot, T_l, T_{l+1})$ have been modeled as exponential semimartingales according to (8) and are $\mathbb{P}_{l+1}$-uniformly integrable martingales, for all $l \in \{k+1, \dots, N-1\}$. By repeatedly applying Lemma A.4, we derive the $\mathbb{P}_{k+1}$-characteristics of X, which have the form

$$\begin{aligned}
b_t^{k+1} &= b_t^N + c_t^N \sum_{j=k+1}^{N-1} \mathrm{D} f^j(t, X_{t-}) \\
&\quad + \int_{\mathbb{R}^d} \left(\prod_{j=k+1}^{N-1} \mathrm{e}^{f^j(t,X_{t-}+x)-f^j(t,X_{t-})} - 1 \right) x F_t^N(\mathrm{d}x) \\
c_t^{k+1} &= c_t^N \\
F_t^{k+1}(\mathrm{d}x) &= \prod_{j=k+1}^{N-1} \mathrm{e}^{f^j(t,X_{t-}+x)-f^j(t,X_{t-})} F_t^N(\mathrm{d}x).
\end{aligned} \tag{12}$$

Now, the forward price process $F(\cdot, T_k, T_{k+1})$ with dynamics

$$F(t, T_k, T_{k+1}) = \mathrm{e}^{f^k(t,X_t)}, \quad t \in [0, T_{k-1}],$$

is a $\mathbb{P}_{k+1}$-local martingale if the following condition holds

$$
\begin{aligned}
\langle \mathrm{D} f^k(t, X_{t-}), b_t^{k+1} \rangle
&= -\frac{\mathrm{d}}{\mathrm{d}t} f^k(t, X_{t-}) - \frac{1}{2} \sum_{i,j=1}^{d} \mathrm{D}^2_{ij} f^k(t, X_{t-}) (c_t^{k+1})^{ij} \\
&\quad - \frac{1}{2} \langle \mathrm{D} f^k(t, X_{t-}), c^{k+1} \mathrm{D} f^k(t, X_{t-}) \rangle \qquad (13) \\
&\quad - \int_{\mathbb{R}^d} \left(\mathrm{e}^{f^k(t, X_{t-}+x) - f^k(t, X_{t-})} - 1 - \langle \mathrm{D} f^k(t, X_{t-}), x \rangle \right) F_t^{k+1}(\mathrm{d}x);
\end{aligned}
$$

cf. Proposition A.1. By replacing (12) into (13) we see, after some straightforward calculations, that the latter is equivalent to the ($\mathbb{DRIFT}$) condition. We can also verify that conditions (A.4) and (A.5) from Proposition A.2 hold for the function f^k that satisfies ($\mathbb{LIP}$) and the process X that satisfies ($\mathbb{INT}$). Indeed, we have

$$
\int_0^{T_N} \| c_t^{k+1} \| \mathrm{d}t = \int_0^{T_N} \| c_t^N \| \mathrm{d}t < C_2,
$$

hence condition (A.5) holds. Moreover, using ($\mathbb{LIP}$) and ($\mathbb{INT}$) we get that

$$
\begin{aligned}
&\int_0^{T_N} \int_{\mathbb{R}^d} (|x|^2 \wedge 1) F_t^{k+1}(\mathrm{d}x)\mathrm{d}t + \int_0^{T_N} \int_{|x|>1} |x| \mathrm{e}^{K^k |x|} F_t^{k+1}(\mathrm{d}x)\mathrm{d}t \\
&\quad = \int_0^{T_N} \int_{\mathbb{R}^d} (|x|^2 \wedge 1) \prod_{j=k+1}^{N-1} \mathrm{e}^{f^j(t, X_{t-}+x) - f^j(t, X_{t-})} F_t^N(\mathrm{d}x)\mathrm{d}t \\
&\qquad + \int_0^{T_N} \int_{|x|>1} |x| \mathrm{e}^{K^k |x|} \prod_{j=k+1}^{N-1} \mathrm{e}^{f^j(t, X_{t-}+x) - f^j(t, X_{t-})} F_t^N(\mathrm{d}x)\mathrm{d}t \\
&\quad \le \int_0^{T_N} \int_{\mathbb{R}^d} \left\{ (|x|^2 \wedge 1) \mathrm{e}^{\sum_{j=k+1}^{N-1} K^j |x|} + 1_{\{|x|>1\}} |x| \mathrm{e}^{\sum_{j=k}^{N-1} K^j |x|} \right\} F_t^N(\mathrm{d}x)\mathrm{d}t \\
&\quad \le \mathrm{const} \cdot \int_0^{T_N} \int_{\mathbb{R}^d} \left\{ |x|^2 1_{\{|x| \le 1\}} + |x| \mathrm{e}^{K|x|} 1_{\{|x|>1\}} \right\} F_t^N(\mathrm{d}x)\mathrm{d}t \\
&\quad < C_1,
\end{aligned}
$$

where the second to last inequality holds because the exponential function is bounded in the unit hypercube and the Lipschitz constants are positive. Hence, condition (A.4) holds as well. Thus, Proposition A.2 yields that the process $f^k(\cdot, X)$ is exponentially special and the forward price process $F(\cdot, T_k, T_{k+1})$ is a $\mathbb{P}_{k+1}$-uniformly integrable martingale.

Therefore, exactly as in the previous steps we can use the $\mathbb{P}_{k+1}$-uniformly integrable martingale $f^k(\cdot, X)$ to define the measure $\mathbb{P}_k$ via

$$\frac{\mathrm{d}\mathbb{P}_k}{\mathrm{d}\mathbb{P}_{k+1}}\bigg|_{\mathcal{F}_\cdot} = \frac{F(\cdot, T_k, T_{k+1})}{F(0, T_k, T_{k+1})} = \frac{\mathrm{e}^{f^k(\cdot, X)}}{\mathrm{e}^{f^k(0, X_0)}}.$$

Then, we can compute the $\mathbb{P}_k$-characteristics of X using Lemma A.4 and consider the 'next' forward price process with dynamics

$$F(\cdot, T_{k-1}, T_k) = \mathrm{e}^{f^{k-1}(\cdot, X^{k-1})}.$$

This procedure produces an arbitrage-free semimartingale model for the forward price process, and thus also for the LIBOR rate, if the ($\mathbb{DRIFT}$) condition holds for each $k \in \bar{\mathcal{K}}$.

Finally, we can easily show that the measures $\mathbb{P}_k$ are indeed forward measures, i.e. that $B(\cdot, T_l)/B(\cdot, T_k)$ is a $\mathbb{P}_k$-martingale for all $1 \leq k, l \leq N$. This follows directly from Proposition III.3.8 in Jacod and Shiryaev [15], using that

$$\frac{B(\cdot, T_l)}{B(\cdot, T_k)} \frac{\mathrm{d}\mathbb{P}_k}{\mathrm{d}\mathbb{P}_{l+1}} = \mathrm{e}^{f^l(\cdot, X)}$$

which is a $\mathbb{P}_{l+1}$-martingale. □

Remark 3.3 Let us point out that, although the true martingale property of the forward price process is not necessary to guarantee the absence of arbitrage, it is required in order to define the forward measures and to construct the model via backward induction. Aside from this, forward measures play a crucial role in term structure models since they allow to derive tractable formulas for interest rate derivatives. Indeed, the major advantage of forward measures for derivative pricing is that we can avoid the numerical computation of multidimensional integrals over joint distributions.

Remark 3.4 We may assume, if desired, that $d \geq N - 1$ in order to ensure there are at least as many driving factors as the number of forward price processes. Moreover, by suitable choices of the functions f^k we may select the components of X driving a certain forward price process. See, for example, Sect. 4.1, where we work with an $N - 1$-dimensional process X and set $f^k(x) = \hat{f}^k(x_k)$, for $x = (x_1, \ldots, x_{N-1})$ and $\hat{f}^k : \mathbb{R} \to \mathbb{R}$, i.e. each forward price process is driven by a different component of the process X.

3.2 Modeling Rates Under the Terminal Measure

Another possibility for constructing a model for the forward LIBOR rates, or equivalently the forward price processes, is to start with the family of forward price processes with respect to the terminal bond price $B(\cdot, T_N)$ in the tenor structure, i.e.

$$F(\cdot, T_k, T_N) = \frac{B(\cdot, T_k)}{B(\cdot, T_N)}$$

for all $k \in \bar{\mathcal{K}}$, and to model them simultaneously under the same measure, typically the terminal forward measure $\mathbb{P}_N$. Similarly to the previous section, the construction is based on specifying exponential semimartingale dynamics for the forward price process of the following functional form

$$F(\cdot, T_k, T_N) = \mathrm{e}^{g^k(\cdot, X^k)}, \tag{14}$$

where g^k are suitable functions and X^k are d-dimensional semimartingales, for $k \in \bar{\mathcal{K}}$.

Consider a collection of $\mathbb{R}^d$-valued semimartingales $X^k = (X^k_t)_{0 \le t \le T_N}$ on $(\Omega, \mathcal{F}, \mathbb{F}, \mathbb{P}_N)$ and a collection of functions $g^k : [0, T_N] \times \mathbb{R}^d \to \mathbb{R}$ for all $k \in \bar{\mathcal{K}}$, which satisfy the following assumptions:

($\mathbb{LIP}'$) The function g^k belongs to $C^{1,2}([0, T_N] \times \mathbb{R}^d)$ and is globally Lipschitz, i.e.

$$|g^k(t, x) - g^k(t, y)| \le \tilde{K}^k |x - y|,$$

for every $t \ge 0$ and any $x, y \in \mathbb{R}^d$, where $\tilde{K}^k > 0$ is a constant.

($\mathbb{INT}'$) The process X^k is an $\mathbb{R}^d$-valued semimartingale with absolutely continuous characteristics $(b^{k,N}, c^{k,N}, F^{k,N})$ under $\mathbb{P}_N$, such that the following conditions hold

$$\int_0^{T_N} \int_{\mathbb{R}^d} \left\{ |x|^2 1_{\{|x| \le 1\}} + |x| \mathrm{e}^{\tilde{K}^k |x|} 1_{\{|x| > 1\}} \right\} F^{k,N}_t(\mathrm{d}x)\mathrm{d}t < \tilde{C}^k_1 \tag{15}$$

and

$$\int_0^{T_N} \|c^{k,N}_t\| \mathrm{d}t < \tilde{C}^k_2, \tag{16}$$

for some constants $\tilde{C}^k_1, \tilde{C}^k_2 > 0$. Recall that the truncation can be chosen the identity.

Theorem 3.5 *Consider $\mathbb{R}^d$-valued semimartingales X^k and functions g^k such that Assumptions ($\mathbb{LIP}'$) and ($\mathbb{INT}'$) are satisfied for each $k \in \bar{\mathcal{K}}$. Assume that the forward price processes are modeled via*

$$F(t, T_k, T_N) = \mathrm{e}^{g^k(t, X^k_t)}, \quad t \in [0, T_k], \tag{17}$$

and the following drift condition *is satisfied*

$$
\begin{aligned}
\langle \mathrm{D}g^k(t, X^k_{t-}), b^{k,N}_t \rangle
&= -\frac{\mathrm{d}}{\mathrm{d}t} g^k(t, X^k_{t-}) - \frac{1}{2} \sum_{i,j=1}^{d} \mathrm{D}^2_{ij} g^k(t, X^k_{t-}) (c^{k,N}_t)^{ij} \\
&\quad - \frac{1}{2} \langle \mathrm{D}g^k(t, X^k_{t-}), c^{k,N}_t \mathrm{D}g^k(t, X^k_{t-}) \rangle \qquad (\mathbb{DRIFT}') \\
&\quad - \int_{\mathbb{R}^d} \Big(\mathrm{e}^{g^k(t, X^k_{t-}+x) - g^k(t, X^k_{t-})} - 1 - \langle \mathrm{D}g^k(t, X^k_{t-}), x \rangle \Big) F^{k,N}_t(\mathrm{d}x),
\end{aligned}
$$

for all $k \in \bar{\mathcal{K}}$. Then, the forward price processes are uniformly integrable martingales with respect to the terminal forward measure $\mathbb{P}_N$, for all $k \in \bar{\mathcal{K}}$. In particular, the model is arbitrage-free and satisfies Axioms ($\mathbb{A}1$) *and* ($\mathbb{A}2$), *as well as Property* ($\mathbb{B}1$).

Proof The proof is simpler compared to the proof of Theorem 3.2 because we work only under the terminal measure $\mathbb{P}_N$. Furthermore, we can work simultaneously with all forward price processes $F(\cdot, T_k, T_N)$ for each $k \in \bar{\mathcal{K}}$. More precisely, for all $k \in \bar{\mathcal{K}}$ the function g^k satisfies ($\mathbb{LIP}'$) and the process X^k satisfies ($\mathbb{INT}'$), hence the process $g^k(\cdot, X^k)$ is an exponentially special semimartingale by Proposition A.2. Using Proposition A.1, we get by virtue of the ($\mathbb{DRIFT}'$) condition that $F(\cdot, T_k, T_N)$ is a $\mathbb{P}_N$-local martingale. Moreover, Proposition A.2 yields that $F(\cdot, T_k, T_N)$ is actually a $\mathbb{P}_N$-uniformly integrable martingale. □

Remark 3.6 In this construction we can use a family of semimartingales X^k, $k \in \bar{\mathcal{K}}$, where each forward price process is driven by a different semimartingale. This is possible because we do not have to perform measure changes as we did in the backward construction, since all forward price processes are modeled under a common measure. Hence, at this stage, we do not need to know the dependence structure between the processes X^k which is necessary when applying Girsanov's theorem. However, for pricing purposes and also for linking the backward and the terminal measure constructions, we revert to a common $\mathbb{R}^d$-valued driving process X for which the dependence structure between its components is obviously fully known. Naturally, the dimension of the process X can be chosen such that each rate is driven by a different component of the process; compare with Remark 3.4.

Remark 3.7 Based on (14), we can immediately deduce the dynamics of the forward price process $F(\cdot, T_k, T_{k+1})$ and the forward LIBOR rate $L(\cdot, T_k)$, for all $k \in \bar{\mathcal{K}}$. Using that

$$
1 + \delta L(\cdot, T_k) = F(\cdot, T_k, T_{k+1}) = \frac{F(\cdot, T_k, T_N)}{F(\cdot, T_{k+1}, T_N)},
$$

we obtain that

$$
1 + \delta L(\cdot, T_k) = F(\cdot, T_k, T_{k+1}) = \mathrm{e}^{g^k(\cdot, X^k) - g^{k+1}(\cdot, X^{k+1})}. \tag{18}
$$

Remark 3.8 Assumptions ($\mathbb{LIP}'$) and ($\mathbb{INT}'$) are sufficient to produce an arbitrage-free family of LIBOR rates, but they are by no means necessary. Indeed, we can weaken them slightly by assuming that the functions g^k satisfy ($\mathbb{LIP}'$) and the processes X^k have finite exponential moments. Then the previous theorem yields an arbitrage-free model that satisfies Axioms ($\mathbb{A}1$) and ($\mathbb{A}2$), but not necessarily ($\mathbb{B}1$). However, as pointed out also in Remark 3.3, the latter is needed to define forward measures, which are very useful because they typically lead to tractable pricing formulas.

Remark 3.9 Let us consider the case where all semimartigales X^k coincide, i.e. $X^k \equiv X$ for all $k \in \bar{\mathcal{K}}$. Then, we can easily link the approach using backward induction presented in Sect. 3.1 and the approach under the terminal measure presented in this section. More precisely, starting from a family of functions g^k, $k \in \bar{\mathcal{K}}$, and a semimartingale X satisfying ($\mathbb{LIP}'$), ($\mathbb{INT}'$) and ($\mathbb{DRIFT}'$), we define

$$f^k(t,x) := g^k(t,x) - g^{k+1}(t,x). \tag{19}$$

The functions f^k obviously satisfy ($\mathbb{LIP}$) with the constants $K^k := \tilde{K}^k + \tilde{K}^{k+1}$. Assume moreover that the semimartingale X satisfies ($\mathbb{INT}$) with K^k as above. Then the model for the terminal forward prices given by (14) can be equivalently written as

$$F(\cdot, T_k, T_{k+1}) = \mathrm{e}^{f^k(\cdot, X)}, \qquad k \in \bar{\mathcal{K}},$$

with f^k given by (19) and all assertions of Theorem 3.2 remain valid.

Conversely, assuming that a model for the forward prices (8) is given via a family of functions f^k, $k \in \bar{\mathcal{K}}$, and a semimartingale X satisfying ($\mathbb{LIP}$), ($\mathbb{INT}$) and ($\mathbb{DRIFT}$), we define

$$g^k(t,x) := \sum_{j=k}^{N-1} f^j(t,x). \tag{20}$$

The functions g^k satisfy condition ($\mathbb{LIP}$) with the constants $\tilde{K}^k := \sum_{j=k}^{N-1} K^j$. Assuming furthermore that the semimartingale X satisfies ($\mathbb{INT}'$) with $\tilde{K}^k$ as above, the model for the forward prices (8) can be equivalently written as

$$F(\cdot, T_k, T_N) = \mathrm{e}^{g^k(\cdot, X)}, \qquad k \in \bar{\mathcal{K}},$$

with g^k defined in (20). This easily follows from the following telescopic product

$$F(\cdot, T_k, T_N) = \frac{B(\cdot, T_k)}{B(\cdot, T_N)} = \prod_{j=k}^{N-1} \frac{B(\cdot, T_j)}{B(\cdot, T_{j+1})} = \prod_{j=k}^{N-1} F(\cdot, T_j, T_{j+1}). \tag{21}$$

Thus, we conclude that Theorem 3.5 is valid for the semimartingale X and the functions g^k, $k \in \bar{\mathcal{K}}$.

3.3 Observations and Ramifications

Next, we discuss further properties of the models constructed in the previous two sections. In particular, we derive conditions such that a LIBOR model is structure preserving and produces non-negative rates. In order to provide a unified treatment of both modeling approaches, we assume that $X^k \equiv X$ in Sect. 3.2, for all $k \in \bar{\mathcal{K}}$.

Lemma 3.10 (i) *If the functions f^k are non-negative for all $k \in \bar{\mathcal{K}}$, then the LIBOR rates in the model* (8) *are non-negative, i.e. Property* ($\mathbb{C}$) *is satisfied.*

(ii) *If the functions g^k are non-negative and such that $g^k(t,x) \geq g^{k+1}(t,x)$ for all $k \in \bar{\mathcal{K}}$ and all $(t,x) \in [0,T_N] \times \mathbb{R}^d$, then the LIBOR rates in the model* (17) *are non-negative, i.e. Property* ($\mathbb{C}$) *is satisfied.*

Proof This follows directly from the relation between forward prices and LIBOR rates, see (3) and (18). □

The second tractability property ($\mathbb{B}$2) states that a LIBOR model is *structure preserving* if the characteristics of the driving process are transformed in a deterministic way under different forward measures, which ensures that the driving processes remain in the same class under all forward measures. In order to formalize the statement, we consider the following assumption.

($\mathbb{E}$) Let $U := \mathbb{R}^d$ (respectively $U := \mathbb{R}^d_+$). The measure $\mathbb{P}_N^{X_{t-}}$ is absolutely continuous with positive Lebesgue density on U, for all $t \in [0,T_N]$.

We say that a LIBOR model is structure preserving if the tuple (β^l, Y^l) defining the change of measure from the forward measure $\mathbb{P}_l$ to $\mathbb{P}_{l-1}$, for $l = N,\dots,1$, via Girsanov's theorem as in Lemma A.4, is deterministic.

Notice that under Assumption ($\mathbb{E}$), β^l is deterministic if and only if Y^l is so; indeed, if $Y^l(t,x) = \mathrm{e}^{f^l(t,X_{t-}+x)-f^l(t,X_{t-})}$ is assumed to be deterministic, the function f^l must satisfy

$$f^l(t,y+x) - f^l(t,y) = h^l(t,x) \tag{22}$$

for every $x, y \in U$, for some function h^l. Taking derivatives with respect to y, we get that

$$\mathrm{D}f^l(t,y+x) = \mathrm{D}f^l(t,y)$$

for every $x, y \in U$, hence $\mathrm{D}f^l(\cdot,y)$ is constant, and thus $f^l(t,\cdot)|_U$ is an affine function. This implies that $\beta_t^l = \mathrm{D}f^l(t,X_{t-})$ is deterministic.

Conversely, assume that the variable $\beta_t^l = \mathrm{D}f^l(t,X_{t-})$ is deterministic. Since the support of $\mathbb{P}_N^{X_{t-}}$ is U, $\mathrm{D}f^l$ is continuous and $\mathbb{P}_N^{X_{t-}}$ has a positive Lebesgue measure on U, we obtain that $\mathrm{D}f^l$ is constant and hence f^l is affine in the second variable. Thus, we conclude that $Y^l(t,x)$ is deterministic.

The next result provides necessary and sufficient conditions for ($\mathbb{B}2$) to be satisfied.

Proposition 3.11 *If the functions f^k and g^k are affine in the second variable for every $k \in \bar{\mathcal{K}}$, then the LIBOR models in* (8) *and* (17) *are structure preserving. Conversely, assume* ($\mathbb{E}$). *If the LIBOR models in* (8) *and* (17) *are structure preserving, then the functions $f^k|_{[0,T_N]\times U}$ and $g^k|_{[0,T_N]\times U}$ are affine in the second variable for every $k \in \bar{\mathcal{K}}$.*

Proof We will concentrate on the model constructed by backward induction, while the other one follows analogously. Following the argumentation in the proof of Theorem 3.2, the characteristics of X under $\mathbb{P}_l$ have the following form

$$\begin{aligned} b_t^l &= b_t^N + c_t^N \sum_{j=l+1}^{N-1} \mathrm{D}f^j(t, X_{t-}) \\ &\quad + \int_{\mathbb{R}^d} \left(\prod_{j=l+1}^{N-1} \mathrm{e}^{f^j(t,X_{t-}+x)-f^j(t,X_{t-})} - 1 \right) x F_t^N(\mathrm{d}x) \\ c_t^l &= c_t^N \\ F_t^l(\mathrm{d}x) &= \prod_{j=l+1}^{N-1} \mathrm{e}^{f^j(t,X_{t-}+x)-f^j(t,X_{t-})} F_t^N(\mathrm{d}x). \end{aligned} \tag{23}$$

Assume that the function $f^k(t,x)$ is affine in x, i.e. there exist $\alpha^k(t)$ and $\beta^k(t)$ such that $f^k(t,x) = \alpha^k(t) + \langle \beta^k(t), x \rangle$, then we can easily deduce that (b^l, c^l, F^l) in (23) is only a deterministic transformation of (b^N, c^N, F^N).

The converse statement is already implied by the arguments preceding this Proposition. □

The statement of Proposition 3.11 can be generalized to allow for more general driving processes. Assumption ($\mathbb{E}$), for instance, can be formulated for more general sets U. As an example, X could be a process that is positive in some coordinate and real or negative in another. On the other hand, processes with fixed jump sizes, such as the Poisson process, require a slightly different approach than in the proof above, taking care of the state space of the process and the support of the jump measure.

The following remark summarizes further interesting properties of LIBOR rates that can be easily deduced from this general modeling framework.

Remark 3.12 If the function f^k is affine in the second argument, i.e.

$$f^k(t,x) = \alpha^k(t) + \langle \beta^k(t), x \rangle, \tag{24}$$

with functions $\alpha^k, \beta^k \in C^1(\mathbb{R}_+)$ and the process is required to satisfy

$$F(\cdot, T_k, T_{k+1}) \geq 1,$$

i.e. to produce non-negative LIBOR rates, then the process X has to be bounded from below.

4 Examples

4.1 LIBOR Market Models

We start by revisiting the class of LIBOR market models in view of the general framework developed in the previous section. We will concentrate on the Lévy LIBOR model of Eberlein and Özkan [10] in order to fix ideas and processes, and as a representative of other LIBOR market models which fit in this framework as well, such as models with local volatility, stochastic volatility or driven by jump-diffusions. See, among many other references, Brigo and Mercurio [5], Schoenmakers [31], Glasserman [12], and Andersen and Piterbarg [1].

We assume that the driving process X is an $\mathbb{R}^{N-1}$-valued semimartingale of the form

$$X = B + \Lambda \cdot L, \tag{25}$$

where L is an $\mathbb{R}^n$-valued time-inhomogeneous Lévy process with characteristic triplet $(0, c^L, F^L)$ under the terminal measure $\mathbb{P}_N$ with respect to the truncation function $h(x) = x$, and $\Lambda = [\lambda(\cdot, T_1), \ldots, \lambda(\cdot, T_{N-1})]$ is an $(N-1) \times n$ volatility matrix where, for every $k \in \bar{\mathcal{K}}$, $\lambda(\cdot, T_k)$ is a deterministic, n-dimensional function. Moreover, $\Lambda \cdot L$ denotes the Itô stochastic integral of Λ with respect to L, while the drift term $B = \int_0^{\cdot} b(s)\mathrm{d}s = \left(\int_0^{\cdot} b(s, T_1)\mathrm{d}s, \ldots, \int_0^{\cdot} b(s, T_{N-1})\mathrm{d}s\right)$ is an $(N-1)$-dimensional stochastic process. We further assume that the following exponential moment condition is satisfied:

($\mathbb{EM}$) Let $\varepsilon > 0$ and $M > 0$, then

$$\int_0^{T_N} \int_{|x|>1} \mathrm{e}^{\langle u, x\rangle} F_s^L(\mathrm{d}x)\mathrm{d}s < \infty \qquad \text{for all } u \in [-(1+\varepsilon)M, (1+\varepsilon)M]^n;$$

while the volatility functions satisfy:

($\mathbb{VOL}$) The volatility $\lambda(\cdot, T_k) : [0, T_N] \to \mathbb{R}_+^n$ is a deterministic, bounded function such that for $s > T_k$, $\lambda(s, T_k) = 0$, for every $k \in \bar{\mathcal{K}}$. Moreover,

$$\sum_{k=1}^{N-1} \lambda^j(s, T_k) \leq M, \tag{26}$$

for every $s \in [0, T_{N-1}]$ and every coordinate $j \in \{1, \dots, n\}$.

The construction of the Lévy LIBOR model will follow the backward induction approach of Sect. 3.1. Define, for all $k \in \bar{\mathcal{K}}$ and $x = (x_1, \dots, x_{N-1}) \in \mathbb{R}^{N-1}$, the functions

$$f^k(t, x) := \log\left(1 + \delta_k L(0, T_k) \mathrm{e}^{x_k}\right) \tag{27}$$

and set

$$F(t, T_k, T_{k+1}) = \mathrm{e}^{f^k(t, X_t)}, \qquad k \in \bar{\mathcal{K}}. \tag{28}$$

Then, it follows easily that

$$F(t, T_k, T_{k+1}) = 1 + \delta_k L(0, T_k) \mathrm{e}^{X_t^k},$$

which coincides with the dynamics of the Lévy LIBOR model of Eberlein and Özkan [10], that are provided by

$$L(t, T_k) = L(0, T_k) \exp\left(\int_0^t b(s, T_k) \mathrm{d}s + \int_0^t \lambda(s, T_k) \mathrm{d}L_s\right). \tag{29}$$

The function f^k is Lipschitz continuous with constant 1, hence condition ($\mathbb{LIP}$) is satisfied with $K^k = 1$ for each $k \in \bar{\mathcal{K}}$. Moreover, thanks to Assumptions ($\mathbb{EM}$) and ($\mathbb{VOL}$), condition ($\mathbb{INT}$) is also satisfied for every $k \in \bar{\mathcal{K}}$. Therefore, an application of Theorem 3.2 yields the drift $B^k = \int_0^{\cdot} b(s, T_k) \mathrm{d}s$ of this model under the terminal measure. More precisely, we have that

$$\partial_{x_k} f^k(t, x) = \frac{\delta_k L(0, T_k) \mathrm{e}^{x_k}}{1 + \delta_k L(0, T_k) \mathrm{e}^{x_k}} =: \ell^k(x_k),$$

and $\partial_{x_j} f^k(t, x) = 0$, for $j \neq k$, while also $\frac{\mathrm{d}}{\mathrm{d}t} f^k(t, x) = 0$. Moreover,

$$\partial_{x_k x_k} f^k(t, x) = \frac{\delta_k L(0, T_k) \mathrm{e}^{x_k}}{(1 + \delta_k L(0, T_k) \mathrm{e}^{x_k})^2},$$

and $\partial_{x_i x_j} f^k(t, x) = 0$, for all $(i, j) \neq (k, k)$. According to Proposition 2.4 in Kallsen [18], the $\mathbb{P}_N$-characteristics (b^N, c^N, F^N) of X are given by

$$
\begin{aligned}
b_t^N &= b(t) \\
c_t^N &= \big\langle \Lambda(t), c_t^L \Lambda(t) \big\rangle \\
F_t^N(A) &= \int_{\mathbb{R}^n} 1_A(\Lambda(t)x) F_t^L(\mathrm{d}x), \qquad A \in \mathcal{B}(\mathbb{R}^{N-1}) \setminus \{0\},
\end{aligned} \tag{30}
$$

hence the ($\mathbb{DRIFT}$) condition from Theorem 3.2 becomes

$$
\begin{aligned}
\ell^k(X_{t-}^k) b(t, T_k) = &-\frac{1}{2} \frac{\ell^k(X_{t-}^k)}{1 + \delta_k L(0, T_k) \mathrm{e}^{X_{t-}^k}} \langle \lambda(t, T_k), c_t^L \lambda(t, T_k) \rangle \\
&- \frac{1}{2} (\ell^k(X_{t-}^k))^2 \langle \lambda(t, T_k), c_t^L \lambda(t, T_k) \rangle \\
&- \sum_{j=k+1}^{N-1} \ell^k(X_{t-}^k) \ell^j(X_{t-}^j) \langle \lambda(t, T_k), c_t^L \lambda(t, T_j) \rangle \\
&- \int_{\mathbb{R}^{N-1}} \Big[\Big(\mathrm{e}^{f^k(X_{t-}^k + x_k) - f^k(X_{t-}^k)} - 1 \Big) \prod_{j=k+1}^{N-1} \Big(\mathrm{e}^{f^j(X_{t-}^j + x_j) - f^j(X_{t-}^k)} \Big) \\
&\qquad - \ell^k(X_{t-1}^k) x_k \Big] F_t^X(\mathrm{d}x).
\end{aligned} \tag{31}
$$

Notice that

$$
\begin{aligned}
\frac{\ell^k(X_{t-}^k)}{1 + \delta_k L(0, T_k) \mathrm{e}^{X_{t-}^k}} + (\ell^k(X_{t-}^k))^2 &= \frac{\delta_k L(0, T_k) \mathrm{e}^{X_{t-}^k} + (\delta_k L(0, T_k) \mathrm{e}^{X_{t-}^k})^2}{(1 + \delta_k L(0, T_k) \mathrm{e}^{X_{t-}^k})^2} \\
&= \frac{\delta_k L(0, T_k) \mathrm{e}^{X_{t-}^k}}{1 + \delta_k L(0, T_k) \mathrm{e}^{X_{t-}^k}} = \ell(X_{t-}^k)
\end{aligned}
$$

and that, for all $j = k, \dots, N-1$,

$$
\begin{aligned}
\mathrm{e}^{f^j(X_{t-}^j + x_j) - f^j(X_{t-}^k)} &= \frac{1 + \delta_j L(0, T_j) \mathrm{e}^{X_{t-}^j + x_j}}{1 + \delta_j L(0, T_j) \mathrm{e}^{X_{t-}^j}} \\
&= \frac{1 + \delta_j L(0, T_j) \mathrm{e}^{X_{t-}^j} + \delta_j L(0, T_j) \mathrm{e}^{X_{t-}^j} (\mathrm{e}^{x_j} - 1)}{1 + \delta_j L(0, T_j) \mathrm{e}^{X_{t-}^j}} \\
&= 1 + \ell^j(X_{t-}^j)(\mathrm{e}^{x_j} - 1).
\end{aligned}
$$

Inserting the above simplifications into (31) yields

$$b(t,T_k) = -\frac{1}{2}\langle \lambda(t,T_k), c_t^L \lambda(t,T_k)\rangle - \sum_{j=k+1}^{N-1} \ell^j(X_{t-}^j)\langle \lambda(t,T_k), c_t^L \lambda(t,T_j)\rangle$$
$$- \int_{\mathbb{R}^N} \Big[(e^{x_k}-1)\prod_{j=k+1}^{N-1}\Big(1+\ell^j(X_{t-}^j)(e^{x_j}-1)\Big) - x_k\Big] F_t^X(dx)$$
$$= -\frac{1}{2}\langle \lambda(t,T_k), c_t^L \lambda(t,T_k)\rangle - \sum_{j=k+1}^{N-1} \ell^j(X_{t-}^j)\langle \lambda(t,T_k), c_t^L \lambda(t,T_j)\rangle$$
$$- \int_{\mathbb{R}^n} \Big[(e^{\langle \lambda(t,T_k),y\rangle}-1)\prod_{j=k+1}^{N-1}\Big(1+\ell^j(X_{t-}^j)(e^{\langle \lambda(t,T_j),y\rangle}-1)\Big) \tag{32}$$
$$- \langle \lambda(t,T_k), y\rangle\Big] F_t^L(dy),$$

where the second equality follows by (30). The equation above now can be recognized as the drift condition of the Lévy LIBOR model; cf. Papapantoleon et al. [30, Eq. (2.7)].

Remark 4.1 The LIBOR market models satisfy Axioms ($\mathbb{A}1$) and ($\mathbb{A}2$), as well as Properties ($\mathbb{B}1$) and ($\mathbb{B}4$) by construction. On the other hand, Properties ($\mathbb{B}2$) and ($\mathbb{B}3$) are not satisfied. Regarding ($\mathbb{B}2$), this follows immediately by Proposition 3.11 (at least for driving processes satisfying ($\mathbb{E}$), which is typically the case), since the functions f^k, $k \in \bar{\mathcal{K}}$, are not affine in the second argument. Moreover, the drift term (32) which contains the random terms $\delta_j L(t,T_j)/(1+\delta_j L(t,T_j))$ implies that the vector of LIBOR rates $(L(\cdot,T_k))_{k\in\bar{\mathcal{K}}}$ considered as a whole is Markovian, but not the single LIBOR rates, because their dynamics depend on the other rates as well. Hence, ($\mathbb{B}3$) does not hold. Finally, Property ($\mathbb{C}$) is obviously satisfied in this model.

4.2 Lévy Forward Price Models

Next, we show that the Lévy forward price models can be easily embedded in our general framework starting from the terminal measure construction; starting from the backward induction approach is even easier. The Lévy forward price models were introduced by Eberlein and Özkan [10, pp. 342–343]; see also Kluge [23] for a detailed construction and Kluge and Papapantoleon [24] for a concise presentation.

We will model the dynamics of the forward price relative to the terminal bond price under the terminal measure $\mathbb{P}_N$, via

$$F(t,T_k,T_N) = e^{g^k(t,X_t^k)}, \tag{33}$$

where the function g^k is of the following affine form

$$g^k(t,x) := \log F(0,T_k,T_N) + x, \tag{34}$$

while the process X^k has the following dynamics

$$X^k := \int_0^{\cdot} b_s^{k,N} \mathrm{d}s + \sum_{i=k}^{N-1} \int_0^{\cdot} \lambda(s, T_i) \mathrm{d}L_s. \tag{35}$$

The driving process L and the volatility functions $\lambda(\cdot, T_i)$ are specified, while the drift term $b^{k,N}$ is determined by the no-arbitrage $(\mathbb{DRIFT}')$ condition. In particular, L is an $\mathbb{R}^n$-valued time-inhomogeneous Lévy process with $\mathbb{P}_N$-local characteristics $(0, c^L, F^L)$ satisfying condition $(\mathbb{EM})$ and the volatility functions satisfy condition $(\mathbb{VOL})$. The function g^k trivially satisfies the $(\mathbb{LIP}')$ condition with constant 1, while the process X^k satisfies the $(\mathbb{INT}')$ condition by virtue of $(\mathbb{EM})$ and $(\mathbb{VOL})$. Therefore, we can apply Theorem 3.5 and, after some computations, the $(\mathbb{DRIFT}')$ condition yields that

$$b_t^{k,N} = -\frac{1}{2} c_t^{k,N} - \int_{\mathbb{R}} (\mathrm{e}^x - 1 - x) F_t^{k,N}(\mathrm{d}x). \tag{36}$$

Moreover, using Kallsen and Shiryaev [20, Lemma 3], the $\mathbb{P}_N$-local characteristics of the stochastic integral process X^k are

$$c_t^{k,N} = \left\langle \sum_{i=k}^{N-1} \lambda(t, T_i), c_t^L \sum_{i=k}^{N-1} \lambda(t, T_i) \right\rangle \tag{37}$$

and

$$F_t^{k,N}(A) = \int_{\mathbb{R}^n} 1_A \left(\sum_{i=k}^{N-1} \langle \lambda(t, T_i), x \rangle \right) F_t^L(\mathrm{d}x), \quad A \in \mathcal{B}(\mathbb{R}). \tag{38}$$

Now, using (33)–(35), we get that the dynamics of the forward price process $F(\cdot, T_k, T_{k+1})$ are provided by

$$\begin{aligned} F(t, T_k, T_{k+1}) &= \frac{F(t, T_k, T_N)}{F(t, T_{k+1}, T_N)} = F(0, T_k, T_{k+1}) \mathrm{e}^{X_t^k - X_t^{k+1}} \\ &= F(0, T_k, T_{k+1}) \exp\left(\int_0^t (b_s^{k,N} - b_s^{k+1,N}) \mathrm{d}s + \int_0^t \lambda(s, T_k) \mathrm{d}L_s \right), \end{aligned}$$

hence the forward price process is driven by its corresponding volatility function and the time-inhomogeneous Lévy process, as specified in the Lévy forward process models. We just have to check that the drift terms coincide as well. Indeed, using (36)–(38), after some straightforward calculations we get that

$$
\begin{aligned}
b_s^{k,N} - b_s^{k+1,N} &= -\frac{1}{2}\big\langle \lambda(s,T_k), c_s^L \lambda(s,T_k)\big\rangle - \sum_{i=k+1}^{N-1} \big\langle \lambda(s,T_k), c_s^L \lambda(s,T_i)\big\rangle \\
&\quad - \int_{\mathbb{R}^n} \Big\{ \Big(\mathrm{e}^{\langle \lambda(s,T_k), x\rangle} - 1\Big) \mathrm{e}^{\sum_{i=k+1}^{N-1} \langle \lambda(s,T_i), x\rangle} - \langle \lambda(s,T_k), x\rangle \Big\} F_s^L(\mathrm{d}x),
\end{aligned}
$$

which is exactly the $\mathbb{P}_N$-drift of the forward price process; compare with Kluge and Papapantoleon [24, Eqs. (19)–(21)].

Remark 4.2 The Lévy forward price model satisfies Axioms ($\mathbb{A}1$) and ($\mathbb{A}2$) as well as Properties ($\mathbb{B}1$) and ($\mathbb{B}4$) by construction. Moreover, it satisfies Properties ($\mathbb{B}2$) and ($\mathbb{B}3$); cf. Proposition 3.11. Property ($\mathbb{C}$) is not satisfied however, i.e. the LIBOR rates can become negative; cf. Remark 3.12.

4.3 *Affine LIBOR Models*

Finally, we examine a class of LIBOR models driven by affine processes, and in particular the affine LIBOR models proposed by Keller-Ressel et al. [21]. Our main reference for the definition and properties of affine processes is Duffie et al. [7].

Let $X = (X_t)_{0 \le t \le T_N}$ be a conservative affine process according to Definitions 2.1 and 2.5 in Duffie et al. [7] with state space $D = \mathbb{R}_+$. We consider a one-dimensional process here only for notational simplicity; the d-dimensional case can be treated in exactly the same manner. Moreover, the state space is restricted to the positive half-line following Keller-Ressel et al. [21], which is necessary in order to produce a model satisfying ($\mathbb{C}$); see also Remark 3.12. We can equally well choose the state space $D = \mathbb{R}$, and then interest rates in the model will also take negative values.

The process X is a semimartingale with absolutely continuous characteristics, and the local characteristics (b^X, c^X, F^X) of X with respect to the truncation function $h(x) := 1 \wedge x$, for $x \in D$, are given as

$$
\begin{aligned}
b_t^X &= \tilde{b} + \beta X_{t-} \\
c_t^X &= 2\alpha X_{t-} \\
F_t^X(\mathrm{d}\xi) &= F^1(\mathrm{d}\xi) + X_{t-} F^2(\mathrm{d}\xi)
\end{aligned}
$$

for some $\tilde{b} > 0$, $\beta \in \mathbb{R}$, $\alpha > 0$ and Lévy measures F^1 and F^2 on $D\backslash\{0\}$ (cf. Theorem 2.12 in Duffie et al. [7]), with

$$
\tilde{b} := b + \int_{\xi>0} h(\xi) F^1(\mathrm{d}\xi) .
$$

Affine processes are characterized by the following property of their moment generating function:

$$\mathbb{E}_{\mathrm{x}}\big[\exp(uX_t)\big] = \exp\big(\phi(t,u) + \psi(t,u)\mathrm{x}\big), \tag{39}$$

for all $(t, u, \mathrm{x}) \in [0, T_N] \times \mathcal{I}_T \times D$, where $\mathbb{E}_{\mathrm{x}}$ denotes the expectation with respect to $\mathbb{P}_{\mathrm{x}}$—a probability measure such that $X_0 = \mathrm{x} \in D$, $\mathbb{P}_{\mathrm{x}}$-a.s. Moreover, the set $\mathcal{I}_T$ is defined by

$$\mathcal{I}_T := \big\{u \in \mathbb{R} : \mathbb{E}_{\mathrm{x}}\big[\mathrm{e}^{uX_{T_N}}\big] < \infty, \ \text{for all } \mathrm{x} \in D\big\}, \tag{40}$$

while (ϕ, ψ) is a pair of deterministic functions $\phi, \psi : [0, T_N] \times \mathcal{I}_T \to \mathbb{R}$. The functions ϕ and ψ are given as solutions to generalized Riccati equations (cf. Theorem 2.7 in Duffie et al. [7]), that is

$$\begin{aligned} \partial_t \phi(t,u) &= F\big(\psi(t,u)\big), \quad \phi(0,u) = 0 \\ \partial_t \psi(t,u) &= R\big(\psi(t,u)\big), \quad \psi(0,u) = u, \end{aligned} \tag{41}$$

where

$$\begin{aligned} F(u) &= bu + \int_{\xi>0} \big(\mathrm{e}^{u\xi} - 1\big) F^1(\mathrm{d}\xi), \\ R(u) &= \alpha u^2 + \beta u + \int_{\xi>0} \big(\mathrm{e}^{u\xi} - 1 - uh(\xi)\big) F^2(\mathrm{d}\xi). \end{aligned} \tag{42}$$

We introduce next the class of *affine forward price models*, where the forward price is an exponentially-affine function of the driving affine process X. In particular, we consider the setting of the terminal measure construction of Sect. 3.2 with

$$g^k(t,x) = \theta^k(t) + \vartheta^k(t)x \quad \text{and} \quad X^k \equiv X. \tag{43}$$

The next result shows that the functions θ^k, ϑ^k are solutions to generalized Riccati equations themselves.

Proposition 4.3 *Let X be an affine process with values in D, $X_0 = 1$ and satisfying* ($\mathbb{INT}'$)*, and g^k, $k \in \bar{\mathcal{K}}$, be a collection of functions given by* (43) *where $\theta^k, \vartheta^k : [0, T_N] \to \mathbb{R}$ are deterministic functions of class C^1. Then, the forward price process given by*

$$F(t, T_k, T_N) = \mathrm{e}^{\theta^k(t) + \vartheta^k(t)X_t}, \quad t \in [0, T_k], \tag{44}$$

is a uniformly integrable martingale, for all $k \in \bar{\mathcal{K}}$, if the functions θ^k and ϑ^k satisfy

$$\begin{aligned} \partial_t \theta^k(t) &= -F\big(\vartheta^k(t)\big), \\ \partial_t \vartheta^k(t) &= -R\big(\vartheta^k(t)\big), \end{aligned} \tag{45}$$

with F and R given by (42).

Proof The process X satisfies ($\mathbb{INT}'$) by assumption, while the functions g^k satisfy ($\mathbb{LIP}'$). Therefore, we can apply Theorem 3.5 and the result follows after straightforward calculations, by inserting the characteristics of X into the ($\mathbb{DRIFT}'$) condition and using that

$$\partial_t g^k(t,x) = \partial_t \theta^k(t) + \partial_t \vartheta^k(t)x, \quad \partial_x g^k(t,x) = \vartheta^k(t), \quad \partial_{xx} g^k(t,x) = 0. \qquad \square$$

Remark 4.4 The affine forward price models given by (44) satisfy Axioms ($\mathbb{A}1$) and ($\mathbb{A}2$) as well as Properties ($\mathbb{B}1$)–($\mathbb{B}3$). Property ($\mathbb{B}4$) is not satisfied and the model has to be calibrated to the initial term structure, similarly to short rate models. Indeed, notice that the initial forward price $F(0, T_k, T_N)$ does not appear in the function $g^k(t,x)$, contrary to the previous two examples.[1] Moreover, these models satisfy Property ($\mathbb{C}$) if and only if the functions θ^k and ϑ^k are non-negative; compare also with Remark 3.12.

The affine LIBOR models introduced by Keller-Ressel et al. [21] can naturally be embedded in this construction. More precisely, we have the following.

Corollary 4.5 *The affine LIBOR models whose dynamics are provided by*

$$F(t, T_k, T_N) = \mathbb{E}_N \left[e^{u_k X_{T_N}} | \mathcal{F}_t \right] = e^{\phi(T_N - t, u_k) + \psi(T_N - t, u_k) X_t},$$

with parameters $u_k \in \mathbb{R}_+$ for $k \in \bar{\mathcal{K}}$, is a special case of the affine forward price models with

$$\theta^k(t) := \phi(T_N - t, u_k) \quad \text{and} \quad \vartheta^k(t) := \psi(T_N - t, u_k),$$

where $\phi(\cdot, u_k)$ and $\psi(\cdot, u_k)$ are solutions to (41).

Appendix A: Semimartingale Characteristics and Martingales

Let $(\Omega, \mathcal{F}, (\mathcal{F}_t)_{t \in [0, T_*]}, \mathbb{P})$ denote a complete stochastic basis and T_* denote a finite time horizon. Let X be an $\mathbb{R}^d$-valued semimartingale on this basis whose characteristics are absolutely continuous, i.e. its local characteristics are given by $(b, c, F; A)$ with $A_t = t$, for some truncation function h; cf. Jacod and Shiryaev [15, Proposition II.2.9]. Moreover, let $f : \mathbb{R}_+ \times \mathbb{R}^d \to \mathbb{R}$ be a function of class $C^{1,2}(\mathbb{R}_+ \times \mathbb{R}^d)$.

[1] We could, of course, use the following affine function $g^k(t,x) = \log F(0, T_k, T_N) + \theta^k(t) + \vartheta^k(t)x$ and the model fits automatically the initial term structure. However, it becomes then difficult to provide models that produce non-negative LIBOR rates.

The process $f(\cdot, X)$ is a real-valued semimartingale which has again absolutely continuous characteristics. Let us denote its local characteristics by (b^f, c^f, F^f) for a truncation function h^f. Then, noting that Itô's formula holds for the function $f \in C^{1,2}(\mathbb{R}_+ \times \mathbb{R}^d)$ and reasoning as in the proof of Corollary A.6 from Goll and Kallsen [13], we have that

$$\begin{aligned}
b_t^f &= \frac{\mathrm{d}}{\mathrm{d}t} f(t, X_{t-}) + \langle \mathrm{D} f(t, X_{t-}), b_t \rangle + \frac{1}{2} \sum_{i,j=1}^{d} \mathrm{D}_{ij}^2 f(t, X_{t-}) c_t^{ij} \\
&\quad + \int_{\mathbb{R}^d} \Big(h^f\big(f(t, X_{t-} + x) - f(t, X_{t-})\big) - \langle \mathrm{D} f(t, X_{t-}), h(x) \rangle \Big) F_t(\mathrm{d}x) \\
c_t^f &= \big\langle \mathrm{D} f(t, X_{t-}), c_t \mathrm{D} f(t, X_{t-}) \big\rangle \qquad \text{(A.1)} \\
F_t^f(G) &= \int_{\mathbb{R}^d} 1_G\big(f(t, X_{t-} + x) - f(t, X_{t-})\big) F_t(\mathrm{d}x), \qquad G \in \mathcal{B}(\mathbb{R} \setminus \{0\}).
\end{aligned}$$

Proposition A.1 *Let X be an $\mathbb{R}^d$-valued semimartingale with absolutely continuous characteristics (b, c, F) and let $f : \mathbb{R}_+ \times \mathbb{R}^d \to \mathbb{R}$ be a function of class $C^{1,2}$ such that the process Y defined by*

$$Y_t := \mathrm{e}^{f(t, X_t)} \qquad \text{(A.2)}$$

is exponentially special. If the following condition holds

$$\begin{aligned}
\langle \mathrm{D} f(t, X_{t-}), b_t \rangle &= -\frac{\mathrm{d}}{\mathrm{d}t} f(t, X_{t-}) - \frac{1}{2} \sum_{i,j=1}^{d} \mathrm{D}_{ij}^2 f(t, X_{t-}) c_t^{ij} \\
&\quad - \frac{1}{2} \big\langle \mathrm{D} f(t, X_{t-}), c_t \mathrm{D} f(t, X_{t-}) \big\rangle \qquad \text{(A.3)} \\
&\quad - \int_{\mathbb{R}^d} \Big(\mathrm{e}^{f(t, X_{t-} + x) - f(t, X_{t-})} - 1 - \langle \mathrm{D} f(t, X_{t-}), h(x) \rangle \Big) F_t(\mathrm{d}x),
\end{aligned}$$

then Y is a local martingale.

Proof The proof follows from Theorem 2.18 in Kallsen and Shiryaev [19]: set $\theta = 1$ and apply the theorem to the semimartingale $f(\cdot, X)$. Indeed, since $f(\cdot, X)$ has absolutely continuous characteristics it is also quasi-left continuous, hence assertions (6) and (1) of Theorem 2.18. yield

$$K^{f(\cdot, X)}(1) = \widetilde{K}^{f(\cdot, X)}(1) = \int_0^{\cdot} \Big(b_t^f + \frac{1}{2} c_t^f + \int_{\mathbb{R}} \big(\mathrm{e}^x - 1 - h^f(x)\big) F_t^f(\mathrm{d}x) \Big) \mathrm{d}t.$$

By definition of the exponential compensator and Theorem 2.19 in Kallsen and Shiryaev [19] it follows that

$$e^{f(\cdot,X)-K^{f(\cdot,X)}(1)} \in \mathcal{M}_{\mathrm{loc}}.$$

Therefore, $e^{f(\cdot,X)} \in \mathcal{M}_{\mathrm{loc}}$ if and only if $K^{f(\cdot,X)}(1) = 0$ up to indistinguishability. Equivalently,

$$b_t^f + \frac{1}{2}c_t^f + \int_{\mathbb{R}} \left(e^x - 1 - h^f(x)\right) F_t^f(\mathrm{d}x) = 0$$

for every t. Inserting the expressions for b^f, c^f and F^f, cf. (A.1), into the above equality yields condition (A.3). □

Proposition A.2 *Let X be an $\mathbb{R}^d$-valued semimartingale with absolutely continuous characteristics (b, c, F) such that*

$$\int_0^{T_*}\int_{\mathbb{R}^d} (|x|^2 \wedge 1) F_t(\mathrm{d}x)\mathrm{d}t + \int_0^{T_*}\int_{|x|>1} |x|e^{K|x|} F_t(\mathrm{d}x)\mathrm{d}t < C_1 \tag{A.4}$$

and

$$\int_0^{T_*} \|c_t\|\mathrm{d}t < C_2, \tag{A.5}$$

for some deterministic constants $C_1, C_2 > 0$. Moreover, let $f : \mathbb{R}_+ \times \mathbb{R}^d \to \mathbb{R}$ be a function of class $C^{1,2}$ and globally Lipschitz with constant $K > 0$ such that

$$|f(t,x) - f(t,y)| \le K|x-y|, \qquad t \ge 0, \ \ x, y \in \mathbb{R}^d.$$

Then, the process $f(\cdot, X)$ is exponentially special, while the process Y defined by (A.2) *and satisfying* (A.3) *is a uniformly integrable martingale.*

Proof The process $f(\cdot, X)$ is exponentially special if and only if

$$1_{\{|x|>1\}}e^x * \nu^f \in \mathcal{V}.$$

Hence, it suffices to show that $1_{\{|x|>1\}}e^x * \nu^f_{T_*} < \infty$, as the integrand is positive. Since f is globally Lipschitz, we have

$$
\begin{aligned}
1_{\{|x|>1\}}\mathrm{e}^{x} * \nu^{f}_{T_*} &= \int_0^{T_*}\int_{|x|>1} \mathrm{e}^{x} F^{f}_{t}(\mathrm{d}x)\mathrm{d}t \\
&\overset{(A.1)}{=} \int_0^{T_*}\int_{\mathbb{R}^d} 1_{\{|f(t,X_{t-}+x)-f(t,X_{t-})|>1\}}\mathrm{e}^{f(t,X_{t-}+x)-f(t,X_{t-})} F_t(\mathrm{d}x)\mathrm{d}t \\
&\leq \int_0^{T_*}\int_{K|x|>1} \mathrm{e}^{K|x|} F_t(\mathrm{d}x)\mathrm{d}t < \infty,
\end{aligned}
$$

which holds by the Lipschitz property and (A.4).

Moreover, if $F = \mathrm{e}^{f(\cdot,X)} \in \mathcal{M}_{\mathrm{loc}}$, applying Proposition 3.1 in Criens et al. [6] it is also a uniformly integrable martingale if the following condition holds:

$$
\int_0^{T_*}\left(c^f_t + \int_{\mathbb{R}^d}\left[(|x|^2 \wedge 1) + |x|\mathrm{e}^{x} 1_{\{|x|>1\}}\right]F^f_t(\mathrm{d}x)\right)\mathrm{d}t < C^f, \tag{A.6}
$$

for some constant $C^f > 0$. We first check the condition for the diffusion coefficient

$$
\int_0^{T_*} c^f_t\,\mathrm{d}t = \int_0^{T_*}\langle \mathrm{D}f(t,X_{t-}), c_t \mathrm{D}f(t,X_{t-})\rangle \mathrm{d}t \leq \int_0^{T_*}\|c_t\||\mathrm{D}f(t,X_{t-})|^2\mathrm{d}t < C^f_1,
$$

which follows from (A.5) and the fact that $\mathrm{D}f(\cdot, X_-)$ is bounded as a consequence of f being globally Lipschitz. As for the jump part, we have that

$$
\begin{aligned}
&\int_0^{T_*}\int_{\mathbb{R}}(|x|^2 \wedge 1)F^f_t(\mathrm{d}x)\mathrm{d}t + \int_0^{T_*}\int_{|x|>1} |x|\mathrm{e}^{x} F^f_t(\mathrm{d}x)\mathrm{d}t \\
&\quad\overset{(A.1)}{=} \int_0^{T_*}\int_{\mathbb{R}^d}(|f(t,X_{t-}+x)-f(t,X_{t-})|^2 \wedge 1)F_t(\mathrm{d}x)\mathrm{d}t \\
&\qquad + \int_0^{T_*}\int_{\mathbb{R}^d} 1_{\{|f(t,X_{t-}+x)-f(t,X_{t-})|>1\}} \\
&\qquad \times |f(t,X_{t-}+x)-f(t,X_{t-})|\,\mathrm{e}^{f(t,X_{t-}+x)-f(t,X_{t-})} F_t(\mathrm{d}x)\mathrm{d}t \\
&\quad\leq \int_0^{T_*}\int_{\mathbb{R}^d}(K^2|x|^2 \wedge 1)F(\mathrm{d}x)\mathrm{d}t + \int_0^{T_*}\int_{K|x|>1} K|x|\mathrm{e}^{K|x|}F^f(\mathrm{d}x)\mathrm{d}t < C^f_2,
\end{aligned}
$$

using again the Lipschitz property and (A.4). □

Next, we provide the representation of Y as a stochastic exponential.

Lemma A.3 *Let X be an $\mathbb{R}^d$-valued semimartingale with absolutely continuous characteristics (B, C, ν) and let $f : \mathbb{R}_+ \times \mathbb{R}^d \to \mathbb{R}_+$ be a function of class $C^{1,2}(\mathbb{R}_+ \times \mathbb{R}^d)$. Define a real-valued semimartingale Y via* (A.2)*. If $Y \in \mathcal{M}_{\mathrm{loc}}$, then it can be written as*

$$Y = \mathcal{E}\big(\mathrm{D}f(\cdot, X_-) \cdot X^c + W(\cdot, x) * (\mu^X - \nu)\big),$$

where X^c is the continuous martingale part of X, μ^X is the random measure of jumps of X with compensator ν and

$$W(\cdot, x) := \mathrm{e}^{f(\cdot, X_- + x) - f(\cdot, X_-)} - 1.$$

Proof Theorem 2.19 in Kallsen and Shiryaev [19] yields that

$$\mathrm{e}^{f(\cdot, X)} = \mathcal{E}\left(f(\cdot, X)^c + (\mathrm{e}^x - 1) * (\mu^f - \nu^f)\right),$$

using that $f(\cdot, X)$ is quasi-left continuous since X is also quasi-left continuous. Here $f(\cdot, X)^c$ denotes the continuous martingale part of $f(\cdot, X)$ and μ^f its random measure of jumps. The result now follows using the form of the local characteristics c^f, F^f of the process $f(\cdot, X)$ in (A.1); see also the proof of Corollary A.6 in Goll and Kallsen [13].

Lemma A.4 *Let X be an $\mathbb{R}^d$-valued semimartingale with absolutely continuous characteristics (b, c, F) with respect to the truncation function h. Let $f : \mathbb{R}_+ \times \mathbb{R}^d \to \mathbb{R}_+$ be a function of class $C^{1,2}(\mathbb{R}_+ \times \mathbb{R}^d)$ and globally Lipschitz. Assume that conditions* (A.3), (A.4) *and* (A.5) *are satisfied.*

Define the probability measure $\mathbb{P}' \sim \mathbb{P}$ via

$$\left.\frac{\mathrm{d}\mathbb{P}'}{\mathrm{d}\mathbb{P}}\right|_{\mathcal{F}} := \mathrm{e}^{f(\cdot, X)}.$$

Then, the $\mathbb{P}'$-characteristics of the semimartingale X are absolutely continuous and provided by (b', c', F'), where

$$\begin{aligned} b'_t &= b_t + c_t \beta_t + \int_{\mathbb{R}^d} (Y_t(x) - 1) h(x) F_t(\mathrm{d}x) \\ c'_t &= c_t \\ F'_t(\mathrm{d}x) &= Y_t(x) F_t(\mathrm{d}x), \end{aligned}$$

with $\beta_t = \mathrm{D}f(t, X_{t-})$ and $Y_t(x) = \mathrm{e}^{f(t, X_{t-} + x) - f(t, X_{t-})}$, for $t \in \mathbb{R}_+$ and $x \in \mathbb{R}^d$.

Proof The result follows directly from the previous lemma and Proposition 2.6 in Kallsen [18]. □

References

1. Andersen, L.B.G., Piterbarg, V.V.: Interest Rate Modeling, vols. 3. Atlantic Financial Press (2010)
2. Belomestny, D., Mathew, S., Schoenmakers, J.: Multiple stochastic volatility extension of the LIBOR market model and its implementation. Monte Carlo Methods Appl. **15**, 285–310 (2009)
3. Bianchetti, M., Morini, M. (eds.): Interest Rate Modelling After the Financial Crisis. Risk Books (2013)
4. Brace, A., Gątarek, D., Musiela, M.: The market model of interest rate dynamics. Math. Financ. **7**, 127–155 (1997)
5. Brigo, D., Mercurio, F.: Interest Rate Models: Theory and Practice, 2nd edn. Springer (2006)
6. Criens, D., Glau, K., Grbac, Z.: Martingale property of exponential semimartingales: a note on explicit conditions and applications to financial models. Preprint (2015). arXiv:1506.08127
7. Duffie, D., Filipović, D., Schachermayer, W.: Affine processes and applications in finance. Ann. Appl. Probab. **13**, 984–1053 (2003)
8. Eberlein, E., Kluge, W.: Exact pricing formulae for caps and swaptions in a Lévy term structure model. J. Comput. Financ. **9**(2), 99–125 (2006)
9. Eberlein, E., Kluge, W.: Valuation of floating range notes in Lévy term structure models. Math. Financ. **16**, 237–254 (2006)
10. Eberlein, E., Özkan, F.: The Lévy LIBOR model. Financ. Stoch. **9**, 327–348 (2005)
11. Filipović, D., Trolle, A.: The term structure of interbank risk. J. Financ. Econom. **109**, 707–733 (2013)
12. Glasserman, P.: Monte Carlo Methods in Financial Engineering. Springer (2003)
13. Goll, T., Kallsen, J.: Optimal portfolios for logarithmic utility. Stoch. Process. Appl. **89**, 31–48 (2000)
14. Grbac, Z., Runggaldier, W.J.: Interest Rate Modeling: Post-Crisis Challenges and Approaches. SpringerBriefs in Quantitative Finance. Springer, Berlin (2015)
15. Jacod, J., Shiryaev, A.N.: Limit Theorems for Stochastic Processes, 2nd edn. Springer (2003)
16. Jamshidian, F.: LIBOR and swap market models and measures. Financ. Stoch. **1**, 293–330 (1997)
17. Jamshidian, F.: LIBOR Market Model with Semimartingales. Working Paper, NetAnalytic Ltd. (1999)
18. Kallsen, J.: A didactic note on affine stochastic volatility models. In: Kabanov, Y, Lipster, R., Stoyanov, J. (eds.) From Stochastic Calculus to Mathematical Finance: The Shiryaev Festschrift, pp. 343–368. Springer (2006)
19. Kallsen, J., Shiryaev, A.N.: The cumulant process and Esscher's change of measure. Financ. Stoch. **6**, 397–428 (2002)
20. Kallsen, J., Shiryaev, A.N.: Time change representation of stochastic integrals. Theory Probab. Appl. **46**, 522–528 (2002)
21. Keller-Ressel, M., Papapantoleon, A., Teichmann, J.: The affine LIBOR models. Math. Financ. **23**, 627–658 (2013)
22. Klein, I., Schmidt, T., Teichmann, J.: No-arbitrage theory for bond markets. In: Advanced modelling in mathematical finance, pp. 381–421. Springer, Switzerland (2016)
23. Kluge, W.: Time-inhomogeneous Lévy processes in interest rate and credit risk models. PhD thesis, Univ. Freiburg (2005)
24. Kluge, W., Papapantoleon, A.: On the valuation of compositions in Lévy term structure models. Quant. Financ. **9**, 951–959 (2009)

25. Ladkau, M., Schoenmakers, J., Zhang, J.: Libor model with expiry-wise stochastic volatility and displacement. Int. J. Portfolio Anal. Manag. **1**, 224–249 (2013)
26. Miltersen, K.R., Sandmann, K., Sondermann, D.: Closed form solutions for term structure derivatives with log-normal interest rates. J. Financ. **52**, 409–430 (1997)
27. Musiela, M., Rutkowski, M.: Continuous-time term structure models: forward measure approach. Financ. Stoch. **1**, 261–291 (1997)
28. Musiela, M., Rutkowski, M.: Martingale Methods in Financial Modelling 2nd edn. Springer (2005)
29. Papapantoleon, A.: Old and new approaches to LIBOR modeling. Stat. Neerl. **64**, 257–275 (2010)
30. Papapantoleon, A., Schoenmakers, J., Skovmand, D.: Efficient and accurate log-Lévy approximations to Lévy-driven LIBOR models. J. Comput. Financ. **15**(4), 3–44 (2012)
31. Schoenmakers, J.: Robust LIBOR Modelling and Pricing of Derivative Products. Chapman & Hall/CRC Press (2005)
32. Wheatley, M.: The Wheatley Review of LIBOR. Technical report, HM Treasury (2012)
33. Wu, L., Zhang, F.: LIBOR market model with stochastic volatility. J. Industr. Manag. Optim. **2**, 199–227 (2006)

Approximate Option Pricing in the Lévy Libor Model

Zorana Grbac, David Krief and Peter Tankov

Abstract In this paper we consider the pricing of options on interest rates such as caplets and swaptions in the Lévy Libor model developed by Eberlein and Özkan (Financ. Stochast. 9:327-348 (2005) [9]). This model is an extension to Lévy driving processes of the classical log-normal Libor market model (LMM) driven by a Brownian motion. Option pricing is significantly less tractable in this model than in the LMM due to the appearance of stochastic terms in the jump part of the driving process when performing the measure changes which are standard in pricing of interest rate derivatives. To obtain explicit approximation for option prices, we propose to treat a given Lévy Libor model as a suitable perturbation of the log-normal LMM. The method is inspired by recent works by Černý, Denkl, and Kallsen (Preprint (2013) [6]) and Ménassé and Tankov (Preprint (2015) [14]). The approximate option prices in the Lévy Libor model are given as the corresponding LMM prices plus correction terms which depend on the characteristics of the underlying Lévy process and some additional terms obtained from the LMM model.

Keywords Libor market model · Caplet · Swaption · Lévy Libor model · Asymptotic approximation

1 Introduction

The goal of this paper is to develop explicit approximations for option prices in the Lévy Libor model introduced by Eberlein and Özkan [9]. In particular, we shall be

Z. Grbac (✉) · D. Krief · P. Tankov
Laboratoire de Probabilités et Modèles Aléatoires,
Université Paris Diderot, Paris, France
e-mail: grbac@math.univ-paris-diderot.fr

D. Krief
e-mail: krief@math.univ-paris-diderot.fr

P. Tankov
e-mail: peter.tankov@polytechnique.org

© Springer International Publishing Switzerland 2016
J. Kallsen and A. Papapantoleon (eds.), *Advanced Modelling in Mathematical Finance*,
Springer Proceedings in Mathematics & Statistics 189,
DOI 10.1007/978-3-319-45875-5_19

interested in price approximations for caplets, whose pay-off is a function of only one underlying Libor rate and swaptions, which can be regarded as options on a "basket" of multiple Libor rates of different maturities.

A full-fledged model of Libor rates such as the Lévy Libor model is typically used for the purposes of pricing and risk management of exotic interest rate products. The prices and hedge ratios must be consistent with the market-quoted prices of liquid options, which means that the model must be calibrated to the available prices / implied volatilities of caplets and swaptions. To perform such a calibration efficiently, one therefore needs explicit formulas or fast numerical algorithms for caplet and swaption prices.

Computation of option prices in the Lévy Libor model to arbitrary precision is only possible via Monte Carlo. Efficient simulation algorithms suitable for pricing exotic options have been proposed in Kohatsu-Higa and Tankov [13] and Papapantoleon et al. [16], however, these Monte Carlo algorithms are probably not an option for the purposes of calibration because the computation is still too slow due to the presence of both discretization and statistical error.

Eberlein and Özkan [9], Kluge [12] and Belomestny and Schoenmakers [1] propose fast methods for computing caplet prices which are based on Fourier transform inversion and use the fact that the characteristic function of many parametric Lévy processes is known explicitly. Since in the Lévy Libor model, the Libor rate L^k is not a geometric Lévy process under the corresponding probability measure $\mathbb{Q}^{T_k}$, unless $k = n$ (see Remark 2 below for details), using these methods for $k < n$ requires an additional approximation (some random terms appearing in the compensator of the jump measure of L^k are approximated by their values at time $t = 0$, a method known as freezing).

In this paper we take an alternative route and develop approximate formulas for caplets and swaptions using asymptotic expansion techniques. Inspired by methods used in Černý et al. [6] and Ménassé and Tankov [14] (see also Benhamou et al. [2, 3] for related expansions "around a Black-Scholes proxy" in other models), we consider a given Lévy Libor model as a perturbation of the log-normal LMM. Starting from the driving Lévy process $(X_t)_{t\geq 0}$ of the Lévy Libor model, assumed to have zero expectation, we introduce a family of processes $X_t^\alpha = \alpha X_{t/\alpha^2}$ parameterized by $\alpha \in (0, 1]$, together with the corresponding family of Lévy Libor models. For $\alpha = 1$ one recovers the original Lévy Libor model. When $\alpha \to 0$, the family X^α converges weakly in Skorokhod topology to a Brownian motion, and the option prices in the Lévy Libor model corresponding to the process X^α converge to the prices in the log-normal LMM. The option prices in the original Lévy Libor model can then be approximated by their second-order expansions in the parameter α, around the value $\alpha = 0$. This leads to an asymptotic approximation formula for a derivative price expressed as a linear combination of the derivative price stemming from the LMM and correction terms depending on the characteristics of the driving Lévy process. The terms of this expansion are often much easier to compute than the option prices in the Lévy Libor model. In particular, we shall see the expansion for caplets is expressed in terms of the derivatives of the standard Black's formula, and the various terms of the expansion for swaptions can be approximated using one of the many swaption approximations for the log-normal LMM available in the literature.

This paper is structured as follows. In Sect. 2 we briefly review the Lévy Libor model. In Sect. 3 we show how the prices of European-style options may be expressed as solutions of partial integro-differential equations (PIDE). These PIDEs form the basis of our asymptotic method, presented in detail in Sect. 4. Finally, numerical illustrations are provided in Sect. 5.

2 Presentation of the Model

In this section we present a slight modification of the Lévy Libor model by Eberlein and Özkan [9], which is a generalization, based on Lévy processes, of the Libor market model driven by a Brownian motion, introduced by Sandmann et al. [17], Brace et al. [4] and Miltersen et al. [15].

Let a discrete tenor structure $0 \leq T_0 < T_1 < \ldots < T_n = T^*$ be given, and set $\delta_k := T_k - T_{k-1}$, for $k = 1, \ldots, n$. We assume that zero-coupon bonds with maturities T_k, $k = 0, \ldots, n$, are traded in the market. The time-t price of a bond with maturity T_k is denoted by $B_t(T_k)$ with $B_{T_k}(T_k) = 1$.

For every tenor date T_k, $k = 1, \ldots, n$, the forward Libor rate L_t^k at time $t \leq T_{k-1}$ for the accrual period $[T_{k-1}, T_k]$ is a discretely compounded interest rate defined as

$$L_t^k := \frac{1}{\delta_k} \left(\frac{B_t(T_{k-1})}{B_t(T_k)} - 1 \right). \tag{1}$$

For all $t > T_{k-1}$, we set $L_t^k := L_{T_{k-1}}^k$.

To set up the Libor model, one needs to specify the forward Libor rates L_t^k, $k = 1, \ldots, n$, such that each Libor rate L^k is a martingale with respect to the corresponding forward measure $\mathbb{Q}^{T_k}$ using the bond with maturity T_k as numéraire. We recall that the forward measures are interconnected via the Libor rates themselves and hence each Libor rate depends also on some other Libor rates as we shall see below. More precisely, assuming that the forward measure $\mathbb{Q}^{T_n}$ for the most distant maturity T_n (i.e. with numéraire $B(T_n)$) is given, the link between the forward measure $\mathbb{Q}^{T_k}$ and $\mathbb{Q}^{T_n}$ is provided by

$$\left.\frac{d\mathbb{Q}^{T_k}}{d\mathbb{Q}^{T_n}}\right|_{\mathcal{F}_t} = \frac{B_t(T_k)}{B_t(T_n)} \frac{B_0(T_n)}{B_0(T_k)} = \prod_{j=k+1}^{n} \frac{1 + \delta_j L_t^j}{1 + \delta_j L_0^j}, \tag{2}$$

for every $k = 1, \ldots, n-1$. The forward measure $\mathbb{Q}^{T_n}$ is referred to as the terminal forward measure.

2.1 The Driving Process

Let us denote by $(\Omega, \mathcal{F}, \mathbf{F} = (\mathcal{F}_t)_{0 \le t \le T^*}, \mathbb{Q}^{T_n})$ a complete stochastic basis and let X be an $\mathbb{R}^d$-valued Lévy process $(X_t)_{0 \le t \le T^*}$ on this stochastic basis with Lévy measure F and diffusion matrix c. The filtration $\mathbf{F}$ is generated by X and $\mathbb{Q}^{T_n}$ is the forward measure associated with the date T_n, i.e. with the numeraire $B_t(T_n)$. The process X is assumed without loss of generality to be driftless under $\mathbb{Q}^{T_n}$.

Moreover, we assume that $\int_{|z|>1} |z| F(dz) < \infty$. This implies in addition that X is a special semimartingale and allows to choose the truncation function $h(z) = z$, for $z \in \mathbb{R}^d$. The canonical representation of X is given by

$$X_t = \sqrt{c} W_t^{T_n} + \int_0^t \int_{\mathbb{R}^d} z(\mu - \nu^{T_n})(ds, dz), \tag{3}$$

where $W^{T_n} = (W_t^{T_n})_{0 \le t \le T_n}$ denotes a standard d-dimensional Brownian motion with respect to the measure $\mathbb{Q}^{T_n}$, μ is the random measure of jumps of X and $\nu^{T_n}(ds, dz) = F(dz)ds$ is the $\mathbb{Q}^{T_n}$-compensator of μ.

2.2 The Model

Denote by $L = (L^1, \ldots, L^n)^\top$ the column vector of forward Libor rates. We assume that under the terminal measure $\mathbb{Q}^{T_n}$, the dynamics of L is given by the following SDE

$$dL_t = L_{t-}(b(t, L_t)dt + \Lambda(t)dX_t), \tag{4}$$

where $b(t, L_t)$ is the drift term and $\Lambda(t)$ a deterministic $n \times d$ volatility matrix. We write $\Lambda(t) = (\lambda^1(t), \ldots, \lambda^n(t))^\top$, where $\lambda^k(t)$ denotes the d-dimensional volatility vector of the Libor rate L^k and assume that $\lambda^k(t) = 0$, for $t > T_{k-1}$.

One typically assumes that the jumps of X are bounded from below, i.e. $\Delta X_t > C$, for all $t \in [0, T^*]$ and for some strictly negative constant C, which is chosen such that it ensures the positivity of the Libor rates given by (4).

The drift $b(t, L_t) = (b^1(t, L_t), \ldots, b^n(t, L_t))$ is determined by the no-arbitrage requirement that L^k has to be a martingale with respect to $\mathbb{Q}^{T_k}$, for every $k = 1, \ldots, n$. This yields

$$\begin{aligned} b^k(t, L_t) = &- \sum_{j=k+1}^{n} \frac{\delta_j L_t^j}{1 + \delta_j L_t^j} \langle \lambda^k(t), c\lambda^j(t) \rangle \\ &+ \int_{\mathbb{R}^d} \langle \lambda^k(t), z \rangle \left(1 - \prod_{j=k+1}^{n} \left(1 + \frac{\delta_j L_t^j \langle \lambda^j(t), z \rangle}{1 + \delta_j L_t^j} \right) \right) F(dz). \end{aligned} \tag{5}$$

The above drift condition follows from (2) and Girsanov's theorem for semimartingales noticing that

$$\begin{aligned} dL_t^k &= L_{t-}^k(b^k(t, L_t)dt + \lambda^k(t)dX_t) \\ &= L_{t-}^k\lambda^k(t)dX_t^{T_k}, \end{aligned}$$

where

$$X_t^{T_k} = \sqrt{c}W_t^{T_k} + \int_0^t \int_{\mathbb{R}^d} z(\mu - \nu^{T_k})(ds, dz) \tag{6}$$

is a special semimartingale with a d-dimensional $\mathbb{Q}^{T_k}$-Brownian motion W^{T_k} given by

$$dW_t^{T_k} := dW_t^{T_n} - \sqrt{c}\left(\sum_{j=k+1}^{n} \frac{\delta_j L_t^j}{1+\delta_j L_t^j}\lambda^j(t)\right) dt \tag{7}$$

and the $\mathbb{Q}^{T_k}$-compensator ν^{T_k} of μ given by

$$\begin{aligned} \nu^{T_k}(dt, dz) &:= \prod_{j=k+1}^{n}\left(1 + \frac{\delta_j L_{t-}^j}{1+\delta_j L_{t-}^j}\langle\lambda^j(t), z\rangle\right)\nu^{T_n}(dt, dz) \\ &= \prod_{j=k+1}^{n}\left(1 + \frac{\delta_j L_t^j}{1+\delta_j L_t^j}\langle\lambda^j(t), z\rangle\right)F(dz)dt \\ &= F_t^{T_k}(dz)dt \end{aligned} \tag{8}$$

with

$$F_t^{T_k}(dz) := \prod_{j=k+1}^{n}\left(1 + \frac{\delta_j L_t^j}{1+\delta_j L_t^j}\langle\lambda^j(t), z\rangle\right)F(dz). \tag{9}$$

Equalities (7) and (8), and consequently also the drift condition (5), are implied by Girsanov's theorem for semimartingales applied first to the measure change from $\mathbb{Q}^{T_n}$ to $\mathbb{Q}^{T_{n-1}}$ and then proceeding backwards. We refer to Kallsen [11, Proposition 2.6] for a version of Girsanov's theorem that can be directly applied in this case. Note that the random terms $\frac{\delta_j L_t^j}{1+\delta_j L_t^j}$ appear in the measure change due to the fact that for each $j = n, n-1, \ldots, 1$ we have

$$d(1+\delta_j L_t^j) = (1+\delta_j L_{t-}^j)\left(\frac{\delta_j L_{t-}^j}{1+\delta_j L_{t-}^j}b^j(t, L_t)dt + \frac{\delta_j L_{t-}^j}{1+\delta_j L_{t-}^j}\lambda^j(t)dX_t\right). \tag{10}$$

We point out that the predictable random terms $\frac{\delta_j L_{t-}^j}{1+\delta_j L_{t-}^j}$ can be replaced with $\frac{\delta_j L_t^j}{1+\delta_j L_t^j}$ in equalities (5), (7) and (8) due to absolute continuity of the characteristics of X.

Therefore, the vector process of Libor rates L, given in (4) with the drift (5), is a time-inhomogeneous Markov process and its infinitesimal generator under $\mathbb{Q}^{T_n}$ is given by

$$\mathcal{A}_t f(x) = \sum_{i=1}^{n} x_i b^i(t,x)\frac{\partial f(x)}{\partial x_i} + \frac{1}{2}\sum_{i,j=1}^{n} x_i x_j (\Lambda(t)c\Lambda(t)^\top)_{ij}\frac{\partial f(x)}{\partial x_i \partial x_j} \tag{11}$$
$$+ \int_{\mathbb{R}^d}\left(f(\operatorname{diag}(x)(\mathbf{1}+\Lambda(t)z)) - f(x) - \sum_{j=1}^{n} x_j(\Lambda(t)z)_j \frac{\partial f(x)}{\partial x_j}\right) F(dz),$$

for a function $f \in C_0^2(\mathbb{R}^n, \mathbb{R})$ and with the function $b^i(t,x)$, for $i = 1, \dots, n$ and $x = (x_1, \dots, x_n) \in \mathbb{R}^n$, given by

$$b^i(t,x) = -\sum_{j=i+1}^{n} \frac{\delta_j x_j}{1+\delta_j x_j}\langle \lambda^i(t), c\,\lambda^j(t)\rangle$$
$$+ \int_{\mathbb{R}^d} \langle \lambda^i(t), z\rangle \left(1 - \prod_{j=k+1}^{n}\left(1 + \frac{\delta_j x_j \langle \lambda^j(t), z\rangle}{1+\delta_j x_j}\right)\right) F(dz).$$

Remark 1 (Connection to the Lévy Libor model of Eberlein and Özkan [9]) The dynamics of the forward Libor rate L^k, for all $k = 1, \dots, n$, in the Lévy Libor model of Eberlein and Özkan [9] (compare also Eberlein and Kluge [8]) is given as an ordinary exponential of the following form

$$L_t^k = L_0^k \exp\left(\int_0^t \tilde{b}^k(s, L_s)ds + \int_0^t \tilde{\lambda}^k(s)d\tilde{Y}_s\right), \tag{12}$$

for some deterministic volatility vector $\tilde{\lambda}^k$ and the drift $\tilde{b}^k(t, L_t)$ which has to be chosen such that the Libor rate L^k is a martingale under the forward measure $\mathbb{Q}^{T_k}$. Here $\tilde{Y}$ is a d-dimensional Lévy process given by

$$\tilde{Y}_t = \sqrt{c}W_t^{T_n} + \int_0^t \int_{\mathbb{R}^d} z(\tilde{\mu} - \tilde{\nu}^{T_n})(ds, dz),$$

with the $\mathbb{Q}^{T_n}$-characteristics $(0, c, \tilde{F})$, where $\tilde{\nu}^{T_n}(ds, dz) = \tilde{F}(dz)ds$. The Lévy measure $\tilde{F}$ has to satisfy the usual integrability conditions ensuring the finiteness of the exponential moments. The dynamics of L^k is thus given by the following SDE

$$dL_t^k = L_{t-}^k \left(b^k(t, L_t)dt + \sqrt{c}\tilde{\lambda}^k(t)dW_t^{T_n} + (e^{\langle \tilde{\lambda}^k(t), z\rangle} - 1)(\tilde{\mu} - \tilde{\nu}^{T_n})(dt, dz)\right)$$
$$= L_{t-}^k \left(b^k(t, L_t)dt + dY_t^k\right),$$

for all k, where Y^k is a time-inhomogeneous Lévy process given by

$$Y_t^k = \int_0^t \sqrt{c}\tilde{\lambda}^k(s)dW_s^{T_n} + \int_0^t \int_{\mathbb{R}^d} (e^{\langle \tilde{\lambda}^k(s), z\rangle} - 1)(\tilde{\mu} - \tilde{\nu}^{T_n})(ds, dz)$$

and the drift $b^k(t, L_t)$ is given by

$$\begin{aligned} b^k(t, L_t) &= \tilde{b}^k(t, L_t) + \frac{1}{2}\langle \tilde{\lambda}^k(t), c\tilde{\lambda}^k(t)\rangle \\ &\quad + \int_{\mathbb{R}^d} (e^{\langle \tilde{\lambda}^k(t), z\rangle} - 1 - \langle \tilde{\lambda}^k(t), z\rangle)\tilde{F}(dz). \end{aligned}$$

3 Option Pricing via PIDEs

Below we present the pricing PIDEs related to general option payoffs and then more specifically to caplets and swaptions. We price all options under the given terminal measure $\mathbb{Q}^{T_n}$.

3.1 General Payoff

Consider a European-type payoff with maturity T_k given by $\xi = g(L_{T_k})$, for some tenor date T_k. Its time-t price P_t is given by the following risk-neutral pricing formula

$$\begin{aligned} P_t &= B_t(T_k)\mathbb{E}^{\mathbb{Q}^{T_k}}[g(L_{T_k}) \mid \mathcal{F}_t] \\ &= B_t(T_n)\mathbb{E}^{\mathbb{Q}^{T_n}}\left[\frac{B_{T_k}(T_k)}{B_{T_k}(T_n)} g(L_{T_k}) \middle| \mathcal{F}_t\right] \\ &= B_t(T_n)\mathbb{E}^{\mathbb{Q}^{T_n}}\left[\prod_{j=k+1}^{n} (1 + \delta_j L_{T_k}^j) g(L_{T_k}) \middle| \mathcal{F}_t\right] \\ &= B_t(T_n)u(t, L_t), \end{aligned}$$

where u is the solution of the following PIDE[1]

[1] A detailed proof of this statement is out of scope of this note. Here we simply assume that Eq. (13) admits a unique solution which is sufficiently regular and is of polynomial growth. The existence of such a solution may be established first by Fourier methods for the case when there is no drift and then by a fixed-point theorem in Sobolev spaces using the regularizing properties of the Lévy kernel for the general case (see De Franco [7, Chap. 7] for similar arguments). Once the existence of a regular solution has been established, the expression for the option price follows by the standard Feynman-Kac formula.

$$\begin{aligned} \partial_t u + \mathcal{A}_t u &= 0 \\ u(T_k, x) &= \tilde{g}(x) \end{aligned} \tag{13}$$

and $\tilde{g}$ denotes the transformed payoff function given by

$$\tilde{g}(x) := \tilde{g}(x_1, \ldots, x_n) = \prod_{j=k+1}^{n} (1 + \delta_j x_j) g(x_1, \ldots, x_n).$$

In what follows we shall in particular focus on two most liquid interest rate options: caps (caplets) and swaptions.

3.2 Caplet

Consider a caplet with strike K and payoff $\xi = \delta_k (L^k_{T_{k-1}} - K)^+$ at time T_k. Note that here the payoff is in fact a $\mathcal{F}_{T_{k-1}}$-measurable random variable and it is paid at time T_k. This is known as *payment in arrears*. There exist also other conventions for caplet payoffs, but this one is the one typically used.

The time-t price of the caplet, denoted by P_t^{Cpl} is thus given by

$$\begin{aligned} P_t^{Cpl} &= B_t(T_k)\delta_k \mathbb{E}^{\mathbb{Q}^{T_k}}[(L^k_{T_{k-1}} - K)^+ \mid \mathcal{F}_t] \\ &= B_t(T_n)\delta_k \mathbb{E}^{\mathbb{Q}^{T_n}}\left[\prod_{j=k+1}^{n} (1 + \delta_j L^j_{T_{k-1}})(L^k_{T_{k-1}} - K)^+ \middle| \mathcal{F}_t \right] \\ &= B_t(T_n)\delta_k u(t, L_t) \end{aligned} \tag{14}$$

where u is the solution to

$$\begin{aligned} \partial_t u + \mathcal{A}_t u &= 0 \\ u(T_{k-1}, x) &= \tilde{g}(x) \end{aligned} \tag{15}$$

with

$$\tilde{g}(x) := (x_k - K)^+ \prod_{j=k+1}^{n} (1 + \delta_j x_j).$$

For the second equality in (14) we have used the measure change from $\mathbb{Q}^{T_k}$ to $\mathbb{Q}^{T_n}$ given in (2).

Remark 2 Noting that the payoff of the caplet depends on one single underlying forward Libor rate L^k, it is often more convenient to price it directly under the corresponding forward measure $\mathbb{Q}^{T_k}$, using the first equality in (14). Thus, one has

$$P_t^{Cpl} = B_t(T_k)\delta_k u(t, L_t),$$

where u is the solution to

$$\begin{aligned} \partial_t u + \mathcal{A}_t^{T_k} u &= 0 \\ u(T_{k-1}, x) &= \tilde{g}(x) \end{aligned} \tag{16}$$

with $\tilde{g}(x) := (x_k - K)^+$ and where $\mathcal{A}^{T_k}$ is the generator of L under the forward measure $\mathbb{Q}^{T_k}$. In the log-normal LMM this leads directly to the Black's formula for caplet prices. However, in the Lévy Libor model the driving process X under the forward measure $\mathbb{Q}^{T_k}$ is not a Lévy process anymore since its compensator of the random measure of jumps becomes stochastic (see (9)). Therefore, passing to the forward measure in this case does not lead to a closed-form pricing formula and does not bring any particular advantage. This is why in the forthcoming section we shall work directly under the terminal measure $\mathbb{Q}^{T_n}$.

3.3 Swaptions

Let us consider a swaption, written on a fixed-for-floating (payer) interest rate swap with inception date T_0, payment dates $T_1, \dots, T_n$ and nominal $N = 1$. We denote by K the swaption strike rate and assume for simplicity that the maturity T of the swaption coincides with the inception date of the underlying swap, i.e. we assume $T = T_0$. Therefore, the payoff of the swaption at maturity is given by $\left(P^{Sw}(T_0; T_0, T_n, K)\right)^+$, where $P^{Sw}(T_0; T_0, T_n, K)$ denotes the value of the swap with fixed rate K at time T_0 given by

$$\begin{aligned} P^{Sw}(T_0; T_0, T_n, K) &= \sum_{j=1}^{n} \delta_j B_{T_0}(T_j) \mathbb{E}^{\mathbb{Q}^{T_j}} \left[L^j_{T_{j-1}} - K \middle| \mathcal{F}_{T_0} \right] \\ &= \sum_{j=1}^{n} \delta_j B_{T_0}(T_j) \left(L^j_{T_0} - K \right) \\ &= \left(\sum_{j=1}^{n} \delta_j B_{T_0}(T_j) \right) (R(T_0; T_0, T_n) - K) \end{aligned}$$

where

$$R(t; T_0, T_n) = \frac{\sum_{j=1}^{n} \delta_j B_t(T_j) L_t^j}{\sum_{j=1}^{n} \delta_j B_t(T_j)} =: \sum_{j=1}^{n} w_j L_t^j \tag{17}$$

is the swap rate i.e. the fixed rate such that the time-t price of the swap is equal to zero. Here we denote

$$w_j(t) := \frac{\delta_j B_t\left(T_j\right)}{\sum_{k=1}^{n} \delta_k B_t\left(T_k\right)} \tag{18}$$

Note that $\sum_{j=1}^{n} w_j(t) = 1$. Dividing the numerator and the denominator in (17) by $B_t(T_n)$ and using the telescopic products together with (1) we see that $w_j(t) = f_j(L_t)$ for a function f_j given by

$$f_j(x) = \frac{\delta_j \prod_{i=j+1}^{n}(1+\delta_i x_i)}{\sum_{k=1}^{n} \delta_k \prod_{i=k+1}^{n}(1+\delta_i x_i)} \tag{19}$$

for $j = 1, \ldots, n$.

Therefore, the swaption price at time $t \leq T_0$ is given by

$$\begin{aligned}
&P^{Swn}(t; T_0, T_n, K) \\
&= B_t(T_0)\mathbb{E}^{\mathbb{Q}^{T_0}}\left[\left(P^{Sw}(T_0; T_0, T_n, K)\right)^+ \middle| \mathcal{F}_t\right] \\
&= B_t(T_0)\mathbb{E}^{\mathbb{Q}^{T_0}}\left[\left(\sum_{j=1}^{n} \delta_j B_{T_0}(T_j)\right)(R(T_0; T_0, T_n) - K)^+ \middle| \mathcal{F}_t\right] \\
&= B_t(T_n)\mathbb{E}^{\mathbb{Q}^{T_n}}\left[\frac{\sum_{j=1}^{n} \delta_j B_{T_0}(T_j)}{B_{T_0}(T_n)}(R(T_0; T_0, T_n) - K)^+ \middle| \mathcal{F}_t\right] \\
&= B_t(T_n)u(t, L_t)
\end{aligned} \tag{20}$$

where u is the solution to

$$\begin{aligned}
\partial_t u + \mathcal{A}_t u &= 0 \\
u(T_0, x) &= \tilde{g}(x)
\end{aligned} \tag{21}$$

with $\tilde{g}(x) := \delta_n f_n(x)^{-1}\left(\sum_{j=1}^{n} f_j(x)x_j - K\right)^+$.

4 Approximate Pricing

4.1 Approximate Pricing for General Payoffs Under the Terminal Measure

Following an approach introduced by Černý et al. [6], we introduce a small parameter into the model by defining the rescaled Lévy process $X_t^\alpha := \alpha X_{t/\alpha^2}$ with $\alpha \in (0, 1)$.

The process X^α is a martingale Lévy process under the terminal measure $\mathbb{Q}^{T_n}$ with characteristic triplet $(0, c, F_\alpha)$ with respect to the truncation function $h(z) = z$, where

$$F_\alpha(A) = \frac{1}{\alpha^2} F(\{z \in \mathbb{R}^d : z\alpha \in A\}, \quad \text{for } A \in \mathcal{B}(\mathbb{R}^d).$$

We now consider a family of Lévy Libor models driven by the processes X^α, $\alpha \in (0, 1)$, and defined by

$$dL_t^\alpha = L_{t-}^\alpha (b_\alpha(t, L_t^\alpha)dt + \Lambda(t)dX_t^\alpha), \tag{22}$$

where the drift b_α is given by (5) with F replaced by F_α. Substituting the explicit form of F_α, we obtain

$$\begin{aligned}
b_\alpha^k(t, L_t) &= -\sum_{j=k+1}^{n} \frac{\delta_j L_t^j}{1+\delta_j L_t^j} \langle \lambda^k(t), c\lambda^j(t)\rangle \\
&\quad + \frac{1}{\alpha} \int_{\mathbb{R}^d} \langle \lambda^k(t), z\rangle \left(1 - \prod_{j=k+1}^{n} \left(1 + \frac{\alpha \delta_j L_t^j \langle \lambda^j(t), z\rangle}{1+\delta_j L_t^j}\right)\right) F(dz) \\
&= -\sum_{j_0=k+1}^{n} \Sigma_{kj_0}(t) \frac{\delta_{j_0} L_t^{j_0}}{1+\delta_{j_0} L_t^{j_0}} \\
&\quad - \sum_{p=1}^{n-k-1} \alpha^p \sum_{j_0=k+1}^{n} \sum_{j_1=j_0+1}^{n} \cdots \sum_{j_p=j_{p-1}+1}^{n} M_t^{p+2}(\lambda^k, \lambda^{j_0}, \ldots, \lambda^{j_p}) \prod_{l=0}^{p} \frac{\delta_{j_l} L_t^{j_l}}{1+\delta_{j_l} L_t^{j_l}} \\
&=: -\sum_{p=0}^{n-k-1} \alpha^p b_p^k(t, L_t)
\end{aligned}$$

where we define

$$\Sigma_{ij}(t) := (\Lambda(t)c\Lambda(t)^\top)_{ij} + \int_{\mathbb{R}^d} \langle \lambda^i(t), z\rangle \langle \lambda^j(t), z\rangle F(dz), \tag{23}$$

for all $i, j = 1, \ldots, n$, and

$$M_t^k(\lambda^1, \ldots, \lambda^k) := \int_{\mathbb{R}^d} \prod_{p=1}^{k} \langle \lambda^p(t), z\rangle F(dz) \tag{24}$$

for all $k = 1, \ldots, n$. We denote the infinitesimal generator of L^α by $\mathcal{A}_t^\alpha$. For a smooth function $f : \mathbb{R}^d \to \mathbb{R}$, the infinitesimal generator $\mathcal{A}_t^\alpha f$ can be expanded in powers of α as follows:

$$\mathcal{A}_t^\alpha f(x) = \sum_{i=1}^n b_\alpha^i(t,x) x_i \frac{\partial f(x)}{\partial x_i} + \frac{1}{2}\sum_{i,j=1}^n \Sigma_{ij}(t) x_i x_j \frac{\partial^2 f(x)}{\partial x_i \partial x_j}$$
$$+ \sum_{k=3}^\infty \sum_{i_1,\dots,i_k=1}^n \frac{\alpha^{k-2}}{k!} x_{i_1}\dots x_{i_k} \frac{\partial^k f(x)}{\partial x_{i_1}\dots\partial x_{i_k}} M_t^k(\lambda^{i_1},\dots,\lambda^{i_k}).$$

Consider now a financial product whose price is given by a generic PIDE of the form (13) with $\mathcal{A}_t$ replaced by $\mathcal{A}_t^\alpha$. Assuming sufficient regularity,[2] one may expand the solution u^α in powers of α:

$$u^\alpha(t,x) = \sum_{p=0}^\infty \alpha^p u_p(t,x). \tag{25}$$

Substituting the expansions for $\mathcal{A}_t^\alpha$ and b_α into this equation, and gathering terms with the same power of α, we obtain an 'open-ended' system of PIDE for the terms in the expansion of u^α.

The zero-order term u_0 satisfies

$$\partial_t u_0 + \mathcal{A}_t^0 u_0 = 0, \quad u_0(T_k, x) = \tilde{g}(x)$$

with

$$\mathcal{A}_t^0 u_0(t,x) = \sum_{i=1}^n b_0^i(t,x) x_i \frac{\partial u_0(t,x)}{\partial x_i} + \frac{1}{2}\sum_{i,j=1}^n \Sigma_{ij}(t) x_i x_j \frac{\partial^2 u_0(t,x)}{\partial x_i \partial x_j} \tag{26}$$

$$b_0^i(t,x) = -\sum_{j=i+1}^n \Sigma_{ij}(t) \frac{\delta_j x_j}{1+\delta_j x_j}. \tag{27}$$

Hence, by the Feynman-Kac formula

$$u_0(t,x) = E^{\mathbb{Q}^{T_n}}\left[\tilde{g}(X_{T_k}^{t,x})\right] \tag{28}$$

where the process $X^{t,x} = (X^{i,t,x})_{i=1}^n$ satisfies the stochastic differential equation

$$dX_s^{i,t,x} = X_s^{i,t,x}\{b_0^i(s, X_s^{i,t,x})ds + \sigma_i dW_s\}, \quad X_t^{i,t,x} = x_i, \tag{29}$$

with W a d-dimensional standard Brownian motion with respect to $\mathbb{Q}^{T_n}$ and σ an $n \times d$-dimensional matrix such that $\sigma\sigma^\top = (\Sigma_{i,j})_{i,j=1}^n$.

To obtain an explicit approximation for the higher order terms $u_1(t,x)$ and $u_2(t,x)$ given above, we consider the following proposition.

[2] See Ménassé and Tankov [14] for rigorous arguments in a simplified but similar setting.

Proposition 3 *Let Y be an n-dimensional log-normal process whose components follow the dynamics*

$$dY_t^i = Y_t^i(\mu_i(t)dt + \sigma_i(t)dW_t),$$

where μ and σ are measurable functions such that

$$\int_0^T (\|\mu(t)\| + \|\sigma(t)\|^2)dt < \infty$$

and for all $y \in \mathbb{R}^n$ and some $\varepsilon > 0$,

$$\inf_{0 \le t \le T} y\sigma(t)\sigma(t)^T y^T \ge \varepsilon \|y\|^2.$$

We denote by $Y^{t,y}$ the process starting from y at time t, and by $Y^{t,y,i}$ the i-th component of this process. Let f be a bounded measurable function and define

$$v(t, y) = \mathbb{E}[f(Y_T^{t,y})].$$

Then, for all $i_1, \ldots, i_m$, the process

$$Y_s^{t,y,i_1} \ldots Y_s^{t,y,i_m} \frac{\partial^m v(Y_s^{t,y})}{\partial y_{i_1} \ldots \partial y_{i_m}}, \quad s \ge t,$$

is a martingale.

The proof can be carried out by direct differentiation for smooth f together with a standard approximation argument for a general measurable f.

Furthermore, we assume the following simplification for the drift terms:

For all $i = 1, \ldots, n-1$ and $p = 1, \ldots, n-k-1$, the random quantities in the terms $b_p^i(t, L_t)$ in the expansion of the drift of the Libor rates under the terminal measure are constant and equal to their value at time t, i.e. for all $j = 1, \ldots, n$:

$$\frac{\delta_j L_s^j}{1 + \delta_j L_s^j} = \frac{\delta_j L_t^j}{1 + \delta_j L_t^j}, \quad \text{for all } s \ge t. \tag{30}$$

This simplification is known as *freezing of the drift* and is often used for pricing in the Libor market models.

Coming back now to the first-order term u_1, we see that it is the solution of

$$\partial_t u_1 + \mathcal{A}_t^0 u_1 + \mathcal{A}_t^1 u_0 = 0, \quad u_1(T_k, x) = 0 \tag{31}$$

with

$$\mathcal{A}_t^1 u_0(t,x) = \sum_{j=1}^{n} b_1^j(t,x) x_j \frac{\partial u_0(t,x)}{\partial x_j} + \frac{1}{6} \sum_{i_1,i_2,i_3=1}^{n} x_{i_1} x_{i_2} x_{i_3} \frac{\partial^3 u_0(t,x)}{\partial x_{i_1} \partial x_{i_2} \partial x_{i_3}} M_t^3(\lambda^{i_1}, \lambda^{i_2}, \lambda^{i_3}) \tag{32}$$

and the drift term

$$b_1^j(t,x) = -\sum_{j_0=j+1}^{n} \sum_{j_1=j_0+1}^{n} M_t^3(\lambda^j, \lambda^{j_0}, \lambda^{j_1}) \frac{\delta_{j_0} x_{j_0}}{1+\delta_{j_0} x_{j_0}} \frac{\delta_{j_1} x_{j_1}}{1+\delta_{j_1} x_{j_1}}. \tag{33}$$

Moreover,

$$\mathcal{A}_t^0 u_1(t,x) = \sum_{i=1}^{n} b_0^i(t,x) x_i \frac{\partial u_1(t,x)}{\partial x_i} + \frac{1}{2} \sum_{i,j=1}^{n} \Sigma_{ij}(t) x_i x_j \frac{\partial^2 u_1(t,x)}{\partial x_i \partial x_j}.$$

We have

Lemma 4 *Consider the model* (22). *Under the simplification* (30), *the first-order term $u_1(t,x)$ in the expansion* (25) *can be approximated by*

$$\begin{aligned} u_1(t,x) \approx{}& \frac{1}{6} \sum_{i_1,i_2,i_3=1}^{n} x_{i_1} x_{i_2} x_{i_3} \frac{\partial^3 u_0(t,x)}{\partial x_{i_1} \partial x_{i_2} \partial x_{i_3}} \int_t^{T_k} M_s^3(\lambda^{i_1}, \lambda^{i_2}, \lambda^{i_3}) ds \\ &- \sum_{j=1}^{n} \sum_{j_0=j+1}^{n} \sum_{j_1=j_0+1}^{n} \frac{\delta_{j_0} x_{j_0}}{1+\delta_{j_0} x_{j_0}} \frac{\delta_{j_1} x_{j_1}}{1+\delta_{j_1} x_{j_1}} x_j \frac{\partial u_0(t,x)}{\partial x_j} \int_t^{T_k} M_s^3(\lambda^j, \lambda^{j_0}, \lambda^{j_1}) ds \\ =:{}& \tilde{u}_1(t,x). \end{aligned} \tag{34}$$

Proof Applying the Feynman-Kac formula to (31), we have,

$$\begin{aligned} u_1(t,x) ={}& \frac{1}{6} \int_t^{T_k} ds \sum_{i_1,i_2,i_3=1}^{n} M_s^3(\lambda^{i_1}, \lambda^{i_2}, \lambda^{i_3}) \mathbb{E}^{\mathbb{Q}^{T_n}} \left[X_s^{t,x,i_1} X_s^{t,x,i_2} X_s^{t,x,i_3} \frac{\partial^3 u_0(s, X_s^{t,x})}{\partial x_{i_1} \partial x_{i_2} \partial x_{i_3}} \right] \\ &+ \int_t^{T_k} ds \sum_{j=1}^{n} \mathbb{E}^{\mathbb{Q}^{T_n}} \left[b_1^j(s, X_s^{t,x}) X_s^{t,x,j} \frac{\partial u_0(s, X_s^{t,x})}{\partial x_j} \right], \end{aligned} \tag{35}$$

with the process $(X_s^{t,x})$ defined by (29). Under the simplification (30), we can apply Proposition 3 to obtain (34). □

Similarly, the second-order term u_2 is the solution of

$$\partial_t u_2 + \mathcal{A}_t^0 u_2 + \mathcal{A}_t^1 u_1 + \mathcal{A}_t^2 u_0 = 0, \quad u_2(T_k, x) = 0 \tag{36}$$

with

$$\mathcal{A}_t^2 u_0(t,x) = \sum_{j=1}^{n} b_2^j(t,x) x_j \frac{\partial u_0(t,x)}{\partial x_j} + \frac{1}{24} \sum_{i_1,i_2,i_3,i_4=1}^{n} x_{i_1} x_{i_2} x_{i_3} x_{i_4} \frac{\partial^4 u_0(t,x)}{\partial x_{i_1} \partial x_{i_2} \partial x_{i_3} x_{i_4}} M_t^4(\lambda^{i_1}, \lambda^{i_2}, \lambda^{i_3}, \lambda^{i_4}) \tag{37}$$

and the drift

$$b_2^j(t,x) = -\sum_{j_0=j+1}^{n} \sum_{j_1=j_0+1}^{n} \sum_{j_2=j_1+1}^{n} M_t^4(\lambda^j, \lambda^{j_0}, \lambda^{j_1}, \lambda^{j_2}) \frac{\delta_{j_0} x_{j_0}}{1+\delta_{j_0} x_{j_0}} \cdot \frac{\delta_{j_1} x_{j_1}}{1+\delta_{j_1} x_{j_1}} \frac{\delta_{j_2} x_{j_2}}{1+\delta_{j_2} x_{j_2}}. \tag{38}$$

Lemma 5 *Consider the model* (22)*. Under the simplification* (30)*, the second-order term $u_2(t,x)$ in the expansion* (25) *can be approximated by*

$$u_2(t,x) \approx \tilde{u}_2(t,x) := \tilde{E}_1 + \tilde{E}_2 + \tilde{E}_3 + \tilde{E}_4, \tag{39}$$

with

$$\tilde{E}_1 := \frac{1}{6} \sum_{i_1,i_2,i_3=1}^{n} x_{i_1} x_{i_2} x_{i_3} \int_t^{T_k} ds M_s^3(\lambda^{i_1}, \lambda^{i_2}, \lambda^{i_3}) \cdot \left[\frac{1}{6} \sum_{i_4,i_5,i_6=1}^{n} \left(\int_s^{T_k} M_v^3(\lambda^{i_4}, \lambda^{i_5}, \lambda^{i_6}) dv \right) \frac{\partial^3 v^{i_4,i_5,i_6}(t,x)}{\partial x_{i_1} \partial x_{i_2} \partial x_{i_3}} - \sum_{j_4=1}^{n} \sum_{j_5=j_4+1}^{n} \sum_{j_6=j_5+1}^{n} \left(\int_s^{T_k} M_v^3(\lambda^{j_4}, \lambda^{j_5}, \lambda^{j_6}) dv \right) \frac{\partial^3 \bar{v}^{j_4,j_5,j_6}(t,x)}{\partial x_{i_1} \partial x_{i_2} \partial x_{i_3}} \right] \tag{40}$$

$$\tilde{E}_2 := -\sum_{j=1}^{n} \sum_{j_0=j+1}^{n} \sum_{j_1=j_0+1}^{n} \frac{\delta_{j_0} x_{j_0}}{1+\delta_{j_0} x_{j_0}} \frac{\delta_{j_1} x_{j_1}}{1+\delta_{j_1} x_{j_1}} x_j \int_t^{T_k} ds M_s(\lambda^j, \lambda^{j_0}, \lambda^{j_1}) \cdot \left[\frac{1}{6} \sum_{i_4,i_5,i_6=1}^{n} \left(\int_s^{T_k} M_v^3(\lambda^{i_4}, \lambda^{i_5}, \lambda^{i_6}) dv \right) \frac{\partial v^{i_4,i_5,i_6}(t,x)}{\partial x_j} - \sum_{j_4=1}^{n} \sum_{j_5=j_4+1}^{n} \sum_{j_6=j_5+1}^{n} \left(\int_s^{T_k} M_v^3(\lambda^{j_4}, \lambda^{j_5}, \lambda^{j_6}) dv \right) \frac{\partial^3 \bar{v}^{j_4,j_5,j_6}(t,x)}{\partial x_j} \right] \tag{41}$$

$$\tilde{E}_3 := \frac{1}{24} \sum_{i_1,i_2,i_3,i_4=1}^{n} x_{i_1} x_{i_2} x_{i_3} x_{i_4} \frac{\partial^4 u_0(t,x)}{\partial x_{i_1} \partial x_{i_2} \partial x_{i_3} \partial x_{i_4}} \int_t^{T_k} ds M_s^4(\lambda^{i_1}, \lambda^{i_2}, \lambda^{i_3}, \lambda^{i_4}) \tag{42}$$

and

$$\tilde{E}_4 := -\sum_{j=1}^{n} \sum_{j_0=j+1}^{n} \sum_{j_1=j_0+1}^{n} \sum_{j_2=j_1+1}^{n} \frac{\delta_{j_0} x_{j_0}}{1+\delta_{j_0} x_{j_0}} \frac{\delta_{j_1} x_{j_1}}{1+\delta_{j_1} x_{j_1}} \frac{\delta_{j_2} x_{j_2}}{1+\delta_{j_2} x_{j_2}} x_j \frac{\partial u_0(t,x)}{\partial x_j} \cdot \int_t^{T_k} M_s^4(\lambda^j, \lambda^{j_0}, \lambda^{j_1}, \lambda^{j_2}) ds \quad (43)$$

where we define

$$v^{i,j,l}(t,x) := x_i x_j x_l \frac{\partial^3 u_0(t,x)}{\partial x_i \partial x_j \partial x_l} \quad (44)$$

for all $i, j, l = 1, \ldots, n$ *and*

$$\bar{v}^{i,j,l}(t,x) := x_i \frac{\delta_j x_j}{1+\delta_j x_j} \frac{\delta_l x_l}{1+\delta_l x_l} \frac{\partial u_0(t,x)}{\partial x_i} \quad (45)$$

for all $i = 1, \ldots, n$, $j = i+1, \ldots, n$ *and* $l = j+1, \ldots, n$.

Proof Once again by the Feynman-Kac formula applied to (36) we have

$$\begin{aligned} u_2(t,x) &= \frac{1}{6} \int_t^{T_k} ds \sum_{i_1,i_2,i_3=1}^{n} M_s^3(\lambda^{i_1}, \lambda^{i_2}, \lambda^{i_3}) \mathbb{E}^{\mathbb{Q}^{T_n}} \left[X_s^{t,x,i_1} X_s^{t,x,i_2} X_s^{t,x,i_3} \frac{\partial^3 u_1(s, X_s^{t,x})}{\partial x_{i_1} \partial x_{i_2} \partial x_{i_3}} \right] \\ &+ \int_t^{T_k} ds \sum_{j=1}^{n} \mathbb{E}^{\mathbb{Q}^{T_n}} \left[b_1^j(s, X_s^{t,x}) X_s^{t,x,j} \frac{\partial u_1(s, X_s^{t,x})}{\partial x_j} \right] \\ &+ \frac{1}{24} \int_t^{T_k} ds \sum_{i_1,i_2,i_3,i_4=1}^{n} M_s^4(\lambda^{i_1}, \lambda^{i_2}, \lambda^{i_3}, \lambda^{i_4}) \\ &\quad \cdot \mathbb{E}^{\mathbb{Q}^{T_n}} \left[X_s^{t,x,i_1} X_s^{t,x,i_2} X_s^{t,x,i_3} X_s^{t,x,i_4} \frac{\partial^4 u_0(s, X_s^{t,x})}{\partial x_{i_1} \partial x_{i_2} \partial x_{i_3} \partial x_{i_4}} \right] \\ &+ \int_t^{T_k} ds \sum_{j=1}^{n} \mathbb{E}^{\mathbb{Q}^{T_n}} \left[b_2^j(s, X_s^{t,x}) X_s^{t,x,j} \frac{\partial u_0(s, X_s^{t,x})}{\partial x_j} \right] \\ &=: E_1 + E_2 + E_3 + E_4 \end{aligned} \quad (46)$$

with the process $(X_s^{t,x})$ given by (29), $b_1^j(s,x)$ by (33) and $b_2^j(s,x)$ by (38).

In order to obtain an explicit expression for $u_2(t,x)$, we apply Proposition 3 combined with the simplification (30) for the drift terms b_1^j and b_2^j above. More precisely, the expressions for the third and the fourth expectation, which are present in the terms E_3 and E_4, follow by a straightforward application of Proposition 3 after using the simplification for b_2^j. We get

$$E_3 \approx \tilde{E}_3 \qquad \text{and} \qquad E_4 \approx \tilde{E}_4$$

with $\tilde{E}_3$ and $\tilde{E}_4$ given by (42) and (43), respectively.

To obtain explicit expressions for E_1 and E_2, firstly we insert the expression for $u_1(s, X_s^{t,x})$ as given by (35). After some straightforward calculations, based again on the application of Proposition 3 and the simplification (30) for b_1^j, which yields

$$E_1 \approx \tilde{E}_1 \qquad \text{and} \qquad E_2 \approx \tilde{E}_2$$

with $\tilde{E}_1$ and $\tilde{E}_2$ given by (40) and (41), respectively.

Collecting the terms above concludes the proof. □

Summarizing, we get the following expansion for the time-t price $P^\alpha(t; g)$ of the payoff $g(L_{T_k})$ when $\alpha \to 0$.

Proposition 6 *Consider the model* (22) *and a European-type payoff with maturity* T_k *given by* $\xi = g(L_{T_k})$. *Assuming* (30), *its time-t price* $P^\alpha(t; g)$ *for* $\alpha \to 0$ *satisfies*

$$P^\alpha(t; g) = P_0(t; g) + \alpha P_1(t; g) + \alpha^2 P_2(t; g) + O(\alpha^3), \tag{47}$$

with

$$P_0(t; g) := B_t(T_n)u_0(t, L_t) =: P^{LMM}(t; g)$$

$$P_1(t; g) := B_t(T_n)u_1(t, L_t) \approx B_t(T_n)\tilde{u}_1(t, L_t)$$

$$P_2(t; g) := B_t(T_n)u_2(t, L_t) \approx B_t(T_n)\tilde{u}_2(t, L_t)$$

where $P^{LMM}(t; g)$ *denotes the time-t price of the payoff* $g(L_{T_k})$ *in the log-normal LMM with covariance matrix* Σ *and the drift given by* (27), $u_0(t, x)$ *is given by* (28) *and* $\tilde{u}_1(t, x)$ *and* $\tilde{u}_2(t, x)$ *by* (34) *and* (39), *respectively.*

4.2 Approximate Pricing of Caplets

Recalling that the caplet price is given by (14), where u is the solution of the PIDE (15), we can approximate this price using the development

$$u^\alpha(t, x) = u_0(t, x) + \alpha u_1(t, x) + \alpha^2 u_2(t, x) + O(\alpha^3)$$

where the zero-order term u_0 satisfies

$$\partial_t u_0 + \mathcal{A}_t^0 u_0 = 0, \quad u_0(T_{k-1}, x) = (x_k - K)^+ \prod_{j=k+1}^{n} (1 + \delta_j x_j)$$

$$\text{with} \quad \mathcal{A}_t^0 u_0 = \sum_{i=1}^{n} b_0^i(t,x) x_i \frac{\partial u_0(t,x)}{\partial x_i} + \frac{1}{2} \sum_{i,j=1}^{n} \Sigma_{ij}(t) x_i x_j \frac{\partial^2 u_0(t,x)}{\partial x_i \partial x_j}$$

$$\text{and} \quad b_0^i(t,x) = -\sum_{j=i+1}^{n} \Sigma_{ij}(t) \frac{\delta_j x_j}{1 + \delta_j x_j}.$$

The solution to the above PDE can be found via the Feynman-Kac formula, where the conditional expectation is computed in the log-normal LMM model with covariation matrix $(\Sigma_{ij})_{i,j=1}^n$ as in Sect. 4.1. Performing a measure change from $\mathbb{Q}^{T_n}$ to $\mathbb{Q}^{T_k}$ and denoting by $P_{BS}(V, S, K)$ the Black-Scholes price of a call option with variance V,

$$P_{BS}(V, S, K) = \mathbb{E}\left[\left(Se^{-\frac{V}{2}+\sqrt{V}Z} - K\right)^+\right], \quad Z \sim N(0,1),$$

we see that the zero-order term is given by

$$u_0(t,x) = P_{BS}(V_{t,T_{k-1}}^{Cpl}, x_k, K) \prod_{j=k+1}^{n} (1 + \delta_j x_j), \tag{48}$$

where

$$V_{t,T}^{Cpl} := \int_t^T \Sigma_{kk}(s) ds. \tag{49}$$

Now, in complete analogy to the case of a general payoff, the first-order term $u_1(t,x)$ and the second-order term $u_2(t,x)$ are given by (35) and (46), respectively, with $u_0(t,x)$ as in (48). Noting that $u_0(t,x)$ depends only on $x_k, x_{k+1}, \ldots, x_n$, the derivatives of $u_0(t,x)$ with respect to $x_1, \ldots, x_{k-1}$ are zero and the sums in (35) and (46) in fact start from the index k. An application of Proposition 3 and simplification (30) thus yields the following proposition, which provides an approximation of the caplet price $P^{Cpl,\alpha}(t; T_{k-1}, T_k, K)$ when $\alpha \to 0$.

Proposition 7 *Consider the model* (22) *and a caplet with strike K and maturity T_{k-1}. Assuming* (30), *its time-t price $P^{Cpl,\alpha}(t; T_{k-1}, T_k, K)$ for $\alpha \to 0$ satisfies*

$$\begin{aligned} P^{Cpl,\alpha}(t; T_{k-1}, T_k, K) = {} & P_0^{Cpl}(t; T_{k-1}, T_k, K) + \alpha P_1^{Cpl}(t; T_{k-1}, T_k, K) \\ & + \alpha^2 P_2^{Cpl}(t; T_{k-1}, T_k, K) + O(\alpha^3), \end{aligned} \tag{50}$$

with

$$P_0^{Cpl}(t; T_{k-1}, T_k, K) := B_t(T_n)\delta_k u_0(t, L_t)$$
$$= B_t(T_n)\delta_k P_{BS}(V^{Cpl}_{t,T_{k-1}}, L_t^k, K) \prod_{j=k+1}^{n} (1 + \delta_j L_t^j)$$

$$P_1^{Cpl}(t; T_{k-1}, T_k, K)$$
$$:= B_t(T_n)\delta_k \left\{ \frac{1}{6} \sum_{i_1,i_2,i_3=k}^{n} L_t^{i_1} L_t^{i_2} L_t^{i_3} \frac{\partial^3 u_0(t,x)}{\partial x_{i_1}\partial x_{i_2}\partial x_{i_3}}\Big|_{x=L_t} \int_t^{T_{k-1}} M_s^3(\lambda^{i_1}, \lambda^{i_2}, \lambda^{i_3})ds \right.$$
$$- \sum_{j=k}^{n} \sum_{j_0=j+1}^{n} \sum_{j_1=j_0+1}^{n} \frac{\delta_{j_0} L_t^{j_0}}{1+\delta_{j_0} L_t^{j_0}} \frac{\delta_{j_1} L_t^{j_1}}{1+\delta_{j_1} L_t^{j_1}} L_t^j \frac{\partial u_0(t,x)}{\partial x_j}\Big|_{x=L_t}$$
$$\left. \cdot \int_t^{T_{k-1}} M_s^3(\lambda^j, \lambda^{j_0}, \lambda^{j_1})ds \right\}$$

$$P_2^{Cpl}(t; T_{k-1}, T_k, K)$$
$$:= B_t(T_n)\delta_k \left\{ \frac{1}{6} \sum_{i_1,i_2,i_3=k}^{n} L_t^{i_1} L_t^{i_2} L_t^{i_3} \int_t^{T_{k-1}} ds M_s^3(\lambda^{i_1}, \lambda^{i_2}, \lambda^{i_3}) \right.$$
$$\cdot \left[\frac{1}{6} \sum_{i_4,i_5,i_6=k}^{n} \left(\int_s^{T_{k-1}} M_v^3(\lambda^{i_4}, \lambda^{i_5}, \lambda^{i_6})dv \right) \frac{\partial^3 v^{i_4,i_5,i_6}(t,x)}{\partial x_{i_1}\partial x_{i_2}\partial x_{i_3}}\Big|_{x=L_t} \right.$$
$$\left. - \sum_{j_4=k}^{n} \sum_{j_5=j_4+1}^{n} \sum_{j_6=j_5+1}^{n} \left(\int_s^{T_{k-1}} M_v^3(\lambda^{j_4}, \lambda^{j_5}, \lambda^{j_6})dv \right) \frac{\partial^3 \bar{v}^{j_4,j_5,j_6}(t,x)}{\partial x_{i_1}\partial x_{i_2}\partial x_{i_3}}\Big|_{x=L_t} \right]$$

$$- \sum_{j=k}^{n} \sum_{j_0=j+1}^{n} \sum_{j_1=j_0+1}^{n} \frac{\delta_{j_0} L_t^{j_0}}{(1+\delta_{j_0} L_t^{j_0}} \frac{\delta_{j_1} L_t^{j_1}}{(1+\delta_{j_1} L_t^{j_1}} L_t^j \int_t^{T_{k-1}} ds M_s(\lambda^j, \lambda^{j_0}, \lambda^{j_1})$$
$$\cdot \left[\frac{1}{6} \sum_{i_4,i_5,i_6=k}^{n} \left(\int_s^{T_{k-1}} M_v^3(\lambda^{i_4}, \lambda^{i_5}, \lambda^{i_6})dv \right) \frac{\partial v^{i_4,i_5,i_6}(t,x)}{\partial x_j}\Big|_{x=L_t} \right.$$
$$\left. - \sum_{j_4=k}^{n} \sum_{j_5=j_4+1}^{n} \sum_{j_6=j_5+1}^{n} \left(\int_s^{T_{k-1}} M_v^3(\lambda^{j_4}, \lambda^{j_5}, \lambda^{j_6})dv \right) \frac{\partial^3 \bar{v}^{j_4,j_5,j_6}(t,x)}{\partial x_j}\Big|_{x=L_t} \right]$$

$$+ \frac{1}{24} \sum_{i_1,i_2,i_3,i_4=k}^{n} L_t^{i_1} L_t^{i_2} L_t^{i_3} L_t^{i_4} \frac{\partial^4 u_0(t,x)}{\partial x_{i_1}\partial x_{i_2}\partial x_{i_3}\partial x_{i_4}}\Big|_{x=L_t} \int_t^{T_{k-1}} M_s^4(\lambda^{i_1}, \lambda^{i_2}, \lambda^{i_3}, \lambda^{i_4})ds$$

$$- \sum_{j=k}^{n} \sum_{j_0=j+1}^{n} \sum_{j_1=j_0+1}^{n} \sum_{j_2=j_1+1}^{n} \frac{\delta_{j_0} L_t^{j_0}}{1+\delta_{j_0} L_t^{j_0}} \frac{\delta_{j_1} L_t^{j_1}}{1+\delta_{j_1} L_t^{j_1}} \frac{\delta_{j_2} L_t^{j_2}}{1+\delta_{j_2} L_t^{j_2}} L_t^j \frac{\partial u_0(t,x)}{\partial x_j}\Big|_{x=L_t}$$
$$\left. \cdot \int_t^{T_{k-1}} M_s^4(\lambda^j, \lambda^{j_0}, \lambda^{j_1}, \lambda^j)ds \right\} .$$

with $V_{t,T_{k-1}}^{Cpl}$ *given by* (49), $u_0(t,x)$ *by* (48), *the terms* $M_s^3(\cdot)$ *and* $M_s^4(\cdot)$ *by* (24) *and* $v^{i_4,i_5,i_6}(t,x)$ *and* $\bar{v}^{j_4,j_5,j_6}(t,x)$ *by* (44) *and* (45), *respectively.*

Remark 8 Recalling that

$$u_0(t,x) = P_{BS}(V_{t,T}^{Cpl}, x_k, K) \prod_{j=k+1}^{n} (1+\delta_j x_j)$$

we see that the functions v and $\bar{v}$ given by

$$v^{i,j,l}(t,x) := x_i x_j x_l \frac{\partial^3 u_0(t,x)}{\partial x_i \partial x_j \partial x_l}$$

for all $i, j, l = k, \ldots, n$ and

$$\bar{v}^{i,j,l}(t,x) := x_i \frac{\delta_j x_j}{1+\delta_j x_j} \frac{\delta_l x_l}{1+\delta_l x_l} \frac{\partial u_0(t,x)}{\partial x_i}$$

for all $i = k, \ldots, n$, $j = i+1, \ldots, n$ and $l = j+1, \ldots, n$, become in fact linear combinations of the terms which are polynomials in x multiplied by derivatives of $P_{BS}(\cdot)$ up to order three.

4.3 Approximate Pricing of Swaptions

Let us consider a swaption defined in Sect. 3.3. For swaption pricing we again use the general result under the terminal measure $\mathbb{Q}^{T_n}$ given in Proposition 6. The price of the swaption $P^{Swn}(t; T_0, T_n, K)$ then satisfies

$$\begin{aligned} P^{Swn}(t; T_0, T_n, K) &= B_t(T_n)(u_0(t, L_t) + \alpha u_1(t, L_t) + \alpha^2 u_2(t, L_t)) + O(\alpha^3) \\ &=: P_0^{Swn}(t; T_0, T_n, K) + \alpha P_1^{Swn}(t; T_0, T_n, K) \\ &\quad + \alpha^2 P_2^{Swn}(t; T_0, T_n, K) + O(\alpha^3), \end{aligned}$$

where the function u_0 satisfies the equation

$$\partial_t u_0 + \mathcal{A}_t^0 u_0 = 0, \qquad u_0(T_0, x) = \tilde{g}(x)$$

with $\tilde{g}(x) = \delta_n f_n(x)^{-1} \left(\sum_{j=1}^{n} f_j(x) x_j - K\right)^+$. We see that the zero-order term $P_0^{Swn}(t; T_0, T_n, K)$ corresponds to the price of the swaption in the log-normal LMM model with volatility matrix $\Sigma(t)$.

The function u_0 related to the swaption price in the log-normal LMM is of course not known in explicit form but one can use various approximations developed in the literature (Jäckel and Rebonato [10], Schoenmakers [18]). To introduce the approximation of Jäckel and Rebonato [10], we compute the quadratic variation of the log swap rate expressed as function of Libor rates:

$$R(t; T_0, T_n) = R(L_t^1, \dots, L_t^n) = \frac{\sum_{j=1}^n \delta_j L_t^j \prod_{k=1}^j (1 + \delta_k L_t^k)}{\sum_{j=1}^n \delta_j \prod_{k=1}^j (1 + \delta_k L_t^k)}.$$

$$\langle \log R(\cdot; T_0, T_n) \rangle_T = \int_0^T \frac{d\langle R(\cdot; T_0, T_n) \rangle_t}{R(t; T_0, T_n)^2} = \int_0^T \sum_{i,j=1}^n \frac{\partial R(L_t)}{\partial L^i} \frac{\partial R(L_t)}{\partial L^j} \frac{d\langle L^i, L^j \rangle_t}{R(t; T_0, T_n)^2}$$

$$= \int_0^T \sum_{i,j=1}^n \frac{\partial R(L_t)}{\partial L^i} \frac{\partial R(L_t)}{\partial L^j} \frac{L_t^i L_t^j \Sigma_{ij}(t) dt}{R(t; T_0, T_n)^2}.$$

The approximation of Jäckel and Rebonato [10] consists in replacing all stochastic processes in the above integral by their values at time 0; in other words, the swap rate becomes a log-normal random variable such that $\log R(t; T_0, T_n)$ has variance

$$V_T^{swap} = \sum_{i,j=1}^n \frac{\partial R(L_0)}{\partial L^i} \frac{\partial R(L_0)}{\partial L^j} \frac{L_0^i L_0^j}{R(0; T_0, T_n)^2} \int_0^T \Sigma_{ij}(t) dt.$$

The function $u_0(0, x)$ can then be approximated by applying the Black-Scholes formula (for simplicity $t = 0$):

$$u_0(0, x) \approx P_{BS}(V_T^{swap}, R(0; T_0, T_n), K).$$

5 Numerical Examples

In this section, we test the performance of our approximation at pricing caplets on Libor rates in the model (4), where X_t is a unidimensional CGMY process (Carr et al. [5]). The CGMY process is a pure jump process, so that $c = 0$, with Lévy measure

$$F(dz) = \frac{C}{|z|^{1+Y}} \left(e^{-\lambda_- z} \mathbf{1}_{\{x<0\}} + e^{-\lambda_+ z} \mathbf{1}_{\{x>0\}} \right) dz .$$

The jumps of this process are not bouded from below but the parameters we choose ensure that the probability of having a negative Libor rate value is negligible. We choose the time grid $T_0 = 5$, $T_1 = 6$, ... $T_5 = 10$, the volatility parameters $\lambda_i = 1$, $i = 1, \dots, 5$, the initial forward Libor rates $L_0^i = 0.06$, $i = 1, \dots, 5$ and the bond price for the first maturity $B_0(T_0) = 1.06^{-5}$. The CGMY model parameters are chosen

Table 1 Price of ATM caplet computed using the analytic approximation together with the 95 % confidence bounds computed by Monte Carlo over 10^6 trajectories

	Case 1	Case 2	Case 3	Case 4
Order 0	0.008684	0.006392	0.003281	0.007112
Order 1	0.008677	0.006361	0.003241	0.006799
Order 2	**0.008677**	**0.006351**	**0.003172**	**0.006556**
MC lower bound	0.008626	0.006306	0.003178	0.006493
MC upper bound	0.008712	0.006361	0.003204	0.006578

according to four different cases described in the following table, which also gives the standard deviation and excess kurtosis of X_1 for each case. Case 1 corresponds to a Lévy process that is close to the Brownian motion (Y close to 2 and λ_+ and λ_- large) and Case 4 is a Lévy process that is very far from Brownian motion:

Case	C	λ_+	λ_-	Y	Volatility (%)	Excess kurtosis
1	0.01	10	20	1.8	23.2	0.028
2	0.1	10	20	1.2	17	0.36
3	0.2	10	20	0.5	8.7	3.97
4	0.2	3	5	0.2	18.9	12.7

We first calculate the price of the ATM caplet with maturity T_1 written on the Libor rate L^1 with the zero-order, first-order and second-order approximation, using as benchmark the jump-adapted Euler scheme of Kohatsu-Higa and Tankov [13]. The first Libor rate is chosen to maximize the nonlinear effects related to the drift of the Libor rates, since the first maturity is the farthest from the terminal date. The results are shown in Table 1. We see that for all four cases, the price computed by second-order approximation is within or at the boundary of the Monte Carlo confidence interval, which is itself quite narrow (computed with 10^6 trajectories).

Secondly, we evaluate the prices of caplets with strikes ranging from 3 to 9 % and explore the performance of our analytic approximation for estimating the caplet implied volatility smile. The results are shown in Fig. 1. We see that in Cases 1, 2 and 3, which correspond to the parameter values most relevant in practice given the value of the excess kurtosis, the second order approximation reproduces the volatility smile quite well (in Case 1 there is actually no smile, see the scale on the Y axis of the graph). In Case 4, which corresponds to very violent jumps and pronounced smile, the qualitative shape of the smile is correctly reproduced, but the actual values are often outside the Monte Carlo interval. This means that in this extreme case the model is too far from the Gaussian LMM for our approximation to be precise. We also note that the algorithm runs in $\mathcal{O}(n^6)$, for the second order approximation, due to the number of partial derivatives that one has to calculate. The algorithm may therefore run slowly, should n become too large.

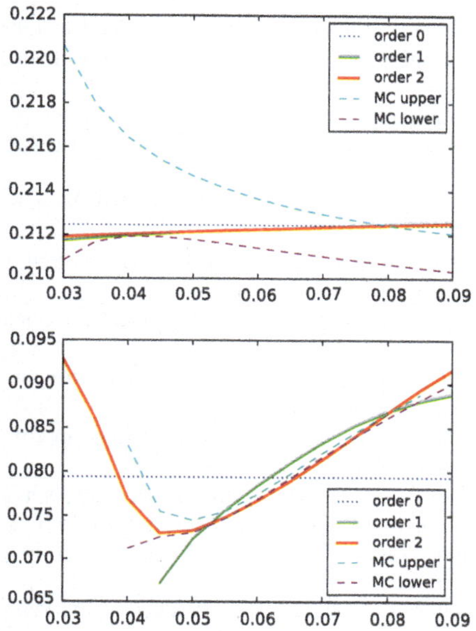

Fig. 1 Implied volatilities of caplets with different strikes computed using the analytic approximation together with the Monte Carlo bound. *Top graphs* Case 1 (*left*) and Case 2 (*right*). *Bottom graphs* Case 3 (*left*) and Case 4 (*right*)

Acknowledgements This research of Peter Tankov is partly supported by the Chair Financial Risks of the Risk Foundation sponsored by Société Générale.

References

1. Belomestny, D., Schoenmakers, J.: A jump-diffusion Libor model and its robust calibration. Quant. Financ. **11**(4), 529–546 (2011)
2. Benhamou, E., Gobet, E., Miri, M.: Smart expansion and fast calibration for jump diffusions. Financ. Stochast. **13**(4), 563–589 (2009)
3. Benhamou, E., Gobet, E., Miri, M.: Time dependent Heston model. SIAM J. Finan. Math. **1**(1), 289–325 (2010)
4. Brace, A., Gatarek, D., Musiela, M.: The market model of interest rate dynamics. Math. Financ. **7**, 127–155 (1997)
5. Carr, P., Geman, H., Madan, D.B., Yor, M.: Self-decomposability and option pricing. Math. Financ. **17**, 31–57 (2007)
6. Černý, A., Denkl, S., Kallsen J.: Hedging in Lévy models and the time step equivalent of jumps. Preprint. (2013). arXiv:1309.7833
7. De Franco, C.: Two studies in risk management: portfolio insurance under risk measure constraint and quadratic hedge for jump processes. Ph. D. thesis, University Paris Diderot (2012)
8. Eberlein, E., Kluge, W.: Calibration of Lévy term structure models. In Fu, M., Jarrow, R.A., Yen, J.-Y., Elliott, R.J. (eds.) Advances in Mathematical Finance: In Honor of D. B. Madan, pp. 147–172. Birkhauser (2007)
9. Eberlein, E., Özkan, F.: The Lévy Libor model. Financ. Stochast. **9**, 327–348 (2005)

10. Jäckel, P., Rebonato, R.: The link between caplet and swaption volatilities in a Brace-Gatarek-Musiela/Jamshidian framework: approximate solutions and empirical evidence. J. Comput. Financ. **6**(4), 41–60 (2003)
11. Kallsen, J.: A didactic note on affine stochastic volatility models. In Kabanov, Y., Lipster, R., Stoyanov, J. (eds) From Stochastic Calculus to Mathematical Finance: The Shiryaev Festschrift, pp. 343–368. Springer (2006)
12. Kluge, W.: Time-Inhomogeneous Lévy Processes in Interest Rate and Credit Risk Models. Ph. D. thesis, University of Freiburg (2005)
13. Kohatsu-Higa, A., Tankov, P.: Jump-adapted discretization schemes for Lévy-driven SDEs. Stochast. Process. Appl. **120**(11), 2258–2285 (2010)
14. Ménassé, C., Tankov, P.: Asymptotic indifference pricing in exponential Lévy models. Applied Mathematical Finance (published online) (2015)
15. Miltersen, K.R., Sandmann, K., Sondermann, D.: Closed form solutions for term structure derivatives with log-normal interest rates. J. Financ. **52**, 409–430 (1997)
16. Papapantoleon, A., Schoenmakers, J., Skovmand, D.: Efficient and accurate log-Lévy approximations to Lévy driven LIBOR models. J. Comput. Financ. **15**(4), 3–44 (2012)
17. Sandmann, K., Sondermann, D., Miltersen, K.R.: Closed form term structure derivatives in a Heath–Jarrow–Morton model with log-normal annually compounded interest rates. In: Proceedings of the Seventh Annual European Futures Research Symposium Bonn, pp. 145–165. Chicago Board of Trade (1995)
18. Schoenmakers, J.: Robust Libor Modelling and Pricing of Derivative Products. CRC Press (2005)

Cointegrated Commodity Markets and Pricing of Derivatives in a Non-Gaussian Framework

Fred Espen Benth

Abstract We analyse cointegration in commodity markets based on continuous-time non-Gaussian stochastic processes. Using simple Lévy-based processes, we propose a cointegrated spot price model in two commodity markets, and derive the implied futures price dynamics using the Esscher transform to introduce a pricing measure. A simple Heath-Jarrow-Morton cointegrated futures price dynamics is introduced motivated from these considerations. We study the question of pricing spread and quanto options in commodity markets, based on a Fourier approach.

Keywords Cointegration · Heath-Jarrow-Morton modeling · Ornstein-Uhlenbeck processes · Lévy processes · Fourier transform · Spread options · Quanto options

1 Introduction

Eberlein and co-authors have, in a series of papers, developed an extensive theory for non-Gaussian interest-rate modelling (see Eberlein et al. [21], Eberlein and Kluge [23], Eberlein and Özkan [24] and Eberlein and Raible [26]). Using Lévy processes, short rate models, Heath-Jarrow-Morton (HJM) dynamics for forward rates as well as LIBOR models have been proposed and analysed in these papers. Moreover, in Eberlein et al. [19, 20], Fourier methods have been studied and applied to price various derivatives (see also the recent survey paper by Eberlein [18]).

In this paper we will look at cointegration in a non-Gaussian framework for commodity futures markets. There are close relationships between fixed-income and commodity markets when it comes to modelling and analysis of derivatives (see Clewlow and Strickland [16], Benth and Koekebakker [9], Eydeland and Wolyniec [28] and Benth et al. [7]). Hence, we adopt the ideas of Lévy-based modelling from fixed-income markets to commodities, and propose an approach where the concept of

F.E. Benth (✉)
Department of Mathematics, University of Oslo,
P.O. Box 1053, 0316 Blindern, Oslo, Norway
e-mail: fredb@math.uio.no

© Springer International Publishing Switzerland 2016

J. Kallsen and A. Papapantoleon (eds.), *Advanced Modelling in Mathematical Finance*,
Springer Proceedings in Mathematics & Statistics 189,
DOI 10.1007/978-3-319-45875-5_20

cointegration can be included. As it turns out, our bivariate spot and futures price dynamics are tailor-made for pricing spread and quanto options, financial derivatives which are important in commodity markets.

Cointegration is a popular concept in econometrics to model the joint price dynamics of two or more assets. Individually, the asset prices may be non-stationary, while a linear combination of them are stationary. Duan and Pliska [17] suggested to model two asset prices by a bivariate exponential Gaussian process, with logarithmic prices consisting of a common non-stationary (drifted Brownian motion) dynamics and, possibly dependent, stationary processes (Ornstein-Uhlenbeck processes). Although the assets are cointegrated under the market probability, Duan and Pliska [17] argued that the equivalent martingale measure will remove the cointegration effect, as this enters into the drift. Hence, in their context the price of a spread option, say, will not be affected by cointegration.

In Benth and Koekebakker [10], the view of Duan and Pliska [17] is questioned in the context of commodity markets, and more specifically in energy markets. The spot is in some markets not tradeable in the classical financial sense (electricity, weather, and freight), and in other markets it may be very illiquid (gas). Hence, the pricing measure Q does not necessarily need to be a martingale measure for the spot. Indeed, for example in electricity markets, any equivalent measure Q can serve as a pricing measure for derivatives contracts (like for example futures), since the spot cannot be traded financially. This opens for the possibility that cointegration may be preserved when going from the market probability to the pricing probability. Benth and Koekebakker [10] demonstrate in a Gaussian modelling framework that this is indeed the case, and in fact the futures price dynamics become cointegrated as well when looking at two contracts with the same time-to-maturity (e.g., using the Musiela parametrization). As a side-remark, a pricing measure Q can be seen as a model of the risk premium in the market.

In this paper we generalize the cointegration analysis in Benth and Koekebakker [10] to Lévy models. In particular, we consider the common non-stationary factor to be a general Lévy process, whereas the stationary factors are assumed to be Volterra-type process which can be represented as stochastic integrals of deterministic kernel functions with respect to Lévy processes. Such models are significantly generalizing Ornstein-Uhlenbeck processes, which are the typical choice for modelling stationary price behaviour in commodity markets (see Benth et al. [7]). After proposing a cointegrated Lévy based spot price model (specified under the market probability), we introduce a pricing measure by the Esscher transform. Next, we derive the futures price dynamics by computing the conditional expected spot price at maturity and we obtain that cointegration is preserved for futures with a fixed time-to-maturity. We remark here that we focus on dynamics which are cointegrated in the sense that the difference of logarithmic prices are stationary, after removing possible seasonality effects. Seasonality is an important ingredient in many commodity markets, for example weather and power markets.

Motivated from the cointegrated spot model and the related futures prices, we discuss the HJM approach (see Heath et al. [31] for their seminal paper on direct forward rate modelling). A no-arbitrage condition is derived, which essentially states a relationship between the initial futures curve, the stationary "volatility" functions and the driving Lévy processes.

As an application of our cointegration analysis, we price two classes of options with some interest in energy markets. Spread options are popular in energy and commodity markets as a tool to hedge price differentials, for example to hedge the difference between the power price and a price of a fuel like coal or gas. Another recently emerged class of derivatives is the so-called *energy quanto options*. Such options are intended to provide a hedge towards price *and* volume risk in the power market, say. The payoff function from an energy quanto option is typically given as a product of a European option (put, say) on the energy price and a European option (put, say) on a temperature index.

In this paper we derive integral expressions for the price of spread and energy quanto options in the context of cointegrated futures price dynamics. Using Fourier methods, we can conveniently express the price in terms of the characteristics of the involved Lévy processes along with the Fourier transforms of the payoff functions. In the spread option case, we can reduce the bivariate feature of the option to the problem of pricing a standard call option using a convenient change of probabilities. For the quanto option we show by conditioning that the price can be reduced into that of two univariate options.

Our results are presented as follows: in the next section we propose a cointegrated bivariate spot price dynamics, with factors driven by Lévy processes. The implied futures price dynamics is derived based on Esscher transform. Section 3 contains a discussion on the HJM approach, inspired by the results from Sect. 2. Pricing of spread and quanto options are presented in Sect. 4.

2 Cointegrated Spot Price Dynamics and the Relation to Futures Prices

Let $(\Omega, \mathscr{F}, P)$ be a complete probability space equipped with a filtration $\{\mathscr{F}_t\}_{t\geq 0}$ and $\mathbf{L} = (L, U_1, U_2)$ be a trivariate Lévy process. We denote the cumulant function, that is the distinguished logarithm of the characteristic function (see Sato [37, Sect. 7]), of $\mathbf{L}$ by $\psi_{\mathbf{L}}(\mathbf{z})$ for $\mathbf{z} \in \mathbb{R}^3$. The marginal cumulants are denoted $\psi(z) := \psi_{\mathbf{L}}(z, 0, 0)$, $\psi_1(z) = \psi_{\mathbf{L}}(0, z, 0)$ and $\psi_2(z) = \psi_{\mathbf{L}}(0, 0, z)$ for $z \in \mathbb{R}$, being the cumulants of L, U_1 and U_2, resp. The joint cumulant of U_1 and U_2, $\psi_{1,2}(z_1, z_2) := \psi_{\mathbf{L}}(0, z_1, z_2)$, $z_1, z_2 \in \mathbb{R}$, will be important. We assume that $\mathbf{L}$ has finite exponential moments of all orders $\theta \in [-K, K]^3$, where $K > 0$ is a constant. Furthermore, for convenience we suppose that the real-valued Lévy processes U_1 and U_2 have zero drift.

Consider two commodities with spot price dynamics S_1, S_2 defined by

$$\ln S_k(t) = s_k + \Lambda_k(t) + L(t) + Y_k(t), k = 1, 2. \tag{1}$$

Here, for $k = 1, 2$, $\Lambda_k(t)$ is a deterministic seasonality function assumed to be bounded and measurable, and

$$Y_k(t) = \int_0^t g_k(t-s)\,dU_k(s), \tag{2}$$

for $g_k \in L^2(\mathbb{R}_+)$. Furthermore, $s_k := \ln S_k(0) - \Lambda_k(0) \in \mathbb{R}$ is the deviation of the initial (logarithmic) spot price from its current seasonal mean.

We note that Y_k is a stationary process in the sense of having a limiting distribution when time approaches infinity. Indeed, for $z \in \mathbb{R}$ (and log denoting the distinguished logarithm),

$$\lim_{t\to\infty} \log \mathbb{E}\left[e^{iz\int_0^t g_k(t-s)\,dU_k(s)}\right] = \lim_{t\to\infty} \int_0^t \psi_k(zg_k(s))\,ds = \int_0^\infty \psi_k(zg_k(s))\,ds.$$

The latter equality holds by dominated convergence, noting that ψ_k is of quadratic growth as U_k is integrable (in fact, exponentially integrable of orders in the interval $[-K, K]$) and driftless and g_k is square integrable on $\mathbb{R}_+$. Defining for $k = 1, 2$

$$\widetilde{Y}_k(t) := \int_{-\infty}^t g_k(t-s)\,dU_k(s), \tag{3}$$

we readily see that this is a stationary process in the strict sense (e.g., in the sense of having a distribution independent of t for all times) with cumulant $\int_0^\infty \psi_k(zg_k(s))\,ds$. Hence, $Y_k(t)$ converges to $\widetilde{Y}_k(0)$ in distribution when $t \to \infty$. Sometimes $\widetilde{Y}_k(t)$ is referred to as a Lévy stationary (LS) process. Note that we have used a two-sided Lévy process U_k in the definition of $\widetilde{Y}_k$ in (3).

As an example, assume that $L(t) = \mu t + \sigma B(t)$ for $\mu, \sigma \in \mathbb{R}$, $(B, U_1, U_2) := (B, W_1, W_2)$, a trivariate Brownian motion and $g_k(x) = \sigma_k \exp(-\alpha_k x)$, $\sigma_k, \alpha_i \in \mathbb{R}_+, k = 1, 2$. The choice of g_k implies that Y_k follows an Ornstein-Uhlenbeck process with speed of mean reversion α_k and volatility σ_k. Under this specification, the spot dynamics in (1) coincide with the cointegration model proposed by Duan and Pliska [17]. Moreover, marginally, the spot prices follow a two-factor dynamics with L modeling the long-term variations while Y_k represents the short term fluctuations. This is a typical commodity spot price model, see for example Lucia and Schwartz [34] for an application to the NordPool electricity market. Remark that one may have cross-dependencies between the short and long term factors (that is, between B and W_k), and between the short term factors (that is, between W_1 and W_2). Benth and Koekebakker [10] proposed a cointegrated spot price model of this class. Indeed, our spot model in (1) extends the Gaussian set-up in Benth and Koekebakker [10] to general Lévy processes.

An extension of the Gaussian two-factor model was proposed in Benth et al. [8] as a model for the EEX spot prices, where the long-term factor L was supposed to be a normal inverse Gaussian (NIG) Lévy process. The short term factor was driven by a stable Lévy process with kernel function $g(x)$ coming from a continuous-time autoregressive moving average process. This is a two-factor extension of the exponential hyperbolic Lévy process suggested by Eberlein and Stahl [27] for German power spot prices. Barndorff-Nielsen et al. [2] propose a related model for energy spot markets, namely a volatility modulated Volterra dynamics.

We observe from (1) that S_1 and S_2 are cointegrated since

$$\ln S_1(t) - \ln S_2(t) = \Lambda_1(t) - \Lambda_2(t) + s_1 - s_2 + Y_1(t) - Y_2(t),$$

which is stationary around the difference of the seasonal means $\Lambda_1(t) - \Lambda_2(t)$. We remark that the notion of cointegration is slightly modified here to take seasonality into account. Strictly speaking, in view of Duan and Pliska [17], it is the *de-seasonalized* logarithmic spot prices which are cointegrated, being equal to $Y_1(t) - Y_2(t)$ which has a limiting distribution. Indeed, we find the cumulant of the last difference to be

$$\begin{aligned}\lim_{t\to\infty} \log \mathbb{E}\left[\exp\left(\mathrm{i}z\Big(\int_0^t g_1(t-s)\,dU_1(s) - \int_0^t g_2(t-s)\,dU_2(s)\Big)\right)\right] \\ = \lim_{t\to\infty} \int_0^t \psi_{1,2}(zg_1(s), -zg_2(s))\,ds \\ = \int_0^\infty \psi_{1,2}(zg_1(s), -zg_2(s))\,ds.\end{aligned}$$

This limit is the cumulant of the random variable $\widetilde{Y}_1(t) - \widetilde{Y}_2(t)$, with $\widetilde{Y}_k$ defined in (3), $k = 1, 2$.

To derive the futures price dynamics from the cointegrated spot model we need a pricing measure Q. Due to market frictions like storage costs (for gas, oil and agriculture, say) or non-tradability (for power, weather and freight, say), the spot dynamics does not need to be a martingale under the pricing measure (see Benth et al. [7] for an argument of this). In this paper we apply a simple Esscher transform to introduce an equivalent pricing measure Q (see Gerber and Shiu [30] and Kallsen and Shiryaev [33] for the Esscher transform in insurance and mathematical finance applications, and Benth et al. [7] for the Esscher transform in energy markets).

Let $\mathbf{a} \in [-K, K]^3$ and note that $\psi_{\mathbf{L}}(-\mathrm{i}\mathbf{a})$ is finite by the exponential integrability condition on $\mathbf{L}$. Consider the stochastic process

$$M(t) = \exp\left(\mathbf{a}'\mathbf{L}(t) - \psi_{\mathbf{L}}(-\mathrm{i}\mathbf{a})\,t\right), t \geq 0, \tag{4}$$

which is a martingale due to the exponential moment condition. For a given time horizon $T < \infty$, we define a probability Q with Radon-Nikodym derivative $dQ/dP|_{\mathcal{F}_T} = M(T)$. This will be our pricing measure.

Following Benth et al. [7] or Kallsen and Shiryaev [33], we find have that $\mathbf{L}$ is a Lévy process with respect to the probability Q, where the cumulant, denoted by $\psi_{\mathbf{L},Q}$, becomes,

$$\psi_{\mathbf{L},Q}(\mathbf{z}) = \psi_{\mathbf{L}}(\mathbf{z} - \mathrm{i}\mathbf{a}) - \psi_{\mathbf{L}}(-\mathrm{i}\mathbf{a}), \tag{5}$$

for $\mathbf{z} \in \mathbb{R}^3$. Indeed, the effect of the probability Q is an exponential tilting of the Lévy measure of $\mathbf{L}$ and a change of drift, which can be related back to the notion of *market price of risk* and eventually the *risk premium* (see Benth et al. [7]).

For $k = 1, 2$, we define the futures price $F_k(t, T)$ at time $t \geq 0$ for a contract delivering commodity $k = 1, 2$ at time $T \geq t$ by

$$F_k(t, T) = \mathbb{E}_Q\left[S_k(T) \mid \mathscr{F}_t\right], \tag{6}$$

if S_k is Q-integrable. Here, we have assumed that the risk-free interest rate is a constant, from now on denoted by r. In order to have that S_k is Q-integrable, we must make sure that the exponential integrability condition of $\mathbf{L}$ is verified. For this purpose, we assume from now on that $g_k(s)$, $k = 1, 2$, are bounded functions on $\mathbb{R}_+$, and that $\mathbf{a}$ in the Esscher transform satisfies the condition that $\mathbf{a} + \mathbf{e}_1 + g_k(s)\mathbf{e}_{k+1} \in [-K, K]^3$ for $s \geq 0$ and $k = 1, 2$.

In the next Proposition we compute the futures price dynamics:

Proposition 1 *For $0 \leq t \leq T$ and $k = 1, 2$ it holds that*

$$F_k(t, T) = \exp\left(\Lambda_k(T) + h_k(T - t) + L(t) + \int_0^t g_k(T - s)\, dU_k(s)\right)$$

where

$$h_k(x) = s_k + \int_0^x \psi_{\mathbf{L},Q}(-\mathrm{i}(\mathbf{e}_1 + g_k(s)\mathbf{e}_{k+1}))\, ds$$

for $x \geq 0$ and $\mathbf{e}_j$, $j = 1, 2, 3$ are the canonical basis vectors in $\mathbb{R}^3$

Proof We show the result for $k = 1$. It holds that

$$\begin{aligned}
F_1(t, T) &= \mathbb{E}_Q\left[\exp\left(\Lambda(T) + s + L(T) + \int_0^T g_1(T - v)\, dU_1(v)\right) \mid \mathscr{F}_t\right] \\
&= \exp(\Lambda(T) + s + L(t) + \int_0^t g_1(T - v)\, dU_1(v)) \\
&\quad \times \mathbb{E}_Q\left[\exp\left(L(T) - L(t) + \int_t^T g_1(T - v)\, dU_1(v)\right) \mid \mathscr{F}_t\right],
\end{aligned}$$

from $\mathscr{F}_t$-measurability. As L and U_1 are Lévy processes under Q, with finite exponential moments of sufficient orders, we find by the independent increment property

$$\begin{aligned}
&\mathbb{E}_Q\left[\exp\left(L(T) - L(t) + \int_t^T g_1(T - v)\, dU_1(v)\right) \mid \mathscr{F}_t\right] \\
&\quad = \mathbb{E}_Q\left[\exp\left(L(T) - L(t) + \int_t^T g_1(T - v)\, dU_1(v)\right)\right] \\
&\quad = \exp\left(\int_0^{T-t} \psi_{\mathbf{L},Q}(-\mathrm{i}, -\mathrm{i}g_1(v), 0)\, dv\right),
\end{aligned}$$

where we have used that $dQ/dP|_{\mathscr{F}_t} = M(t)$. □

We remark that the above result is a special case of Proposition 3.1 in Benth et al. [5] (with $n = 2$ and the limiting specification $T_1 = T_2$ in their notation).

Note that using the Musiela parametrization $x = T - t$, that is, expressing the futures prices in terms of *time-to-maturity*, we find that with $G_k(t, x) := F_k(t, t + x)$,

$$\ln G_1(t, x) - \ln G_2(t, x) = \Lambda_1(t, x) - \Lambda_2(t, x) + h_1(x) - h_2(x) \\ + \int_0^t g_1(t + x - s)\, dU_1(s) - \int_0^t g_2(t + x - s)\, dU_2(s). \tag{7}$$

For fixed $x \geq 0$, the last two terms are stationary in the sense of having a limiting distribution. Indeed, we find

$$\begin{aligned} &\lim_{t\to\infty} \log \mathbb{E}\left[\exp\left(\mathrm{i}z(\int_0^t g_1(t + x - s)\, dU_1(s) - \int_0^t g_2(t + x - s)\, dU_2(s))\right)\right] \\ &\quad = \lim_{t\to\infty} \int_x^{t+x} \psi_{1,2}(zg_1(s), -zg_2(s))\, ds \\ &\quad = \int_x^{\infty} \psi_{1,2}(zg_1(s), -zg_2(s))\, ds. \end{aligned}$$

The difference of the h-functions

$$h_1(x) - h_2(x) = \int_0^x \psi_{\mathbf{L},Q}(-\mathrm{i}(\mathbf{e}_1 + g_1(s)\mathbf{e}_2)) - \psi_{\mathbf{L},Q}(-\mathrm{i}(\mathbf{e}_1 + g_2(s)\mathbf{e}_3))\, ds,$$

is, of course, constant over time t, and therefore we can conclude that G_1 and G_2 are cointegrated (around the seasonal function). Cointegration holds both under Q and under P. Note that F_1 and F_2 are not cointegrated, as the integral involving the market price of risk will vary with t. Hence, for *roll-over* futures we have cointegration, whereas for futures with fixed maturity this property does not hold.

3 A Cointegrated HJM Futures Price Dynamics

Motivated by the analysis of cointegrated spot prices and their futures price dynamics, we discuss the HJM approach. As our goal is pricing of derivatives (on futures), it is reasonable to follow the HJM framework and model the futures price dynamics directly under a pricing measure Q. Thus, we assume that we have given a complete probability space $(\Omega, \mathcal{F}, Q)$ equipped with a filtration $\{\mathcal{F}_t\}_{t\geq 0}$.

We preserve the notation from the previous Section, and let $F_k(t, T), k = 1, 2$ denote the futures price at time $0 \leq t \leq T$ for two contracts delivering commodity 1 and 2, respectively, at time T. In the Musiela parametrization, we let $G_k(t, x) := F_k(t, t + x), k = 1, 2$, with $x = T - t \geq 0$ being the *time-to-maturity* of the futures contracts. We define the dynamics of the two futures prices under the Musiela parametrization as follows: for $k = 1, 2$,

$$G_k(t, x) = \exp\left(\Lambda_k(t + x) + h_k(x) + L(t) + \int_0^t g_k(t + x - s)\, dU_k(s)\right), \quad (8)$$

where $\Lambda_k:\mathbb{R}_+ \to \mathbb{R}$ is a measurable function, modelling the seasonal variations in the futures prices. Furthermore, $\mathbf{L} = (L, U_1, U_2)$ is a trivariate Q-Lévy process with finite exponential moments of all orders $\theta \in [-K, K]^3$ for a constant $K > 0$, where the cumulant is denoted by $\psi_{\mathbf{L},Q}(\mathbf{z}), \mathbf{z} \in \mathbb{R}^3$. The functions $g_k:\mathbb{R}_+ \to \mathbb{R}$ are assumed square integrable, that is, $g_k \in L^2(\mathbb{R}_+)$. Moreover, assume that $h_k : \mathbb{R}_+ \to \mathbb{R}$ are measurable functions. Observe that $G_k(0, x) = \exp(\Lambda_k(x) + h_k(x))$, or, $h_k(x) = \ln G_k(0, x) - \Lambda_k(x)$. Hence, $h_k(x)$ is the initial futures curve (on a logarithmic scale) less the seasonality function. In particular, after choosing $x = 0$, we find that $h_k(0) = \ln G_k(0, 0) - \Lambda_k(0) = \ln S_k(0) - \Lambda_k(0)$. It is therefore natural to assume that $h_k(0) = s_k$, where we recall s_k from (1). Remark also that the functions Λ_k are measuring the seasonality effect on the futures prices, and therefore we use the argument $t + x$ as this will refer to the actual date of delivery.

In order to have an arbitrage-free model, the futures prices $t \mapsto F_k(t, T), t \leq T$, for $k = 1, 2$ must be Q-martingales. With our specific model, we can easily formulate a no-arbitrage condition on the involved parameters:

Proposition 2 *Suppose that the functions $g_k(s)$ are such that $\mathbf{e}_1 + g_k(s)\mathbf{e}_{k+1} \in [-K, K]^3$ for $s \geq 0$ and $k = 1, 2$. The processes $t \mapsto F_k(t, T), k = 1, 2$, are Q-martingales if and only if*

$$h_k(x) = s_k + \int_0^x \psi_{\mathbf{L},Q}(-\mathrm{i}(\mathbf{e}_1 + g_k(v)\mathbf{e}_{k+1}))\, dv. \quad (9)$$

Proof It follows from the exponential moment condition of $\mathbf{L}$ and the assumption on g_k that $t \mapsto F_k(t, T), t \leq T$, are integrable processes. By the independent increment property and adaptedness of the Q-Lévy process $\mathbf{L}$, we find for $s \leq t \leq T$,

$$\begin{aligned}\mathbb{E}_Q\left[F_k(t, T) \mid \mathcal{F}_s\right] &= \exp(\Lambda_k(T) + h_k(T - t))\, \mathbb{E}_Q\left[\exp\left(L(t) + \int_0^t g_k(T - v)\, dU_k(v)\right) \mid \mathcal{F}_s\right] \\ &= \exp\left(\Lambda_k(T) + h_k(T - t) + L(s) + \int_0^s g_k(T - v)\, dU_k(s)\right) \\ &\quad \times \mathbb{E}_Q\left[\exp\left(L(t) - L(s) + \int_s^t g_k(T - v)\, dU_k(v)\right)\right] \\ &= F_k(s, T) \exp\left(h_k(T - t) - h_k(T - s) + \int_s^t \psi_{\mathbf{L},Q}(-\mathrm{i}(\mathbf{e}_1 + g_k(T - v)\mathbf{e}_k))\, dv\right).\end{aligned}$$

Hence, after a change of variables, we have that F_k are Q-martingales if and only if

$$h_k(T - t) - h_k(T - s) + \int_{T-s}^{T-t} \psi_{\mathbf{L},Q}(-\mathrm{i}(\mathbf{e}_1 + g_k(v)\mathbf{e}_k))\, dv = 0,$$

for all $s \leq t \leq T$. The claim follows. □

Condition (9) in the Proposition above gives a precise relationship between h_k and g_k, and the Lévy process $\mathbf{L}$, that ensures an arbitrage-free dynamics. If we have specified the initial futures curves from data, say, then we have implicitly specified $h_k(x)$, since $h_k(x) = \ln G_k(0, x) - \Lambda_k(x)$. Then the no-arbitrage condition puts restrictions on the freedom to model g_k and $\mathbf{L}$. On the other hand, if we have specified $\mathbf{L}$ and g_k, then there is no flexibility in the modelling of h_k. Note also that the assumption on g_k implies that it is bounded by K, and furthermore that L must have an exponential moment of order 1 and thus $K \geq 1$.

Observe that we have specified $G_k(t, x)$ in (8) directly as an exponential Lévy driven process, and not as a solution to a stochastic differential equation. In the original HJM approach (see Heath et al. [31]), the forward rate dynamics is modelled as a stochastic differential equation, whereas under the Musiela parametrization the forward rates will follow a particular stochastic partial differential equation (see Carmona and Tehranchi [14] for an analysis of such stochastic partial differential equations in the Gaussian framework and Peszat and Zabczyk [36] for the extension to the Lévy case). In particular, the no-arbitrage condition for such forward rate models links the drift of the dynamics to the noise. For example, in a simple linear Lévy-based forward rate dynamics, Eq. (3.83) in Eberlein [18] states the HJM drift condition in terms of the cumulant of the Lévy process and the volatility. In our exponential model, the function h_k plays the role of the drift, and the no-arbitrage condition becomes slightly simpler in our context as it is the futures prices that must be martingales and not the discounted zero-coupon bond prices implied from the forward *rate* dynamics.

A direct modelling approach as we propose here was advocated in equity markets by Eberlein and Keller [23] and Barndorff-Nielsen [1] as a statistically convenient way to specify a Lévy-driven stochastic dynamics. One avoids ad-hoc hypothesis on the jumps of the Lévy process to ensure positivity of the prices, which comes for free in (8). Moreover, a simple stochastic model for the logarithmic returns is readily available from the direct model, given in terms of increments of Lévy processes.

Consider the case of $\mathbf{L}$ being a trivariate Brownian motion. Then, $\psi_{\mathbf{L},Q}(\mathbf{z}) = -\frac{1}{2}\mathbf{z}'C\mathbf{z}$ for $\mathbf{z} \in \mathbb{R}^3$ with C being a symmetric, positive definite 3×3-matrix with 1's on the diagonal (C is the correlation matrix). The no-arbitrage condition now becomes

$$\begin{aligned} h_k(x) &= s_k + \frac{1}{2}\int_0^x (\mathbf{e}_1 + g_k(v)\mathbf{e}_k)' C(\mathbf{e}_1 + g_k(v)\mathbf{e}_k)\,dv \\ &= s_k + \frac{1}{2}\left(x(\mathbf{e}_1'C\mathbf{e}_1) + 2(\mathbf{e}_1'C\mathbf{e}_k)\int_0^x g_k(v)\,dv + (\mathbf{e}_k'C\mathbf{e}_k)\int_0^x g_k^2(v)\,dv\right) \\ &= s_k + \frac{1}{2}x + \mathrm{corr}(L, U_k)\int_0^x g_k(v)\,dv + \frac{1}{2}\int_0^x g_k^2(v)\,dv, \end{aligned}$$

as expected.

In the bivariate futures price model in (8), L plays the role of a common non-stationary factor, while $Y_k(t, x) := \int_0^t g_k(t + x - s)\, dU_k(s)$ is the stationary factor. A simple specification of L is $L(t) = \mu t + \sigma B(t)$, a drifted Brownian motion (here $\mu, \sigma \in \mathbb{R}$ and B is a standard Brownian motion). Letting $g_k(x) = \sigma_k \exp(-\alpha_k x)$ for $x \geq 0$ and parameters $\sigma_k, \alpha_k \in \mathbb{R}_+$ yields that Y_k becomes an Ornstein-Uhlenbeck process. Further assuming U_k to be a Brownian motion leads us back to Benth and Koekebakker [10]. There, a cointegrated HJM dynamics is proposed stated as a simple stochastic differential equation. In our proposed futures price dynamics we have used only two factors for the sake of simplicity. Observe that one can easily incorporate more factors to explain the price dynamics.

We have that

$$\ln G_1(t, x) - \ln G_2(t, x) - (\Lambda_1(t + x) - \Lambda_2(t + x))$$
$$= h_1(x) - h_2(x) + \int_0^t g_1(t + x - s)\, dU_1(s) - \int_0^t g_2(t + x - s)\, dU_2(s). \tag{10}$$

The cumulant of the difference of the two stochastic integrals on the right hand side is (with $z \in \mathbb{R}$)

$$\psi_Q(t, z) := \log \mathbb{E}_Q \left[\exp \left(\mathrm{i}z \left(\int_0^t g_1(t + x - s)\, dU_1(s) - \int_0^t g_2(t + x - s)\, dU_2(s) \right) \right) \right]$$
$$= \int_x^{t+x} \psi_{\mathbf{L},Q}(z(g_1(s)\mathbf{e}_2 - g_2(s)\mathbf{e}_3))\, ds.$$

Since, $\psi_Q(t, z) \to \int_x^\infty \psi_{\mathbf{L},Q}(z(g_1(s)\mathbf{e}_2 - g_2(s)\mathbf{e}_3))\, ds$ when $t \to \infty$, we have that the HJM futures price model is cointegrated in the sense that the futures price dynamics for fixed time to maturity x is cointegrated.

Note that we do not need to have the same time to maturity in order for two futures contracts to be cointegrated. Indeed, consider $\ln G_1(t, x) - \ln G_2(t, y)$ for $x, y \in \mathbb{R}_+$. We have

$$\ln G_1(t, x) - \ln G_2(t, y) - (\Lambda_1(t + x) - \Lambda_2(t + y))$$
$$= h_1(x) - h_2(y) + \int_0^t g_1(t + x - s)\, dU_1(s) - \int_0^t g_2(t + y - s)\, dU_2(s).$$

We find, with obvious modification of the notation,

$$\lim_{t \to \infty} \psi_Q(t, z) = \int_0^\infty \psi_{\mathbf{L},Q}(z(g_1(x + s)\mathbf{e}_2 - g_2(y + s)\mathbf{e}_3))\, ds,$$

which shows stationarity of the last difference of stochastic integrals, and thus cointegration. In particular, choosing $x = 0$, any futures with fixed time to maturity will be cointegrated with the spot. If $x = y = 0$, we find the spot dynamics as

$$S_k(t) = G_k(t,0) = \exp\left(\Lambda_k(t) + s_k + L(t) + \int_0^t g_k(t-s)\,dU_k(s)\right)$$

which leads us back to the dynamics considered in the previous Section. Hence, not unexpectedly, the HJM model produces a cointegrated (under Q) spot price model as well.

4 Pricing of Spread and Quanto Options Using Fourier Methods

In this Section we will analyse the price of spread and quanto options, two classes of derivatives which are popular for risk management in energy markets. Spread and quanto options are examples of options where the payoff is a function of two underlying assets, in our case being two futures contracts. Throughout the Section we consider the futures prices $F_k(t,T), t \leq T$ under a pricing measure Q defined by the cointegrated dynamics (8) in Sect. 3 (or, essentially equivalently, the one derived from the cointegrated spot model in Sect. 2):

$$F_k(t,T) = \exp\left(\Lambda_k(T) + h_k(T-t) + L(t) + \int_0^t g_k(T-s)\,dU_k(s)\right).$$

We recall that $\mathbf{L} = (L, U_1, U_2)$ is a trivariate Q-Lévy processes, with cumulant denoted by $\psi_{\mathbf{L},Q}(\mathbf{z}), \mathbf{z} \in \mathbb{R}^3$, which satisfies an exponential integrability condition of orders $\theta \in [-K, K]^3$. From here on we invoke the assumption on g_k of Proposition 2.

In our study of the price of spread and quanto options, we shall apply the Fourier approach. Recall that for a function $f \in L^1(\mathbb{R})$, we define its Fourier transform (following Folland [29]) by

$$\widehat{f}(y) = \int_{\mathbb{R}} f(x)\mathrm{e}^{-\mathrm{i}xy}\,dx,\ y \in \mathbb{R}. \tag{11}$$

Further, if $\widehat{f} \in L^1(\mathbb{R})$, the Fourier inversion formula tells that (see Folland [29])

$$f(x) = \frac{1}{2\pi}\int_{\mathbb{R}} \widehat{f}(y)\mathrm{e}^{\mathrm{i}xy}\,dy,\ \text{a.e. } x \in \mathbb{R}. \tag{12}$$

We now study the price of spread and quanto options.

4.1 Spread Options

We consider a spread option with exercise time $\tau > 0$ written on two cointegrated forward contracts F_1 and F_2 with delivery at time $T \geq \tau$, that is, an option with payoff

$$\max\left(F_1(\tau, T) - F_2(\tau, T), 0\right),$$

at the exercise time τ. This is indeed an exchange-option first analysed by Margrabe [35] in the case of a bivariate geometric Brownian motion dynamics for the two underlying assets. In energy markets, spread options are written on the difference between the energy price and the price of fuel, for example gas (called the *spark spread*) or coal (called the *dark spread*). Such options provide a protection for power generators against too high fuel prices. There are also *geographical spread* options traded, which are written on the price differential across different power markets, for example between the Nordic NordPool and the German EEX market. On the NYMEX market, options on price differences between various blends of oil are traded (so-called *crack spreads*). We refer to Carmona and Durrleman [13] for an extensive discussion of various spread options and their use.

Spread option pricing formulas using Fourier methods have been derived by several authors under various model assumptions (see Eberlein et al. [19], Benth et al. [4], Barndorff-Nielsen et al. [3] and Benth and Zdanowicz [12]). Here we will derive the spread option price for the cointegrated futures price dynamics.

From the no-arbitrage theory, the price of the spread is defined as

$$V_{\text{spread}} = \mathrm{e}^{-r\tau}\mathbb{E}_Q\left[\max\left(F_1(\tau, T) - F_2(\tau, T), 0\right)\right]. \tag{13}$$

Introduce the notation $H_k(\tau, T) := \exp(\Lambda_k(T) + h_k(T - \tau))$ and we see that

$$V_{\text{spread}} = \mathrm{e}^{-r\tau}\mathbb{E}_Q\left[\mathrm{e}^{L(\tau)}\max\left(H_1(\tau, T)\exp\left(\int_0^\tau g_1(T-s)\,dU_1(s)\right)\right.\right.$$
$$\left.\left.-H_2(\tau, T)\exp\left(\int_0^\tau g_2(T-s)\,dU_2(s)\right), 0\right)\right]. \tag{14}$$

We apply a combination of Esscher transform and Fourier methods, along the lines of the abovementioned papers, to derive an expression for V_{spread} (see also Eberlein et al. [25, Theorem. 5.1] for spread option pricing in a general semimartingale framework).

First, we show that $L(\tau)$ can be factorized out in the spread option price:

Proposition 3 *Define the probability $Q^L \sim Q$ with density*

$$\frac{dQ^L}{dQ}|_{\mathcal{F}_t} = \exp(L(t) - \psi_{\mathrm{L},Q}(-\mathrm{i}, 0, 0)\,t)$$

for $t \leq \tau$. *Then* (U_1, U_2) *is a* Q^L*-Lévy process on* $t \leq \tau$ *with cumulant* $\psi^L_{1,2}(x, y) := \psi_{\mathbf{L},Q}(-\mathrm{i}, x, y) - \psi_{\mathbf{L},Q}(-\mathrm{i}, 0, 0)$ *for* $x, y \in \mathbb{R}$, *and it holds that*

$$V_{spread} = e^{-(r-\psi_{\mathbf{L},Q}(-\mathrm{i},0,0))\tau}\mathbb{E}_{Q^L}\left[\max\left(H_1(\tau, T)\exp\left(\int_0^\tau g_1(T-s)\,dU_1(s)\right)\right.\right.$$
$$\left.\left.-H_2(\tau, T)\exp\left(\int_0^\tau g_2(T-s)\,dU_2(s)\right), 0\right)\right].$$

Proof As L has an exponential moment of order 1 by assumption, $t \mapsto \exp(L(t) - \psi_{\mathbf{L},Q}(-\mathrm{i}, 0, 0)t)$ is a martingale with expectation 1, and thus Q^L is an equivalent probability to Q with the given density process. Indeed, this is an Esscher transform which preserves the independent and stationary increment properties of (U_1, U_2) (see Benth et al. [7]) A straightforward calculation shows for $x, y \in \mathbb{R}$,

$$\begin{aligned}
&\mathbb{E}_{Q^L}\left[\exp\left(\mathrm{i}(xU_1(1) + yU_2(1))\right)\right] \\
&\quad = \mathbb{E}_Q\left[\exp\left(\mathrm{i}(-\mathrm{i})L(1) + xU_1(1) + yU_2(1))\right)\right]\mathrm{e}^{-\psi_{\mathbf{L},Q}(-\mathrm{i},0,0)} \\
&\quad = \exp\left(\psi_{\mathbf{L},Q}(-\mathrm{i}, x, y) - \psi_{\mathbf{L},Q}(-\mathrm{i}, 0, 0)\right).
\end{aligned}$$

Hence, (U_1, U_2) is a Q^L-Lévy process with $\psi^L_{1,2}(x, y)$ as cumulant. By changing probability from Q to Q^L in the expectation expressing V_{spread}, the result follows. □

By the Lévy-Khintchine representation we can represent $\psi_{\mathbf{L},Q}(\mathbf{z})$, $\mathbf{z} \in \mathbb{R}^3$ as

$$\psi_{\mathbf{L},Q}(\mathbf{z}) = \mathrm{i}\mathbf{z}'\mathbf{b} - \frac{1}{2}\mathbf{z}'C\mathbf{z} + \int_{\mathbb{R}^3}(\mathrm{e}^{\mathrm{i}\mathbf{z}'\mathbf{u}} - 1 - \mathrm{i}\mathbf{z}'\mathbf{u}1_{|\mathbf{u}|\leq 1})\,\ell_Q(d\mathbf{u})$$

where $\mathbf{b} \in \mathbb{R}^3$ is the drift, $C \in \mathbb{R}^{3\times 3}$ is a symmetric positive definite matrix and $\ell_Q(d\mathbf{u})$ is the Lévy measure. Then, after a little algebra (using $\mathbf{x} = (0, x, y) \in \mathbb{R}^3$)

$$\begin{aligned}
\psi^L_{1,2}(x, y) &= \mathrm{i}\mathbf{x}'\left(\mathbf{b} + C\mathbf{e}_1 + \int_{\mathbb{R}^3}(\mathrm{e}^{\mathbf{e}_1'\mathbf{u}} - 1)\mathbf{u}1_{|\mathbf{u}|\leq 1}\,\ell_Q(d\mathbf{u})\right) - \frac{1}{2}\mathbf{x}'C\mathbf{x} \\
&\quad + \int_{\mathbb{R}^3}(\mathrm{e}^{\mathrm{i}\mathbf{x}'\mathbf{u}} - 1 - \mathrm{i}\mathbf{x}'\mathbf{u}1_{|\mathbf{u}|\leq 1})\,\mathrm{e}^{\mathbf{e}_1'\mathbf{u}}\,\ell_Q(d\mathbf{u}).
\end{aligned} \tag{15}$$

We see that the Esscher transform exponentially tilts the Lévy measure by $\exp(u_1)$, while the covariance operator is unchanged. The drift of (U_1, U_2) will be altered by a contribution from both the Lévy measure and the covariance operator when going from Q to Q^L.

Consider the expectation operator in the price V_{spread} in Proposition 3. We find

$$
\begin{aligned}
&\mathbb{E}_{Q^L}\left[\max\left(H_1(\tau,T)\mathrm{e}^{\int_0^\tau g_1(T-s)\,dU_1(s)} - H_2(\tau,T)\mathrm{e}^{\int_0^\tau g_2(T-s)\,dU_2(s)},0\right)\right] \\
&\quad = H_2(\tau,T)\mathbb{E}_{Q^L}\left[\exp(\int_0^\tau g_2(T-s)\,dU_2(s))\right. \\
&\qquad \left.\times \max\left(\frac{H_1(\tau,T)}{H_2(\tau,T)}\mathrm{e}^{\int_0^\tau g_1(T-s)\,dU_1(s)-\int_0^\tau g_2(T-s)\,dU_2(s)} - 1,0\right)\right] \\
&\quad = H_2(\tau,T)\exp\left(\int_0^\tau \psi_{1,2}^L(0,-\mathrm{i}g_2(T-s))\,ds\right) \\
&\qquad \times \mathbb{E}_{Q_2}\left[\max\left(\frac{H_1(\tau,T)}{H_2(\tau,T)}\mathrm{e}^{\int_0^\tau g_1(T-s)\,dU_1(s)-\int_0^\tau g_2(T-s)\,dU_2(s)} - 1,0\right)\right],
\end{aligned}
$$

where the probability $Q_2 \sim Q^L$ has density process

$$
\frac{dQ_2}{dQ^L}|_{\mathscr{F}_t} = \exp\left(\int_0^t g_2(T-s)\,dU_2(s) - \int_0^t \psi_{1,2}^L(0,-\mathrm{i}g_2(T-s))\,ds\right),
$$

for $t \leq \tau$. Note that by the boundedness of g_2 and the exponential integrability assumption on U_2, this is a martingale. Again, we have applied an Esscher transform to simplify the expectation functional for the spread option price. This time, the Esscher transform is not time-homogeneous, and hence the Lévy property of U_1 and U_2 is not preserved. However, we can still compute an expression for the cumulant of the involved random variables, which is exactly what we need in order to exploit Fourier techniques, as we investigate next.

So far, we have found that

$$
\begin{aligned}
V_{\text{spread}} &= \mathrm{e}^{\int_0^\tau (\psi_{1,2}^L(0,-\mathrm{i}g_2(T-s))+\psi_{L,Q}(-\mathrm{i},0,0)-r)\,ds} H_2(\tau,T) \\
&\quad \times \mathbb{E}_{Q_2}\left[\max\left(\frac{H_1(\tau,T)}{H_2(\tau,T)}\mathrm{e}^{\int_0^\tau g_1(T-s)\,dU_1(s)-\int_0^\tau g_2(T-s)\,dU_2(s)} - 1,0\right)\right]. \qquad (16)
\end{aligned}
$$

Thus, what remains to compute V_{spread} is effectively to price a call option. We have the following Lemma, which is a version of a result that can be traced back to the seminal paper by Carr and Madan [15] on Fourier methods applied to option pricing.

Lemma 1 *Let C be some positive constant. If X is a random variable with cumulant ψ_X where $\mathbb{E}[\exp(\alpha X)] < \infty$ for a given $\alpha > 1$. Then*

$$
\mathbb{E}[\max(C\exp(X)-1,0)] = \frac{1}{2\pi}\int_{\mathbb{R}} \frac{C^{\alpha+\mathrm{i}y}}{((\alpha-1)+\mathrm{i}y)(\alpha+\mathrm{i}y)}\exp(\psi_X(y-\mathrm{i}\alpha))\,dy.
$$

Proof Define $f_\alpha(x) := \mathrm{e}^{-\alpha x}\max(C\mathrm{e}^x - 1, 0)$. Thus, $f_\alpha(x) = 0$ whenever $x < -\ln C$ and $f_\alpha(x) = C\exp(-(\alpha-1)x) - \exp(-\alpha x)$ for $x \geq -\ln C$. Hence, $f_\alpha \in L^1(\mathbb{R})$, and its Fourier transform becomes

$$\widehat{f_\alpha}(y) = \frac{C^{\alpha+\mathrm{i}y}}{(\alpha-1+\mathrm{i}y)(\alpha+\mathrm{i}y)}.$$

Since $|\widehat{f_\alpha}(y)| \sim y^{-2}$ for $|y|$ large, $\widehat{f_\alpha} \in L^1(\mathbb{R})$, and from the Fourier inversion formula we find

$$\max(Ce^x - 1, 0) = \frac{1}{2\pi}\int_{\mathbb{R}} \widehat{f_\alpha}(y)\mathrm{e}^{\mathrm{i}(y-\mathrm{i}\alpha)x}\,dy.$$

Since $\mathbb{E}[|\exp((\alpha+\mathrm{i}y)X)|] = \mathbb{E}[\exp(\alpha X)] < \infty$ by assumption, we find from Fubini's Theorem the assertion of the Lemma. □

To price our spread option, choose now $C = H_1(\tau,T)/H_2(\tau,T)$ and define $X = \int_0^\tau g_1(T-s)\,dU_1(s) - \int_0^\tau g_2(T-s)\,dU_2(s)$. We find the cumulant of X with respect to Q_2 to be

$$\begin{aligned}\psi_X(x) &= \log \mathbb{E}_{Q_2}\left[\exp\left(\mathrm{i}x(\int_0^\tau g_1(T_s)\,dU_1(s) - \int_0^\tau g_2(T-s)\,dU_s(s))\right)\right]\\ &= \log \mathbb{E}_{QL}\left[\exp\left(\mathrm{i}x\int_0^\tau g_1(T-s)\,dU_1(s) + \mathrm{i}(-\mathrm{i}-x)\int_0^\tau g_2(T-s)\,dU_2(s)\right)\right]\\ &\quad - \int_0^\tau \psi^L_{1,2}(0, -\mathrm{i}g_2(T-s))\,ds\\ &= \int_0^\tau \psi^L_{1,2}(xg_1(T-\tau+s), -(\mathrm{i}+x)g_2(T-\tau+s))\\ &\quad - \psi^L_{1,2}(0, -\mathrm{i}g_2(T-\tau+s))\,ds.\end{aligned}$$

We wrap up to conclude with the following:

Proposition 4 *Let $\psi_{\mathbf{L},Q}$ be the cumulant of the Q-Lévy process $\mathbf{L} = (L, U_1, U_2)$ and suppose that there exists a constant $\alpha > 1$ such that $\mathbf{e}_1 + \alpha g_1(s)\mathbf{e}_2 + (1-\alpha)g_2(s)\mathbf{e}_3 \in [-K, K]^3$ for all $s \geq 0$. Then*

$$\begin{aligned}V_{spread} &= \exp\left(\int_0^\tau \psi_{\mathbf{L},Q}(-\mathrm{i}, 0, -\mathrm{i}g_2(T-\tau+s)) - r\,ds\right) H_2(\tau,T)\\ &\quad\times \frac{1}{2\pi}\int_{\mathbb{R}} \frac{(H_1(\tau,T)/H_2(\tau,T))^{\alpha+\mathrm{i}y}}{((\alpha-1)+\mathrm{i}y)(\alpha+\mathrm{i}y)} \exp(\psi_X(y-\mathrm{i}\alpha))\,dy,\end{aligned}$$

with

$$\begin{aligned}\psi_X(x) = \int_0^\tau &\psi_{\mathbf{L},Q}(-\mathrm{i}, xg_1(T-\tau+s), -(\mathrm{i}+x)g_2(T-\tau+s))\\ &- \psi_{\mathbf{L},Q}(-\mathrm{i}, 0, -\mathrm{i}g_2(T-\tau+s))\,ds,\end{aligned}$$

for $x \in \mathbb{R}$ and $H_k(\tau,T) = \exp(\Lambda_k(T) + h_k(T-\tau)), k = 1, 2$.

Proof From the definition of $\psi^L_{1,2}$ in Proposition 3, we find that

$$\psi^L_{1,2}(0, -\mathrm{i}g_2(T+\tau-s)) = \psi_{\mathbf{L},Q}(-\mathrm{i}, 0, -\mathrm{i}g_2(T+\tau-s)) - \psi_{\mathbf{L},Q}(-\mathrm{i}, 0, 0),$$

which implies the expression for ψ_X. Recalling the formula in (16), the result follows from Lemma 1 after noting that the condition on g_1 and g_2 implies that $\mathbb{E}[\exp(\alpha X)] < \infty$ for the given definition of X. □

Observe that since $\Lambda_k(T) = \Lambda_k(\tau + (T - \tau))$, we have that H_k becomes a function of τ and $T - \tau$. Therefore, the spread option price is indeed depending on the exercise time τ and time left until delivery from exercise, $T - \tau$. On the other hand, after a simple change of variables,

$$\int_0^\tau \psi_{\mathbf{L},Q}(-\mathrm{i}, 0, -\mathrm{i}g_2(T - \tau + s)) - r\,ds = \int_{T-\tau}^T \psi_{\mathbf{L},Q}(-\mathrm{i}, 0, -\mathrm{i}g_2(s)) - r\,ds,$$

and likewise for the integral expressing ψ_X. Hence, the option price V_{spread} can also be viewed as a function which depends on $T - \tau$ and T, i.e., time left to delivery from exercise and the delivery date.

In order to compute the price V_{spread} in practice, we must evaluate two integrals over a time segment where the integrands are linear combinations of $\psi_{\mathbf{L},Q}$ with arguments involving $g_1(s)$ and $g_2(s)$. Additionally, there is a third integration of the cumulants and the Fourier transform of a dampened call payoff function over the real line. Note that this latter inverse Fourier transform consists of only a one-dimensional integral, although we have two underlying futures contracts involved in the spread option, which depend on altogether three processes. We pay for this reduction of dimension by a complex structure of ψ_X, which is an integral that is hardly possible to compute analytically in most interesting cases. Thus, numerical integration must be performed to compute ψ_X. We remark in passing that Hurd and Zhou [32] has developped a Fourier approach to pricing spread options with non-zero strikes.

4.2 *Energy Quanto Options*

Producers and retailers operating in the energy market face both price and volume risk. The demand for power varies with temperature, and hence a producer may experience losses incurred by both low demand and low prices in a period of cold summer weather and little use of air-conditioning. In the German market, say, power generators based on fossil fuels will need to reduce their production in periods with a lot of wind. Wind mills will then cover the demand, as well as lower the prices. Both these examples show the need for risk management tools where one can hedge both price and volume risk. Energy quanto options offer such a tool.

Quanto options in energy typically have a payoff which is a product of two European options, for example a product of two put options. As discussed in Benth et al. [11], most energy quanto options are written on an energy price like gas or power and an index of temperature in a given location. Temperature determines to a large extent demand, and is used as a measure for the volume risk. The index can be heating-degree days (HDD) or cooling-degree days (CDD) over a given time period, to indicate the demand for power to heat or cool, resp. As shown by Benth et al. [11], many energy quanto options can be re-phrased as options written on two futures contracts, a futures on energy and a futures on a weather factor. For example, an energy quanto option can have the payoff

$$\max(K_1 - F_{\text{power}}(\tau, T), 0) \times \max(K_2 - F_{\text{CDD-temp}}(\tau, T), 0),$$

i.e., paying the product of two put options on power and CDD temperature futures at exercise time $\tau \leq T$, where the futures "deliver" at time T. Such a quanto option gives a protection against too low prices combined with too low CDD, the latter meaning little demand for air-conditioning cooling. We note that there is a market for weather futures (temperature, precipitation, hurricanes) at the Chicago Mercantile Exchange (CME) (see Benth and Šaltytė Benth [6] for a discussion and mathematical analysis of this market). Hence, one can actually trade in CDD temperature futures.

In Benth et al. [11], a "Black-Scholes" -type of formula is derived for the price of an energy quanto option (call-call payoff) when the two underlying futures follow a bivariate geometric Brownian motion. Further, they present an empirical analysis of various energy quanto options applied to gas and temperatures in New York and Chicago areas.

We now proceed with an analysis of the price of energy quanto options based on the cointegration model presented in the previous Section. This extends the analysis in Benth et al. [11] in two directions. Firstly, we allow for more realistic Lévy-based pricing models, and secondly, dependency between the two futures are not only modelled by correlation, but also by a common non-stationary factor. It is reasonable to assume that energy prices co-move with temperature index futures, as the demand for energy is directly linked to temperature, and temperature futures are traded to hedge energy price and volume risk.

We have the following general result, which will be useful when pricing a general quanto option:

Proposition 5 *Let X, Y and Z be three random variables where Z is independent of X and Y. Assume $f, g \in L^1(\mathbb{R})$ with $f(X+Z)g(Y)$ being integrable. If $\widehat{f}, \widehat{g} \in L^1(\mathbb{R})$, then*

$$\mathbb{E}[f(X+Z)g(Y)] = \frac{1}{4\pi^2} \int_{\mathbb{R}} \int_{\mathbb{R}} \widehat{f}(x)\widehat{g}(y) \exp\left(\psi_Z(x) + \psi_{X,Y}(x,y)\right) dx\, dy$$

where ψ_Z and $\psi_{X,Y}$ are the cumulants of Z and (X, Y), resp.

Proof By the tower property of conditional expectation,

$$\begin{aligned} \mathbb{E}[f(X+Z)g(Y)] &= \mathbb{E}\left[\mathbb{E}[f(X+Z)g(Y) \mid X, Y]\right] \\ &= \mathbb{E}\left[g(Y)\mathbb{E}[f(X+Z) \mid X, Y]\right]. \end{aligned}$$

We find from the Fourier inversion formula (12) and Fubini's Theorem that

$$
\begin{aligned}
\mathbb{E}[f(X+Z)\,|\,X,Y] &= \frac{1}{2\pi}\int_{\mathbb{R}} \widehat{f}(x)\mathbb{E}[\mathrm{e}^{\mathrm{i}x(X+Z)}\,|\,X,Y]\,dx \\
&= \frac{1}{2\pi}\int_{\mathbb{R}} \widehat{f}(x)\mathbb{E}[\mathrm{e}^{\mathrm{i}xZ}]\mathrm{e}^{\mathrm{i}xX}\,dx \\
&= \frac{1}{2\pi}\int_{\mathbb{R}} \widehat{f}(x)\exp(\psi_Z(x))\mathrm{e}^{\mathrm{i}xX}\,dx,
\end{aligned}
$$

where we have used the independence of Z in the second equality. Again appealing to Fubini's Theorem and the Fourier inversion formula (12), we get

$$
\begin{aligned}
\mathbb{E}[f(X+Z)g(Y)] &= \frac{1}{2\pi}\int_{\mathbb{R}} \widehat{f}(x)\exp(\psi_Z(x))\mathbb{E}[g(Y)\mathrm{e}^{\mathrm{i}xX}]\,dx \\
&= \frac{1}{4\pi^2}\int_{\mathbb{R}}\int_{\mathbb{R}} \widehat{f}(x)\exp(\psi_Z(x))\widehat{g}(y)\mathbb{E}[\mathrm{e}^{\mathrm{i}xX+\mathrm{i}yY}]\,dx\,dy \\
&= \frac{1}{4\pi^2}\int_{\mathbb{R}}\int_{\mathbb{R}} \widehat{f}(x)\widehat{g}(y)\exp(\psi_Z(x)+\psi_{X,Y}(x,y))\,dx\,dy.
\end{aligned}
$$

This concludes the proof. □

Let for $k=1,2$, $p_k:\mathbb{R}_+\to\mathbb{R}_+$ be two measurable functions. Suppose we want to price an energy quanto option with payoff $p_1(F_1(\tau,T))p_2(F_2(\tau,T))$ at exercise time $\tau\le T<\infty$. We suppose that $p_1(F_1(\tau,T))p_2(F_2(\tau,T))\in L^1(Q)$ and the no-arbitrage price is

$$
V_{\text{quanto}} = \mathrm{e}^{-r\tau}\mathbb{E}_Q\left[p_1(F_1(\tau,T))p_2(F_2(\tau,T))\right]. \tag{17}
$$

To this end, introduce the measurable functions $f,g:\mathbb{R}\to\mathbb{R}$

$$
\begin{aligned}
f(u) &= p_1\left(\exp\left(\Lambda_1(T)+h_1(T-\tau)+u\right)\right), \\
g(v) &= p_2\left(\exp\left(\Lambda_2(T)+h_2(T-\tau)+v\right)\right),
\end{aligned}
$$

where we assume that p_1 and p_2 are such that $f,g,\widehat{f},\widehat{g}\in L^1(\mathbb{R})$. Set $X=L(\tau)$, $Z=\int_0^\tau g_1(T-s)\,dU_1(s)$ and $Y=L(\tau)+\int_0^\tau g_2(T-s)\,dU_2(s)$. In order to have Z independent of (X,Y), we assume that the Lévy process U_1 is independent of (L,U_2). From a practical perspective, this means that the short-term stationary factor of the energy futures is independent of its long-term factor, and of the short-term stationary factor of the temperature index futures. This is admittedly a rather restrictive assumption. However, one may alternatively consider a factorization of the Lévy processes in terms of dependent and independent processes which opens for more general models. To avoid technicalities, we refrain from following this path further.

It follows that $\psi_Z(x)=\int_0^\tau \psi_{\mathbf{L},Q}(0,xg_1(T-\tau+s),0)\,ds$, where $x\in\mathbb{R}$. Furthermore,

$$\psi_{X,Y}(x,y) = \log \mathbb{E}_Q\left[\exp\left(\mathrm{i}xL(\tau) + \mathrm{i}y(L(\tau) + \int_0^\tau g_2(T-s)\,dU_2(s))\right)\right]$$
$$= \int_0^\tau \psi_{\mathbf{L},Q}(x+y, 0, yg_2(T-\tau+s))\,ds.$$

By Proposition 5 we then obtain a Fourier expression for V_{quanto} in terms of the Fourier transforms of the payoff functions given by f and g, and the cumulants ψ_Z and $\psi_{X,Y}$.

If the payoff functions p_1 and p_2 are of call or put type, we do not have that f and g are integrable. In this case we do an exponential dampening of f and g as described in the proof of Lemma 1. This leads to some simple modifications of the pricing formula above. We also remark that Eberlein et al. [19] state neat sufficient conditions on the payoff function and the cumulant (or the distribution) of the involved random variables, allowing for a Fourier representation of the expectation operator.

Acknowledgements Financial support from the project "Finance, Insurance, Energy, Weather and Stochastics (FINEWSTOCH)", funded by the Norwegian Research Council, is gratefully acknowledged. Part of this work was done when the author visited the Center of Advanced Study (CAS) in Oslo. Two anonymous referees and the editors Jan Kallsen and Antonis Papapantoleon are thanked for their careful reading and critics of an earlier version of this paper.

References

1. Barndorff-Nielsen, O.E.: Processes of normal inverse Gaussian type. Financ. Stoch. **2**, 41–68 (1998)
2. Barndorff-Nielsen, O.E., Benth, F.E., Veraart, A.E.D.: Modelling energy spot prices by volatility modulated Lévy-driven Volterra processes. Bernoulli **19**(3), 803–845 (2013)
3. Barndorff-Nielsen, O.E., Benth, F.E., Veraart, A.E.D.: Cross-commodity modelling by multivariate ambit fields fields. In: Aid, R., Ludkovski, M., Sircar, R. (eds.) Commodities, pp. 109–148. Springer, Energy and Environmental Finance (Fields Institute Communications) (2015)
4. Benth, F.E., Di Nunno, G., Khedher, A., Schmeck, M.: Pricing of spread options on a bivariate jump market and stability to model risk. Appl. Math. Financ. **22**(1), 28–62 (2015)
5. Benth, F.E., Kallsen, J., Meyer-Brandis, T.: A non-Gaussian Ornstein-Uhlenbeck process for electricity spot price modeling and derivatives pricing. Appl. Math. Financ. **14**, 153–169 (2007)
6. Benth, F.E., Šaltytė Benth, J.: Modeling and Pricing in Financial Markets for Weather Derivatives. World Scientific, Singapore (2013)
7. Benth, F.E., Šaltytė Benth, J., Koekebakker, S.: Stochastic Modeling of Electricity and Related Markets. World Scientific, Singapore (2008)
8. Benth, F.E., Klüppelberg, C., Müller, G., Vos, L.: Futures pricing in electricity markets based on stable CARMA spot models. Energy Econ. **44**, 392–406 (2014)
9. Benth, F.E., Koekebakker, S.: Stochastic modeling of financial electricity contracts. Energy Econ. **30**(3), 1116–1157 (2008)
10. Benth, F.E., Koekebakker, S.: Pricing of forwards and other derivatives in cointegrated commodity markets. Energy Econ. **52**, 104–117 (2015)
11. Benth, F.E., Lange, N., Myklebust, T.Aa: Pricing and hedging quanto options in energy markets. J. Energy Markets. **8**(1), 1–35 (2015)
12. Benth, F.E., Zdanowicz, H.: Pricing and hedging of energy spread options and volatility modulated Volterra processes. Int. J. Theor. Appl. Financ. **19**(1), 1650002 (2016)

13. Carmona, R., Durrleman, V.: Pricing and hedging spread options. SIAM Rev. **45**(4), 627–685 (2003)
14. Carmona, R., Tehranchi, M.: Interest Rate Models: an Infinite Dimensional Stochastic Analysis Perspective. Springer, Berlin, Heidelberg (2006)
15. Carr, P., Madan, D.: Option valuation and using the fast Fourier transform. J. Comp. Financ. **2**(4), 61–73 (1999)
16. Clewlow, L., Strickland, C.: Energy Derivatives: Pricing and Risk Management. Lacima Publications, London (2000)
17. Duan, J.-C., Pliska, S.R.: Option valuation with cointegrated asset prices. J. Econ. Dynam. Control **28**, 727–754 (2004)
18. Eberlein, E.: Fourier based valuation methods in mathematical finance. In Benth, F.E., Kholodnyi, V., Laurence, p. (eds.) Quantitative Energy Finance, pp. 85–114. Springer (2014)
19. Eberlein, E., Glau, K., Papapantoleon, A.: Analysis of Fourier transform valuation formulas and applications. Appl. Math. Financ. **17**, 211–240 (2010)
20. Eberlein, E., Glau, K., Papapantoleon, A.: Analyticity of the Wiener-Hopf factors and valuation of exotic options in Lévy models. In: Di Nunno, G., Øksendal, B. (eds.) Advanced Mathematical Methods of Finance, pp. 223—245. Springer (2011)
21. Eberlein, E., Jacod, J., Raible, S.: Lévy term structure models: no-arbitrage and completeness. Financ. Stoch. **9**, 67–88 (2005)
22. Eberlein, E., Keller, U.: Hyperbolic distributions in finance. Bernoulli **1**, 281–299 (1995)
23. Eberlein, E., Kluge, W.: Valuation of floating range notes in Lévy term structure models. Math. Financ. **16**, 237–254 (2006)
24. Eberlein, E., Özkan, F.: The Lévy Libor model. Financ. Stoch. **9**, 327–348 (2005)
25. Eberlein, E., Papapantoleon, A., Shiryaev, A.N.: Esscher transform and the duality principle for multidimensional semimartingales. Ann. Appl. Probab. **19**, 1944–1971 (2009)
26. Eberlein, E., Raible, S.: Term structure models driven by general Lévy processes. Math. Financ. **9**, 31–53 (1999)
27. Eberlein, E., Stahl, G.: Both sides of the fence: a statistical and regulatory view of electricity risk. Energy Power Risk Manag. **8**(6), 34–38 (2003)
28. Eydeland, A., Wolyniec, K.: Energy and power risk management: new developments in modeling, pricing and hedging. John Wiley & Sons, Hoboken, New Jersey (2003)
29. Folland, G.B.: Real Analysis. John Wiley & Sons, New York (1984)
30. Gerber, H.U., Shiu, E.S.W.: Option pricing by Esscher transforms. Trans. Soc. Actuaries, **46**, 99–191 (1994) (with discussion)
31. Heath, D., Jarrow, R., Morton, A.: Bond pricing and the term structure of interest rates: a new methodology for contingent claim valuation. Econometrica **60**, 77–105 (1992)
32. Hurd, T.R., Zhou, Z.: A Fourier transform method for spread option pricing. SIAM J. Finan. Math. **1**, 142–157 (2009)
33. Kallsen, J., Shiryaev, A.: The cumulant process and Esscher's change of measure. Financ. Stoch. **6**, 397–428 (2002)
34. Lucia, J., Schwartz, E.S.: Electricity prices and power derivatives: evidence from the Nordic Power Exchange. Rev. Deriv. Res. **5**(1), 5–50 (2002)
35. Margrabe, W.: The value of an option to exchange one asset for another. J. Financ. **33**, 177–186 (1978)
36. Peszat, S., Zabczyk, J.: Stochastic Partial Differential Equations with Lévy Noise. Cambridge University Press, Cambridge (2007)
37. Sato, K.-I.: Lévy Processes and Infinitely Divisible Distributions. Cambridge University Press, Cambridge (1999)

The manufacturer's authorised representative in the EU is Springer Nature Customer Service Centre GmbH, Europaplatz 3, 69115 Heidelberg, Germany. If you have any concerns regarding our products, please contact ProductSafety@springernature.com

Printed and bound by CPI Group (UK) Ltd, Croydon, CR0 4YY

30/09/2026

02225230-0018